KENDALL'S

MUSCLES

TESTING AND FUNCTION WITH POSTURE AND PAIN

SIXTH EDITION

KENDALL'S

MUSCLES

TESTING AND FUNCTION WITH POSTURE AND PAIN

SIXTH EDITION

Vincent M. Conroy, PT, DScPT

Brian N. Murray, Jr., PT, DPT, OCS

Quinn T. Alexopulos, MSN, RN, CNL

Jordan McCreary, MD, BS (Biology)

Wolters Kluwer

Philadelphia • Baltimore • New York • London
Buenos Aires • Hong Kong • Sydney • Tokyo

Senior Acquisitions Editor: Matt Hauber
Development Editors: Robin Levin Richman and Amy Millholen
Editorial Coordinator: Sean Hanrahan
Editorial Assistant: Parisa Saranj
Marketing Manager: Phyllis Hitner
Production Project Manager: Barton Dudlick and Catherine Ott
Manager, Graphic Arts & Design: Stephen Druding
Art Director, Illustration: Jennifer Clements
Manufacturing Coordinator: Margie Orzech
Prepress Vendor: Lumina Datamatics Ltd

Sixth Edition

9 8 7 6 5 4 3 2 1

Printed in Mexico

Library of Congress Cataloging-in-Publication Data

Names: Conroy, Vincent M., author. | Murray, Brian (Brian N.), Jr., author.
Title: Kendall's muscles : testing and function with posture and pain / Vincent M. Conroy, Brian Murray.
Other titles: Muscles.
Description: 6th edition. | Philadelphia, PA : Wolters Kluwer, [2024] | Preceded by Muscles : testing and function with posture and pain / Florence Peterson Kendall ... [et al.]. 5th ed. c2005. | Includes bibliographical references and index. | Summary: "Kendall's Muscles: Testing and Function, with Posture and Pain, 6th Edition, transforms this landmark Physical Therapy classic to prepare you for unparalleled clinical success in today's practice. Timeless coverage of manual muscle testing, evaluation, and treatment meets the latest evidence-based practices, engaging imagery, and dynamic digital resources to create a powerful resource you will reference for years to come. The extensively revised 6th Edition of this proven classic details normal and abnormal range of motion, developmental factors and environmental influences on posture, length tests and stretching exercises, entrapment and compression syndromes, scoliosis, and more, incorporating extensive updates and enhancements to help you develop clinical confidence and ensure safe, effective practice throughout your career. Expanded ergonomics and body mechanics content familiarizes you with the most current perspectives in these critical areas. Considerations for special populations alert you to specific approaches for pediatric and geriatric populations, as well as patients with comorbidities. Photographs and illustrations clarify the latest technologies and clinical advances in muscle testing in a clear, user-friendly format. Case studies enhance your clinical application capabilities with realistic patient scenarios. Classic Kendall boxes detail key assessment, process, and treatment content in the trademark Kendall approach for unparalleled understanding at a glance. Historical Notes enrich your understanding of important links between past and current practices. References and Suggested Readings point you to the most current research literature for further study. "-- Provided by publisher.
Identifiers: LCCN 2022039475 | ISBN 9781975159894 (hardback) | ISBN 9781975159900 (ebook)
Subjects: MESH: Muscles--physiology | Posture--physiology | Physical Examination--methods | Musculoskeletal Diseases--diagnosis | Musculoskeletal Diseases--therapy | BISAC: MEDICAL / Allied Health Services / Physical Therapy | MEDICAL / Education & Training
Classification: LCC QP321 | NLM WE 504 | DDC 612.7/4--dc23/eng/20220919
LC record available at https://lccn.loc.gov/2022039475

shop.lww.com

QUADM1122

Dedicated to Florence and Henry Kendall, our families, students, and mentors who have contributed to "the process" of gaining knowledge and to all who have assisted with or will benefit from this publication.

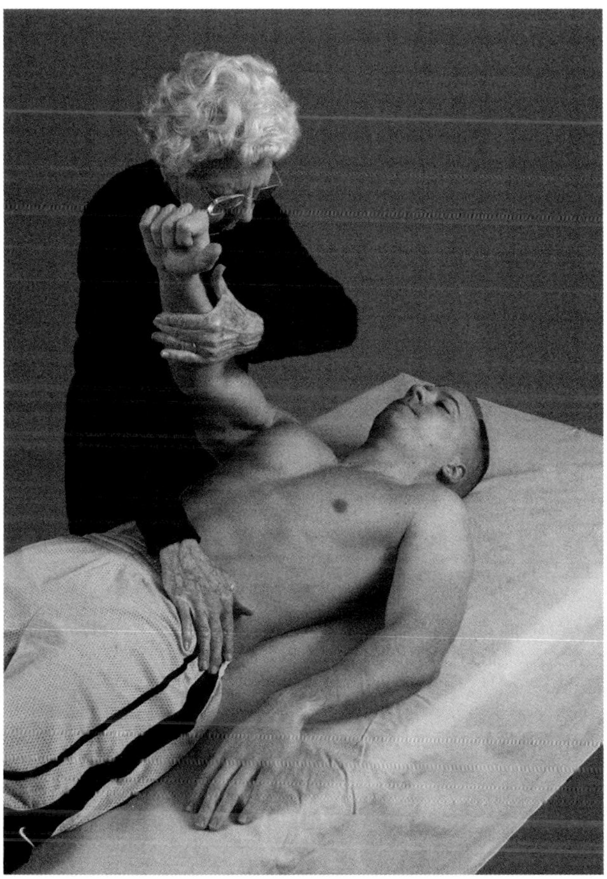

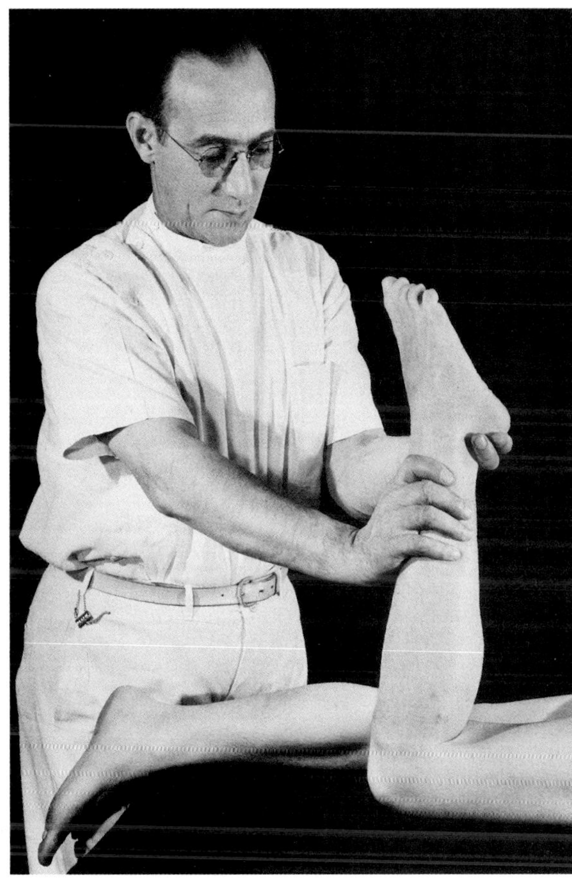

FOREWORD TO THE SIXTH EDITION

It is an honor to be asked to write this foreword for the 6th edition of *Kendall's Muscles: Testing and Function with Posture and Pain.* My connections with this work and with Florence P. Kendall are many.

Florence and her husband, Henry O. Kendall, PT, were both charter members of the Maryland Chapter of the American Physical Therapy Association (APTA) in 1939 and Florence served as its first president. They published the first edition of this book in 1949, setting the standard for physical therapy muscle testing. In 1956, they helped start the physical therapy program at the University of Maryland. That program became the Department of Physical Therapy and Rehabilitation Science and was Florence's academic home since its inception.

One of the highlights of my 15 years as chair of that department was working with Florence on the 5th edition of this textbook. I often worked on the book with Florence at her daughter Susie's house in Severna Park, MD, where Florence lived. Florence would start our work session by serving cookies and tea, or a meal followed by ice cream, but then she got down to business. She made me feel like part of her family and I loved working with her.

Having worked closely with Florence Kendall on her last publication, I know that she would be so very happy that the new coauthors (one of whom teaches in our program) have continued to improve upon the work about which she was so passionate. This new edition preserves the original Kendall philosophy of using basics while introducing several new enhancements. The book's structure has been reorganized to provide better continuity and flow while preserving the lucid and easily accessible approach of previous editions. The content has been modified to eliminate redundancy in the illustrations and descriptions. New graphs, tables, full-color photographs, figures, and evidence-based examples of clinical diagnoses and interventions have been added.

I feel privileged to write this foreword in honor of Florence and in recognition of the excellent contributions of the new coauthors. This latest edition will continue as the comprehensive text on examination, evaluation, and diagnostic processes of the musculoskeletal system for students, clinicians, and faculty.

Mary M. Rodgers, PT, PhD, FAPTA, FASB, FISB
Professor Emerita
Department of Physical Therapy &
Rehabilitation Science
University of Maryland School of Medicine

KENDALL FAMILY HISTORY

Kendall's Muscles: Testing and Function with Posture and Pain is more than a textbook. It is the culmination of a lifetime of hands-on research and set the standard for the fledgling field of physical therapy. This is the first edition published in 17 years, and the first without Florence P. Kendall as an author. Her career spanned more than 75 years. She has been hailed as the mother of physical therapy.

Florence P. Kendall

Born Della May Anna Florence Peterson to Swedish immigrants on a farm in Warman, Minnesota, on May 5, 1910, she was one of 11 surviving children. Florence received her bachelor's of science in physical education from the University of Minnesota in 1930. After graduation, she taught high school physical education in Minnesota for one year. The following year, she took a position at Walter Reed Army Medical Center in 1931 to study physical therapy. When her position was eliminated, she was hired at Children's Hospital in Baltimore, Maryland.

On May 5, 1933, Florence attended a lecture presented by Children's Hospital's staff physical therapist Henry Otis Kendall. That was the day they first met. She told her family that she found a four-leaf clover on that day and knew that moment was special. She believed everything happened for a reason. Florence began working for Kendall the same year, and in 1935, Florence and Henry were married.

The series of events that brought Henry O. Kendall to physical therapy began with his service in World War I. As a young soldier, he was sent to run through the battlefields and trigger landmines ahead of the advancing troop. He did not step away in time and was blinded by shrapnel. He lost one eye and was set to have the other removed when a nurse at the combat hospital in France spoke up and asked that his other eye be spared. In time, he did regain 20% vision in his remaining eye, but was legally blind.

Kendall was one of many soldiers blinded in the war. Programs were developed across Europe and eventually the United States to train these men in skills that could be performed without sight. Kendall was sent to Evergreen, U.S. Army General Hospital No 7, in Baltimore, Maryland. Evergreen, which later became the Red Cross Institute for the Blind, was the country's first blind veteran rehabilitation center. There he was taught anatomy and massage, which prepared him for the newly evolving field of physical therapy (see following page).

In the 1930s and 1940s, Florence and Henry wrote and produced material on the care of patients with polio. They were instrumental in establishing physical therapy as a licensed profession in the State of Maryland and in the founding of the American Physical Therapy Association. Florence served as the first president of the Maryland chapter.

In 1949, the Kendalls published the first edition of *Muscles: Testing and Function*. *Posture and Pain* was then published in 1952, the same year the Kendalls opened one of the first private physical therapy practices in the country. Their books set the standard for assessment and treatment of painful conditions caused by disease and poor alignment. During this time, they served on the faculty of Johns Hopkins University and the University of Maryland to train students in anatomy and proper body mechanics.

In 1971, following the publication of the second edition of *Muscles: Testing and Function*, which included the material from *Posture and Pain*, Henry retired. He died in 1979. Florence continued their work by traveling all

Henry O. Kendall is the third student from the left. Anatomy class at Evergreen, U.S. Army General Hospital (later known as The Red Cross Institute for the Blind), approximately 1918. With permission of Florence Kendall.

over the country to give workshops to teach the fundamentals of physical therapy.

Florence authored three more editions of *Muscles: Testing and Function* in her lifetime. She was extremely proud of the fifth edition, published in 2005, the same year she celebrated her 95th birthday. The entire world of physical therapy appeared to celebrate this milestone with her. Ever the perfectionist, before the latest edition hit the shelves, she was already revising.

Florence Kendall was my grandmother. In our family, she is called Amma; *Muscles: Testing and Function* to this day is known as "The Book." The pages of The Book are like a family album, as the majority of the models are her children and grandchildren. Some of her models, particularly examples of painful conditions and poor alignment, were not relatives. On more than one occasion, she approached complete strangers on the beach to ask if they would consent to be photographed for The Book. Her daughters were mortified. Even on vacation, Florence could never stop seeing the world as a physical therapist would.

She taught with a hands-on approach, which had been critical for Henry's initial training, as he learned to evaluate without sight. As the licensing requirements for physical therapy required higher levels of education, Florence emphasized the need to go "back to basics": to effectively treat a patient, you must always start with the patient. You must perform your own assessment to determine the best individualized patient care.

Even in her final days, she continued to treat patients in need. One of her nurses shared that her husband was having severe back problems. He came to see if she could help. From her bed, she was able to determine that he needed a lift in one shoe to correct his alignment and decrease his pain.

Florence P. Kendall knew how to make an exit. She died on January 28, 2006, on the heels of that year's American Physical Therapy Association convention. Her funeral was delayed so the physical therapy community could pay their respects. She planned her entire funeral and asked that a picture of Henry O. Kendall be displayed to honor his role in her remarkable career.

If by any chance you have taken the time to read this small section of The Book, our family thanks you. What is to you a textbook is the legacy left by two people who came together to improve the quality of life for patients and along the way created the standards of a profession.

Quinn Tyler Alexopulos, MSN, RN, CNL

PREFACE

Muscles: Testing and Function with Posture and Pain has served the academic and clinical realms of the healthcare arena since the initial publications of *Muscles: Testing and Function* (1949) and *Posture and Pain* (1952). The assessment of muscle performance and its relationship to posture, pain, impairment, and functional movement continue to shape the clinician's problem-solving process when interacting with persons in need of assistance. The straightforward simplicity of the narrative and the practical approach are what continue to make this text the gold standard in the field. Much of the narrative and imaging remains intact throughout the 6th edition, a reflection of the enduring quality of groundbreaking work the Kendalls and other authors have contributed. The new edition continues to hold to this purpose.

STRUCTURE AND ORGANIZATION

While keeping the original goal of previous editions, this edition has modifications to the basic format to provide the reader with improved continuity and flow of the content being discussed. Each chapter is divided into five or six sections, with Chapters 1 and 2 offering section titles specific to Fundamental Concepts and Posture. Chapters 3, 6, and 7 are organized consistently by the following section heads: Innervation, two regional sections outlining Muscles and Muscle Tests, Clinical Findings, and Intervention. Chapter 4 is organized with the following section heads: Innervation, Intervertebral Joints and Motions, Muscles and Muscle Tests, Clinical Findings, and Intervention. Chapter 5 follows a similar organization as Chapters 3, 6, and 7 and includes an extra section dedicated to the pelvis.

Finally, key to the clarity and accessibility of previous editions and carried on in this one, content on muscles and muscle testing is carefully presented where images most easily illustrate it; nothing is left to chance in the layout and balance of all information and the illustrations and photographs that reflect it.

NEW TO THIS EDITION

We retained content when emphasis was required and eliminated redundancies when appropriate. Beyond the overall structure, specific modifications to chapters include the following:

- **Graphs and Tables:** *New graphs, charts, and tables*
- **Full-Color Art:** New full-color photographs and figures
- **Evidence-Based Practice:** Specific evidence-based examples of clinical diagnoses and interventions.
- **References:** Plentiful new and updated references that highlight evidence-based best practices
- **Classic Kendall** and **Historical Notes:** We have added more of these descriptive pieces, which will continue to shine the light on and allow the reader to benefit from Henry and Florence Kendall's knowledge and pioneering expertise.

Specific chapter changes are as follows:

- **Chapter 1**, to which we moved pertinent and foundational information that was previously in Chapter 2, continues to address fundamental concepts that are expanded upon in subsequent chapters. Additional content, new color photographs, tables, and updated evidence-based references are provided when and where indicated.
- **Chapter 2** content that focuses on adult posture, sample diagnoses, and interventions, as well as the section on child posture, is now in Appendix A.
- **Chapter 3** has new color photographs to support the narrative.
- **Chapter 4** speaks to the regions of the Neck and Back and includes content from the previous edition's Chapter 5 to align with anatomical and clinical organization.
- **Chapter 5** content from the previous edition's Chapter 4 is now located in Chapter 5. Content is

added to provide more detail about the pelvic floor and muscles that act upon it.

- **Chapters 6** and **7** have selected new and updated full-color photographs. The clinical findings and intervention sections are expanded with relevant examples.

Appendices include the following:

- **Appendix A**, an updated discussion on child posture assessment and intervention.
- **Appendix B**, updated information on classic diagnoses that were highlighted in previous editions.
- **Appendix C** is a new table to provide the reader with examples of standardized evaluative techniques that can prove useful in the clinic.
- **Appendix D**, the Spinal Segment Distribution to Nerves and Muscles charts, provides insight and guidance to how the spinal level distributions are attributed to muscles throughout the text.

ACKNOWLEDGMENTS

This textbook and the instruction received by Florence Kendall has been embedded as a component of my practice as a clinician for over 30 years. To the Kendall family, Thank You for your trust! The invitation to coauthor this edition is an honor that I will relish for a lifetime, and I am grateful to be a part of this project. Every project comes to fruition only through the dedication and commitment of a talented team. I would be remiss if my heartfelt Thank You did not get extended to several individuals. Quinn and Jordon, your ever-present reminders helped to keep the Kendall philosophy ingrained in this edition. Brian, your attention to detail was and remains unwavering and I will always appreciate your ability to meet on a moment's notice. I look forward to future collaborations. Meg, BJ, Mackenzie, Maddie, and Mia, Thank You for sharing a very intelligent and witty Brian with me for the length of this project. To the folks at Wolters Kluwer – Matt, Robin, Amy, Sean, Fred, Julie, and others – Thank You for your guidance and patience. To the previous authors, thank you for beginning and fostering the continuation of this publication. To my wife Susan, my daughter Julia, my extended and not so extended family, thank you for your love, support, and understanding throughout the duration of this project.

—Vinnie

I had no idea what I was getting myself into when this team came together at our first meeting years ago. And that was before we had to adapt to the development of a pandemic that would drastically impact our lives and the work we had been doing thus far. None of the past years of work would have been possible, however, without the support and encouragement of my wife, Meg. I would like to thank her and our children for their support and patience during many early and late-day meetings, allowing me to take on this challenge. To Vinnie Conroy, thank you for sharing your time, experience, and knowledge with me. To the Conroy family, thank you for sharing Vinnie! To my mom and sisters, thank you for being that needed ear during this process. To the team at Wolters Kluwer and Robin Richman, thank you for your guidance and willingness to educate me in real time on best practices. A special word of thanks to Laura Schmitt for presenting me with this unique opportunity. And thank you, especially, to Florence Kendall, Quinn Alexopulos, Jordan McCreary, and the extended Kendall family for accepting me as part of this truly special undertaking

—Brian

First and foremost, this book is dedicated to Florence Peterson Kendall (Amma). I would like to thank Vinnie and Brian (and Robin) for their respect of the original work and for carefully reorganizing the text in a way that makes it an even stronger resource for the field of physical therapy. I also want to thank the team at Wolters Kluwer, especially Matt, for their persistence and faith that there was still an audience for this book. I believe Amma would say she was "tickled" to see the new edition. Finally, I want to dedicate this book to my incredible family for maintaining and honoring the Kendall legacy. It has been a lifelong dream to work on The Book. I'm grateful that Jordan and I have had the opportunity. I hope that my daughters, Daphne and Freya, get to see this book continue in print for many years to come.

—Quinn

CONTENTS IN BRIEF

DETAILED CONTENTS

FUNDAMENTAL CONCEPTS

1

OUTLINE

INTRODUCTION

The underlying philosophy of this book is that there is a continuing need to "get back to basics." This philosophy is especially pertinent in this era of time-limited treatments, advancing technology, and the need to justify the initiation or continuation of care by practitioners.

Muscle function, body mechanics, and simple treatment procedures do not change. With respect to musculoskeletal problems, the underlying purposes of treatment have been, and continue to be, to restore and maintain appropriate flexibility, extensibility, range of motion, good postural alignment, and muscle balance.

It is essential for practitioners to choose and effectively perform tests that aid in solving problems, whether to provide a differential diagnosis, establish or change treatment procedures, improve function, or relieve pain. Of paramount importance for students and clinicians is the ability to think critically, to apply objectivity, and to use the caution and care needed for appropriate objective tests and measurements so that clinical reasoning can be both accurate and meaningful.

The role of prevention of musculoskeletal problems, as predicted in previous editions of this text, has become an increasingly important issue. Health practitioners play an effective role in promoting wellness if they are aware of adverse effects of muscle imbalance, faulty alignment, and improper exercise.

A thorough understanding of musculoskeletal assessment and those painful muscle conditions associated with poor posture enables practitioners to develop safe and effective home programs for their patients. The costs to society for treatment of common problems, such as low back pain (LBP), have reached a critical point. Many cases of LBP are related to postures that can be addressed, corrected, and alleviated.

Manual Muscle Testing

Muscle testing is an integral part of physical examination. It provides information, not obtained by other procedures, that is useful in differential diagnosis, prognosis, and treatment of neuromuscular and musculoskeletal disorders.

Muscle *length testing* is used to determine whether the muscle length is limited or excessive. A muscle that is too short will limit normal range of motion and a muscle that is elongated may allow too much range of motion that leads to instability of adjacent joints and tissues. When stretching is indicated, tight muscles should be stretched in a manner that is not injurious to the structures or to the body as a whole. Range of motion should be increased to permit normal joint function unless restriction of motion is a desired outcome for the sake of stability (1, 2).

Muscle *strength testing* is used to determine the capability of muscles or muscle groups to function in movement as this has an impact on their ability to provide stability and support (3).

Muscle Weakness and Muscle Imbalance

Muscle Weakness. Many conditions are characterized by **muscle weakness**. Some show definite patterns of muscle involvement; others show intermittent weakness without any apparent pattern. In some cases, weakness is symmetrical; in others, asymmetrical. The site or level of peripheral lesion may be determined because the muscles distal to the site of the lesion will show weakness or paralysis. Careful testing and accurate recording of test results will reveal the characteristic findings and aid in diagnosis.

Many factors influence weakness and the return of muscle strength. Weakness may be due to nerve involvement, disuse atrophy, stretch weakness, pain, or fatigue. Return of muscle strength may be due to recovery following a disease process, return of nerve impulse, after trauma and repair, hypertrophy of unaffected muscle fibers, muscular development resulting from exercises to overcome disuse atrophy, or return of strength after stretch and strain have been relieved.

Muscle weakness should be treated in accordance with the basic cause of weakness. If due to lack of use, then exercise; if due to overwork and fatigue, then rest; if due to stretch and strain, then relief of stretch and strain before the stress of additional exercise is thrust upon the weak muscle.

Muscle Imbalance. Conditions frequently show patterns of muscle imbalance. Some patterns are associated with handedness, some with habitually poor posture. Muscle imbalance may also result from occupational or recreational activities in which there is persistent use of certain muscles without adequate exercise of opposing muscles. Imbalance that affects postural alignment is an important factor in many painful conditions (4–6).

Muscle imbalance can distort postural alignment and set the stage for undue stress and strain on joints, ligaments, and muscles. Manual muscle testing is the tool of choice to determine the extent of imbalance.

Muscle Functions

Every muscle is a prime mover in some specific action. When any one muscle is paralyzed, stability of the adjacent joints and structures is impaired and normal movement is lost. Some of the most dramatic evidence of muscle function comes from observing the effects of loss of the ability to contract as seen in paralyzed muscles, or the effect of excessive shortening as seen in a muscle contracture and the resultant deformity.

The muscle testing described in this book is directed toward examination of individual muscles insofar as is practical. The overlap of muscle actions, as well as the interdependence of muscles in movement, is well recognized by those involved in muscle testing. Because of this close relationship in functions, accurate testing of individual muscles requires strict adherence to the fundamental principles of muscle testing and rules of procedure.

Fundamental components of manual muscle testing are test performance and evaluation of muscle strength and length. To become proficient in these procedures, one must possess a comprehensive and detailed knowledge of human anatomy and muscle function. This knowledge must include an understanding of joint motion because length and strength tests are described in terms of joint movements and positions. It must also include knowledge of muscular innervations, the agonistic and antagonistic actions of muscles, and their role in fixation and compensation. In addition, it requires the ability to palpate the muscle or its tendon, to distinguish between normal and atrophied contour, and to recognize abnormalities of position or movement.

Evidence-Based Objectivity and Reliability in Muscle Testing

Muscle testing measures must be objective. With the high cost of medical care, the economics of reimbursement requires documentation that treatment is necessary and improvement has resulted from treatment. The more gradual the improvement, the more important the numbers become so that even minimal changes can be documented.

Many advocate the use of instrumentation to eliminate the subjective component of manual muscle tests (7). How do new problems and variables introduced by instruments affect the accuracy, reliability, and validity of muscle tests? To date, the research is conflicting.

Instrumentation, Machines, and Devices

The value of objective measurements obtained using modern instrumentation must be weighed against their limited usefulness, cost, and complexity. Length tests, if performed with precision, can provide objective data using simple devices such as goniometers to measure angles and rulers or tape measures to measure distance. Strength tests cannot rely on these devices. Objectivity is based on the examiner's ability to palpate and observe the tendon or muscle response in very weak muscles. Additionally, an examiner can visualize a tendon that becomes prominent (i.e., a trace grade), movement of a segment in the horizontal plane (i.e., a poor grade), and a segment being held in an antigravity position (i.e., a fair grade). Even the fair+ grade, which is based on holding the antigravity position against slight pressure by the examiner, is easy to identify. For these grades of strength, mechanical devices are not applicable or necessary to ensure objectivity. However, instrumentation may play a role in the assessment of the muscle grades of good and normal as well as strength above normal.

Isokinetic Machines and Electromyography. Under controlled research conditions, isokinetic machines can help in obtaining valuable information. At present, however, their usefulness in the clinic is limited. Problems occur both in testing muscle strength and in exercising. One problem with machines is providing adequate stabilization to control variables and to ensure the standardization of testing techniques. Tests by machines lack specificity, and compensation occurs. In addition to the high cost of the machines, setting up patients is time-consuming; both are important factors when comparing cost-effectiveness of the testing procedures with that of skilled manual muscle testing.

Electromyography (EMG) is another important research tool, but its usefulness in muscle strength testing is questionable. According to Gregory Rash, "EMG data cannot tell us how strong the muscle is, if one muscle is stronger than another muscle, if the contraction is a concentric or eccentric contraction, or if the activity is under voluntary control by the individual" (8).

Handheld Devices. Handheld devices (HHDs), such as dynamometers, measure the amount of force exerted by the examiner and countered by the subject. HHDs are more readily available in today's clinics and provide objective data regarding the amount of force that is used during manual muscle strength testing. The problem with an HHD is that it comes between the examiner and the segment being tested. It also interferes with use of the examiner's hand. The examiner's hand must not be encumbered for positioning the segment, controlling the specific direction of pressure, or applying pressure with the fingers, palm, or whole hand as needed.

A review of the literature regarding dynamometers reveals some of the problems associated with the use of these devices. Reliability across testers, examiner strength, and gender differences have all been found to limit intertester reliability when using handheld dynamometry. A study of intertester reliability concluded that "the handheld dynamometer shows limited reliability when used by two or more testers" (9). Two studies have demonstrated good intratester reliability using handheld dynamometers (10, 11). However, "hand-held dynamometers … may underestimate a patient's true maximal isometric strength, due to difficulties in stabilization of the device" (12). Bohannon reports that although no formal determination of minimal clinically important difference has been described for handheld dynamometry, the report provided relevant information.

Examiner strength presents another variable in handheld dynamometer reliability. Work by Marino et al. identified examiner strength as the reason for discrepancy between two examiners testing hip abductor strength (13). The examiner's strength affects the stability of the handheld dynamometer when used with stronger subjects (11). This problem was also related to gender differences by Mulroy et al. The subject's maximal knee extension force, measured by a handheld dynamometer, was accurate only for the male examiner testing female patients (14). (The validity of measurements obtained by HHD, like the reliability of the measurements, is dependent on the tester having adequate strength to hold stably against the effort of the tested individual. Without such strength, the maximum force the tester can measure is limited by his or her own strength.)

Hands, the Preferred Tools of Measurement

After a decade of scientific review, Newton and Waddell concluded that the "judgment of the clinician appears to be more accurate in determining effort of the patient, than evaluating the results from the machines" (15).

As tools, our hands are very sensitive, fine-tuned instruments. One hand of the examiner positions and stabilizes the segment adjacent to the segment being tested. The other hand determines the pain-free range of motion, guides the tested segment into precise test position, and gives the appropriate amount of pressure to determine strength. All the while, this instrument we call the hand is hooked up to the most marvelous computer ever created—the human mind—which can store valuable and useful information on the basis of which judgments about evaluation and treatment can be made. Such information contains objective data that are obtained without sacrificing the art and science of manual muscle testing to the demand for objectivity.

HISTORICAL NOTE

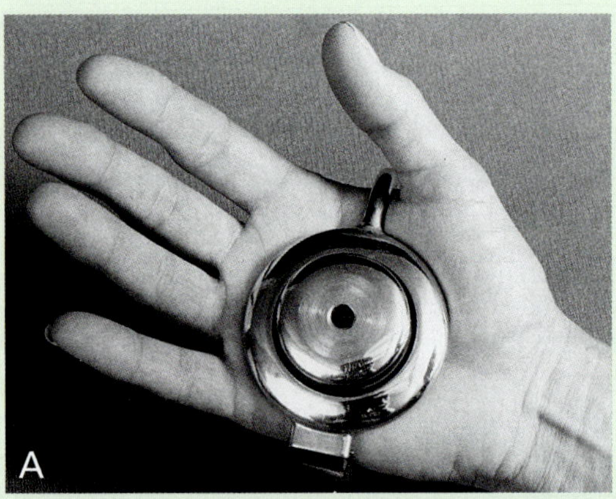

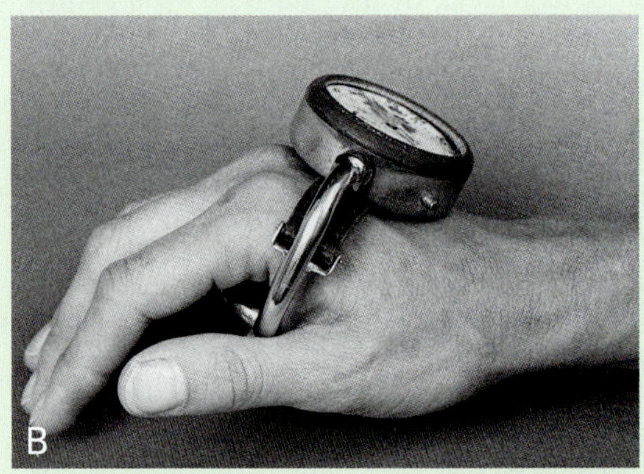

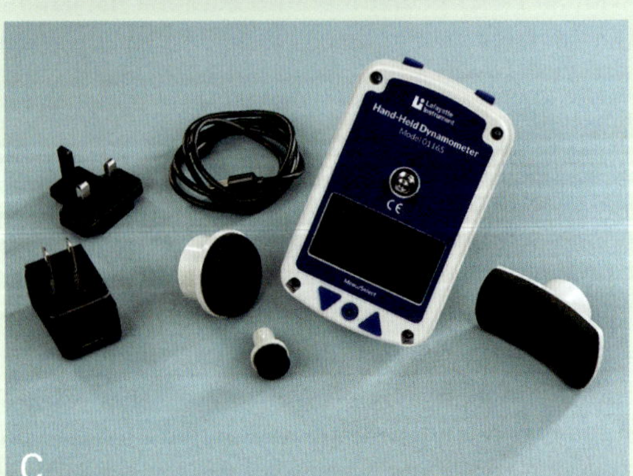

In 1941, while engaged in a research study for the Foundation for Infantile Paralysis, Florence Kendall designed an HHD to measure the force applied by the examiner during manual muscle testing. The Foundation turned over the design to Dr. W. Beasley in Washington, DC, who made a prototype. One year later, this device was presented at a symposium on polio. ***A*** shows the pressure-sensitive pad in the palm of the hand from which force was transmitted to the gauge on the dorsum of the hand, shown in ***B*** (Lippincott, Williams, & Wilkins, © Copyright 2005.) This may have been one of the first handheld dynamometers. ***C***. Modern manual muscle tester (C/O Lafayette Instrument Company, Inc.)

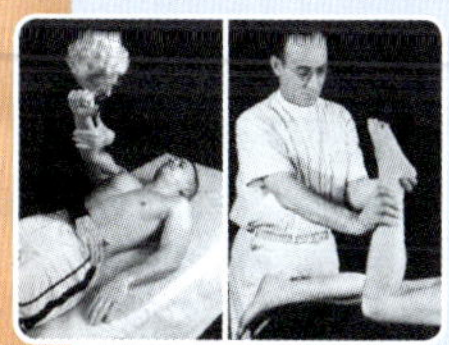

CLASSIC KENDALL

One of the unique features of this text is the preservation of more than half a century of postural analyses and the careful evaluation of muscle balance as it relates to function and pain. Many of the photographs provide outstanding historic examples of postural faults that are genuine rather than posed.

It is essential that every practitioner develop effective problem-solving skills that will result in choosing and performing appropriate and accurate tests to provide meaningful data for the establishment of a successful treatment plan. Anatomy has not changed, but time constraints in some current practice settings have resulted in testing "shortcuts" that can lead to an incorrect diagnosis.

The Kendalls were early pioneers in performing clinical research as part of their continual quest for knowledge regarding how muscle length and weakness relate to painful conditions. A study performed in the early 1950s compared hundreds of "normal" subjects—cadets, physicians, physical therapists, and student nurses (age range, 18–40 years)—with patients who had LBP. That study led to a better understanding of common muscle imbalances in the general population as compared to those in patients with LBP. In addition, it helped to define the differences in these imbalances between males and females. The data from this clinical study are included in the following table.

Male (% [*n*])				Female (% [*n*])		
100 LBP Patients	36 Physicians	275 Cadets	Case Findings	307 Student Nurses	50 Physical Therapists	100 LBP Patients
58% (58)	25% (9)	5% (14)	Weakness in "upper" anterior abdominals	44% (135)	52% (26)	81% (81)
69% (69)	31% (11)	33% (91)	Weakness in "lower" anterior abdominals	79% (243)	72% (36)	96% (96)
71% (71)	45% (16)	10% (28)	Limitation of forward flexion	5% (15)	10% (5)	48% (48)
71% (71)	77% (28)	26% (72)	Right gluteus medius weakness	40% (123)	76% (38)	90% (90)
15% (15)	3% (1)	5% (14)	Left gluteus medius weakness	5.5% (17)	10% (5)	6% (6)
0% (0)	0% (0)	0.3% (1)	Bilateral gluteus medius weakness	5.5% (17)	0% (0)	12% (12)

SECTION I
BODY SYSTEMS

POSITIONAL REFERENCE

Body Segments

Posture is a composite of the positions of all the joints of the body at any given moment, and static postural alignment is best described in terms of the positions of the various joints and body segments. This chapter provides basic information about anatomical positions, axes, planes, and movements of joints. This information is essential when analyzing postural alignment (Figure 1-1).

Anatomical Position

The anatomical position of the body is an erect posture with face forward, arms at sides, palms forward, and fingers and thumb in extension. This is the position of reference for definitions and descriptions of body planes and axes (Figure 1-2).

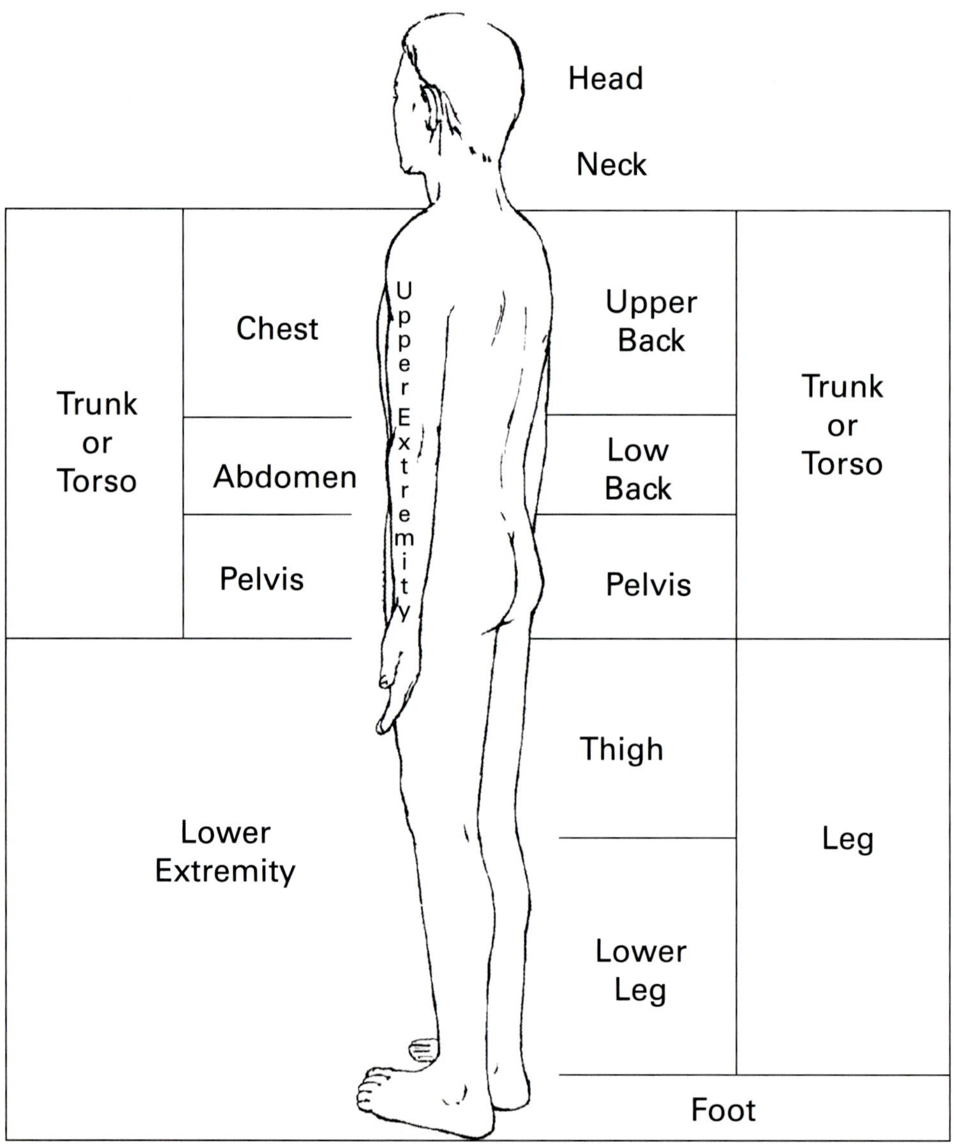

FIGURE 1-1: Common terminology to describe anatomical positions, axes, planes, and movements of joints.

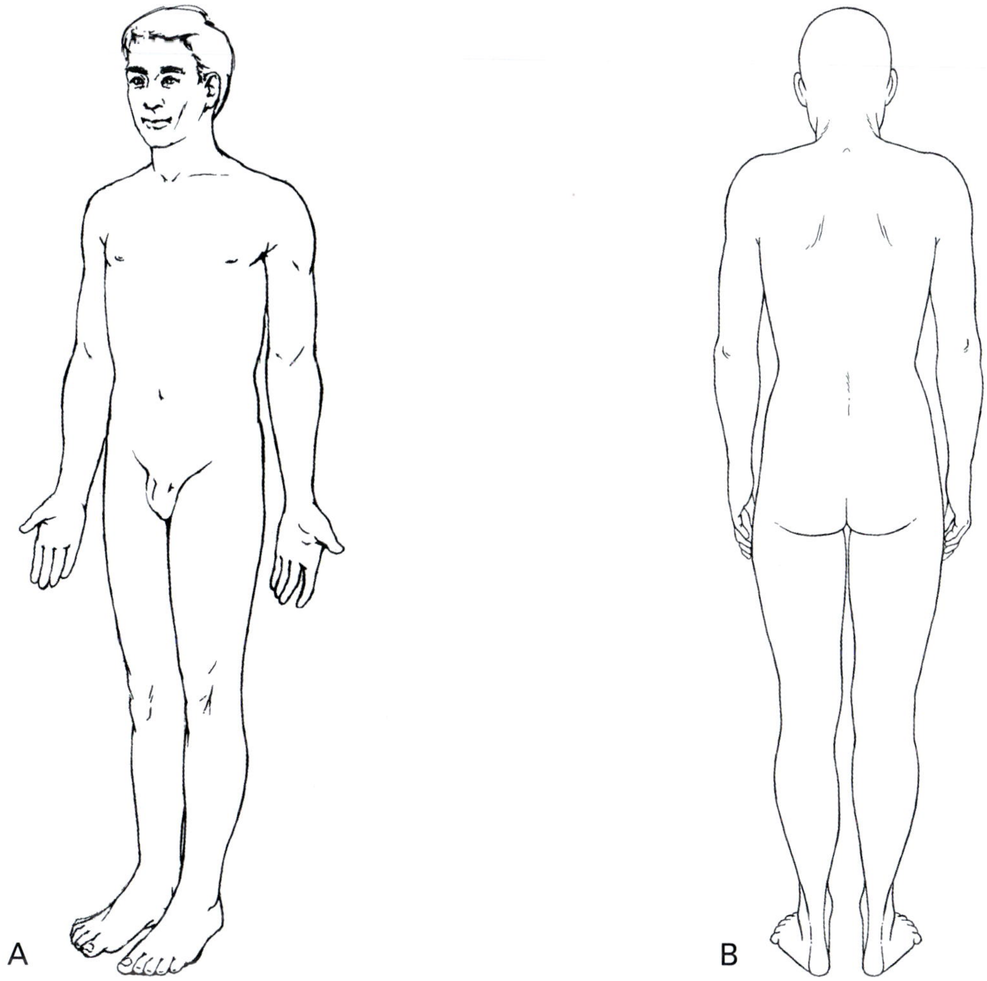

FIGURE 1-2: Anatomical position of reference for definitions and descriptions of body planes and axes: **A**, Front; **B**, Zero position (back).

Axes

Axes are lines, real or imaginary, about which movement takes place. Related to the planes of reference seen in the following text are three basic types of axes at right angles to each other (16):

1. A *sagittal (anterior–posterior) axis* lies in the sagittal plane and extends horizontally from front to back. The movements of abduction and adduction take place about this axis in a coronal plane.
2. A *coronal (medial–lateral) axis* lies in the coronal plane and extends horizontally from side to side. The movements of flexion and extension take place about this axis in a sagittal plane.
3. A *longitudinal (vertical or superior–inferior) axis* extends in a cranial-caudal direction. The movements of medial and lateral rotation take place about this axis in a transverse plane.

Exceptions to these general definitions occur with respect to movements of the scapula, clavicle, and thumb.

Planes

The three basic planes of reference are derived from the dimensions in space and are at right angles to each other (Figure 1-3) (16):

1. A *sagittal plane* is vertical and extends from front to back, deriving its name from the direction of the sagittal suture of the skull. It may also be called an antero-posterior plane. The median sagittal plane, or *midsagittal*, divides the body into right and left portions.
2. A *coronal plane* is vertical and extends from side to side, deriving its name from the direction of the coronal suture of the skull. It is also called the frontal or lateral plane, and it divides the body into an anterior and a posterior portion.
3. A *transverse plane* is horizontal and divides the body into upper (cranial) and lower (caudal) portions.

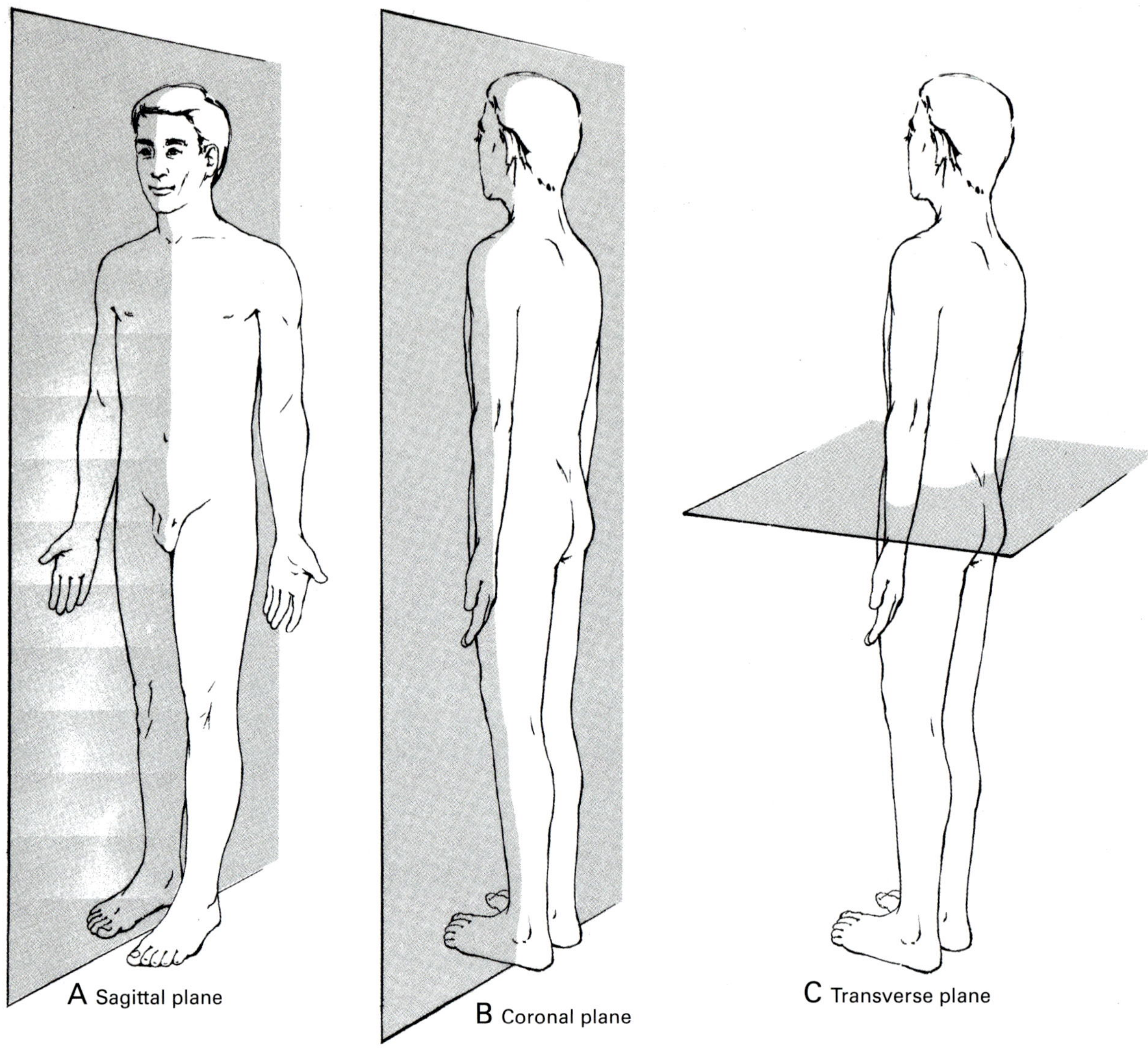

FIGURE 1-3: Three basic planes of reference: **A**, Sagittal; **B**, Coronal; **C**, Transverse.

Movements

Movements in the Coronal Plane

Movements of abduction, adduction, and lateral flexion occur within the coronal plane about a sagittal axis. The sagittal axis is positioned at a right angle to the coronal plane.

Abduction and Adduction. Abduction is movement away from and adduction is movement toward the midsagittal plane of the body for all joints of the extremities except the thumb, fingers, and toes (Figures 1-4 and 1-5) (16). The movements of the thumb are referenced to the plane of the palm of the hand. Abduction and adduction of digits 2–5 are movements away from and toward the axial line that extends through the third digit. For the toes, the axial line extends through the second digit.

Lateral Flexion. Lateral flexion denotes abduction and adduction movement dedicated to the spine, resulting in positional changes of the head, neck, and trunk.

Movements in the Sagittal Plane

Movements of flexion and extension occur within the sagittal plane about a coronal axis. The coronal axis is positioned at a right angle to the sagittal plane.

Flexion and Extension. At an early stage, the limbs of the embryo are directed ventrally, the flexor surfaces medially, and the great toes and thumbs cranially. With further development, the limbs rotate 90° at their girdle articulation so that the thumbs turn laterally and the flexor surfaces of the upper extremities ventrally, while the great toes turn medially and the flexor surfaces of the lower extremities dorsally. As a result of this 90° rotation of the limbs in opposite directions, movement that approximates

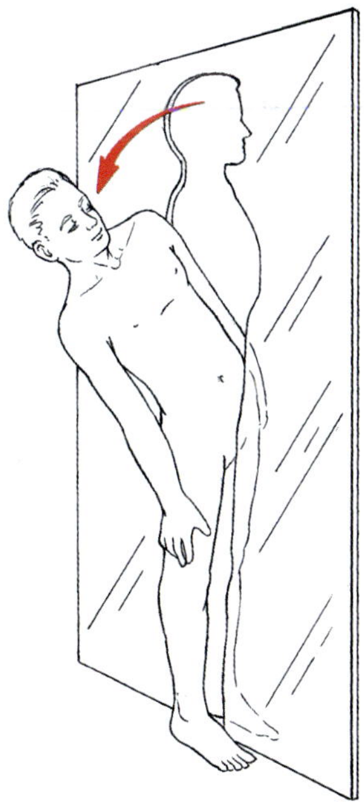

FIGURE 1-4: The body bending within the coronal plane.

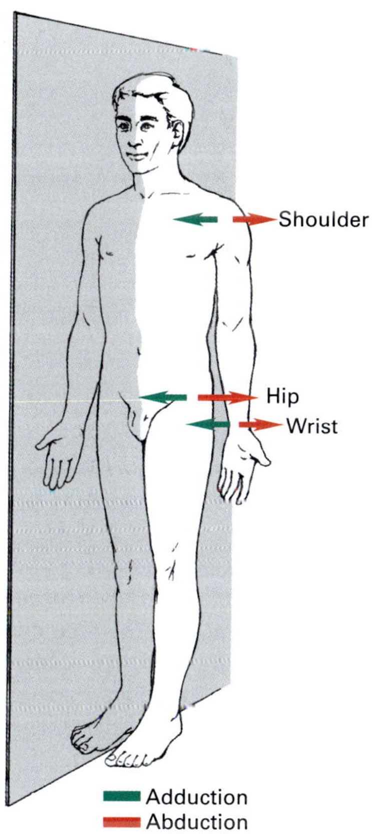

FIGURE 1-5: Points of abduction and adduction on the body within the sagittal plane.

the hand and the anterior surface of the forearm is termed flexion, because it is performed by flexor muscles. Movement that approximates the foot and anterior surface of the leg is termed extension, because it is performed by extensor muscles. See Figures 1-6 and 1-7.

Flexion is the movement of bending forward (i.e., in an anterior direction) for joints of the spine, upper extremity, and hip, and movement in the posterior direction for the knee, ankle, and joints of the toes.

Extension is movement in the direction opposite to flexion (i.e., in a posterior direction) for the joints of the spine, upper extremity, and hip, and movements in an anterior direction for the knee, ankle, and joints of the toes, returning the body to the anatomical position. The difference occurs because the developmental pattern of the lower extremities differs from that of the upper extremities.

Hyperextension is the term used to describe excessive movement in the direction of extension, as in hyperextension of the knees. It is also used in reference to the increased lumbar curvature as in a hyperlordosis with anterior pelvic tilt, or an increased cervical curvature as in a forward head position. In such instances, the range of motion through which the lumbar or cervical spine moves is not excessive, but the position of extension is greater than desirable from a postural standpoint.

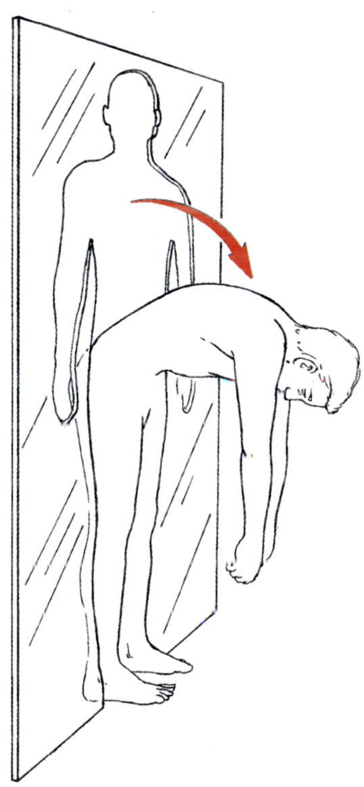

FIGURE 1-6: The body bending within the sagittal plane.

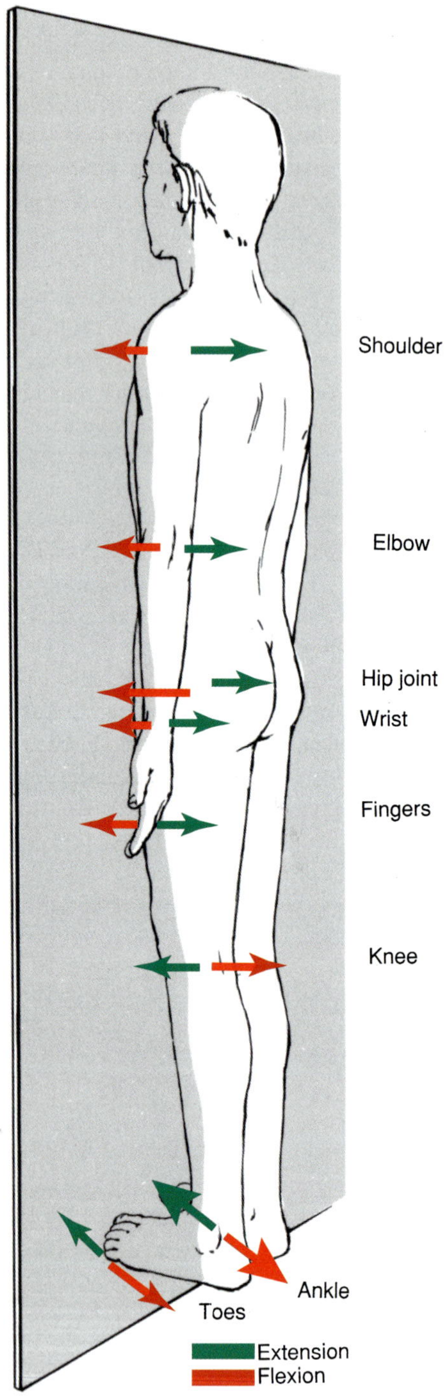

FIGURE 1-7: Points of flexion and extension on the body within the coronal plane.

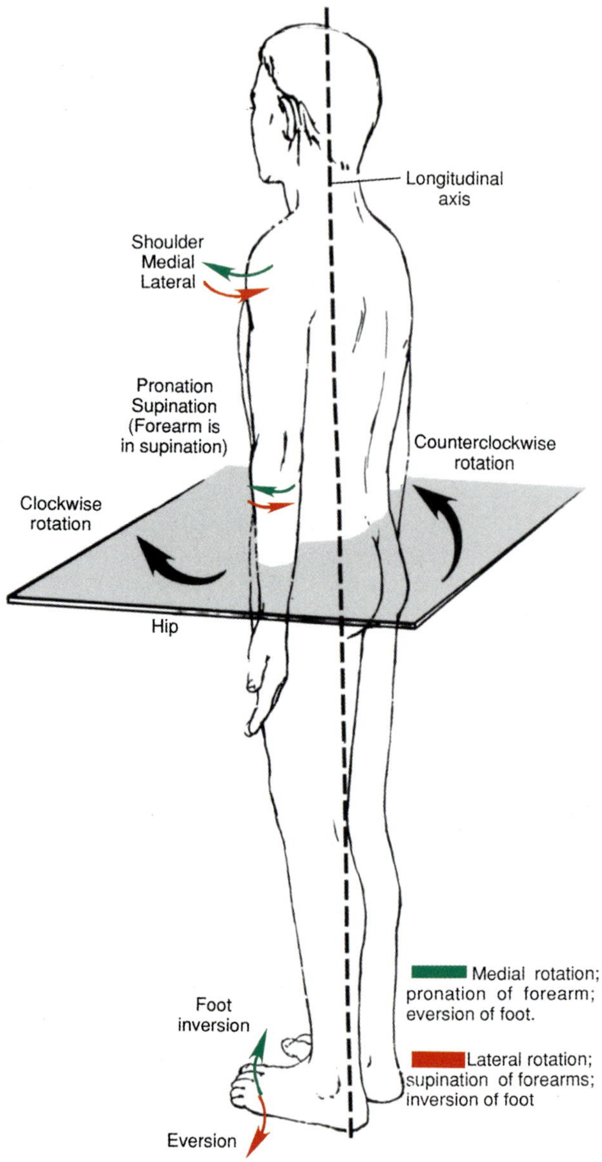

FIGURE 1-8: Points of rotation about a longitudinal axis within a transverse plane.

Movements in the Transverse Plane

Rotation. Rotation refers to movement around a longitudinal axis, in a transverse plane, for all areas of the body except the scapula and clavicle. A *longitudinal* axis is vertical, extending in a cranial-caudal direction.

In the extremities, rotation occurs about the anatomical axis except in the case of the femur, which rotates about a mechanical axis. In the extremities, the anterior surface of the extremity is used as a reference area. Rotation of the anterior surface toward the midsagittal plane of the body is *medial* rotation, and that away from the midsagittal plane is *lateral* rotation (Figure 1-8).

Because the head, neck, thorax, and pelvis rotate about longitudinal axes in the midsagittal area, rotation cannot be named in reference to the midsagittal plane. Rotation of the head and thorax is described as right or left rotation. Rotation of the pelvis may also be described as clockwise or counterclockwise. With the transverse plane as a reference and 12 o'clock at midpoint anteriorly, *clockwise* rotation occurs when the left side of the thorax or pelvis is more forward than the right, and *counterclockwise* rotation occurs when the right side is more forward.

Tilt. Tilt describes certain movements of the scapula and pelvis. The pelvis may tilt in an anterior or posterior direction about a coronal axis. Anterior tilt of the pelvis results in extension of the lumbar spine and posterior tilt results in flexion (flattening) of the lumbar spine.

The pelvis may also tilt laterally, moving about a sagittal axis. Lateral tilt of the pelvis is termed high on one side or low on the other and will be accompanied by lateral flexion of the lumbar spine to either the ipsilateral or contralateral side.

Because the pelvis moves at times as a unit, a tilt may be viewed as an anterior, posterior, or lateral tilt. An anterior or posterior tilt would occur in the sagittal plane and a lateral tilt would occur in the coronal plane. Rotation of the pelvis would occur in the transverse plane as seen in Figure 1-8.

With the scapula in a neutral position, there may be anterior tilt but not posterior tilt, except that return from anterior tilt may be referred to as such.

Combination Movements

Circumduction. Circumduction is a movement that successively combines flexion, abduction, extension, and adduction in which the segment being moved describes a cone. The proximal end of the extremity forms the apex of the cone, serving as a pivot, and the distal end circumscribes a circle. Such movements are possible only in ball-and-socket, condyloid, and saddle type joints.

MUSCULOSKELETAL SYSTEM

The musculoskeletal system is composed of striated **muscles**, various types of **connective tissues**, and the **skeleton**. This system provides the essential components for strength, flexibility, and stability in weight bearing.

The bones of the skeleton are joined together by **ligaments**, which are strong, fibrous bands (sheets) of connective tissue. They are flexible but not extensible. Some ligaments limit motion to such an extent that the joint is immovable; some allow freedom of movement. Ligaments are classified as **capsular**, **extracapsular**, and **intracapsular**. They contain nerve endings that are important in reflex mechanisms and in the perception of movement and position. Ligaments may differ from the standpoint of mechanical function. For example, a collateral ligament is an extracapsular type that remains taut throughout the range of joint motion, whereas a cruciate ligament (as in the knee joint) becomes slack during some movements and taut during others.

Skeletal muscle fibers are classified primarily into two types: type I (red slow twitch) and type II (white fast twitch). The two types of fibers are intermingled in most muscles. Usually, however, one type predominates, with the predominant type depending on the contractile properties of the muscle as a whole. Type I fibers seem to predominate in some postural muscles, such as the erector spinae and soleus. Type II fibers often predominate in limb muscles, where rapid, powerful forces are needed. Variability does occur, however, in these ratios in the population, especially as related to development and aging. Skeletal muscles constitute approximately 40% of body weight and are attached to the skeleton by aponeuroses, fasciae, or tendons.

Aponeuroses are sheets of dense connective tissue and are glistening white in color. They furnish the broad origins for the latissimus dorsi muscles. The external and internal oblique muscles are attached to the linea alba by means of aponeuroses. The palmaris longus inserts into and tenses the palmar aponeurosis.

Fascia is of two types: **superficial**, which lies beneath the skin and permits free movement of the skin, and **deep**, which envelopes, invests, and separates muscles. Some deep fasciae furnish attachments for muscles. For example, the iliotibial tract is a strong band of deep fasciae that provides attachments for the tensor fasciae latae into the tibia and for the gluteus maximus into the femur and tibia. The thoracolumbar fascia furnishes attachment for the transversus abdominis.

Tendons are white, fibrous bands that attach muscles to bones. They have great tensile strength but are practically inelastic and resistant to stretch. Tendons have few blood vessels but are supplied with sensory nerve fibers that terminate in organs of Golgi near the musculotendinous junction. In injuries that involve a severe stretch, the muscle most likely is affected, and sometimes the tendinous attachment to the bone is affected. For example, the peroneus brevis attachment at the base of the fifth metatarsal may be disrupted in an inversion injury of the foot. Tendons can also rupture. When the Achilles tendon ruptures, there is retraction of the gastrocnemius and soleus muscles with spasm and acute pain.

Joints: Definitions and Classification

A **joint** is defined as a union or junction between two or more bones and can be classified into three general morphologic types: fibrous joints, cartilaginous joints, and synovial joints (17, 18). Joints are supported by fibrous, cartilaginous, or synovial tissue and are typically named according to the bones that comprise the given articulation.

For some joints, the bones are held so close together that no appreciable motion occurs. They provide great stability. Some joints provide stability in one direction and freedom of motion in the opposite direction, and some provide freedom of motion in all directions.

Joints that provide little or no movement are those that hold the two sides of the body together. The sagittal suture of the skull is considered to be an immovable joint, held together by a strong fibrous membrane. The symphysis pubis is considered to be slightly movable and is held together by strong **fibrocartilaginous** membranes.

TABLE 1-1 Classification of Joints

Tissue	Joint		Movement	Example
Fibrous	Synarthrosis	Syndesmosis	Immovable	Tibiofibular (distal)
		Sutura	Immovable	Suture of skull
		Gomphosis	Immovable	Tooth in bony socket
Cartilaginous	Amphiarthrosis	Synchondrosis	Slightly movable	First sternocostal
		Symphysis	Slightly movable	Symphysis pubis
Synovial	Diarthrosis	Spheroid or ball-and-socket	All joint movements	Shoulder (2) and hip
		Ginglymus	Flexion and extension	Elbow
		Modified ginglymus	Flexion, extension, and slight rotation	Knee and ankle
		Ellipsoid or condyloid	All except rotation and opposition	Metacarpopha-langeal and metatar-sophalangeal
		Trochoid or pivot	Supination, pronation, and rotation	Atlantoaxial and radioulnar
		Reciprocal-reception or saddle	All except rotation	Calcaneocuboid and carpometacarpal
		Plane or gliding	Gliding	Head of fibula with lateral condyle of tibia
		Combined ginglymus and gliding	Flexion, extension, and gliding	Temporomandibular

Most joints fall into the category of freely movable joints held together by synovial membranes. The elbow and knee joints are essentially hinge joints. The structure of the joint surfaces and the strong lateral and medial ligaments limit sideways movements, and anterior (elbow) and posterior (knee) ligaments and muscles limit extension. Hence, there is stability and strength in the extended position. In contrast, the shoulder joints are movable in all directions and have less stability (Table 1-1).

Gross Structure of Muscle

The gross structure of muscle helps to determine muscle action and affects the way that a muscle responds to stretching. Muscle fibers are arranged in bundles called **fasciculi**. The arrangement of fasciculi and their attachments to tendons varies structurally. Two main divisions are found in gross structure: fusiform (or spindle) and pennate. A third arrangement, fan-shaped, is probably a modification of the other two but has a distinct clinical significance. See Figure 1-9 A–C.

Fusiform fibers are arranged essentially parallel to the line from origin to insertion and the fasciculi terminate at both ends of the muscle in flat tendons. **Pennate** fibers are inserted obliquely into the tendon or tendons that extend the length of the muscle on one side (i.e., unipennate) or through the belly of the muscle (i.e., bipennate).

Other structural examples include flat (external abdominal oblique, e.g.), circular (orbicularis oculi, e.g.), quadrate (quadratus femoris, e.g.), and multihead/bellied (biceps brachii).

In all probability, the long fusiform muscle is the most vulnerable to stretch. The joint motion is in the same direction as the length of the fiber, and each longitudinal component is dependent on every other one. The pennate muscles are probably the least vulnerable to stretch, both because the muscle fiber is oblique to the direction of joint motion and because the fibers and fasciculi are short and parallel and, thereby, are not dependent on other segments for continuity in action.

The fan-shaped muscle has advantages and disadvantages of the fusiform and the pennate muscles. It might be thought of as a group of muscles arranged side by side to form a fan-shaped unit. Each segment is independent in that it has its own origin with a common insertion. For example, in the fan-shaped pectoralis major, the clavicular part may be unaffected but the sternal part paralyzed in a spinal cord lesion.

According to *Gray's Anatomy,* the "arrangement of fasciculi is correlated with the power of the muscles. Those with comparatively few fasciculi, extending the length of the muscle, have a greater range of motion but not as much power. Penniform muscles, with a large number of fasciculi distributed along their tendons, have greater power but smaller range of motion" (17).

Types of Muscle Contraction

All types of contractions will be utilized throughout the procedures of manual muscle testing. Understanding the normal tonic phase of muscle and muscle groups will be useful during the posture screen and evaluation. The *tonic* phase of muscle refers to its resting state. During this phase it is stated that muscle rests in a slight state of contraction.

Phasic contractions of muscle include *isotonic* and *isometric* contractions. **Isotonic** muscle contractions involve a

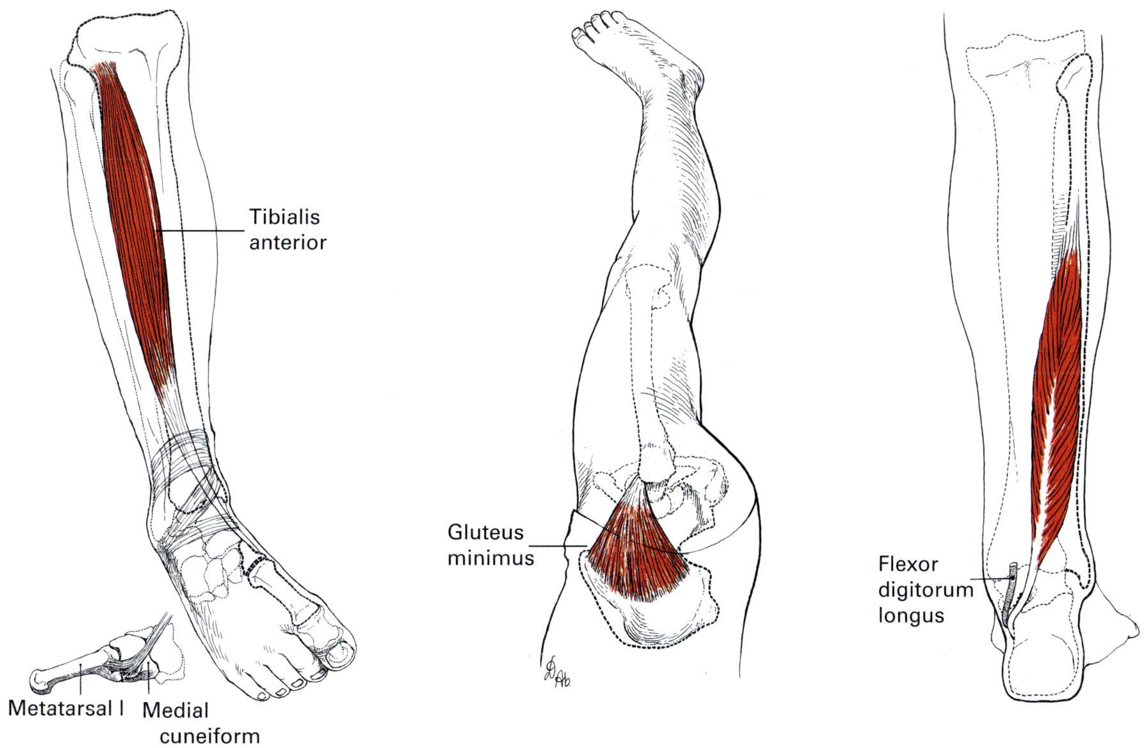

FIGURE 1-9: Gross structure of muscles: **A**, Fusiform; **B**, Fan-shaped; **C**, Pennate.

change in length of muscle that results in an action about a given joint, thus producing movement. An **isometric** contraction involves little to no change in the length of muscle that typically does not generate an action about a given joint, therefore does not produce movement. During functional tasks, muscle actions will produce movement by transitioning through these phases utilizing each type of contraction in an orchestrated fashion to complete a given task.

NERVOUS SYSTEM

Nervous system function is imperative for the control and activation of muscles allowing for the independent completion of task performance. The nervous system detects internal and external environmental changes that impact the body, then works in tandem with other systems to respond accordingly (18, 19).

The nervous system is composed of two primary divisions: the central nervous system (CNS) and the peripheral nervous system (PNS). Subdivisions of the PNS include the somatic, autonomic, and enteric systems. The former two will be discussed in brief detail below.

The somatic nervous system is considered to be a voluntary system that allows for conscious control of movement. Motor neurons innervate skeletal muscle and sensory neurons provide feedback from the environment through receptors in the skin, tendons, and joints via feedback and feedforward mechanisms.

The autonomic nervous motor system is an involuntary system that mitigates visceral function. Motor neurons innervate cardiac muscle, smooth muscle found in glands, blood vessels, and other organs of the body. The autonomic system is further divided into parasympathetic and sympathetic systems. The parasympathetic system is responsible for the maintenance of or return to a level of homeostasis while the sympathetic system activates in response to stress.

This text will focus on the transition of the CNS to the PNS and their integration with muscle function. Types of peripheral nerves include the spinal nerves and the cranial nerves.

Spinal Nerves

There are 31 pairs of **spinal nerves**. Spinal nerves are mixed nerves that originate from the spinal cord. Each of the 31 pairs of spinal nerves arises from the spinal cord by two spinal **nerve roots**. The **ventral root** is composed of motor neurons and the **dorsal root** is composed of sensory neurons. These nerve roots unite as they approach the intervertebral foramen to form the spinal nerve. A spinal **segment** is the component of the spinal cord that gives rise to each pair of spinal nerves. Each spinal nerve contains motor and sensory fibers numbered from a single spinal segment.

Shortly after the spinal nerve exits through the foramen, it divides into a **dorsal primary ramus** and a **ventral primary ramus.** The dorsal rami are directed posteriorly, and the sensory and motor fibers innervate the skin and extensor muscles of the neck and trunk. The ventral rami, except those in the thoracic region, contain the nerve fibers that form the plexuses.

Nerve Plexuses: Definitions

The term **plexus** comes from the Latin word plectere meaning braid. A **nerve plexus** results from the dividing, reuniting, and intertwining of nerves into a complex network. When describing the origins, components, and terminal branches of a plexus, the terms **nerves, roots,** and **cord** are used with dual meanings. There are spinal nerves and peripheral nerves, roots of the spinal nerves and roots of the plexus, and the spinal cord and cords of the plexus. To avoid confusion, appropriate modifying words are used in the descriptions that follow.

Plexus illustrations have been included with the appropriate chapters: cervical with neck, brachial with upper extremity, and both lumbar and sacral with lower extremity. Trunk muscles receive innervation directly from the thoracic nerves, plus a branch from the lumbar plexus.

Peripheral nerves emerge from the plexuses at various regions of the plexus with terminal branches extending from the medial, lateral, and posterior cords. As a result of the interchange of fibers within the plexus, peripheral nerves contain fibers from at least two and, in some instances, as many as five spinal segments.

The **spinal cord** lies within the vertebral column, extending from the first cervical vertebra to the level of the second lumbar vertebra.

Segmental Distributions

For anatomists and clinicians, the determination of spinal segment distribution to peripheral nerves and muscles has proven to be a difficult task. The pathway of the spinal nerves is obscured by the intertwining of the nerve fibers as they pass through the nerve plexuses. Since it is almost impossible to trace the course of an individual nerve fiber through the maze of its plexus, information regarding spinal segment distribution has been derived mainly from clinical observation. The use of this empirical method has resulted in a variety of findings regarding the segmental origins of these nerves and the muscles they innervate. An awareness of possible variations is important in the diagnosis and the location of a nerve lesion. To focus attention on the range of variations that exists, the Kendalls tabulated information from six well-known sources. (Please see the series of charts in Appendix D.)

OTHER BODY SYSTEMS

While the musculoskeletal and nervous systems are the primary focus of manual muscle testing and postural evaluations, an astute clinician recognizes that other systems play roles in normal muscle function. These systems include the circulatory, respiratory, integumentary, digestive, lymphatic, endocrine, and renal systems. Table 1-2 provides limited examples of how these systems can impact muscle function.

TABLE 1-2 Body Systems and Their Impact on Muscle Function

System	Components	Potential Impact on Muscle Function	Presentation
Circulatory	Heart, great vessels, arteries, veins, capillaries	Decreased nutrition, metabolite evacuation and O_2 supply contracted state decreases blood flow	Decreased strength, atrophy, pain
Respiratory	Lungs, airway, pulmonary vessels	Decrease in O_2 supply, limit of CO_2 evacuation	Decreased strength, atrophy, pain
Integumentary	Skin, hair	Scarring	Limit available range of motion
Digestive	Alimentary canal	Decreased nutrition, impaired glycogen storage in muscle	Decreased strength, fatigue
Renal	Kidneys, bladder, ureters, urethra	Edema, increased weight of extremity	Edema
Lymphatic	Lymph vessels, glands, spleen	Edema, increased weight of extremity	Edema
Endocrine	Pancreas, thyroid, adrenal glands	Change in blood sugar, decreased glycogen storage in muscle, hypo/hyperthyroid myopathies, proximal muscle weakness/ wasting in Cushing syndrome, etc.	Decreased muscle strength, atrophy, decreased coordination

SECTION II
MANUAL TESTING (PROCEDURE, OBJECTIVITY, GRADING)

MUSCLE STRENGTH TESTS

Kendall Classification for Muscles

Test for Strength

Class I. One-joint muscles that actively shorten (i.e., concentric contraction) through range to completion of joint motion and exhibit maximal strength at completion of range (i.e., short and strong).

Examples: Triceps, medial and lateral heads; deltoid; pectoralis major; three one-joint thumb muscles; gluteus maximus; iliopsoas; and soleus.

Class II. Two-joint and multi-joint muscles that act like one-joint muscles by actively shortening over both or all joints simultaneously and exhibiting maximal strength at completion of range (i.e., short and strong).

Examples: Sartorius, tibialis anterior and posterior, and peroneus longus, brevis, and tertius.

Class III. Two-joint muscles that shorten over one-joint and lengthen over the other to provide midrange of the overall muscle length for maximal contraction and strength (as represented by the length–tension curve).

Examples: Rectus femoris, hamstrings, and gastrocnemius.

Class IV. Two-joint or multi-joint muscles that physiologically act in one direction but are prevented from overshortening by the coordinated action of synergic muscles.

Example of Two-Joint Muscle: The biceps act to flex the shoulder joint and the elbow joint. If acting to flex both joints simultaneously, the muscle would become overshortened. To prevent this, the shoulder extensors, as synergists, extend the shoulder joint, thereby lengthening the biceps over the shoulder joint when the elbow is maximally flexed by the biceps.

Example of Multi-Joint Muscle: If acting in one direction by flexing the wrists and fingers simultaneously, the finger flexors and extensors would overshorten and become actively insufficient. Nature, however, prevents this from happening. In forceful flexion of fingers, such as when making a fist, the flexors shorten over the finger joints but are prevented from shortening over their entire length by the synergic action of wrist extensors that hold the wrist in moderate extension, thereby lengthening the flexors over wrist joint for them to forcefully shorten over the finger joints.

Passive Insufficiency

As defined by O'Connell and Gardner:

> Passive insufficiency of a muscle is indicated whenever a full range of motion of any joint or joints that the muscle crosses is limited by that muscle's length, rather than by the arrangement of ligaments or structures of the joint itself (20).

As defined by Kendall:

> Passive insufficiency. Shortness of a two-joint (or multi-joint) muscle; the length of the muscle is not sufficient to permit *normal elongation* over both joints simultaneously, e.g., short hamstrings.

> **NOTE:** *By both definitions, the term* ***passive insufficiency*** *refers to lack of muscle length. In contrast, the term* ***active insufficiency*** *refers to lack of muscle strength.*

Active Insufficiency

As defined by O'Connell and Gardner:

> If a muscle which crosses two or more joints produces simultaneous movement at all of the joints that it crosses, it soon reaches a length at which it can no longer generate a useful amount of force. Under these conditions, the muscle is said to be *actively insufficient.* An example of such insufficiency occurs when one tries to achieve full hip extension with maximal knee flexion. The two-joint hamstrings are incapable of shortening sufficiently to produce a complete range of motion of both joints simultaneously (20).

As defined by Kendall:

> Active insufficiency. The inability of a Class III or IV two-joint (or multi-joint) muscle to generate an effective force when placed in a fully shortened position. The same meaning is implied by the expression "the muscle has been put on a slack".

The two definitions above only apply to two-joint or multi-joint muscles. However, the statement that one-joint muscles exhibit their greatest strength at completion of range of motion has appeared in all five editions of Kendall's *Muscles: Testing and Function.* Knowing where the muscle exhibits its greatest strength in relation to the range of motion is of utmost importance for determining test position. After careful analysis, it is evident that there are four classifications.

Strength Testing Procedures

The order in which muscles are tested is largely a matter of choice, but it generally is arranged to avoid frequent and unnecessary changes of position for the subject (Box 1-1). Muscles that are closely related in position or action tend to appear in a testing order in sequence in order to distinguish test differences. *As a general rule, length testing precedes strength testing.* When the specific order of tests is important, it is so indicated in the text.

Terms Used in Description of Muscle Strength Tests

Descriptions of the muscle tests in Chapters 4 through 7 are presented under the headings of *Patient*, *Fixation*, *Test*, and *Pressure.* This chapter discusses each of these topics in detail to point out its particular significance in relation to accurate muscle testing.

Patient

In the description of each muscle test, this heading is followed by the position in which the patient is placed to accomplish the desired test. The position is important in relation to the test in two respects. First, insofar as is practical, the position of the body should permit function against gravity for all muscles in which gravity is a factor in grading. Second, the body should be placed in such a position that the segments not being tested will remain as stable as possible. (This point is discussed further under *Fixation.*)

In all muscle testing, the comfort of the patient and the intelligent handling of affected muscles are important factors. In some instances, the comfort of the patient or the condition of the affected muscles will necessitate some modification of the test position. For example, insisting on an antigravity position may result in absurd positioning of a patient. Side-lying, which offers the best test position for several muscles, may be uncomfortable and result in strain of other muscles.

Fixation

This heading refers to the stability of the body or body part, which is necessary to insure an accurate test of a muscle or muscle group. Stabilization (i.e., holding steady or holding down), support (i.e., holding up), and counterpressure (i.e., equal and opposite pressure) are all included under fixation. Fixation will be influenced by the firmness of the table, body weight of the patient, positioning and technique/skill of the clinician, and in some tests, the muscles that furnish fixation.

Adequate fixation depends to a great extent on the positioning of the patient and firmness of the examining table, which offers much of the necessary support. Testing and grading of strength will not be accurate if the table on which the patient lies has a thick, soft pad or soft mattress that "gives" as the examiner applies pressure.

Body weight may furnish the necessary fixation. Because the weight of the body is an important factor in

BOX 1-1

Basic Rules of Procedure That Apply to Muscle Strength Testing

1. Place the subject in a position that offers the best fixation of the body as a whole (usually supine, prone, or side-lying).
2. Stabilize the segment proximal to the tested segment or, as in the case of the hand, adjacent to the tested segment. Stabilization is necessary for specificity in testing.
3. Place the segment to be tested in precise antigravity test position, whenever appropriate, to help elicit the desired muscle action and aid in grading.
4. Use test movements in the horizontal plane when testing muscles that are too weak to function against gravity. Use test movements in antigravity positions for most trunk muscle tests in which body weight offers sufficient resistance.
5. Apply pressure directly opposite to the line of pull of the muscle or the muscle segment being tested. Like the antigravity position, the direction of pressure helps to elicit the desired muscle action.
6. Apply pressure gradually but not too slowly, allowing the subject to "get set and hold." Apply uniform pressure; avoid localized pressure that can cause discomfort.
7. Use a long lever whenever possible, unless contraindicated. The length of the lever is determined by the location of the pressure along the lever arm. Better discrimination of strength for purposes of grading is obtained through use of a long lever.
8. Use a short lever if the intervening muscles do not provide sufficient fixation for use of a long lever.

offering stability, the horizontal position, whether supine, prone, or side-lying, offers the best fixation for most tests. In the extremities, the body segment that is proximal to the tested segment must be stable.

The examiner may stabilize the proximal segments in tests of finger, wrist, toe, and foot muscles, but in other tests the body weight should help to stabilize the proximal segment. In some instances, the examiner may offer fixation in addition to the weight of the proximal segment. There may be a need to hold a segment firmly down on the table so that the pressure applied on the distal segment (plus the weight of that segment) does not displace the weight of the proximal segment. In rotation tests, it is necessary for the examiner to apply counterpressure to ensure exact test performance.

In some tests, muscles furnish fixation. The muscles that furnish fixation do not cross the same joint or joints as the muscle being tested. The muscles that stabilize the scapula during arm movements and the pelvis during leg movements are referred to as **fixation muscles.** They do not enter directly into the test movement, but they do stabilize the movable scapula to the trunk or the pelvis to the thorax and, thereby, make it possible for the tested muscle to have a firm origin from which to pull. In the same way, anterior abdominal muscles fix the thorax to the pelvis as anterior neck flexors act to lift the head forward in flexion from a supine position.

Muscles that have an antagonistic action give fixation by preventing excessive joint movement. This principle is illustrated by the fixation that the lumbricals and interossei provide in restricting hyperextension at the metacarpophalangeal joint during finger extension. In the presence of weak lumbricals and interossei, the pull of a strong extensor digitorum results in hyperextension of these joints and passive flexion of the interphalangeal joints. This hyperextension does not occur, however, and the fingers can be extended normally if the examiner prevents hyperextension of the metacarpophalangeal joints by fixation equivalent to that of the lumbricals and interossei.

When the fixation muscles are either too weak or too strong, the examiner can simulate the normal stabilization by assisting or restricting movement of the segment in question. The examiner must be able to differentiate between the normal action of these muscles in fixation and the abnormal actions that occur when compensation or muscle imbalance is present.

Test

In muscle testing, weakness must be distinguished from restriction of range of motion. Frequently, a muscle cannot complete the normal range of joint motion. It may be that the muscle is too weak to complete the movement, or it may be that the range of motion is restricted because of shortness of the muscles, capsule, or ligamentous structures. The examiner should passively carry the segment through the range of motion to determine whether any restriction exists. If no restriction is present, then failure by the subject to hold the test position may be interpreted as weakness unless joint or tendon laxity is present.

When testing one-joint muscles in which the ability to hold the segment at completion of range of motion is expected, the examiner must distinguish between muscle weakness and tendon insufficiency. For example, the quadriceps may be strong but unable to fully extend the knee because the patellar tendon or quadriceps tendon has been stretched.

Muscle examinations should consider superimposed factors like relaxed, unstable joints. The degree of actual muscle weakness is difficult to judge in such cases. From the standpoint of function, the muscle is weak and should be so graded. When the muscle exhibits a strong contraction, however, it is important to recognize this as the potential for improvement. In a muscle that fails to function because of joint instability rather than because of weakness of the muscle itself, treatment should be directed at correcting the joint problem and relieving strain on the muscle. For example, it is not uncommon for the deltoid muscle to show a "fullness" of contraction throughout the muscle belly yet fail to lift the weight of the arm. Such a muscle should be protected from strain by application of an adequate support for the express purpose of allowing the joint structures to shorten to their normal position. Failure to distinguish between real and apparent muscle weakness resulting from joint instability may deprive a patient of adequate follow-up treatment.

Test Position. Test position is the position in which the segment is placed by the examiner and held (if possible) by the patient. It is the position used for the purpose of evaluating strength for most muscles because the segment and relevant musculature is aligned such that accurate testing of that muscle's strength be can be assessed.

The **optimal test position** is at the completion of range for one-joint muscles and for two- or multi-joint muscles that act like one-joint muscles. The optimal test position for other two- or multi-joint muscles is at mid-range of overall length, in accordance with the length-tension principle.

Use of the test position also enables the examiner to detect compensatory movements. When muscle weakness exists, other muscles immediately act in an attempt to hold a position resembling the test position. The visible shift from the test position indicates a compensatory movement.

Placing the segment in the test position expedites grading the muscle strength. As the effort is made to hold the test position, the ability or inability to hold the position against gravity is at once established. If it fails to hold, the examiner tests for strength below the fair grade. If the position is held, the examiner then applies pressure to grade above fair (Table 1-3).

TABLE 1-3 Key to Muscle Grading

	Function of Muscle	Muscle Grades and Symbols				
No Movement	No contraction felt or seen in the muscle	Zero	0	0	0	0
	Tendon becomes prominent or feeble contraction felt in muscle with no visible movement	Trace	T	1	T	
Supported in the Horizontal Plane*	Movement through partial range of motion	Poor−	P−	2−	1	+
	Movement through complete range of motion for the muscle being tested	Poor	P	2	2	
	Holds against slight pressure in test position**	Poor+	P+	2+	3	
Tests in the Antigravity Position	Moves through partial range of motion against gravity	Poor+	P+	2+	3	
	Gradual release from test position occurs	Fair−	F−	3−	4	
	Holds test position (no added pressure)	Fair	F	3	5	++
	Holds test position against slight pressure	Fair+	F+	3+	6	
	Holds test position against slight to moderate pressure	Good−	G−	4−	7	
	Holds test position against moderate pressure	Good	G	4	8	+++
	Holds test position against moderate to strong pressure	Good+	G+	4+	9	
	Holds test position against strong pressure	Normal	N	5	10	++++

*Support of the part being tested should ideally be provided by a firm, smooth surface that minimizes resistance to movement in the horizontal plane, such as a powder board.

**Testing for a Poor+ grade in the horizontal plane requires that the muscle being tested (1) be able to move the part through the muscle's range of motion without resistance (Poor grade), then (2) be able to hold against slight pressure in the test position where it exhibits greatest strength (e.g., Class I and II muscles should be tested at completion of range, while Class III and IV muscles should be tested at midrange of overall length of muscle.)

According to the Key, the highest test movement grade in the antigravity position is a 3, or Poor+. Test movements for lateral trunk flexors, upper and lower abdominal musclels, and back extensors are exceptions. See individual tests for grading of these muscles.

Testing of the muscles of the fingers and toes does not depend on gravity. See Chapter 6.

Test Movement. Test movement is a movement of a segment in a specified direction and through a specific arc of motion. For strength tests of extremity muscles that are too weak to act against gravity (i.e., muscles that grade in the range of poor), tests are done in the horizontal plane. Test movement is also used when testing the trunk lateral flexors, upper abdominal flexors, back extensors, quadratus lumborum, serratus anterior (in standing), and gastrocnemius.

Test movement may be used for certain muscles, such as those that cross hinge joints, but it is not practical when a test requires a combination of two or more joint positions or movements. It is difficult for a patient to assume the exact position through verbal instruction or imitating a movement demonstrated by the examiner. For accurate testing, the examiner should place the segment in precisely the desired test position.

Pressure

The term **pressure** is used throughout this text to refer to the external force that is applied by the examiner to determine the strength of the muscle holding in the test *position* (i.e., for grades of F+ or better).

The term **resistance** refers to the external force that opposes the test *movement.* The resistance may be the force of gravity or a force that is supplied by the examiner. Resistance may vary according to body weight (i.e., back extensor test), arm position (i.e., upper abdominal test), or leg positions (i.e., lower abdominal test). Occasionally, the examiner may offer resistance. An example of this is the traction the examiner provides in the quadratus lumborum test.

The placement, direction, and amount of pressure/resistance are important factors when testing for strength above the grade of fair. In the descriptions of muscle tests, pressure is specified as against or in the direction of. *Against* refers to the position of the examiner's hand in relation to the patient; *in the direction of* describes the direction of the force that is applied directly opposite to the line of pull of the muscle or its tendon.

In some of the illustrations of muscle tests, the examiner's hand has been held extended for the purpose of indicating, photographically, that the direction of pressure is perpendicular to the palmar surface of the hand. Pressure should be applied only in the direction indicated. (It is not necessary that the extended hand position be imitated during routine muscle testing.) An extended hand is not appropriate when applying pressure in a test that includes a rotation component.

Just as the direction of the pressure is an important component of accurate test performance, the *amount* of pressure is the determining factor in grading strength above fair (see *Grading*).

The *place* at which the pressure is applied depends on muscle insertions, strength of intervening muscles, and leverage. As a general rule, pressure is applied near the distal end of the segment on which the muscle is inserted. For example, pressure is applied near the distal end of the forearm during the biceps test. Exceptions to this rule occur when pressure on the bone of insertion does not provide adequate leverage to obtain discrimination for grading or when a patient cannot tolerate application of pressure at that site for some reason.

Both the length of the lever and the amount of pressure are closely related with respect to grading above fair. Using a long lever gives the examiner a mechanical advantage and allows more sensitive grading of muscle strength. Test results might be more indicative of the lack of strength of the examiner than of the subject if the examiner did not have the advantage of leverage.

When testing strong muscles like hip abductors, it is necessary to use a long lever (i.e., placing pressure just proximal to the ankle). When testing hip adductors, however, it is necessary to use a shorter lever, with pressure just above the knee joint, to avoid strain on the anteromedial area of that joint.

Pressure must be applied *gradually* to determine the degree of strength above fair in muscles. The patient must be allowed to *get set and hold* the test position against the examiner's pressure. The examiner cannot gauge the degree of strength unless pressure is applied gradually, because slight pressure that is applied suddenly can break the pull of a strong muscle. Grading strength involves a subjective evaluation based on the amount of pressure applied. Differences in strength are so apparent, however, that an observer who understands grading can estimate the strength with a high degree of accuracy while watching the examiner apply pressure.

Compensatory Presentations

Compensation results from one or more muscles attempting to compensate for the lack of strength in another muscle or group of muscles. Compensation is a good indication that the tested muscle is weak, that adequate fixation has not been applied, or that the subject has not been given adequate instruction concerning how to perform the test. Muscles that normally act together in movements may act in compensation. These include fixation muscles, agonists, and antagonists.

Compensation by fixation muscles occurs specifically in relation to movements of the shoulder joint and the hip joint. Muscles that move the scapula may produce a secondary movement of the arm; muscles that move the pelvis may produce a secondary movement of the thigh. These compensatory movements appear similar to—but are not—movements of the shoulder or hip joint.

The close relationship of muscles determines their action in compensation, assistance, and stabilization during tests of individual muscles. The grouping of muscles according to joint action, as seen in the charts in Chapters 6 and 7, has been done to aid the examiner in understanding the allied action of muscles.

True abduction of the hip joint is accomplished by hip abductors with normal fixation by the lateral trunk muscles. When the hip abductors are weak, apparent abduction may occur by the compensatory action of lateral trunk muscles. The pelvis is hiked up laterally, the leg is raised from the table, but no true hip joint abduction occurs.

Antagonists may produce movements similar to test movements. If finger flexors are weak, action of the wrist extensors may produce passive finger flexion by the tension placed on flexor tendons.

Compensation by other **agonists** results in either a movement of the segment in the direction of the stronger agonist or a shift of the body in a way that favors the pull of that agonist. For example, during the gluteus medius test in side-lying, the thigh will tend to flex if the tensor fasciae latae is attempting to compensate for a weak gluteus medius, or the trunk may rotate back so that the tensor fasciae latae can hold a position that appears to be the desired test position.

For accurate muscle examinations, no compensations should be permitted. The position or movement described as the test should be done without shifting the body or turning the segment. Such secondary movements allow other muscles to substitute for the weak or paralyzed muscle.

An experienced examiner who is aware of the ease with which normal muscles perform the tests will readily detect compensations. When test position is employed instead of test movement, even an inexperienced examiner can detect the sudden shift of the body or the segment that results from an effort to compensate for the muscle weakness.

Presentation of Weakness, Shortness, and Contracture

Included with the descriptions of the muscles in this text is a discussion of the loss of movement or the position of deformity that results from muscle weakness or muscle shortness.

Weakness is used as an overall term that covers a range of strength from zero to fair in non-weight-bearing muscles but also includes fair+ in weight-bearing muscles. Weakness will result in loss of movement if the muscle cannot contract sufficiently to move the part through partial or complete range of motion.

A contracture or shortness will result in loss of motion if the muscle cannot be elongated through its full range of motion. **Contracture** refers to a degree of shortness that results in a marked loss of range of motion. **Shortness** refers to a degree of shortness that results in slight to moderate loss of range of motion.

A fixed deformity usually does not exist as a result of weakness unless contractures develop in the stronger opponents. In the wrist, for example, a fixed deformity will not develop as a result of wrist extensor weakness unless the opposing flexors maintain the position of wrist flexion.

A state of **muscle imbalance** exists when a muscle is weak and its antagonist is strong. The stronger of the two opponents tends to shorten, and the weaker of the two tends to elongate. Either weakness or shortness can cause faulty alignment. Weakness permits a position of deformity, but shortness creates a position of deformity.

In some regions of the body, positions of deformity may develop as a result of weakness even though the opposing muscles do not become contracted. A hyperkyphotic position of the upper back may result from weakness of the upper back muscles regardless of whether the anterior trunk muscles become contracted. A position of pronation of the foot may exist if the inverters are weak because the body weight in standing will distort the bony alignment. If opposing fibular (peroneal) muscles become contracted, a fixed deformity will result.

The word **tight** has two meanings. It may be used interchangeably with the term **short**, or it may be used to mean **taut**, in which case it may be applied to either a short or a stretched muscle. On palpation, hamstrings that are short and drawn taut will feel tight. Hamstrings that are stretched and drawn taut will also feel tight. From the standpoint of prescribing treatment, it is very important to recognize the difference between stretched muscles and shortened muscles. In addition, some muscles are short and remain in what appears to be a state of semi-contraction. On palpation, they feel firm or even rigid without being drawn taut. For example, posterior neck and upper trapezius muscles often are tight in people with bad posture of the upper back, head, and shoulders.

RANGE-OF-JOINT MOTION AND MUSCLE LENGTH TESTING

The phrases range of joint motion (range of motion (ROM)) and range of muscle length (extensibility) have specific meanings. **Range of joint motion** refers to the number of degrees of motion that are present in a joint. Descriptions of joints and the joint measurement charts include references to normal ranges of joint motion. **Range of muscle length**, also expressed in terms of degrees of joint motion, refers to the length of the muscle.

For muscles that pass over one joint only, the range of joint motion and range of muscle length will measure the same. Both may be normal, limited, or excessive. In some instances, when measuring range of joint motion, it is necessary to allow the muscle to be slack over one joint to determine the full range of joint motion in the other. For example, when measuring the range of knee joint flexion, the hip is flexed to allow the rectus femoris to be slack over the hip joint and permit full range of joint motion at the knee. When measuring range of hip joint flexion, the knee is flexed to allow the hamstrings to be slack over the knee joint and permit full range of joint motion at the hip.

Measuring Joint Motion and Muscle Length

It is easier and more accurate to use a measuring device that permits the stationary arm of the caliper to rest on the table and the examiner to place the movable arm in line with or parallel to the axis of the humerus or femur, as the case may be. The fulcrum will be shifted to permit this change, but the angle will remain the same—as if the stationary arm were held parallel to the table along the trunk in line with the shoulder joint or hip joint.

Correlation Between Joint Range and Muscle Length

An interesting correlation exists between the total range of joint motion and the range of muscle length chosen as a standard for hamstring and hip flexor length tests. In each case, the muscle length adopted as a standard is approximately 81% of the total range of joint motion of the two joints over which the muscles pass. The following are the joint ranges considered to be normal:

Hip—30° extension, 125° flexion, for a total of 155°
Knee—0° (10°) extension, 135° flexion, for a total of 135° (145)°
Total of both joints—290°

Hip flexor length test used as a standard: Supine, with the low back and sacrum flat on the table, hip joint extended, and hip flexors elongated 155° over the hip joint. With the knee flexed over the end of the table at an angle of 80°, the two-joint hip flexors are elongated 80° over the knee joint, for a total of 215°. Thus, 215° divided by 290° is 78.18%, and range of muscle length is 81% of total joint range.

Hamstring length test used as a standard: Supine, with the low back and sacrum flat on the table and straight-leg raising to an 80° angle with table. Hamstrings are elongated 155° over the knee by full extension and 80° over the hip joint by the straight-leg raising, for a total of 235°. Thus, 235° divided by 290° equals 81%, and range of muscle length is 81% of total joint range.

Muscle Length Tests

Muscle length tests are performed to determine whether the range of muscle length is normal, limited, or excessive. Muscles that are excessive in length are usually weak and can allow adaptive shortening of opposing muscles; muscles that are too short are usually strong and maintain opposing muscles in a lengthened position.

Muscle length testing consists of movements that increase the distance between origin and insertion, thereby elongating muscles in directions opposite to those of the muscle actions.

Accurate muscle length testing usually requires that the bone of origin be in a fixed position while the bone of insertion moves in the direction of lengthening the muscle. Length tests use passive or active-assisted movements to determine the extent to which a muscle can be elongated.

Grading Muscle Strength

Grades represent an examiner's assessment of the strength or weakness of a muscle or a muscle group. In manual muscle testing, grading is based on a system in which the ability to hold the tested segment in a given position against gravity establishes a grade referred to as fair or the numerical equivalent (depending on the grading symbols being used). The grade of fair is the most objective grade because the pull of gravity is a constant factor.

For grades above fair, pressure is applied in addition to the resistance offered by gravity. A **break test** is a muscle strength test to determine the maximal effort exerted by a subject who is performing an isometric contraction as the examiner applies a gradual buildup of pressure to the point that the effort by the subject is overcome. It is used in determining grades of fair+ through good+.

No effort is made to break the subject's hold if the examiner has determined that the strength is normal. To continue exerting force to make the muscle yield by performing a break test is unnecessary and may even be injurious.

The symbols used in grading vary and include the use of words, letters, numbers, or other signs. To avoid listing the equivalents each time this text refers to a grade, the symbols are used in the descriptions of grades below.

Gravity is a form of resistance that is basic to manual muscle testing. It is used in tests of the trunk, neck, and extremity muscles. It is a factor, however, only in approximately 60% of the extremity muscles. It is not required in tests of finger and toe muscles because the mass of the segment is so small in comparison with the strength of the muscle that the effect of gravity on the segment is negligible. Another example is supination and pronation of the forearm. These are movements of rotation in which the effect of gravity is also not a significant factor.

Testing muscles that are very weak involves movements in the horizontal plane on a supporting surface where the resistance by gravity is decreased. To avoid use of phrases such as gravity-lessened, gravity-decreased, or gravity-minimized, the text and the *Key to Muscle Grading* will refer to movements in the horizontal plane.

Detailed grading of muscle strength is more important in relation to prognosis than to diagnosis. The extent of involvement may be determined by such simple grading as zero, weak, and normal. On the other hand, more precise grading helps to establish the rate and degree of return of muscle strength and is also useful in determining a prognosis. A muscle might appear to be weak for months, although the record shows that it has progressed from poor to fair during this same period.

Accuracy in grading depends on many factors: the stable position of the patient, the fixation of the segment proximal to the segment being tested, the precision of the test position, and the direction and amount of pressure. The amount of pressure varies with the age and the size of the patient, the segment being tested, and the leverage. If one extremity is unaffected, the examiner may use the strength in the unaffected extremity as an index for the patient's normal strength when testing the affected extremity.

An examiner must build a basis for comparison of test results through experience in muscle testing. Such experience is necessary when testing both paralytic and normal individuals. For many, however, experience in muscle testing is insufficient or has been limited to the examination of patients with disease or injury. As a result, these examiners' idea of normal strength tends to be a measure of what appears to be good functional recovery following weakness or injury.

To gain experience, according to the Kendalls, an examiner must make an effort to test individuals of all ages and body types as well as those with good and faulty postures. If it is not possible to examine a large number of normal individuals, an effort should be made to examine the trunk and unaffected extremities in cases involving only one or two extremities.

Testing and grading procedures are modified during examination of infants and children to the age of 5 or 6 years. The ability to determine a child's muscle strength up to the grade of fair is usually not difficult, but grading strength above fair depends on the understanding and cooperation of the child in holding against resistance or pressure. Young children seldom cooperate in strong test movements. Very often, tests must be recorded as "apparently normal," which indicates that although the strength may, in fact, be normal, one cannot be sure.

Grading Symbols

Robert W. Lovett, M.D., introduced a method of testing and grading muscle strength using gravity as resistance (21). A description of the Lovett system was published in 1932 and listed the following definitions:

Gone—no contraction felt.

Trace—muscle can be felt to tighten but cannot produce movement.

Poor—produces movement with gravity eliminated but cannot function against gravity.

Fair—can raise the segment against gravity.

Good—can raise the segment against outside resistance as well as against gravity.

Normal—can overcome a greater amount of resistance than a good muscle.

The symbols used may vary, but the movement and weight factors set forth by Lovett form the basis of most current muscle testing. The Kendalls introduced the use of numbers for computing the amount of change in muscle strength when doing research with patients recovering from poliomyelitis. They had used the word and letter symbols previously and, for the most part, it was possible to translate grades from one scale to the other.

The Kendalls believed it was in the best interest of those who engage in manual muscle testing that an effort be made to standardize (as much as possible) the descriptions of the tests and the symbols used. Numerals are commonly used and such use is needed for research that involves muscle test grades.

Table 1-3 is basically the same as the Lovett system, but with added definitions for the minus and plus grades. The poor+ grade provides for movement in the horizontal plane and for partial arc against gravity.

In this text, the normal minus (N–) grade has been eliminated, and the scale has been changed from 0 to 10. Leaving zero as 0 and trace as T, the word and letter symbols translate directly as indicated in Table 1-3. No movement is involved with the 0 and T grades, and the numerals 1–10 refer to test movement and test position grades.

Grades above Fair

Standardization of muscle testing techniques related to grading strength above fair requires a specific place in the arc of motion where the segment is held by the subject as manual pressure is applied.

Muscle strength is not constant throughout the range of motion, and in manual muscle testing, it is not practical to try to grade the strength at various points in the arc of motion (see *Grading* for the place in the arc used as the position for grading).

If the test position is used, the segment is placed in the specific position by the examiner and then pressure is applied. For there to be standardization of testing techniques and grading when test movement is used, the movement must proceed to the same place in the arc of motion as that established as the test position. For this reason, the movement factor is omitted in Table 1-3 when defining grades above fair.

Normal Grade

The grade of **normal** means that the muscle can hold the test position against gravity and strong pressure. This grade is not intended to indicate the maximum strength of the subject, but rather the maximum pressure that the examiner applies to determine what might be termed as full strength of the muscle. In terms of judgment, it might be described as strength that is adequate for ordinary functional activities. To become competent in judging this full strength, an examiner should test normal individuals of various ages, sizes, and genders.

It should be noted that the term *normal* has a variety of meanings. It may mean average, typical, natural, or standard. If one adheres to the definition used when muscle grading, a grade of poor will be recorded for a small child who cannot lift the head in flexion from a supine position. Knowing that it is natural for small children to exhibit weakness of the anterior neck muscles, an examiner might say this child's neck is normal, using normal in the sense that it is natural. On administering a leg-lowering test for abdominal strength in a large group of adolescent children and finding that a grade of fair+ or good– is the average strength for the group, one might say that this grade of strength is normal for this age. Thus, we have three different uses of *normal* applied rather freely in muscle testing: standard, natural, and average.

Because normal is defined as a standard when used in the scale of grading, grades of strength should relate to that standard, and appropriate terms other than normal should be used in the interpretation of results. One of the advantages of using numerical grades is that it leaves the term *normal* free for use in the interpretation of those grades. In the following discussion, this term will be employed in this manner.

Most grades are based on adult standards, so it is necessary to acknowledge what is normal for children of a given age. This is particularly true regarding the strength of the anterior neck and anterior abdominal muscles. The size of the head and trunk in relation to the lower extremities as well as the long span and normal protrusion of the abdominal wall affect the relative strength of these muscles. Anterior neck muscles may grade about poor+ in a 3-year-old child, about fair in a 5-year-old child, and gradually increase up to the standard of performance for adults by 10 or 12 years of age. Many adults will exhibit no more than fair+ strength. This need not be interpreted as neurogenic, however, because it usually is associated with faulty posture of the head and upper back.

The prime example of a standard that is an infant rather than an adult accomplishment is that of toe flexor strength. In general, children have more strength in their toe flexors than many adults do. It is not uncommon to find that women who have worn high heels and rather narrow-toed shoes have weakness of toe flexors in which the grade is no more than fair–. With the standard being the ability to flex the toes and hold against strong resistance or pressure, the adult must be graded against that standard; however, this weakness of the toe flexors should not be interpreted as being normal for age. One becomes so accustomed to toe flexor weakness among adults that a degree of weakness might be assumed to be normal in the sense that "normal" is "average." Marked weakness of the toe flexors is almost invariably associated with some degree of disability of the foot. However, the term *normal* should not apply to such weakness unless one is ready to accept the disability itself as being normal.

This toe flexor weakness represents a loss of strength from childhood to adulthood, and it should be regarded as an unnatural, acquired weakness. This type of weakness may be present in other muscles as a result of stretch and strain associated with occupational or recreational activities or faulty posture. Acquired weakness usually does not drop below the grade of fair, but fair and fair+ grades of strength might be interpreted as neurogenic if one were not aware that such degrees of weakness can result from stretch and strain of the muscles.

Good Grade

The grade of **good** means that the muscle can hold the test position against moderate pressure.

Fair Grade

The grade of **fair** indicates that a muscle can hold the segment in test position against the resistance of gravity but cannot hold if even slight pressure is added. In tests such as those for the triceps and quadriceps, the examiner should avoid a locked (closed-packed) position of the joint that could give undue advantage to a muscle that was slightly less than fair in strength.

When considering a fair grade, the question arises of whether the strength to hold the test position is equivalent to the strength required to move through range of motion to the test position. With some exceptions, the general rule is that the test movement can be performed if the test position can be held.

In some muscle tests, the bone on which the muscle is inserted moves from a position of suspension in the vertical plane toward the horizontal plane. The quadriceps, deltoid, and hip rotators tested in the sitting position and the triceps and shoulder rotators tested in the prone position compose this group. The leverage exerted by the weight of the segment increases as the segment moves toward completion of the arc, and the muscle strength required to hold the test position against gravity usually is sufficient to perform the test movement against gravity.

In a few tests, the bone on which the muscle is inserted moves from a horizontal position toward a vertical position, and less strength is required to hold the test position than is needed to perform the test movement. This occurs during tests of hamstrings when tested by knee flexion in the prone position and tests of the elbow flexors when examined in the supine position.

Poor Grade

The ability to move through a partial arc of motion in the horizontal plane is graded as **poor–**. The grade of **poor** means that the muscle is capable of completing the range of motion in the horizontal plane. The grade of **poor+** denotes the ability to move in the horizontal plane to completion of the range of motion against resistance or to hold the completed position against pressure. It also means that the muscle is capable of moving through a partial arc of motion in the antigravity position.

The ranges of strength within the grade of poor are significant enough to deserve these subclassifications for purposes of more definitive grading. The ability to perform the full range of motion in the horizontal plane is not close to the ability to perform the test against gravity for most muscles. Adding pressure or resistance to the element of movement in the horizontal plane provides the added force that approaches that of gravity in the antigravity position.

Testing for the various grades of poor is justified and meaningful when used appropriately. In the rehabilitation of persons with severe neuromuscular and musculoskeletal involvement, the minute but visible changes that show improvement are very important. Maintaining a record of these significant changes, however slight, is important to the morale and continuing motivation of the patient and is necessary in determining his or her progress and the justification for continued care. These small changes can be as meaningful to a patient as the gains achieved by a recovering athlete at the later stages of rehabilitation.

The overall grade of poor can be assigned without unnecessary changes of position required for testing in the horizontal plane. If it has been determined that the muscle does not grade a fair minus (F–) by the test in the antigravity position but does grade more than a trace (which can be established in almost any position), then the overall grade of poor can be assigned without any need for further testing.

In practice, clinicians must realize frequent changes of position or repetition of a test in various positions may be tiring for a patient and time-consuming for the examiner. Patients should not be subjected to unnecessary procedures during examination if the results obtained are not meaningful.

Tests in the horizontal plane include several variables. The partial range of motion for the poor– grade is not specific, because there is no indication of where in the arc of motion the partial range should be. It may be at the beginning of the range of motion, within the midrange, or near the end.

With respect to partial arc of motion in antigravity position for a poor+ grade, it may mean starting from the suspended (i.e., vertical) position for quadriceps. For the hamstrings, it may mean that in the prone position, the subject can flex the last few degrees required to bring the leg to the vertical position.

When testing hip extensors or hip flexors in the side-lying position, a horizontal movement through the range of motion furnishes a means to obtain an objective grade of poor. The surface of the table, smooth or rough, changes the amount of friction and resistance. The strength of the hip adductors (if the underneath leg is being tested) may make a material difference in the results of the flexor and extensor tests. If the adductors are paralyzed, the full weight of the extremity will rest on the table and make flexion and extension difficult. If the adductors are strong, they will tend to raise the extremity so that the full weight does not rest on the table, thereby reducing the friction, and the flexion and extension movements will thus be made easier.

Trace Grade

The grade of **trace** means that a contraction can be appreciated by palpation and/or observation in a muscle or tendon as it becomes slightly prominent; however, no movement of the segment is visible. Trace grades can be determined in almost any position.

When testing very weak muscles, the examiner usually moves the segment into test position and tries to help the patient feel the movement and elicit a muscle response. The examiner should be sure that the movement starts from a relaxed position. If the segment is carried to the beginning of the range of motion and slight tension is put on the muscle, there may be a rebound or springing back, which can be confused with active movement.

Zero Grade

The grade of **zero** is assigned when no evidence of any muscle contraction is visible or palpable.

Key to Muscle Grading

According to Table 1-3, the highest test movement grade in the antigravity position is a 3, or poor+. Test movements for lateral trunk flexors, upper and lower abdominal muscles, and back extensors are exceptions. See individual tests for grading of these muscles. Testing of the muscles of the fingers and toes does not depend on gravity.

SECTION III
NERVE AND MUSCLE CHARTS

MUSCLE INNERVATIONS, SPINAL NERVES, AND MUSCLE CHARTS

The recording of test results is an important component of muscle examinations. Records are valuable from the standpoints of diagnosis, treatment, prognosis, and progress. An examination performed without recording the details can be of value at the moment, but one has an obligation to the patient, to the institution (if one is involved), and to oneself to record the findings.

Charts used for recording the findings of muscle examinations should permit complete tabulation of test results. In addition, the arrangement of the information should facilitate its interpretation. There are two charts in this category provided by Florence Kendall: one for the neck, diaphragm, and upper extremity and the other for the trunk and lower extremity. These charts have been designed especially for use as an aid in the differential diagnosis of lesions of the spinal nerves (Box 1-2). The motor involvement, as determined by manual muscle tests, can aid in determining whether a lesion of the nerve exists at the root, plexus, or peripheral level. The charts may also be useful in determining the level of a spinal cord lesion.

Use of the *Spinal Nerve and Muscle Charts* is illustrated by the case studies in Chapters 6 and 7.

In the upper and lower extremity charts, the names of the muscles appear in the left column and are grouped, as indicated by heavy black lines, according to their innervations, which are listed to the left of the muscle names. The space between the column of muscle names and the nerves is used to record the grade of muscle strength. Following is a brief description of a few sections of these extremity charts.

Peripheral Nerves

Peripheral nerves and their segmental origins are listed across the top of the center of the chart and follow the order of proximal-distal branching insofar as possible. For the peripheral nerves that arise from cords of the brachial plexus, the appropriate cord is indicated. The key at the top of the charts explains the abbreviations used.

Below this section, in the body of the chart, the dots indicate the peripheral nerve supply to each muscle (see Appendix D for sources of material for this section.)

BOX 1-2

Use of Charts in Differential Diagnosis

- Muscle strength grades are recorded in the column to the left of the muscle names. The grade symbols may be numerals or letters. Grades can be translated as indicated on the *Key to Grading Symbols.*
- After the grades have been recorded, the nerve involvement is plotted, when applicable, by circling the dots under peripheral supply or outlining the numbers under spinal segment distribution that corresponds with each involved muscle (see Chapters 6 and 7).
- The involvement of peripheral nerves and/or parts of the plexus is ascertained from the encircled dots by following the vertical lines upward to the top of the chart or the horizontal to the left margin. When evidence of involvement at spinal segment level exists, the level of the lesion may be indicated by a heavy black line drawn vertically to separate the involved from the uninvolved spinal segments.
- As a rule, muscles graded as good (i.e., 8) or above may be considered as not being involved from a neurological standpoint. This degree of weakness may be the result of factors such as inactivity, stretch weakness, or lack of fixation by other muscles. It should be remembered, however, that a grade of good might indicate a deficit of a spinal segment that minimally innervates the muscle.
- Weakness with grades of fair or less may be the result of inactivity, disuse atrophy, immobilization, or neurological problems. Faulty posture of the upper back and shoulders may cause weakness of the middle and lower trapezius. It is not uncommon to find bilateral weakness of these muscles with grades as low as fair–. A neurological problem with involvement of the spinal accessory nerve is unlikely in cases of isolated weakness of these muscles, unless there is also involvement of the upper trapezius.

Spinal Segment

In this section, a number denotes the spinal segment origin of nerve fibers innervating each of the muscles listed in the left column (see Appendix D for sources for material for this section).

In the accompanying spinal nerve and muscle charts and subsequent text, distribution is indicated by numbers. Major distribution is indicated by a number in bold type, a small distribution by a number in regular type, and a possible or infrequent distribution by a number in parenthesis.

Sensory

On the right side of the charts are diagrams showing the dermatomes and the distribution of cutaneous nerves for the upper extremity on one and for the trunk and lower extremity on the other (17, 22).

It is possible to use the illustrations for charting areas of sensory involvement by shading or using a colored pencil to outline the areas of the involvement for any given patient. Only drawings of the right extremity are used on the extremity charts, but labeling can indicate, when necessary, that the recorded information pertains to the left side.

Neck, Diaphragm, and Upper Extremity

See Figure 1-10 for grading muscle strength of the neck, diaphragm, and upper extremity.

Trunk and Lower Extremity

See Figure 1-11 for grading muscle strength of the trunk and lower extremity.

Name Date

PERIPHERAL NERVES

KEY ⟶ D. = Dorsal Prim. Ramus; V. = Vent. Prim. Ramus; P.R. = Plexus Root; S.T. = Superior Trunk; P. = Posterior Cord; L. = Lateral Cord; M. = Medial Cord

		MUSCLE STRENGTH GRADE	MUSCLE	Cervical (D.; Cervical 1-8, T. 1)	Cervical (V.; 1-8)	Cervical (V.; 1-4)	Phrenic (V.; 3, 4, 5)	Long Thor. (P.R.; 5, 6, 7, (8))	Dor. Scap (P.R.; 4, 5)	N. to Subcl. (S.T.; 5, 6)	Suprascap (S.T.; 4, 5, 6)	U. Subscap. (P.; (4), 5, 6, (7))	Thoracodor (P.; (5), 6, 7, 8)	L. Subscap. (P.; 5, 6, (7))	Lat. Pect. (L.; 5, 6, 7)	Med. Pect. (M.; (6), 7, 8, T. 1)	Axillary (P.; 5, 6)	Musculocu. (L.; (4), 5, 6, 7)	Radial (P.; 5, 6, 7, 8, T. 1)	Median (L.M.; 5, 6, 7, 8, T. 1)	Ulnar (M.; 7, 8, T. 1)	C1	C2	C3	C4	C5	C6	C7	C8	T1
			SPINAL SEGMENT																											
			HEAD & NECK EXTENSORS	•																		1	2	3	4	5	6	7	8	1
Cervical Nerves			INFRAHYOID MUSCLES			•																1	2	3						
Cervical Nerves			RECTUS CAP ANT. & LAT.			•																1	2							
Cervical Nerves			LONGUS CAPITIS			•																1	2	3	(4)					
Cervical Nerves			LONGUS COLLI		•																		2	3	4	5	6	(7)		
Cervical Nerves			LEVATOR SCAPULAE			•			•															3	4	5				
Cervical Nerves			SCALENI (A. M. P.)		•																			3	4	5	6	7	8	
Cervical Nerves			STERNOCLEIDOMASTOID			•																(1)	2	3						
Cervical Nerves			TRAPEZIUS (U. M. L.)			•																	2	3	4					
Cervical Nerves			DIAPHRAGM				•																	3	4	5				
Brachial Plexus	Root		SERRATUS ANTERIOR					•																		5	6	7	8	
Brachial Plexus	Root		RHOMBOIDS MAJ & MIN						•																4	5				
Brachial Plexus	Trunk		SUBCLAVIUS							•																5	6			
Brachial Plexus	Trunk		SUPRASPINATUS								•														4	5	6			
Brachial Plexus	Trunk		INFRASPINATUS								•														(4)	5	6			
Brachial Plexus	P Cord		SUBSCAPULARIS									•		•												5	6	7		
Brachial Plexus	P Cord		LATISSIMUS DORSI										•														6	7	8	
Brachial Plexus	P Cord		TERES MAJOR											•												5	6	7		
Brachial Plexus	L		PECTORALIS MAJ (UPPER)												•											5	6	7		
Brachial Plexus	M&L		PECTORALIS MAJ (LOWER)												•	•											6	7	8	1
Brachial Plexus	M&L		PECTORALIS MINOR													•											(6)	7	8	1
Axil.			TERES MINOR														•									5	6			
Axil.			DELTOID														•									5	6			
Musculo-cutan			CORACOBRACHIALIS															•									6	7		
Musculo-cutan			BICEPS															•								5	6			
Musculo-cutan			BRACHIALIS															•								5	6			
Radial			TRICEPS																•								6	7	8	1
Radial			ANCONEUS																•									7	8	
Radial	Lat.M		BRACHIALIS (SMALL PART)																•							5	6			
Radial	Lat.M		BRACHIORADIALIS																•							5	6			
Radial	Lat.M		EXT CARPI RAD L																•							5	6	7	8	
Radial	Lat.M		EXT CARPI RAD B																•							5	6	7	8	
Radial	Post Inter		SUPINATOR																•							5	6	(7)		
Radial	Post Inter		EXT DIGITORUM																•								6	7	8	
Radial	Post Inter		EXT DIGITI MINIMI																•								6	7	8	
Radial	Post Inter		EXT CARPI ULNARIS																•								6	7	8	
Radial	Post Inter		ABD POLLICIS LONGUS																•								6	7	8	
Radial	Post Inter		EXT POLLICIS BREVIS																•								6	7	8	
Radial	Post Inter		EXT POLLICIS LONGUS																•								6	7	8	
Radial	Post Inter		EXT INDICIS																•								6	7	8	
Median			PRONATOR TERES																	•							6	7		
Median			FLEX CARPI RADIALIS																	•							6	7	8	
Median			PALMARIS LONGUS																	•							(6)	7	8	1
Median			FLEX DIGIT SUPERFICIALIS																	•								7	8	1
Median	A Inter		FLEX DIGIT PROF I & II																	•								7	8	1
Median	A Inter		FLEX POLLICIS LONGUS																	•							(6)	7	8	1
Median	A Inter		PRONATOR QUADRATUS																	•								7	8	1
Median			ABD POLLICIS BREVIS																	•							6	7	8	1
Median			OPPONENS POLLICIS																	•							6	7	8	1
Median			FLEX POLL BREV (SUP. H)																	•							6	7	8	1
Median			LUMBRICALES I & II																	•							(6)	7	8	1
Ulnar			FLEX CARPI ULNARIS																		•							7	8	1
Ulnar			FLEX DIGIT. PROF. III & IV																		•							7	8	1
Ulnar			PALMARIS BREVIS																		•							(7)	8	1
Ulnar			ABD DIGITI MINIMI																		•							(7)	8	1
Ulnar			OPPONENS DIGITI MINIMI																		•							(7)	8	1
Ulnar			FLEX DIGITI MINIMI																		•							(7)	8	1
Ulnar			PALMAR INTEROSSEI																		•								8	1
Ulnar			DORSAL INTEROSSEI																		•								8	1
Ulnar			LUMBRICALES III & IV																		•							(7)	8	1
Ulnar			ADDUCTOR POLLICIS																		•								8	1
Ulnar			FLEX POLL BREV. (DEEP H.)																		•								8	1

SENSORY

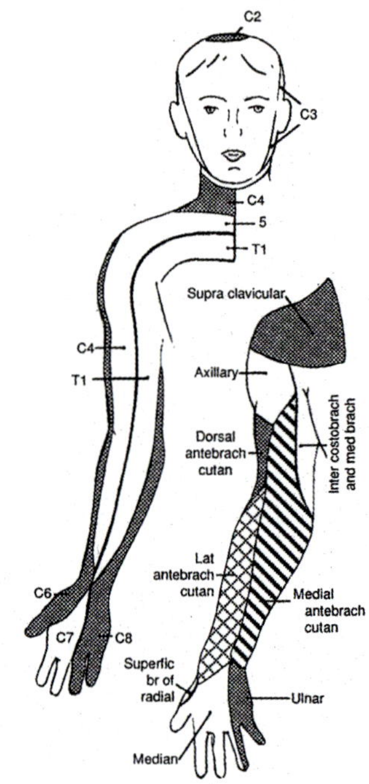

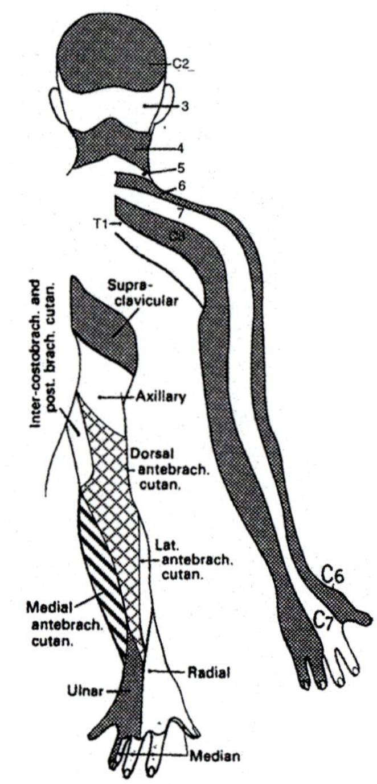

Dermatomes redrawn from Keegan and Garrett Anat Rec 102. 409. 437. 1948
Cutaneous Distribution of peripheral nerves redrawn from Gray's *Anatomy of the Human Body*. 28th ed

FIGURE 1-10: Grading muscle strength: neck, diaphragm, and upper extremity. © 1993 Florence P. Kendall.

Name Date

		MUSCLE STRENGTH GRADE	MUSCLE	D. T1-12, L1-5, S1-3	V. T1, 2, 3, 4	V. T5, 6	V. T7, 8	V. T9, 10, 11, 12	V. Iliohypogastric T12 L1	V. Ilioinguinal T(12) L1	V. Lumb. Plex. T(12) L1, 2, 3, 4	P. Femoral L(1) 2, 3, 4	A. Obturator L(1) 2, 3, 4	P. Sup. Glut. L4, 5, S1	P. Inf. Glut. L5, S1, 2	V. Sac. Plex. L4, 5, S1, 2, 3	P. Sciatic L4, 5, S1, 2	A. Sciatic L4, 5, S1, 2, 3	P. C. Peroneal L4, 5, S1, 2	A. Tibial L4, 5, S1, 2, 3	L1	L2	L3	L4	L5	S1	S2	S3
			ERECTOR SPINAE	●																	1	2	3	4	5	1	2	3
Thoracic Nerves			SERRATUS POST SUP		●																							
Thoracic Nerves			TRANS THORACIS		●	●	●																					
Thoracic Nerves			INT INTERCOSTALS		●	●	●	●																				
Thoracic Nerves			EXT INTERCOSTALS		●	●	●	●																				
Thoracic Nerves			SUBCOSTALES		●	●	●	●																				
Thoracic Nerves			LEVATOR COSTARUM		●	●	●	●																				
Thoracic Nerves			OBLIQUUS EXT ABD			(●)	●	●																				
Thoracic Nerves			RECTUS ABDOMINIS			●	●	●																				
Thoracic Nerves			OBLIQUUS INT ABD				●	●	●	(●)											1							
Thoracic Nerves			TRANSVERSUS ABD				●	●	●	(●)											1							
Thoracic Nerves			SERRATUS POST INF					●																				
Lumb. Plexus			QUAD LUMBORUM								●										1	2	3					
Lumb. Plexus			PSOAS MINOR								●										1	2						
Lumb. Plexus			PSOAS MAJOR								●										1	2	3	4				
Femoral			ILIACUS									●									(1)	2	3	4				
Femoral			PECTINEUS									●	(●)									2	3	4				
Femoral			SARTORIUS									●										2	3	(4)				
Femoral			QUADRICEPS									●										2	3	4				
Obturator	Ant.		ADDUCTOR BREVIS										●									2	3	4				
Obturator	Ant.		ADDUCTOR LONGUS										●									2	3	4				
Obturator	Ant.		GRACILIS										●									2	3	4				
Obturator	Post		OBTURATOR EXT										●										3	4				
Obturator	Post		ADDUCTOR MAGNUS										●					●				2	3	4	5	1		
Gluteal	Sup		GLUTEUS MEDIUS											●										4	5	1		
Gluteal	Sup		GLUTEUS MINIMUS											●										4	5	1		
Gluteal	Sup		TENSOR FAS LAT											●										4	5	1		
Gluteal	In.		GLUTEUS MAXIMUS												●										5	1	2	
Sacral Plexus			PIRIFORMIS													●									(5)	1	2	
Sacral Plexus			GEMELLUS SUP													●									5	1	2	
Sacral Plexus			OBTURATOR INT													●									5	1	2	
Sacral Plexus			GEMELLUS INF													●								4	5	1	(2)	
Sacral Plexus			QUADRATUS FEM													●								4	5	1	(2)	
Sciatic	P		BICEPS (SHORT H)														●								5	1	2	
Sciatic	Tibial		BICEPS (LONG H)															●							5	1	2	3
Sciatic	Tibial		SEMITENDINOSUS															●						4	5	1	2	
Sciatic	Tibial		SEMIMEMBRANOSUS															●						4	5	1	2	
Common Peroneal	Deep		TIBIALIS ANTERIOR																●					4	5	1		
Common Peroneal	Deep		EXT HALL LONG																●					4	5	1		
Common Peroneal	Deep		EXT DIGIT LONG																●					4	5	1		
Common Peroneal	Deep		PERONEUS TERTIUS																●					4	5	1		
Common Peroneal	Deep		EXT DIGIT BREVIS																●					4	5	1		
Common Peroneal	Sup		PERONEUS LONGUS																●					4	5	1		
Common Peroneal	Sup		PERONEUS BREVIS																●					4	5	1		
Tibial	Tibial		PLANTARIS																	●				4	5	1	(2)	
Tibial	Tibial		GASTROCNEMIUS																	●						1	2	
Tibial	Tibial		POPLITEUS																	●				4	5	1		
Tibial	Tibial		SOLEUS																	●					5	1	2	
Tibial	Tibial		TIBIALIS POSTERIOR																	●				(4)	5	1		
Tibial	Tibial		FLEX DIGIT LONG																	●					5	1	(2)	
Tibial	Tibial		FLEX HALL LONG																	●					5	1	2	
Tibial	Med Pl		FLEX DIGIT BREVIS																	●				4	5	1		
Tibial	Med Pl		ADDUCTOR HALL																	●				4	5	1		
Tibial	Med Pl		FLEX HALL BREVIS																	●				4	5	1		
Tibial	Med Pl		LUMBRICALIS I																	●				4	5	1		
Tibial	Lat Plant		ABD DIGITI MIN																	●						1	2	
Tibial	Lat Plant		QUAD PLANTAE																	●						1	2	
Tibial	Lat Plant		FLEX DIGITI MIN																	●						1	2	
Tibial	Lat Plant		OPP. DIGITI MIN																	●						1	2	
Tibial	Lat Plant		ADDUCTORS HALL																	●						1	2	
Tibial	Lat Plant		PLANT INTEROSSEI																	●						1	2	
Tibial	Lat Plant		DORSAL INTEROSSEI																	●						1	2	
Tibial	Lat Plant		LUMB II,III,IV																	●				(4)	(5)	1	2	

KEY
- D Dorsal Primary Ramus
- V Ventral Primary Ramus
- A Anterior Division
- P Posterior Division

SENSORY

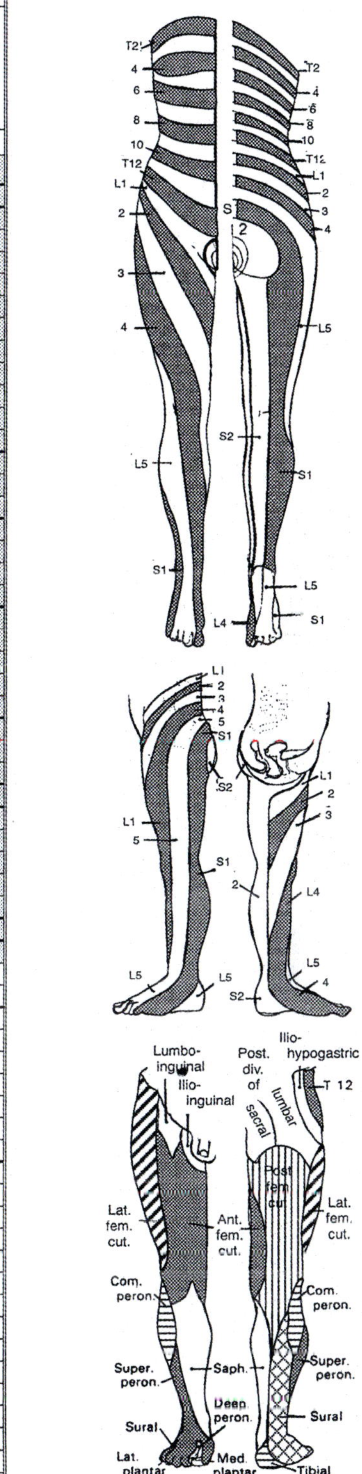

Dermatomes redrawn from Keegan and Garrett Anat Rec 102. 409. 437. 1948
Cutaneous Distribution of peripheral nerves redrawn from *Gray's Anatomy of the Human Body*. 28th ed

FIGURE 1-11: Grading muscle strength: trunk and lower extremity.

SECTION IV
CLINICAL FINDINGS

SYSTEM PATHOLOGIES

Numerous neurological and musculoskeletal conditions arise from **noninvasive** trauma that can cause compression or tension (i.e., traction) on a nerve or tissues containing nerve endings. The trauma may be sudden or gradual, with the latter type resulting from intrinsic mechanical barriers, maintained positions, or repetitive movements. Involvement may vary from being widespread throughout an extremity to being localized to a single nerve branch. Noninvasive trauma may be transitory or result in permanent deficits.

Body tissues are subject to **invasive** trauma in many areas of the body, as well. Invasive trauma may be accidental, such as lacerations, piercing wounds, injections of medications, or nerves cut or injured during surgery. Invasive trauma may also be caused by necessary procedures, such as a nerve resection or rhizotomy.

Mechanical Causes of Pain

Pain—whether it is perceived in the muscle, the joint, or the nerve itself—is a response of the nerve. Regardless of where the stimulus may arise, it is conducted by the nerves and interpreted by the brain. The mechanical factors that give rise to pain must, therefore, directly affect the nerve fibers. Two such factors need to be considered in problems of faulty body mechanics: pressure and tension.

Pressure or **compression** on nerve root, trunk, nerve branches, or nerve endings may be caused by some adjacent, firm structure, such as bone, cartilage, fascia, scar tissue, or taut muscle. Pain resulting from ligamentum flavum hypertrophy or a protruded disk exemplifies nerve root pressure. Scalenus anticus syndrome, in cases of arm pain, and piriformis syndrome, in cases of sciatica, are examples of peripheral nerve irritation.

Tension on structures containing nerve endings that are sensitive to deformation, as found in stretch or strain of muscles, tendons, or ligaments, can cause slight or excruciating pain, depending on the severity of the strain. Forces within the body that exert an injurious tension resulting in strain of soft tissue usually arise from a prolonged distortion of bony alignment or from a sudden muscle pull.

Distribution of pain along the course of the involved nerve and the areas of cutaneous sensory disturbance aid in determining the site of the lesion. Pain may be localized below the level of direct involvement or be widespread because of reflex or referred pain. In a root lesion, pain tends to extend from the origin of the nerve to its periphery, and cutaneous sensory involvement is on a dermatomal basis.

Peripheral nerve involvement is often distinguished by pain below the level of the lesion. Most peripheral nerves contain both sensory and motor fibers. Symptoms of pain or tingling usually appear in the cutaneous areas that are supplied by the nerve before numbness or weakness becomes apparent. Numerous muscles are supplied by nerves that are purely motor to the muscle, however, and symptoms of weakness can appear without previous or concurrent symptoms of pain or tingling.

Nerve Compression and Tension

Trauma may also result from an **external force causing compression** on a nerve. *External force causing compression is exemplified by:*

- Radial, median, or ulnar nerve (or some combination of these), as in "Saturday night palsy" from an arm hanging over the back of bench or a chair.
- Radial or median nerve (or both) from crutch paralysis.
- Radial, median, and ulnar nerves from a tourniquet.
- Median nerve from various sleeping positions (e.g., supine, with arm overhead; side-lying on the arm in adduction) (23).
- Ulnar nerve from trauma to the elbow.
- Ulnar or median nerve from sudden or repeated trauma to the hypothenar or thenar eminence.
- Anterior interosseous nerve from (forearm) armband sling (24).
- Brachial plexus from a strap over the shoulder.
- Fibular (peroneal) nerve by a cast, adhesive strapping, or garter producing pressure over the head of the fibula or by prolonged sitting with the legs crossed and one knee resting on the other.
- A transitory external compressive force is exemplified by a bump on the elbow (commonly referred to as hitting the "funny bone," which is so named because it is the distal end of the humerus), which compresses the ulnar nerve. The bruise hurts and causes tingling into the fourth and fifth digits, but the symptoms do not persist.
- Trauma by an external force causing *tension* on nerves can occur to the brachial plexus, such as from an accident or a manipulation that puts excessive traction on the plexus. The long thoracic nerve is susceptible to stretch from carrying a heavy bag with a strap over the shoulder.

- Internal forces created in areas of the body where there is a close association between nerves and firm skeletal structures may result in neural compression or tension. Under ordinary conditions, a groove or a tunnel may protect nerves along their path, but in cases of injury or inflammation with swelling or scar tissue, the confined area becomes a source of entrapment.

Internal compression is exemplified by pressure on:

- Spinal nerve root from calcium deposits in the foramen.
- Suprascapular nerve as it passes under the ligament and through the scapular notch (25–28).
- Brachial plexus from a cervical rib (see posture in relation to cervical rib).
- Brachial plexus from the coracoid process and a tight pectoralis minor (23, 29).
- Axillary nerve in the quadrilateral space (27, 30).
- Median nerve, as in carpal tunnel syndrome.
- Nerve to (usually) the fourth toe, as in Morton's neuroma.

Internal tension on a nerve is exemplified by:

- Suprascapular nerve as it passes through the scapular notch, being subject to stretch with displacement of shoulder and scapula (31).
- Fibular (peroneal) nerve, secondary to spasm in the tensor fasciae latae, with resultant traction on the iliotibial band to its insertion.
- Fibular (peroneal) nerve, secondary to traction on the leg, by inversion of the foot (23, 28).

Nerve Impingement

In this text, the term **impingement** is used with reference to nerve irritation associated with muscles.

During the 1930s, there was great reluctance to speak about the possibility that in addition to bone and other firm structures, muscles might play a role in causing irritation to the nerves. In a 1934 article regarding the piriformis muscle, Albert H. Freiberg stated that "pressure of a muscle belly upon the trunk of the sciatic nerve can be productive of pain and tenderness [but] must be looked upon as unproved at present" (32). Freiberg was cautious and almost apologetic about suggesting that the muscle could play that kind of a role.

During that same era, one of the original authors of *Muscles: Testing and Function*, Henry O. Kendall, rather courageously offered such explanations for several clinical entities. Most instances were related to muscles that were pierced by a peripheral nerve and in which movement and alteration of muscle length were factors in causing a friction type of irritation to the nerve. Symptoms of pain or discomfort could be elicited by stretching the muscle, by having the muscle actively contract, or by repetitive movements.

Explaining peripheral nerve pain on the basis of pressure or friction by muscles remains a controversial issue with respect to certain syndromes, notably the piriformis (27, 33). However, the concept is well recognized in regard to nerve involvement with numerous muscles.

Under normal conditions and through normal range of motion, it may be presumed that a muscle will not cause irritation to a nerve that lies in close proximity to it or that pierces it. A muscle that is drawn taut, however, becomes firm and has the potential for exerting a compressive or a friction force. A muscle that has developed adaptive shortness moves through less range and becomes taut before reaching normal length; a stretched muscle moves through more than normal range before becoming taut. A taut muscle, especially a weight-bearing muscle, can cause friction on a nerve during repetitive movements.

In mild cases, the symptoms may be discomfort and dull ache rather than sharp pain when the muscles contract or are elongated. Sharp pain may be elicited by vigorous movements but tends to be intermittent, because the subject finds ways to avoid the painful movements.

Recognizing this phenomenon in the early stages can increase the likelihood of counteracting or preventing the more painful or disabling problems that develop later. Physical therapists who deal with stretching and strengthening exercises have the opportunity to observe early signs of impingement among their patients.

The axillary nerve emerges with the posterior circumflex humeral artery through the quadrangular (quadrilateral) space that is bounded posteriorly by the teres major, latissimus dorsi, long head of the triceps, and humerus. When stretching a tight teres major, a patient may complain of a shooting pain in the area of cutaneous sensory distribution of the axillary nerve (upper lateral cutaneous nerve of the arm). The assumption is that the axillary nerve is being compressed or stretched against the tight teres major. The pain that results from direct irritation to the nerve is in contrast to the discomfort that is often associated with the usual stretching of tight muscles (see cutaneous nerve distribution and teres syndrome).

The femoral nerve pierces the psoas major muscle. During assisted stretching exercises, a patient with tight iliopsoas muscles may complain of pain along the anteromedial aspect of the leg in the area of cutaneous sensory distribution of the saphenous nerve (see cutaneous nerve distribution).

The greater occipital nerve emerges from the suboccipital triangle and superficially pierces the trapezius muscle and fascia. Movements of the head and neck in the direction of contracting or stretching the trapezius may elicit pain in the area of the back of the head and the cervical region (see occipital headache). Other examples include:

- Supinator with radial nerve (27, 34).
- Pronator with median nerve (27, 31, 34).
- Flexor carpi ulnaris with ulnar nerve (23).
- Lateral head of the triceps with radial nerve (27, 34).
- Trapezius with greater occipital nerve (23).

- Scalenus medius with C5 and C6 root of the plexus and long thoracic nerve (23).
- Coracobrachialis with musculocutaneous nerve (27, 31).

Muscle Spasm

Spasm is an involuntary contraction of a muscle or a segment within a muscle that can result from painful nerve stimulation. Irritation of root, plexus, or peripheral nerve branch levels may cause spasm of a number of muscles, whereas spasm caused by irritation of the nerve endings within a muscle may be limited to the muscle involved or be widespread because of reflex pain mechanisms (35).

Treatment of muscle spasm depends on the type of spasm. Relief of spasm resulting from initial nerve irritation of the root, trunk, or peripheral branch must depend on relief of such nerve irritation. Aggressive treatment of the muscle or muscles in spasm will tend to aggravate the symptoms. For example, avoid the use of heat, massage, and stretching of the hamstring muscles in cases of acute sciatica. Rigid immobilization of the extremity is also contraindicated.

Protective spasm may occur secondary to injury of underlying structures, such as a ligament or bone. This protective guarding, such as often occurs following a back injury, prevents movement and further irritation of the injured structure. Protective spasm should be treated by application of a protective support to relieve the muscles of this extraordinary function. Muscle spasm tends to subside rapidly and pain diminishes when a support is applied. As the muscles relax, the support maintains the function of protection to permit healing of whatever underlying injury gave rise to the protective muscle response (36).

Besides the relief from restriction of motion, the support gives added relief by putting pressure on the muscles in spasm. The positive response to direct pressure on the muscle distinguishes this type of spasm from that caused by initial nerve irritation. In the low back, where protective muscle spasm frequently occurs, a brace with a lumbar pad, or a corset with posterior stays bent to conform to the contour of the low back, may be used for both immobilization and pressure.

In most instances, one may assume that the underlying disturbance is severe enough to require the use of a support for at least a few days to permit healing. However, it is not uncommon to find, when the acute onset of pain is caused by a sudden exaggeration of movement, that a rigid posture persists because of the patient's fear of movement rather than because of the continued need for a protective reaction. Because of this possibility, it is often useful to apply heat and gentle massage as a diagnostic aid in determining the extent of protective reaction.

Segmental muscle spasm is an involuntary contraction of the uninjured segment of a muscle as a result of an injury to the muscle. The contraction of this part puts tension on the injured part, and a condition of strain is present. Pain associated with tension within the muscle may be outlined by the margins of the muscle or be widespread because of reflex or referred pain mechanisms. Treatment requires immobilization in a position that relieves tension on the affected muscle. A positive response may also be obtained by gentle, localized massage to the area in spasm.

Muscle spasm associated with tendon injury differs from the above when the tension is exerted on the tendon rather than on a part of the muscle. Tendons contain many nerve endings that are sensitive to stretch, and pain associated with tendon injury tends to be severe.

Adaptive Shortening

Adaptive shortening is tightness that results from the muscle remaining in a shortened position. Unless the opposing muscle is able to pull the segment back to neutral position or some outside force is exerted to lengthen the short muscle, the muscle in question will remain in a shortened condition.

Shortness represents a slight to moderate decrease in muscle length and results in a corresponding restriction of range of motion. It is considered to be reversible, but stretching movements must be done gradually to avoid damaging the tissue structures. A period of several weeks is usually necessary for restoration of mobility in muscles exhibiting moderate tightness.

Wheelchair users and those who frequently maintain a sedentary, seated position may develop adaptive shortening in the one-joint hip flexors (iliopsoas). Prolonged sitting with the knees partially extended places the foot in a position of plantar flexion and may result in adaptive shortening of the soleus. Women who wear high-heeled shoes much of the time may also develop adaptive shortening of the soleus. Such shortness can affect both balance and standing alignment.

Stretch Weakness

Stretch weakness is defined as weakness that results from muscles remaining in an elongated condition, however slight, beyond the neutral physiological rest position but not beyond the normal range of muscle length. The concept relates to the duration rather than the severity of the faulty alignment. (It does not refer to overstretch, which means beyond the normal range of muscle length.)

Many cases of stretch weakness have responded to treatment that supported the muscles in a favorable position, although the muscles had been weak or partially paralyzed for a long time even several years after the onset of the initial problem. Return of strength in such instances indicates that damage to the muscles was not irreparable.

A familiar example of stretch weakness superimposed on a normal muscle is the foot drop that may develop in a bedridden patient as a result of bed linens holding the foot in plantar flexion. Weakness in the dorsiflexors results from the continuous stretch on these muscles even though there is no neurological involvement.

Stretch weakness superimposed on muscles affected by anterior horn cell involvement was seen numerous times in patients with poliomyelitis (see example, Appendix B).

Stretch weakness superimposed on a lesion of the CNS has been observed in patients with multiple sclerosis, especially with regard to the wrist extensors and ankle dorsiflexors. Stretching opposing muscles that have become shortened and applying a support in the form of a cock-up splint for the wrist or an orthosis for the ankle have resulted in improved strength and functional ability.

Stretch weakness of a less dramatic nature is frequently seen in cases of occupational and postural strain. The muscles most often affected have been one-joint muscles: gluteus medius and minimus, iliopsoas, hip external rotators, abdominal muscles, and middle and low trapezius.

Muscles exhibiting stretch weakness should not be treated by stretching or movement through the full range of joint motion in the direction of elongating the weak muscles. The condition has resulted from continuous stretching, and it responds to immobilization in a physiological rest position for a sufficient period of time to allow recovery to occur. Realignment of the segment, bringing it into a neutral position, and use of supportive measures to help restore and maintain such alignment until weak muscles recover strength are important factors in treatment. Any opposing tightness that tends to hold the segment out of alignment must be corrected to relieve tension on the weak muscles. Faulty occupational positions that impose continuous tension on certain muscles must also be adjusted or corrected. Care must be taken not to overwork a muscle that has been subjected to a prolonged tension stress. As the muscles improve in strength and become capable of maintaining the gain, the patient is expected to use the muscles by working to maintain proper muscle balance and good alignment.

SECTION V TREATMENT FUNDAMENTALS

See Box 1-3 for guidelines for clinicians.

STABILITY OR MOBILITY

In the treatment of conditions affecting joints and muscles, one must determine the overall objectives of treatment based on whether **stability** or **mobility** is the desired outcome for optimal function. Joint structures are designed so that along with greater mobility there is less stability, and along with greater stability there is less mobility.

It is generally accepted that along with growth from childhood to adulthood, a tightening up of the ligamentous structures occurs, along with a corresponding decrease in flexibility of the muscles. This change affords greater stability and strength for adults than for children.

Individuals with ligamentous laxity may demonstrate less stability than an individual with less flexibility. A knee that goes into hyperextension, for example, is not mechanically as stable for weight bearing as one that is held in neutral extension.

Lack of stability of the spine in a flexible individual can lead to problems when work requires prolonged sitting, standing, or the need to lift and carry heavy objects. Muscles do not succeed in functioning for both *movement* and the *support* normally afforded by the ligaments. When symptoms occur, they will appear as fatigue first and only later as pain. A young adult with excellent strength but excessive spinal flexibility may require a back support to relieve painful symptoms.

From a mechanical standpoint, two types of faults relate to *alignment* and *mobility*: undue compression on articulating surfaces of bone and undue tension on bones, ligaments, or muscles. Eventually, two types of bony changes may occur: Excessive compression produces an eroding effect on the articulating surface, whereas tension may result in bony hypertrophy at the point of connective tissue or muscle attachments.

Lack of mobility is closely associated with persistent faulty alignment as a factor in causing undue compression. When mobility is lost, stiffness occurs, and a certain alignment remains constant. This may be a result of the restriction of motion by tight muscles or of the inability of weak muscles to move the segment through the arc of motion. Muscle tightness is a constant factor, tending to maintain the segment in faulty alignment regardless of the position of the body. Muscle weakness is a less constant factor because changing the body position can bring about a change in the alignment of the segment. With normal movement in joints, wear and tear on the joint surfaces tends to be distributed; however, with limitation of range, the wear will take place only on the joint surfaces that represent the arc of use. If the segment that is restricted by muscle tightness is protected against any movement that may cause strain, the other parts that must compensate for such restriction can become susceptible to strain.

BOX 1-3

Guidelines for the Clinician

- Be guided by the age-old adage: "Thou shalt do no harm."
- Obtain the patient's confidence and cooperation.
- Listen carefully to the patient.
- Observe posture, body language, and spontaneous movements that provide helpful diagnostic clues.
- Apply basic knowledge of anatomy, physiology, and body mechanics in musculoskeletal evaluations and treatments of patients.
- Consider whether the occupational or recreational activities of the patient alleviate or aggravate existing conditions.
- Educate your patients; help them to understand the nature of their problems.
- Be guided by the patient's reaction to previous treatments.
- Be patient with your patients. It often takes more than one session to overcome anxiety and guarding against pain.
- Start treatments in a gentle manner.
- Remember that it is essential to obtain patient relaxation before attempting to stretch tight muscles. Stretching that is too vigorous will retard rather than hasten recovery.
- Understand that muscles weakened as a result of injury or disease must be handled with more care than a muscle exhibiting normal function.
- When applying traction, use a firm but gentle grasp. Avoid pinching, twisting, or pulling the skin over the segment that is being held.
- Expect that favorable responses to treatment will progress gradually, based on the patient's tolerance to pain or discomfort.
- Avoid the attitude of "more is better." Reactions to treatment are often delayed, so one may not know until the next day that the previous treatment was "too much."
- Avoid application of heat over areas with impaired sensation or circulation, and over muscles exhibiting stretch weakness.
- Recognize that continuation of treatment is contraindicated if any of the following symptoms appear: swelling, redness, abnormal temperature of the region, marked tenderness, loss of range of motion, or persistent pain.
- Involve the patient in setting treatment goals and in planning a home treatment program.
- Be accountable. Document your assessment, evaluation, treatment plan, and follow-up care.

Excessive joint mobility results in tension on the ligaments that normally limits range of motion and can result in undue compression on the margins of the articulating surfaces when the excessive range is long-standing or frequent.

ROLE OF MUSCLES IN SUPPORTING THE SKELETON

In addition to their role in movement, muscles have an important role in supporting the skeletal structures. A muscle must be long enough to permit normal mobility of the joints yet short enough to contribute effectively to joint stability.

When range of motion is limited because of tight muscles, treatment consists of the use of various modalities and procedures to promote muscle relaxation and optimal length. Stretching exercises should be gradual, and although they may cause mild discomfort, they should not cause pain.

When range of motion is excessive, care must be taken to avoid overstretching. If the patient has excessive mobility, with or without pain, in many instances it is prudent to apply a support that can allow the affected structures to tighten. It may or may not be necessary to add specific exercises, because many muscles that are weakened by stretching recover with normal activity when overstretching is avoided.

TREATMENT OPTIONS

Following are treatment options which may be employed in the care of patients presenting with impairments of posture and pain (see Table 1-4). This is not intended to be an exhaustive, all-encompassing list. Greater detail may be provided in later chapters as appropriate.

TABLE 1-4 Treatment Options Used for Impairments of Posture and Pain

Treatment	Definition	Effects
Traction	• A force used therapeutically to produce elongation or stretch of joint structures and/or muscles. Properly applied, the force pulls in the direction perpendicular to the articulating surfaces of a joint. Traction may be applied: • Manually, • using a mechanical traction device, • static weights, or positional distraction.	• Relief of pain and spasm • Reduction/prevention of adhesions • Stretching of tight musculature • Improved circulation
Massage	**Manual therapy** techniques are skilled hand movements and skilled passive movements of joints and soft tissue. Techniques may include: • manual lymphatic drainage, • manual traction, • massage, • mobilization/manipulation, and • passive range of motion.	• Improve circulation • Promote relaxation of muscles • Loosen scar tissue • Stretch tight muscles/fasciae • Relieve edema/swelling • Improve tissue extensibility • Increase range of motion • Mobilize/manipulate soft tissue and joints • Modulate pain • Reduce restriction
Exercise (19, 31)	*Therapeutic exercise* is the systematic performance or execution of planned physical movements or activities intended to enable the patient or client to remediate or prevent impairments of body functions and structures, enhance activities and participation, reduce risk, optimize overall health, and enhance fitness and well-being. Therapeutic exercise may include aerobic and endurance conditioning and reconditioning; agility training; body mechanics training; breathing exercises; coordination exercises; developmental activities training; muscle lengthening; movement pattern training; neuromotor development activities training; neuromuscular education or reeducation; perceptual training; range of motion exercises and soft tissue stretching; relaxation exercises; and strength, power, and endurance exercises.	• Enhance bone density • Enhance breathing • Enhance/maintain physical performance • Improve safety • Increase aerobic capacity/endurance • Increase muscle strength, power, and endurance • Enhance postural control and relaxation • Increase sensory awareness • Increase tolerance of activity • Prevent or remediate impairments in body functions and structures, activity limitations, and participation restrictions to improve physical function • Enhance health, wellness, and fitness • Reduce complications, pain, restriction, and swelling • Reduce risk and increase safety during activity performance • Strengthen weak muscles • Lengthen short muscles for the purpose of restoring the elasticity on which normal muscle function depends • Increase endurance • Improve coordination • Restore function • Stimulate circulation
Supports	• Correction of alignment faults associated with weakness often requires supportive measures. • Inclusive of orthotic/prosthetic positioning technologies intended to improve positioning and function	• Immobilize • Minimize pain • Correct faulty alignment • Relieve strain on weak muscles • Facilitate function • Restrict movement
	Modalities	
Electrical Stimulation	• A technique used to elicit a muscle contraction using electrical impulses	• Modulate pain • Reeducate muscle • Manage Edema • Decrease inflammation
Heat (32)	• A form of thermal energy transfer from a heating agent to a body segment	• Pain and muscle spasm relief • Decrease joint stiffness • Increased extensibility of collagen tissue • Increased blood flow • Resolution of inflammatory infiltrates (32) • Facilitate stretch
Cold (32)	A form of thermal energy transfer from a body segment to a cooling agent	• Reduce pain and swelling/edema • Inhibit spasticity • Facilitate muscular contraction for various forms of neurogenic weakness • Muscle reeducation

REFERENCES

1. Page P. Current concepts in muscle stretching for exercise and rehabilitation. Int J Sports Phys Ther. 2012;7(1):109–119.
2. Weppler, CH, Magnusson, SP. Increasing muscle extensibility: A matter of increasing length or modifying sensation? Phys Ther. 2010;90(3):438–449, doi: 10.2522/ptj.20090012.
3. Henricson EK, Abresch RT, Cnaan A, et al. The cooperative international neuromuscular research group Duchenne natural history study: glucocorticoid treatment preserves clinically meaningful functional milestones and reduces rate of disease progression as measured by manual muscle testing and other commonly used clinical trial outcome measures. Muscle Nerve. 2013;48(1): 55–67. doi: 10.1002/mus.23808.
4. Morton DP, Callister R. Influence of posture and body type on the experience of exercise-related transient abdominal pain. J Sci Med Sport. 2010;13(5):485–488. doi: 10.1016/j.jsams.2009.10.487. PMID: 20022301.
5. Alizadehkhaiyat O, Roebuck MM, Makki AT, Frostick SP. Postural alternations in patients with subacromial impingement syndrome. Int J Sports Phys Ther. 2017;12(7):1111–1120. doi: 10.26603/ijspt20171111.
6. Lee SP, Souza RB, Powers CM. The influence of hip abductor muscle performance on dynamic postural stability in females with patellofemoral pain. Gait Posture. 2012;36(3):425–429. doi: 10.1016/j.gaitpost.2012.03.024. Epub 2012 May 16. PMID: 22607792.
7. Dunn JC, Iversen MD. Interrater reliability of knee muscle forces obtained by hand-held dynamometer from elderly subjects with degenerative back pain. J Geriatr Phys Ther. 2003;26(3):23–29. Accessed September 17, 2021. https://search-ebscohost-com.proxy-hs.researchport.umd.edu/login.aspx?direct=true&db=rzh&AN=106734482&site=eds-live
8. Rash G. Electromyography Fundamentals. http://www.gcmas.org. Accessed 8/03, 2003.
9. Rheault W, Beal J, Kubick K, Novack T, Shepley J. Intertester reliability of the hand-held dynamometer for wrist flexion and extension. Arch Phys Med Rehabil. 1989;70:909.
10. Surburg P, Suomi R, Poppy W. Validity and reliability of a hand-held dynamometer applied to adults with mental retardation. Arch Phys Med Rehabil. 1992;73(6):535–539.
11. Wadsworth C R K, Sear M, Harrold J, Nielsen D. Intrarater reliability of manual muscle testing and hand-held dynametric muscle testing. Phys Ther. 1987;67(9):1342–1347.
12. Brinkman JR. Comparison of a hand-held to a fixed dynamometer in tracking strength change. [Abstract R226] In: Abstracts of papers accepted for presentation at 67th Annual Conference of American Physical Therapy Association, June 14–16, 1992. Phys Ther. 1992;72(6) Suppl.
13. Marino M, Nicholas J, Gleim G, Rosenthal P, Nicholas J. The efficacy of manual assessment of muscle strength using a new device. Am J Sports Med. 1982;10(6):360–364.
14. Mulroy SJ, Lassen KD, Chambers SH, Perry J. The ability of male and female clinicians to effectively test knee extension strength using manual muscle testing. Orthop Sports Phys Ther. 1997;26(4):192–199.
15. Newton M, Waddell G. Trunk strength testing with iso-machine. Part 1: Review of a decade of scientific evidence. Spine. 1993;18(7):801–811.
16. Norkin C, Levangie P. Joint Structure and Function. Philadelphia: F.A. Davis, 1992.
17. Goss CM, ed. Gray's Anatomy of the Human Body, 28th ed. Philadelphia: Lea & Febiger; 1966: 380–381.
18. Moore KL, Agur AMR, Dalley AF II. Clinically Oriented Anatomy, 8th ed. Wolters Kluwer; 2018. Accessed February 19, 2021. http://search.ebscohost.com/login.aspx?direct=true&db=cat01362a&AN=hshs.004959764&site=eds-live.
19. Huang, J. Dynamic activity of human brain task-specific networks. Sci Rep. 2020;10:7851. doi: 10.1038/s41598-020-64897-2.
20. O'Connell A, Gardner E. Understanding the scientific basis of human motion. Baltimore: Williams & Wilkins; 1972.
21. Legg AT. Physical therapy in infantile paralysis. In: Mock, ed. Principles and practice of physical therapy. Vol II. Hagerstown, MD: WF Prior; 1932:45.
22. Keegan J, Garrett F. The segmental distribution of the cutaneous nerves in the limbs of man. Anat Rec. 1948;102:409–437.
23. Sunderland S. Nerve and Nerve Injuries, 2nd ed. New York: Churchill Livingstone; 1978.
24. O'Neill DB, Zarins B, Gelberman RH, Keating TM, Louis D. Compression of the anterior interosseous nerve after use of a sling for dislocation of the acromioclavicular joint. J Bone Joint Surg [AM]. 1990;72-A(7):1100.
25. Post M, Mayer JM. Suprascapular nerve entrapment. Clin Orthop Relat Res. 1987;223:126–135.
26. Hadley MN, Sonntag VKH, Pittman HW. Suprascapular nerve entrapment. J Neurosurg. 1986;64:843–848.
27. Dawson DM, Hallett M, Millender LH. Entrapment Neuropathies, 2nd ed. Boston: Little, Brown; 1990.
28. Conway S, Jones H. Entrapment and compression neuropathies. In: Tollison C, ed. Handbook of chronic pain management. Baltimore: Williams & Wilkins; 1989; 433, 437, 438.
29. Kendall HO, Kendall FP, Boynton DA. Posture and Pain. Baltimore: Williams & Wilkins; 1952.
30. Cahill BR. Quadrilateral space syndrome. In: Omer GE, Spinner M. Management of peripheral nerve problems. Philadelphia: WB Saunders; 1980: 602–606.
31. Sunderland S. Nerve Injuries and Their Repair: A Critical Appraisal. London: Churchill Livingstone; 1991: 161.
32. Freiberg AH, Vinke TH. Sciatica and sacro-iliac joint. J Bone Joint Surg [AM]. 1934;16:126–136.
33. Jankiewicz JJ, Henrikus WL, Houkom JA. The appearance of the prirformis muscle syndrome in computed tomography and magnetic resonance imaging. Clin Orthop Relat Res. 1991;262:207.
34. Spinner M. Management of nerve compression lesions of the upper extremity. In: Omer G, Spinner M, eds. Management of peripheral nerve problems. Philadelphia: WB Sanders; 1980.
35. Hirayama J, Yamagata M, Ogata S, Shimizu K, Ikeda Y, Takahashi K. Relationship between low-back pain, muscle spasm and pressure pain thresholds in patients with lumbar disc herniation. Eur Spine J. 2006;15(1):41–47. doi: 10.1007/s00586-004-0813-2.
36. Jeon ET, Jung JH, Moon JH, et al. The effects of spinal support device on pain and extensibility of the hamstrings in patients with non-specific low back pain. J Phys Ther Sci. 2017;29(8): 1301–1304. doi: 10.1589/jpts.29.1301.

POSTURE

2

OUTLINE

SECTION I
FUNDAMENTALS OF POSTURE

INTRODUCTION

Ideal posture is a good habit that contributes to the well-being of the individual. The structure and function of the body provide the potential for attaining and maintaining ideal posture. Conversely, faulty posture is an alternative habit that, unfortunately, is all too common (1). Postural faults have their origin in the diminished or misuse of capacities afforded by the body, not in the structure and function of the normal body. If faulty posture were merely an aesthetic problem, the concerns about it might be limited to those regarding appearance. However, postural faults that persist can give rise to discomfort, pain, impaired function, or disability (1–5). The range of effects, from discomfort to incapacitating disability, is often related to the severity and persistence of the faults.

Discussion of the importance of ideal posture springs from a recognition of the prevalence of postural problems and associated painful conditions. This text attempts to define the concepts of ideal posture, to analyze postural faults, to present treatment options, and to discuss some of the developmental factors and environmental influences that affect posture. The objective is to help decrease the incidence of postural faults that result in painful conditions.

Cultural patterns of modern civilization add to the stresses on the basic structures of the human body by imposing increasingly specialized activities. It is necessary to recognize common compensatory strategies employed in everyday life to achieve optimal function. The high incidence of postural faults in adults is related to the process of aging and highly specialized or repetitive patterns of activity (1, 3, 6). Correction of the existing conditions depends on understanding the underlying influences and implementing a program of positive and preventive educational measures. Both require an understanding of the mechanics of the body and its response to the stresses and strains imposed on it.

Inherent in the concept of good body mechanics are the inseparable qualities of alignment and muscle balance. Examination and treatment procedures are directed toward restoration and preservation of ideal body mechanics in posture and movement. Education and therapeutic exercises to strengthen weak muscles and stretch tight muscles are the chief means by which muscle balance is restored. Ideal body mechanics requires that range-of-joint motion be adequate but not excessive. Normal flexibility is an attribute; excessive flexibility is not. A basic principle regarding joint movements can be summarized as follows: the more flexibility, the less stability; the more stability, the less flexibility. A problem arises, however, because skilled performance in a variety of sport, dance, and acrobatic activities requires excessive flexibility and muscle length. *Although "the more, the better" may apply to improving performance, it may adversely affect the well-being of the performer.*

The following definition of posture was included in a report by the Posture Committee of the American Academy of Orthopedic Surgeons (7). It is so well stated that it bears repeating:

> Posture is usually defined as the relative arrangement of the parts of the body. Good posture is that state of muscular and skeletal balance which protects the supporting structures of the body against injury or progressive deformity, irrespective of the attitude (erect, lying, squatting, or stooping) in which these structures are working or resting. Under such conditions the muscles will function most efficiently, and the optimum positions are afforded for the thoracic and abdominal organs. Poor posture is a faulty relationship of the various parts of the body which produces increased strain on the supporting structures and in which there is less efficient balance of the body over its base of support.

POSTURE AND PAIN

Complaints of pain related to faulty body mechanics are so common that most adults have examples of their own that impact their daily life. Painful lower backs have been the most frequent complaints, although cases of neck, shoulder, arm, foot, and knee pain have become increasingly prevalent (1, 3, 5, 8, 9).

When discussing pain in relation to postural faults, questions are often asked about why many cases of faulty posture exist without symptoms of pain, and why seemingly mild postural defects give rise to symptoms of mechanical and muscular strain. The answer to both depends on the constancy of the fault and the individual it affects.

A posture may appear to be very faulty, yet the individual may be flexible and the position of the body may change readily, thus limiting the amount of time they spend in the position. Alternatively, a posture may appear to be good, but stiffness or muscle tightness may so limit mobility that the position of the body cannot change readily. The lack of mobility, which is not apparent as an alignment fault but which is detected in tests for flexibility and muscle length, may be the more significant factor.

Basic to an understanding of pain in relation to posture is the concept that the cumulative effects of constant or repeated small stresses over long periods of time can give rise to the same kind of difficulties that occur with a sudden, severe stress. Cases of postural pain are extremely variable in the manner of onset and in the severity of symptoms. In some cases, only acute symptoms appear, usually as a result of an unusual stress or injury. Other cases have an acute onset and develop chronically painful symptoms. Still others exhibit chronic symptoms that later become acute.

Symptoms associated with an acute onset can be focal or widespread. Measures to relieve pain are indicated for these patients. Only after acute symptoms have subsided can tests for underlying faults in alignment and muscle balance be done and specific therapeutic measures be instituted.

Important differences exist between treatment of an acutely painful condition and that of a chronic one. A given procedure may be recognized and accepted as therapeutic if it is applied at the proper time. Applied at the wrong time, this same procedure may be ineffective or even harmful.

Just like an injured neck, shoulder, or ankle, an injured back may need support. Nature's way of providing protection is by protective muscle spasm or muscle guarding, in which the back muscles hold the back rigid to prevent painful movements. Muscles can become secondarily involved, however, when they are overburdened by the work of protecting the back. Use of an appropriate support to immobilize the back can temporarily relieve the muscles of this function, permit healing of the underlying injury, and lead to a decrease in pain.

Immobilization is often a necessary expedient for the relief of pain, but prolonged immobilization can lead to adaptive shortening of the surrounding soft tissue. Stiffness of the body part is not a desirable outcome. The patient should understand that a transition from the acute stage to the recovery stage requires moving from immobilization to restoration of normal motion. Continuing use of a support that should have been discarded can perpetuate a problem that might otherwise resolve. See Box 2-1 for the principles of alignment, joints, and muscles.

THE STANDARD POSTURE

Posture is a composite of the positions of all the joints of the body at any given moment, and static postural alignment is best described in terms of the positions of the various joints and body segments. This chapter provides basic but essential information for analyzing postural alignment. One must have a clear understanding of the anatomical position as described previously in Chapter 1. Posture may also be described in terms of muscle balance. This chapter describes the muscle balance or imbalance associated with static postural positions.

As is true in all testing, there must be a standard when evaluating postural alignment. It is essential that the standard be met if the whole system of posture training that is built around it is to be sound. Basmajian and DeLuca states that "among mammals, man has the most economical of antigravity mechanisms once the upright posture is attained. The expenditure of muscular energy for what seems to be a most awkward position is actually extremely economical" (10).

In the *standard position*, the spine presents the normal curves and the bones of the lower extremities are in ideal alignment for weight bearing. The neutral position of the pelvis is conducive to good alignment of the abdomen and trunk and of the extremities below. The chest and thoracic spine are in a position that favors optimal function of the respiratory organs. The head is erect and in a well-balanced position that minimizes stress on the neck musculature (Figure 2-1).

The body contour in illustrations of standard posture shows the relationship of skeletal structures to surface outline in ideal alignment. Variations occur in body type and size, and the shape and proportions of the body are factors in weight distribution. Variations in contour are correlated, to some degree, with variations in skeletal alignment (11, 12). This is essentially true regardless of body build. An experienced observer should be able to estimate the position of the skeletal structures by observing the contours of the body (13, 14).

BOX 2-1

Principles of Alignment, Joints, and Muscles

Evaluating and treating postural problems requires an understanding of the basic principles relating to alignment, joints, and muscles:

- Faulty alignment results in undue stress and strain on bones, joints, ligaments, and muscles.
- Joint positions indicate which muscles appear to be elongated and which appear to be shortened.
- A relationship exists between alignment and muscle test findings if posture is habitual.
- Muscle shortness holds the origin and insertion of the muscle closer together.
- Adaptive shortening can develop in muscles that remain in a shortened condition.
- Muscle weakness allows separation of the origin and insertion of the muscle.
- Stretch weakness can occur in one-joint muscles that remain in an elongated condition.

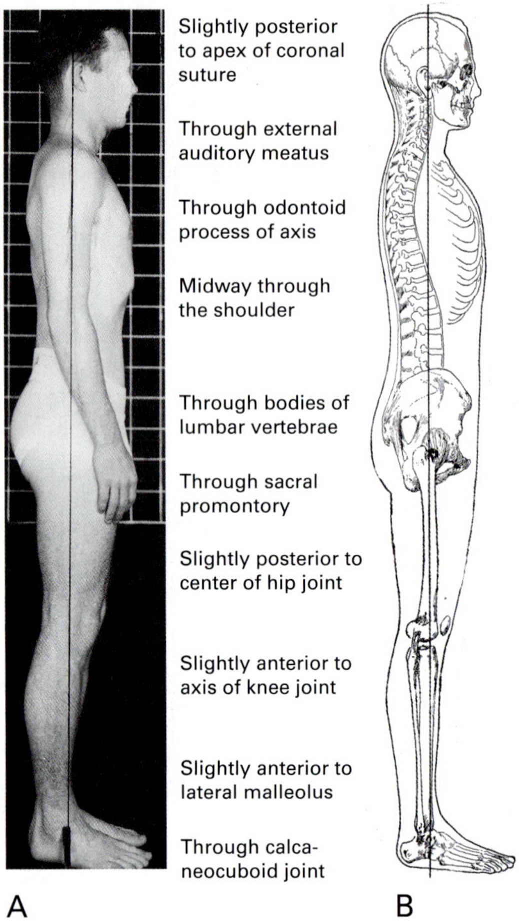

FIGURE 2-1: Postural examination landmark alignment.

The intersection of the midsagittal and coronal planes of the body forms a line that is analogous to the *gravity line* (15). Around this line, the body is hypothetically in a position of equilibrium. Such a position implies a balanced distribution of weight and a stable position of each joint.

Various machines are available for use in evaluating postural alignment. These complicated machines, which may be cost prohibitive to most clinics, often introduce variables that are difficult to control. Whitmore and Berman (1996) noted that "[c]ommercially available movement/posture evaluation systems require extensive data collection procedures, rigid camera calibrations, and referencing points" (16). Fortunately, accurate postural examinations can be done with simple equipment at minimal cost. See Table 2-1 for body regions, postural views, and lines of reference.

Head and Neck

The ideal alignment of the head and neck is one in which the head is in a well-balanced position that is maintained with minimal muscular effort. The head is not tilted upward or downward, and it is not tilted sideways or rotated. The chin is not retracted.

Good alignment of the thoracic spine is essential for good alignment of the head and neck; faulty alignment of the thoracic spine adversely affects the alignment of the head and neck. If the thoracic spine slumps into a rounded position when sitting or standing, a compensatory change will occur in the position of the head and neck.

If the head position were to remain fixed with the neck held in normal lordosis as the thoracic spine flexed into hyperkyphosis, the head would be inclined forward and downward. However, "eyes seek eye level," and the head must be raised from that position by extending the cervical spine. In normal extension of the cervical spine, there is an approximation of the occiput and the seventh cervical vertebra. As the head is raised to seek eye level, the distance between the occiput and the seventh cervical vertebra is reduced remarkably. Compared with the separation between the two points in ideal alignment, there may be as much as 2 or 3 inches of difference between the two positions.

The forward head position is one in which the neck extensors are in a shortened position and are strong, and the potential exists for the development of adaptive shortening in these muscles. The anterior vertebral neck flexors are in an elongated position and give evidence of weakness when tested for strength. See Figures 2-2 through 2-7.

TABLE 2-1 Body Regions, Postural Views, and Lines of Reference

Body Region	Sagittal View	Posterior View
	Line of Reference	*Line of Reference*
Head and Neck	• Lobe of the ear (through the external auditory meatus)	• Midline of the head Cervical spinous processes
Thoracic Spine		• Thoracic spinous processes
Upper Extremity	• Midway through the glenohumeral joint	
Pelvis and Lumbar Spine	• Posterior to the axis of the hip joint	• Sacral crest, Lumbar spinous processes
Hips and Knees	• Anterior to the axis of the knee joint	

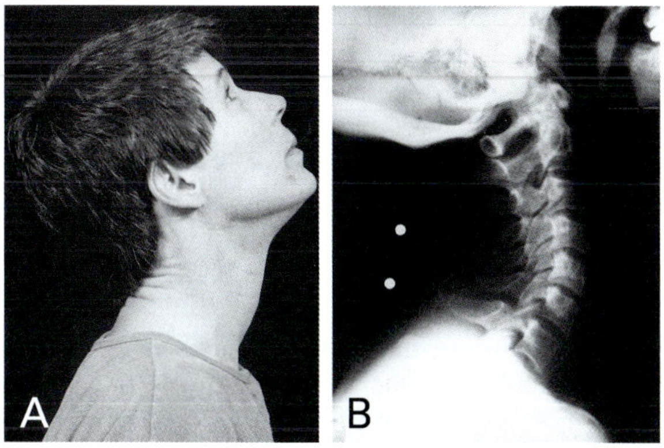

FIGURE 2-2: Subject with normal flexibility was photographed and x-rayed in five neck positions. Markers were placed at the hairline and at C7. The photograph shows cervical spine extension by tilting the head in a posterior direction. Note the approximation of the markers on the x-ray.

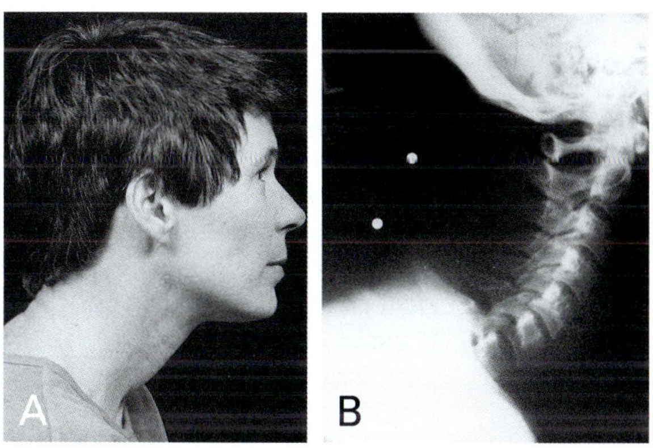

FIGURE 2-3: Cervical spine extension in a typical forward-head posture. Note the similarity in the curve and the positions of the markers to those in Figure 2-2. Often, this slumped posture is mistakenly referred to as flexion of the lower cervical spine and extension of the upper cervical spine. However, the extension is more pronounced in the lower than in the upper cervical region.

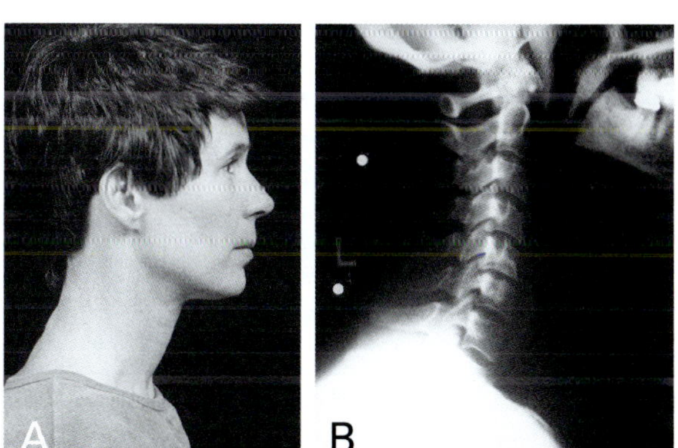

FIGURE 2-4: Good alignment of the cervical spine.

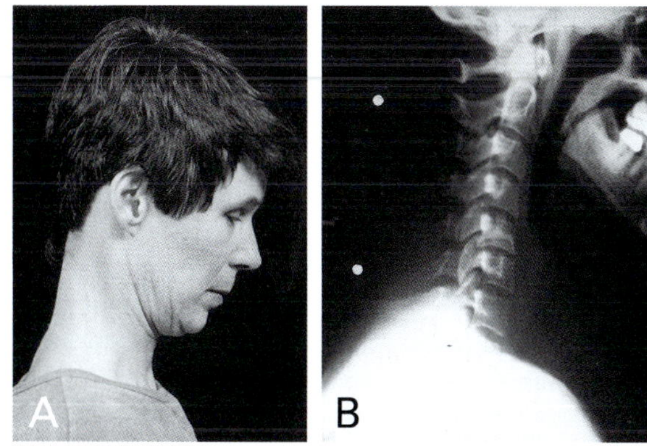

FIGURE 2-5: Flexion (flattening) of the cervical spine by tilting the head in an anterior direction.

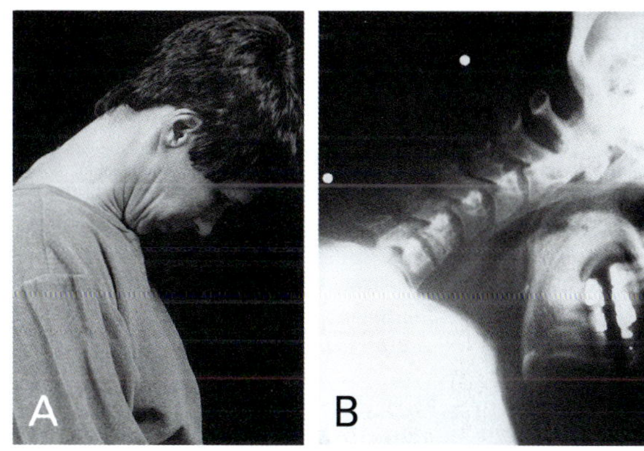

FIGURE 2-6: Flexion of both the cervical spine and the upper thoracic spine occurs when the chin is brought toward the chest.

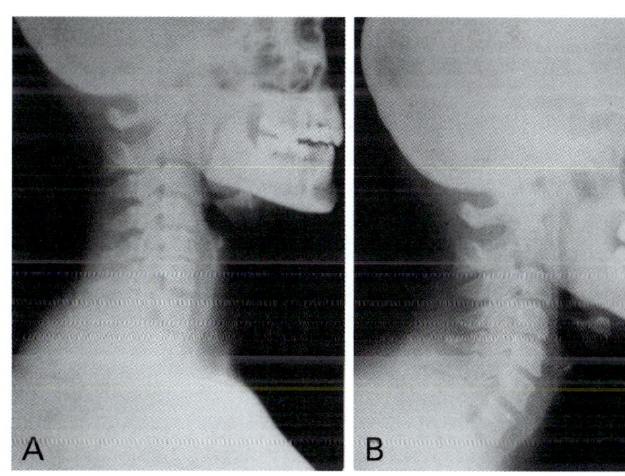

FIGURE 2-7: Cervical spine, good and faulty positions. For the x-ray in **A**, the subject sat erect, with the head and upper trunk in good alignment. For the x-ray in **B**, the same subject sat in a typically slumped position, with a round upper back and a forward head. As illustrated, the cervical spine is in extension.

Thoracic Spine

In ideal alignment, the thoracic spine curves slightly in a posterior direction (kyphosis). Just as the positions of the head and neck are affected by the position of the thoracic spine, the thoracic spine is affected by the positions of the lumbar spine and pelvis. With the pelvis and lumbar spine in ideal alignment, the thoracic spine can assume the ideal position. If a normally flexible individual assumes a position of increased lordosis of the lumbar spine (i.e., increased anterior curve), the thoracic spine tends to straighten, decreasing the normal posterior curve. On the other hand, habitual positions and repetitive activities may give rise to the development of a lordotic-kyphotic posture, in which one tends to compensate for the other. For example, in a sway-back posture, the position of increased posterior curvature of the thoracic spine compensates for a forward deviation of the pelvis (see Figure 2-17 later in this chapter).

Upper Extremity

The position of the arm and shoulder depends on the positions of the scapulae and thoracic spine. In ideal alignment, the scapulae lie flat against the thoracic spine approximately between the second and seventh thoracic vertebrae, and approximately 4 inches apart (more or less depending on the size of the individual). Faulty positions of the scapulae adversely affect the position of the shoulder, and malalignment of the glenohumeral joint can predispose to injury and chronic pain. The elbow will typically rest in slight flexion, the radioulnar joint in slight pronation, the wrist in neutral, and fingers (metacarpophalangeal and interphalangeal joints) in slight relaxed flexion.

Pelvis and Lumbar Spine

The relationship of the pelvis to the line of reference is determined to a great extent by the relationship of the pelvis to the hip joints. Because the sagittal view line of reference represents the plane passing slightly posterior to the axis of the hip joint of the viewed side, the pelvis will be intersected at the acetabula. However, these points of reference are not sufficient to establish the position of the pelvis, because the pelvis can tilt either anteriorly or posteriorly about the axis through the hip joint.

It is therefore necessary to define the **neutral position of the pelvis** in the standard posture. The neutral position used as the standard in this text is one in which the anterior-superior iliac spine is in the same horizontal plane and in which the anterior-superior iliac spines and the symphysis pubis are in the same vertical plane. From the standpoint of the action of muscles attached to the anterior iliac spine and the symphysis pubis, opposing groups of muscles have an equal mechanical advantage in a straight line of pull. The rectus abdominis, with its attachment on the pubis, extends upward to the sternum, and the rectus femoris, sartorius, and tensor fasciae latae, with their attachments on the anterior iliac spines, extend downward to the thigh.

Because of the structural variations of the pelvis, it is not practical to describe a neutral position on the basis of a specific anterior point and a specific posterior point being in the same horizontal plane. The anterior-superior iliac spines and the posterior-superior iliac spines are approximately in the same plane, however. In neutral position of the pelvis, there is a normal anterior curve (lordosis) in the low back. In anterior tilt of the pelvis, there is a hyperlordosis. In posterior tilt of the pelvis, there is a flat back.

Without minimizing the importance of proper foot positions that establish the base of support, it may be said that the position of the pelvis is the key to good or faulty postural alignment. The muscles that maintain good alignment of the pelvis anteriorly, posteriorly, and laterally are of utmost importance in maintaining good overall alignment. Imbalance between muscles that oppose each other in the standing position changes the alignment of the pelvis and adversely affects the posture of the body parts both above and below.

Hips and Knee

The standard sagittal-view line of reference through the lower extremities passes slightly posterior to the center of the hip joint and slightly anterior to the axis of the knee joint and represents a stable position of the hip and knee joints.

If the center of the hip and knee joints coincides with the line of gravity, there is an equal tendency for those joints to flex or to extend. However, this on-center position of the joint is not a stable one for weight bearing. The slightest force exerted in either direction will cause it to move off center unless it is stabilized by constant muscular effort. If the body must call on muscular effort to maintain a stable position, energy is expended unnecessarily.

If the hip joint and knee joint moved freely in extension as well as in flexion, there would be no stability, and constant effort would be required to resist movement in both directions. A stable off-center position for a joint is dependent on limitation of joint motion in one direction. For the hip and knee, extension is limited. Ligamentous structures, strong muscles, and tendons are the restraining forces preventing hyperextension. Stability in the standing position is obtained by this normal limitation of joint motion.

Exercises or manipulations that tend to hyperextend the knee or hip joint or that excessively stretch muscles such as hamstrings should be scrutinized carefully. The normal restraining influence of the ligaments and muscles helps to maintain good postural alignment with a minimum of muscular effort. When muscles and ligaments fail to offer adequate support, the joints exceed their normal range, and posture can become faulty with respect to the positions of knee and hip hyperextension.

Ankle

The standard line of reference passes slightly anterior to the lateral malleolus and approximately through the apex of the arch, designated laterally by the calcaneocuboid joint. In the standard position, dorsiflexion at the ankle with the knee extended is approximately 10°. This means that standing barefoot with feet in a position of slight out-toeing and with knees straight, the lower leg cannot sway forward on the foot more than about 10°. Forward deviation of the body (dorsiflexion at the ankle) is checked by the restraining tension of strong posterior muscles and ligaments. However, this element of restraint is materially altered with changes in heel height that place the ankle in varying degrees of plantar flexion, and it is appreciably altered if the knees are flexed.

Feet

In the standard posture, the position of the feet is one in which the heels are separated approximately 3 inches and the forepart of the feet are separated so that the angle of out-toeing is approximately 8–10° from the midline on each side, making a total of 20° or less. This position of the feet refers only to the static and barefoot position. Both elevation of the heels and motion affect the position of the feet.

In establishing a standard position of the feet—and in determining where, if at all, out-toeing should occur—it is necessary to consider the foot in relation to the rest of the lower extremity. The out-toeing position cannot occur at the knee, because there is no rotation in extension. In ideal alignment, the axis of the extended knee joint is in a frontal plane. With the knee joint in this plane, out-toeing cannot take place from the knee-joint level. There can be a position of out-toeing as a result of external rotation of the hip. In this case, however, the entire extremity would be externally rotated, and the degree of out-toeing would be exaggerated. This makes the question of whether there should be rotation of the foot into an out-toeing position dependent on the relationship of the foot to the ankle joint. The ankle joint permits dorsiflexion and plantarflexion primarily; it does not permit appreciable rotation. The ankle joint is not positioned entirely in the frontal plane. According to anatomists, it is in a slightly oblique plane. The line of obliquity is such that it extends from slightly anterior at the medial malleolus to slightly posterior at the lateral malleolus. The angle at which the axis of the ankle joint deviates from the frontal plane suggests that the foot is normally in a position of slight out-toeing in relation to the lower leg.

The foot is not a rigid structure. Movements of the subtalar and transverse tarsal joints permit pronation and supination of the foot as well as abduction and adduction of the forefoot. The combination of pronation and forefoot abduction is seen as eversion of the foot and the combination of supination and forefoot adduction as inversion. Passive or active movements of the foot and ankle reveal that the foot tends to move outward as it moves upward and to move inward as it moves downward.

In the standing position, the foot is not fully dorsiflexed on the leg, nor is it in full eversion. However, the person who stands with flexed knees and marked out-toeing of the feet will be in dorsiflexion and eversion—a position that results in stress and strain on the foot and leg. When influenced by shoes with heels, the standing position represents varying degrees of plantar flexion of the foot, dependent on the heel height. As heel height increases, the tendency toward a parallel position of the feet, or in-toeing, also increases.

The relationship of heel height to out-toeing or in-toeing of the foot is analogous to the position of the foot in standing, walking, and running. When standing barefoot, a slight degree of out-toeing is natural. Standing with heels raised or walking fast, the feet tend to become parallel. As speed increases from walking to sprinting, the heels do not contact the ground, and the weight is borne on the anterior part of the foot entirely. There is then a tendency for the print of the forefoot to show in-toeing.

SECTION II
ALIGNMENT

In the study of body mechanics, **plumb lines** represent the vertical planes. With the anatomical position of the body as the basis, positions and movements are defined in relation to these planes. **Body mechanics** analyzes the static and dynamic forces acting on the body. It is not an exact science, but to the extent that it is possible and meaningful, standards and precision must be incorporated in its study. The ideal alignment of the body is the standard.

When viewing a standing posture, a plumb line is used as a line of reference. The plumb line is a cord with a plumb bob attached to provide an absolute vertical line. The point in line with which a plumb line is suspended must be a standard fixed point. Because the only fixed point in the standing posture is at the base, where the feet are in contact with the floor, the point of reference must be at the base. A movable point is not acceptable as a standard. The position of the head is not stationary; therefore, using the lobe of the ear as a point in line with which to suspend a plumb line is not appropriate.

The plumb line test is used to determine whether the points of reference of the individual being tested are in the same alignment as the corresponding points in the standard posture. The deviations of the various points of reference from the plumb line reveal the extent to which the subject's alignment is considered faulty.

For the purpose of testing, subjects step up to a suspended plumb line. In posterior view, they stand with the feet equidistant from the line. In lateral view, a point just in front of the lateral malleolus is in line with the plumb line.

Deviations from the plumb alignment are described as slight, moderate, or marked rather than in terms of inches or degrees. During routine examinations, it is not practical to determine exactly how much each point of reference deviates from the plumb line.

The standing position may be regarded as a composite alignment of a subject from four views: anterior, posterior, sagittal right, and sagittal left side.

With ideal alignment as the standard, the positions of the head, neck, upper extremities, thoracic spine, lumbar spine, pelvis, and lower extremities are described and illustrated on the following pages.

IDEAL ALIGNMENT: ANTERIOR VIEW

Figure 2-8 shows an anterior view of the ideal alignment.

Head: Neutral position, neither tilted nor rotated.

Cervical Spine: Straight, no lateral flexion or rotation.

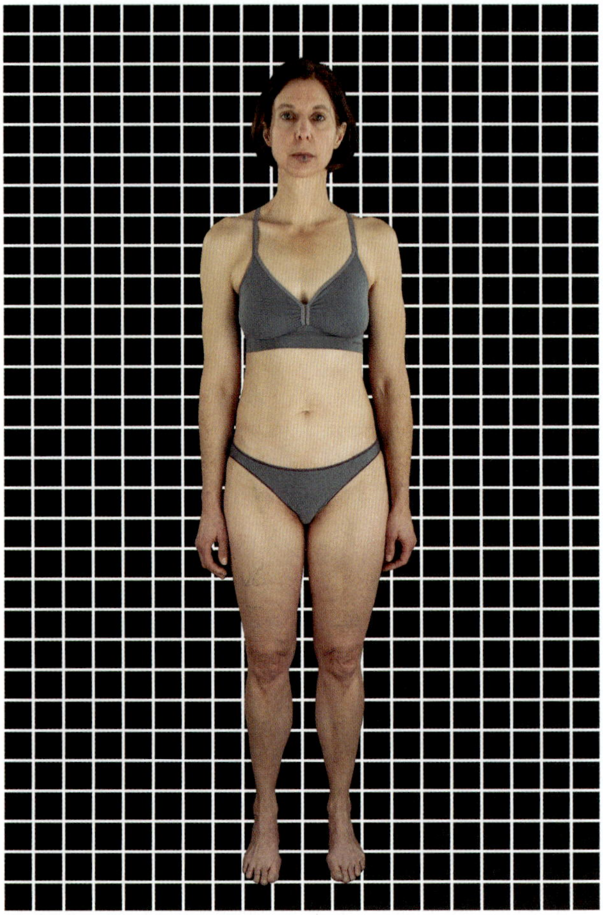

FIGURE 2-8: Ideal alignment, anterior view.

Shoulder: Level, not elevated or depressed.

Elbow: Slight carrying angle.

Thoracic Spine: Straight, no lateral flexion or rotation.

Lumbar Spine: Straight, no lateral flexion or rotation.

Pelvis: Level, both anterior-superior iliac spines in the same transverse plane.

Hip Joints: Neutral position, not adducted, abducted or rotated.

Knees Joints: Neutral position, no genu valgum or varum.

Ankle Joints: Neutral, no excessive inversion or eversion.

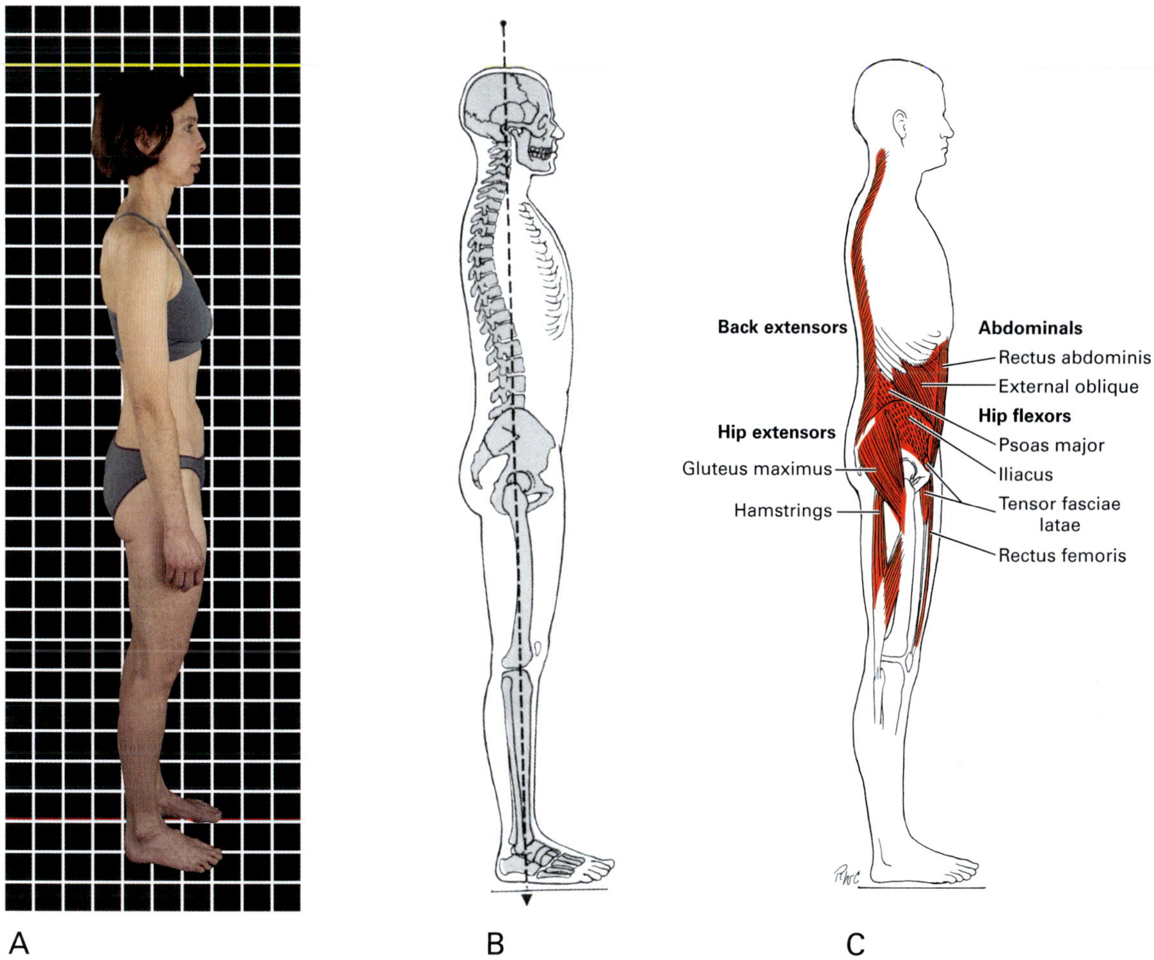

FIGURE 2-9: A, B, C, Ideal alignment, sagittal view.

IDEAL ALIGNMENT: SAGITTAL VIEW

In sagittal view (Figure 2-9), the standard line of reference in the drawings and the plumb line in the photographs represent a projection of the gravity line in the coronal plane. This plane hypothetically divides the body into anterior and posterior sections of equal weight. These sections are not symmetrical, and no line of division is obvious on the basis of anatomical structures.

Head: Neutral position, not tilted forward or back

Cervical Spine: Normal curve, slightly convex anteriorly.

Scapulae: Flat against the posterior thorax.

Thoracic Spine: Normal curve, slightly convex posteriorly.

Lumbar Spine: Normal curve, slightly convex anteriorly.

Pelvis: Neutral position, anterior-superior spines in the same vertical plane as the symphysis pubis.

Hip Joints: Neutral position, neither flexed nor extended.

Knee Joints: Neutral position, neither flexed nor hyperextended.

Ankle Joints: Neutral position, leg vertical and at a right angle to the sole of the foot.

In sagittal view, the anterior and posterior muscles attached to the pelvis maintain it in ideal alignment. Anteriorly, the abdominal muscles pull upward, and the hip flexors pull downward. Posteriorly, the back muscles pull upward, and the hip extensors pull downward. Thus, the anterior abdominal and hip extensor muscles work together to tilt the pelvis posteriorly; the low back and hip flexor muscles work together to tilt the pelvis anteriorly.

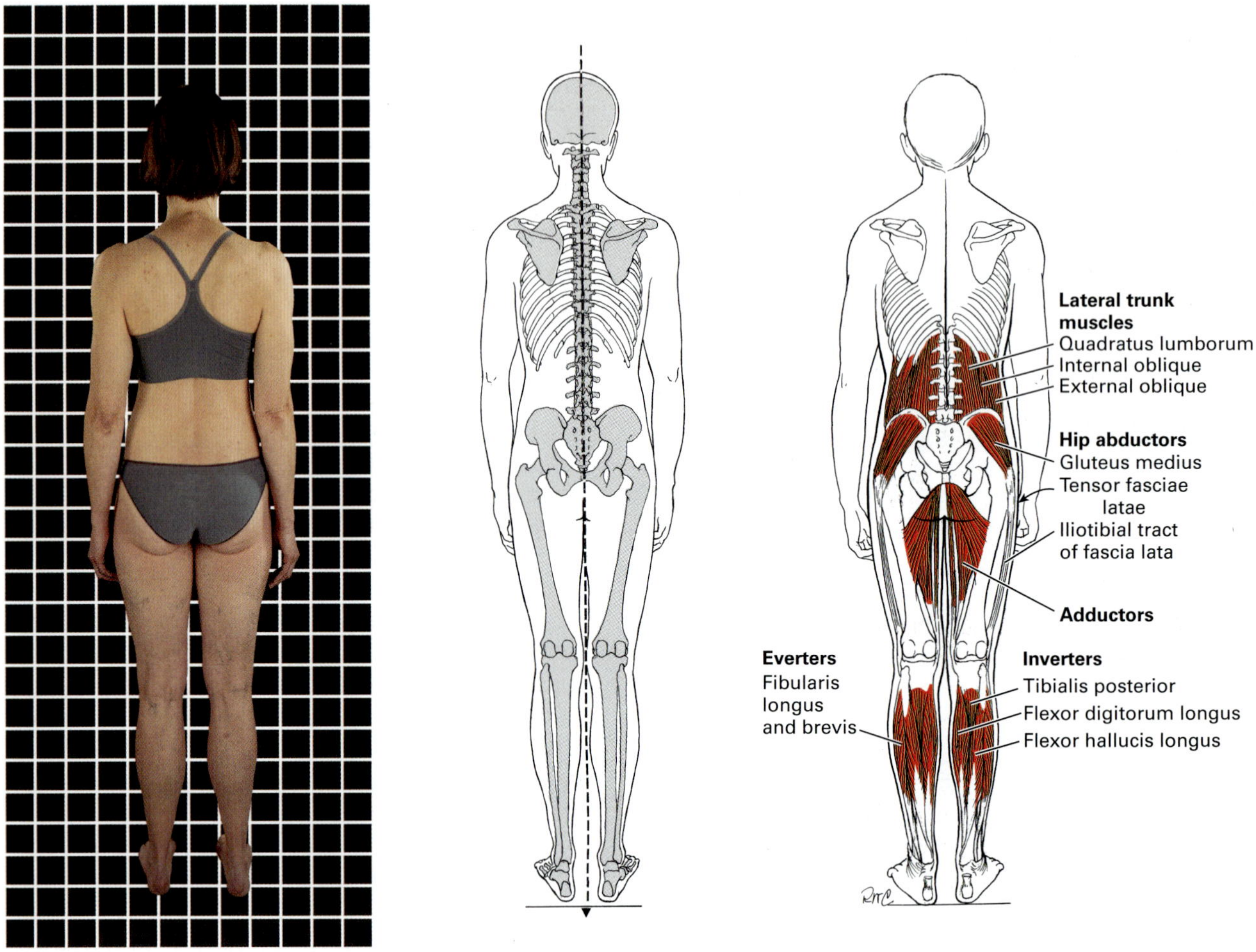

FIGURE 2-10: **A**, **B**, **C**, Ideal alignment, posterior view.

IDEAL ALIGNMENT: POSTERIOR VIEW

As shown in Figure 2-10, in a posterior view, the standard line of reference in the drawings and the plumb line in the photographs represent a projection of the gravity line in the midsagittal plane. Beginning midway between the heels, it extends upward midway between the lower extremities, through the median sacral crest, spine, sternum, and skull. The right and left halves of the skeletal structures are essentially symmetrical, and by hypothesis, the two halves of the body exactly counterbalance (17).

Head: Neutral position, neither tilted nor rotated.

Cervical Spine: Straight.

Shoulders: Level, not elevated or depressed.

Scapulae: Neutral position, medial borders essentially parallel and approximately 3 to 4 inches apart.

Thoracic and Lumbar Spines: Straight.

Pelvis: Level, both posterior-superior iliac spines in the same transverse plane.

Hip Joints: Neutral position, not adducted or abducted.

Lower Extremities: Straight, no genu valgum or genu varum.

Feet: Parallel or slight out-toeing. Outer malleolus and outer margin of the sole of the foot in same vertical plane so that the foot is not pronated or supinated. Tendo calcaneus should be vertical when seen in posterior view.

Shoulders and Scapulae

Shoulders and Scapulae, Good Position: The subject in Figure 2-11 illustrates a good position of the shoulders and scapulae. The scapulae lie flat against the thorax, and no angle or border is unduly prominent. Their position is not distorted by unusual muscular development or misdirected efforts at postural correction.

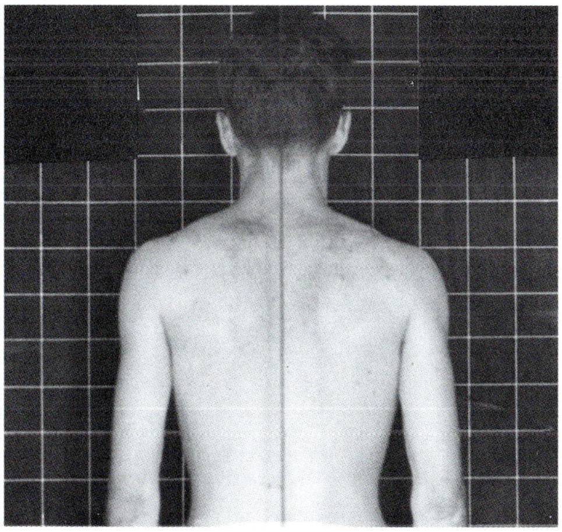

FIGURE 2-11: Shoulder and scapulae, good position.

SITTING POSTURE

Maintaining ideal alignment of the body in the sitting position can reduce or even prevent pain associated with posture-related problems (Figure 2-12). Part A shows ideal alignment, requiring the least expenditure of muscle energy. Part B shows the low back in a lordosis. This posture is mistakenly regarded as a correct position. The back muscles fatigue because it takes effort to maintain this position. Part C is a familiar slumped position that results in strain from lack of support for the low back and results in very faulty positions of the upper back, neck, and head.

People are usually advised to sit with their feet flat on the floor. If the knees are crossed, they should alternate so that they are not always crossed in the same manner. Some people, especially those with poor circulation in the legs, should avoid sitting with their knees crossed.

Some people may be comfortable in a chair with a pad in the lumbar area. Others may experience discomfort and even pain from such a lumbar support. Certain people find that a contoured pad in the sacroiliac area, or a chair that is rounded to conform to the body in that area, will enable them to sit comfortably.

There is no one correct chair. The height and depth of the chair must be appropriate to the individual. The chair should be of a height that allows the feet to rest comfortably on the floor to avoid pressure on the back of the thighs. In a chair that is too deep from front to back, either the individual's back will be unsupported or undue pressure will be placed against the lower leg. Hips and knees should be approximately at 90° angles and the back of the chair should incline approximately 10°.

Not all chairs are conducive to ideal sitting posture. So-called "posture chairs," which support the back only in the lumbar region, tend to increase the lumbar curve and are often undesirable. Sitting for long periods of time in a swivel chair that tilts back at too great an angle may contribute to a very faulty position of the upper back and head.

If the chair has armrests that are too high, the shoulders will be elevated. If the armrests are too low, the upper extremities will not have adequate support. With proper armrests it should be possible to pull the chair close to the desk. Whenever practical, tools and desk equipment should be placed within reach to avoid undue stretch or torsion.

When sitting for hours at a time, it is necessary to shift positions since a sitting position keeps the hips, the knees, and usually the back in flexion. Simple extension movements and occasionally standing up can alleviate the stress and strain associated with prolonged sitting positions (18).

In an automobile, it is important that the seat is comfortable. Pain and fatigue in the neck and shoulder region can often be attributed to the head being held in a forward head or laterally flexed position while driving.

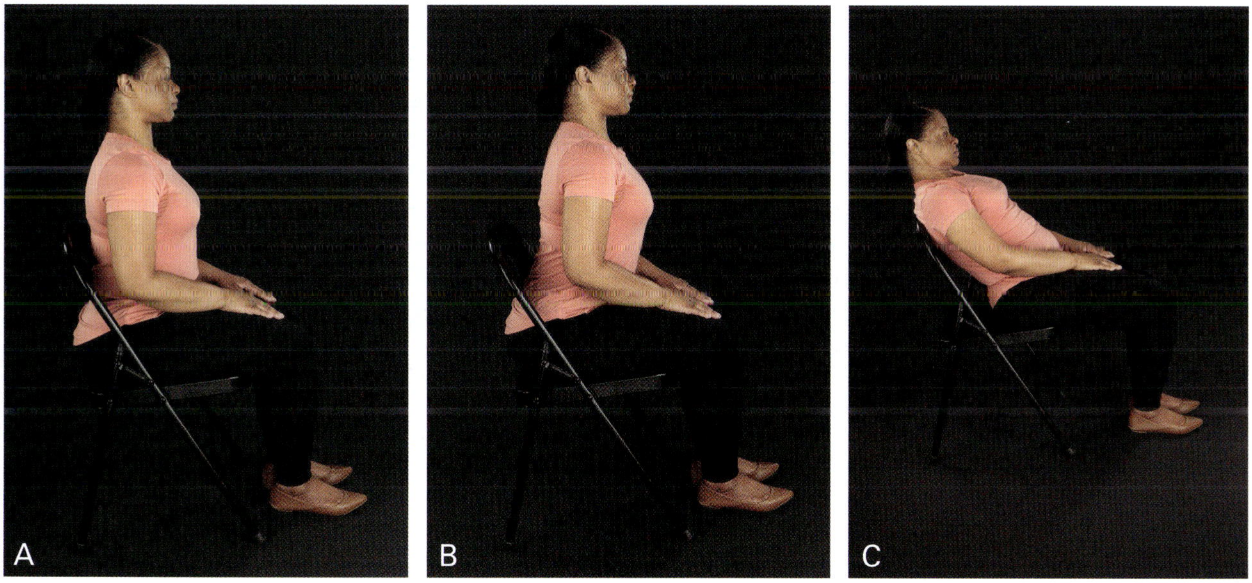

FIGURE 2-12: Sitting posture and alignment: **A**, Good alignment; **B**, The low back in lordosis; **C**, Slumped position.

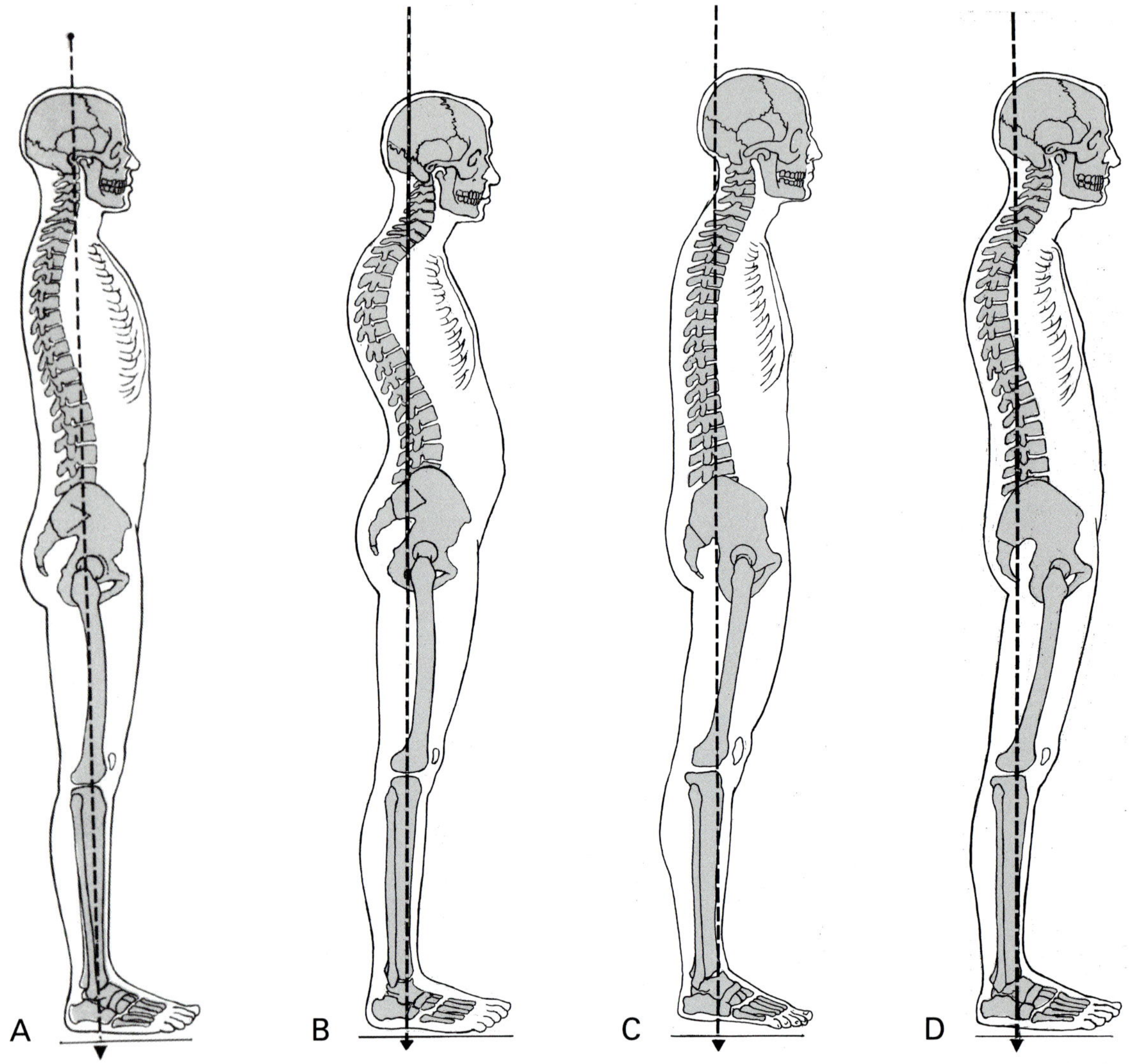

FIGURE 2-13: Four types of postural alignment: **A**, Ideal alignment; **B**, Kyphotic-lordotic posture; **C**, Flat-back posture; **D**, Sway-back posture.

TYPES OF VARIABLE POSTURAL ALIGNMENT

See Figure 2-13. The normal curves of the spine consist of a curve that is convex anteriorly in the cervical region (lordosis), a curve that is convex posteriorly in the thoracic region (kyphosis), and a curve that is convex anteriorly in the lumbar region (lordosis). This describes the neutral position of the spine. When there is a normal curve in the lumbar spine, the pelvis can attain a neutral position more readily. In Part A, the bony prominences at the front of the pelvis are in a neutral position, as indicated by the anterior-superior iliac spines and the symphysis pubis being in the same vertical plane.

In a faulty postural position, the pelvis may be in an anterior, posterior, or lateral tilt. Any tilting of the pelvis involves simultaneous movements of the low back and hip joints. In **anterior pelvic tilt**, as shown in Part B, the pelvis tilts forward, decreasing the angle between the pelvis and the thigh anteriorly, resulting in flexion of the hip joint; the low back arches forward, creating an increased anterior curve (hyperlordosis) in the lumbar spine. In **posterior pelvic tilt**, as shown in Parts C and D, the pelvis tilts backward, the hip joints extend, and the lumbar spine flattens. In **lateral pelvic tilt**, one hip is higher than the other, and the spine curves with convexity toward the lower side.

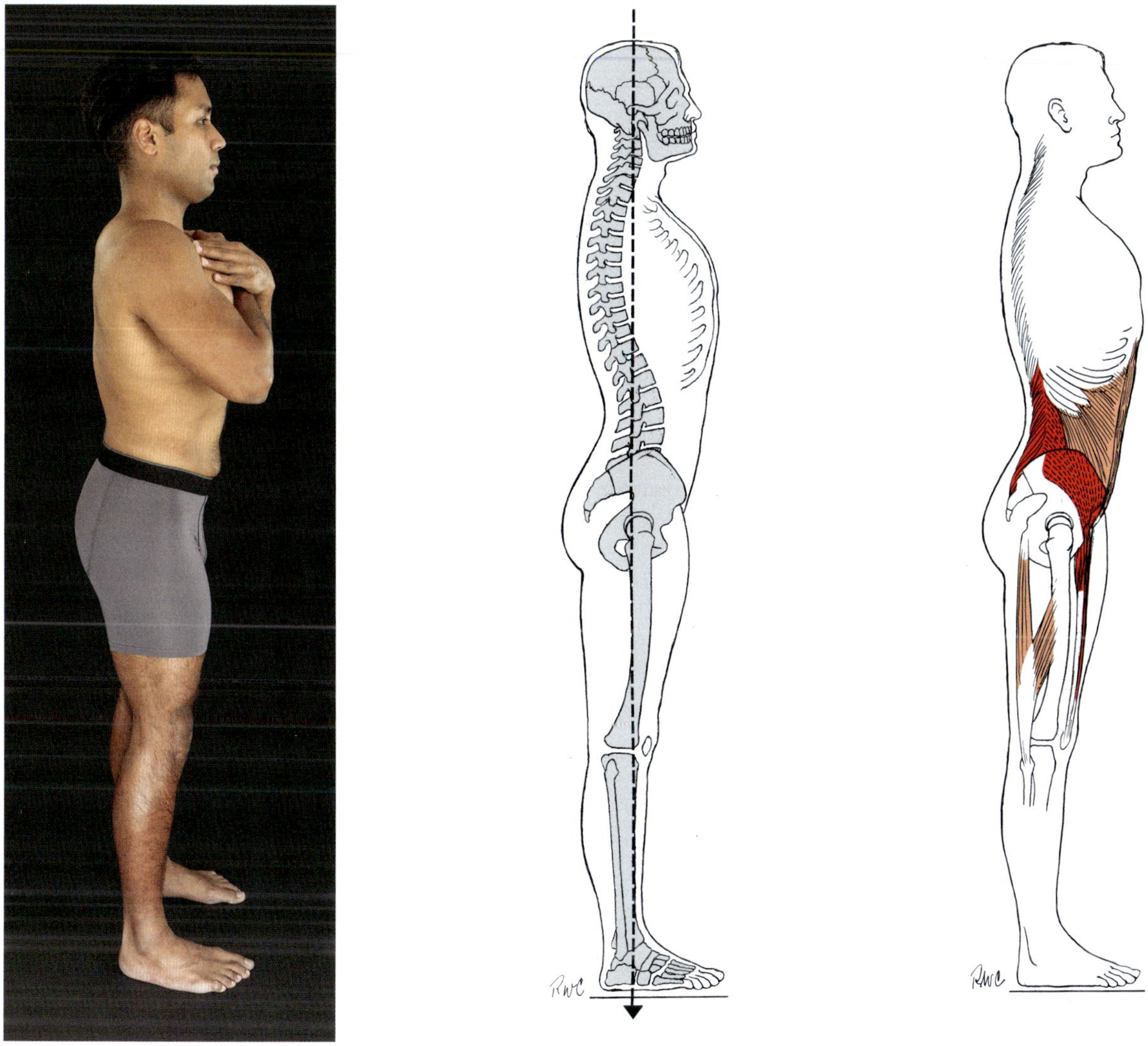

A B C

FIGURE 2-14: A, B, C, Lordotic posture.

Lordotic Posture

See Figure 2-14 for the lordotic posture.

Head: Neutral position.

Cervical Spine: Normal curve (lordosis).

Thoracic Spine: Normal curve (kyphosis).

Lumbar Spine: Hyperextended (hyperlordosis).

Pelvis: Anterior tilt.

Knee Joints: Slightly hyperextended.

Ankle Joints: Slightly plantar flexed.

Elongated and Weak: Anterior abdominals. Hamstring muscles are somewhat elongated but may or may not be weak.

Short and Strong: Low back and hip flexor muscles.

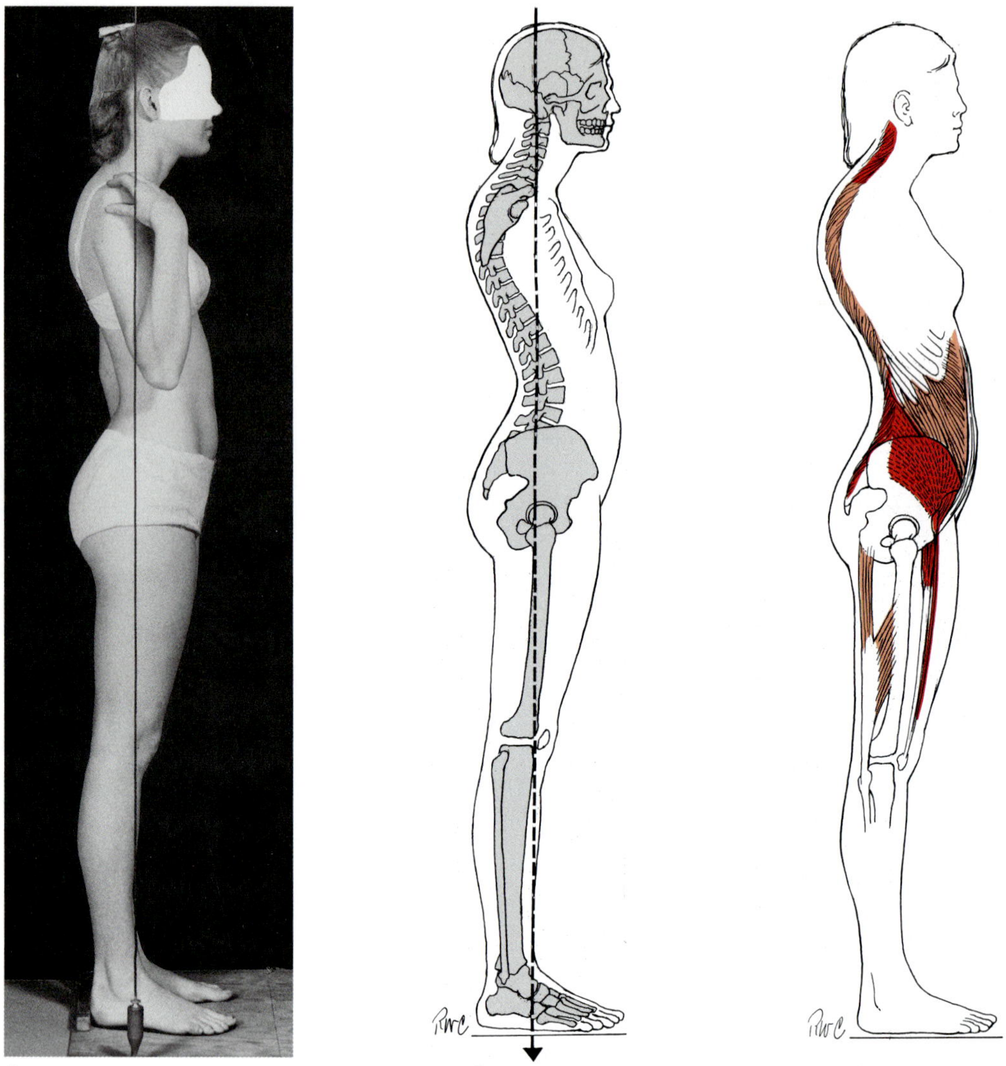

A B C

FIGURE 2-15: A, B, C, Kyphotic-lordotic posture.

Kyphotic-Lordotic Posture

Figure 2-15 shows the kyphotic-lordotic posture.

Head: Forward.

Cervical Spine: Hyperlordosis.

Scapulae: Abducted.

Thoracic Spine: Increased flexion (**hyperkyphosis**).

Lumbar Spine: Hyperextended (hyperlordosis).

Pelvis: Anterior tilt.

Hip Joints: Flexed.

Knee Joints: Slightly hyperextended.

Ankle Joints: Slight plantar flexion because of backward inclination of the leg.

Elongated and Weak: Neck flexors, erector spinae of the thoracic region, external oblique. Hamstrings are slightly elongated but may or may not be weak.

The rectus abdominis is not necessarily elongated, because the depressed position of the chest (due to thoracic hyperkyphosis) offsets the effect of the anterior pelvic tilt.

Hip flexors are in a shortened position in both the sitting posture and the lordotic posture in standing (as illustrated above). However, low back muscles may or may not be tight. In sitting the back may flatten. This combination of circumstances may explain why low back muscle shortness is less prevalent than hip flexor shortness in this type of posture.

Short and Strong: Neck extensors and hip flexors. The low back is strong and may or may not develop shortness.

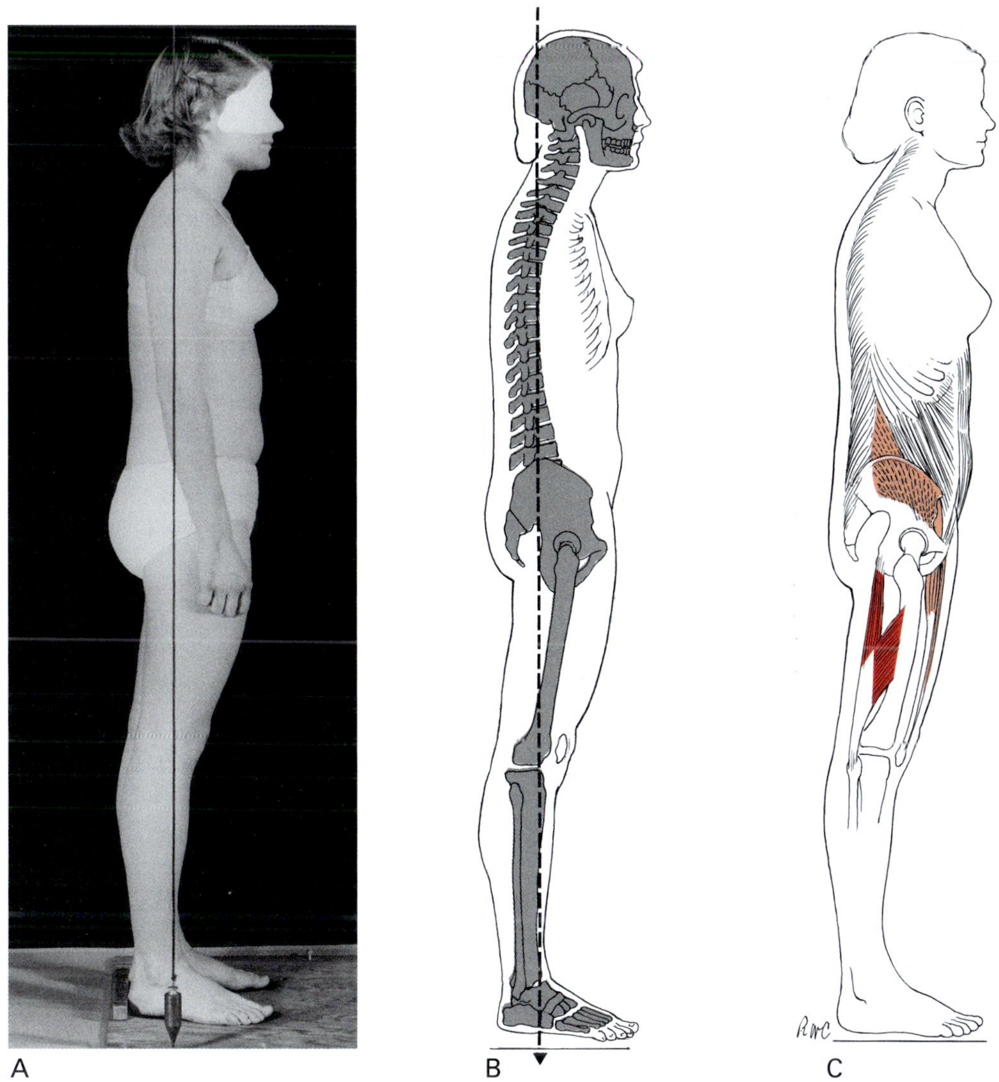

FIGURE 2-16: A, B, C, Flat-back posture.

Flat-Back Posture

The flat back posture is illustrated in Figure 2-16.

Head: Forward.

Cervical Spine: Upper cervical extension, lower cervical flexion.

Thoracic Spine: Upper part, increased flexion; lower part, straight.

Lumbar Spine: Flexed (straight).

Pelvis: Posterior tilt.

Hip Joints: Extended.

Knee Joints: Extended.

Ankle Joints: Slight plantar flexion.

Elongated and Weak: One-joint hip flexors.

Short and Strong: Hamstrings.

Frequently, abdominal muscles are strong. Although back muscles are slightly elongated when the normal anterior curve is eliminated, they are not weak. Sometimes, knees are slightly flexed rather than hyperextended along with the flat-back posture.

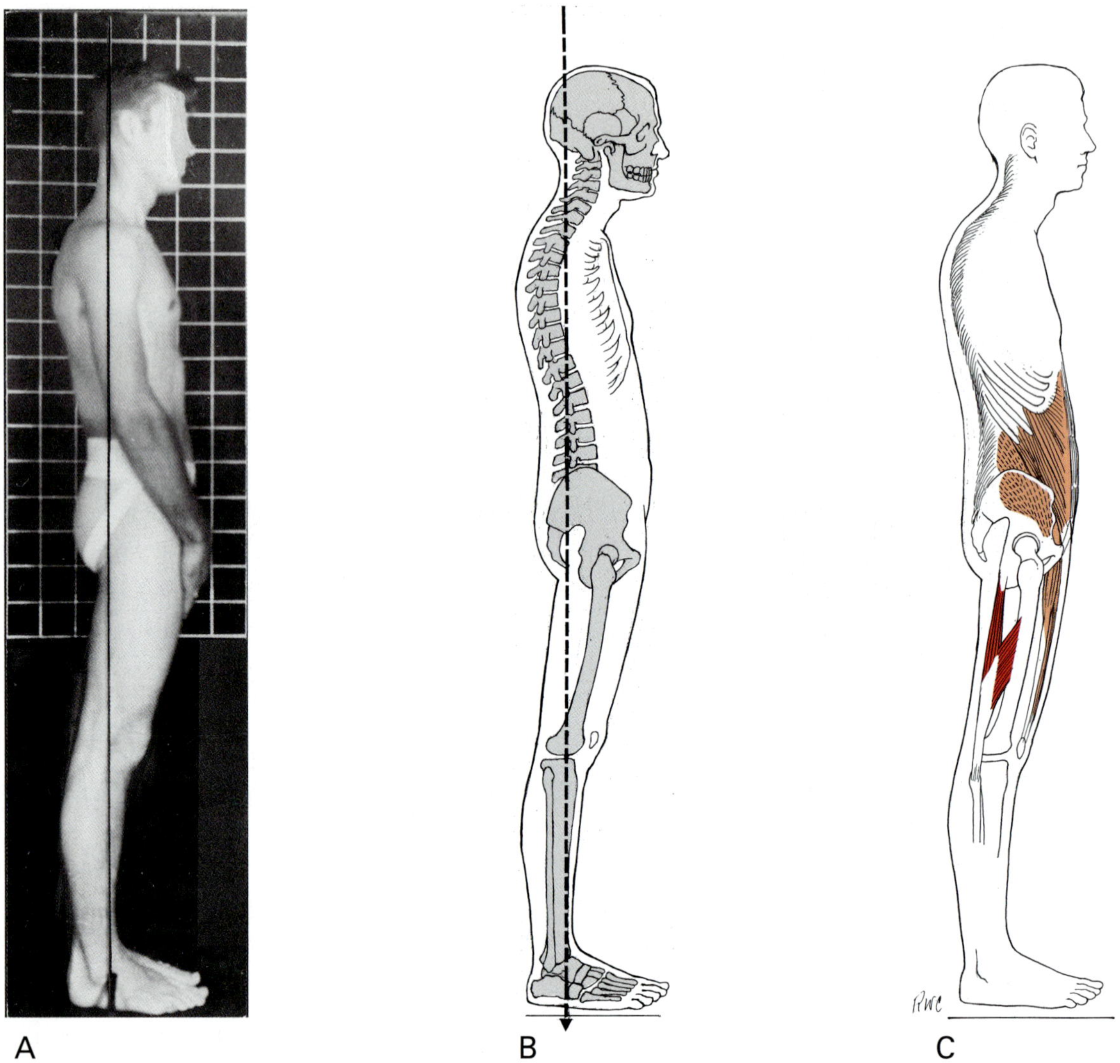

FIGURE 2-17: **A**, **B**, **C**, Sway-back posture.

Sway-Back Posture

Head: Forward.

Cervical Spine: Slightly extended.

Thoracic Spine: Increased flexion (long kyphosis) with posterior displacement of the upper trunk.

Lumbar Spine: Flexion (flattening) of the lower lumbar area.

Pelvis: Posterior tilt.

Hip Joints: Hyperextended with anterior displacement of the pelvis.

Knee Joints: Hyperextended.

Ankle Joints: Neutral. Knee joint hyperextension usually results in plantar flexion of the ankle joint, but that does not occur here because of anterior deviation of the pelvis and thighs.

Elongated and Weak: One-joint hip joint flexors, external oblique, upper back extensors, neck flexors.

Short and Strong: Hamstrings, upper fibers of the internal oblique.

Strong but Not Short: Low back muscles.

In Figure 2-17, the pelvis is in posterior tilt and sways forward in relation to the stationary feet, causing the hip joint to extend. The effect is equivalent to extending the leg backward with the pelvis stationary. With posterior pelvic tilt, the lumbar spine flattens. Hence, there is no lordosis, although the long curve in the thoracolumbar region (caused by the backward deviation of the upper trunk) is sometimes mistakenly referred to as a lordosis. (The term **sway-back posture** is an appropriate label and requires that the word *sway-back* not be used synonymously with *lordosis*.)

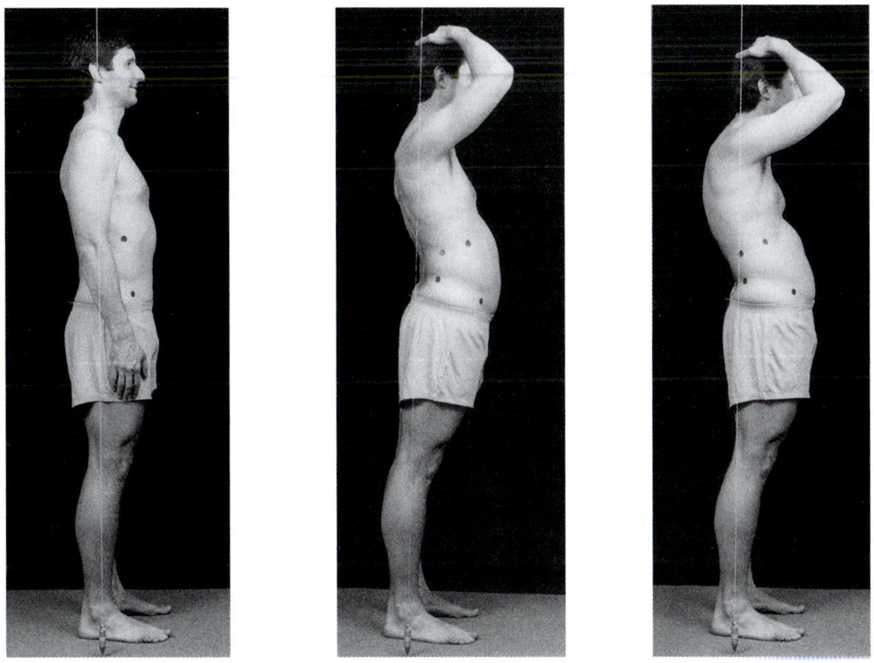

A Good posture alignment B Lordotic posture C Sway-back posture

FIGURE 2-18: Measurements of postural alignment: **A**, Dots representing the external oblique are 6 inches apart with the subject in good alignment. **B**, Dots representing the external oblique are 7 inches apart with the subject in a lordotic posture. **C**, Dots representing the external oblique are 71/2 inches apart with the subject in sway-back posture.

ABDOMINAL MUSCLES IN RELATION TO POSTURE

Transversus Abdominis in Relation to Posture

Despite its limited role in producing isolated joint movement, the transversus abdominis plays a pivotal role in supporting the lumbar spine during task performances, including sitting, lifting, and standing. Its origin from the thoracolumbar fascia allows for this stabilization. Normal length and tone of this muscle, along with that of the erector spinae, diaphragm, and pelvic floor muscles, has been shown to decrease isolated loading of spinal segments, control postural flexion in sitting, and promote core stabilization during movement of the trunk and limbs (19).

Faulty postures that can occur as a result of decreased transversus abdominis activation include increased thoracic kyphosis, decreased lumbar lordosis, or a combination of these two postures while an individual is in standing or sitting. Transversus abdominis activation assessment and training will be discussed further in Chapter 5.

External Oblique in Relation to Posture

The muscles that hold the pelvis in posterior tilt during leg lowering are chiefly the rectus abdominis and external oblique. In many instances, abdominal strength is normal on the trunk-raising test, but the muscles grade very weak on the leg-lowering test. Because the rectus must be strong to perform the trunk curl, the inability to keep the low back flat during leg lowering cannot be attributed to that muscle. It is logical to attribute the lack of strength to the external oblique, not to the rectus. Furthermore, the postural deviations that exist in persons who show weakness on the leg-lowering test are associated with elongation of the external oblique.

Two types of posture exhibit this weakness: anterior tilt (lordotic posture) and anterior displacement of the pelvis with posterior displacement of the thorax (sway-back posture). The lateral fibers of the external oblique extend diagonally from the posterolateral rib cage to the anterolateral pelvis. By this line of pull, they are in a position to help maintain good alignment of the thorax in relation to the pelvis or to restore the alignment when there is displacement. See Figure 2-18A, B, C.

The difference in grades between the trunk-raising test and the leg-lowering test is often very marked. Examination frequently reveals leg-lowering grades of only fair (5) to fair+ (6) in persons who can perform many curled trunk sit-ups. It becomes very clear in such situations that the trunk-raising exercise does not improve the ability to hold the low back flat during leg lowering. Indeed, it appears that repeated and persistent trunk-flexion exercises may contribute to continued weakness of the lateral fibers of the external oblique.

The type of postural deviation that occurs depends to a great extent on associated muscle weakness. In the anterior tilt, or *lordotic posture*, there is often *hip flexor tightness* along with the abdominal weakness; in the *sway-back posture*, there is *hip flexor weakness*, specifically of the iliopsoas.

See Figure 2-18A, B, C. Note the similarity between the lordotic and the sway-back curves in the back. Without careful analysis of the differences in plumb alignment and pelvic tilt, the sway-back curve might be referred to as lordosis, which it is not.

Good Postural Alignment: Pelvis is in neutral position.

Lordotic Posture: Pelvis is in anterior tilt.

Sway-Back Posture: Pelvis is in posterior tilt.

IDEAL POSTURE OF FEET, KNEES, AND LEGS

Ideal Posture of Feet and Knees

Ideal Alignment of Feet and Knees: The patellae face directly forward, and the feet are neither pronated nor supinated (Figure 2-19).

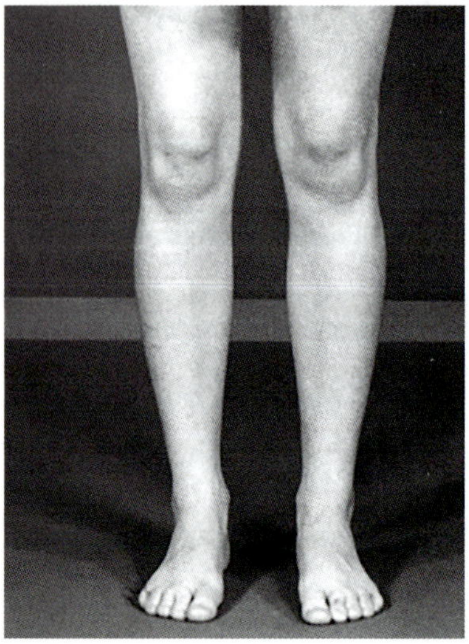

FIGURE 2-19: Ideal alignment of feet and knees.

Common Posture of Knees and Legs

Knees, Ideal Alignment: In good alignment of the knees, as in this side view (Figure 2-20), the plumb line passes slightly anterior to the axis of the knee joint.

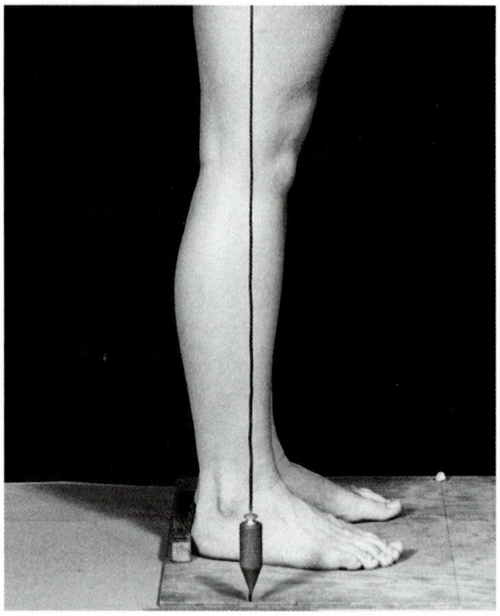

FIGURE 2-20: Ideal alignment of knees.

Ideal Alignment of Legs and Feet: Normal arch height, neutral subtalar and transverse tarsal position, slight (normal = approximately 4–7°) out-toeing (Figure 2-21).

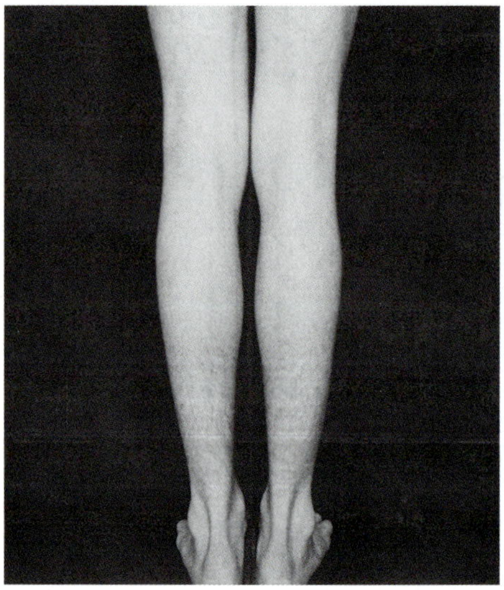

FIGURE 2-21: Ideal alignment of legs and feet.

SECTION III
POSTURAL EXAMINATION

Postural examination consists of assessing alignment in standing or sitting. Based upon that initial assessment, tests for flexibility, muscle length, and/or muscle strength may be indicated.

PROCEDURE FOR POSTURAL EXAMINATION

Equipment

The equipment required to perform a postural examination includes adequate space and lighting, frames of reference, and standard positioning cues. Individuals being evaluated must be attired in clothing that is comfortable and provides a balance of preserving modesty while exposing anatomical landmarks that are necessary to align with the points of reference.

Vertical Reference

Items used as a vertical reference in the clinic include the vertical portion of a door frame, corner of a treatment room, grid poster, laser level, and a plumb line.

Horizontal Reference

Items used as a horizontal reference in the clinic include the horizontal portion of a window frame, level treatment table that is height adjustable, grid poster, and a laser level.

Standard Foot Position

Standardizing the position of the foot within the examination environment is critical for minimizing error with intersession reliability. Floor grout lines or tape aligned perpendicular to the vertical reference is a low-cost, useful method to standardize foot position during a posture examination.

Other Equipment

Rulers, washable markers, goniometers, and gravity-dependent inclinometers are useful tools for quantifying normal and variant positions between anatomical landmarks. Mirrors, still photographs, video recordings, and the use of motion capture applications on cell phones can provide real-time feedback to individuals during a posture examination and subsequent training sessions.

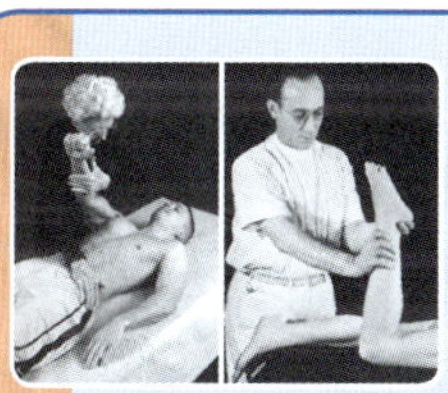

CLASSIC KENDALL

Equipment Used by the Kendalls

The equipment used by the Kendalls consisted of posture boards, plumb line, folding ruler with spirit level, set of six blocks, a marking pencil, and appropriate clothing.

Posture Boards

These are boards on which footprints have been drawn. Footprints may be painted on the floor of the examining room, but the posture boards have the advantage of being portable.

Plumb Line

This line is suspended from an overhead bar, and the plumb bob is hung in line with the point on the posture board that indicates the standard base point (i.e., anterior to the lateral malleolus in sagittal view, midway between the heels in back view).

Folding Ruler with Spirit Level

This is used to measure the difference in level of the posterior iliac spines. It also may be used to detect any differences in shoulder level. A background with squares (as shown in many of the photographs) is a more practical aid in detecting differences in shoulder level.

Set of Six Blocks

These blocks measure 4 inches by 10 inches and are of the following thicknesses: 1/8, 1/4, 3/8, 1/2, 3/4, and 1 inch. They are used for determining the amount of lift needed to level the pelvis laterally.

(continued)

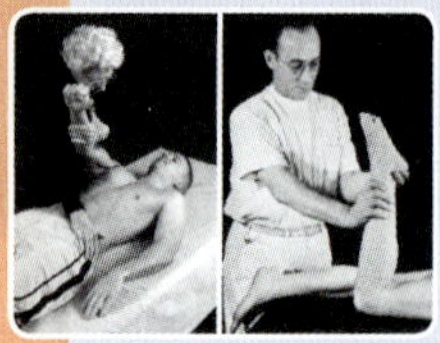

CLASSIC KENDALL *(continued)*

Marking Pencil

This is used for marking the spinous processes to observe the position of the spine in cases of lateral deviation.

Appropriate Clothing

Clothing, such as a two-piece bathing suit for girls or swim trunks for boys, should be worn by subjects for a postural examination. Such an examination of schoolchildren is unsatisfactory when children are clothed in ordinary gym suits. In hospital clinics, gowns or other suitable garments should be provided.

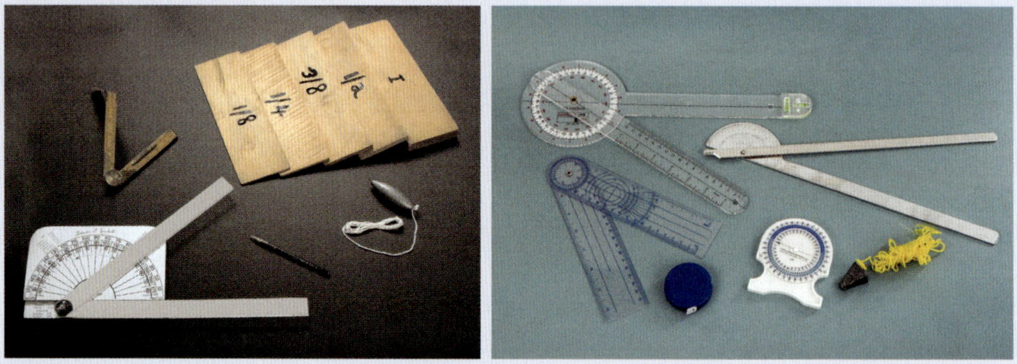

A B

FIGURE 2-22: Equipment for postural examination: **A**, Classic Kendall; **B**, Modern.

The equipment in Figure 2-22A consist of (left to right) protractor and caliper, folding ruler with spirit level, set of blocks, plumb line, and marking pencil. Figure 2-22B is the modern, currently used equipment: Top row left to right: large goniometer, large metal goniometer. Bottom row left to right: small goniometer, tape measure, bubble inclinometer, plumb line (weight with yellow string).

Figure 2-23 shows the posture boards with footprints on which the subject stands for alignment tests.

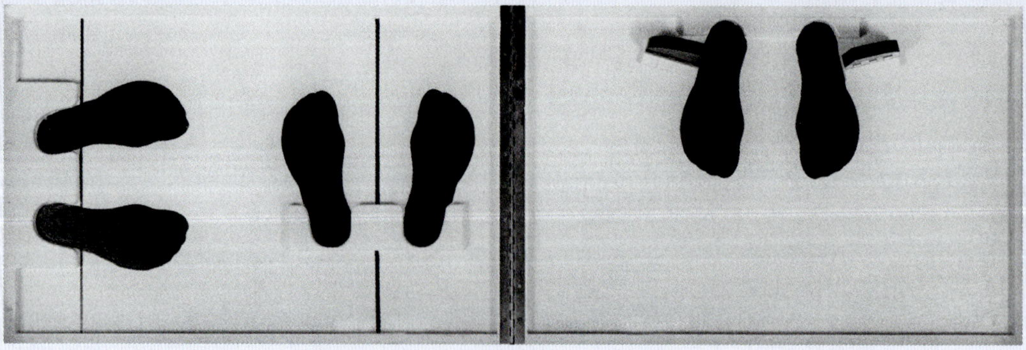

A B C

FIGURE 2-23: Posture boards with footprints on which the subject stands for alignment tests: **A**, Sagittal view; **B**, Posterior view; **C**, Anterior view.

Approach to Postural Assessment

Subjects stand on the posture boards with their feet in the position indicated by the footprints. If posture boards are not available, foot position must be standardized in the clinic for consistency of successive assessment.

Anterior View

Observe the position of the feet, knees, and legs. Toe positions, appearance of the longitudinal arch, alignment regarding pronation or supination of the foot, rotation of the femur as indicated by position of the patella, genu valgum, or genu varum should be noted. Any rotation of the head or abnormal appearance of the ribs should also be noted. Findings are recorded in the medical record.

Sagittal View

With the plumb line hung in line with a point just anterior to the lateral malleolus, the relationship of the body as a whole to the plumb line is noted and recorded under a heading such as Plumb Alignment. It should be observed from both the right and the left sides for the purpose of detecting rotation faults. Descriptions such as the following may be used in recording findings: "Body anterior from ankles up," "Pelvis and head anterior," "Good except lordosis," or "Upper trunk and head posterior."

Segmental alignment faults may be noted with or without the plumb line. Observe whether the knees are in good alignment, hyperextended, or flexed. Note the position of the pelvis as seen from the sagittal view and whether the anteroposterior curves of the spine are normal or exaggerated. Also note the head position (forward or tilted up or down), the chest position (whether normal, depressed, or elevated), and the contour of the abdominal wall. Findings are recorded on the chart under a heading such as Segmental Alignment.

Posterior View

With the plumb line hung in line with a point midway between the heels, the relationship of the body or parts of the body to the plumb line are expressed as good or as deviations toward the right or left. These findings are recorded on the chart (Figure 2-24), under the heading Note.

From the standpoint of segmental alignment, one should note the alignment of the tendo calcaneus, postural adduction or abduction of the hips, relative height of the posterior iliac spines, lateral pelvic tilt, lateral deviations of the spine, and positions of the shoulders and the scapulae. For example, a lateral pelvic tilt may result from one foot being pronated or one knee being habitually flexed, allowing a dropping of the pelvis on that side in standing.

TESTS AND INTERPRETATIONS

Test for Flexibility and Muscle Length

Findings regarding flexibility and muscle length are recorded on the chart in the space provided. Forward bending is designated as Normal, Limited, or Normal+, with the number of inches from or beyond the toes also being recorded (see charts regarding normal for various ages). In Figure 2-24, the Postural Examination chart, *Bk* indicates back, *H.S.* indicates hamstrings, and *G.S.* indicates gastroc-soleus.

Forward bending may be checked in the standing or sitting position, but the authors consider the test in the sitting position to be more indicative of flexibility. If flexibility is normal when sitting and limited when standing, there is usually some rotation or lateral tilt of the pelvis, resulting in rotation of the lumbar spine that in turn restricts the flexion in the standing position.

Findings regarding the arm overhead elevation tests may be recorded as normal or limited. If limited, findings may be further recorded as slight, moderate, or marked.

Trunk extension is the movement of backward bending, and it may be done in the standing position to help differentiate the flexibility of the back from the strength of the back muscles as done in the prone position. Normally, the back should arch in the lumbar region. If hyperextension is limited, the subject may try to simulate backward bending by flexing the knees and leaning backward. Knees should be kept straight during this test.

Lateral flexion movements are used to test the lateral flexibility of the trunk. The length of the left lateral trunk muscles permits range of motion for trunk bending toward the right, and vice versa. In other words, if flexibility of the trunk toward the right is limited, it should be interpreted as some muscle tightness of the left lateral trunk muscles—unless, of course, there is the element of limited spinal motion because of ligamentous or joint tightness.

Among other things, variations among individuals in length of the torso and in space between the ribs and iliac crest make for differences in flexibility. It is impractical to try to measure the degree of lateral flexion. Range of motion is considered to be normal when the rib cage and iliac crest are closely approximated in side-bending. Most people can bring their fingertips to about the level of the knee when bending directly sideways.

Name .... Doctor ....

Diagnosis .... Date of 1st Ex. ....

Onset .... Date of 2nd Ex. ....

Occupation .... Height .... Weight ....

Handedness .... Age .... Sex .... Leg length: Left .... Right ....

PLUMB ALIGNMENT

Side view: Lt .... Rt. ....

Back view: Deviated lt. .... Deviated rt. ....

SEGMENTAL ALIGNMENT

Feet	Hammer toes	Hallux valgus	Low ant. arch	Ant. foot varus
	Pronated	Supinated	Flat long. arch	Pigeon toes
	Med. rotat.	Lat. rotat.	Knock-knees	
Knees	Hyperext.	Flexed	Bowlegs	Tibial torsion
Pelvis	Leg in postural add.	Rotation	Tilt	Deviation
Low back	Lordosis	Flat	Kyphosis	Operation
Up. back	Kyphosis	Flat	Scap. abducted	Scap. elevated
Thorax	Depressed chest	Elevated chest	Rotation	Deviation
Spine	Total curve	Lumbar	Thoracic	Cervical
Abdomen	Protruding	Scars	....	....
Shoulder	Low	High	Forward	Med. rotated
Head	Forward	Torticollis	Lateral Tilt	Rotation ....

TESTS FOR FLEXIBILITY AND MUSCLE LENGTH

Forward bending .... Bk. .... H.S. ...... G.S. ....

Arm overhead elevation: Lt. .... Rt. ....

Hip flexors: Lt. .... Rt. ....

Tensor fas. lata.: Lt. .... Rt. ....

Trunk extension: ....

Trunk lat. flex.: To lt. .... To rt. ....

MUSCLE STRENGTH TESTS

L		R
	Mid. trapezius	
	Low. trapezius	
	Back extensors	
	Glut. medius	
	Glut. maximus	
	Hamstrings	
	Hip flexors	
	Tib. posterior	
	Toe flexors	

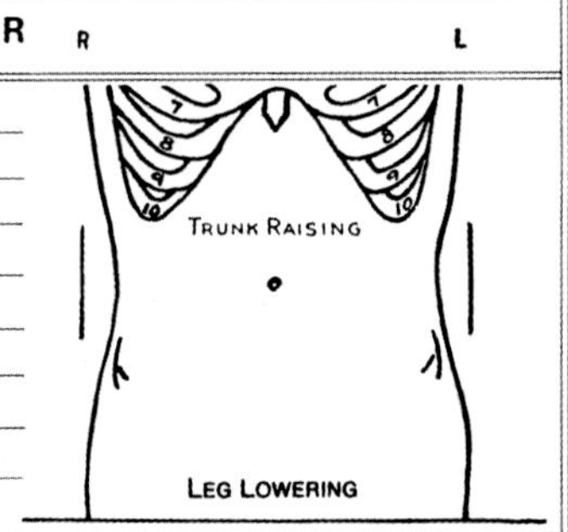

SHOE CORRECTION

Left		Right
	(Wide Heel) Inner wedge (Narrow heel)	
	Level heel raise	
	Metatarsal support	
	Longitudinal support	

NOTES: ....

....

....

....

....

....

TREATMENT

....

....

....

....

Exercises:

Bk. Lying	Pel. tilt and breath.	....
	Pel. tilt and leg sl.	....
	Head and sh. raising	....
	Shoulder add. stretch	....
	Straight leg-raise	....
	Hip flex. stretch	....
Sd. Lying	Stretch .... tensor	....
Sitting	Forward bending	
	To stretch low bk.	....
	To stretch h. s.	....
	Wall-sitting	
	Middle trapezius	....
	Lower trapezius	....
Standing	Foot and knee ex.	....
	Wall-standing	....

Other Exercises: ....

....

....

....

....

....

....

....

....

Support: ....

....

....

FIGURE 2-24: Postural examination chart.

Muscle Strength Tests

The essential muscle tests during postural examinations are described in Chapters 5–7. They include tests of the upper, lower, and oblique abdominals as well as the lateral trunk flexors, back extensors, middle and lower trapezius, serratus anterior, gluteus medius, gluteus maximus, hamstrings, hip flexors, soleus, and toe flexors.

With problems of anteroposterior deviations in postural alignment, it is especially important to test the abdominal muscles, back muscles, hip flexors and extensors, and soleus. With problems of lateral deviation of the spine or lateral tilt of the pelvis, it is especially important to test the oblique abdominal muscles, lateral trunk flexors, and gluteus medius.

Interpretation of Findings

In the usual case of faulty posture, the pattern of faulty body mechanics determined by the alignment test will be confirmed by the muscle tests if both procedures have been accurate. At times, however, there may be an apparent discrepancy in test findings. This inconsistency may result from:

- the effects of an old injury or disease altering the alignment pattern, particularly as related to handedness patterns;
- the effects of a recent illness or injury superimposed on an established pattern of imbalance; and
- a child with a lateral curvature of the spine in a transition stage between a C-curve and an S-curve.

Except in flexible children, postural faults seen at the time of examination will usually correspond with the habitual faults of the individual. With children, it is necessary and advisable to do repeated tests of alignment and to obtain information regarding their habitual posture from the parents and teachers who see them frequently. It is also advisable to keep photographic records of posture to attain a worthwhile evaluation of postural changes in growing children. See Figure 2-24.

SECTION IV
CLINICAL FINDINGS

HANDEDNESS: EFFECT ON POSTURE

Parts A and B in Figure 2-25 illustrate typical patterns of posture as related to handedness. Part A shows the pattern typical of right-handed individuals. The right shoulder is lower than the left, the pelvis is deviated slightly toward the right, and the right hip appears to be slightly higher than the left. Usually, there is a slight deviation of the spine toward the left, and the left foot is more pronated than the right. The right gluteus medius is usually weaker than the left.

Handedness patterns related to posture may begin at an early age. The slight deviation of the spine toward the side opposite to the higher hip may appear as early as 8 or 10 years of age. There tends to be a compensatory low shoulder on the side of the higher hip. In most cases, the low shoulder is less significant than the high hip. Usually, shoulder correction tends to follow correction of lateral pelvic tilt, but the reverse does not necessarily occur.

Part B shows the opposite pattern, which is typical of left-handed individuals. Usually, however, the depressed shoulder is not as marked as in right-handed individuals. See also Acquired Postural Weakness for handedness patterns of acquired postural weakness.

SAGITTAL AND POSTERIOR VIEWS

Figure 2-26, Part A is an example of posture that appears to be good in back view but is very faulty in sagittal view. The sagittal-view posture shows marked segmental faults, but the anterior and posterior deviations compensate for each other so that the plumb alignment is quite good. The contour of the abdominal wall almost duplicates the curve of the low back.

Part B shows a posture that is faulty in both side and back views. The back view shows a marked deviation of the body to the right of the plumb line, a high right hip, and a low right shoulder. The sagittal view shows that the plumb alignment is worse than the segmental alignment. The knees are posterior, and the pelvis, trunk, and head are markedly anterior. Segmentally, the anterior and posterior curves of the spine are only slightly exaggerated. The knees, however, are quite hyperextended.

The result in this subject is so much forward deviation of the trunk and head that the posture is most unstable and requires a good deal of muscular effort to maintain balance. The anterior part of the foot shows evidence of strain. An individual with this type of fault might appear as someone with good posture when fully clothed.

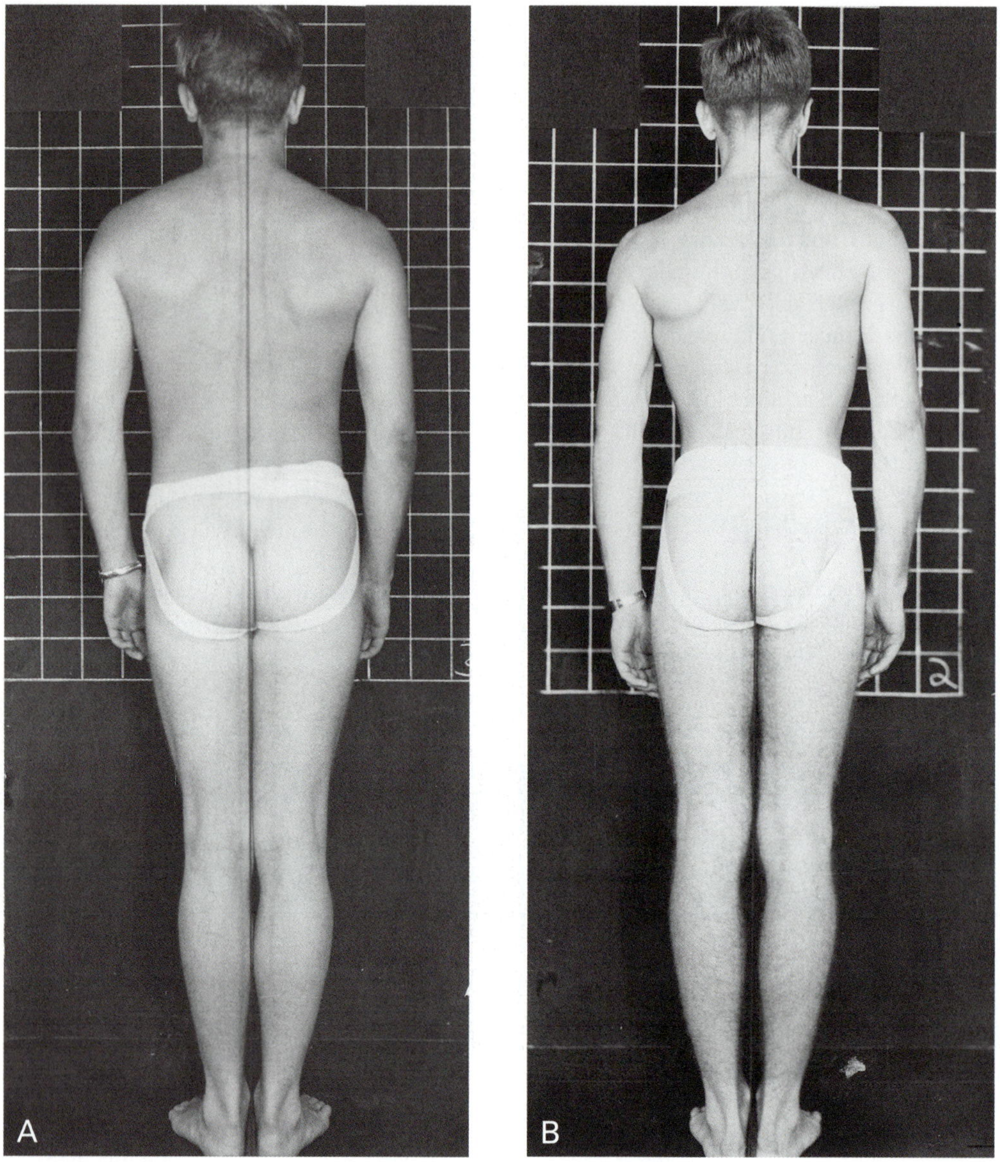

Right-handed Left-handed

FIGURE 2-25: Handedness pattern: **A**, Right-handed; **B**, Left-handed.

Shoulders and Scapulae

Scapulae, Abducted and Slightly Elevated: In Figure 2-27, in this subject, both scapulae are abducted, with the left one more than the right. Both are also slightly elevated. The abduction and elevation are accompanied by forward shoulders and round upper back.

Shoulder Elevated, Scapulae Adducted: In Figure 2-28, in this subject, both shoulders are elevated, with the right slightly higher than the left. The scapulae are adducted. The upper trapezius and other shoulder elevators are tight.

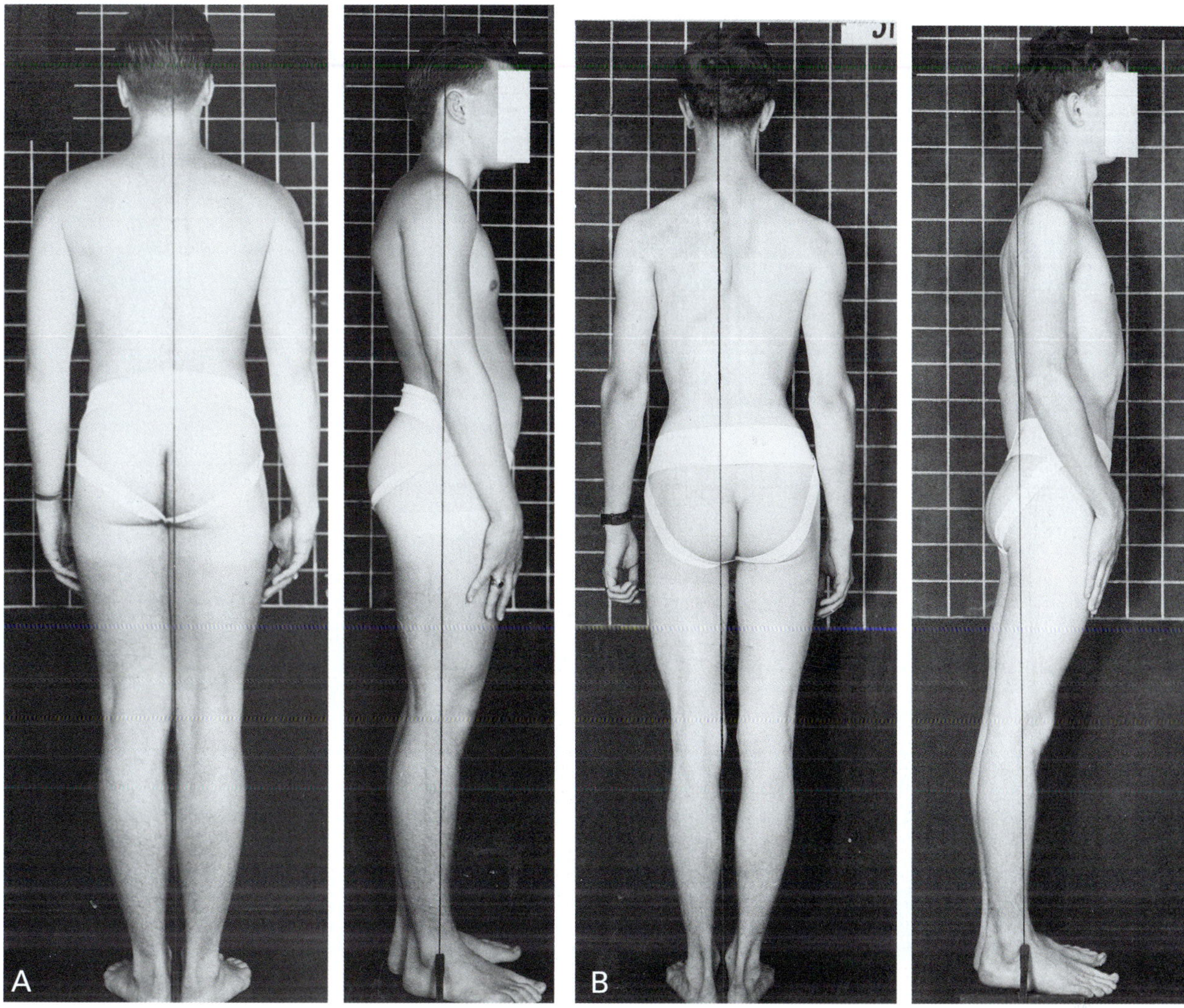

Back Side

FIGURE 2-26: Faulty posture (side and back views): **A**, Posture appears to be good in back view, but is faulty in side view; **B**, Posture is faulty in both back and side views.

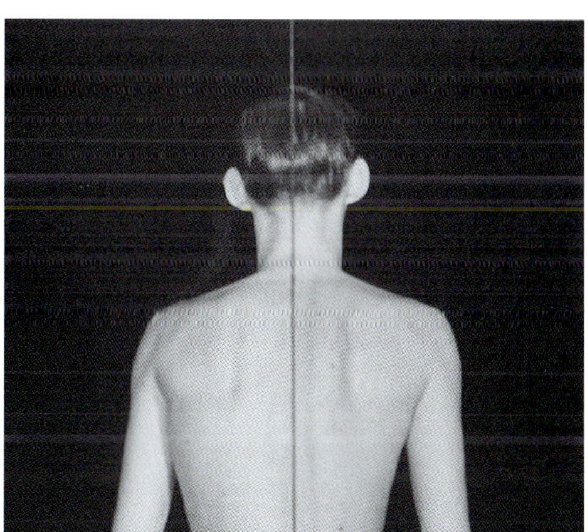

FIGURE 2-27: Scapulae, abducted and slightly elevated.

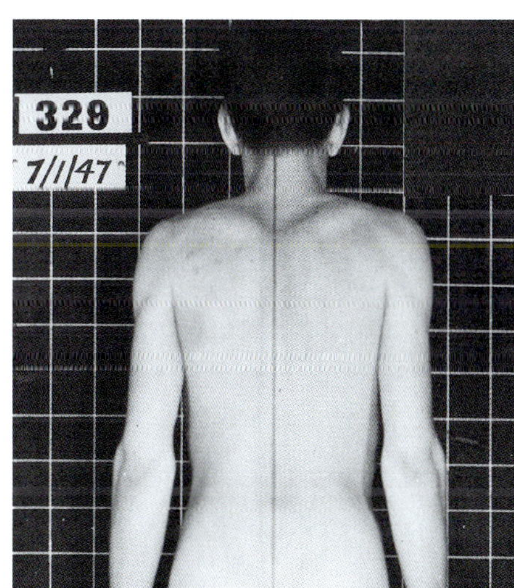

FIGURE 2-28: Shoulder elevated, scapulae adducted.

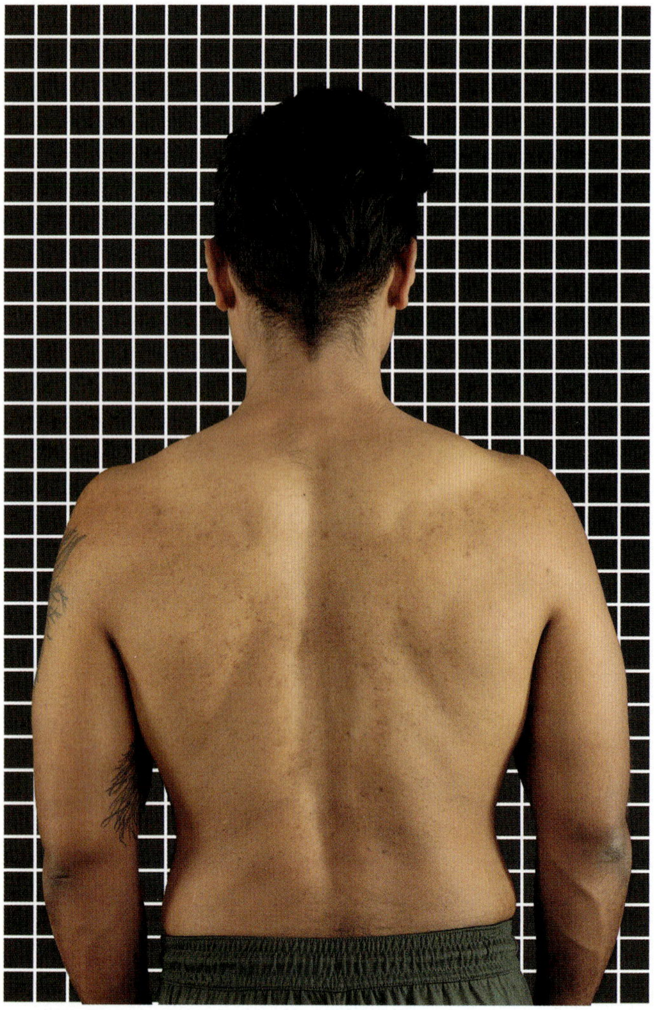

FIGURE 2-29: Shoulder suppressed, scapula abducted.

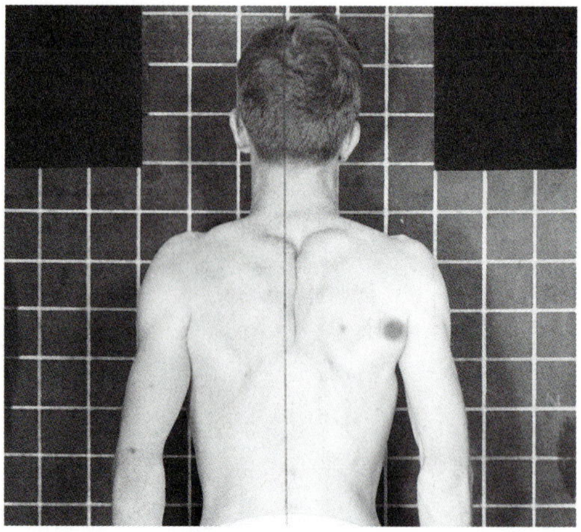

FIGURE 2-30: Scapulae, adducted and elevated.

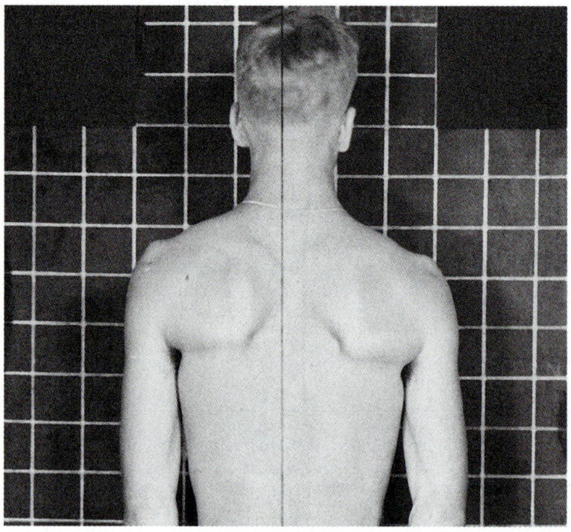

FIGURE 2-31: Scapulae, abnormal appearance.

Shoulders Depressed, Scapulae Abducted: In Figure 2-29, the shoulders of this subject slope downward sharply, accentuating their natural broadness. The marked abduction of the scapulae also contributes to this effect. Exercises to strengthen the trapezius muscles, especially the upper part, are needed to correct the faulty posture of the shoulders.

Scapulae, Adducted and Elevated: In Figure 2-30, the subject's scapulae are completely adducted and considerably elevated. The position illustrated appears to be held by voluntary effort, but if this habit persists, the scapulae will not return to the normal position when the subject tries to relax. This position is a potential result of engaging in the military brace posture that requires standing with the chin up, chest out, and shoulders back.

Scapulae, Abnormal Appearance: This subject in Figure 2-31 shows abnormal hypertrophy of some of the scapular muscles with a faulty position of the scapulae. The teres major and rhomboids are clearly visible and form a V at the inferior angle. The scapula is rotated so that the axillary border is more nearly horizontal than normal. The appearance suggests weakness of either the serratus anterior, or the trapezius, or both.

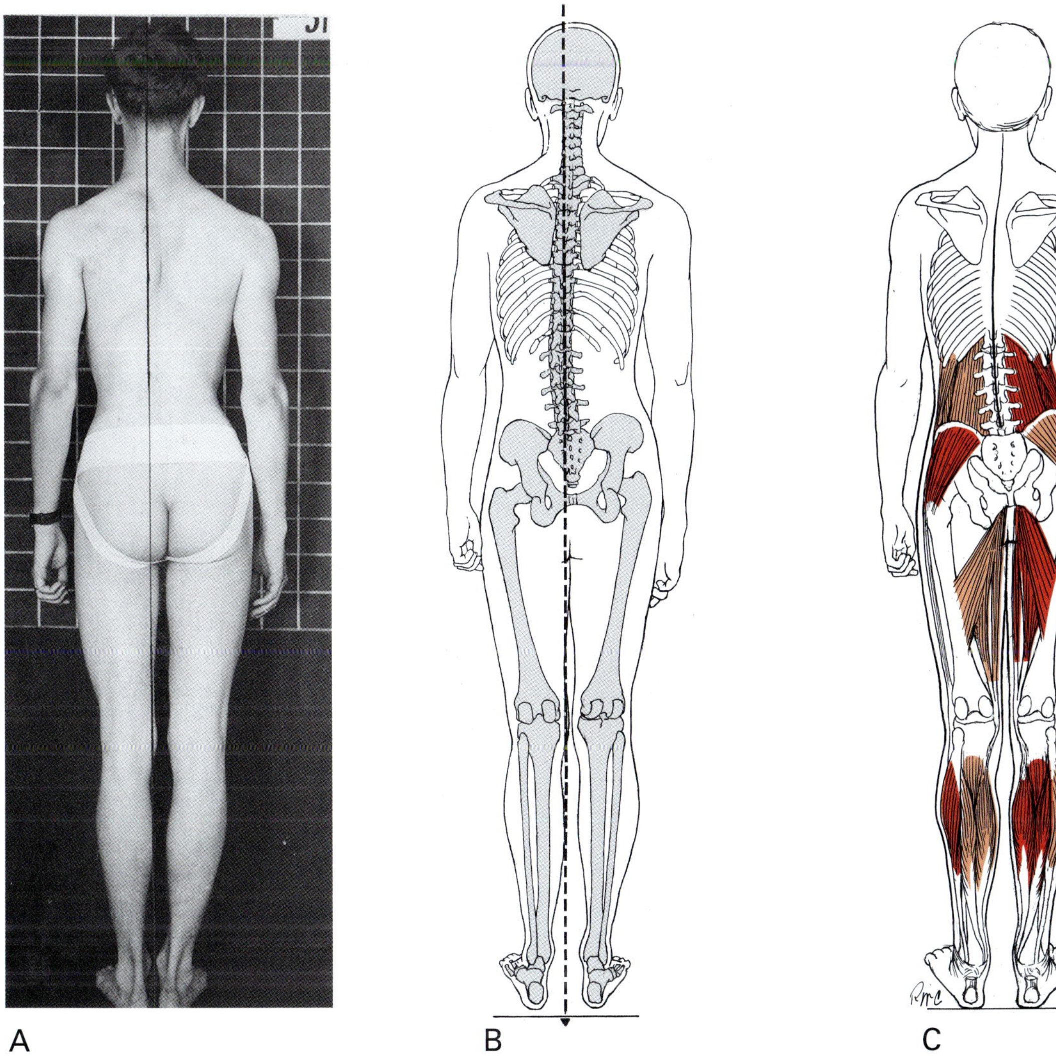

FIGURE 2-32: **A**, **B**, **C**, Faulty alignment, posterior view.

Faulty Alignment: Posterior View

Head: Erect, neither tilted nor rotated. (It appears to be slightly tilted and rotated right in Figure 2-32 because of deviations in the thoracic spine.)

Cervical Spine: Straight.

Shoulder: Right low.

Scapulae: Adducted, right slightly depressed.

Thoracic and Lumbar Spines: Thoracolumbar curve convex toward the left.

Pelvis: Lateral tilt, high on the right.

Hip Joints: Right adducted and slightly medially rotated, left abducted.

Lower Extremities: Straight, demonstrating neither genu valgum or genu varum.

Feet: In Part A, the photograph, the right is slightly pronated, as seen in the alignment of the tendo calcaneus. The left is in a position of slight postural pronation by virtue of the deviation of the body toward the right.

Elongated and Weak: Left lateral trunk muscles, right hip abductors (especially posterior gluteus medius), left hip adductors, right fibularis longus and brevis, left tibialis posterior, left flexor hallucis longus, left flexor digitorum longus. The right tensor fasciae latae may or may not be weak.

Short and Strong: Right lateral trunk muscles, left hip abductors, right hip adductors, left fibularis longus and brevis, right tibialis posterior, right flexor hallucis longus, right flexor digitorum longus. The left tensor fasciae latae is usually strong, and there may be tightness in the iliotibial band. The right leg is in postural adduction, and the position of the hip gives the appearance of a longer right leg. This posture is typical of right-handed individuals.

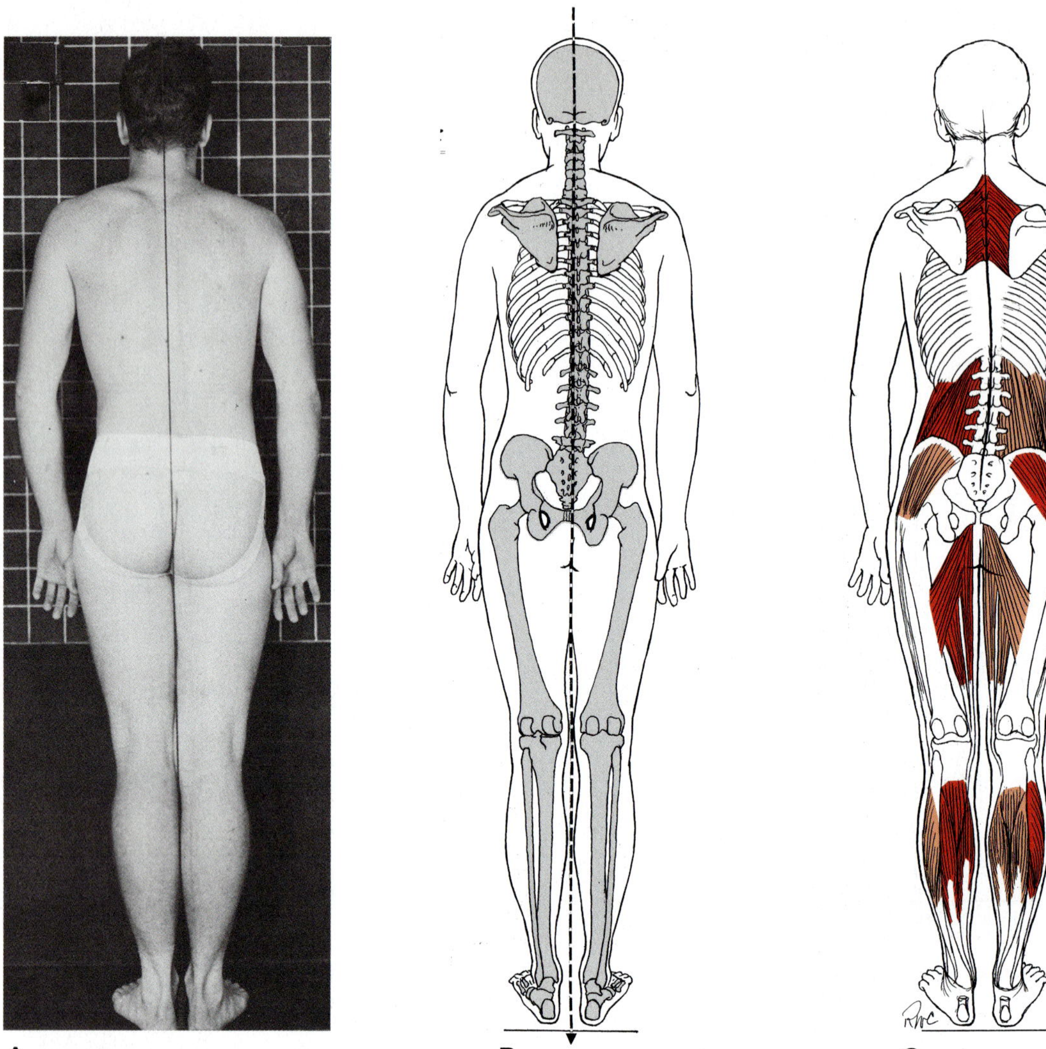

A B C

FIGURE 2-33: A, B, C, Faulty alignment, posterior view that shows the pronation of feet and medial rotation of the femur.

Figure 2-33 shows the pronation of feet and medial rotation of the femur.

Head: Erect, neither tilted nor rotated.

Cervical spine: Straight.

Shoulder: Elevated and adducted.

Shoulders Joints: Medially rotated, as indicated by position of the hands facing posteriorly.

Scapulae: Adducted and elevated.

Thoracic and Lumbar Spines: Slight thoracolumbar curve convex toward the right.

Pelvis: Lateral tilt, higher on the left.

Hip Joints: Left adducted and slightly medially rotated, right abducted.

Lower Extremities: Straight, demonstrating neither genu valgum or genu varum.

Feet: Slightly pronated.

Elongated and Weak: Right lateral trunk muscles, left hip abductors (especially posterior gluteus medius), right hip adductors, right tibialis posterior, right flexor hallucis longus, right flexor digitorum longus, left fibularis longus and brevis.

Short and Strong: Left lateral trunk muscles, right hip abductors, left hip adductors, left tibialis posterior, left flexor hallucis longus, left flexor digitorum longus, right fibularis longus and brevis. With the elevation and adduction of the scapulae, the rhomboids are in a shortened position.

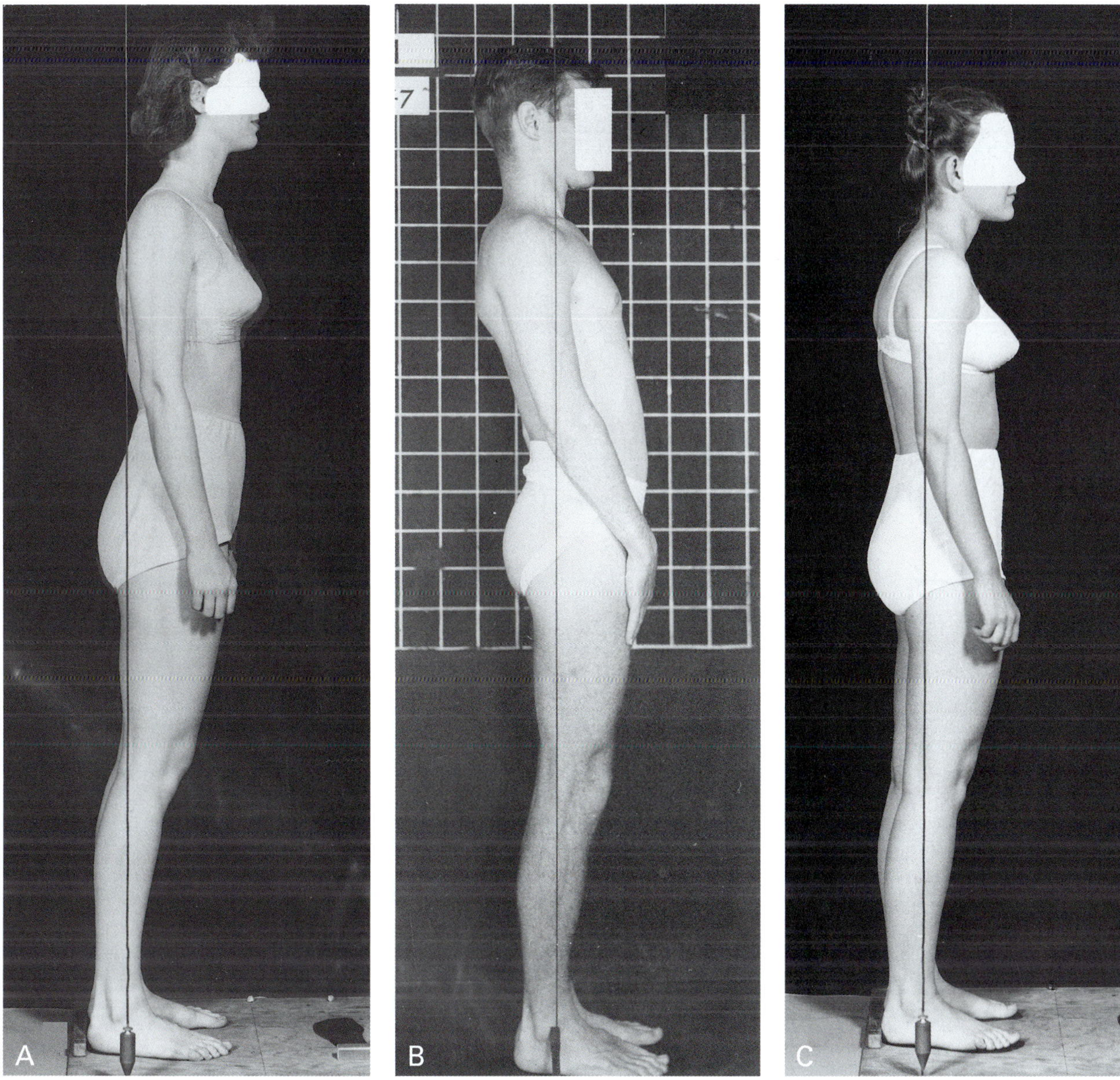

FIGURE 2-34: Faulty plumb alignment, side view.

Faulty Plumb Alignment: Sagittal View

Figure 2-34, Part A shows a marked anterior deviation of the body in relation to the plumb line, with the body weight being carried forward over the balls of the feet. It is seen most frequently among tall and slender individuals. Subjects who habitually stand this way may exhibit strain on the anterior part of the foot, with calluses under the ball of the foot and even under the great toe. Metatarsal arch supports may be indicated along with correction of the overall alignment. The ankle joint is in slight dorsiflexion because of the forward inclination of the leg and the slight flexion of the knee. Posterior muscles of the trunk and lower extremities tend to remain in a state of constant contraction, and the alignment must be corrected to achieve effective relaxation of these muscles.

Part B shows a marked posterior deviation of the upper trunk and head. The knees and pelvis are displaced anteriorly to counterbalance the posterior thrust of the upper part of the body.

Part C shows a rotation to the left (counterclockwise rotation) of the body from the ankles to the cervical region. The deviation of the body from the plumb line appears to be different when viewed from the right and left sides in subjects who have such rotation. The body is anterior to the plumb line as seen from the right, but appears to be in fairly good alignment from the left. From both sides, however, the head appears to be forward.

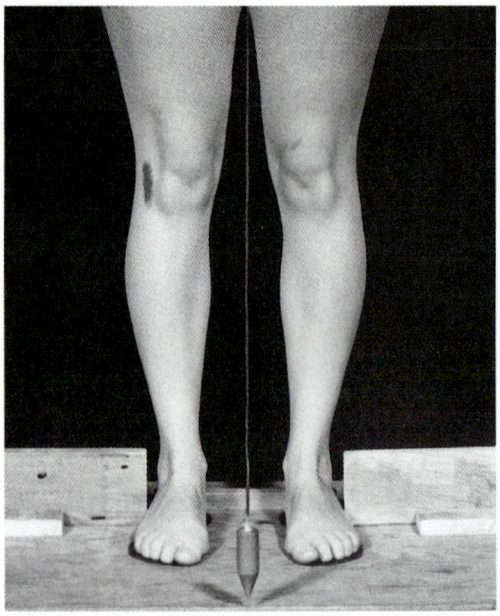

FIGURE 2-35: Pronation of feet and medial rotation of femur.

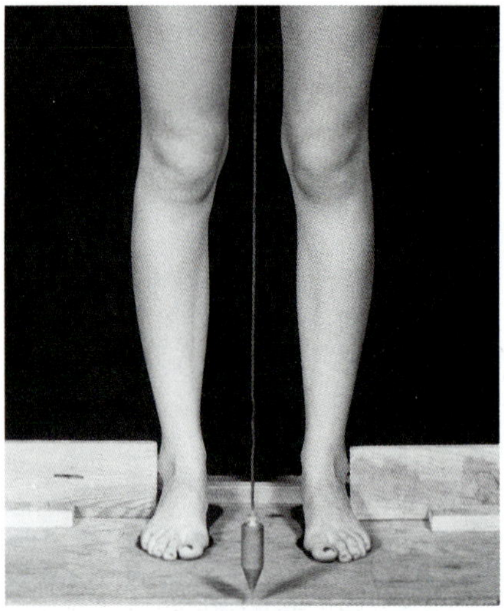

FIGURE 2-37: Good alignment of feet; faulty alignment of knees.

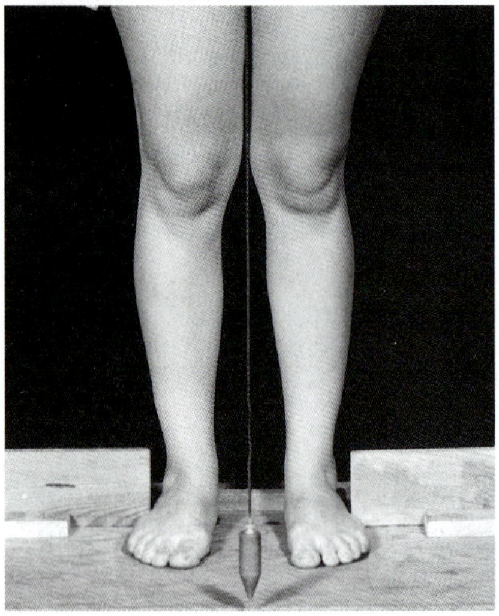

FIGURE 2-36: Pronation of feet and genu valgum.

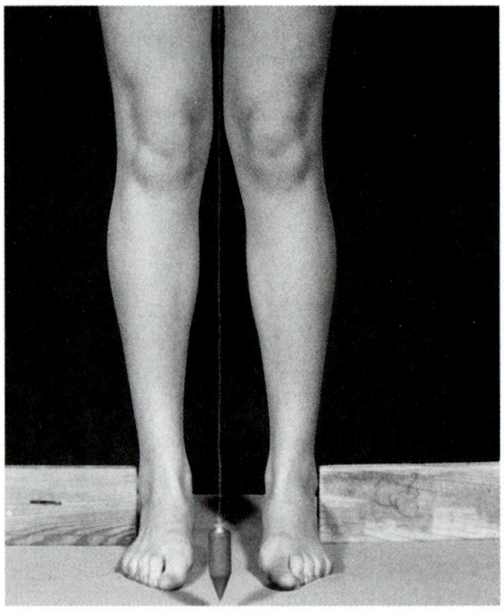

FIGURE 2-38: Supinated feet.

FAULTY POSTURE OF FEET, KNEES, AND LEGS

Faulty Posture of Feet and Knees

Pronation of Feet and Medial Rotation of Hips: The distance between the lateral malleolus and the foot board (located lateral to each lateral malleolus) indicates a moderate pronation of the feet, and the position of the patellae inward indicates a moderate degree of medial rotation of the hips (Figure 2-35).

Pronation of Feet and Genu Valgum: The feet are moderately pronated; there is slight genu valgum but no medial or lateral rotation of the hips (Figure 2-36).

Feet Good, Knees Faulty: The alignment of the feet is very good, but medial rotation of the hips is indicated by the position of the patellae (pointing inward). This fault is harder to correct by use of shoe corrections than one in which pronation accompanies the medial rotation (Figure 2-37).

Supinated Feet: The weight is borne on the lateral borders of the feet, and the long arches are higher than normal. The perpendicular foot board touches the lateral malleolus but is not in contact with the lateral border of the sole of the foot. It appears as if an effort were being made to invert the feet because the anterior tibial muscles are so prominent. However, the position shown is the natural posture of this subject's feet (Figure 2-38).

Faulty Posture of Knees and Legs

Lateral Rotation of the Legs: Lateral rotation of the legs, as seen in Figure 2-39, is the result of lateral rotation at the hip joint. This position is more typical of boys than of girls. It may or may not have serious effects, although

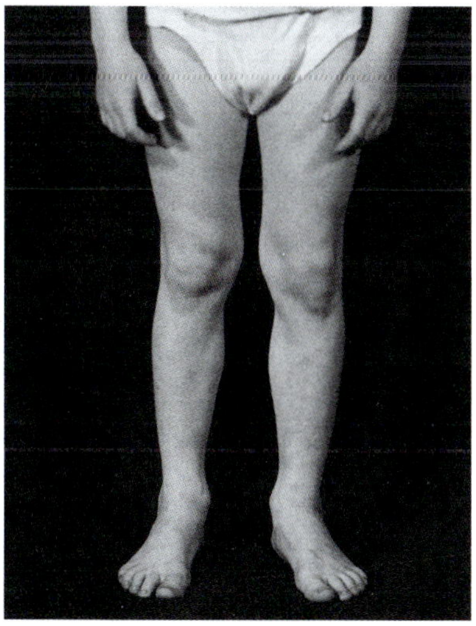

FIGURE 2-39: Lateral rotation of the legs.

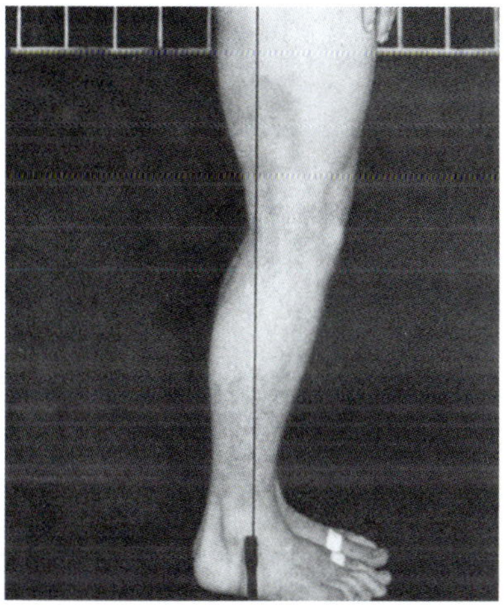

FIGURE 2-40: Knee flexion, moderate.

persistence of such a pattern in walking as well as in standing puts undue strain on the longitudinal arches.

Knee Flexion, Moderate: Flexion of the knees is seen less frequently than hyperextension in cases of faulty posture (Figure 2-40). The flexed position requires constant muscular effort by the quadriceps. Knee flexion in standing may result from hip flexor tightness. When hip flexors are tight, there must be compensatory alignment faults of the knees, the low back, or both. Attempting to reduce a lordosis by flexing the knees in standing is not an appropriate solution when hip flexor stretching is needed.

Knee Hyperextension: With hyperextension of the knee, the ankle joint is in plantarflexion (Figure 2-41). Compensatory forward trunk lean may be noticed to decrease the ground reaction forces experienced at the knee.

Genu Varum: Figure 2-42 shows a mild degree of genu varum (structural bowlegs). (Clinically, imaging would be indicated to distinguish between structural and postural genu varum.)

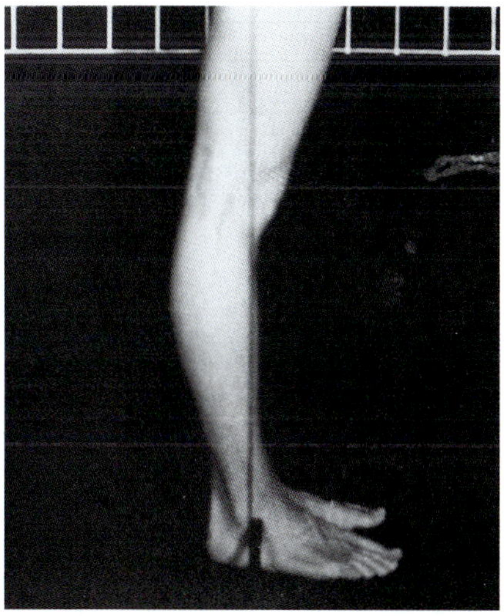

FIGURE 2-41: Knee hyperextension.

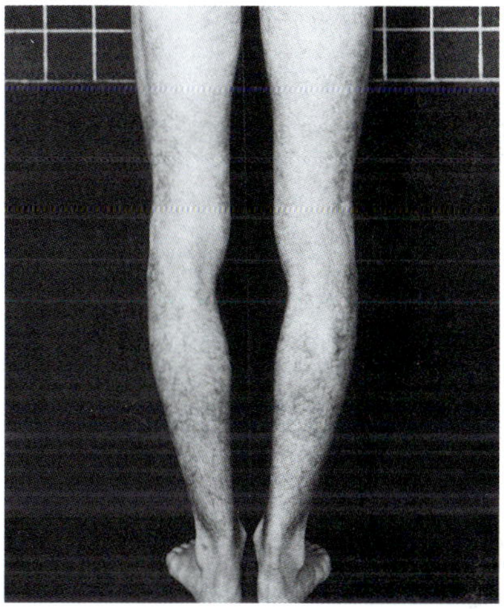

FIGURE 2-42: Genu varum.

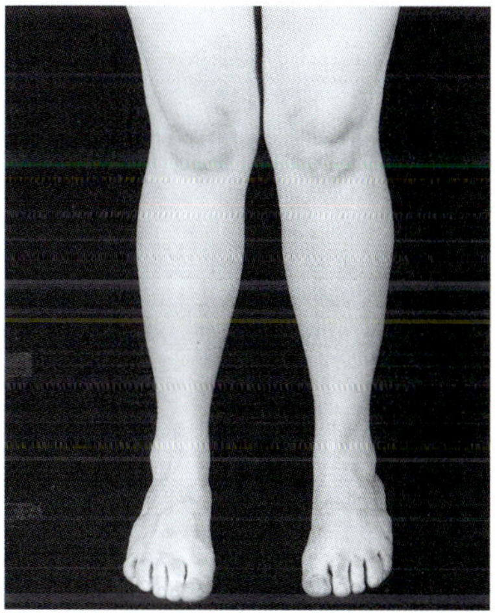

FIGURE 2-43: Genu valgum.

Genu Valgum: Figure 2-43 shows a moderate degree of genu valgum (structural knock-knees). Clinically, imaging would be indicated to distinguish between structural and postural genu valgum.

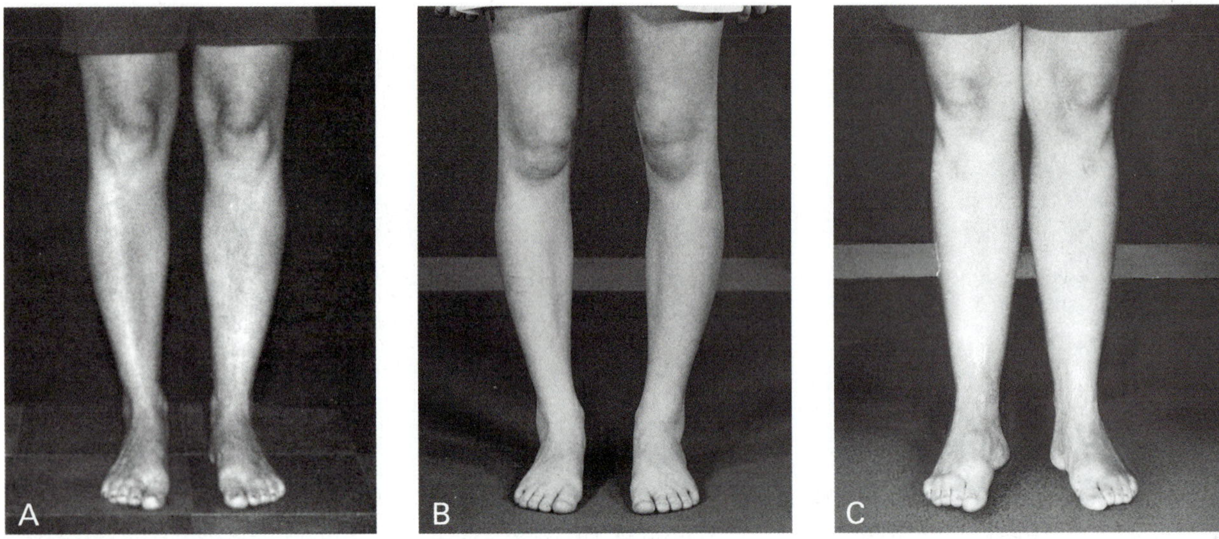

Ideal Alignment | Postural Genu Varum (Bowlegs) | Postural Genu Valgum (Knock-Knees)

FIGURE 2-44: Photographs of postural bowlegs and knock-knees: **A**, Ideal alignment; **B**, Postural bowlegs; **C**, Postural knock-knees.

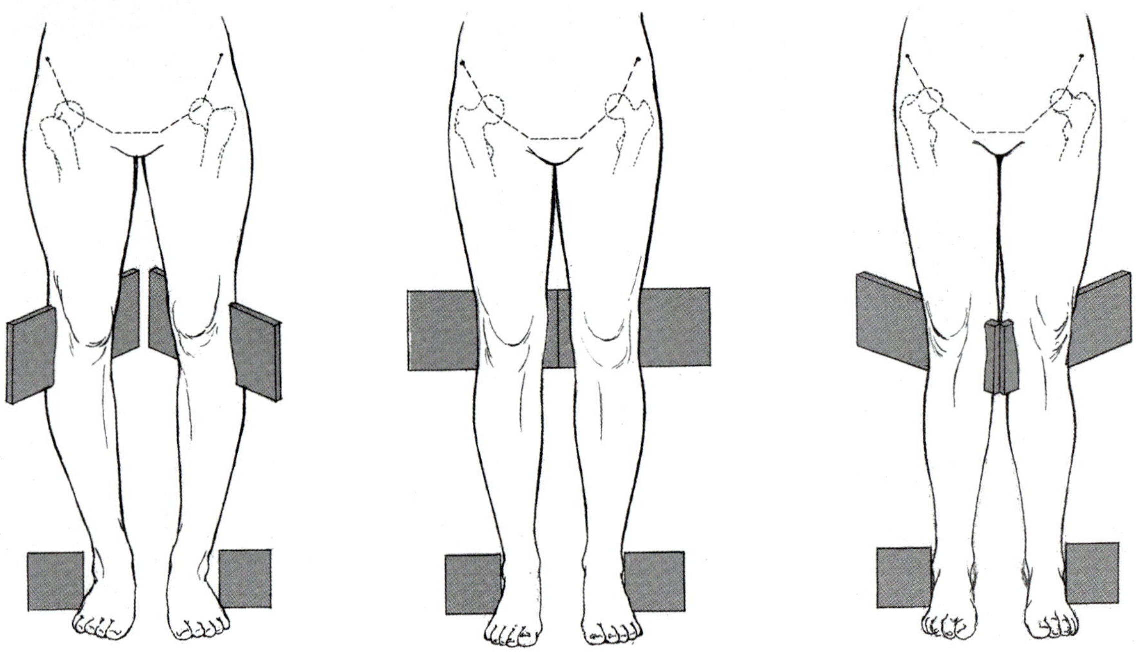

A B C

FIGURE 2-45: Illustrations of postural bowlegs and knock-knees (see Figure 2-44).

Postural Bowlegs and Knock-Knees

See Figures 2-44 and 2-45 for photographs and illustrations of postural bowlegs and knock-knees.

Ideal Alignment: In ideal alignment (A), the hips are neutral in rotation, as evidenced by the position of the patellae facing directly forward. The axis of the knee joint is in the coronal plane, and flexion and extension occur in the sagittal plane. The feet are in good alignment.

Postural Genu Varum: Postural genu varum (B) results from a combination of medial rotation of the hips, pronation of the feet, and hyperextension of the knees. When hips medially rotate, the axis of motion for flexion and extension of the knee is oblique to the coronal plane. From this axis, hyperextension occurs in a posterolateral direction, resulting in a separation at the knees and apparent bowing of the legs.

Postural Genu Valgum: Postural genu valgum (C) results from a combination of lateral rotation of the hips, supination of the feet, and hyperextension of the knees. With lateral rotation, the axis of the knee joint is oblique to the coronal plane, and hyperextension results in adduction at the knees.

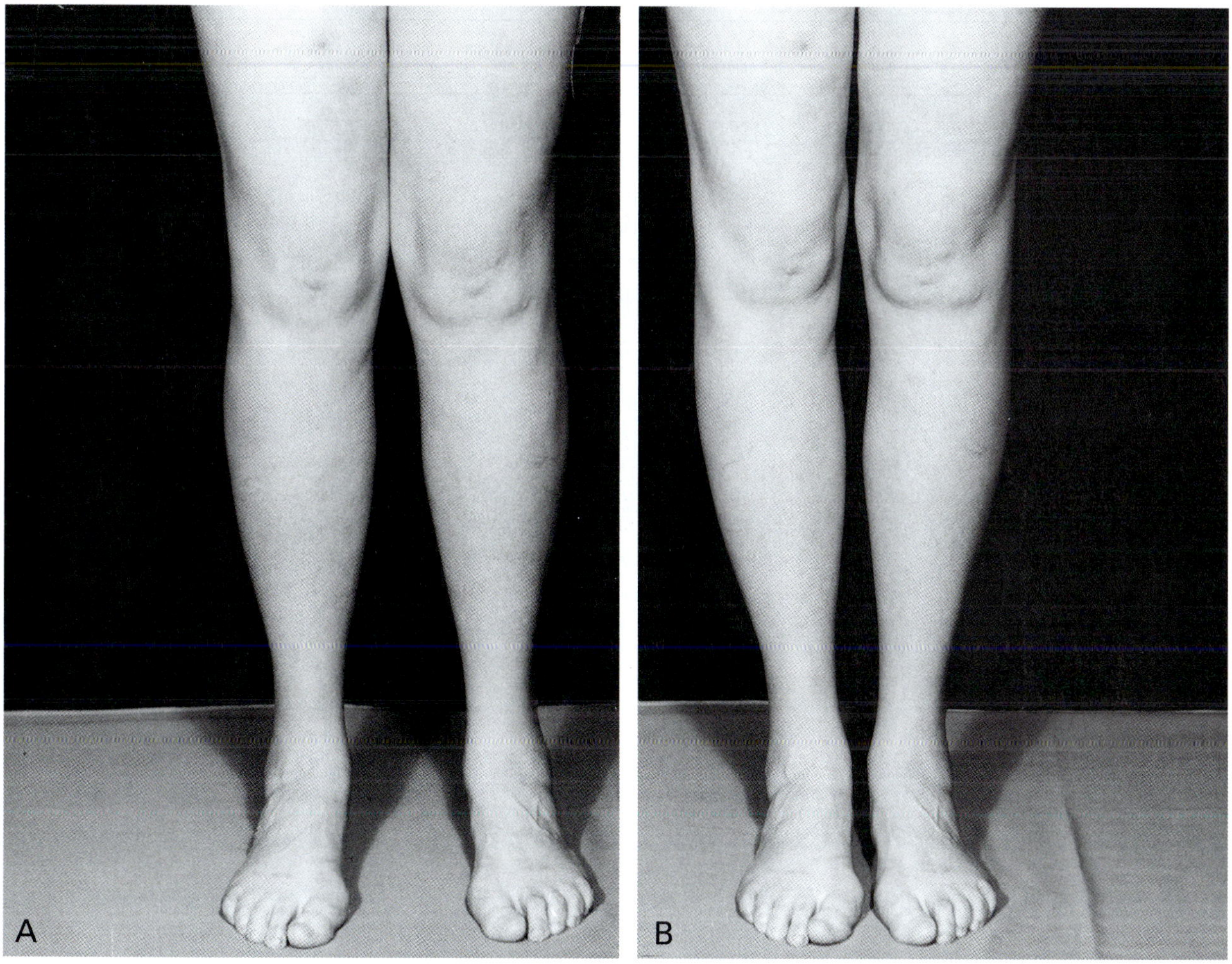

FIGURE 2-46: Mechanism of postural bowing compensatory for genu valgum (knock-knees).

Postural Bowing Compensatory for Genu Valgum

Mechanism of Postural Bowing Compensatory for Genu Valgum: In Figure 2-46, Part A shows the position of genu valgum the subject exhibits when the knees are in good anteroposterior alignment. Part B shows that by hyperextending the knees, the subject is able to produce enough postural bowing to accommodate for the 4-inch separation of her feet shown in Part A.

See Figure 2-46B for the extent of postural bowing that can be produced by hyperextension in an individual without genu valgum.

It is common to observe compensations when genu valgum persists. At times, individuals will hide the genu valgum position by flexing one knee and hyperextending the other so that the knees can be close together. Rotation faults may result if the same knee is habitually flexed while the other is hyperextended.

The appearance of postural genu varum and genu valgum may also result from the combination of knee flexion with rotation (not illustrated). With lateral rotation and slight flexion, the legs will appear to be slightly bowed, and with medial rotation and slight flexion, there will appear to be a position of genu valgum. These variations associated with flexion are of less concern than those associated with hyperextension because flexion is a normal movement but hyperextension is an abnormal movement.

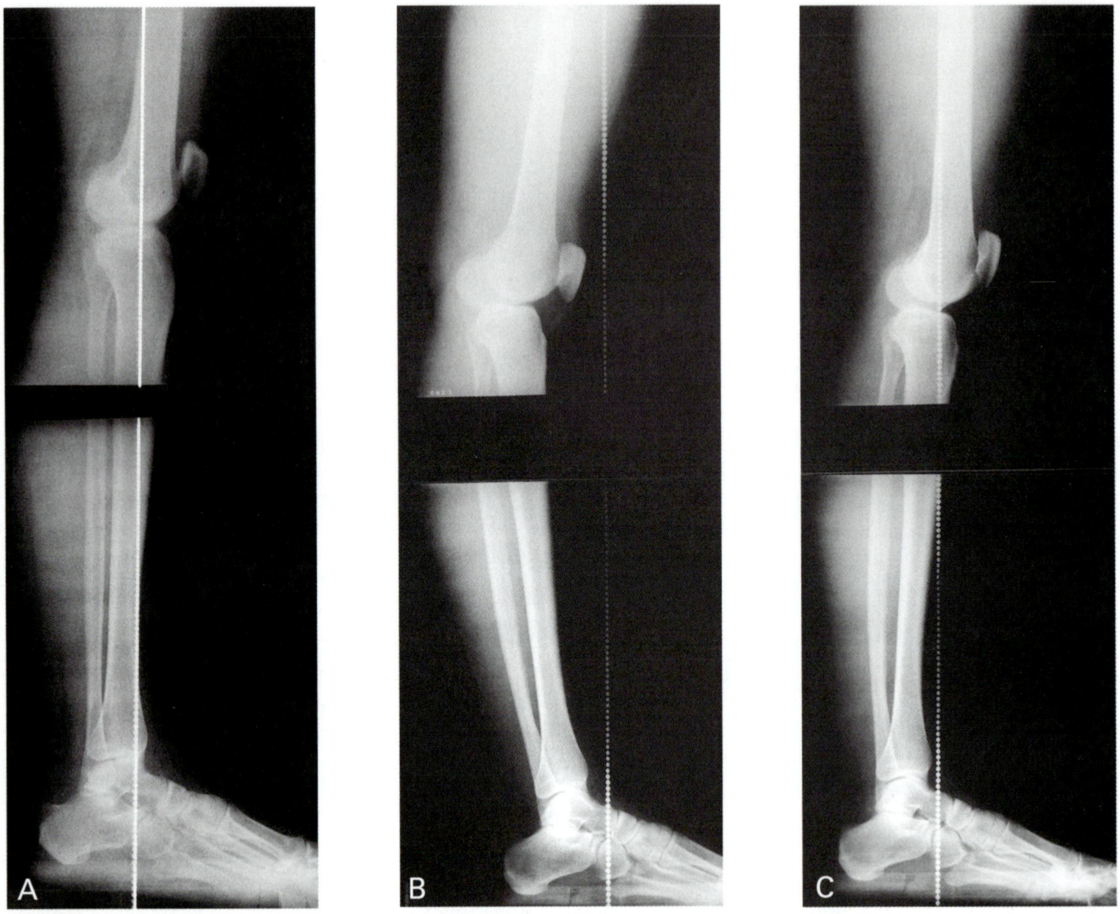

FIGURE 2-47: Radiographs of legs in good and faulty alignment.

Radiographs of Legs in Good and Faulty Alignment

For each part in Figure 2-47, a beaded metal plumb line was suspended beside the subject when the radiograph was taken. Two radiographic films were in position for the single exposure. The above illustration shows the relationship of the plumb line to the bones of the foot and lower leg, with the subject standing in a position of good alignment.

Radiograph B shows a subject who had a habit of standing in hyperextension. The plumb line was suspended in line with the standard base point while the radiograph was obtained. Note the change in position of the patella and the anterior compression of the knee joint.

Radiograph C shows the same subject depicted by the center figure. As an adult, she attempted to correct her hyperextension fault. The alignment through the knee joint and hip are very good, but the tibia and fibula show evidence of posterior bowing. (Compare with the good alignment of these bones as seen in Radiograph A.

See Table 2-2 for a summary of good and faulty posture.

TABLE 2-2 Summary of Good and Faulty Posture

Good Posture	Part	Faulty Posture
In standing, the longitudinal arch has the shape of a half dome. Barefoot or in shoes without heels, the feet out-toe slightly. In shoes with heels, the feet are parallel. In walking with or without heels, the feet are parallel, and the weight is transferred from the heel along the outer border to the ball of the foot. In sprinting, the feet are parallel or in-toe slightly. The weight is on the balls of the feet and toes, because the heels do not come in contact with the ground.	Feet	Low longitudinal arch or flat foot. Low metatarsal arch, usually indicated by calluses under the ball of the foot. Weight borne on the inner side of the foot (pronation). "Ankle rolls in." Weight borne on the outer border of the foot (supination). "Ankle rolls out." Out-toeing while walking or while standing in shoes with heels ("slue-footed"). In-toeing while walking or standing ("pigeon-toed").
Toes should be straight (i.e., neither curled downward nor bent upward). They should extend forward in line with the foot and not be squeezed together or overlap.	Toes	Toes extended at the metatarsophalangeal joint and flexed at interphalangeal joints so that the weight rests on the tips of the toes (hammer toes). This fault is often associated with wearing shoes that are too short. Great toe adducts inward toward the midline of the foot (hallux valgus – "Bunion"). This fault is often associated with wearing shoes that are too narrow and pointed at the toes.
Legs are straight up and down. Kneecaps face straight ahead when feet are in good position. In sagittal view, the knees are straight (i.e., neither bent forward nor locked backward).	Knees and legs	Knees touch when feet are apart (genu valgus). Knees are apart when feet touch (genu varum). Knee curves slightly backward (genu recurvatum). Knee bends slightly forward; that is, it is not as straight as it should be (flexed knee). Kneecaps face slightly toward each other (medially rotated hips).
Ideally, the body weight is borne evenly on both feet, and the hips are level. One side is not more prominent than the other as seen from front or back, nor is one hip more forward or backward than the other as seen from the side. The spine does not curve toward the left or toward the right. (A slight deviation to the left in right-handed individuals and to the right in left-handed individuals is not uncommon. Also, a tendency toward a slightly low right shoulder and slightly high right hip is frequently found in right-handed people, and vice versa in left-handed people.)	Hips, pelvis, and spine (posterior view)	One hip is higher than the other (lateral pelvic tilt). Sometimes, the deviation is not significant but a sideways sway of the body has made it more prominent. (Tailors and dressmakers often notice a lateral tilt, because the hemline of skirts or the length of trousers must be adjusted to the difference.) The hips are rotated so that one is more forward than the other (anterior or posterior rotation).
The front of the pelvis and the thighs are in a straight line. The buttocks are not prominent in the back but slope slightly downward. The spine has four natural curves. In the neck and lower back, the curves are lordotic; in the upper back and lowest part of the spine (sacral region), there is a kyphosis. The sacral curve is a fixed curve, whereas the other three are flexible.	Spine and pelvis (sagittal view)	The lower back arches forward too much (hyperlordosis). The pelvis tilts anteriorly too much. The front of the thigh forms an angle with the pelvis when this tilt is present. The normal lordosis in the lower back has straightened. The pelvis tips posteriorly as in sway-back and flat-back postures. Increased kyphosis in the upper back (round upper back). Hyperlordosis in the neck. Almost always accompanied by round upper back and seen as a forward head. Lateral curve of the spine (scoliosis) toward one side (C curve) or both sides (S curve).
In children up to approximately 10 years of age, the abdomen normally protrudes somewhat. In older children and adults, the abdomen should be flat.	Abdomen	Entire abdomen protrudes. Lower part of the abdomen protrudes; the upper part is pulled in.
A good position of the chest is one in which it is slightly up and slightly forward (with the back remaining in good alignment). The chest appears to be in a position approximately halfway between that of a full inspiration and a forced expiration	Chest	Depressed ("hollow-chest") position. Lifted and held up too high, brought about by arching the back. Ribs more prominent on one side than on the other. Lower ribs flaring out or protruding.

(continues on the next page)

TABLE 2-2 (*continued*)

Good Posture	Part	Faulty Posture
Arms hang relaxed at the sides with palms facing toward the body. Elbows are slightly bent, so the forearms hang slightly forward. Shoulders are level, and neither one is more anterior or posterior than the other when seen from the side. Shoulder blades lie flat against the rib cage. They are neither too close together nor too wide apart. In adults, a separation of about 4 inches is average.	Arms and shoulders	Holding the arms stiffly in any position anterior, posterior, or out from the body. Shoulders internally rotated, so palms face posteriorly. One shoulder higher than the other. Both shoulders hiked-up. One or both shoulders drooping forward or sloping. Shoulders rotated either to the right or left. Shoulder blades adducted too hard. Shoulder blades too far apart. Shoulder blades too prominent, standing out from the rib cage (winged scapulae).
Head is held erect, in a position of good balance.	Head	Chin up too high. Head protruding forward. Head laterally flexed or rotated to one side.

SECTION V
INTERVENTION

FAULTY POSTURE: ANALYSIS AND TREATMENT

See Table 2-3 for the analysis and treatment of faulty posture.

TABLE 2-3 Analysis of and Treatment for Faulty Posture

Postural Fault	Anatomical Position of Joints	Muscles in Shortened Position	Muscles in Lengthened Position	Treatment Procedures
Forward head	Cervical spine hyperextension	Cervical spine extensors Upper trapezius and levator	Cervical spine flexors	Stretch cervical spine extensors, if short, by trying to flatten the cervical spine. Strengthen cervical spine flexors, if weak. A forward head position is usually the result of a faulty upper back posture. If neck muscles are not tight posteriorly, the head position will usually correct as the upper back is corrected. Strengthen the thoracic spine extensors. Do deep breathing exercises to help stretch the intercostals and the upper parts of abdominal muscles. Stretch the pectoralis minor. Stretch the shoulder adductors and internal rotators, if short. Strengthen the middle and lower trapezius. Use shoulder support when indicated to help stretch the pectoralis minor and relieve strain on the middle and lower trapezius (see exercises and supports, pp. 74, 143, 306). Stretch low back muscles, if tight. Strengthen abdominals by posterior pelvic tilt exercises and, if indicated, by trunk curl. Avoid sit-ups, because they shorten hip flexors. Stretch hip flexors, when short. Strengthen hip extensors, if weak. Instruct regarding proper body alignment. Depending on the degree of lordosis and extent of muscle weakness and pain, use support (corset) to relieve strain on abdominals and help correct the lordosis.
Kyphosis and depressed chest	Thoracic spine flexion Intercostal spaces diminished	Upper and lateral fibers of internal oblique Shoulder adductors Pectoralis minor Intercostals	Thoracic spine extensors Middle trapezius Lower trapezius	
Forward shoulders	Scapulae abducted and (usually) elevated	Serratus anterior Pectoralis minor Upper trapezius	Middle trapezius Lower trapezius	
Lordotic posture	Lumbar spine hyperextension Pelvis, anterior tilt Hip joint flexion	Lower back erector spinae Internal oblique (upper) Hip flexors	Abdominals, especially external oblique (lateral) Hip extensors	

(continues on the next page)

TABLE 2-3 (*continued*)

Postural Fault	Anatomical Position of Joints	Muscles in Shortened Position	Muscles in Lengthened Position	Treatment Procedures
Flat-back posture	Lumbar spine flexion	Anterior abdominals	Lower back erector spinae	Low back muscles are seldom weak, but if they are, do exercises to strengthen them and restore the normal anterior curve. Tilt the pelvis forward, bringing the low back into an anterior curve. *Avoid* prone hyperextension, because it increases posterior pelvic tilt and stretches hip flexors (see p. 49).
	Pelvis, posterior tilt			Instruct in proper body alignment. If the back is painful and in need of support, apply a corset that holds the back in a normal anterior lumbar curve.
	Hip joint extension	Hip extensors	Hip flexors (one-joint)	Strengthen hip flexors to help produce a normal anterior lumbar curve. Stretch hamstrings, if tight.
Sway-back posture (pelvis displaced forward, upper trunk backward)	Lumbar spine position depends on level of posterior displacement of upper trunk	Upper anterior abdominals, especially upper rectus and internal oblique	Lower anterior abdominals, especially external oblique	Strengthen lower abdominals (stress external oblique). Stretch arms overhead and do deep breathing to stretch tight intercostals and upper abdominals. Instruct in proper body alignment. Wall-standing exercise is particularly useful.
	Pelvis, posterior tilt Hip joint extension	Hip extensors	Hip flexors (one-joint)	Stretch hamstrings, if tight. Strengthen hip flexors, if weak, using alternate hip flexion in the sitting position or alternate leg raising from the supine position. *Avoid* double leg-raising exercises because of strain on the abdominals.
Slight left C-curve thoracolumbar scoliosis	Thoracolumbar spine: lateral flexion, convex toward left	Right lateral trunk muscles	Left lateral trunk muscles	*If present without lateral pelvic tilt,* stretch the right lateral trunk muscles, if short, and strengthen the left lateral trunk muscles, if weak. *If present with lateral pelvis tilt, see* below for additional treatment procedures. Correct faulty habits that tend to increase the lateral curve: *Avoid* sitting on left foot in manner that thrusts the spine toward the left. *Avoid* lying on the left side, propped up on an elbow, to read or write.
	Opposite for right C-curve.			
		Left psoas major	Right psoas major	If weak, exercise the right iliopsoas in Sitting position (see p. 438).
Prominent or high right hip	Pelvis, lateral tilt, high on right	Right lateral trunk muscles	Left lateral trunk muscles	Stretch the right lateral trunk muscles, if short. Strengthen the left lateral trunk muscles, if weak
	Right hip joint, adducted Left hip joint, abducted	Left hip abductors and fascia lata Right hip adductors	Right hip abductors, especially the gluteus medius Left hip adductors	Stretch the left lateral thigh muscles and fascia, if short. Specific exercises to strengthen the right gluteus medius are not required to correct slight postural weakness; functional activity will suffice if the alignment is corrected and maintained. The subject should: Stand with weight evenly distributed over both feet, with the pelvis level. Avoid standing with weight on the right leg, causing the right hip to be in postural adduction. Temporarily use a straight raise on the heel of the left shoe (usually 3/16 inch) or a pad on the inside heel of the shoe and in bedroom slippers.
	Opposite for posture with right C-curve and high left hip.			

FAULTY LEG, KNEE, AND FOOT POSITIONS: ANALYSIS AND TREATMENT

See Table 2-4 for the analysis and treatment of faulty leg, knee, and foot positions.

TABLE 2-4 Analysis of and Treatment for Faulty Leg, Knee, and Foot Positions

Postural Fault	Anatomical Position of Joints	Muscles in Shortened Position	Muscles in Lengthened Position	Treatment Procedures
Hyperextended knee	Knee hyperextension Ankle plantar flexion	Quadriceps Soleus	Popliteus Hamstrings at knee, short head	Instruct regarding overall postural correction, with emphasis on avoiding knee hyperextension. In those with hemiplegia, use a short leg brace with a right-angle stop.
Flexed knee	Knee flexion Ankle dorsiflexion	Popliteus Hamstrings at knee	Quadriceps Soleus	Stretch the knee flexors, if tight. Perform overall postural correction. Knee flexion may be secondary to hip flexor shortness. Check length of the hip flexors; stretch, if short.
Medially rotated femur (often associated with pronation of foot, see below)	Hip joint, medial rotation	Hip medial rotators	Hip lateral rotators	Stretch the hip medial rotators, if tight. Strengthen the hip lateral rotators, if weak. Young children should avoid sitting in reverse tailor fashion (i.e., W position) (see below for correction of any accompanying pronation).
Knock-knee (Genu valgum)	Hip joint adduction Knee joint abduction	Fascia lata Lateral knee joint structures	Medical knee joint structures	Use an inner wedge on heels, if feet are pronated. Stretch the fascia lata, if indicated
Postural bowlegs	Hip joint medial rotation Knee joint hyperextension Foot pronation	Hip medial rotators Quadriceps Foot evertors	Hop lateral rotators Popliteus Tibialis posterior and long toe flexors	Perform exercises for overall correction of foot, knee, and hip positions. Avoid knee hyperextension. Strengthen the hip lateral rotators. Use inner wedges on heels to correct foot pronation.
		Stand with feet straight ahead and about 2 inches apart. Relax the knees into an "easy" position (i.e., neither stiff nor bent). Tighten the muscles that lift the arches of the feet, rolling the weight slightly toward the outer borders of the feet. Tighten the buttocks muscles to rotate the legs slightly outward (until the kneecaps face directly forward).		
Pronation	Foot eversion	Peroneals and toe extensors	Tibialis posterior and long toe flexors	Use inner wedges on heels. (Usually 1/8 inch on wide heels, and 1/16 inch on medium heels.) Perform overall correction of posture of feet and knees. Use exercises to strengthen the inverters. Instruct in proper standing and walking.
Supination	Foot inversion	Tibialis	Peroneals	Use outer wedge on heels. Perform exercise for peroneals.
Hammer toes and low metatarsal arch	Metatarsophalangeal joint hyperextension Proximal interphalangeal joint flexion	Toe extensors	Lumbricales	Stretch metatarsophalangeal joints by flexion; stretch interphalangeal joints by extension. Strengthen lumbricales by metatarsophalangeal joint flexion. Use a metatarsal pad or bar.

ACQUIRED POSTURAL WEAKNESS

See Box 2-2 for information on how muscles show evidence of acquired postural weakness.

BOX 2-2

Evidence of Acquired Postural Weakness in Muscles

The following muscles tend to show evidence of acquired postural weakness:

- Toe flexors (brevis and lumbricales)
- Middle and lower trapezius
- Upper back extensors
- Anterior abdominal muscles (as tested by leg-lowering test)
- Anterior neck muscles

Right-handed individuals:

- Left lateral trunk muscles
- Right hip abductors
- Right hip lateral rotators
- Right peroneus longus and brevis
- Left tibialis posterior
- Left flexor hallucis longus
- Left flexor digitorum longus

Left-handed individuals, but the pattern is not as common as that occurring in right-handed individuals:

- Right lateral trunk muscles
- Left hip abductors
- Left hip lateral rotators
- Left peroneus longus and brevis
- Right tibialis posterior
- Right flexor hallucis longus
- Right flexor digitorum longus

CORRECTIVE EXERCISES

The exercises that follow are designed to help correct some common postural faults. Additional corrective exercises are located at the end of chapters that follow. Specific exercises are done to improve muscle balance and restore good posture. To be effective, they may be prescribed to be completed every day for a period of weeks, plus daily practice in assuming and maintaining good posture (Figure 2-48) until it becomes a habit (Figure 2-49 A-E).

While working to correct muscle imbalance, it may be advisable to avoid the following exercises:

- Bilateral straight leg raises in supine position
- Sit-ups to a long-sit position with the feet held firmly down
- Supine bicycling with the majority of body weight through the upper back
- Reaching forward to touch toes in standing or sitting with the knees in full extension
- Prone press-ups (for those who have a lumbar hyperlordosis)

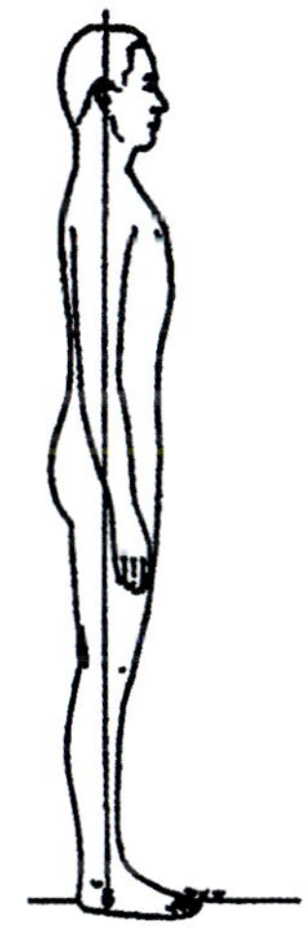

FIGURE 2-48: Ideal posture (goal of corrective exercises).

Posterior Neck Stretch
In back-lying position, bend knees and place feet flat on floor. With elbows bent and hands up beside head, tilt pelvis to flatten low back. Press head back, with chin down and in, trying to flatten neck.

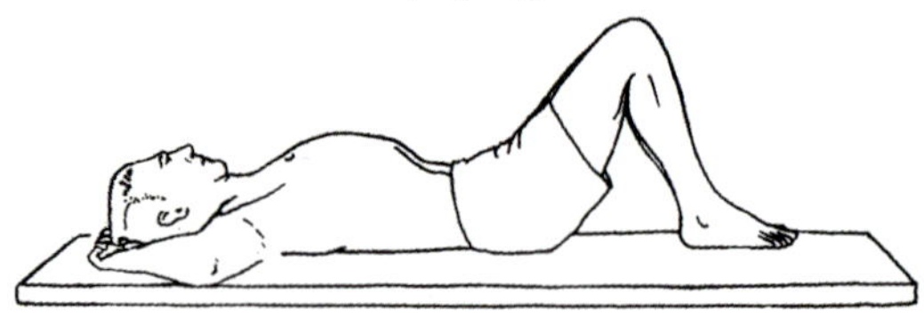

A

Shoulder Adductor Stretching
With knees bent and feet flat on floor, tilt pelvis to flatten low back. Hold the back flat, place both arms overhead, and try to reach arms to the table with elbows straight. Bring upper arms as close to sides of head as possible. (Do NOT allow the back to arch.)

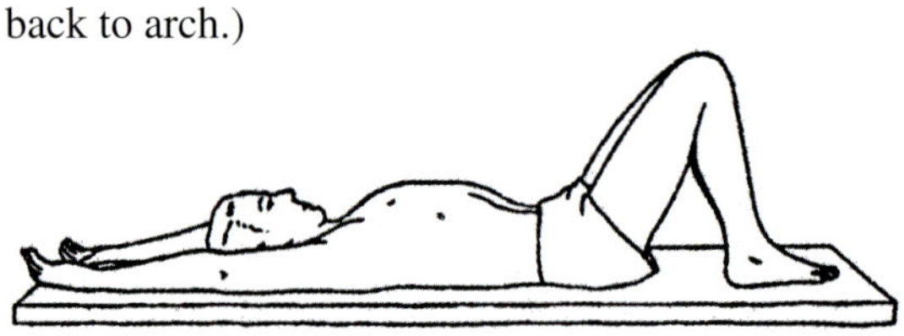

B

Wall-Standing Postural Exercise
Stand with back against a wall, heels about 3 inches from wall. Place hands up beside head with elbows touching wall. If needed, correct feet and knees as in above exercise below, then tilt to flatten low back against wall by *pulling up and in with the lower abdominal muscles*. Keep arms in contact with wall and move arms slowly to a diagonally overhead position.

C

Wall-Sitting Postural Exercise
Sit on a stool with back against wall. Place hands up beside head. Straighten upper back, press head back with chin down and in, and pull elbows back against wall. Flatten low back against wall by *pulling up and in with lower abdominal muscles*. Keep arms in contact with wall and slowly move arms to a diagonally overhead position.

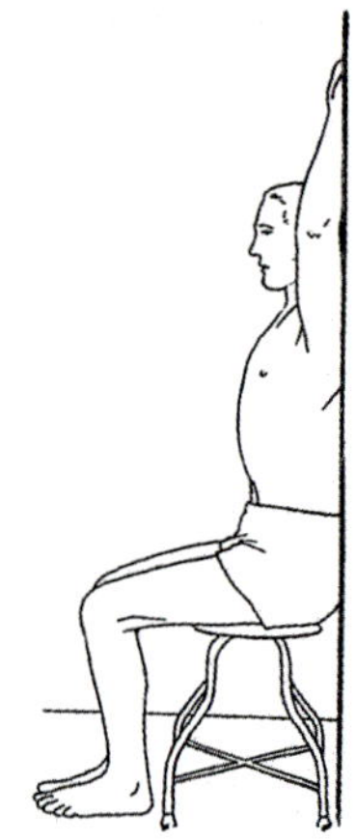

D

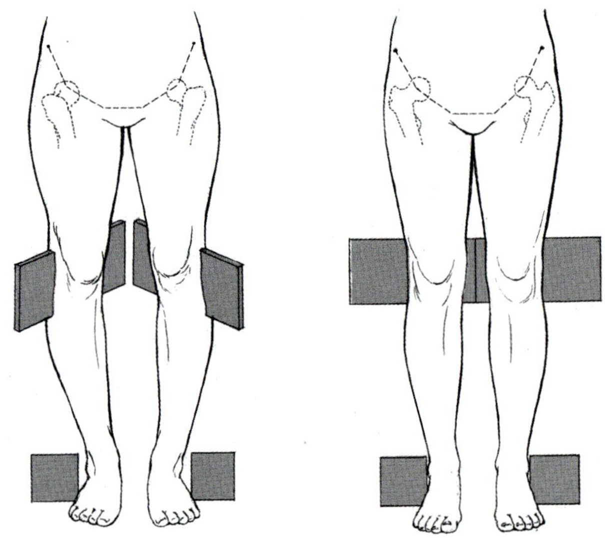

E

Correction of Pronation, Hyperextension and Internal Rotation
Stand with feet about 4 inches apart and toeing-out slightly. Relax knees into an "easy" position, i.e., neither stiff nor bent. Tighten buttock muscles to rotate legs slightly outward (until kneecaps face directly forward). Tighten muscles that lift the arches of the feet, rolling the weight slightly toward outer borders of feet.

FIGURE 2-49: A-E are exercises for correcting some common postural faults. © 2006 by Florence P. Kendall and Patricia G. Provance.

Interventions for Abdominal Weakness

The type of exercise indicated for strengthening the obliques depends on what other muscles are involved and what postural problems are associated with the weakness. The manner in which movements are combined in exercises determines whether they will be therapeutic for the individual. For example, alternate leg raising along with pelvic tilt exercises would be contraindicated in cases of hip flexor shortness but would be indicated in cases of hip flexor weakness.

To correct anterior pelvic tilt, posterior pelvic tilt exercises are indicated. The movement should be done by the external oblique, not by the rectus or by the hip extensors.

The effort must be made to pull upward and inward with the abdominal muscles, making them very firm, particularly in the area of the lateral external oblique fibers.

To exercise the external oblique in cases of sway-back posture, the same effort should be made to pull upward and inward with the lower abdominal muscles, but the pelvic tilt is not emphasized. This type of faulty posture already has a posterior pelvic tilt along with the hip flexor weakness. Contracting the lateral fibers of the external oblique in standing must be accompanied by *straightening, not flexing*, the upper back because these muscles act to shift the thorax forward and the pelvis back by the diagonal line of pull. Properly done, this movement brings the chest up and forward and restores the normal anterior curve in the low back. Additional examples of abdominal strengthening exercises can be found in Chapter 5.

When properly done, the wall-sitting and the wall-standing exercises stress the use of the muscles of the lower abdomen and lateral fibers of the external oblique. Expressions such as "make the lower abdomen cave in" or "hide the tummy under the chest," or in the vernacular of the military, "suck in your gut" are all used to encourage the subject to exert strong effort in the exercise.

Proper exercise of abdominal muscles should be a part of preventive medicine and physical fitness programs. Good strength in these muscles is essential to the maintenance of good posture, but one must avoid overdoing both the trunk curl (i.e., early stage sit-ups) and the pelvic tilt exercises. The normal anterior curve in the low back should not be obliterated in the standing posture.

REFERENCES

1. Karahan A, Bayraktar N. Determination of the usage of body mechanics in clinical settings and the occurrence of low back pain in nurses. Int J Nurs Stud 2004; 41:67–75.
2. Sharma L, Song J, Felson D, Cahue S, Shamiyeh E, Dunlop D. The role of knee alignment in disease progression and functional decline in knee osteoarthritis. J Am Med Assoc 2001; 286(2): 188–195.
3. Hales T, Sauter S, Peterson M, et al. Musculoskeletal disorders among visual display terminal users in a telecommunications company. Ergonomics 1994; 37(10): 1603–1621.
4. Elahi S, Cahue S, Felson D, Engelman L, Sharma L. The association between varus-valgus alignment and patellofemoral osteoarthritis. Arthritis Rheum 2000; 43(8): 1874–1880.
5. Marcus M, Gerr F, Monteilh C, et al. A prospective study of computer users: II. Postural risk factors for musculoskeletal symptoms and disorders. Am J Ind Med 2002; 41: 236–249.
6. Zhang Y, Brenner E, Duysens J, Verschueren S, Smeets JBJ. Effects of aging on postural responses to visual perturbations during fast pointing. Front Aging Neurosci 2018; 10: 401. Published 2018 Dec 4. doi:10.3389/fnagi.2018.00401.
7. American Academy of Orthopaedic Surgeons. Posture and its relationship to orthopaedic disabilities. A report of the Posture Committee of the American Academy of Orthopaedic Surgeons 1947.
8. Matheson G, Clement D, McKenzie D, Taunton J, Lloyd-Smith D, Macintyre J. Stress fractures in athletes: A study of 320 cases. Am J Sports Med 1987; 15(1): 46–58.
9. Macintyre J, Taunton J, Clement D, Lloyd-Smith D, McKenzie D, Morrell R. Running injuries: A clinical study of 4173 cases. Clin J Sport Med 1991; 1(2): 81–87.
10. Basmajian J, DeLuca D. Muscles Alive. 5th Ed. Baltimore: Williams & Wilkins, 1985, pp. 255, 414.
11. Levine D, Whittle MW. The effects of pelvic movement on lumbar lordosis in the standing position. J Orthop Sports Phys Ther 1996; 24(3): 130–135.
12. McLean I, Gillan G, Ross J, Aspden R, Porter R. A comparison of methods for measuring trunk list. Spine 1996; 21(14): 1667–1670.
13. Fedorak C, Nigel A, Marshall J, Paull H. Reliability of the visual assessment of cervical and lumbar lordosis. How good are we? Spine 2003; 28(16): 1857–1859.
14. Griegel-Morris P, Larson K, Mueller-Klaus K, Oatis C. Incidence of common postural abnormalities in the cervical, shoulder, and thoracic regions and their association with pain in two age groups of healthy subjects. Phys Ther 1992; 72(6): 425–431.
15. Soderberg G. Kinesiology: Application to Pathological Motion. Baltimore: Lippincott Williams & Wilkins, 1997.
16. Whitmore M, Berman A. The Evaluation of the Posture Video Analysis Tool (PVAT). NASA Technical Paper 3659. Lockhead Martin Engineering & Science Services, Houston Texas. 1996.
17. Norkin C, Levangie P. 1992; Philadelphia. Posture. Joint Structure & Function. Baltimore: Williams and Wilkins; 1992: 428–432.
18. Tsuboi Y, Oka T, Nakatsuka K, Isa T, Ono R. Effectiveness of workplace active rest programme on low back pain in office workers: a stepped-wedge cluster randomised controlled trial. BMJ Open 2021; 11(6): e040101. Published 2021 Jun 25. doi:10.1136/bmjopen-2020-040101119.
19. Gaudreault N, Benoît-Piau J, van Wingerden JP, Stecco C, Daigle F, Léonard G. An investigation of the association between transversus abdominis myofascial structure and activation with age in healthy adults using ultrasound imaging. Int J Sports Phys Ther 2021; 16(4): 1093–1103. Published 2021 Aug 1. doi:10.26603/001c.25168.

2 POSTURE

HEAD AND FACE 3

OUTLINE

INTRODUCTION

Figure 3-1 portrays a sagittal section of the skull at approximately the center of the left orbit, except that the complete eyeball is shown. The muscles illustrated are the deep facial muscles and head muscles, mainly those of the tongue, the pharyngeal area, and the eyeball.

The left hemisphere of the brain has been reflected upward to show its inferior surface and the cranial nerve roots. Lines, numbered according to the respective cranial nerves, connect the nerve roots to the corresponding nerve trunks in the lower part of the drawing. Nerve roots I, II, and VIII are sensory and are shown in white. The motor and mixed nerves are shown in yellow with one exception: Because the motor part of cranial nerve V is such a tiny branch, it is shown in yellow; the rest of cranial nerve V is shown in white.

A side view of the superficial muscles of the head and neck is seen in Figure 3-2, and cranial nerves and the muscles they innervate are listed below the illustration. Facial muscles are called the muscles of expression. The facial nerve, through its many branches, innervates most of the facial muscles. Numerous muscles may act together to create movement (e.g., as in a grimace), or movement may occur in a single area (e.g., as in raising an eyebrow). Loss of function of the facial muscles interferes with the ability to communicate feelings through facial expressions and to speak clearly.

A smile, a frown, a look of surprise—expressions such as these are created by the actions of muscles that insert directly into the skin. Because of the unique insertions of facial muscles, tests of these muscles differ from other manual muscle tests that require test position and fixation for the subject and pressure or resistance by the examiner. Instead, the subject is asked to imitate facial expressions while looking at photographs of a person performing the test movements or while watching the examiner perform them. Grading the strength of muscles is essentially a subjective estimate by the examiner of how well the muscle functions on a scale of zero, trace, poor, fair, good, and normal. Common practice is to grade these muscles as simply absent, impaired, or intact. Facial and eye muscle tests are illustrated and named on the following pages.

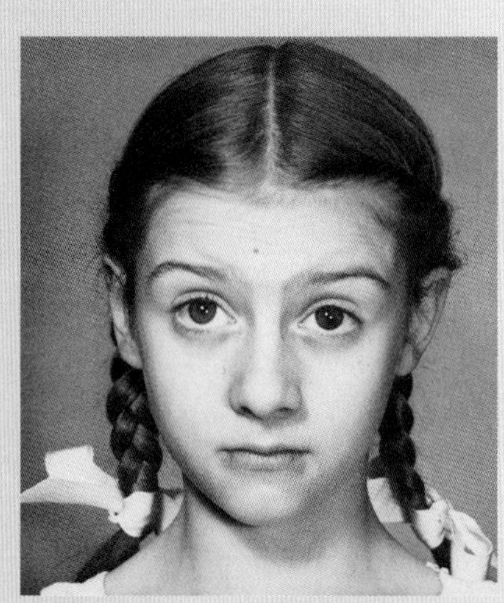

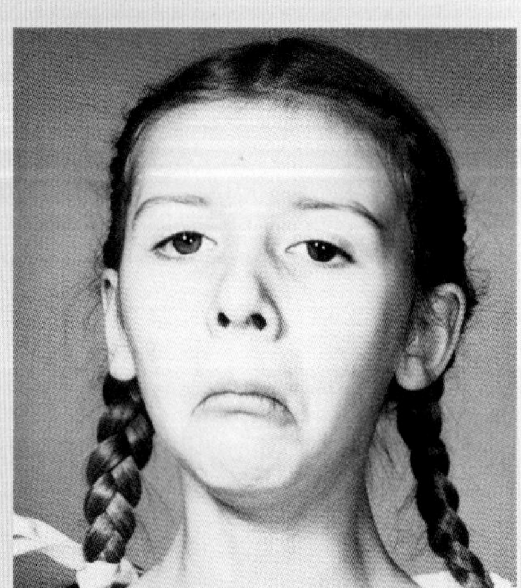

SECTION I
INNERVATION

CRANIAL NERVES AND DEEP FACIAL MUSCLES

Cranial nerves and deep facial muscles are shown in Figure 3-1.

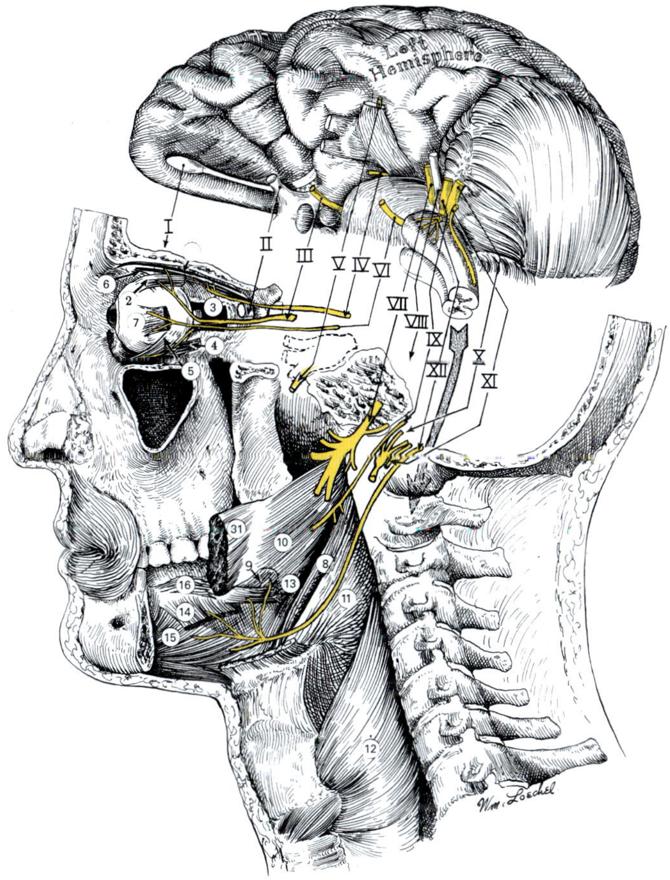

FIGURE 3-1: Cranial nerves and deep facial muscles.

I	**Olfactory nerve** (sensory)	
II	**Optic nerve** (sensory)	
III	**Oculomotor nerve**	
	Levator palpebrae superioris	(1)
	Rectus superior	(2)
	Rectus medialis	(3)
	Rectus inferior	(4)
	Obliquus inferior	(5)
IV	**Trochlear nerve**	
	Obliquus superior	(6)
V	**Trigeminal nerve, mandibular branch**	
	Masseter	(17)
	Temporalis	(18)
	Anterior digastric	(19)
VI	**Abducens nerve**	
	Rectus lateralis	(7)
VII	**Facial nerve**	
	Occipitalis	(20)
	Auricularis posterior	(21)
	Posterior digastric	(22)
	Stylohyoid	(23)
	Auricularis superior	(24)
	Auricularis anterior	(25)
	Frontalis	(26)
	Corrugator supercilia	(27)
	Orbicularis oculi	(28)
	Levator labii superioris	(29)
	Zygomaticus major and minor	(30)
	Buccinator	(31)
	Risorius	(32)
	Orbicularis oris	(33)

CERVICAL NERVES AND SUPERFICIAL FACIAL AND NECK MUSCLES

Cervical nerves and superficial facial and neck muscles are shown in Figure 3-2.

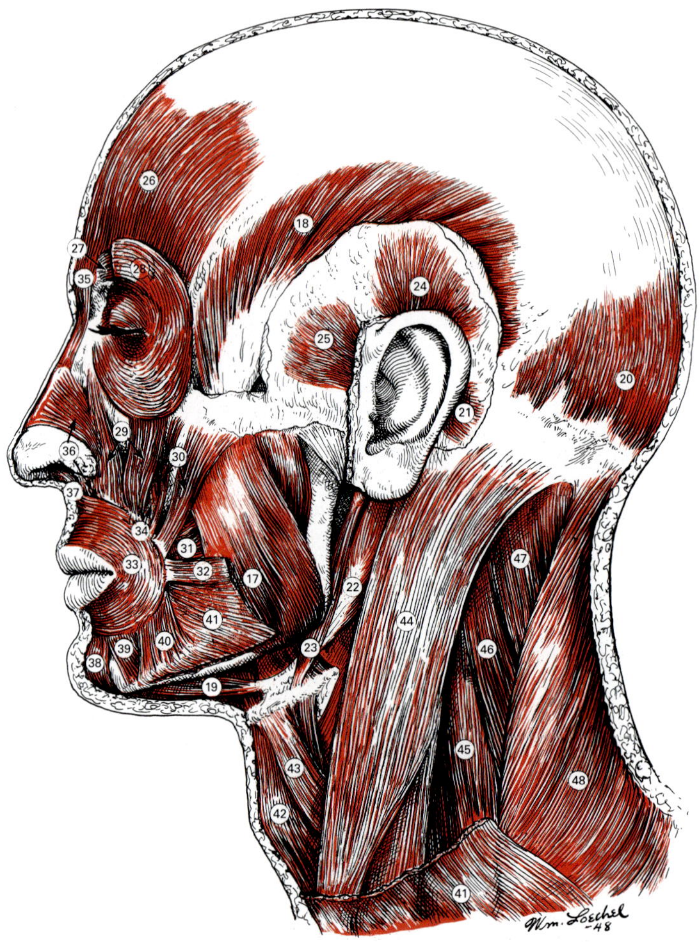

FIGURE 3-2: Cervical nerves and superficial facial and neck muscles.

	Levator anguli oris	(34)
	Procerus	(35)
	Nasalis	(36)
	Depressor septi nasi	(37)
	Mentalis	(38)
	Depressor labii inferioris	(39)
	Depressor anguli oris	(40)
	Platysma	(41)
VIII	**Vestibulocochlearis nerve** (sensory)	
IX	**Glossopharyngeal nerve**	
	Stylopharyngeus	(8)
X	**Vagus nerve**	
XI	**Accessory nerve** (spinal portion)	
	Sternocleidomastoid	(44)
	Trapezius	(48)
XII	**IX, X, and XI Pharyngeal plexus**	
	Palatoglossus	(9)
	Constrictor pharyngis superior	(10)
	Constrictor pharyngis medius	(11)
	Constrictor pharyngis inferior	(12)
XIII	**Hypoglossal nerve**	
	Styloglossus	(13)
	Hyoglossus	(14)
	Genioglossus	(15)
	Tongue intrinsics	(16)
Miscellaneous from cervical nerves		
	Sternohyoid	(42)
	Omohyoid	(43)
	Scalenus medius	(45)
	Levator scapulae	(46)
	Splenius capitis	(47)

CRANIAL NERVES AND MUSCLES

Figure 3-3 lists all the cranial nerves and the specific muscles they innervate. A column is provided in which to record the strength of the muscles that can be tested. On the right side of the page are drawings of the head that show the areas of distribution for the cutaneous nerves.

This chart was designed primarily as a reference sheet and secondarily as a form on which to record the results of tests involving the facial muscles. Because of this dual purpose, the chart contains some material that would not be included on a form solely intended for recording test results. All the cranial nerves (sensory, motor, or mixed) are listed and some muscles are included that cannot be tested (individually or in groups) by voluntary movements.

CRANIAL NERVE AND MUSCLE CHART

Name   Date

		MUSCLE STRENGTH GRADE	SENSORY OR MOTOR TO:		I Olfactory S	II Optic S	III Oculomotor M	IV Trochlear M	V Trigeminal S & M	VI Abducens M	VII Facial S & M	VIII Vestibulocochlear S	IX Glossopharyngeal S & M	X Vagus S & M	XI Accessory M	XII Hypoglossal M
I	NOSE	S	SENSORY—SMELL		•											
II	EYE	S	SENSORY—SIGHT			•										
III	EYELID		LEVATOR PALPEBRAE SUPERIORIS				•									
	EYE		RECTUS SUPERIOR				•									
			OBLIQUUS INFERIOR				•									
			RECTUS MEDIALIS				•									
			RECTUS INFERIOR				•									
IV	EYE		OBLIQUUS SUPERIOR					•								
V	→	S	SENSORY—FACE & INT. STRUCTURES OF HEAD						•							
	EAR		TENSOR TYMPANI						•							
	PALATE		TENSOR VELI PALATINI						•							
	MASTICATION		MASSETER						•							
			TEMPORALIS						•							
			PTERYGOIDEUS MEDIALIS						•							
			PTERYGOIDEUS LATERALIS						•							
	S. HYOID		MYLOHYOIDEUS						•							
			ANTERIOR DIGASTRIC						•							
VI	EYE		RECTUS LATERALIS							•						
VII	TONGUE	S	SENSORY—TASTE, ANTERIOR ⅔ TONGUE								•					
	→	S	SENSORY—EXTERNAL EAR								•					
	EAR		STAPEDIUS								•					
	S. HYOID		POSTERIOR DIGASTRIC								•					
			STYLOHYOIDEUS								•					
	SCALP		OCCIPITALIS								•					
	EAR		INTRINSIC EAR MUSCLES	POST. AURICULAR BRANCH							•					
			AURICULARIS POSTERIOR								•					
			AURICULARIS ANTERIOR								•					
			AURICULARIS SUPERIOR	TEMPORAL BRANCH							•					
	SCALP		FRONTALIS								•					
	EYEBR.		CORRUGATOR SUPERCILII	TEMP. & ZYGO. BRANCH							•					
	EYELID		ORBICULARIS OCULI								•					
	NOSE		PROCERUS								•					
			DEPRESSOR SEPTI								•					
			NASALIS TRANSVERSE & ALAR								•					
	MOUTH		ZYGOMATICUS MAJOR & MINOR								•					
			LEVATOR LABII SUPERIORIS	BUCCAL BRANCH							•					
			BUCCINATOR								•					
			ORBICULARIS ORIS								•					
			LEVATOR ANGULI ORIS								•					
			RISORIUS								•					
			DEPRESSOR ANGULI ORIS								•					
			DEPRESSOR LABII INFERIORIS	MANDIBULAR BRANCH							•					
	CHIN		MENTALIS								•					
	NECK		PLATYSMA	CERVICAL BRANCH							•					
VIII	EAR	S	SENSORY—HEARING & EQUILIBRIUM									•				
IX	TONGUE	S	SENSORY—POSTERIOR ⅓ TONGUE										•			
	PHARYNX	S	SENSORY—PHARYNX, FAUCES, SOFT PALATE										•			
			STYLOPHARYNGEUS										•			
		—	STRIATED MUSCLES—PHARYNX										•			
X	→	—	STRIATED MUSCLES—SOFT PALATE, PHARYNX & LARYNX											•		
	→	—	INVOLUNTARY MUSCLES—ALIMENTARY TRACT											•		
	→	—	INVOLUNTARY MUSCLES—AIR PASSAGES											•		
	→	—	INVOLUNTARY CARDIAC MUSCLE											•		
	→	S	SENSORY—AURICULAR											•		
	→	S	SENSORY—ALIMENTARY TRACT											•		
	→	S	SENSORY—AIR PASSAGES											•		
	→	S	SENSORY—ABDOMINAL VISCERA & HEART											•		
XI	NECK		TRAPEZIUS & STERNOCLEIDOMASTOID												•	
	PALATE		LEVATOR VELI PALATINI												•	
	→		STRIATED MUSCLES—SOFT PALATE, PHARYNX, & LARYNX												•	
XII	TONGUE		STYLOGLOSSUS													•
			HYOGLOSSUS													•
			GENIOGLOSSUS													•
			TONGUE INTRINSICS													•

SENSORY

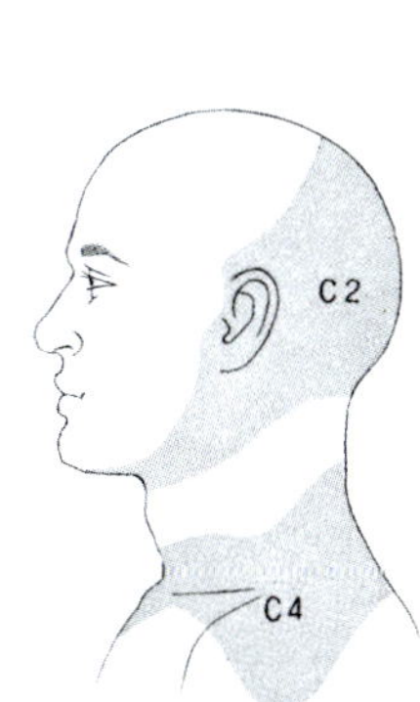

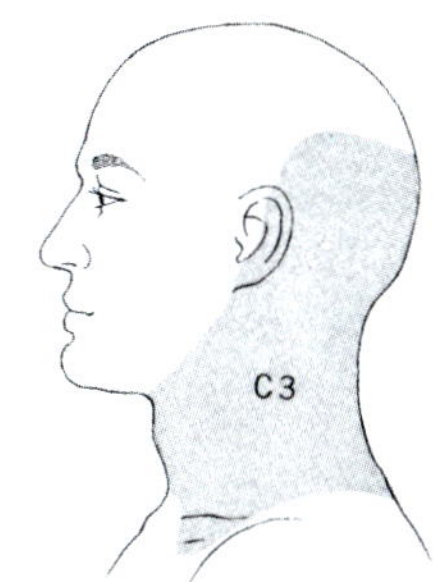

DERMATOMES

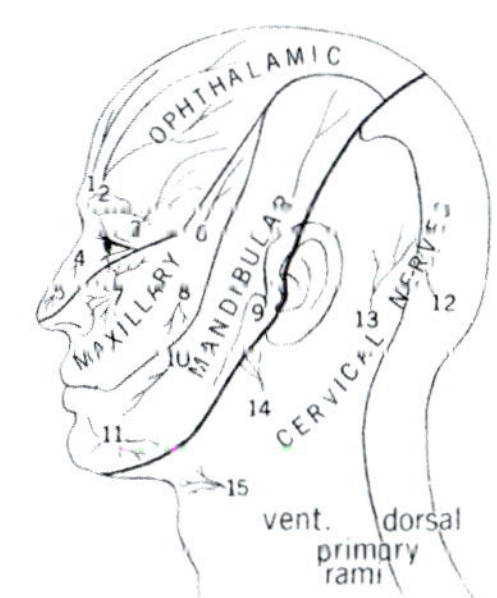

CUTANEOUS DISTRIBUTION OF CRANIAL NERVES

Ophthalamic
1. Supratrochlear N.
2. Supraorbital N.
3. Lacrimal N.
4. Infratrochlear N.
5. Nasal N.

Maxillary
6. Zygomatico-temporal N.
7. Infraorbital N.
8. Zygomatico-facial N.

Mandibular
9. Auriculotemporal N.
10. Buccal N.
11. Mental N.

Cervical Nerves
12. Greater Occipital N.
13. Lesser Occipital N.
14. Great Auricular N.

Redrawn from *Gray's Anatomy of the Human Body*, 28th ed.

 Revised 1970   HENRY O. & FLORENCE P. KENDALL & GLADYS E. WADSWORTH

FIGURE 3-3: Cranial nerve and muscle chart. © 1993 Florence P. Kendall.

SECTION II
FACIAL AND EYE MUSCLES

FACIAL AND EYE MUSCLES

Facial and eye muscles are shown in Table 3-1.

TABLE 3-1 Facial and Eye Muscles

Muscles/*Nerves*	Origin	Insertion	Action and Page Reference
Buccinator/*Facial*	Alveolar processes of the maxilla, buccinator ridge of the mandible, and pterygomandibular ligament	Orbicularis oris at angle of the mouth	Compresses the cheeks
Corrugator supercilii *Facial*	Medial end of the supercilliary arch	Deep surface of skin above middle of the orbital arch	Draws the eyebrow downward and inward, with vertical wrinkles in forehead; "frowning muscle"
Depressor anguli oris/*Facial*	Oblique line of the mandible	Angle of the mouth, blending with adjacent muscle	Depresses angle of the mouth
Depressor labii inferioris/*Facial*	Oblique line of the mandible	Integument of the lower lip, blending with the orbicularis oris	Draws the lower lip downward and slightly sideways, as in expressions of irony
Depressor septi nasi/*Facial*	Incisive fossa of the maxilla	Ala and septum of the nose	Draws ala of the nose downward to close the nose
Frontalis/*Facial*	Galea aponeurotica	Muscles and skin of the eyebrow and root of the nose	Raises the eyebrows and wrinkles the forehead, as in expressions of surprise or fright
Levator anguli oris/*Facial*	Canine fossa of the maxilla	Angle of the mouth, blending with the orbicularis oris	Depresses the nasolabial furrow, as in expressions of contempt or disdain
Levator labii superioris/*Facial*	Lower margin of orbit	Orbicularis of the upper lip	Moves the upper lip upward and forward
Levator labii superioris alaeque nasi/*Oculomotor*	Root of nasal process of the maxilla	Greater alar cartilage, skin of the nose, and lateral part of the upper lip	Raises and protrudes the upper lip
Levator palpebrae superioris/*Oculomotor*	Inferior surface of lesser wing of the sphenoid	Skin of the eyelid, tarsal plate of the upper eyelid, orbital wall, and medial and lateral expansion of aponeurosis of the insertion	Raises the upper eyelids
Masseter/*Trigeminal*	*Superficial portion:* Zygomatic process of the maxilla and lower border of the zygomatic arch	Angle and ramus of the mandible	Temporomandibular joint (TMJ) elevation
	Profundus portion: Posterior ⅓ of inferior border and medial surface of the zygomatic arch	Superior ½ of ramus and lateral surface of coronoid process of mandible	
Mentalis/*Facial*	Incisive fossa of the mandible	Skin of the chin	Raises and protrudes the lower lip and wrinkles skin on the chin, as in pouting
Nasalis, alar portion/*Facial*	Maxilla	Ala of the nose	Enlarges the nostrils

TABLE 3-1 (*continued*)

Muscles/Nerves	Origin	Insertion	Action and Page Reference
Nasalis, transverse portion/*Facial*	Above and lateral to incisive fossa of the maxilla	By aponeurosis with nasalis on the opposite side	Depresses cartilaginous part of the nose
Obliquus inferior oculi/*Oculomotor*	Orbital plate of the maxilla	External part of the sclera, between the rectus superior and rectus lateralis and posterior to equator of the eyeball	Directs the cornea upward and outward
Obliquus superior oculi/*Trochlear*	Above medial margin of the optic foramen	Into sclera between the rectus superior and rectus lateralis and posterior to equator of the eyeball	Directs the cornea downward and outward
Orbicularis oculi/*Facial*	Nasal part of the frontal bone, frontal process of the maxilla, and anterior surface of the medial palpebral ligament	Muscle fibers surround the circumference of the orbit, spread downward on the cheek, and blend with adjacent muscular or ligamentous structures	*Palpebral part:* Closes eyes gently *Orbital part:* Stronger closing
Orbicularis oris/*Facial*	Numerous strata of muscular fibers surrounding the mouth; derived in part from other facial muscles	Into skin and mucous membrane of the lips, blending with other muscles	Closes lips and protrudes them forward
Platysma/*Facial*	Fascia covering superior portion of the pectoralis major and deltoid	Lower border of the mandible, the posterior fibers blending with muscles about angle and lower part of the mouth	Retracts and depresses angle of the mouth
Procerus/*Facial*	Fascia covering lower part of the nasal bone and upper part of the lateral nasal cartilage	Into skin over lower part of the forehead between eyebrows	Pulls inner angle of the eyebrows downward and produces transverse wrinkles over bridge of the nose
Pterygoideus lateralis/*Trigeminal*	*Superior head:* Lateral surface of the great wing of the sphenoid and the infratemporal crest *Inferior head:* Lateral surface of the lateral pterygoid	Depression, anterior part of condyle of the mandible, and anterior margin of articular disk of temporomandibular articulation	Opens the jaws TMJ depression, lateral deviation, and protrusion
Pterygoideus medialis/*Trigeminal*	Medial surface of the lateral pterygoid plate, pyramidal process of the palatine bone, and tuberosity of the maxilla	Interior and posterior part of medial surface of the ramus and angle of the mandibular foramen	Closes the jaw TMJ elevation
Recti superior, inferior, and medialis/*Oculomotor* Rectus lateralis/*Abducent*	Fibrous ring that surrounds superior, medial, and inferior margins of the optic foramen	Into sclera, anterior to equator of the eyeball at the site implied by each name	Movement of the eye in the direction indicated by the name of the muscle
Risorius/*Facial*	Fascia over the masseter	Into skin at angle of the mouth	Retracts the angle of the mouth
Temporalis/*Trigeminal*	Temporal fossa and fascia	Coronoid process and anterior border of ramus of the mandible	Closes the jaws TMJ Elevation
Zygomaticus major/*Facial*	Zygomatic bone in front of the temporal process	Angle of the mouth, blending with the adjacent muscles	Draws angle of the mouth up and out, as in a smile
Zygomaticus minor/*Facial*	Zygomatic bone, malar surface	Orbicularis oris of the upper lip	Deepens the nasolabial furrow, as in expressions of sadness

FACIAL AND EYE MUSCLES TESTS

Frontalis

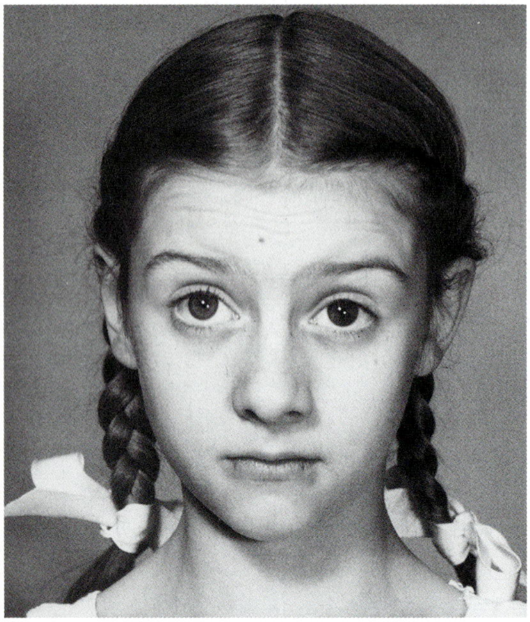

FIGURE 3-4

Test: Raise the eyebrows, wrinkling the forehead, as in an expression of surprise or fright.

Corrugator Supercilii

FIGURE 3-6

Test: Draw the eyebrows together, as in frowning.

Nasalis, Alar Portion

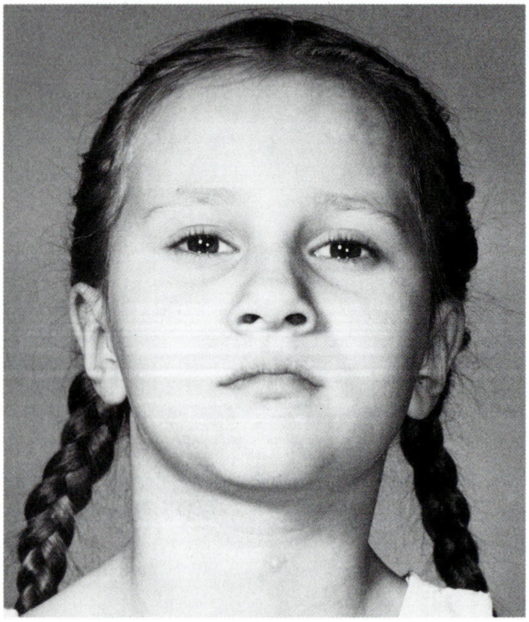

FIGURE 3-5

Test: Widen the apertures of the nostrils, as in forced or difficult breathing.

Depressor Septi and Transverse Portion Nasalis

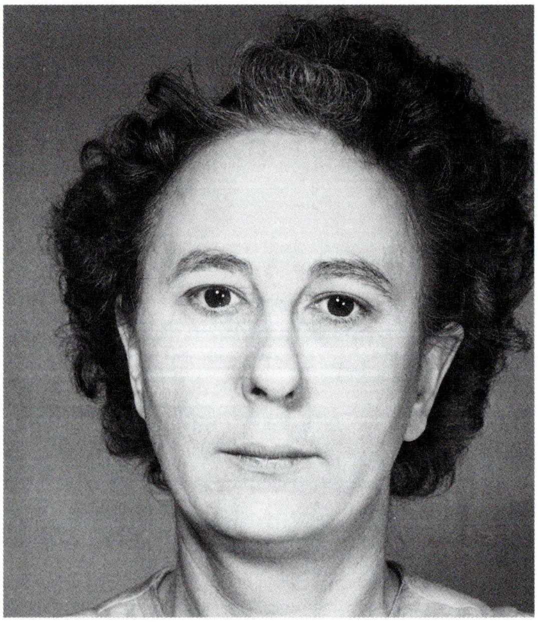

FIGURE 3-7

Test: Draw the point of the nose downward, narrowing the nostrils.

Procerus (Pyramidalis Nasi)

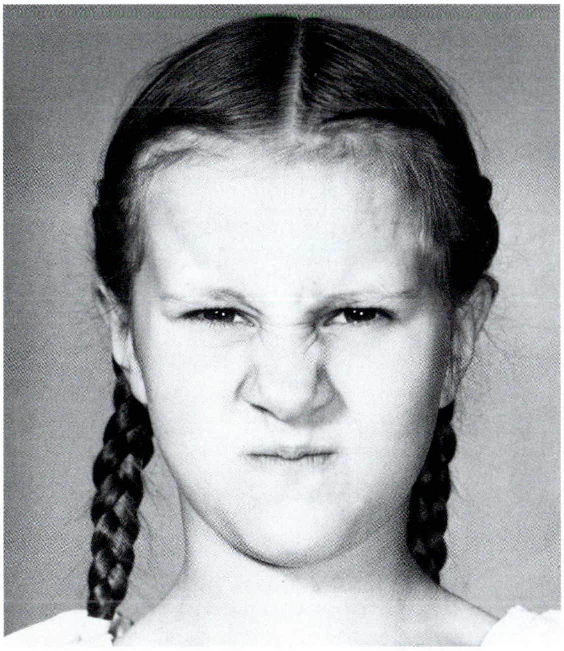

FIGURE 3-8

Test: Pull the skin of the nose upward, forming transverse wrinkles over the bridge of the nose.

Risorius

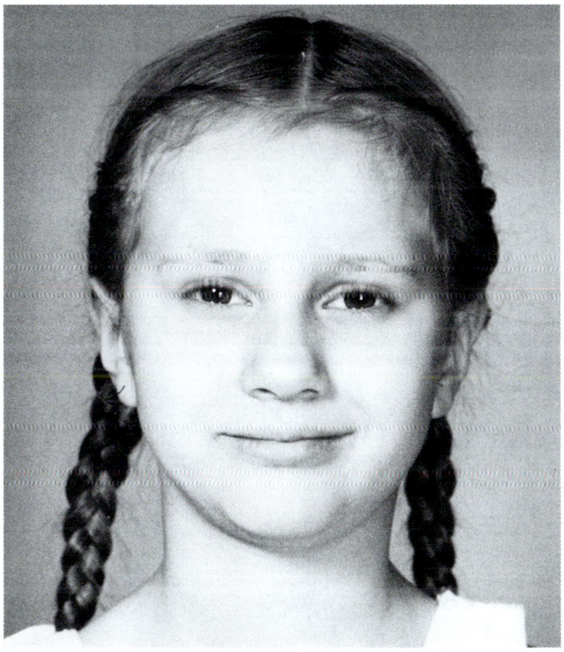

FIGURE 3-9

Test: Draw the angle of the mouth backward.

Levator Anguli Oris

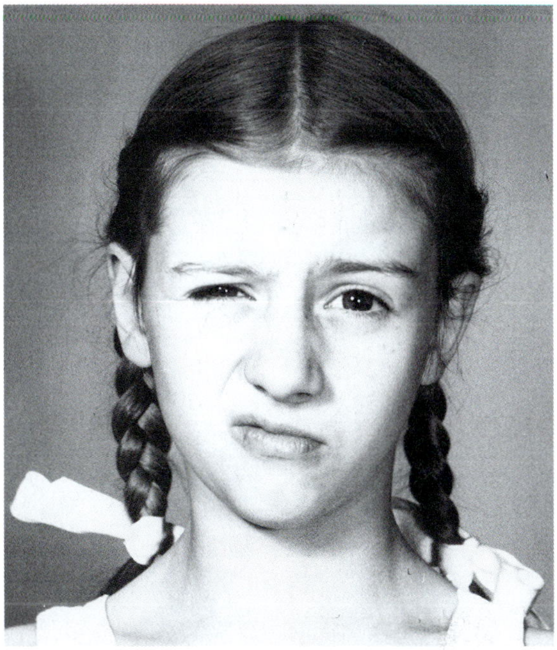

FIGURE 3-10

Test: Draw the angle of the mouth straight upward, deepening the furrow from the side of the nose to the side of the mouth, as in sneering. Suggest that the patient try to show the "eye" (canine) tooth, first on one side and then on the other.

Zygomaticus Major

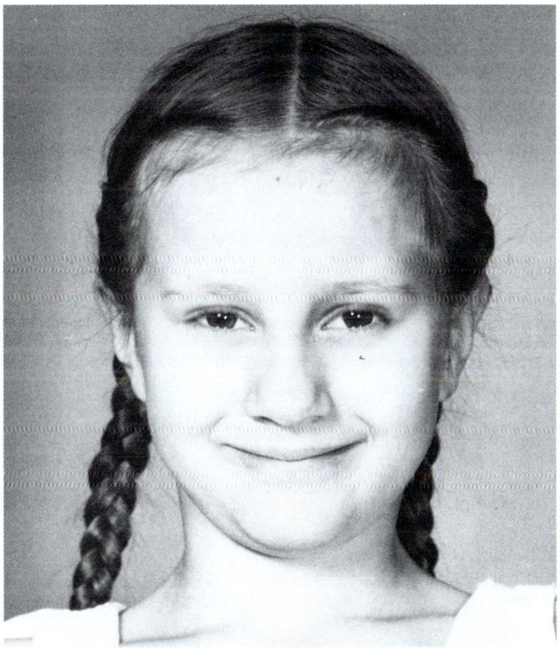

FIGURE 3-11

Test: Draw the angle of the mouth upward and outward, as in smiling.

Levator Labii Superioris

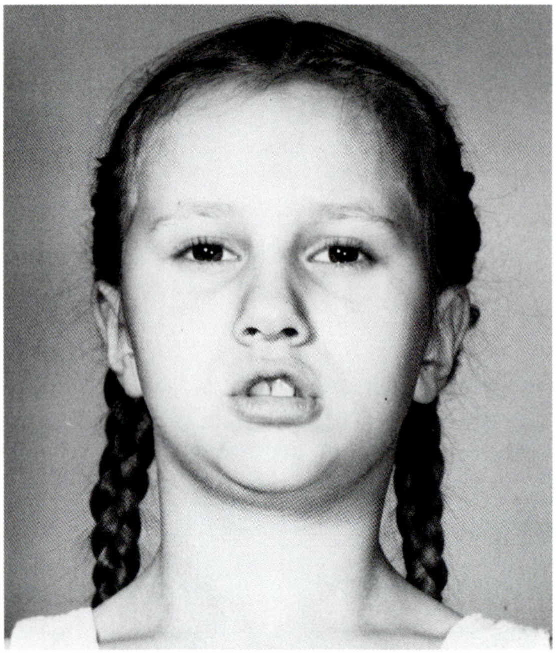

FIGURE 3-12

Test: Raise and protrude the upper lip, as if to show the upper gums.

Orbicularis Oris

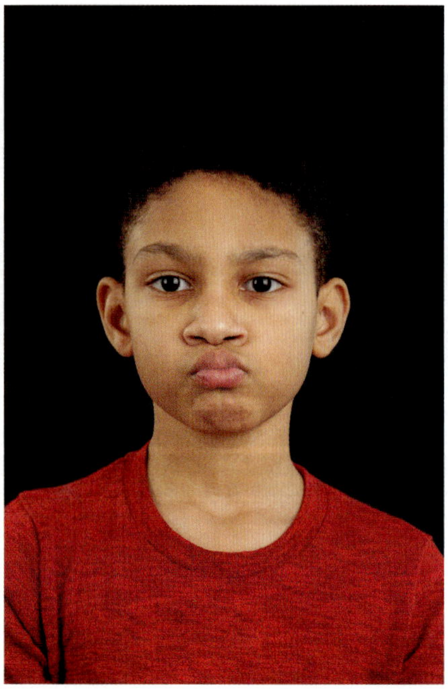

FIGURE 3-13

Test: Close and protrude the lips, as in whistling.

Depressor Labii Inferioris and Platysma

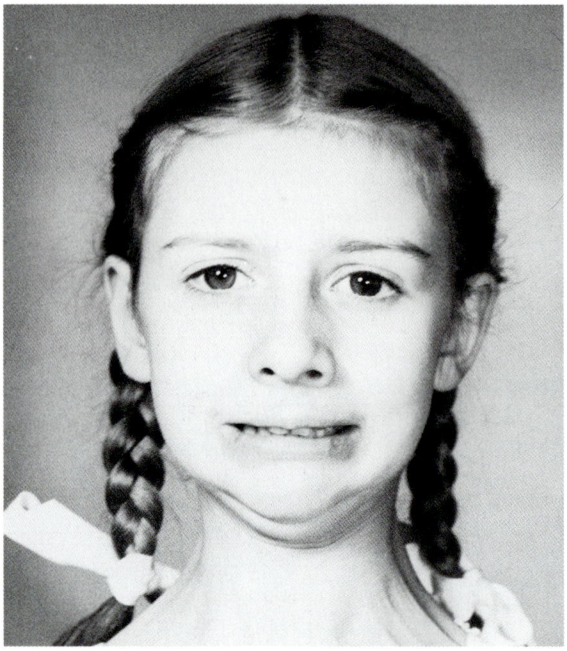

FIGURE 3-14

Test: Draw the lower lip and angle of the mouth downward and outward, tensing the skin over the neck.

Buccinator

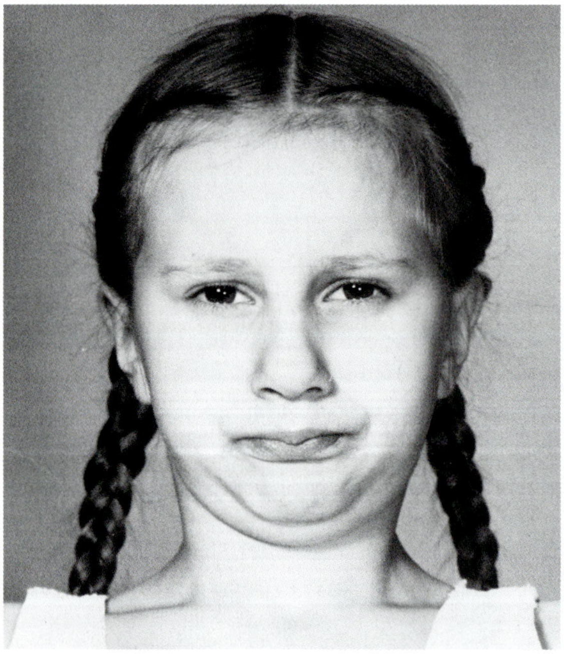

FIGURE 3-15

Test: Press the cheeks firmly against the side teeth and pull back the angle of the mouth, as in blowing a trumpet. (Drawing the chin backward, as seen in this illustration, is not part of the action of the buccinator muscle.)

Mentalis

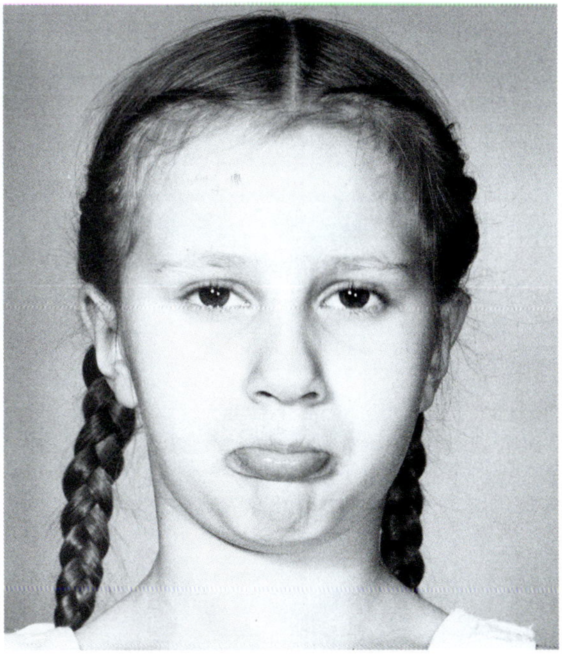

FIGURE 3-16

Test: Raise the skin of the chin. As a result, the lower lip will protrude somewhat, as in pouting.

Lateral Pterygoid

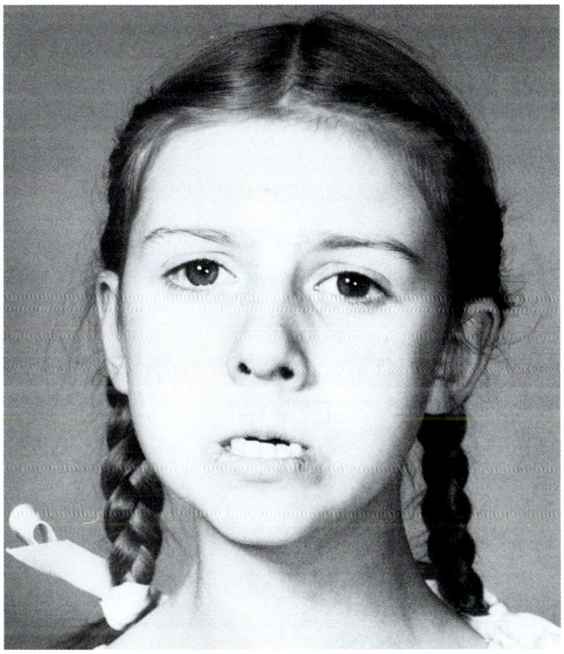

FIGURE 3-17

Test: Open the mouth slightly. (The lateral pterygoid contributes to depression of the temporomandibular joint (TMJ).) Protrudes and then laterally deviates the TMJ.

Depressor Anguli Oris

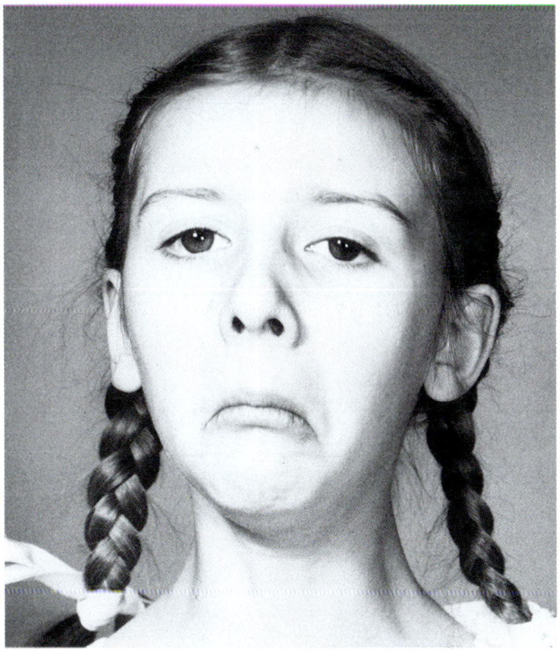

FIGURE 3-18

Test: Draw down the angles of the mouth.

Temporalis, Masseter, and Pterygoideus Medialis

FIGURE 3-19

Test: Elevate the mandible and bite firmly with the mouth slightly open to show that the teeth are being clenched.

Suprahyoid Muscles

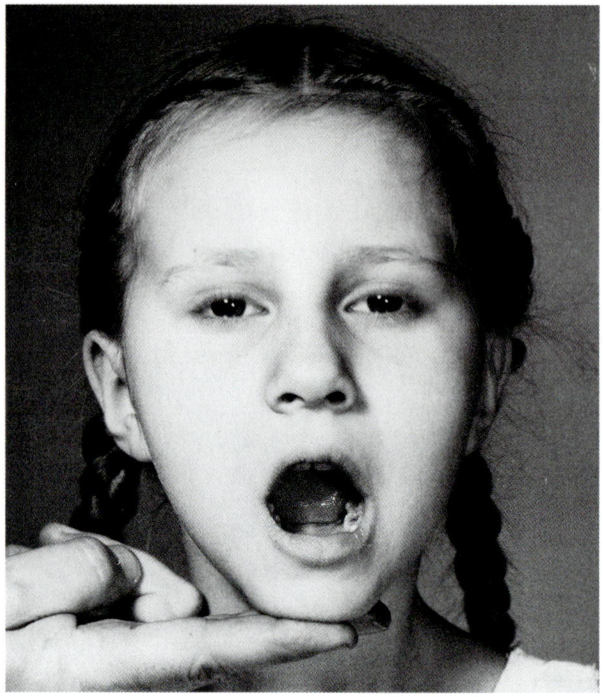

FIGURE 3-20

Test: Depress the mandible against resistance provided by the examiner. During action of the suprahyoid muscles, the infrahyoid muscles furnish fixation of the hyoid bone. (For origins, insertions, actions, and innervations and for an illustration, see Chapter 4.)

Infrahyoid Muscles

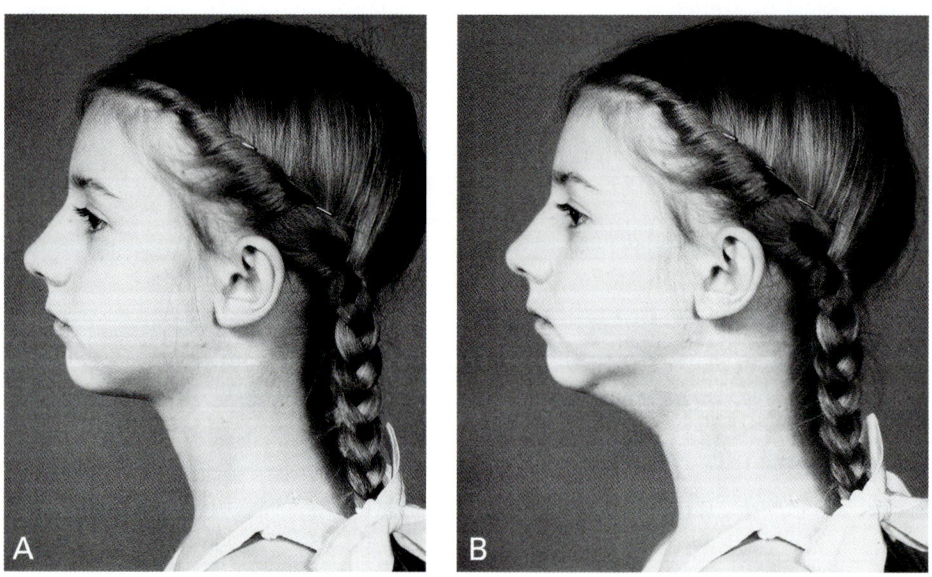

FIGURE 3-21

Test: Begin with a relaxed starting position, as shown at left. Then, depress the hyoid bone, as illustrated at the right. (For origins, insertions, actions, and innervations of infrahyoid muscles and for an illustration, see Chapter 4.)

Orbicularis Oculi

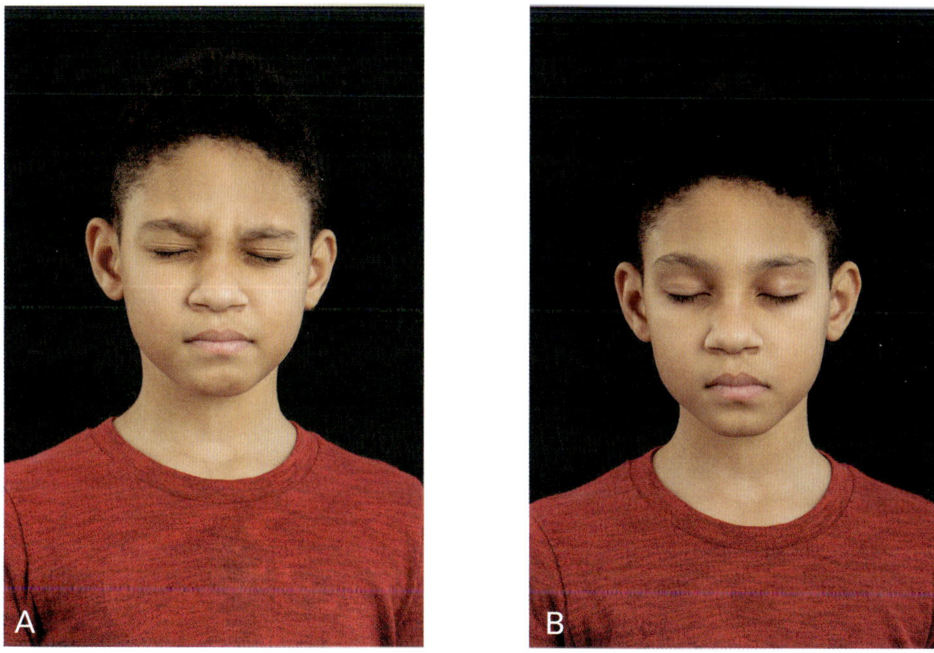

FIGURE 3-22

Test, Orbital Part: Close the eyelid firmly, forming wrinkles that radiate from the outer angle.

Test, Palpebral Part: Close the eyelid gently.

Rectus Medialis Oculi and Rectus Lateralis Oculi

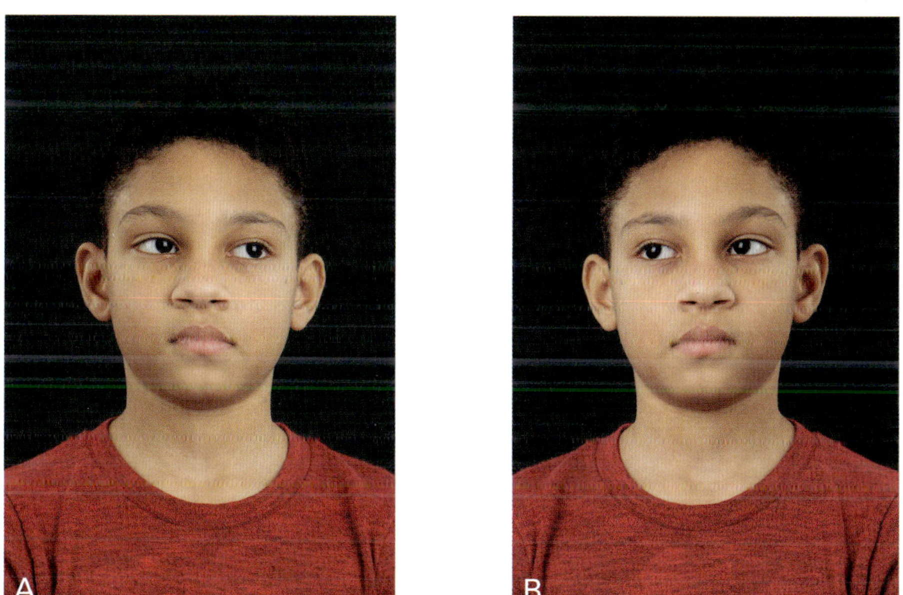

FIGURE 3-23

Test, Right Rectus Medialis and Left Rectus Lateralis: Look horizontally toward the left (as illustrated).

Test, Left Rectus Medialis and Right Rectus Lateralis: Look horizontally toward the right.

Levator Palpebrae Superioris et al.

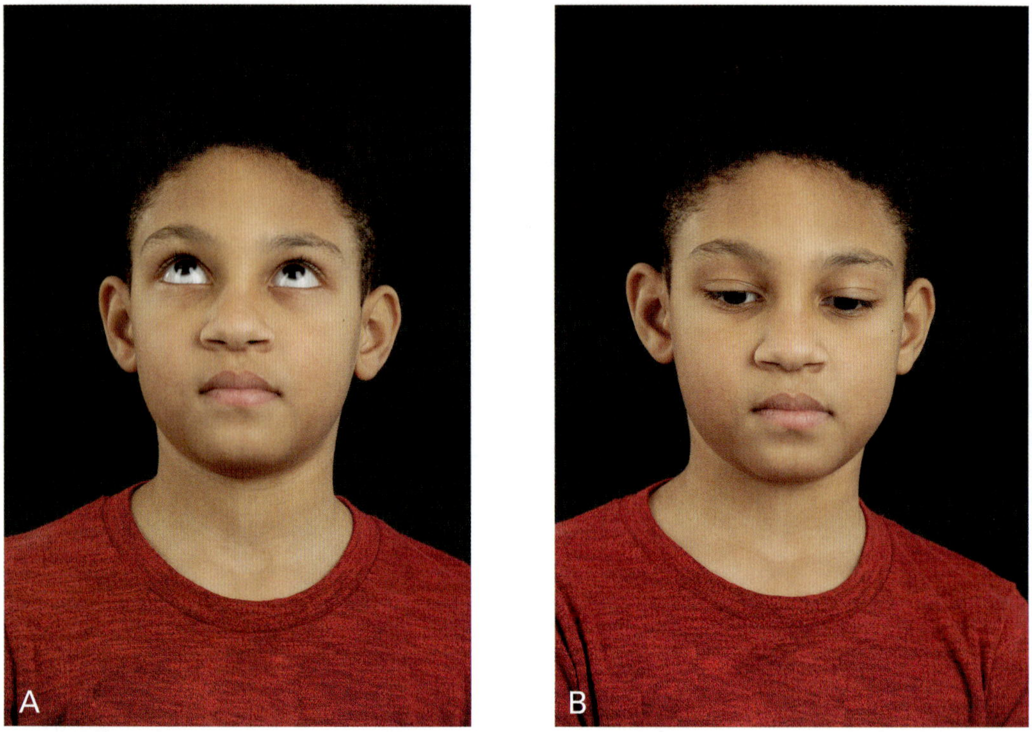

FIGURE 3-24

Test, Levator Palpebrae Superioris: Raise the upper eyelid.

Test, Rectus Superior and Obliquus Inferior: Look straight upward toward the brow.

Test, Rectus Inferior and Obliquus Superior: Look straight downward toward the mouth.

SECTION III

MUSCLES OF MASTICATION, TONGUE, AND PHARYNX

MUSCLES OF DEGLUTITION

The muscles of deglutition include the tongue, soft palate, fauces, suprahyoid, infrahyoid, larynx, and pharynx. Their origin, insertion, action, innervation (motor and sensory), and role are described in Table 3-2.

TABLE 3-2 Muscles of Deglutition

Muscle	Origin	Insertion	Action	Innervation: Motor	Innervation: Sensory	Role in Deglutition
TONGUE						
Sup. longitudinal	Intrinsic	Intrinsic	Shortens tongue Raises sides and tip of Tongue	Hypoglossal XII	General sensation: Ant. ⅔— Trigeminal V Post ⅓— Glossopharyngeal IX Base—Vagus X Special sensation (taste) Ant. ⅔— Facial VII Post. ⅓— Glossopharyngeal IX Base—Vagus X	Bolus Preparation During this phase, the tongue and buccinator muscles keep the food between the molar teeth where it is crushed and ground by the action muscles of mastication. Alternate side-to-side movements and twisting of the tongue, performed chiefly by the intrinsic muscles and by the styloglossi acting unilaterally, aid in mixing the food with saliva and in sorting larger particles from the sufficiently ground portion which is ready to be rolled into a bolus and swallowed.
Transverse	Intrinsic	Intrinsic	Lengthens and narrows tongue			
Vertical	Intrinsic	Intrinsic	Flattens and broadens tongue			
Inf. longitudinal	Intrinsic	Intrinsic	Shortens tongue Turns tip of tongue downward			
Genioglossus	Mental spine	Tongue and body of hyoid	Depresses tongue; protrudes and retracts tongue; elevates hyoid	Hypoglossal XII		
Hyoglossus	Greater horn of hyoid	Tongue	Depresses and pulls tongue posteriorly			
Styloglossus	Styloid process	Tongue	Elevates and pulls tongue posteriorly			
Palatoglossus	Aponeurosis of soft palate	Tongue	Elevates and pulls tongue posteriorly, narrows fauces	Pharyngeal plexus IX, X, XI		
SOFT PALATE						Voluntary Stage The tongue depressor muscles contract and form a groove in the posterior portion of the dorsum of the tongue which cradles the bolus. A movement initiated by the intrinsic muscles raises the anterior portion and then the posterior portion of the tongue to the hard palate. This sequential movement dislodges the bolus and squeezes it toward the fauces. In turn, the base of the tongue is elevated and pulled posteriorly mainly by the action of the styloglossi muscles forcing the bolus through the fauces into the pharynx. Occurring simultaneously with this elevation of the base of the tongue is a moderate elevation of the hyoid bone and the larynx.
Tensor veli palatini	Scaphoid fossa, spine of sphenoid, lateral auditory tube	Aponeurosis of soft palate	Tenses soft palate	Trigeminal V	Trigeminal V Glossopharyngeal IX	
Levator veli palatini	Petrous portion, temporal bone: medial auditory tube	Soft palate	Elevates soft palate	Pharyngeal plexus IX, X, XI		
Uvulae	Posterior nasal spine: aponeurosis of palate	Uvula	Shortens soft palate			
FAUCES						
Palatoglossus	See above			Pharyngeal plexus IX, X, XI	Glossopharyngeal IX	
Palatopharyngeus	Aponeurosis of soft palate	Posterior thyroid cartilage:	Narrows fauces:			
		Posterolateral pharynx	Elevates larynx and pharynx			
SUPRAHYOID						Involuntary (Reflex) Stage As the bolus passes through the fauces to the pharynx, branches of cranial nerves V, IX, and X are stimulated producing impulses in the afferent limb of the swallow reflex. Upon reaching the brainstem, these impulses are transmitted across synapses to efferent fibers of cranial nerves IX, X, and XI completing the reflex arc and effecting the following automatic events.
Digastric Ant. belly	Inferior border of mandible near symphysis	Intermediate tendon to body and cornu of hyoid	Elevates and pulls hyoid anteriorly Assists in depressing the mandible	Trigeminal V		
Post. belly	Mastoid process		Elevates and pulls hyoid posteriorly	Facial VII		
Mylohyoid	Mylohyoid line of mandible	Body of hyoid and median raphe	Elevates hyoid and tongue: depresses mandible	Trigeminal V		
Geniohyoid	Median ridge of mandible	Body of hyoid	Elevates hyoid and tongue: depresses mandible	Ansa cervicalis C1, 2		
Stylohyoid	Styloid process of temporal bone	Body of hyoid	Elevates and pulls hyoid posteriorly	Facial VII		

(continues on the next page)

TABLE 3-2 (*continued*)

Muscle	Origin	Insertion	Action	Innervation: Motor	Innervation: Sensory	Role in Deglutition
INFRAHYOID						
Thyrohyoid	Oblique line of thyroid cartilage	Greater horn of hyoid	Elevates the thyroid cartilage; depresses the hyoid	Ansa cervicalis C1, 2		The soft palate is elevated and brought into contact with the posterior pharyngeal wall by the contraction of the tensor and levator veli palatini muscles. This action closes off the nasopharynx ensuring passage of the bolus into the lumen of the laryngopharynx. This passage is facilitated when the lumen is expanded by the elevation of the pharyngeal wall and the cranial and anterior movement of the hyoid bone and the larynx. When the last of the bolus leaves the oral cavity, the oropharynx opening is closed by contraction of the palatopharyngeal muscles and descent of the soft palate.
Sternohyoid	Manubrium sterni; medial end of clavicle	Body of hyoid, inf. border	Depresses hyoid			
Sternothyroid	Manubrium sterni; costal cartilage of 1st rib	Oblique line of thyroid cartilage	Depresses thyroid cartilage	Ansa cervicalis C1, 2, 3		
Omohyoid-Sup. belly	Superior border of scapula near scapular notch	Intermediate tendon by fascia to clavicle	Depresses the hyoid			
Inf. belly	Intermediate tendon by fascia to clavicle	Body of hyoid, inf. border				
LARYNX						
Aryepiglottic	Apex of arytenoid cartilage	Lateral margin of epiglottis	Assists in closing inlet of larynx			The cranial movement of the thyroid cartilage toward the hyoid bone and of these two structures, in turn, toward the base of the tongue results in tilting the epiglottis posteriorly. The weight of the bolus as it contacts the anterior surface of the epiglottis assists in increasing this posterior tilt. The change of position of the epiglottis aids in directing the bolus material around the sides of the larynx through the piriform sinuses and over the tip of the epiglottis into the hypopharynx. It also aids in preventing foodstuffs from entering the larynx. The major mechanism for protecting the larynx, however, is the concurrent, sphincter-like closure of the laryngeal inlet to the vestibule and the closure of the vestibular and vocal folds of the glottis.
Thyroepiglottic	Medial surface of thyroid cartilage	Lateral margin of epiglottis	Assists in closing inlet of larynx			
Thyroarytenoid	Medial surface of thyroid cartilage	Muscular process of arytenoid cartilage	Assists in closing glottis; shortens vocal folds			
Arytenoid-Oblique	Base of one arytenoid cartilage	Apex of opposite arytenoid cartilage	Assist in closing glottis by adducting arytenoid cartilages			
Transverse	Posterior surface and lateral border of one arytenoid cartilage	Posterior surface and lateral border of opposite arytenoid cartilage		Vagus X Mainly accessory XI, cranial rootwa	Vagus X	
Lat. cricoarytenoid	Upper border of arch of cricoid cartilage	Muscular process of arytenoid cartilage	Adducts and medially rotates arytenoid cartilage assisting in closing glottis			
Vocalis	Medial surface of thyroid cartilage	Vocal process of arytenoid cartilage	Regulates tension of vocal folds			
Post. cricoarytenoid	Posterior surface of lamina of cricoid cartilage	Muscular process of arytenoid cartilage	Abducts arytenoid cartilage widening glottis			
Cricothyroid-Straight Oblique	Anterior and lateral part of arch of cricoid cartilage	Anterior border, inferior horn of thyroid cartilage; Lower border of lamina of thyroid cartilage	Elevates cricoid arch and elongates vocal folds			
PHARYNX						
Salpingopharyngeus	Auditory tube	Pharyngeal wall	Elevates pharynx	Pharyngeal plexus IX, X, XI		Occurring simultaneously with the above events is a sequential contraction of the superior, middle, and inferior constrictors which strips the pharynx forcing the bolus toward the esophagus. Horizontally oriented fibers found between the inferior constrictor and the esophagus have been named the cricopharyngeus muscle. This muscle acts as a sphincter and functionally is related more to the esophagus than to the pharynx. It relaxes when the bolus reaches the caudal extent of the hypopharynx permitting the foodstuff to enter the esophagus.
Palatopharyngeus	See above					
Stylopharyngeus	Styloid process	Posterior border of thyroid cartilage; posterolateral wall of pharynx	Elevates pharynx and larynx	Glossopharyngeal IX		
Superior constrictor	Medial pterygoid plate; pterygomandibular raphe; mandible	Pharyngeal tubercle pharyngeal raphe	Constrict, sequentially, nasopharynx, oropharynx, laryngopharynx		Pharyngeal plexus IX and X	
Middle constrictor	Horns of hyoid	Pharyngeal raphe		Pharyngeal plexus IX, X, XI		
Inferior constrictor	Thyroid and cricoid cartilages	Pharyngeal raphe				
Cricopharyngeus	Arch of cricoid cartilage	Arch of cricoid cartilage	Acts as sphincter to prevent air entering esophagus; relaxes during swallowing			

TEMPOROMANDIBULAR JOINT MOVEMENTS

Temporomandibular joint (TMJ) movements include rotation and translation. The rotational movements include **depression and elevation** of the mandible (i.e., opening and closing the mouth) and **lateral** excursion (deviation) of the mandible (i.e., side-to-side movements). The translational movements include **protrusion** (i.e., movement in an anterior direction) and **retrusion** of the mandible (i.e., movement in a posterior direction) (1). According to Bourban, two primary movements of the TMJ are rotation about a mediolateral axis and translation along the anteroposterior and superoinferior axes (2). Rotation occurs first, and then translation, as the mandibular condyle moves anteriorly and inferiorly on the temporal bone. Closing of the mouth is initiated with posterior translation of the mandible to approximately 2/3 of the maximal opening. The combined translatory and rotatory movements that occur during opening of the mouth are reversed for closing to the resting position (3).

In normal opening and closing, the movements of each TMJ are synchronous so that the mandible does not deviate toward either side. Asymmetrical lateral excursion involves sliding the mandible to one side (4). However, natural mastication requires asymmetrical deviation to complete the act of chewing.

CRANIAL NERVE SCREENING/TESTING

See Table 3-3 for a description of the cranial nerve screening/testing.

TABLE 3-3 Cranial Nerve Screening/Testing

Cranial Nerve	Function	Test Procedure
I-Olfactory	Smell	Identify different odors
II-Optic	Visual acuity	Visual field examination
III-Oculomotor	Eye movement; pupillary reflex	Superior, inferior, and medial eye movement: pupil reaction to light
IV-Trochlear	Eye movement	Medial inferior/superior/oblique eye movements
V-Trigeminal	Sensory to regions of face Chewing	Sensory examination to regions of face (Vi, V2, V3); TMJ motion and strength (V3); corneal reflex (V1)
VI-Abducent	Eye movement	Lateral movement of the eye
VII-Facial	Muscle of facial expression	Open eyes, smile, open and close mouth
VIII-Vestibular cochlear	Balance and hearing	Balance examination; hearing examination
IX-Glossopharyngeal	Taste and swallowing	Gag reflex; ability to swallow
X-Vagus	Taste, pharynx, and larynx control	Gag reflex; ability to swallow; "ahh"
XI-Spinal accessory	Actions of SCM and trapezius	Resisted actions of SCM and trapezius
XII-Hypoglossal	Actions of tongue	Stick tongue out straight/move to right and left

Sternocleidomastoid (SCM)

SECTION IV CLINICAL FINDINGS

CASE STUDIES

The following include two charts with recordings of muscle test findings in two cases of Bell's palsy (i.e., facial paralysis).

Case 1, Bell's Palsy

In this case, the onset of paralysis occurred 1 week before the first examination. As noted in Figure 3-25, 3 muscles graded zero, 10 muscles graded trace, and 2 muscles graded poor. At a second examination 3 weeks later, all the muscles graded good. Approximately 3 weeks after that second examination, all the muscles graded normal, except for three that still graded good.

This case is an example of those patients with facial paralysis who experience a fairly rapid recovery. Sometimes recovery occurs within a few days or a week; in other cases, such as this one, recovery occurs within a 2-month period (5).

On first examination, the orbicularis oculi, which acts to close the eyelid and squeeze it shut, graded poor, and the frontalis, which raises the eyebrow and wrinkles the forehead, graded trace. In some cases of facial paralysis, however, the orbicularis oculi may be slower to respond than the frontalis. In such cases, exercising the frontalis is discouraged, because it acts in opposition to the orbicularis oculi. The reason for this can be illustrated as follows: Raise the eyebrow by contracting the frontalis. Then, with the fingertips placed either on or just above the eyebrow, keep the eyebrow held upward. Now, try to close the eyelid gently, and then try to squeeze the eyelid tightly shut. The difficulty in doing both (and especially the latter) is readily demonstrated.

CRANIAL NERVE AND MUSCLE CHART

Name *Case #1* Date 1 week after onset

		MUSCLE STRENGTH GRADE (Feb. 27)	SENSORY OR MOTOR TO: Left		Mar. 20	Apr. 13	I Olfactory S	II Optic S	III Oculomotor M	IV Trochlear M	V Trigeminal S & M	VI Abducens M	VII Facial S & M	VIII Vestibulocochlear S	IX Glossopharyngeal S & M	X Vagus S & M	XI Accessory M	XII Hypoglossal M
I	NOSE	S	SENSORY—SMELL				•											
II	EYE	S	SENSORY—SIGHT					•										
III	EYELID		LEVATOR PALPEBRAE SUPERIORIS						•									
	EYE		RECTUS SUPERIOR						•									
			OBLIQUUS INFERIOR						•									
			RECTUS MEDIALIS						•									
			RECTUS INFERIOR						•									
IV	EYE		OBLIQUUS SUPERIOR							•								
V	→	S	SENSORY—FACE & INT. STRUCTURES OF HEAD								•							
	EAR		TENSOR TYMPANI								•							
	PALATE		TENSOR VELI PALATINI								•							
	MASTICATION		MASSETER								•							
			TEMPORALIS								•							
			PTERYGOIDEUS MEDIALIS								•							
			PTERYGOIDEUS LATERALIS								•							
	S. HYOID		MYLOHYOIDEUS								•							
			ANTERIOR DIGASTRIC								•							
VI	EYE		RECTUS LATERALIS									•						
VII	TONGUE	S	SENSORY—TASTE, ANTERIOR ⅔ TONGUE										•					
	→	S	SENSORY—EXTERNAL EAR										•					
	EAR		STAPEDIUS										•					
	S. HYOID		POSTERIOR DIGASTRIC										•					
			STYLOHYOIDEUS										•					
	SCALP		OCCIPITALIS	POST. AURICULAR BRANCH									•					
	EAR		INTRINSIC EAR MUSCLES										•					
			AURICULARIS POSTERIOR										•					
			AURICULARIS ANTERIOR	TEMPORAL BRANCH									•					
			AURICULARIS SUPERIOR										•					
	SCALP	T	FRONTALIS		G	N							•					
	EYEBR.	T	CORRUGATOR SUPERCILII	TEMP. & ZYGO. BRANCH	G	N							•					
	EYELID	P	ORBICULARIS OCULI		G	N							•					
	NOSE	P	PROCERUS	BUCCAL BRANCH	G	N							•					
		–	DEP. SEPTI & NAS, TRANS.		–	–							•					
		T	NASALIS, ALAR		G	N							•					
	MOUTH	T	ZYGOMATICUS MAJOR		G	N							•					
			LEVATOR LABII SUPERIORIS		G	N							•					
		T	BUCCINATOR		G	N							•					
		T	ORBICULARIS ORIS		G	G							•					
		T	LEVATOR ANGULI ORIS		G	G							•					
		T	RISORIUS	MANDIBULAR BRANCH	G	N							•					
		T	DEPRESSOR ANGULI ORIS		G	N							•					
		O	DEPRESSOR LABII INFERIORIS		G	N							•					
	CHIN	T	MENTALIS		G	G							•					
	NECK	O	PLATYSMA	CERVICAL BRANCH	G	N							•					
VIII	EAR	S	SENSORY—HEARING & EQUILIBRIUM											•				
IX	TONGUE	S	SENSORY—POSTERIOR ⅓ TONGUE												•			
	PHARYNX	S	SENSORY—PHARYNX, FAUCES, SOFT PALATE												•			
			STYLOPHARYNGEUS												•			
		—	STRIATED MUSCLES - PHARYNX												•			
X	→	—	STRIATED MUSCLES—SOFT PALATE, PHARYNX & LARYNX													•		
	→	—	INVOLUNTARY MUSCLES—ALIMENTARY TRACT													•		
	→	—	INVOLUNTARY MUSCLES—AIR PASSAGES													•		
	→	—	INVOLUNTARY CARDIAC MUSCLE													•		
	→	S	SENSORY—AURICULAR													•		
	→	S	SENSORY—ALIMENTARY TRACT													•		
	→	S	SENSORY—AIR PASSAGES													•		
	→	S	SENSORY—ABDOMINAL VISCERA & HEART													•		
XI	NECK		TRAPEZIUS & STERNOCLEIDOMASTOID														•	
	PALATE		LEVATOR VELI PALATINI														•	
	→		STRIATED MUSCLES—SOFT PALATE, PHARYNX, & LARYNX														•	
XII	TONGUE		STYLOGLOSSUS															•
			HYOGLOSSUS															•
			GENIOGLOSSUS															•
			TONGUE INTRINSICS															•

SENSORY

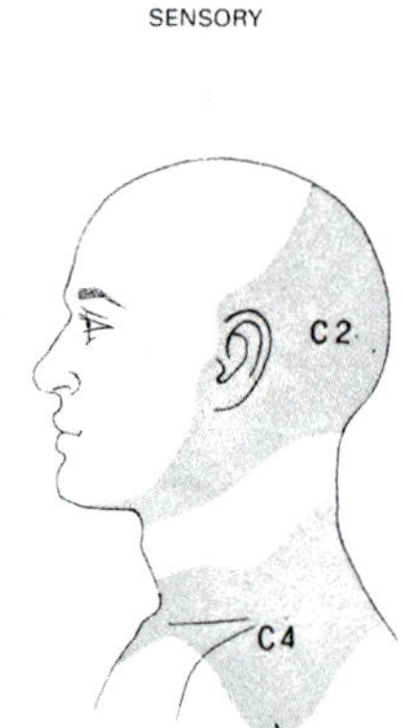

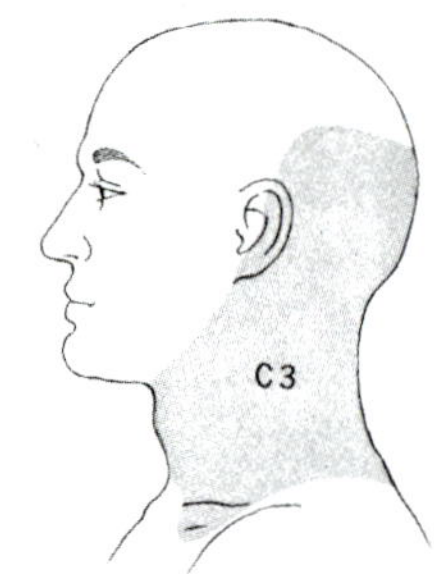

DERMATOMES

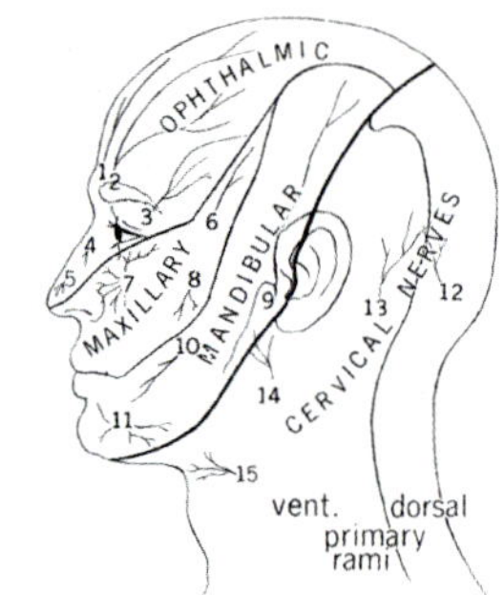

CUTANEOUS DISTRIBUTION OF CRANIAL NERVES

Ophthalmic
1. Supratrochlear N.
2. Supraorbital N.
3. Lacrimal N.
4. Infratrochlear N.
5. Nasal N.

Maxillary
6. Zygomatico-temporal N.
7. Infraorbital N.
8. Zygomatico-facial N.

Mandibular
9. Auriculotemporal N.
10. Buccal N.
11. Mental N.

Cervical Nerves
12. Greater Occipital N.
13. Lesser Occipital N.
14. Great Auricular N.

Redrawn from *Gray's Anatomy of the Human Body*, 28th ed

FIGURE 3-25: Case 1: Bell's palsy, 1. © 2005 Florence P. Kendall.

Case 2, Bell's Palsy

In this case of facial paralysis, which was first seen for examination 3 weeks after onset, no evidence of any muscle function was observed except for a slight action in the corrugator (Figure 3-26). This case showed very little change during the first 3½ months. By the end of 6 months, however, most of the muscles graded either fair or better. By the end of 8 months, further improvement was noted. By the end of 9½ months, approximately 1/3 of the muscles graded fair, and all others graded either good or normal. This case shows the slow but gradual improvement that occurs in some instances (5).

This patient was fitted with a very small, plastic hook that was contoured to fit in the corner of the mouth and was attached by a rubber band to the sidepiece of her glasses. (The sidepiece, or temple, is the part of the frame that extends from the lens both to and over the ear.) She was instructed in how to give herself light massage—upward on the affected side and downward and toward the mouth on the unaffected side. At times, transparent Scotch tape was used to hold up the side of the mouth and cheek. When the patient was not using the hook or the tape, she was advised to make a habit, when sitting, of resting her right elbow on a table or arm of a chair and placing her right hand with the palm under her right chin and the fingers along her cheek to hold the right side of her face upward. Also, when she was speaking, smiling, or laughing, she was to use the hand to push the affected side toward the right and upward to compensate for the weakness as well as to prevent the unaffected side from distorting the mouth in that direction. In addition, she was taught how to exercise the facial muscles by assisting the weak side and restraining the stronger side.

TEMPOROMANDIBULAR DISORDERS

Disorders of the TMJ can lead to headache, facial pain, and limitations of TMJ depression. The muscles usually involved in such disorders are the pterygoids, masseters, and temporalis (6). Integrity and function of the digastric, infra- and suprahyoid muscles, along with noncontractile tissue including associated ligaments and articular discs also need to be assessed. Conservative physical therapy treatment may be sufficient to relieve pain (7–12). Various dental devices may be used to help realign or exercise these muscles (13).

Name Case #2

Date 3 weeks after onset

		MUSCLE STRENGTH GRADE	SENSORY OR MOTOR TO: Right		I Olfactory S	II Optic S	III Oculomotor M	IV Trochlear M	V Trigeminal S & M	VI Abducens M	VII Facial S & M	VIII Vestibulocochlear S	IX Glossopharyngeal S & M	X Vagus S & M	XI Accessory M	XII Hypoglossal M
I	NOSE	S	SENSORY—SMELL		•											
II	EYE	S	SENSORY—SIGHT			•										
III	EYELID		LEVATOR PALPEBRAE SUPERIORIS				•									
III	EYE		RECTUS SUPERIOR				•									
III	EYE		OBLIQUUS INFERIOR				•									
III	EYE		RECTUS MEDIALIS				•									
III	EYE		RECTUS INFERIOR				•									
IV	EYE		OBLIQUUS SUPERIOR					•								
V	⟶	S	SENSORY—FACE & INT. STRUCTURES OF HEAD						•							
V	EAR		TENSOR TYMPANI						•							
V	PALATE		TENSOR VELI PALATINI						•							
V	MASTICATION		MASSETER						•							
V	MASTICATION		TEMPORALIS						•							
V	MASTICATION		PTERYGOIDEUS MEDIALIS						•							
V	MASTICATION		PTERYGOIDEUS LATERALIS						•							
V	S. HYOID		MYLOHYOIDEUS						•							
V	S. HYOID		ANTERIOR DIGASTRIC						•							
VI	EYE		RECTUS LATERALIS							•						
VII	TONGUE	S	SENSORY—TASTE, ANTERIOR ⅔ TONGUE								•					
VII	⟶	S	SENSORY—EXTERNAL EAR								•					
VII	EAR		STAPEDIUS								•					
VII	S. HYOID		POSTERIOR DIGASTRIC								•					
VII	S. HYOID		STYLOHYOIDEUS		[11-3-61]	[12-11-61]	[2-28-62]	[4-17-62]	[6-6-62]		•					
VII	SCALP	[8-22-61]	OCCIPITALIS	POST. AURICULAR BRANCH							•					
VII	EAR		INTRINSIC EAR MUSCLES	POST. AURICULAR BRANCH							•					
VII	EAR		AURICULARIS POSTERIOR	POST. AURICULAR BRANCH							•					
VII	EAR		AURICULARIS ANTERIOR	TEMPORAL BRANCH							•					
VII	EAR		AURICULARIS SUPERIOR	TEMPORAL BRANCH							•					
VII	SCALP	O	FRONTALIS	TEMPORAL BRANCH	T	T	P+	F	F		•					
VII	EYEBR.	P	CORRUGATOR SUPERCILII	TEMP. & ZYGO. BRANCH	P	–	G–	G	G		•					
VII	EYELID	O	ORBICULARIS OCULI	TEMP. & ZYGO. BRANCH							•					
VII	NOSE	O	PROCERUS	BUCCAL BRANCH	O	P	G–	F	G		•					
VII	NOSE	O	DEP. SEPTI & NAS, TRANS.	BUCCAL BRANCH	–	–	–	–	–		•					
VII	NOSE	O	NASALIS, ALAR	BUCCAL BRANCH	O	?	F	F	F		•					
VII	MOUTH	O	ZYGOMATICUS MAJOR	BUCCAL BRANCH	P–	P	G–	G	G		•					
VII	MOUTH	O	LEVATOR LABII SUPERIORIS	BUCCAL BRANCH	?	?	F	F	G		•					
VII	MOUTH	O	BUCCINATOR	BUCCAL BRANCH	–	–	F–	F	F		•					
VII	MOUTH	O	ORBICULARIS ORIS	BUCCAL BRANCH	–	T	F	F–	F		•					
VII	MOUTH	O	LEVATOR ANGULI ORIS	BUCCAL BRANCH	T	?	G–	G	G		•					
VII	MOUTH	O	RISORIUS	BUCCAL BRANCH / MANDIBULAR BRANCH	P–	P	F+	G	G		•					
VII	MOUTH	O	DEPRESSOR ANGULI ORIS	MANDIBULAR BRANCH	?	–	F	F–	F		•					
VII	MOUTH	O	DEPRESSOR LABII INFERIORIS	MANDIBULAR BRANCH	?	–	P+	F–	G		•					
VII	CHIN	O	MENTALIS	MANDIBULAR BRANCH	O	?	F+	G	N		•					
VII	NECK	O	PLATYSMA	CERVICAL BRANCH	T	–	F+	G	G		•					
VIII	EAR	S	SENSORY—HEARING & EQUILIBRIUM									•				
IX	TONGUE	S	SENSORY—POSTERIOR ⅓ TONGUE										•			
IX	PHARYNX	S	SENSORY—PHARYNX, FAUCES, SOFT PALATE										•			
IX	PHARYNX		STYLOPHARYNGEUS										•			
IX	PHARYNX	—	STRIATED MUSCLES - PHARYNX										•			
X	⟶	—	STRIATED MUSCLES—SOFT PALATE, PHARYNX & LARYNX											•		
X	⟶	—	INVOLUNTARY MUSCLES—ALIMENTARY TRACT											•		
X	⟶	—	INVOLUNTARY MUSCLES—AIR PASSAGES											•		
X	⟶	—	INVOLUNTARY CARDIAC MUSCLE											•		
X	⟶	S	SENSORY—AURICULAR											•		
X	⟶	S	SENSORY—ALIMENTARY TRACT											•		
X	⟶	S	SENSORY—AIR PASSAGES											•		
X	⟶	S	SENSORY—ABDOMINAL VISCERA & HEART											•		
XI	NECK		TRAPEZIUS & STERNOCLEIDOMASTOID												•	
XI	PALATE		LEVATOR VELI PALATINI												•	
XI	⟶		STRIATED MUSCLES—SOFT PALATE, PHARYNX, & LARYNX												•	
XII	TONGUE		STYLOGLOSSUS													•
XII	TONGUE		HYOGLOSSUS													•
XII	TONGUE		GENIOGLOSSUS													•
XII	TONGUE		TONGUE INTRINSICS													•

SENSORY

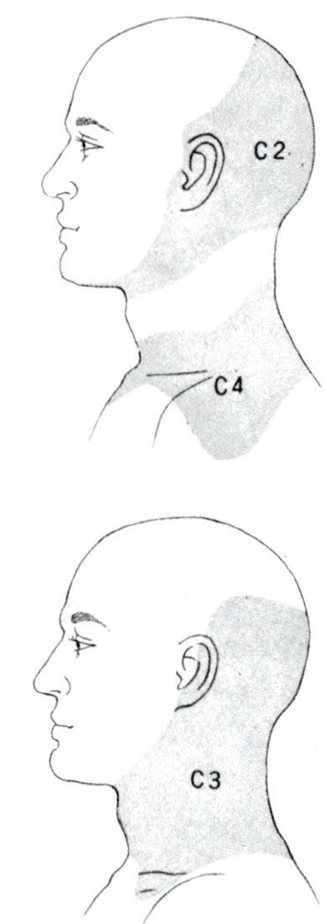

DERMATOMES

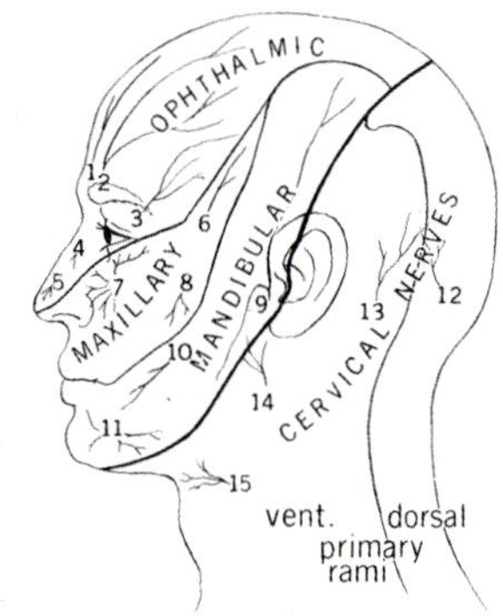

CUTANEOUS DISTRIBUTION OF CRANIAL NERVES

Ophthalmic
1. Supratrochlear N.
2. Supraorbital N.
3. Lacrimal N.
4. Infratrochlear N.
5. Nasal N.

Maxillary
6. Zygomatico-temporal N.
7. Infraorbital N.
8. Zygomatico-facial N.

Mandibular
9. Auriculotemporal N.
10. Buccal N.
11. Mental N.

Cervical Nerves
12. Greater Occipital N.
13. Lesser Occipital N.
14. Great Auricular N.

Redrawn from *Gray's Anatomy of the Human Body*, 28th ed.

FIGURE 3-26: Case 2: Bell's palsy, 2. © 2005 Florence P. Kendall.

SECTION V
INTERVENTION

BELL'S PALSY

It is well documented that the majority of individuals diagnosed with Bell's palsy will regain normal function without intervention (14, 15). For the remaining individuals, a variety of treatment options are available, including medication management with nonsteroidal anti-inflammatory drugs or corticosteroids, therapeutic exercise, massage, acupuncture, laser therapy, electrical stimulation, and eye care to prevent corneal complications (14). There is minimal evidence to support the efficacy of these interventions in restoring the function of the muscles affected by this diagnosis (16).

TEMPOROMANDIBULAR DISORDERS

Treatment interventions for temporomandibular joint disorder (TMD) depends upon the presenting symptoms and the suspected cause (17). Interventions may be provided by a variety of professions within the healthcare arena independently or through interprofessional collaboration. TMD may be managed through nonconservative or conservative means or in combination. While some cases warrant nonconservative intervention such as surgery, there is evidence to support that the majority of patients will achieve satisfying outcomes through conservative interventions (18, 19). Examples of these interventions include posture retraining, manual therapy, active and passive range of motion, and dental appliances (17–19)

REFERENCES

1. Neumann DA. Kinesiology of the Musculoskeletal System: Foundations for Physical Rehabilitation. St. Louis: Mosby, 3rd edition; 2017. pp. 443–449.
2. Bourban B. Musculoskeletal analysis: The temporomandibular joint and cervical spine. In: Scully R, Barnes M, eds. Physical Therapy. Philadelphia: JB Lippincott; 1989.
3. Rocabado M. Arthrokinematics of the temporomandibular joint. Dent Clin North Am. 1983; 27: 573–594.
4. Yustin D, Rieger M, McGuckin R. Determination of the existence of hinge movements of the temporomandibular joint during normal opening by cine-MRI and computer digital addition. J Prosthodont. 1993; 2: 190–195.
5. Baugh RF, Basura GJ, Ishii LE, Schwartz SR, Drumheller CM, Burkholder R, Deckard NA, Dawson C, Driscoll C, Gillespie MB, Gurgel RK, Halperin J, Khalid AN, Kumar KA, Micco A, Munsell D, Rosenbaum S, Vaughan W. Clinical practice guideline: Bell's palsy. Otolaryngol Head Neck Surg. 2013; 149(3 Suppl): S1–S27. doi:10.1177/0194599813505967. PMID: 24189771.
6. Travell J. Temporomandibular joint pain referred from muscles of the head and neck. J Prosthet Dent. 1960; 10(4): 745–763.
7. Urbański P, Trybulec B, Pihut M. The application of manual techniques in masticatory muscles relaxation as adjunctive therapy in the treatment of temporomandibular joint disorders. Int J Environ Res Public Health. 2021; 18(24): 12970. doi:10.3390/ijerpH182412970. PMID: 34948580; PMCID: PMC8700844.
8. De Resende CMBM, de Oliveira Medeiros FGL, de Figueiredo Rêgo CR, Bispo ASL, Barbosa GAS, de Almeida EO. Short-term effectiveness of conservative therapies in pain, quality of life, and sleep in patients with temporomandibular disorders: A randomized clinical trial. Cranio. 2019; 15: 1–9.
9. Pessoa DR, Costa DR, Prianti BDM, Delpasso CA, Arisawa ELS, Nicolau RA. Association of facial massage, dry needling, and laser therapy in temporomandibular disorder: Case report. Codas. 2018; 30: e20170265.
10. De la Serna PD, Plaza-Manzano GP, Cleland J, Fernández-De-Las-Peñas C, Martín-Casas P, Díaz-Arribas MJ. Effects of cervico-mandibular manual therapy in patients with temporomandibular pain disorders and associated somatic tinnitus: A randomized clinical trial. Pain Med. 2019; 21: 613–624.
11. Tuncer A, Ergun N, Tuncer AH, Karahan S. Effectiveness of manual therapy and home physical therapy in patients with temporomandibular disorders: A randomized controlled trial. J Bodyw Mov Ther. 2013; 17: 302–308.
12. Wieckiewicz M, Boening K, Wiland P, Shiau Y-Y, Paradowska-Stolarz A. Reported concepts for the treatment modalities and pain management of temporomandibular disorders. J Headache Pain. 2015; 16: 106.
13. Grace E, Sarlani E, Reid B, Read B. The use of an oral exercise device in the treatment of muscular TMD. J Craniomandib Pract. 2002; 20(3): 204–208.
14. Marson AG, Salinas R. Bell's palsy. West J Med. 2000; 173(4): 266–268. doi:10.1136/ewjm.173.4.266.
15. Holland NJ, Bernstein JM. Bell's palsy. BMJ Clin Evid. 2014; 2014: 1204. Published 2014 Apr 9.
16. Burelo-Peregrino EG, Salas-Magaña M, Arias-Vázquez PI, Tovilla-Zarate CA, Bermudez-Ocaña DY, López-Narváez ML, Guzmán-Priego CG, González-Castro TB, Juárez-Rojop IE. Efficacy of electrotherapy in Bell's palsy treatment: A systematic review. J Back Musculoskelet Rehabil. 2020; 33(5): 865–874. doi:10.3233/BMR-171031. PMID: 32144972.
17. Wright EF, North SL. Management and treatment of temporomandibular disorders: A clinical perspective. J Man Manip Ther. 2009; 17(4): 247–254. doi:10.1179/106698109791352184.
18. Butts R, Dunning J, Pavkovich R, Mettille J, Mourad F. Conservative management of temporomandibular dysfunction: A literature review with implications for clinical practice guidelines (Narrative review part 2). J Bodyw Mov Ther. 2017; 21(3): 541–548. doi:10.1016/j.jbmt.2017.05.021.
19. Shaffer SM, Brismée JM, Sizer PS, Courtney CA. Temporomandibular disorders. Part 2: Conservative management. J Man Manip Ther. 2014; 22(1): 13–23. doi:10.1179/2042618613Y.0000000061.

NECK AND BACK

4

OUTLINE

INTRODUCTION

The cervical spine and the muscles of the neck form a remarkable structure that provides for movement of the head in all directions and stability in various positions. The neck supports the weight of the head in the upright position.

The standard (also referred to as normal) position of the head is one in which the head is level in all planes as the eyes seek eye level. The cervical spine is in a position of lordosis, and the thoracic spine is in a position of kyphosis.

With common atypical postures, the alignment of the head does not change, but the alignment of the neck changes in response to altered thoracic spine positions. If the thoracic spine is straight, the cervical spine may be straight. If the thoracic spine becomes hyperkyphotic, the cervical spine extension increases to the extent that a marked kyphosis may result in a position of full neck extension with the head maintaining a level position (see Figures 4-25 B and D later in this chapter). Chronic problems of the neck may result from faulty posture of the thoracic spine. As seen in the radiographs in Figure 4-25, the extension occurs in the lower cervical area with the upper vertebrae maintaining a level position for support of the head.

Along with many attributes, the neck and back are vulnerable to stress and serious injury. Occupational or recreational activities may demand positions of the head and spine that may result in malalignment and muscle imbalance. Emotional stress can also cause an acute onset of pain with spasm of the neck and back muscles (see examples of incorrect and corrected positions in a workplace situation, Figure 4-68 later in the chapter).

This chapter presents basic evaluation and treatment procedures that address faulty postures, muscle imbalance, and painful conditions of the neck and back. Additionally, this chapter examines the role muscles play in movement and support of the neck and back. Muscles acting in synergy to produce functional and coordinated movements about the hip, pelvis, back, and neck will be explored. The trunk can be defined in two parts: the thoracic spine and the ribcage constitute the upper part; the lumbar spine and the pelvis constitute the lower part. (The pelvis will be discussed in greater detail in Chapters 5 and 7.)

It is interesting to note that muscles that act in unison for certain movements act in opposition to each other in support of good alignment. For example, in the prone position during the movement of spine extension, hip extensors assist by stabilizing the pelvis to the femur. In the supine position during the movement of spine flexion, the hip flexors act to stabilize the pelvis. On the other hand, in support of good postural alignment in standing, the hip extensors act with the abdominal muscles, and the hip flexors act with the back extensors.

Photographs and line drawings will clearly illustrate the differences between the normal movements that occur during testing and changes that occur when there is imbalance between muscles that normally act in unison. In many cases, because of the interaction of some trunk muscles, group tests are more useful than tests of individual muscles.

In relation to the trunk, one of the major concerns is the painful low back. A high percentage of the adult population will suffer from low back pain at one time or another. For many, restoring good postural alignment and muscle balance is the treatment of choice (see section on The Low Back Enigma).

SECTION I
INNERVATION

NECK

Spinal Cord and Nerve Roots

Figure 4-1 shows the spinal cord and nerve roots.

Spinal Nerve and Muscles of the Neck and Diaphragm

Figure 4-2 indicates the spinal nerves and muscles of the neck and diaphragm.

Cervical Plexus

The **cervical plexus** (Figure 4-3) is formed by the ventral primary rami of spinal nerves C1 through C4, with a small contribution from C5. Peripheral nerves arising from the cervical plexus innervate most of the anterior and lateral muscles of the neck and supply sensory innervation to part of the head as well as to much of the neck.

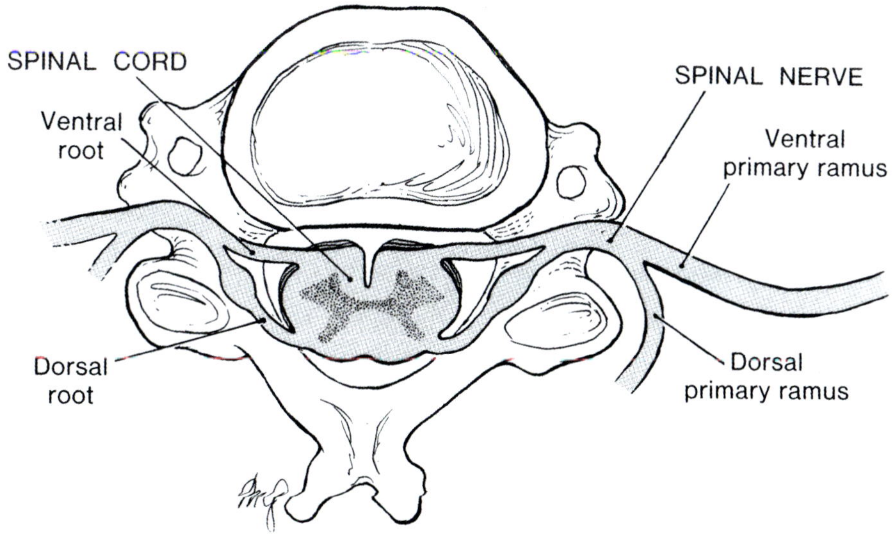

FIGURE 4-1: Spinal cord and nerve roots.

Name | Date

MUSCLE STRENGTH GRADE

PERIPHERAL NERVES

KEY →
- D. = Dorsal Prim. Ramus
- V. = Vent. Prim. Ramus
- P.R. = Plexus Root
- S.T. = Superior Trunk
- P. = Posterior Cord
- L. = Lateral Cord
- M. = Medial Cord

	MUSCLE	Cervical	Cervical	Cervical	Phrenic	Long Thor.	Dor. Scap	N. to Subcl.	Suprascap	U. Subscap.	Thoracodor	L. Subscap.	Lat. Pect.	Med. Pect.	Axillary	Musculocu.	Radial	Median	Ulnar
	Key	D.	V.	V.	V.	P.R.	P.R.	S.T.	S.T.	P.	P.	P.	L.	M.	P.	L.	P.	L.M.	M.
	Cervical	1-8	1-8	1-4	3, 4, 5	5, 6, 7, (8)	4, 5	5, 6	4, 5, 6	(4), 5, 6, (7)	(5), 6, 7, 8	5, 6, (7)	5, 6, 7	(6), 7, 8	5, 6	(4), 5, 6, 7	5, 6, 7, 8	5, 6, 7, 8	7, 8
	T.	1												1			1	1	1
	HEAD & NECK EXTENSORS	●																	
Cervical nerves	INFRAHYOID MUSCLES			●															
Cervical nerves	RECTUS CAP ANT. & LAT.			●															
Cervical nerves	LONGUS CAPITIS			●															
Cervical nerves	LONGUS COLLI		●																
Cervical nerves	LEVATOR SCAPULAE			●			●												
Cervical nerves	SCALENI (A. M. P.)		●																
Cervical nerves	STERNOCLEIDOMASTOID			●															
Cervical nerves	TRAPEZIUS (U. M. L.)			●															
Cervical nerves	DIAPHRAGM				●														

SPINAL SEGMENT

MUSCLE	C1	C2	C3	C4	C5	C6	C7	C8	T1
HEAD & NECK EXTENSORS	1	2	3	4	5	6	7	8	1
INFRAHYOID MUSCLES	1	2	3						
RECTUS CAP ANT. & LAT.	1	2							
LONGUS CAPITIS	1	2	3	(4)					
LONGUS COLLI		2	3	4	5	6	(7)		
LEVATOR SCAPULAE			3	4	5				
SCALENI (A. M. P.)			3	4	5	6	7	8	
STERNOCLEIDOMASTOID	(1)	2	3						
TRAPEZIUS (U. M. L.)		2	3	4					
DIAPHRAGM			3	4	5				

FIGURE 4-2: Spinal nerves and muscles: Neck and diaphragm. © 2005 Florence P. Kendall.

Innervation to Muscles of the Back

Innervation to the trunk muscles does not include an intervening plexus between the spinal cord and the peripheral nerves like the cervical, brachial, lumbar, and sacral plexus. Innervation to the deep back muscles is provided by corresponding dorsal primary within the segmental range of a given muscle's origin and insertion. Muscles of the anterior abdominal wall, superficial back, and chest are provided primarily by ventral primary rami and will be discussed in greater detail in Chapters 5 and 6, respectively.

Spinal Nerve and Muscles of the Trunk

Figure 4-4 shows the spinal nerves and muscles of the trunk.

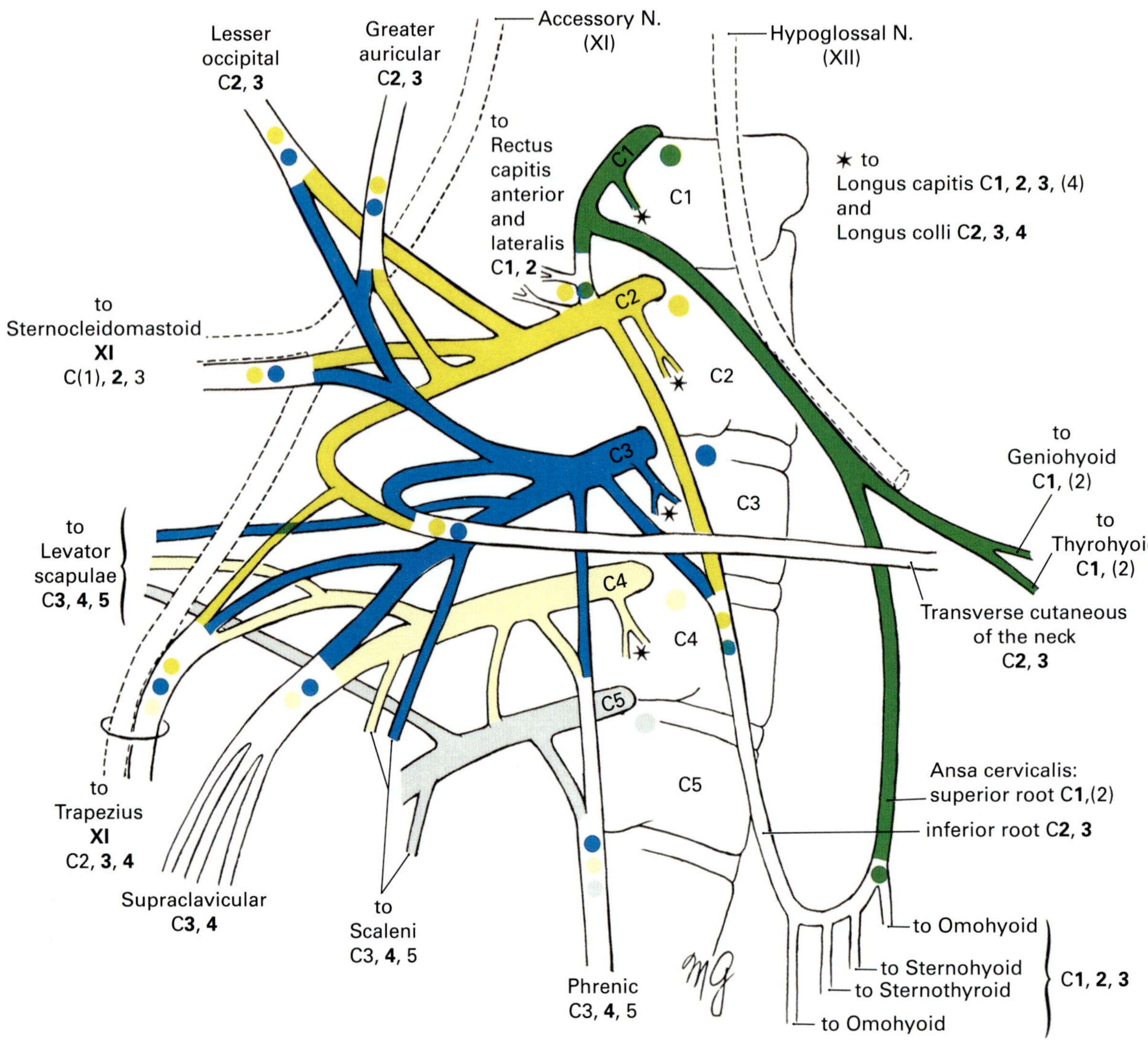

FIGURE 4-3: Cervical plexus.

Name Date

	MUSCLE STRENGTH GRADE	MUSCLE	D. T1-12, L1-5, S1-3	V. T1, 2, 3, 4	V. T5, 6	V. T7, 8	V. T9, 10, 11, 12	V. Iliohypogastric T12 L1	V. Ilioinguinal T(12) L1	V. Lumb. Plex. T(12) L1, 2, 3, 4	P. Femoral L(1) 2, 3, 4	A. Obturator L(1) 2, 3, 4	P. Sup. Glut. L4, 5, S1	P. Inf. Glut. L5, S1, 2	V. Sac. Plex. L4, 5, S1, 2, 3	P. Sciatic L4, 5, S1, 2	A. Sciatic L4, 5, S1, 2, 3	P. C. Peroneal L4, 5, S1, 2	A. Tibial L4, 5, S1, 2, 3	L1	L2	L3	L4	L5	S1	S2	S3
		ERECTOR SPINAE	●																	1	2	3	4	5	1	2	3
Thoracic Nerves		SERRATUS POST SUP		●																							
		TRANS THORACIS		●	●	●																					
		INT INTERCOSTALS		●	●	●	●																				
		EXT INTERCOSTALS		●	●	●	●																				
		SUBCOSTALES		●	●	●	●																				
		LEVATOR COSTARUM		●	●	●	●																				
		OBLIQUUS EXT ABD			(●)	●	●																				
		RECTUS ABDOMINIS			●	●	●																				
		OBLIQUUS INT ABD				●	●	●	(●)											1							
		TRANSVERSUS ABD				●	●	●	(●)											1							
		SERRATUS POST INF					●																				
Lumb. Plexus		QUAD LUMBORUM								●										1	2	3					
		PSOAS MINOR								●										1	2						
		PSOAS MAJOR								●										1	2	3	4				

KEY
D Dorsal Primary Ramus
V Ventral Primary Ramus
A Anterior Division
P Posterior Division

SENSORY

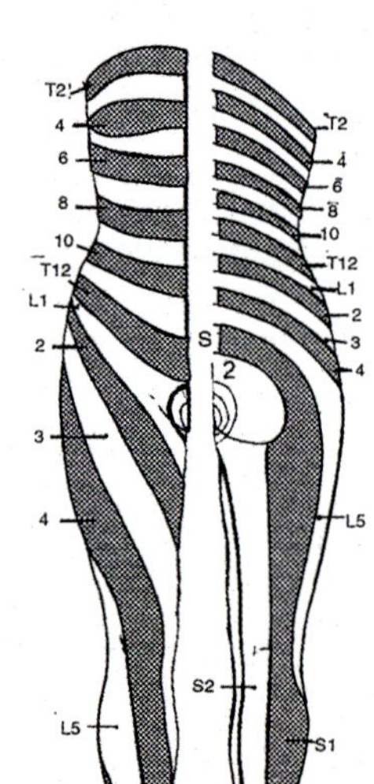

FIGURE 4-4: Spinal nerves and muscles of the trunk and lower extremity. © 2005 Florence P. Kendall.

SECTION II
INTERVERTEBRAL JOINTS AND MOTIONS

DEFINITIONS

The following definitions are essential for understanding the functions of the trunk muscles.

The **trunk** is comprised of the **thorax**, the **abdomen**, and the **pelvis.** The thoracic spine and lumbar spine are located within the thoracic and abdominal region, respectively. The term **trunk raising** may be used to describe raising the trunk against gravity from various positions: from prone, trunk raising backward; from side-lying, trunk raising sideways; and from supine, trunk raising forward. The term may also apply, in standing, to raising the trunk from positions of forward bending, side bending, or backward bending to the erect position.

The thorax is **elevated** (chest lifted upward and forward) by extending the thoracic spine, bringing the rib cage out of a slumped position. The thorax is **depressed** when sitting or standing in a slumped position, or it may be pulled downward by the action of certain abdominal muscles.

The trunk is joined to the thighs at the hip joints. The movement of hip flexion can be achieved by moving the anterior thigh toward the pelvis, as in forward leg raising, or by tilting the pelvis forward toward the thigh, as in the sit-up movement.

JOINTS OF THE VERTEBRAL COLUMN

Intervertebral joints include the bilateral synovial joints of the vertebral arches, where the inferior facets of one vertebra join with the superior facets of the adjacent vertebra, and the fibrous joints between successive vertebral bodies united by intervertebral fibrocartilaginous disks.

The joints between the first two vertebrae of the column are exceptions to the general classification. The **atlanto-occipital joint**, which is between the condyles of the occipital bone and the superior facets of the atlas, is classified as a **condyloid joint**. The movements permitted are flexion and extension with very slight lateral motion. The **atlanto-axial joint** is composed of three joints. The lateral two fit the general description of the joints of the vertebral column. The third, a median joint is formed by the articulation between the odontoid process (dens) of the axis and the fovea dentis of the atlas. This joint is classified as a **trochoid joint** and permits rotation.

Movement between two adjacent vertebrae is slight, and this movement is determined by the slope of the facets and by the flexibility of the intervertebral disks. The range of motion of the column as a whole, however, is considerable, and includes flexion, extension, lateral flexion, and rotation.

MOVEMENTS OF THE VERTEBRAL COLUMN

The normal lordosis of the spine in the cervical region forms a slightly extended position.

Flexion

Cervical spine flexion is movement of the head anteriorly toward the chest, decreasing the normal lordotic curve. Movement may continue to the point of straightening the cervical spine (i.e., the end range of normal flexion), and in some instances, movement may progress to the point that the spine curves convexly backward (i.e., a position of kyphosis). Gore et al., using cervical radiographs, reported cervical kyphosis as a normal variant in asymptomatic individuals (1). Harrison et al. used radiographs to look at the stresses produced by different cervical postures and found stresses in the region of cervical kyphosis were 6–10 times greater than those in the regions of cervical lordosis (2).

In the **thoracic region**, flexion of the spine is movement in the direction of *increasing the normal kyphosis*. In normal flexion, the spine curves convexly backward, producing a continuous, gently rounded contour throughout the thoracic area.

In the **lumbar region**, flexion of the spine is movement in the direction of *decreasing the normal lordosis*. It progresses to the point of straightening or flattening the lower back. Normally, the lumbar spine should not curve convexly backward, but excessive flexion in the low back is not uncommon. Certain types of activities or exercises (e.g., knee-bent sit-ups) can cause flexion beyond the normal range and make the back vulnerable to strain from heavy lifting movements.

From a supine position, normal flexion will allow enough curling of the trunk to lift the scapulae from the supporting surface. The area of the seventh cervical vertebra will be lifted upward approximately 8–10 inches (Figure 4-5).

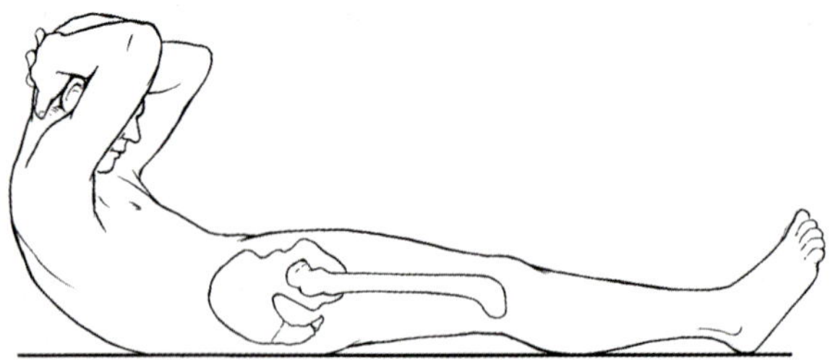

FIGURE 4-5: Normal flexion from a supine position.

Extension

Extension of the spine is movement of the head and trunk in a backward direction, while the spine moves in the direction of curving convexly forward. Cervical spine extension is the return of the head and cervical spine to the anatomical position from a flexed position.

In the **thoracic region**, extension is movement of the spine in the direction of *decreasing the normal backward curve* by straightening the upper back. Movement may progress to, but normally not beyond, the straight (or flat) position.

In the **lumbar region**, extension is movement in the direction of *increasing the normal lordosis* (Figure 4-6). It occurs by bending the trunk backward or by tilting the pelvis forward. As indicated by the photographs in Figures 4-7–4-9, the range of extension is highly variable, making it difficult to establish a standard for the purpose of measurements. Furthermore, these variations may exist without complaints of pain or disability, making it difficult to determine to what extent limited or excessive motion constitutes a disability. Too often, assessment of back extension is inaccurate or arbitrary.

From a prone position, normal extension will allow the head and chest to be raised enough to lift the xiphoid process of the sternum approximately 2–4 inches from the table (Figure 4-8).

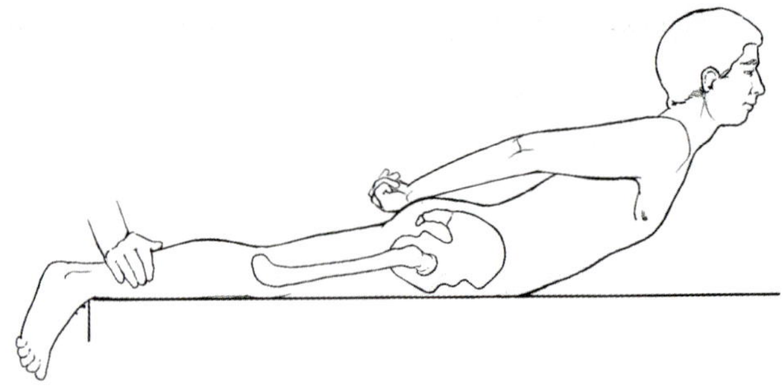

FIGURE 4-6: Extension in the lumbar region.

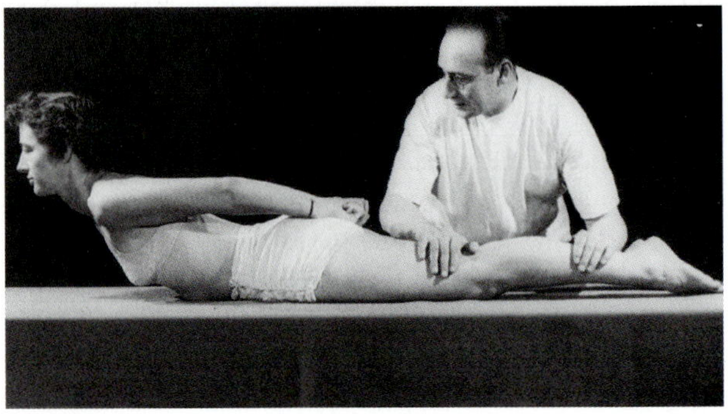

FIGURE 4-7: Less-than-average back extension range of motion but normal muscle strength.

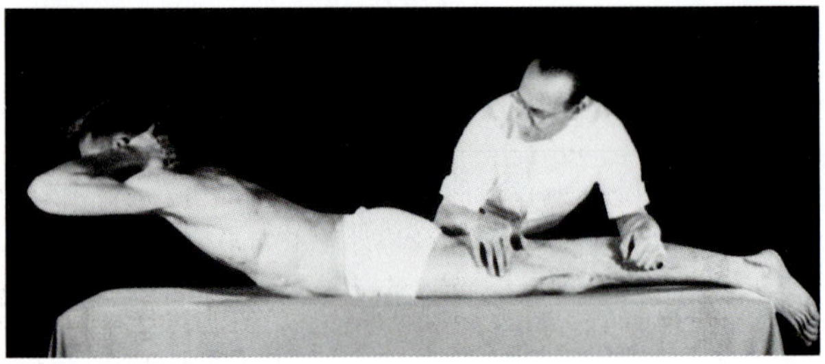

FIGURE 4-8: Average back extension range of motion, with anterosuperior-iliac spines in contact with the table.

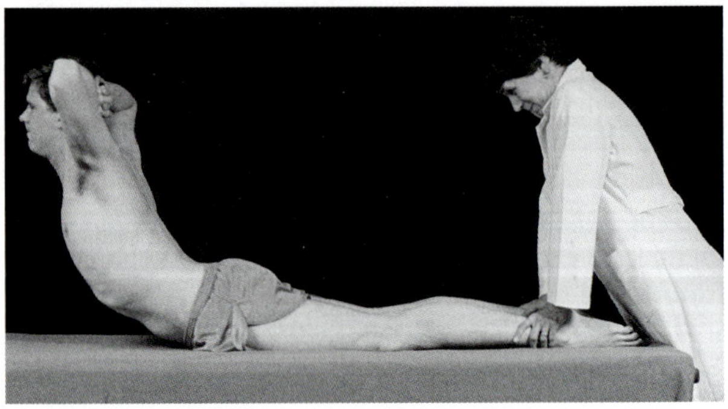

FIGURE 4-9: Excessive range of motion in back extension plus hip joint extension that raises the anterosuperior-iliac spines from the table. This subject is a diver and also has excessive flexion of the back.

Hyperextension

Cervical spine **hyperextension** is movement in the direction of increasing the normal lordosis. It may occur by tilting the head back, bringing the occiput toward the seventh cervical vertebra. It also may occur in sitting or standing by slumping into a rounded thoracic spine and assuming a forward-head position, bringing the seventh cervical vertebra toward the occiput.

Hyperextension of the spine (Figure 4-9) is movement beyond the normal range of motion in extension; it also may refer to a position greater than the normal anterior curve. Hyperextension may vary from slight to extreme. Excessive extension in the standing position is obtained by anterior pelvic tilt and is a position of lordosis. It is important to note that the range of back extension as seen in testing does not automatically translate into the same degree of lordosis in standing. Other factors, such as hip flexor length and abdominal muscle strength, also affect the position of the lumbar spine.

Lateral Flexion

Lateral flexion and rotation are described separately, although they occur in combination and are not considered to be pure movements.

Lateral flexion of the spine occurs in the coronal plane and is observed as movement in which the head and trunk bend toward one side while the spine curves convexly toward the opposite side. A convex curve toward the right is the equivalent of lateral flexion toward the left. From a standing position with the feet approximately 4 inches apart, the body erect, and arms at the sides, normal lateral flexion (i.e., bending directly sideways) will allow the fingertips to reach approximately to the level of the knee.

Consistent with the geometry of the cervical facets, lateral flexion occurs primarily between the occiput and C1, and between C1 and C2 (3). When observing lateral flexion, it is important to stabilize the thoracic and lumbar spines and to ensure that the observed motion is cervical spine lateral flexion and not elevation of the shoulder (4).

Lateral flexion varies according to the region of the spine. It is most free in the cervical and lumbar regions and is restricted in the thoracic region by the rib cage.

Rotation

Rotation is movement in a transverse plane. It is most free in the thoracic region and is slight in the lumbar region. Cervical rotation occurs about a vertical axis between C2 and C7 (5). Due to the coronal and oblique orientation of the cervical facet joints, cervical rotation is combined with lateral flexion. Rotation in the cervical region permits an approximately 90° range of motion of the head and is referred to as rotation of the face toward the right or the left. Rotation of the thorax and lumbar spine follows the same convention.

NECK RANGE OF MOTION

It is important to maintain good neck range of motion. Hence, it is advisable to establish and justify a means by which measurements can be taken to determine the range of motion of the neck in relation to established standards.

Various methods have been employed to measure the range of motion of the cervical spine: radiographs, goniometers, electrogoniometers, inclinometers, tape measures, cervical range-of-motion devices, ultrasound and digital optoelectronic instrumentation, as well as simple estimations of observable motion (6). The broad assortment of instruments and the lack of uniform procedures that have been in both reliability and descriptive studies have contributed to the wide range of published norms for active and passive neck range of motion. However, Table 4-1 provides examples of three sources that do support each other.

TABLE 4-1 Cervical Range of Motion: Comparison of "Norms"

Cervical Movements	Palmer and Eppler 2nd ed. (1998) (8)	Clarkson 2nd ed. (2000) (9)	Reese and Bandy (2002) (10)
Flexion	Cervical 0° to 45°	0° to 45°	0° to 45°–50°
Extension	Cervical 0° to 45°	0° to 45°	0° to 45°–75°
Lateral Flexion	0° to 45°–60°	0° to 45°	0° to 45°
Rotation	0° to 60°–75°	0° to 60°	0° to 80°

Taking measurements of a large number of people is not the answer because too many variables exist. Dvorak et al. found "significant differences both between genders and age decades" (7). In addition, variances exist between individuals of different body types.

It is essential that the subject be placed as close to the ideal postural alignment of the thoracic and cervical spines as possible before taking range-of-motion measurements. Starting with a forward head position will limit movement in every plane.

If the thoracic spine is rigid in a position of kyphosis, soft tissue mobilization and gentle stretching of the tight neck extensors may only be palliative although still worthwhile. If the posture of the thoracic spine is habitually faulty but the person is able to assume a normal alignment, efforts should be directed toward maintaining optimal alignment. Temporary use of a support to help correct postural deviations of the shoulder and thoracic spine may be beneficial.

TRUNK RANGE OF MOTION

Flexion and Extension

Forward bending and backward bending are used to assess the range of motion in flexion and extension of the spine. Several variations of these tests exist.

Range of Motion in Trunk Flexion

The forward-bending, *long-sitting* position involves hip joint flexion along with back flexion. One must try to disregard the hip joint movement when observing the contour of the back (see later the section on Normal Range of Motion).

Range of motion and contour of the back may also be observed by having a subject bend forward from the *standing* position. As a test position, however, this has certain disadvantages. If the pelvis is not level or is rotated, the plane of forward bending will be altered, and the test will not be as satisfactory as one in the long-sitting position in which the pelvis is level and rotation is better controlled.

To assess flexion of the back without associated hip joint flexion, place a subject in the supine position, resting on the forearms with elbows bent at right angles and arms close to the body (Figure 4-10). If the subject can flex the spine in this position with the pelvis flat on the table (i.e., no hip flexion), the range of motion is considered to be good.

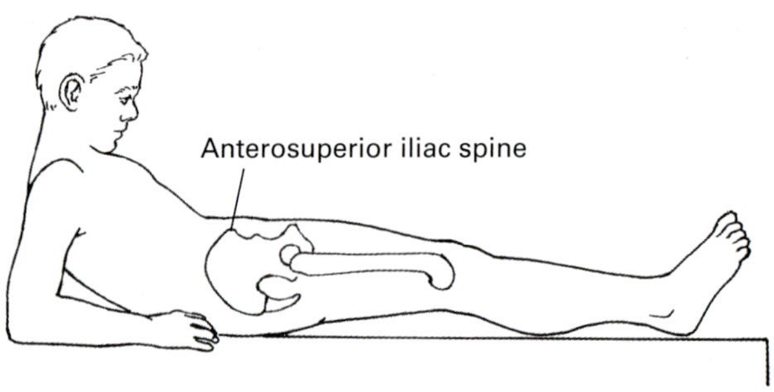

FIGURE 4-10: Assessing flexion of the back without associated hip joint flexion.

Sometimes it is necessary to ascertain the range of back flexion passively. With the subject in the supine position, the examiner lifts the upper trunk in flexion to completion of the subject's range of motion. The subject must relax for the examiner to obtain complete flexion.

Range of Motion in Trunk Extension

Because the low back muscles are seldom weak, the range of back extension may be determined by the active strength test in the prone position. Whether the range of motion is normal, limited, or excessive, the subject is capable of moving through the existing range. The anterosuperior-iliac spines should not be lifted from the table during back extension, because doing so adds hip extension to the back extension range of motion (Figure 4-11).

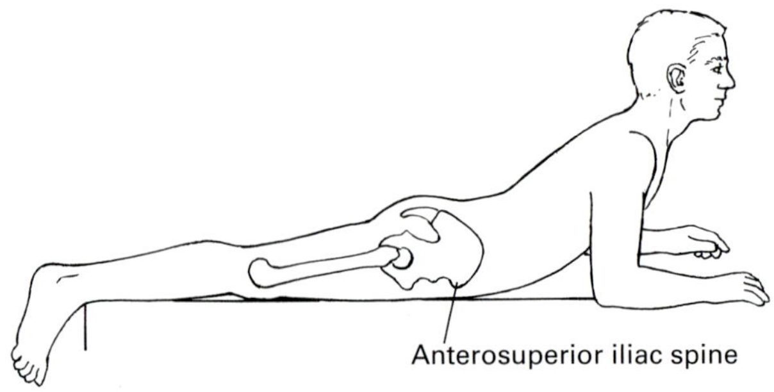

FIGURE 4-11: Assessing range of back extension by the active strength test in the prone position.

Back extension often is checked in the *standing* position. The test is useful as a gross evaluation, but it is not very specific. Swaying forward at the hips is almost a necessity for balance when bending backward, but doing so adds the element of hip extension to the test, or the knees must bend somewhat if the hip does not extend.

Similar to the test to determine range of motion in spine flexion, a test can be done to determine the range in spine extension. The subject lies prone on a table, resting on the forearms with elbows bent at right angles and arms close to the body. If the subject can extend the spine enough to prop up on the forearms with the pelvis flat on the table (i.e., anterosuperior-iliac spines on the table), the range of motion in extension is considered to be good.

Sometimes it is necessary to determine the amount of passive back extension with the subject prone on the table by lifting the subject up in extension through the available range of motion.

Scapular instability and, specifically, serratus anterior weakness can interfere with the back extension test, as seen in Figure 4-12.

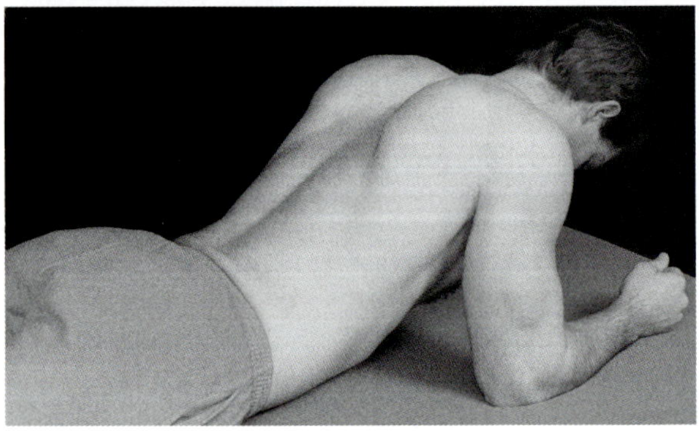

FIGURE 4-12: Scapular instability and, specifically, serratus anterior weakness can interfere with the back extension test.

> **NOTE:** *Push-ups should not be done by individuals who exhibit this type of weakness.*

MOVEMENTS OF THE SPINE AND PELVIS

Backward bending in the standing position (Figure 4-13) requires that the pelvis and thighs be displaced forward for balance. Spine extension must be distinguished from backward bending. How far the spine will bend backward depends on the available range of motion in the spine and the length of the abdominal muscles. How far the body will bend backward depends on the length of the hip flexors in addition to the preceding.

FIGURE 4-13: Backward bending in the standing position.

The subject in Figure 4-14 is not trying to touch fingertips to the floor, which would require more hip joint flexion, but has fully flexed the spine. Flexion is normal, as denoted by the straight lumbar spine, and a smooth, continuous curve is in the thoracic region (see excessive flexion and limited lumbar flexion).

FIGURE 4-14: Full flexion of the spine without touching the floor.

Lateral flexion of the spine depends on the available range of motion in the spine and the length of the opposite lateral trunk flexors (Figure 4-15). How far the body can bend sideways depends on the length of the opposite hip abductors in addition to the above. To use side bending for measuring lateral flexion, the pelvis must be level and the feet a standardized distance apart.

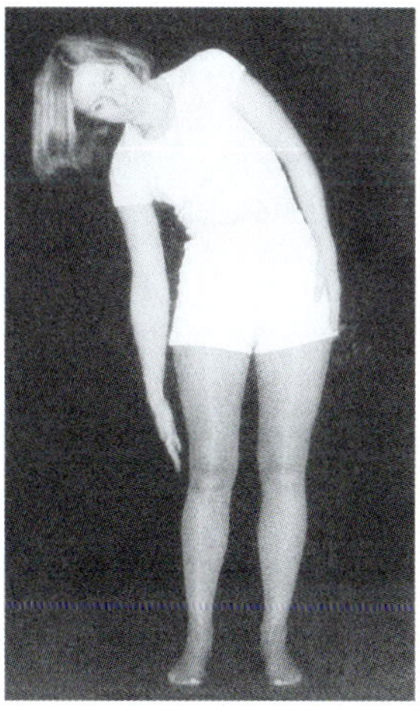

FIGURE 4-15: Lateral flexion of the spine.

The subject in Figure 4-16 has a high hip on the right. If this subject were to do side bending with a measurement taken of the distance from the fingertips to the floor, the measurement would be less on the right than on the left. If these measurements were then read as lateral flexion of the spine, it would be recorded—incorrectly—as lateral flexion more limited toward the right than toward the left. By virtue of the high hip on the right, the spine is already in lateral flexion, so the shoulder and arm will not move downward as far as would occur if the pelvis were level.

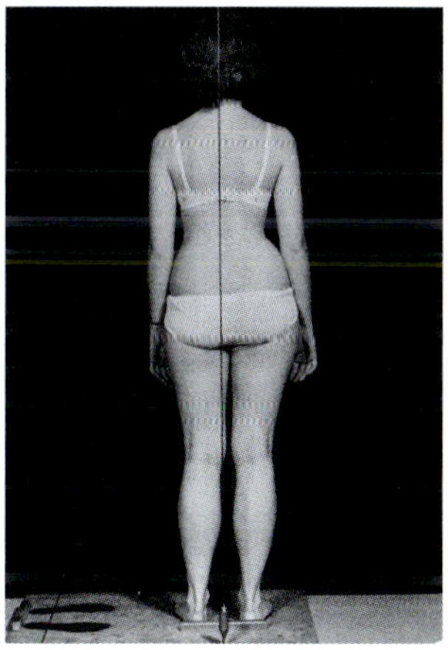

FIGURE 4-16: With a high hip on the right, the spine is already in lateral flexion.

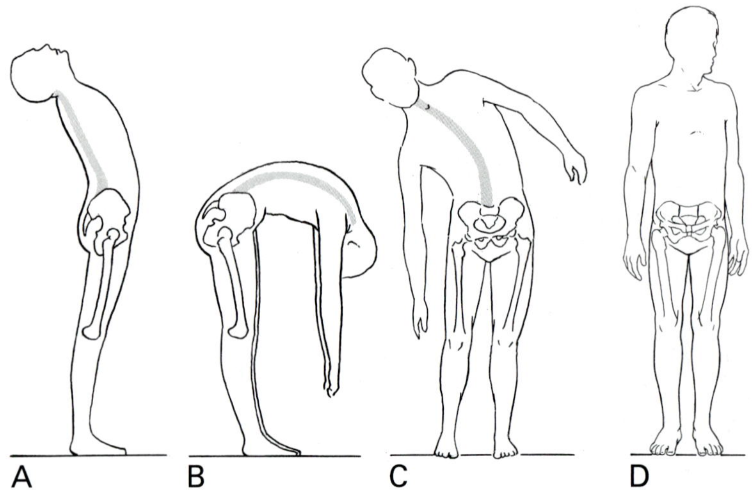

FIGURE 4-17: **A**, Extension; **B**, Flexion; **C**, Lateral flexion; **D**, Rotation.

Accurate measurements of spine extension and flexion as well as of lateral flexion should not include movements in the hip joints, which do occur in the bending movements illustrated in Figure 4-17. Various devices have been developed with the hope that meaningful objective measurements can be obtained. Goniometers, inclinometers, flexible rulers, tape measures, and radiographs have been used in an effort to establish a suitable method for measurement. Without first defining normal flexion of the lumbar spine, however, measurements may not be meaningful.

Pelvis

The neutral position of the pelvis is one in which the anterosuperior-iliac spines are in the same transverse plane and the spines and the symphysis pubis are in the same vertical plane. An **anterior pelvic tilt** is a position of the pelvis in which the vertical plane through the anterosuperior-iliac spines is anterior to a vertical plane through the symphysis pubis. A **posterior pelvic tilt** is a position of the pelvis in which the vertical plane through the anterosuperior-iliac spines is posterior to a vertical plane through the symphysis pubis. In a standing position, an anterior pelvic tilt is associated with hyperextension of the lumbar spine and flexion of the hip joints, whereas a posterior pelvic tilt is associated with flexion of the lumbar spine and extension of the hip joints.

In **lateral pelvic tilt**, the pelvis is not level from side to side; rather, one anterosuperior spine is higher than the other. In standing, a lateral tilt is associated with lateral flexion of the lumbar spine and with adduction and abduction of the hip joints. For example, in a lateral tilt of the pelvis in which the right side is higher than the left, the lumbar spine is laterally flexed resulting in a curve convex toward the left. The right hip joint is in adduction and the left in abduction (see Figure 4-18).

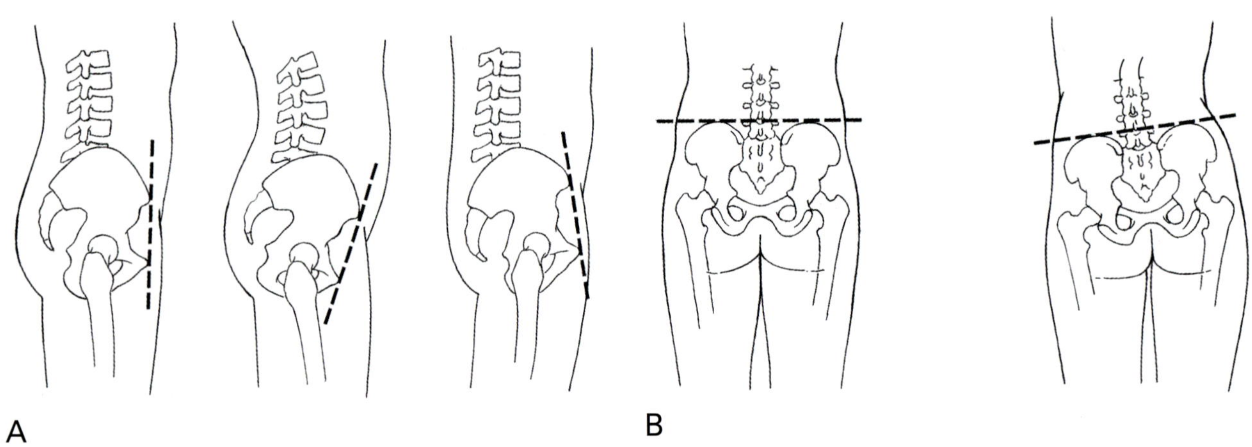

FIGURE 4-18: **A**, Neutral position, anterior tilt, and posterior tilt; **B**, Neutral position and lateral tilt.

The pelvis is in the neutral position, and the lumbar spine is in a normal anterior curve (Figure 4-19).

The pelvis is in a posterior tilt of 10°, and the lower back is flat, that is, normal flexion (Figure 4-20).

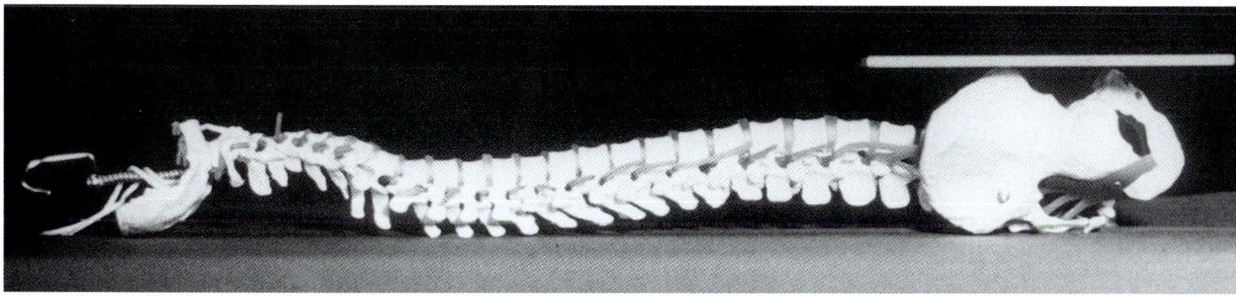

FIGURE 4-19: The pelvis is in the neutral position and the lumbar spine is in a normal anterior curve.

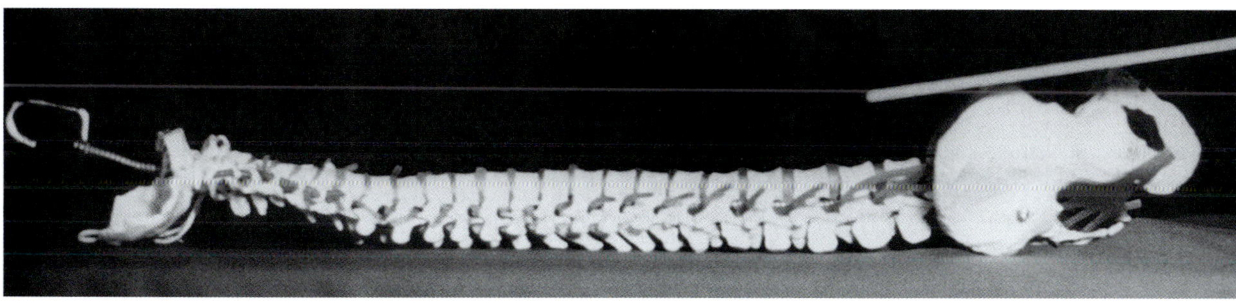

FIGURE 4-20: The pelvis is in a posterior tilt of 10° and the lower back is flat (i.e., normal flexion).

SECTION III
MUSCLES AND MUSCLE TESTS

NECK MUSCLES

Origins and Insertions

Table 4-2 shows the origins and insertions of specific muscles illustrated in Figures 4-21 and 4-22.

Actions and Nerves

Table 4-3 lists the specific actions of muscles and their respective innervations.

TABLE 4-2 Cervical Muscle Origin and Insertion

Muscle	Origin	Insertion
Rectus capitis posterior minor	Tubercle on posterior arch of atlas	Medial part of inferior nuchal line of occipital bone
Rectus capitis posterior major	Spinous process of axis	Lateral part of inferior nuchal line of occipital bone
Obliquus capitis superior	Superior surface of transverse process of atlas	Between superior and inferior nuchal lines of occipital bone
Obliquus capitis inferior	Apex of spinous process of axis	Inferoposterior part of transverse process of atlas
Longus capitis	Anterior tubercles of transverse processes of third through sixth cervical vertebrae	Interior surface of basilar part of occipital bone
Longus colli	*Superior oblique portion:* Anterior tubercles of transverse processes of third through fifth cervical vertebrae *Interior oblique portion:* Anterior surface of bodies of first two or three thoracic vertebrae. *Vertical portion:* Anterior surface of bodies of first three thoracic and last three cervical vertebrae	Tubercle on anterior arch of atlas Anterior tubercles of transverse processes of fifth and sixth cervical vertebrae Anterior surface of bodies of second through fourth cervical vertebrae
Rectus capitis anterior	Root of transverse process; anterior surface of atlas	Interior surface of basilar part of occipital bone
Rectus capitis lateralis	Superior surface of transverse process of atlas	Inferior surface of jugular process of occipital bone
Platysma	Fascia covering superior parts of pectoralis major and deltoid	Inferior margin of mandible; skin of lower part of face and corner of mouth
Sternocleidomastoid	*Medial or sternal head:* Cranial part of manubrium sterni *Lateral or clavicular head:* Medial ⅓ of clavicle	Lateral surface of mastoid process; lateral ½ of superior nuchal line of occipital bone
Scalenus anterior	Anterior tubercles of transverse processes of third through sixth cervical vertebrae	Scalene tubercle and cranial crest of first rib
Scalenus medius	Posterior tubercles of transverse processes of second through seventh cervical vertebrae	First rib, cranial surface between tubercle and subclavian groove
Scalenus posterior	By two or three tendons from posterior tubercles of transverse processes of last two or three cervical vertebrae	Outer surface of second rib
Trapezius, upper	External occipital protuberance, medial ⅓ of superior nuchal line, ligamentum nuchae, and spinous process of seventh cervical vertebra	Lateral ⅓ of clavicle; acromium process of scapula

TABLE 4-3 Cervical Muscle Actions and Innervations

	Acting Bilaterally			Acting Unilaterally		
				Rotation Toward		
Muscle	Extension	Flexion	Lateral Flexion	Same Side	Opposite Side	Nerves
Rectus capitis posterior minor	X					Suboccipital
Rectus capitis posterior major	X			X		Suboccipital
Obliquus capitis superior	X		X			Suboccipital
Obliquus capitis inferior				X		Suboccipital
Longus capitis		X		X		Cervical, 1–3
Longus colli		X	X	X		Cervical, 2–7
Rectus capitis anterior		X		X		Cervical, 1, 2
Rectus capitis lateralis			X			Cervical, 1, 2
Platysma		X				Facial
Sternocleidomastoid	X	X	X		X	Accessory and cervical, 1, 2
Scalenus anterior		X	X		X	Cervical, lower
Scalenus medius			X		X	Cervical, lower
Scalenus posterior			X		X	Cervical, 6–8
Trapezius, upper	X		X		X	Cranial, 1 Cervical, 3, 4

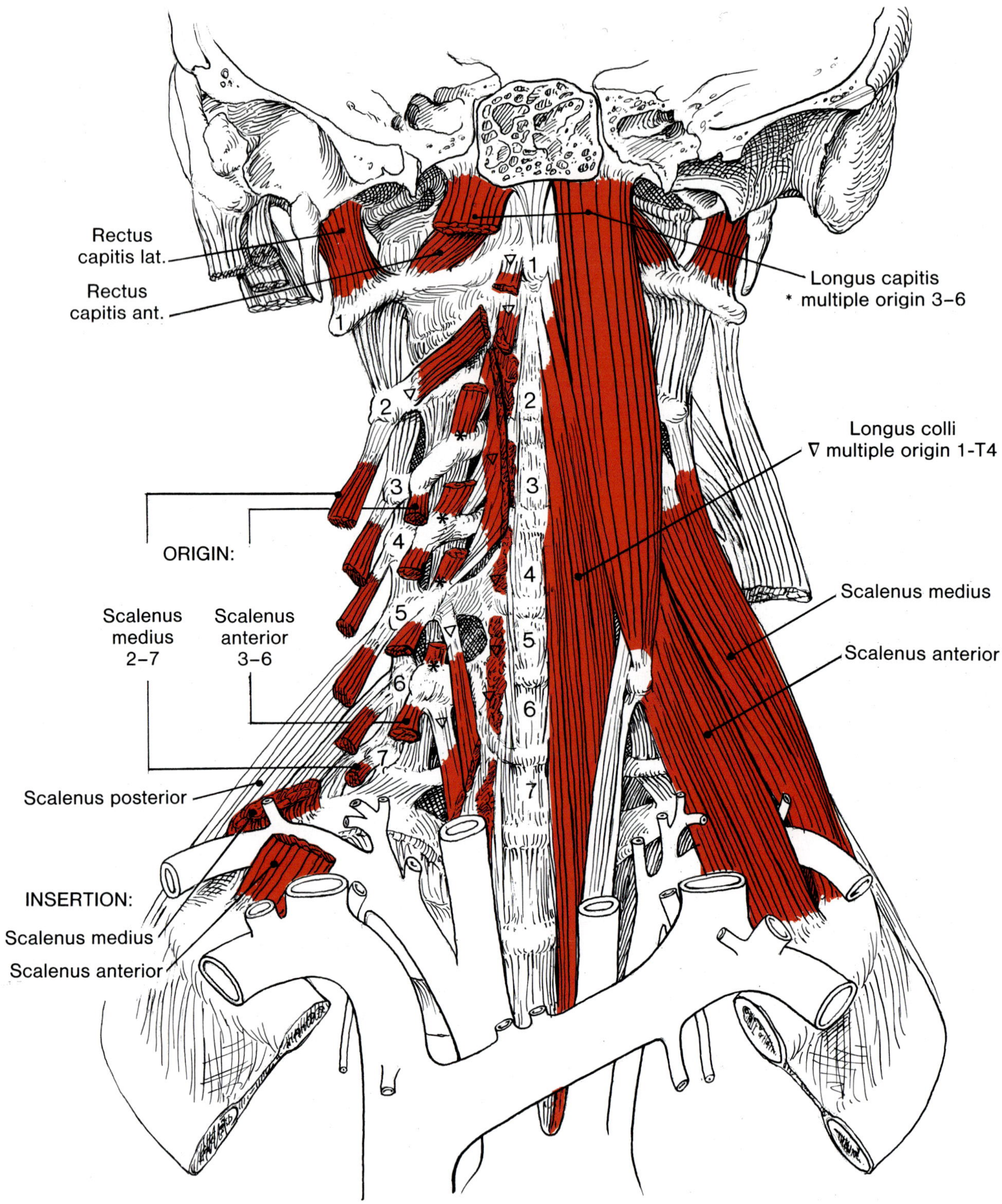

FIGURE 4-21: Anterior and lateral neck muscles.

Anterior and Lateral Neck Muscles

See Tables 4-2 and 4-3 for the origins, insertions, actions, and nerves of the muscles shown in Figure 4-21 (11).

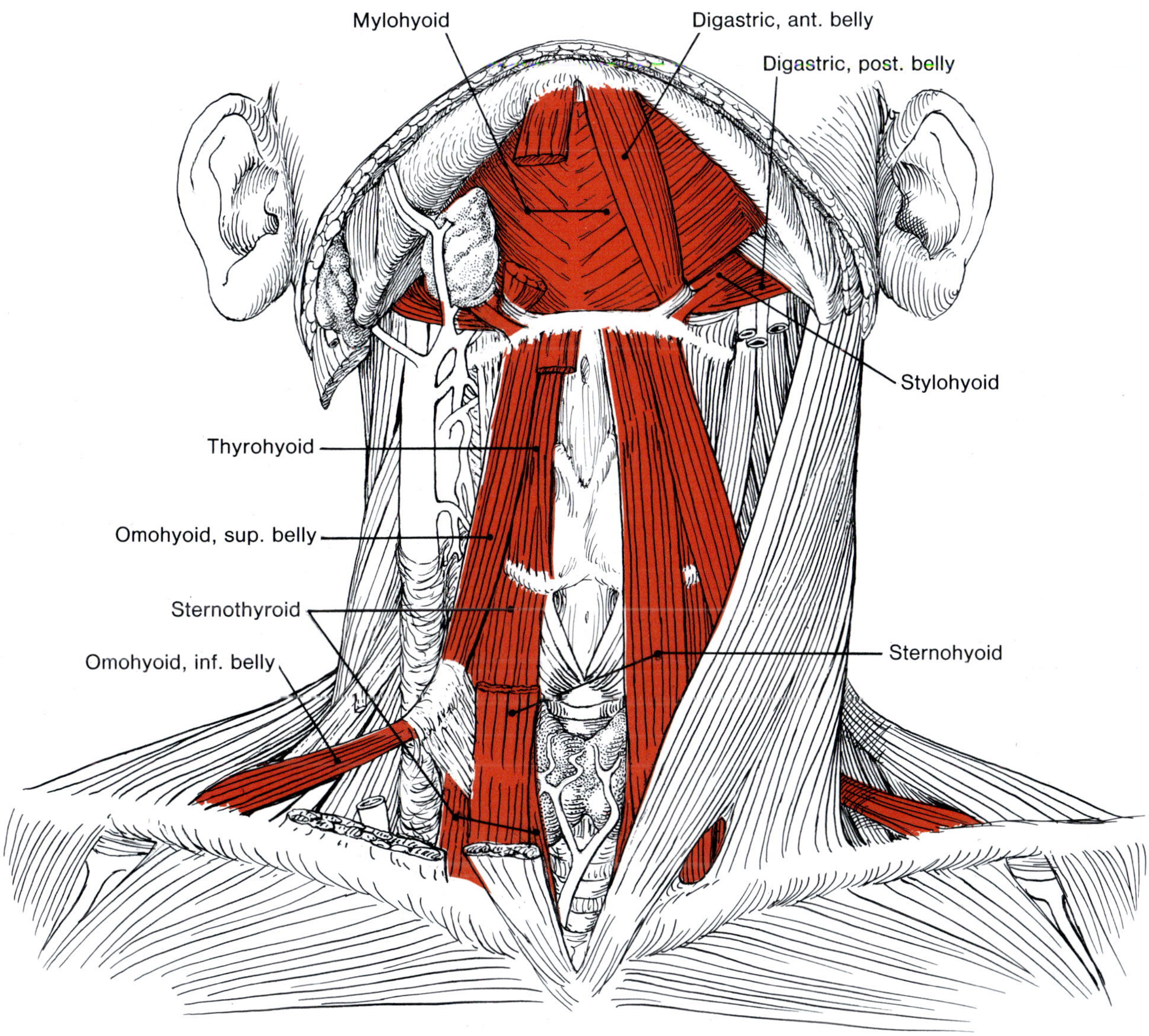

FIGURE 4-22: Suprahyoid and infrahyoid muscles.

Neck Extensors

Suprahyoid and Infrahyoid Muscles

Figure 4-22 illustrates the suprahyoid and infrahyoid muscles. Also see Table 3-2 in Chapter 3 for the origins, insertions, actions, nerves, and roles in deglutition of these muscles (11).

Cervical Spine Extension and Flexion

A subject with normal flexibility was photographed and x-rayed in five neck positions (Figure 4-23A–J). Markers were placed at the hairline and over C7.

Cervical spine extension by tilting the head in a posterior direction (Figure 4-23A). Note the approximation of the markers in Figure 4-23B.

Cervical spine extension in a typical forward-head posture (Figure 4-23C, D). Note the similarity in the curve and the positions of the markers to those in Figure 4-23A. Often, this slumped posture is mistakenly referred to as flexion of the lower cervical spine and extension of the upper cervical spine. However, the extension is more pronounced in the lower than in the upper cervical region.

Good alignment of the cervical spine is illustrated in Figure 4-23 E and F.

Flexion (flattening) of the cervical spine by tilting the head in an anterior direction (Figure 4-23 G and H).

Flexion of both the cervical spine and the upper thoracic spine occurs when the chin is brought toward the chest (Figure 4-23 I and J).

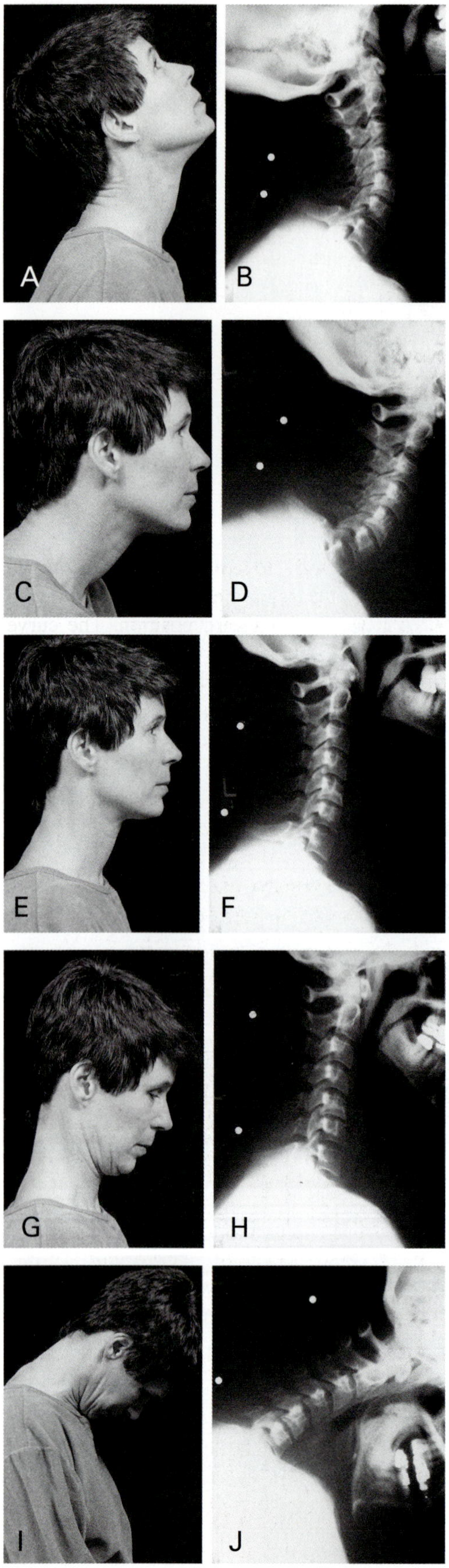

FIGURE 4-23: Cervical spine extension and flexion.

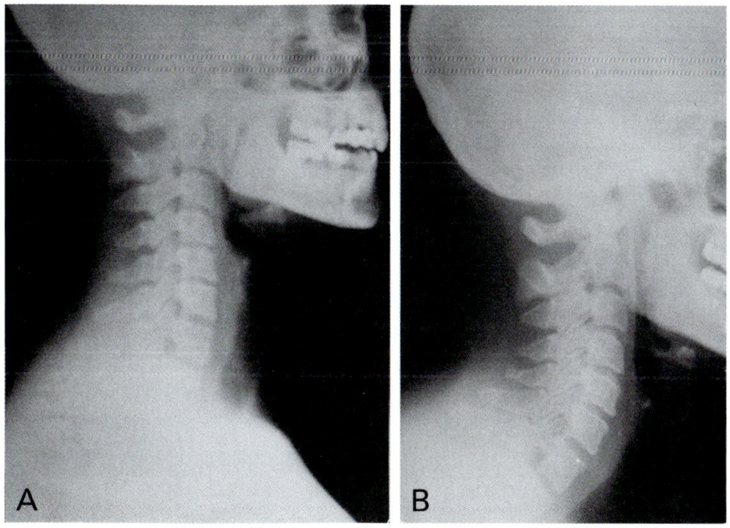

FIGURE 4-24: Cervical spine: good and faulty positions.

Faulty Head and Neck Positions

Cervical Spine, Good and Faulty Positions: For Figure 4-24A, the subject sat erect, with the head and upper trunk in good alignment. For Figure 4-24B, the same subject sat in a typically slumped position, with a round upper back and a forward head. As illustrated, the cervical spine is in extension.

Cervical Spine, Extension: In Figure 4-25, Part A, the head tilts backward, the cervical spine is hyperextended, and the chest and shoulders are elevated.

Cervical Spine, Straight (Flexed): In Figure 4-25, Part B, the head is in slight anterior tilt, the scapulae are prominent, and the upper back is straight.

Forward Head with Attempted Correction: In Figure 4-25, Part C, the subject apparently is trying to correct what is basically a forward position. The curve of the neck begins in a typical way in the lower cervical region, but a sharp angulation occurs at approximately the sixth cervical vertebra. Above this level, the curve seems to be very much decreased. The chin is pressed against the front of the throat. This distorted rather than corrected position of the neck results from a failure to correct the related faulty position of the upper trunk.

Forward Head, Marked: In Figure 4-25, Part D, the subject shows an extremely faulty alignment of the neck and thoracic spine. The degree of deformity in the thoracic spine suggests an epiphysitis. This patient was treated for pain in the posterior neck and occipital region.

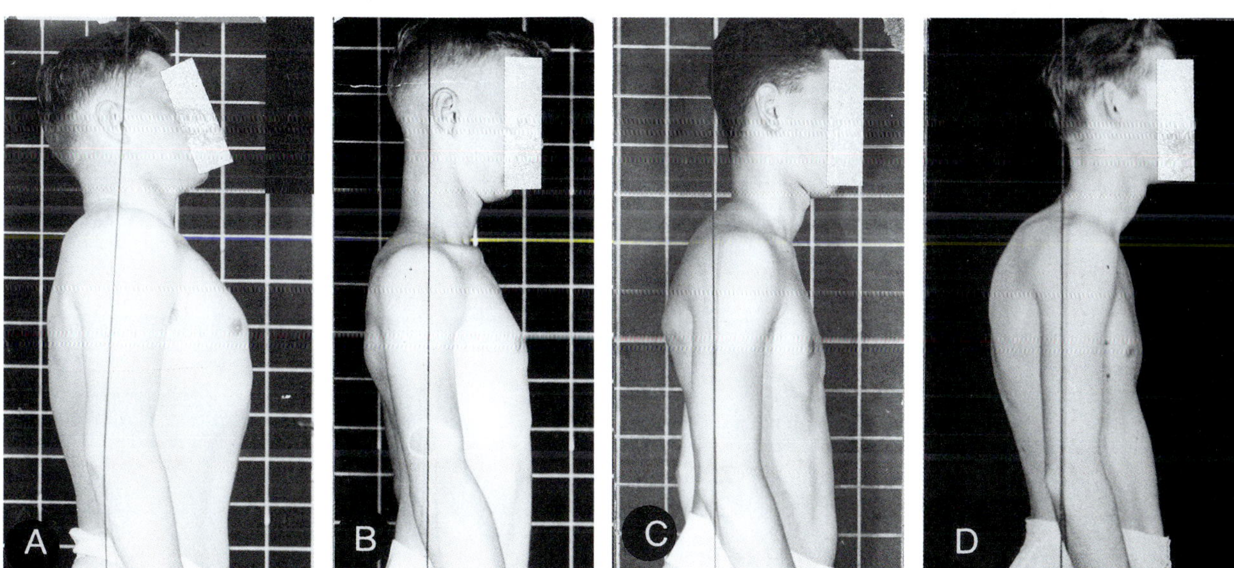

FIGURE 4-25: Faulty head and neck positions.

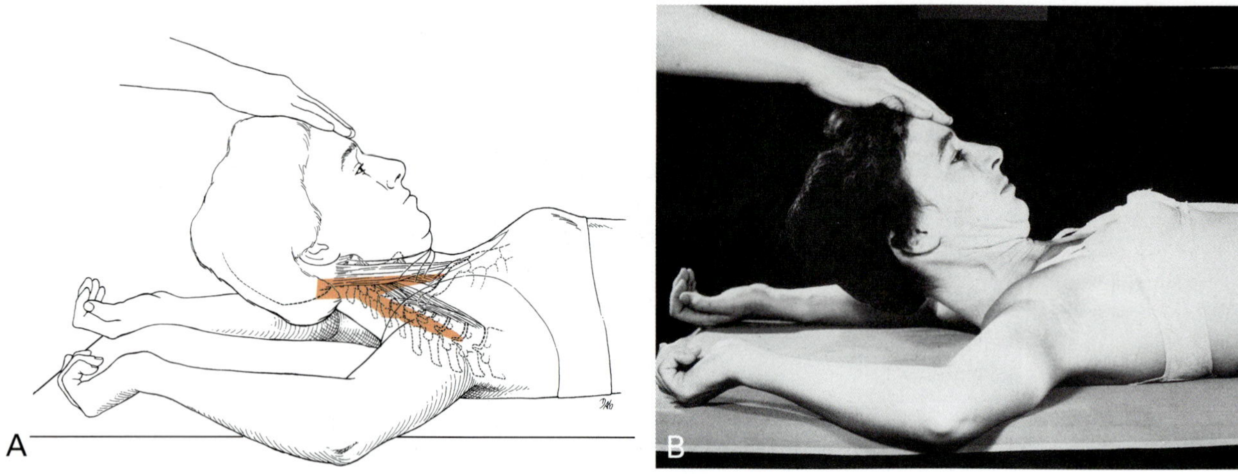

FIGURE 4-26: **A**, **B**, Anterior next flexors.

NECK MUSCLE TESTS

Anterior Neck Flexors

Patient: Supine, with the elbows bent and the hands overhead, resting on the table (Figure 4-26).

Fixation: Anterior abdominal muscles must be strong enough to give anterior fixation from the thorax to the pelvis before the head can be raised by the neck flexors. If the abdominal muscles are weak, the examiner can provide fixation by exerting firm, downward pressure on the thorax. Children approximately 5 years of age and younger should have fixation of the thorax provided by the examiner.

Test: Flexion of the cervical spine by lifting the head from the table, with the chin depressed and approximated toward the sternum.

Pressure: Against the forehead in a posterior direction (see Grading, to follow).

Modified Test: In cases of marked weakness, have the patient make an effort to flatten the cervical spine on the table, approximating the chin toward the sternum.

Pressure: Against the chin in the direction of neck extension.

> **NOTE:** *The anterior vertebral flexors of the neck are the longus capitis, longus colli, and rectus capitis anterior. In this movement, they are aided by the sternocleidomastoid, anterior scaleni, suprahyoids, and infrahyoids. The platysma will also attempt to aid when the flexors are very weak.*

Weakness: Hyperextension of the cervical spine, resulting in a forward-head position.

Contracture: A neck flexion contracture is rarely seen except unilaterally, as in **torticollis**.

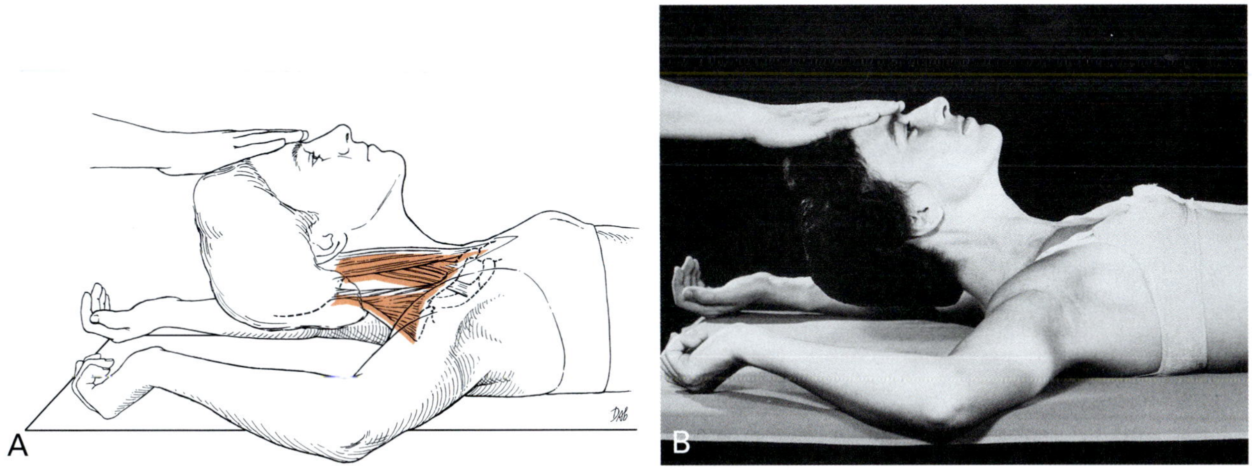

FIGURE 4-27: Error in testing neck flexors.

Error in Testing Neck Flexors

If the anterior vertebral neck flexors are weak and the sternocleidomastoid muscles are strong, an individual can raise the head from the table (as illustrated in Figure 4-27A, B) and hold it against pressure. This is not an accurate test for the neck flexors, however, because the action is accomplished chiefly by the sternocleidomastoids aided by the anterior scaleni and the clavicular portions of the upper trapezius.

Grading: Because most grades of 10 are based on adult standards, it is necessary to acknowledge when a grade of less than 10 is normal for children of a given age. This is particularly true regarding the strength of the anterior neck and abdominal muscles. The size of the head and trunk in relation to the lower extremities as well as the long span and normal protrusion of the abdominal wall affect the relative strength of these muscles.

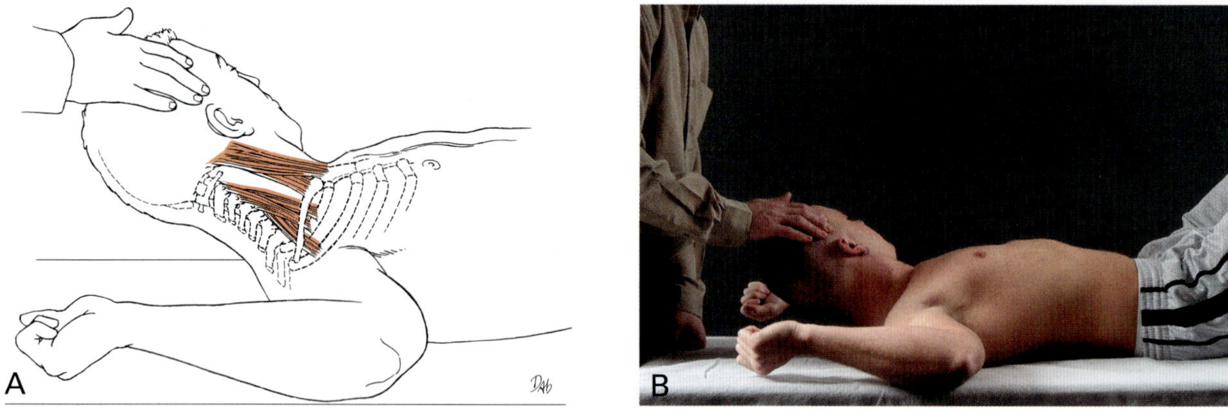

FIGURE 4-28: Anterolateral neck flexors.

Anterolateral Neck Flexors

The muscles acting in this test are chiefly the sternocleidomastoid and scaleni.

Patient: Supine, with elbows bent and hands beside the head, resting on table (Figure 4-28A, B)

Fixation: If the anterior abdominal muscles are weak, the examiner can provide fixation by exerting firm, downward pressure on the thorax.

Test: Anterolateral neck flexion.

Pressure: Against the temporal region of the head in an obliquely posterior direction.

NOTE: *With neck muscles just strong enough to hold but not strong enough to flex completely, a patient can lift the head from the table by raising the shoulders. A patient will do this especially during tests for the right and left neck flexors by taking some weight on the elbow or hand to push the shoulder from the table. To avoid this, keep the patient's shoulder flat against the table.*

Contracture and Weakness: A contracture of the right sternocleidomastoid produces a right torticollis. The face is turned toward the left, and the head is tilted toward the right. Thus, a right torticollis produces a cervical scoliosis that is convex toward the left with the left sternocleidomastoid elongated and weak.

Contracture of the left sternocleidomastoid, with weakness of the right, produces a left torticollis with a cervical scoliosis that is convex toward the right.

In a patient with habitually faulty posture and forward head, the sternocleidomastoid muscles remain in a shortened position and tend to develop shortness.

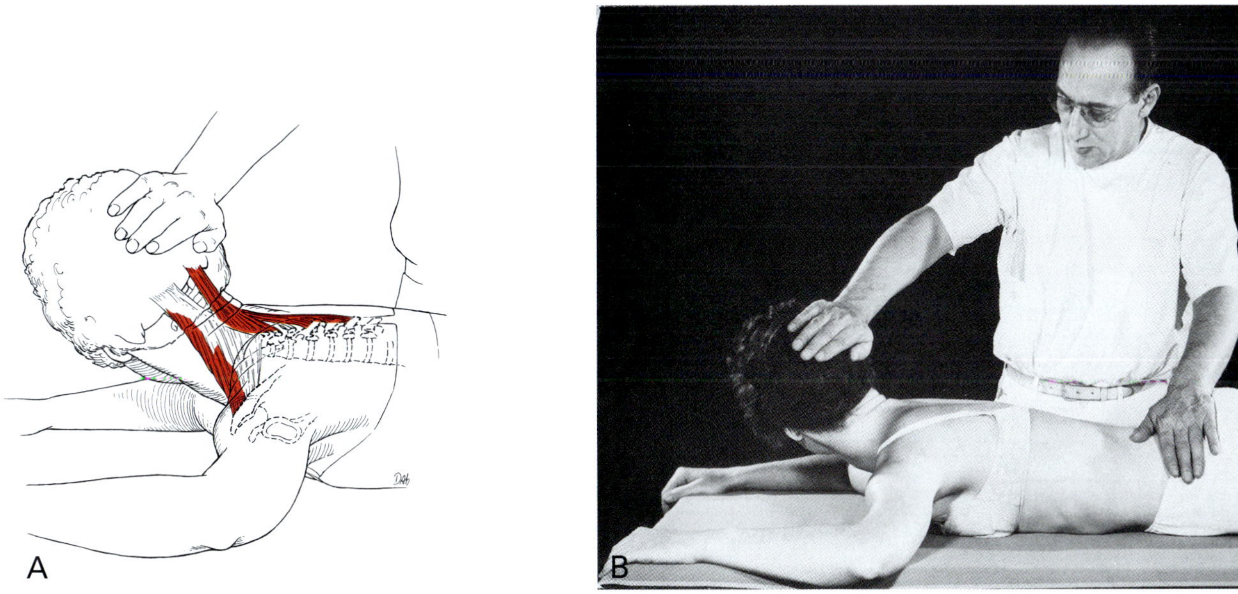

FIGURE 4-29: Posterolateral neck extensors.

Posterolateral Neck Extensors

The muscles acting in this test (Figure 4-29A, B) are chiefly the splenius capitis and cervicis, semispinalis capitis and cervicis, and cervical erector spinae.

Patient: Prone, with elbows bent and hands overhead, resting on the table.

Fixation: None necessary.

Test: Posterolateral neck extension, with the face turned toward the side being tested (see *Note*).

Pressure: Against the posterolateral aspect of the head in an anterolateral direction.

Shortness: The right splenius capitis and left upper trapezius are usually short, along with the sternocleidomastoid, in a left torticollis. The opposite muscles are short in a right torticollis.

NOTE: *The upper trapezius, which is also a posterolateral neck extensor, is tested with the face turned away from the side being tested.*

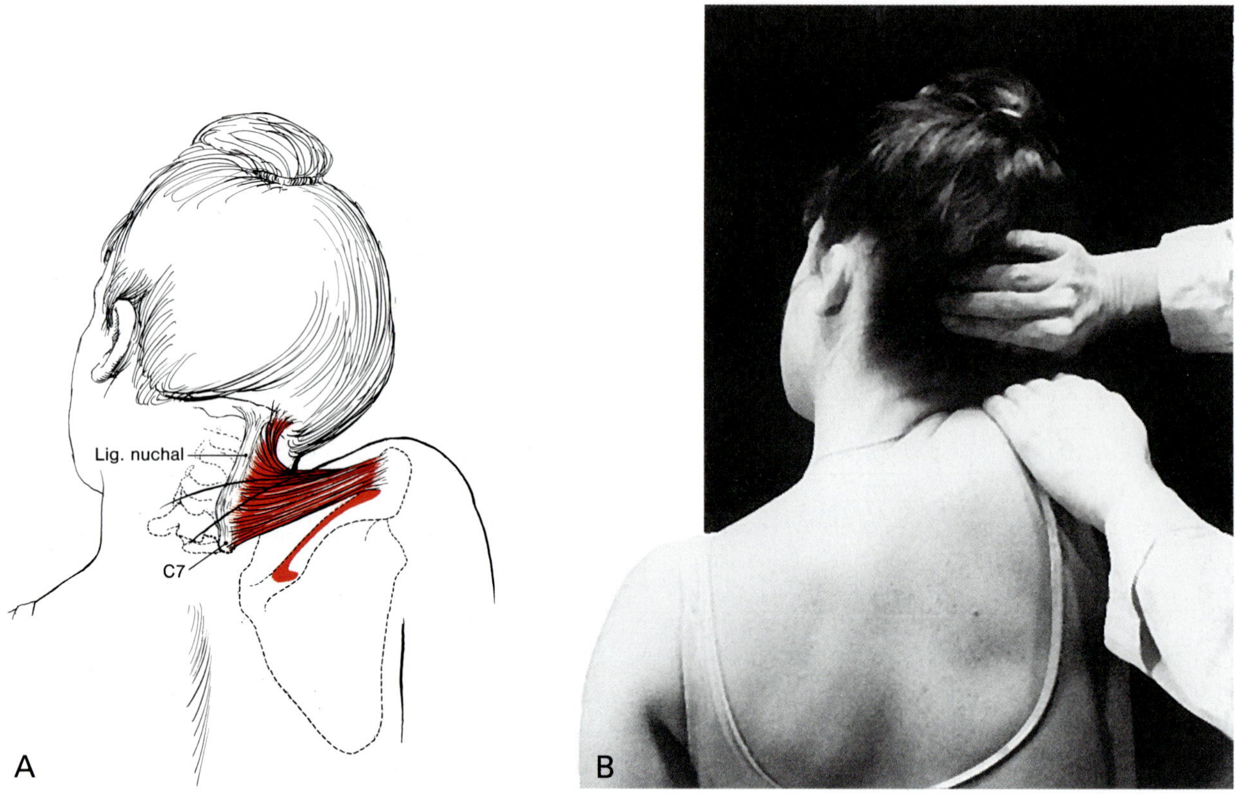

FIGURE 4-30: **A**, **B**, Upper trapezius.

Upper Trapezius

Patient: Sitting.

Fixation: None necessary.

Test: Elevation of the acromial end of the clavicle and scapula and posterolateral extension of the neck, bringing the occiput toward the elevated shoulder with the face turned in the opposite direction.

The upper trapezius can be differentiated from the other elevators of the scapula because it is the only one that elevates the acromial end of the clavicle and the scapula. It also upwardly rotates the scapula as it elevates, in contrast with the straight elevation that occurs when all elevators contract, as in shrugging the shoulders (Figure 4-30).

Pressure: Against the shoulder in the direction of depression, and against the head in the direction of flexion anterolaterally.

Weakness: Unilaterally, weakness decreases the ability to approximate the acromion and the occiput. Bilaterally, weakness decreases the ability to extend the cervical spine (e.g., to raise the head from a prone position).

Shortness: Results in an elevation of the shoulder girdle (commonly seen in prize-fighters and swimmers). In a faulty posture with forward head and hyperkyphosis, the cervical spine is in extension, and the upper trapezius muscles are in a shortened position.

Contracture: Unilateral contracture is frequently seen in cases of torticollis. For example, the right upper trapezius is usually contracted along with the right sternocleidomastoid and scaleni.

TRUNK MUSCLES

Trunk muscles consist of back extensors that extend the thoracic and lumbar spines, lateral flexors that flex to the right or left, anterior abdominals that bend it forward or tilt the pelvis posteriorly, and combinations of these muscles that rotate the trunk in a clockwise or a counterclockwise manner. All of these muscles play a role in stabilizing the trunk, but the back extensors are the most important in this regard. The loss of stability that accompanies paralysis or marked weakness of the back muscles offers dramatic evidence of their importance. Fortunately, marked weakness of these muscles seldom occurs.

A "weak back," as frequently used in connection with low back pain, mistakenly suggests a weakness of the low back muscles. The feeling of weakness that occurs with a painful back is associated with the faulty posture the body assumes, and it is often caused by weakness of the abdominal muscles. Persons who have faulty posture with roundness of the thoracic spine may exhibit weakness in the upper back extensors but have normal strength in those of the low back.

Testing back muscles is less complicated than testing abdominal muscles, and in the field of exercise, few errors occur regarding back exercises. Figure 4-31, Table 4-4, and the following text provide information in detail about the origins, insertions, and actions of these muscles. This information is essential to understanding the functions of these important trunk muscles.

Anteroposterior: Low back muscles oppose anterior abdominal muscles.

Lateral: Lateral trunk muscles oppose each other.

Rotary: Muscles that produce rotation oppose those that produce the contralateral rotation (i.e., left rotators oppose right rotators).

Trunk Muscles Attached to the Pelvis

With the pelvis pivoting on the femora, the opposing groups of muscles act not only in straight anteroposterior opposition but also combine their pulls to tilt the pelvis anteriorly or posteriorly as well as laterally.

There are four main groups of muscles in **anteroposterior opposition**:

1. The erector spinae, quadratus lumborum, and other back muscles attached to the posterosuperior part of the pelvis exert an upward pull posteriorly.
2. The anterior abdominals, especially the rectus abdominis with its insertion on the symphysis pubis and the external oblique with its attachment on the anterior iliac crest, exert an upward pull anteriorly.
3. The gluteus maximus and hamstrings, with attachments on the posterior ilium, sacrum, and ischium, exert a downward pull posteriorly.
4. The hip flexors, including the rectus femoris, tensor fasciae latae, and sartorius, with attachments on the anterior, superior, and inferior spines of the ilium, and the iliopsoas, with attachment on the lumbar spine and inner surface of the ilium, exert a downward pull anteriorly.

The low back muscles act with the hip flexors (especially the psoas, with its direct pull from the lumbar spine to the femur) to tilt the pelvis anteriorly. They are opposed in action by the combined pull of the anterior abdominals, pulling upward anteriorly, and the hamstrings and gluteus maximus, pulling downward posteriorly, to level the pelvis from a position of anterior tilt.

There are two main groups of pelvic muscles in **lateral opposition**:

1. The hip abductors (mainly the gluteus minimus and medius), which arise from the lateral surface of the pelvis, pull downward on the pelvis when the lower extremity is fixed as in standing.
2. The lateral trunk muscles, attached to the lateral crest of the ilium, pull upward laterally on the pelvis.

Hip abductors on one side and lateral trunk muscles on the contralateral side combine in action to tilt the pelvis laterally: Right abductors pull downward on the right side of the pelvis as left lateral trunk muscles pull upward on the left side, and vice versa. These actions are assisted by hip adductors on the same side as the lateral trunk muscles. In combination, right hip abductors, left hip adductors, and left lateral trunk muscles oppose left hip abductors, right hip adductors, and right lateral trunk muscles.

Neck and Back Extensors

Figure 4-31 illustrates the neck and back extensor muscles. Table 4-4 lists the origin, insertion, and innervation of specific extensor muscles.

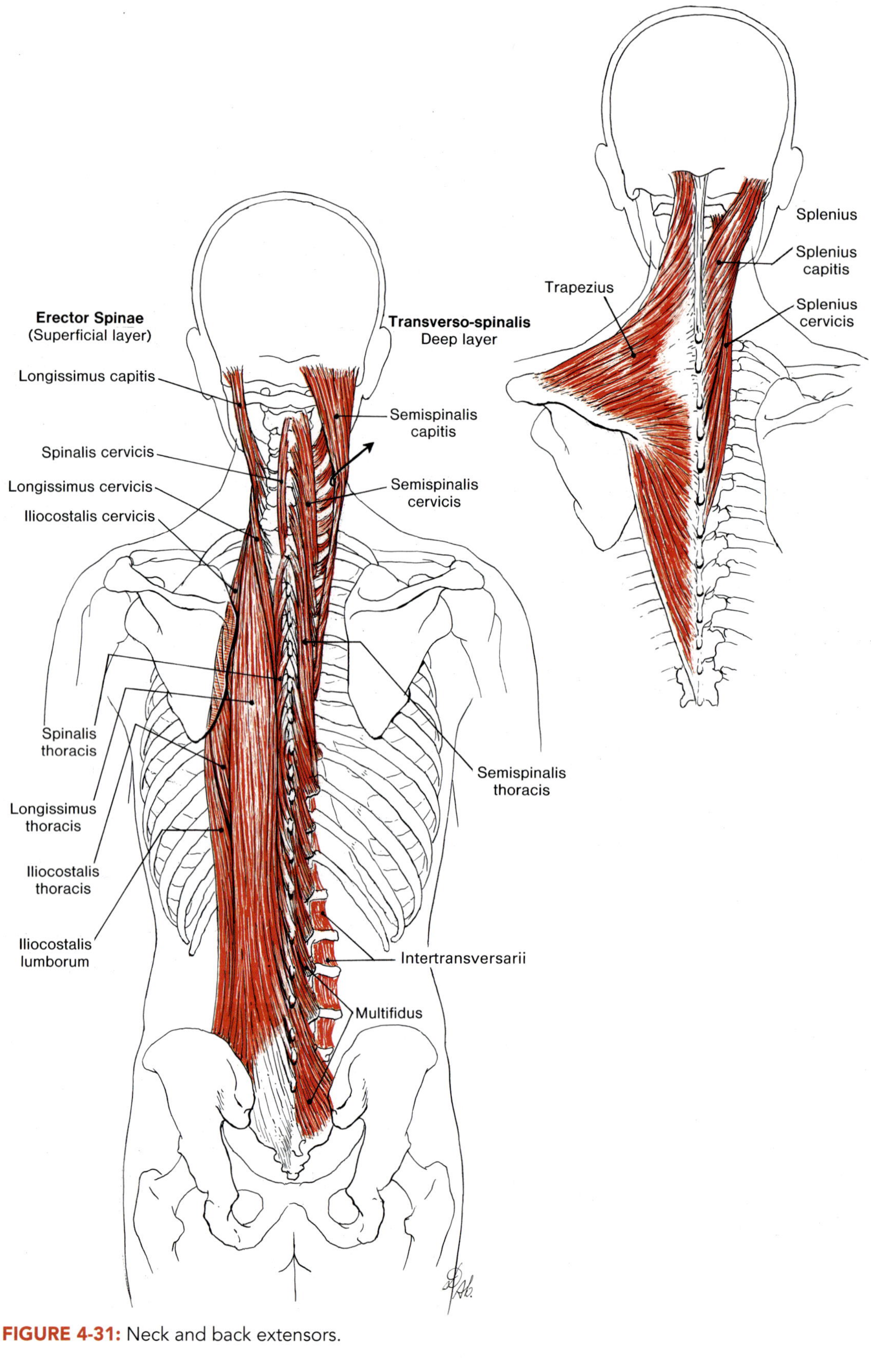

FIGURE 4-31: Neck and back extensors.

TABLE 4-4 Deep Back Muscle Origin, Insertion, and Action

Muscles/Nerves	Origin	Insertion	Action
Erector spinae (superficial) iliocostalis lumborum/*Spinal*	Common origin from anterior surface of broad tendon attached to medial crest of the sacrum, spinous processes of lumbar and 11th and 12th thoracic vertebrae, posterior part of medial lip of iliac crest, supraspinous ligament, and lateral crests of sacrum	By tendons into inferior borders of angles of lower six or seven ribs	Extension of vertebral column in lower thoracic area; draws ribs downward.
Iliocostalis thoracis/*Spinal*	By tendons from upper borders of angles of lower six ribs	Cranial borders of angles of upper six ribs and dorsum of transverse process of seventh cervical vertebra	Extension and lateral flexion of vertebral column in upper thoracic area; draws ribs downward.
Iliocostalis cervicis/*Spinal*	Angles of third through sixth ribs	Posterior tubercles of transverse processes of fourth through sixth cervical vertebrae	Extension of vertebral column in upper thoracic and lower cervical areas.
Longissimus thoracis/*Spinal*	In lumbar region, blended with iliocostalis lumborum, posterior surfaces of transverse and accessory processes of lumbar vertebrae, and anterior layer of thoracolumbar fascia	By tendons into tips of transverse processes of all thoracic vertebrae and by fleshy digitations into lower 9 or 10 ribs between tubercles and angles	Extension and lateral flexion of vertebral column in thoracic area; draws rib downward.
Longissimus cervicis/*Spinal*	By tendons from transverse processes of upper four or five thoracic vertebrae	By tendons into posterior tubercles of transverse processes of second through sixth cervical vertebrae	Extension and lateral flexion of vertebral column in cervical area; draws ribs downward.
Longissimus capitis/*Cervical*	By tendons from transverse processes of upper four or five thoracic vertebrae and articular processes of lower three or four cervical vertebrae	Posterior margin of mastoid process deep to splenius capitis and sternocleidomastoid	Extension, lateral flexion and rotation of cervical spine; turning the head to face toward the same side.
Spinalis thoracis/*Spinal*	By tendons from spinous processes of first two lumbar and last two thoracic vertebrae	Spinous processes of upper four to eight (variable) thoracic vertebrae	Extension of vertebral column in thoracic area.
Spinalis cervicis/*Spinal*	Ligamentum nuchae, lower part; spinous process of seventh cervical vertebra and, sometimes, from spinous process of first and second thoracic vertebrae.	Spinous process of axis and, occasionally, into the spinous processes of C3 and C4	Extension of vertebral column in upper cervical area.
Spinalis capitis/*Spinal*	Inseparably connected with semispinalis capitis	Same as semispinalis capitis	Same as semispinalis capitis.
Transversospinalis (deep) First layer Semispinalis thoracis/*Spinal*	Transverse processes of lower 6–10 thoracic vertebrae	By tendons into spinous processes of first four thoracic and last two cervical vertebrae	Extension of vertebral column and rotation toward opposite side in thoracic area.
Semispinalis cervicis/*Spinal*	Transverse processes of upper five or six thoracic vertebrae	Cervical spinous processes of second through fifth cervical vertebrae	Extension of vertebral column and rotation toward opposite side in upper thoracic and cervical areas.
Semispinalis capitis/*Cervical*	Tips of transverse processes of upper six or seven thoracic and seventh cervical vertebrae and articular processes of fourth through sixth cervical vertebrae.	Between superior and inferior nuchal lines of occipital bone	Extension of neck and rotation of head toward opposite side.

(continues on the next page)

TABLE 4-4 (*continued*)

Muscles/Nerves	Origin	Insertion	Action
Second layer Multifidi/*Spinal*	*Sacral region:* Posterior surface of sacrum, medial surface of posterior iliac spine and postero-sacroiliac ligaments. *Lumbar, thoracic, and cervical regions:* Transverse processes of L5 through C4.	Spanning two to four vertebrae, inserted into spinous process of one of vertebra above from last lumbar to axis (second cervical vertebra)	Extension of vertebral column and rotation toward opposite side.
Third layer Rotatores/*Spinal*	Transverse processes of vertebrae	Base of spinous process of vertebra above	Extension of vertebral column and rotation toward opposite side.
Interspinales/*Spinal*	Placed in pairs between spinous processes of contiguous vertebrae *Cervical:* six pairs *Thoracic:* two or three pairs; between first and second (second and third) and 11th and 12th vertebrae. *Lumbar:* four pairs		Extension of vertebral column.
Intertransversarii anterior and posterior/*Spinal*	Small muscles placed between transverse processes of contiguous vertebrae in cervical, thoracic, and lumbar regions		Lateral flexion of vertebral column.
Splenius cervicis/*Cervical*	Spinous processes of third through sixth thoracic vertebrae	Posterior tubercles of transverse processes of first two or three cervical vertebrae	Extension, lateral flexion and rotation of neck, turning face toward same side. Both sides acting together, extension of neck.
Splenius capitis/*Cervical*	Caudal ½ of ligamentum nuchae, spinous process of seventh cervical vertebrae, and spinous process of first three or four thoracic vertebrae	Occipital bone inferior to lateral ⅓ of superior nuchal line; mastoid process of temporal bone	Extension, lateral flexion, and rotation of neck, turning face toward same side. Both sides acting together, extension of neck.

Back Extensors and Hip Extensors

For back extensors to raise the trunk from a prone position, the hip extensors must fix the pelvis in extension on the thigh. Normally, extension of the hip joints and extension of the lumbar spine are initiated simultaneously, not as two separate movements. The illustrations on the next page show the variations that occur depending on the strength of the two primary muscle groups.

If slight tightness exists in the hip flexors, there is no range of extension in the hip joint, and all the movement in the direction of raising the leg backward is accomplished by lumbar spine hyperextension and pelvic tilt.

For hip extensors to raise the extremity backward from a prone position through the few degrees of true hip joint extension (≈10°), the back extensors must stabilize the pelvis to the trunk (Figure 4-32).

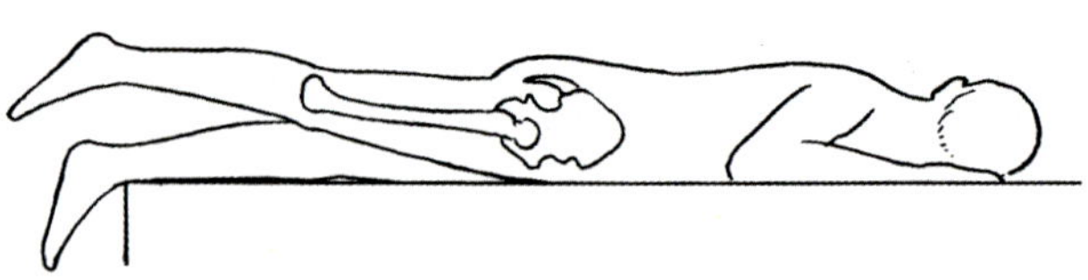

FIGURE 4-32: Hip extension, prone position.

A subject with strong back extensor muscles and strong hip extensor muscles can raise the trunk in extension (Figure 4-33).

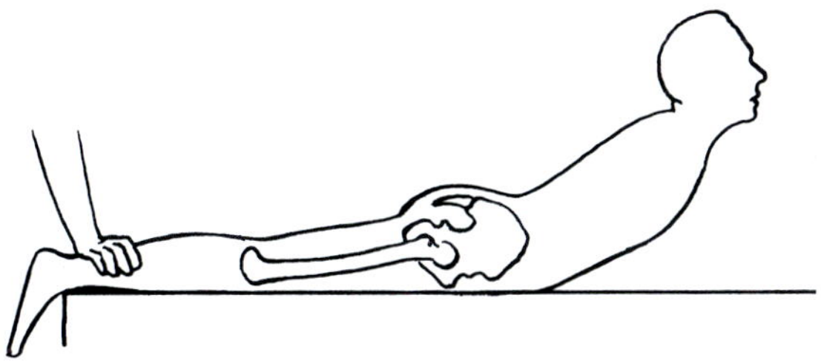

FIGURE 4-33: Strong back and hip extensor muscles.

Raising the extremity higher is accomplished by hyperextension of the lumbar spine and anterior tilting of the pelvis (Figure 4-34). In this latter movement, the back extensors are assisted by hip flexors on the opposite side that help to tilt the pelvis anteriorly.

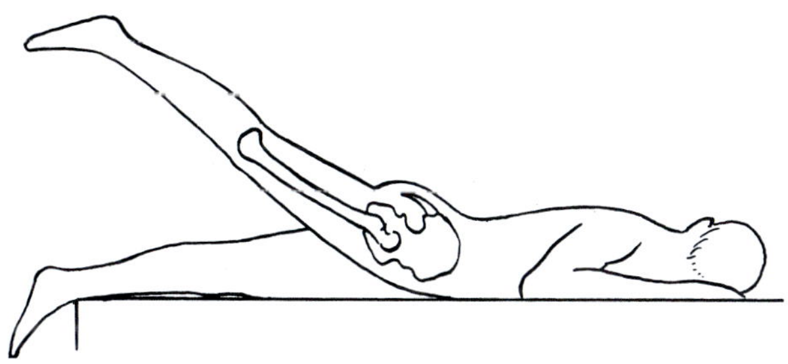

FIGURE 4-34: Raising the extremity higher.

A subject with strong back extensor muscles and markedly weak or paralyzed hip extensor muscles can hyperextend the lumbar spine (Figure 4-35). However, the trunk cannot be lifted high from the table.

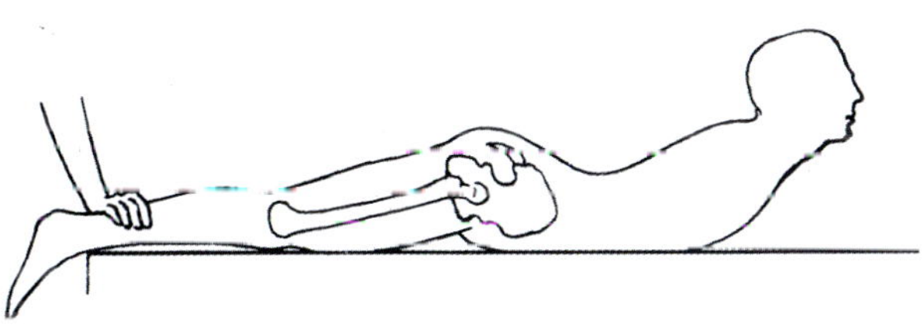

FIGURE 4-35: Strong back extensors and weak or paralyzed hip extensors.

In an effort to lift the extremity, the back muscles contract to fix the pelvis on the trunk, but with little or no strength in the hip extensors, the thigh cannot be extended on the pelvis (Figure 4-36). The unopposed pull of the back muscles results in hyperextension of the back, and the hip joint is passively drawn into flexion despite the effort to extend it.

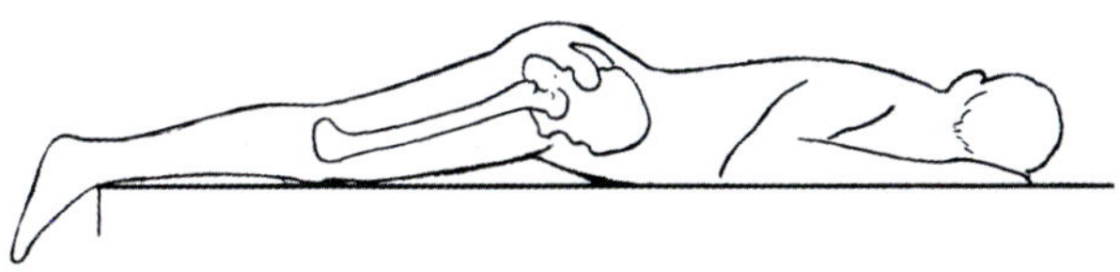

FIGURE 4-36: Minimal strength in hip extensors while trying to lift the extremity.

A subject with weak or paralyzed back extensor muscles and strong hip extensor muscles cannot raise the trunk in extension (Figure 4-37). The hip extensors, in their action to fix the pelvis, are unopposed: The pelvis tilts posteriorly, and the lumbar spine flexes.

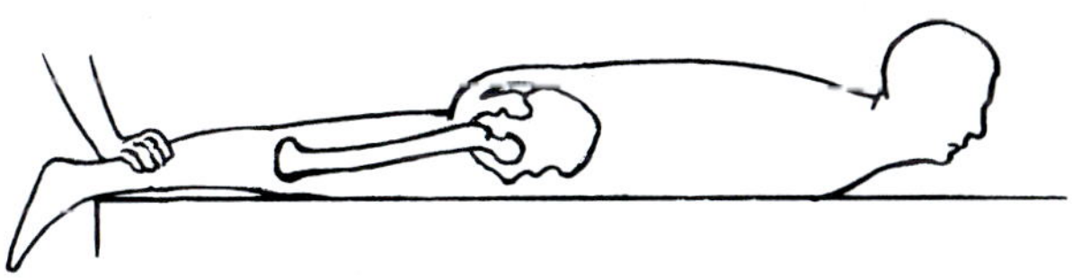

FIGURE 4-37: Weak or paralyzed back extensors and strong hip extensors.

In an effort to lift the extremity, the hip extensors contract (Figure 4-38). The extremity cannot be lifted, however, because the back muscles are unable to stabilize the pelvis. The pelvis tilts posteriorly because of the pull of the hip extensors and the weight of the extremity, instead of tilting anteriorly as it would if the back extensors were normal.

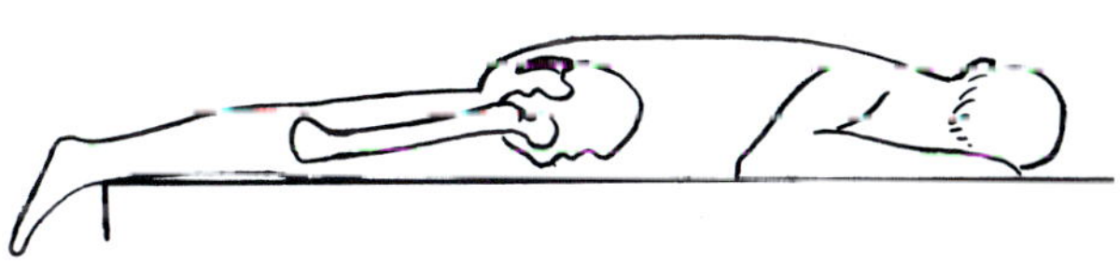

FIGURE 4-38: Hip extensors contract in an effort to lift the extremity.

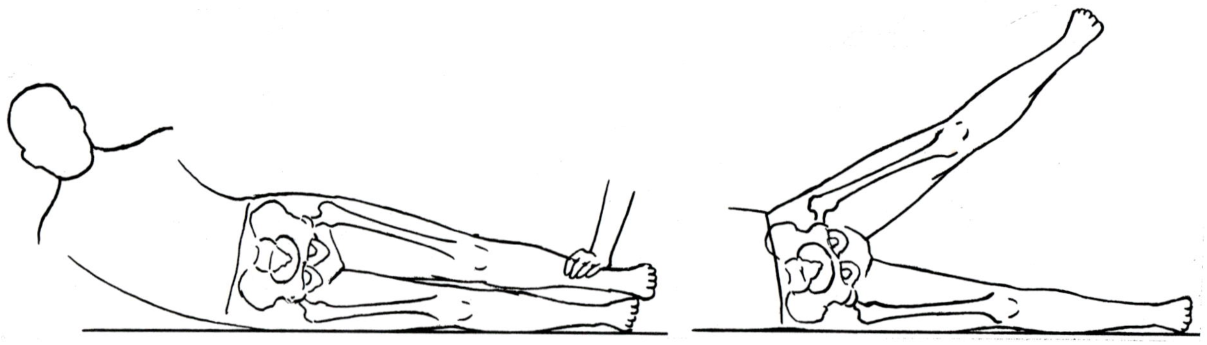

FIGURE 4-39: Strong lateral trunk muscles and strong hip abductor muscles.

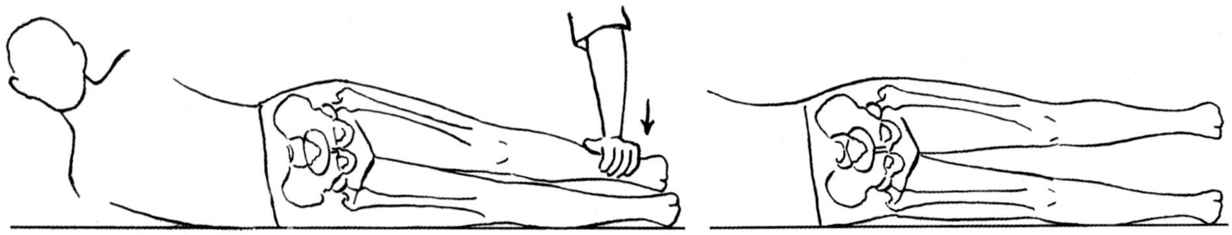

FIGURE 4-40: Strong lateral trunk muscles and paralyzed hip abductor muscles.

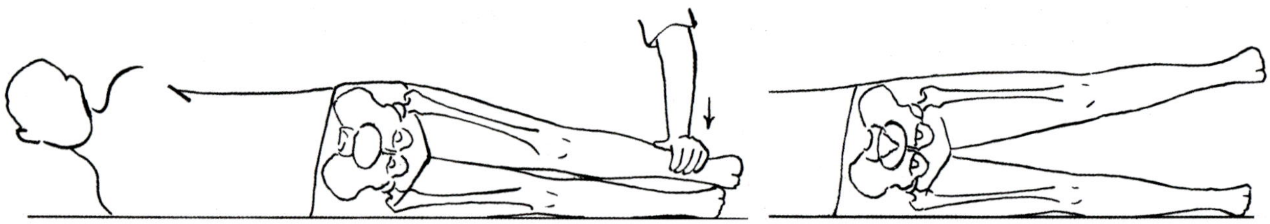

FIGURE 4-41: Weak lateral trunk muscles and strong hip abductor muscles.

Lateral Trunk Flexors and Hip Abductors

Strong Lateral Trunk Muscles and Strong Hip Abductor Muscles

Lateral trunk flexion through the subject's full range of motion (Figure 4-39, left).

Hip abduction through the subject's full range of motion (Figure 4-39, right).

Strong Lateral Trunk Muscles and Paralyzed Hip Abductor Muscles

The subject can laterally flex the trunk (Figure 4-40, left), but the underneath shoulder will scarcely be raised from the table. The pelvis will be drawn upward as the head is raised laterally, and the iliac crest and costal margin will be approximated.

In attempting to raise the extremity in abduction, the movement that occurs is elevation of the pelvis by the lateral trunk muscles (Figure 4-40, right). The extremity may be drawn upward into the position as illustrated, but the hip joint is not abducted. In fact, the thigh has dropped into a position of adduction and is held there by the joint structure rather than by action of the hip muscles.

Weak Lateral Trunk Muscles and Strong Hip Abductor Muscles

The subject cannot raise the trunk in true lateral flexion (Figure 4-41, left). Under certain circumstances, the patient may be able to raise the trunk from the table laterally even though the lateral trunk muscles are quite weak. If the trunk can be held rigid, the hip abductor muscles may raise the trunk in abduction on the thigh. The rib cage and iliac crest will not be approximated laterally as they are when the lateral trunk muscles are strong. By decreasing the pressure providing fixation for the hip abductors, the examiner can make it necessary for the lateral abdominals to attempt initiation of the movement.

The extremity can be lifted in hip abduction, but without fixation by the lateral abdominal muscles, it cannot be raised high off the table. Because of the weakness of the lateral trunk muscles, the weight of the extremity tilts the pelvis downward (Figure 4-41, right).

BACK MUSCLE TESTS

Forward-Bending Test for Length of Posterior Muscles

Equipment: Same as for hamstring length test, plus a ruler. The ruler is used to measure the distance of the fingertips either from or beyond the base of the big toe. This measurement is used only as a record to show the overall forward bending; it in no way indicates where limitation or excessive motion has taken place.

Starting Position: Sitting with legs extended (long-sitting) and feet at, or slightly below, right angles.

Reason: To standardize the position of the feet and knees.

Test Movement: Reach forward, with knees straight, and try to touch the fingertips to the base of the big toe or beyond, reaching as far as the range of muscle length permits.

Reason: Both the back and hamstrings will elongate to their maximum.

Normal length of back, hamstring, and gastroc-soleus muscles (Figure 4-42).

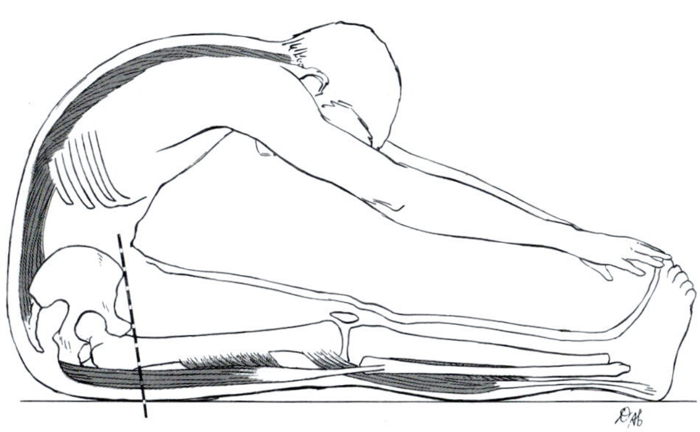

FIGURE 4-42: Normal length of back, hamstring, and gastroc-soleus muscles.

The ability to touch fingertips to toes is a desirable accomplishment for most adults (Figure 4-43). This subject shows hamstring length and back flexibility within normal limits.

FIGURE 4-43: Touching fingertips to toes.

Normal Range of Motion in Forward Bending: Normal length of the hamstrings permits the pelvis to flex toward the thigh to the extent that the angle between the sacrum and the table is approximately 80°. Normal flexion of the lumbar spine permits the spine to flatten. Normal flexion of thoracic spine permits an increase in the posterior convexity, which is seen as a smooth, continuous curve in this area. The average adult will be able to touch fingertips to toes in forward bending with knees straight if the flexibility of the back and the length of the hamstrings are both normal (see Figure 4-42).

Variations in Forward Bending:

Hamstrings and back, both normal.
Hamstrings and back, both excessively flexible.
Hamstrings tight, low back excessively flexible.
Hamstrings normal, upper back excessively flexible.
Hamstrings and back, both tight.
Hamstrings excessive in length, low back tight.

In forward bending, excessive hamstring length permits excessive flexion of the pelvis toward the thigh (hip joint flexion). This subject also has excessive flexion in the mid-back (i.e., thoracolumbar) area (Figure 4-44).

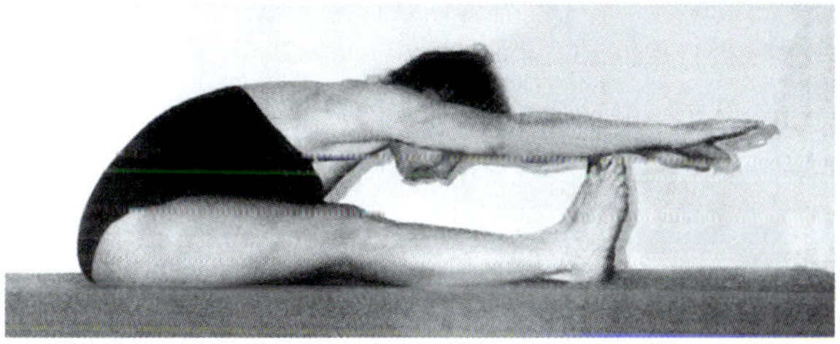

FIGURE 4-44: Excessive hamstring length and excessive flexion in the thoracolumbar area (mid-back).

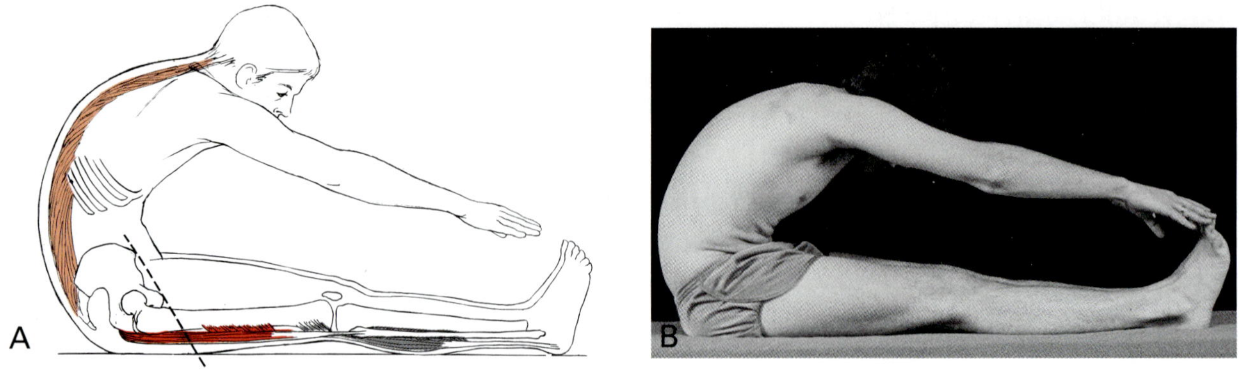

FIGURE 4-45: Short hamstrings, excessively long back muscles.

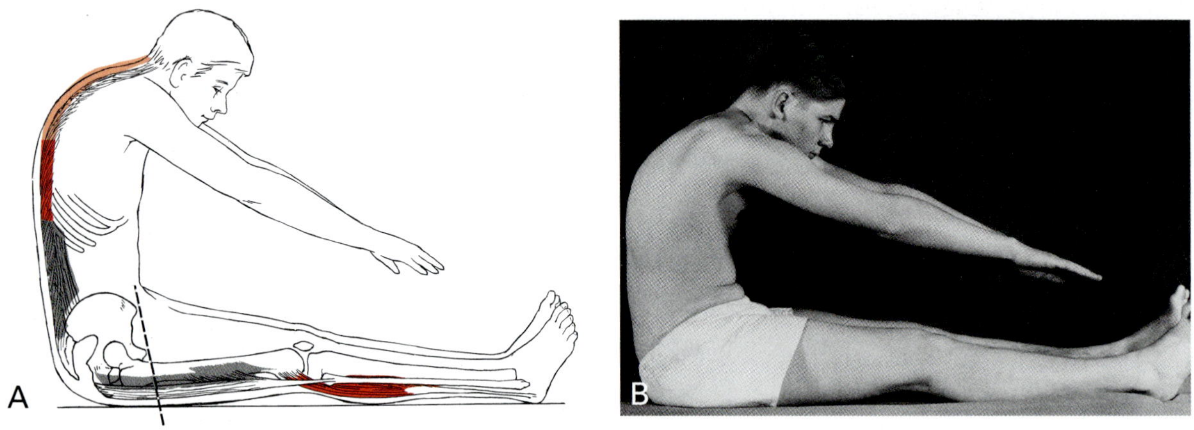

FIGURE 4-46: Short gastric-soleus muscles and limited flexibility in the mid-back.

Variations in Length of Posterior Muscles

Excessive length of back muscles, short hamstrings, and normal length of the gastroc-soleus (Figure 4-45A).

Excessive flexibility of the back overcompensates for shortness of the hamstrings (Figure 4-45B).

Excessive length of the upper back muscles, slight shortness of the muscles in the mid-back and in gastroc-soleus (Figure 4-46A). Hamstrings and low back are normal in length.

This subject is unable to touch his toes because of shortness in the gastroc-soleus, and slight limitation of flexibility in the mid-back area (Figure 4-46B). The upper back shows some excessive flexion.

Normal length of the upper back muscles and short lower back, hamstring, and gastroc-soleus muscles (Figure 4-47).

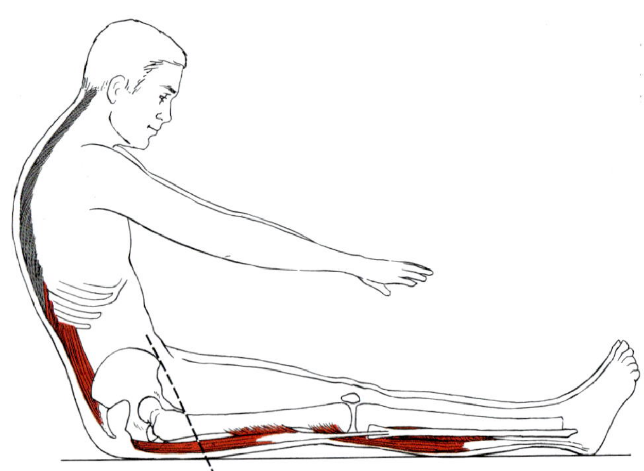

FIGURE 4-47: Normal length of upper back, short lower back, hamstring, and gastroc-soleus muscles.

Normal length of the upper back muscles and contracture of lower back muscles with paralysis and excessive length of extremity muscles (Figure 4-48).

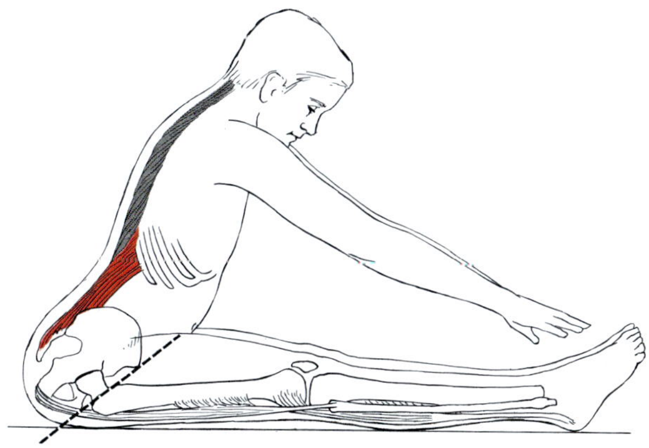

FIGURE 4-48: Normal length of upper back muscles and contracture of lower back muscles with paralysis, plus excessive length of extremity muscles.

Back Extensors: Testing and Grading

In the trunk extension test for the back extensors, the erector spinae muscles are assisted by the latissimus dorsi, quadratus lumborum, and trapezius (Figure 4-49). In the prone position, the low back will assume a normal anterior curve.

To avoid false interpretations of the test results, it may be necessary to perform some preliminary tests. It is not necessary to do so routinely, however, because close observation of the subject in a prone position and of the movements taking place during trunk extension will indicate if preliminary tests for length of hip flexors and strength of the hip extensors are needed.

Patient: Prone, with hands clasped behind the buttocks (or behind the head).

Fixation: Hip extensors must give fixation of the pelvis to the thighs. The examiner stabilizes the legs firmly on the table.

Test Movement: Trunk extension to the subject's full range of motion.

Resistance: Gravity. Hands behind the head, or hands behind the lower back.

Grading: The ability to complete the movement and hold the position with hands behind the head or behind the back may be considered as normal strength. The low back muscles are seldom weak, but if there appears to be weakness, then hip flexor tightness and/or hip extensor weakness must be ruled out first. Actual weakness can usually be determined by having the examiner raise the subject's trunk in extension (to the subject's maximum range) and then asking the subject to hold the completed test position. Inability to hold this position will indicate weakness. Weakness is best described as slight, moderate, or marked based on the judgment of the examiner.

If the range of motion appears to be limited, a second person should hold the legs down (or legs should be held down with straps) while the examiner passively raises the subject's trunk in extension to that individual's completion of spine extension.

If the hip extensors are weak, it is possible that the examiner can stabilize the pelvis firmly in the direction of posterior tilt toward the thighs, provided that the legs are also firmly held down by another person or by straps. Alternatively, the subject may be placed at the end of the table, with the trunk in a prone position and legs hanging down with knees bent as needed. The examiner then stabilizes the pelvis and asks subject to raise the trunk in extension and hold it against pressure. In the presence of tight hip flexors, the back will assume a degree of extension (i.e., lordosis) commensurate with the amount of hip flexor tightness. In other words, the low back will be in extension before beginning the trunk extension movement. In such a case, the subject will be limited in the height to which the trunk can be raised, and the mistaken interpretation may be that the back muscles are weak.

A similar situation may arise if the hip extensor muscles are weak. For strong extension of the back, the hip extensors must stabilize the pelvis toward the thighs.

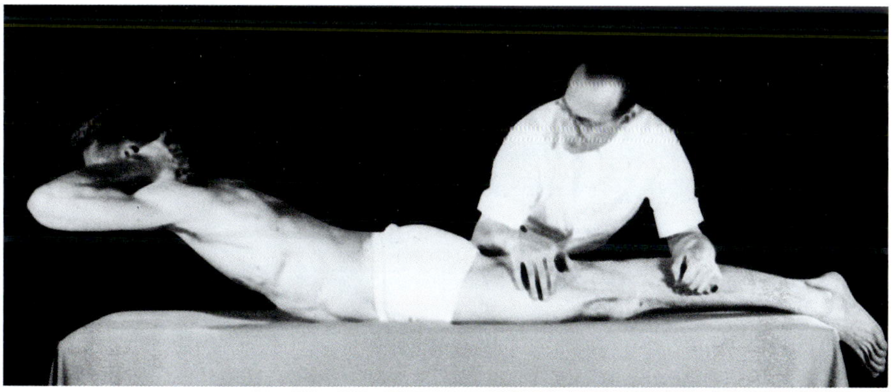

FIGURE 4-49: Testing and grading back extensors.

If the hip extensors cannot provide this stabilization, the pelvis will be pulled upward by the back extensors into a position of back extension. Again, as in the case of hip flexor tightness, if the back is already in some extension before the trunk-raising movement is started, the trunk will not be raised as high off the table as it would be if the pelvis were fixed in extension on the thighs.

Weakness: Bilateral weakness of the back extensor muscles may result in a lumbar kyphosis and an increased thoracic kyphosis. Unilateral weakness results in a lateral curvature with convexity toward the weak side.

Contracture: Bilateral contracture of the low back muscles results in a hyperlordosis. Unilateral contracture results in a scoliosis with convexity toward the opposite side.

Strong Back Extensors, Misdiagnosed

Weakness of the Gluteus Maximus

Lying prone on a table, this subject exhibits a normal anterior curve in the lower back (Figure 4-50).

The moment that back extension is initiated, the curve in the lower back increases because of weakness in the gluteus maximus (Figure 4-51).

When extension is continued, the subject can raise the trunk higher, but not to completion of the range of motion (Figure 4-52).

Holding the pelvis in the direction of posterior pelvic tilt, in the manner provided by a strong gluteus maximus, enables the subject to complete the full range of motion (Figure 4-53).

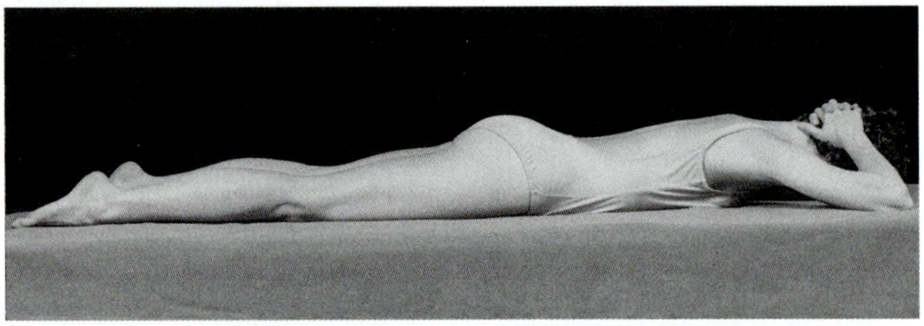

FIGURE 4-50: Normal anterior curve in the lower back.

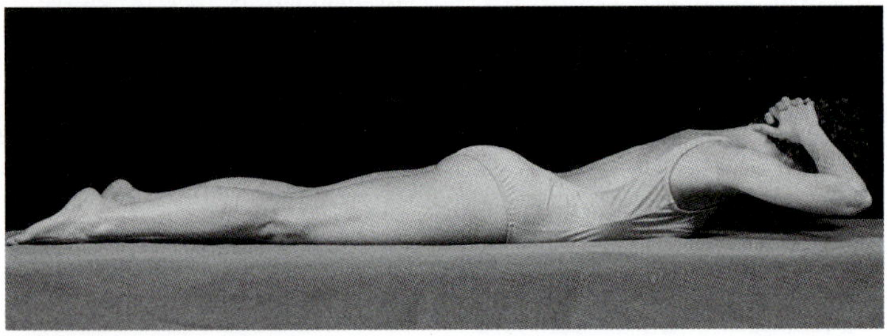

FIGURE 4-51: Weakness in the gluteus maximus.

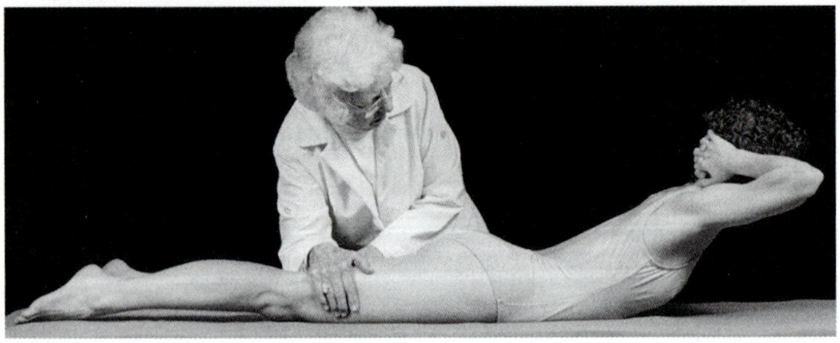

FIGURE 4-52: Back extension without completion of range of motion.

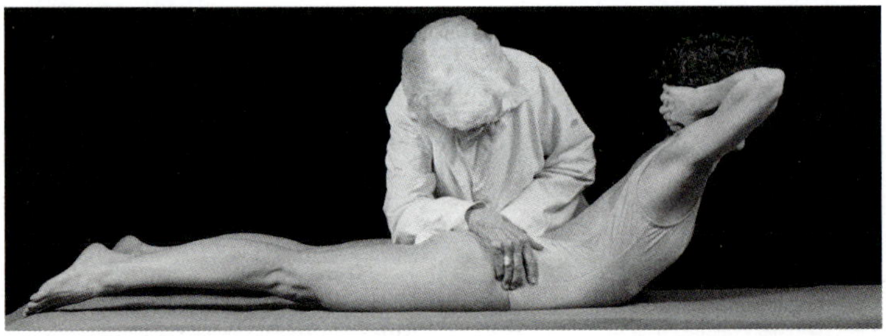

FIGURE 4-53: Strong gluteus maximus.

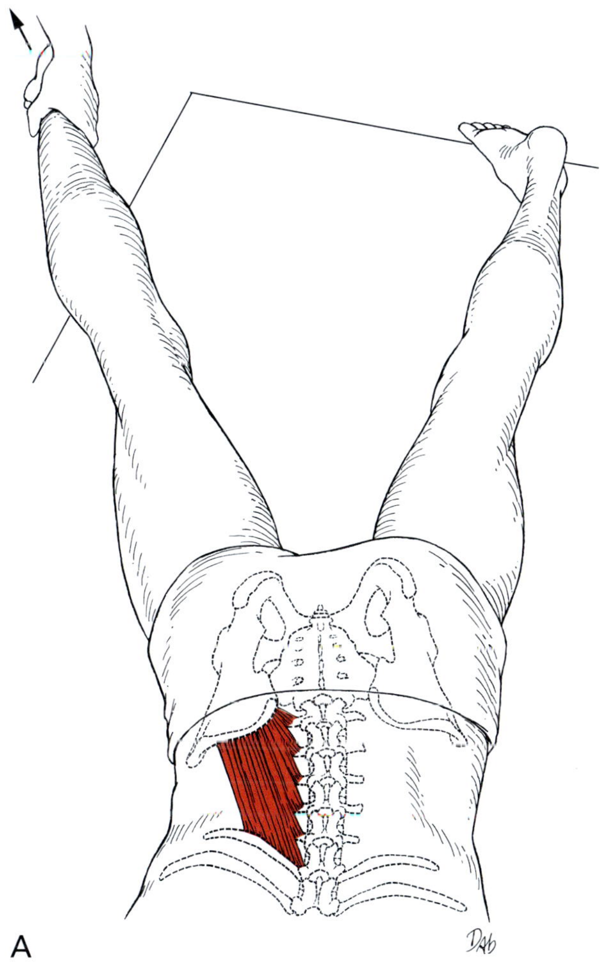

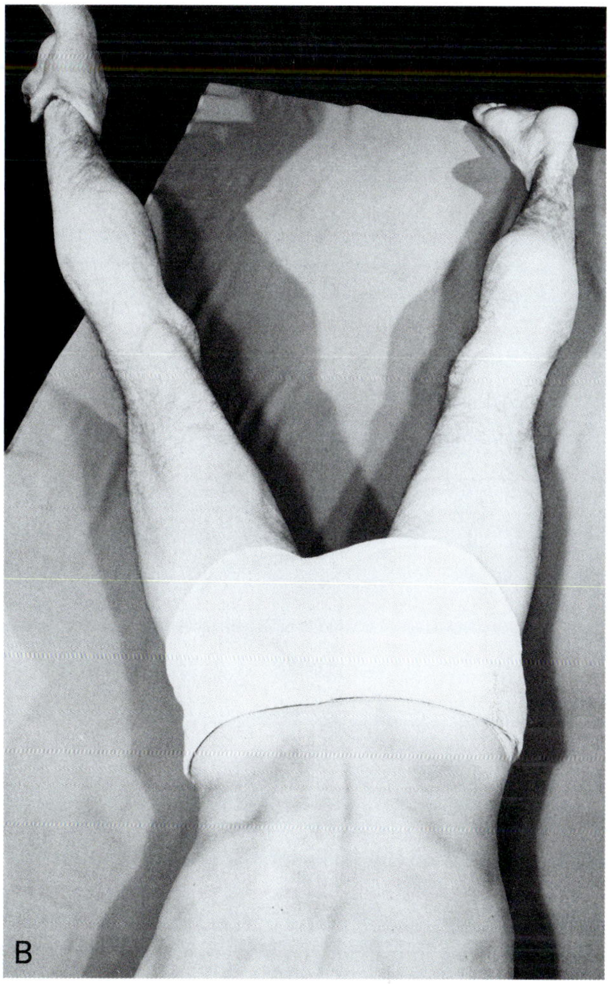

FIGURE 4-54: A, B, Quadratus lumborum.

Quadratus Lumborum

Figure 4-54 shows testing of the quadratus lumborum.

Origin: Iliolumbar ligament, iliac crest. Occasionally from upper borders of the transverse processes of the lower three or four lumbar vertebrae.

Insertion: Inferior border of the last rib and transverse processes of the upper four lumbar vertebrae.

Action: Assists in extension, laterally flexes the lumbar vertebral column, and depresses the last rib. Bilaterally, when acting together with the diaphragm, fixes the last two ribs during respiration.

Nerve: Lumbar plexus, T12, L**1**, **2**, **3**.

Patient: Prone.

Fixation: By muscles that hold the femur firmly in the acetabulum.

Test Movement: Lateral elevation of the pelvis. The extremity is placed in slight extension and in the degree of abduction that corresponds with the line of fibers of the quadratus lumborum.

Resistance: Given in the form of traction on the extremity, directly opposing the line of pull of the quadratus lumborum. If the hip muscles are weak, pressure may be given against the posterolateral iliac crest opposite to the line of pull of the muscle.

The quadratus lumborum acts with other muscles in lateral trunk flexion. It is difficult to palpate this muscle because it lies deep beneath the erector spinae. Although the quadratus lumborum enters into the motion of elevation of the pelvis in the standing position or in walking, the standing position does not offer a satisfactory position for testing. Elevation of the right side of the pelvis in standing, for example, depends as much (if not more) on the downward pull by the abductors of the left hip joint as it does on the upward pull of the right lateral abdominals.

The test should not be considered as limited to action of the quadratus lumborum but as giving the most satisfactory differentiation that can be obtained.

Grading: Grading the strength of this muscle numerically is not recommended. Simply record whether it appears to be weak or strong.

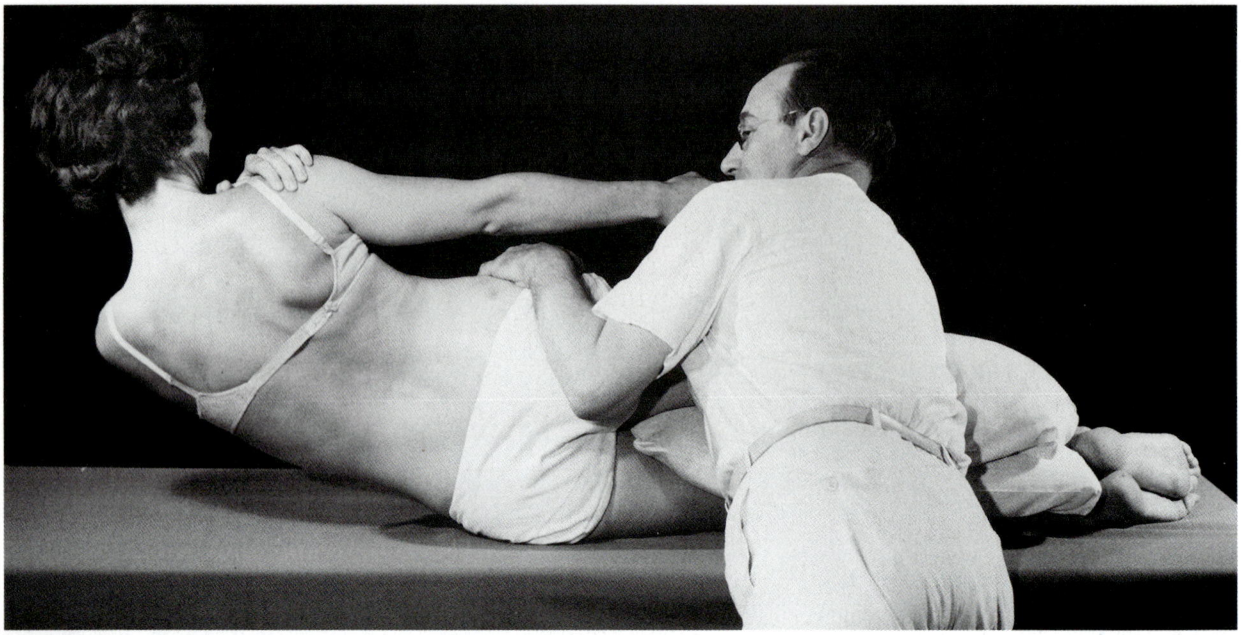

FIGURE 4-55: Testing and grading lateral trunk flexors.

Lateral Trunk Flexors: Testing and Grading

Before testing the lateral trunk muscles, one should test the strength of the hip abductors, adductors, and lateral neck flexors and the range of motion in lateral flexion.

Raising the trunk sideways is a combination of lateral trunk flexion and hip abduction (the latter being produced by downward tilting of the pelvis on the thigh). The lateral trunk muscles entering into the movement are the lateral fibers of the external and internal obliques, the quadratus lumborum, the latissimus dorsi, the rectus abdominis, and the erector spinae on the side being tested.

Patient: Side-lying, with a pillow between the thighs and legs and with the head, upper trunk, pelvis, and lower extremities in a straight line (Figure 4-55). The top arm is extended down along the side, and the fingers are closed so that the patient will not hold onto the thigh and attempt to assist with the hand. The underarm is forward across the chest, with the hand holding the upper shoulder to rule out assistance by pushing up with the elbow.

Fixation: Hip abductors must fix the pelvis to the thigh. The opposite adductors also help to stabilize the pelvis. The lower extremities must be held down by the examiner to counterbalance the weight of the trunk, but they must not be held so firmly as to prevent the superior lower extremity from moving slightly downward to accommodate for the downward displacement of the pelvis on that side. If the pelvis is pushed upward or is not allowed to tilt downward, the subject will be unable to raise the trunk sideways even if the lateral abdominal muscles are strong.

Test Movement: Trunk raising directly sideways without rotation.

Resistance: The body weight offers sufficient resistance.

Normal (10) Grade[1]: The ability to raise the trunk laterally from a side-lying position to a point of maximum lateral flexion.

Good (8) Grade: Same as above, except the underneath shoulder is approximately 4 inches up from table.

Fair (5) Grade: Same as above, except the underneath shoulder is approximately 2 inches up from table (see tests and grades in cases of marked weakness of lateral trunk muscles).

NOTE: *Tests of the lateral trunk muscles may reveal an imbalance in the oblique muscles. In sideways trunk raising, if the legs and the pelvis are held steady (i.e., not permitted to twist forward or backward from the direct side-lying position), the thorax may be rotated forward or backward as the trunk is laterally flexed. A forward twist of the thorax denotes a stronger pull by the external oblique; a backward twist denotes a stronger pull by the internal oblique. If the back hyperextends as the patient raises the trunk, the quadratus lumborum and latissimus dorsi show a stronger pull, indicating that the anterior abdominal muscles cannot counterbalance this pull to keep the trunk in a straight line with the pelvis.*

The test for strength of the lateral trunk flexors is important in cases of scoliosis.

[1] See numerical equivalents for word symbols used in *The Key to Muscle Grading*, Chapter 1.

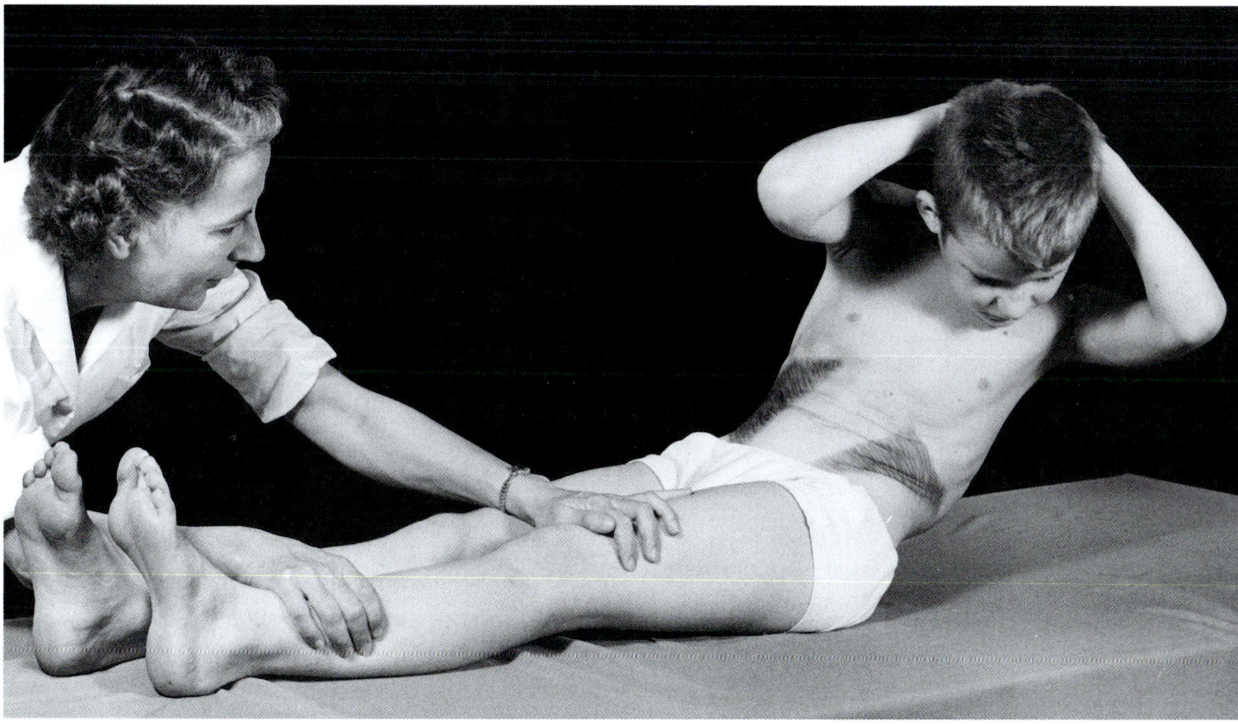

FIGURE 4-56: Testing and grading of oblique trunk flexors.

Oblique Trunk Flexors: Testing and Grading

Raising the trunk obliquely forward combines trunk flexion and rotation. It is accomplished by action of the rectus abdominis and by the external oblique on one side combined with the internal oblique on the opposite side (Figure 4-56).

Patient: Supine. (For upper extremity position, see the discussion of grades below.)

Fixation: An assistant stabilizes the lower extremities as the examiner places the patient in the test position. (The examiner is not shown in this photograph.)

Test: The patient clasps hands behind the head. The examiner places the patient into the precise test position of trunk flexion and rotation and then asks the patient to hold that position. If the muscles are weak, the trunk will derotate and extend. There may be increased flexion of the pelvis on the thighs in an effort to hold the extended trunk up from the table.

Resistance: None in addition to the weight of the trunk. Resistance is varied by the position of the arms.

Normal (10) Grade[2]: Ability to hold the test position with hands clasped behind the head.

Good (8) Grade: Same as above, except with arms folded across the chest.

Fair+ (6) Grade: Same as above, except with arms extended forward (see illustration of arm positions).

Fair (5) Grade: Ability to hold the trunk in enough flexion and rotation to raise both scapular regions from the table (see tests and grades in cases of marked weakness of the oblique trunk muscles).

NOTE: *The test for muscle strength of the oblique abdominal muscles is important in cases of scoliosis.*

[2] See numerical equivalents for word symbols used in *The Key to Muscle Grading*, Chapter 1.

SECTION IV
CLINICAL FINDINGS

NECK

Muscle dysfunctions associated with pain in the posterior neck are essentially of two types: one associated with muscle tightness and the other with muscle strain. Both types are quite prevalent. Symptoms and indications for treatment differ according to the underlying fault. The one associated with muscle tightness usually has a gradual onset of symptoms, whereas the one associated with muscle strain usually has an acute onset.

Tightness of Posterior Neck Muscles

Neck pain and headaches associated with tightness in the posterior neck muscles are most often found in patients with a forward head (cervical hyperlordosis) and a rounded thoracic spine (hyperkyphosis). As shown in the sections on Cervical Spine Extension and Flexion and Faulty Head and Neck Positions, the compensatory head position associated with a slumped, rounded thoracic spine results in hyperextension of the cervical spine.

The faulty mechanics associated with this condition chiefly consist of undue compression posteriorly on the articulating facets and posterior surfaces of the bodies of the vertebrae, stretch weakness of the anterior vertebral neck flexors, and tightness of the neck extensors, including the upper trapezius, splenius capitis, and semispinalis capitis.

Headaches associated with this muscle tightness are essentially of two types: occipital and tension. The greater occipital nerve, which is both sensory and motor, supplies the semispinalis and splenius capitis muscles. It pierces the semispinalis capitis and the trapezius near their attachments to the occipital bone. This nerve also innervates the scalp posteriorly up to the top of the head. In the occipital headache, there usually is pain and tenderness on palpation in the area where the nerve pierces the muscles as well as pain in the scalp in the area supplied by the nerve. In a tension headache, in addition to the faulty postural position of the head and neck and the tightness of the posterior neck muscles, an element of stress is also involved. This makes the condition tend to fluctuate with times of increased or decreased stress. In any event, the tight muscles usually respond to treatment that helps these muscles to relax.

Symptoms in addition to pain may occur with tension headaches: "Occasionally, muscle contraction headaches will be accompanied by nausea, vomiting and blurred vision, but there is no preheadache syndrome as with migraine" (12). Another source states that this forward-head position has been found "to cause an alteration in the rest position of the mandible, upper thoracic respiration with subsequent hyperactivity of the respiratory accessory muscles, and mouth breathing with a loss of the rest position of the tongue ... and may lead to eventual osteoarthrosis and remodeling of the temporomandibular joint" (13).

On palpation, the posterior muscles are tight. Movements of the neck are often limited in all directions except in extension. Pain may be of lower intensity when the patient is recumbent, but it tends to be present regardless of the position the patient assumes.

Unilateral tightness in posterolateral neck muscles is increasingly common as a result of people holding a telephone on the shoulder or habitually side bending the cervical spine as in when talking on a cell phone. In these positions, the shoulder may be elevated, and the head tilted toward the same side. The scapular muscle that is the most direct opponent of the upper trapezius is the lower trapezius, which acts to depress the scapula posteriorly. The most direct opponent of the upper trapezius acting to depress the shoulder and the shoulder girdle directly downward in the coronal plane is the latissimus dorsi. Tests of the strength of this muscle often reveal weakness on the side of the elevated shoulder, and exercises to strengthen this muscle are indicated, along with other exercises to stretch the lateral neck flexors (see the latissimus exercise and exercises to stretch the lateral neck on p. 143).

Upper Trapezius Strain

The upper trapezius is that part of the trapezius muscle extending from the occiput to the lateral one-third of the clavicle and the acromion process of the scapula. A strain of this muscle results in pain, usually acute, in the posterolateral region of the neck.

The stress that gives rise to this condition is often a combination of tension on and contraction of the muscle. Stretching sideways to reach for an object while tilting the head in the opposite direction can cause such an attack (e.g., someone on the floor reaching to recover an object that rolled under a desk, or someone sitting in the front of a car reaching to recover an object from the backseat). The abduction of the arm requires scapular fixation by action of the trapezius, and the sideways tilt of the head puts tension on the muscle. The muscle develops a knot or cramp which is better described as a segmental spasm in the muscle.

Cervical Nerve Root Compression

Arm pain caused by cervical nerve root compression is basically a neurological problem. Faulty posture of the cervical spine may act as a contributory factor when the onset is not associated with sudden trauma. Extension of the cervical spine as seen in a typical forward-head position produces undue compression on the facets and posterior surfaces of the bodies of the cervical vertebrae.

BACK

The Low Back Enigma

The causes of many painful conditions of the low back remain obscure. Low back pain, which is one of the most common types of pain, continues to puzzle the experts. Despite the amount of information now available through modern technology, signs and symptoms largely are used as the basis for determining conservative (i.e., nonsurgical) treatment. Even when these signs and symptoms are objective, examiners often disagree about their clinical significance. In many cases, interpretation of these signs and symptoms is still not adequate to generate a conclusive diagnosis. DeRosa and Porterfield state that "at present, identifying with any certainty the exact tissues involved in most low back pain is virtually impossible" (14).

The inability to establish a definitive diagnosis has contributed to various systems of treatment—supported by evidence of success. Treatment may come in various forms: bed rest and medication, successful mobilization (i.e., manipulation), immediate application of a support that provides immobilization, or gentle treatment that employs various pain-relieving modalities and procedures. Clinical observation demonstrates that a high percentage (as much as 80%) of cases recover within 2 weeks, with or without treatment. In view of these statistics, it is not surprising that there is a high success rate, regardless of the approach or system of treatment. In the minds of those who have been relieved of severe pain, however, there is no doubt that treatment has helped.

Regardless of the approach to treatment, there are numerous references to the need for postural correction in the literature. Sometimes immediate care involves correction of postural alignment, but lasting correction and prevention of future problems are even more important aspects of care. This is the area of treatment with which this text is primarily concerned.

Correction of postural faults involves examination of alignment and tests for muscle length and strength. Preservation of ideal alignment depends on establishing and maintaining good muscle balance. This was the basic thesis as stated by the original authors of this text in the pamphlet *Study and Treatment of Muscle Imbalance in Cases of Low Back and Sciatic Pain* (1936) and in *Posture and Pain* (1952) (15, 16).

The mechanics of the low back are inseparable from that of the overall posture, but especially from that of the pelvis and the lower extremities. As a result, evaluation of faulty posture must include examination of the entire body. Although symptoms and faults often appear in the same area, the faults may not be limited to those areas where symptoms appear. Pain manifested in the leg, for example, may be caused by an underlying problem in the back. A mechanical or functional strain causing muscular imbalance in one part of the body may soon result in compensatory changes in other parts. Conversely, the symptoms appearing in the low back may be caused by underlying faulty mechanics of the feet, legs, or pelvis.

An imbalance may begin with abdominal muscle weakness or strain resulting from surgery or obesity. Among women, pregnancy may be the cause. Low back pain has often followed childbearing, and patients have received complete relief of pain by treatment to strengthen the abdominal muscles and correct faulty posture.

In adults, very few activities require strenuous use of the abdominal muscles, but most activities tend to strengthen the back muscles. An important factor to consider as a predisposing cause of back muscle shortening and abdominal muscle relaxation is that the erector spinae muscles are numerous and short and are attached to a strong bony framework. The abdominal muscles, however, are long, with strong fascial attachments but without a supporting bony framework. In addition, abdominal muscles bear the strain of the weight of the abdominal viscera and, for women, of the muscle stretch and strain accompanying pregnancy.

This section focuses on evaluating the findings of tests for postural alignment, range of motion, muscle length, and muscle strength. It does not put labels on most types of painful low back conditions other than to name the problems of postural alignment and muscle imbalance associated with anterior, posterior, and lateral pelvic tilt. The problems of lower extremity pain are those associated with a tight or a stretched tensor fasciae latae and iliotibial band; with sciatic pain associated with protrusion of a disk; or with a stretched piriformis, with pain and weakness in the region of the posterior gluteus medius and with knee and foot problems in which faulty alignment and muscle imbalance are important factors.

Low Back Pain

Conditions addressed in this category include lumbosacral strain and facet slipping. Lumbosacral strain may be postural in origin. Facet slipping is not primarily considered to be a postural problem, but is an associated problem of alignment and muscle imbalance that often affects this condition.

Lumbosacral Strain

Lumbosacral strain is the most common type of low back problem. The word *strain*, however, which denotes an injurious tension, does not cover the mechanical faults that are present. Essentially, there are two problems: **undue compression** on bony structures, especially in weight-bearing (i.e., standing or sitting), and **undue tension** on muscles and ligaments in weight-bearing and during movement.

A back may have good postural alignment in weight-bearing, but if the low back muscles are tight, they will be subjected to undue tension in a sudden or unguarded attempt to bend forward. Acute muscle strain may follow.

A back may have very faulty alignment, such as a hyperlordosis, without tightness of the low back muscles. Movement may not cause a strain, but standing for any length of time may give rise to pain. Compressive stress resulting from faulty alignment, if marked or constant, may evoke painful symptoms. This type of posture is more common among women than among men. The fault is often associated with weakness of the abdominal muscles. The onset of symptoms usually is gradual rather than acute, and symptoms often remain more or less chronic. Pain is less if the person is active than if the person is standing still, and it is relieved by recumbency or sitting.

In those with a combination of faulty alignment and muscle tightness, both position and movement may give rise to pain. This pain tends to be constant, although it may vary in intensity with changes of position. Stresses that would not be excessive under ordinary circumstances may give rise to pain; an apparently inconsequential act may cause an acute onset of pain.

Facet Slipping

The joints or facets that connect one vertebra with another may show abnormal deviations of position or movement. Conceivably, this may occur at the limit of range in flexion or in hyperextension. As a fault in hyperextension, it may result from a sudden movement in that direction or from a severe, persistent lumbar lordosis; the latter has been seen on radiographs (17, 18). The vertebral interspaces are diminished, and the hyperlordosis is so marked that the force of compression has caused the joint structures to give way and permit the overriding of one facet on another.

The suddenness of onset, acuteness of pain, and absence of previous neuromuscular symptoms suggest the facet may be the cause of some cases of acute low back pain. Usually, these incidents are only of momentary duration, and as such, they are not confirmed by radiography. The diagnosis is established, necessarily, based on subjective rather than objective findings.

The movement of the body and the direction of stress denote the direction of the facet's contribution. Most often, it occurs during flexion, and the patient reports being unable to stand up straight. When the stress results from hyperextension movements, reported symptoms may present as muscle spasm, or it may involve excessive motion at the facet. The faults of alignment and mobility that result in excessive joint motion are the basic factors to be considered in correcting or preventing faults of this type.

Anterior Pelvic Tilt

Kyphotic-Lordotic Posture

Four groups of muscles support the pelvis in anteroposterior alignment. The low back extensors pull upward on the pelvis posteriorly, the hamstrings pull downward posteriorly, the abdominal muscles pull upward anteriorly, and the hip flexors pull downward anteriorly. With good muscle balance, the pelvis is maintained in good alignment. With muscle imbalance, the pelvis tilts anteriorly or posteriorly. With anterior pelvic tilt, the low back arches forward into a position of hyperlordosis. In this position, there is undue compression posteriorly on the vertebrae and the articulating facets, and undue tension on the anterior longitudinal ligament in the lumbar area.

The muscle imbalances that are associated with an anterior tilt may include all or some of the following: weak anterior abdominal muscles, tight hip flexor muscles (chiefly the iliopsoas), tight low back muscles, and weak hip extensor muscles.

Figure 4-57 shows these muscle imbalances. Figure 4-57A shows a marked lordosis. The lordosis shown in Figure 4-57B would also be marked if the subject were to assume an erect posture. When all four muscle groups are involved, correction of the anterior pelvic tilt requires strengthening of the anterior

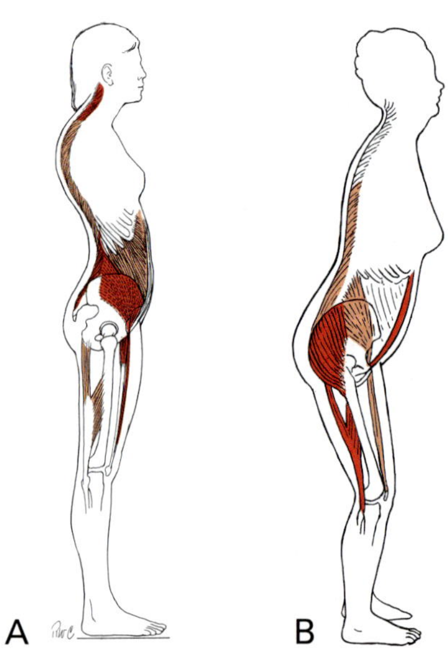

FIGURE 4-57: A, Kyphotic/lordotic posture; **B**, Hip flexion with the trunk inclined forward.

abdominal muscles and hip extensors and stretching of the tight low back and hip flexor muscles. Any one of the above may be a primary factor, but the tight low back and weak hip extensor muscles are least likely to be the primary cause.

Frank Ober stated, "It is well known that a lordotic spine may be a painful spine, but this, of course, is not true in every case" (19). Fahrni and Trueman have emphasized the common association of increased lumbar lordosis and low back pain (20). Some individuals with a hyperlordosis complain of low back pain, whereas others with a more severe lordosis may not complain of any pain. A lordosis may be habitual, but if the muscles of the back are flexible enough that position can be changed from time to time, symptoms may not develop. A back so tight that the hyperlordotic position is fixed, however, tends to be a painful back regardless of the body position.

The best index in regard to a painful low back is not the degree of lordosis or other mechanical defect visible on examination of alignment. Rather, the extent of muscle tightness maintains a fixed anteroposterior alignment, and the extent of muscle weakness allows the faulty position to occur and to persist.

Weak Anterior Abdominal Muscles

Weakness of anterior abdominal muscles allows the pelvis to tilt forward. These muscles are incapable of exerting the upward pull on the pelvis that is needed to help maintain a good alignment. As the pelvis tilts forward, the low back is drawn into a position of hyperlordosis. This may result in complaints of pain across the low back.

During the early stages, this pain may be described as fatigue; later, it may be described as an ache, which may or may not progress to being acutely painful. Pain is usually worse at the end of the day and is relieved by recumbency to such an extent that, after a night's rest, the individual may be free of symptoms.

Tight One-Joint Hip Flexors (Chiefly Iliopsoas)

Tight one-joint hip flexors cause an anterior tilt of the pelvis in standing. The low back goes into a hyperlordosis as the subject stands erect. Occasionally, a subject inclines forward from the hips, avoiding an erect position that would result in a marked lordosis.

This subject (Figure 4-58) had marked lightness in the hip flexors, which limited hip joint extension. The subject also had limitation of back extension. To push up from the table, movement had to take place at the knee joint. As an exercise, this movement would not be appropriate for this subject.

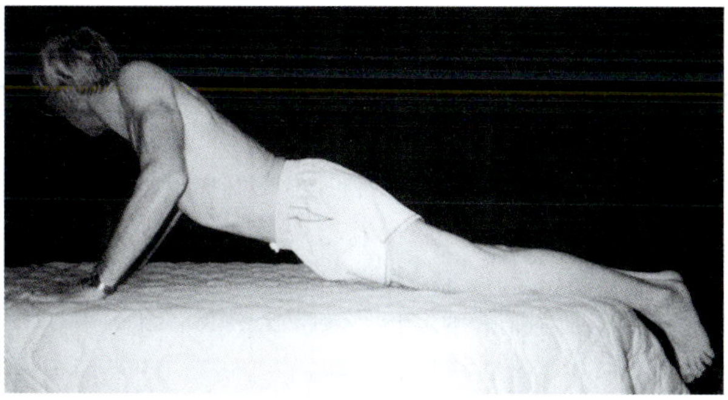

FIGURE 4-58: Marked lightness in hip flexors, which limits hip joint extension.

Tight Two-Joint Hip Flexors

Tightness causes a hyperlordosis in the kneeling position (Figure 4-59). When someone complains that only the kneeling position causes pain in the low back, it is important to examine for two-joint hip flexor shortness (see hip flexor length test). Discussion regarding tightness of two-joint hip flexors can be found in Chapter 7.

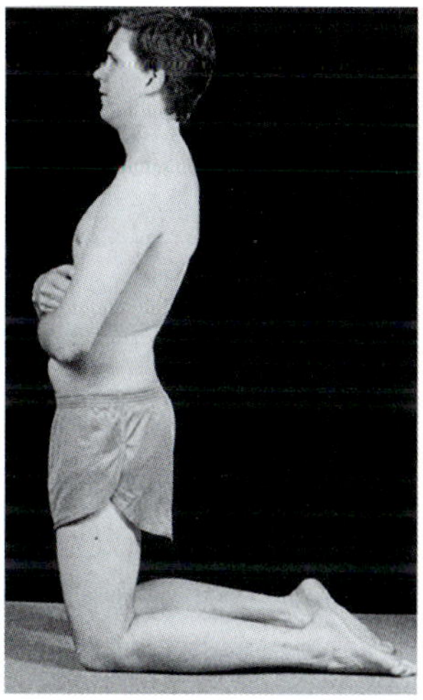

FIGURE 4-59: Tightness of the two-joint hip flexors causes lordosis in the kneeling position.

Tight Low Back Muscles

Tight low back muscles cause an anterior tilt of the pelvis and can hold the low back in a position of lordosis. These muscles cross over joints of the vertebral column, but they do not cross over another joint at which the muscles can give in to the tightness. Regardless of body position, the low back will remain in a degree of extension that corresponds to the degree of tightness of these muscles. In forward bending, the low back remains in an anterior curve (lordosis) and does not straighten.

For cases in which tightness of the low back muscles is a primary factor, pain may be chronic but often has an acute onset. Pain is increased by—and tends to have its onset in—movement rather than standing or sitting positions.

Posterior Pelvic Tilt

Two types of posture exhibit posterior pelvic tilt, hip joint extension, and weakness of the iliopsoas muscle: flat-back posture and sway-back posture. The **flat-back posture**, as the name implies, is a straight back in both the lumbar and the thoracic areas, except that some degree of flexion in the upper thoracic area accompanies the forward head position. The **sway-back posture** is one in which there is posterior displacement (i.e., swaying back) of the upper trunk and anterior displacement (i.e., swaying forward) of the pelvis. A long kyphosis extends into the upper lumbar region, and the lower lumber region is flattened. The posterolateral fibers of the external oblique are elongated. Both postures have been previously discussed in greater detail in Chapter 2 (Figure 2-17).

Excessive Flexion (Hyperflexion)

Excessive flexion (i.e., hyperflexion) of the lumbar spine is not uncommon. It is seen as a kyphosis of the low back in sitting, but it rarely appears as a kyphosis in standing (Chapter 7, Figure 7-32B). In most cases of excessive low back flexion, the back extensor muscles are not weak, but the hamstrings are often tight.

Some individuals with excessive flexion in sitting may stand in a hyperlordotic position. Certain exercises promote excessive flexion of the low back as they strengthen and tend to shorten the hip flexors. Notably, the curled-trunk sit-up from a knee-bent position creates a demand for complete curling of the trunk, including the lumbar spine, and exercises the iliopsoas in hip joint flexion almost to completion of the range of motion.

LIFTING

Much has been written about how to lift and the conditions in the workplace that need to be corrected. The weight of the object to be lifted, the frequency and duration of lifting, and the level from which an object must be lifted are all variables that affect the lifter. However, because of the many variables involved in lifting, there cannot be a single correct way to lift. Some points of agreement, however, relate to the lifter and to the object being lifted:

1. Stand as close to the object as possible.
2. Stand with the feet apart and one foot slightly in front of the other.
3. Bend the knees.
4. Begin the lift slowly, without jerking.
5. Avoid twisting in the forward-bent position.

There is also agreement that lifting from floor height is more challenging than lifting from above floor level. If this is not an option, an assistive device should be used, if possible.

Opinions differ regarding whether to squat (maintaining the lumbar spine's lordosis) or stoop (allowing the lordosis to decrease) when lifting. Squatting involves moderate knee bending; stooping involves bending forward from the hips or the waist (or both) and slight knee bending.

The squat lift has been advocated as a means of placing the load more on the legs and reducing the load on the back. The squat position can place the quadriceps at a mechanical disadvantage. Furthermore, individuals with knee dysfunction may have difficulty lifting from a squat position. Some may tolerate this position but lack the necessary quadriceps strength for a job that requires this type of lifting repetitively.

In many instances, the squat lift is not an option. In these situations, the mechanics of lifting are important, but the *body mechanics of the lifter* are even more important. The decision regarding how to lift must take into consideration the ability of the lifter. Of major concern are the mobility, stability, and strength of the lifter. In the general population, the mobility of the low back varies widely, ranging from excessive to limited. Excessive flexion and excessive extension both represent potential problems related to lifting. Low back stiffness presents the challenge of undue strain elsewhere (if not in the low back itself).

In forward bending, some people exhibit *excessive flexion* (i.e., hyperflexion) in which the lumbar spine curves convexly in a posterior direction and assumes a position of lumbar kyphosis. Although the low back muscles remain strong, the posterior ligaments are stretched and the back is vulnerable to strain when lifting. When this condition exists, the treatment of choice is the use of a support that prevents excessive flexion when lifting. The alternative is to attempt to hold the back in a neutral position by strong co-contraction of the back and abdominal muscles.

Some people exhibit excessive extension in which the lumbar spine curves convexly in an anterior direction and assumes a position of hyperlordosis. Referring to the work of Fahrni, Pope et al. stated that "as the lumbar lordosis increases, the plane of the L5 and S1 disks becomes more vertical and subject to greater shear and cyclic torsional forces, while nonlordotic segments are subject to compressive forces" (20, 21). Referring to the work of Farfan, Pope et al. also stated that "[B]ending and torsional loads are of particular interest, since the bulk of experimental findings suggest that these, and not the compressive loads, are the most damaging to the discs" (21, 22).

When it is advocated that the back be held in a neutral position (the normal lordosis) during lifting, the question arises about precisely what muscles must come into play to hold that exact position. If the low back muscles contract unopposed, the lordosis and anterior pelvic tilt increase, and the potential for overwork of the muscles and injury to the low back also increase and predispose the subject to injury. Chaffin, referring to work by Poulson and Jorgensen, and Tichauer et al., stated that "lumbar muscles (like all skeletal muscles) suffer ischemic pain when statically contracted for prolonged periods of moderate to heavy loading" (23–25).

The opposing force that prevents an increase in the curve must be provided by the anterior abdominal muscles (most specifically the lower abdominals). Tests and exercises specific to these muscles should be applied. Weakness of the lower abdominal muscles is a common finding among otherwise strong individuals that impacts lifting ability. Strengthening the abdominal muscles, however, can affect more than merely the stability of the back. Pope et al. found that "intradiscal pressure fell when abdominal pressure was increased. Thus in the standing posture intradiscal pressure is decreased coincident with increased abdominal muscular activity" (21).

Figure 4-60 shows a weight lifter who developed a backache and had to stop lifting until he built up strength in his abdominal muscles. He then returned to weight lifting and demonstrated the manner in which he would pick up a heavy object from the floor. For those with weakness of the abdominal muscles who continue weight lifting, it is advisable to use a support that provides abdominal and back stabilization.

Many individuals will exhibit a flat low back in the forward-bent position. Flexion of the lumbar spine is movement in the direction of straightening the low back, and a flat low back represents normal flexion. When the low back flexes to—*but not beyond*—the point of flattening, stability is afforded by this limitation of motion, just as there is stability at the knee joint if it does not hyperextend.

The potential for strain of the low back muscles and ligaments exists with hyperflexion, and the potential for ischemic pain with the lordotic back; disk problems may result from either (26). From the standpoint of prevention, one must assess how some exercises adversely affect the body in relation to the challenges of lifting. The knee-bent sit-up is conducive to excessive low back flexion as well as to overdevelopment and shortening of the hip flexors. For many adolescents, the legs are long in relation to the trunk and there is a tendency for tightness in the hamstrings. Forward bending to reach to or beyond the toes often results in excessive back flexion. Press-ups in the prone position that emphasize back extension to the point of fully extending the elbows encourage excessive range of motion in extension.

With an emphasis on maintaining or restoring good body mechanics and muscle balance, or on compensating for deficits by means such as necessary bracing, fewer problems of low back pain will occur from lifting.

FIGURE 4-60: Weight lifters with weak abdominal muscles should use a support that provides abdominal and back stabilization.

SECTION V
INTERVENTION

NECK

Tightness of Posterior Neck Muscles

The patient should use a pillow that permits a comfortable position of the neck. The patient should *not* sleep without a pillow, because the head will drop back in extension of the neck. On the other hand, the use of too high a pillow should be discouraged, because this can result in an increased forward-head position. A commercially available or home-made cervical pillow can provide the needed comfort and keep the neck in good position. The pillow should be flattened in the center to provide support both posteriorly and laterally.

Treatment may consist of heat, soft tissue mobilization, and stretching. Soft tissue mobilization should be gentle and relaxing at first, then progress to deeper kneading. Stretching of the tight muscles must be very gradual, using both active and assisted movements. The patient should actively try to stretch the posterior neck muscles by efforts to flatten the cervical spine (i.e., pulling the chin down and in). This action compares with the effort to flatten the lumbar spine in cases of hyperlordosis and may be done in the supine, sitting, or standing position but not in the prone position. *Exercises that hyperextend the cervical spine are contraindicated.*

Because the faulty head position is usually compensatory to a thoracic hyperkyphosis, which in turn may result from postural deviations of the low back or pelvis, treatment frequently must begin with correction of the associated deviations. Treatment for the neck may need to begin with exercises to strengthen the lower abdominal muscles and with use of a good abdominal support that permits the patient to assume a better upper back and chest position.

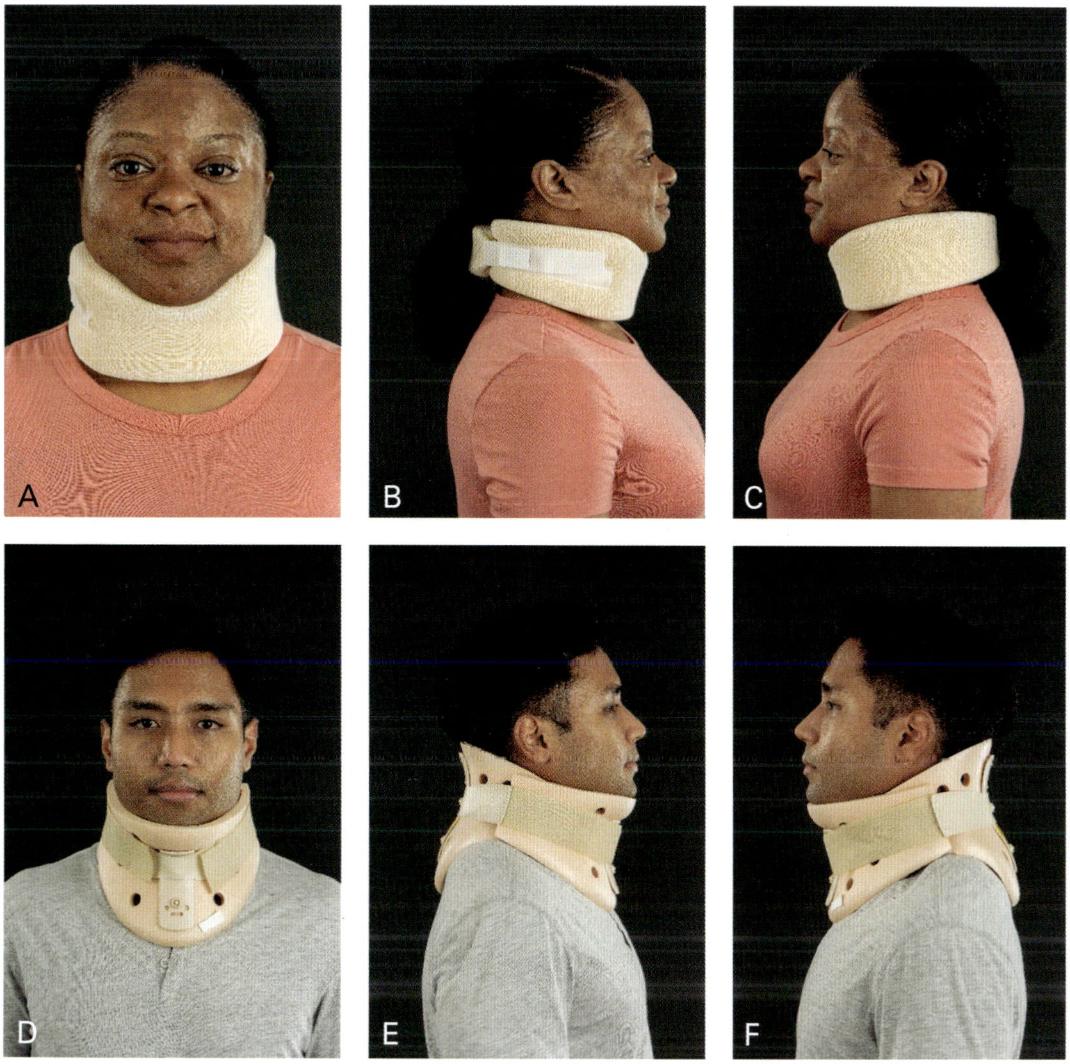

FIGURE 4-61: Soft cervical collar: **A**, Front; **B** and **C**, Side; **D**, Philly collar, Front; **E** and **F**, Philly collar, Side.

Upper Trapezius Strain

Application of heat followed by gentle passive and/or active stretching may be indicated. If a muscle spasm is present, that area should be prioritized. Soft tissue mobilization should be employed, starting with a gentle, kneading technique before increasing as tolerated.

Either an improvised collar or a sling (or both) may be used if the condition remains very painful and does not respond favorably to the soft tissue mobilization (Figure 4-61).

A simple collar can be made from a small towel folded lengthwise to the correct width. The towel is wrapped securely around the neck and then held in place by a strip of strong tape. The collar can be made more firm by placing a strip of cardboard inside the towel. The collar may be needed for only 2 or 3 days.

Cervical Nerve Root Compression

When the condition is acute, significant relief may be obtained by the use of moist heat to relieve protective muscle spasm, gentle soft tissue mobilization to help relax the muscles, and low-level manual or mechanical traction to relieve compression. The use of a collar is often necessary in the early stages. It can provide appropriate support to help immobilize the cervical spine, prevent hyperextension, and transmit the weight of the head to the shoulder girdle. When symptoms are subacute or chronic, treatment should also include exercises to correct any muscle imbalance and underlying postural faults. Conservative treatment may be adequate, or it may be an adjunct to surgical measures.

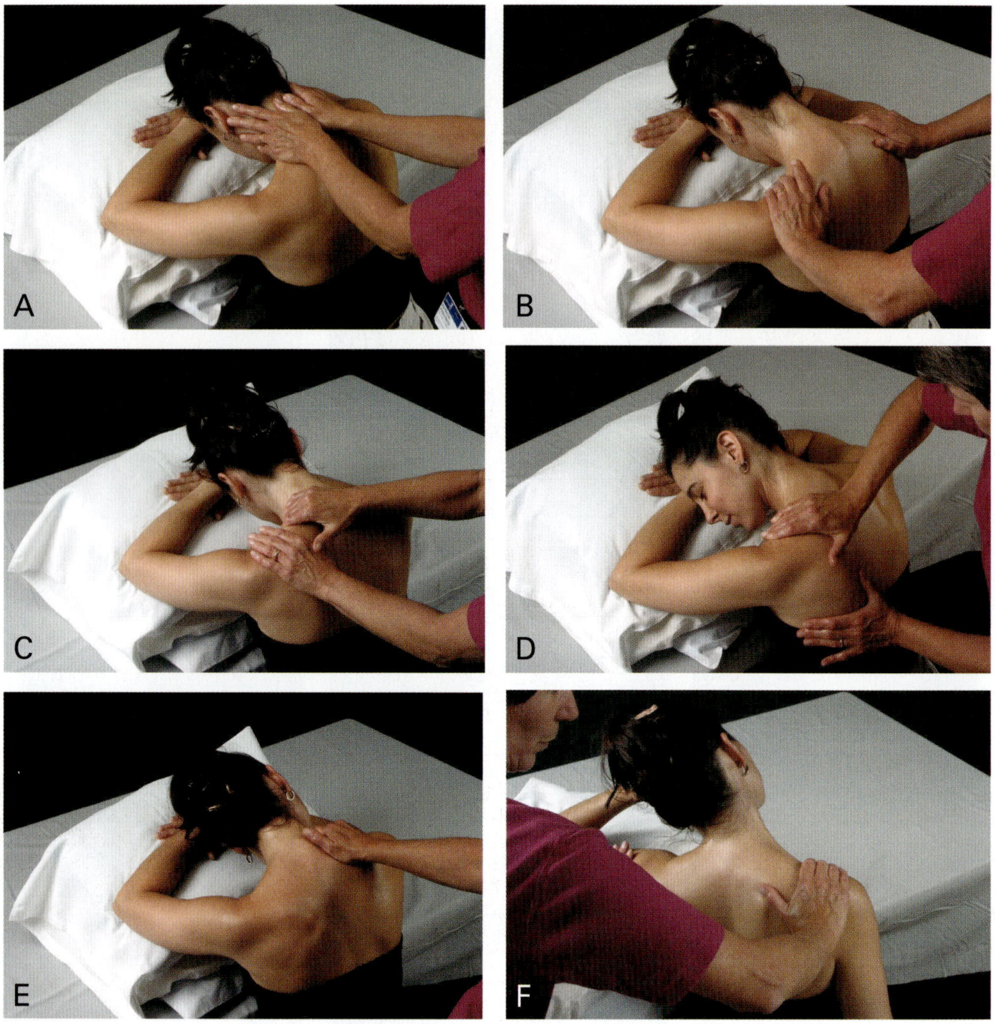

FIGURE 4-62: Massage of neck muscles.

Soft Tissue Mobilization of Neck Muscles

Soft tissue mobilization is an important modality in the treatment of painful neck conditions. The soothing effects of this technique can help relax tense muscles. Neck extensor muscles often become short from a faulty forward head posture. Soft tissue mobilization, along with appropriate exercises, can be used to help relax and stretch the tight muscles and restore normal range of motion (see Chapter 1).

Figure 4-62A–F shows the various positions for effectively applying soft tissue mobilization to help relieve neck tension and stiffness. The subject is seated on a stool beside the treatment table. Pillows on the table are adjusted to a level that is comfortable for the subject when leaning forward to rest the head on the hands. Soft tissue mobilization is applied to posterior and lateral neck muscles, chiefly the upper trapezius (see Flexion and Extension of the neck, and Rotation and Lateral Flexion).

Start the soft tissue mobilization at the occipital attachments of the upper trapezius (Figure 4-62A). Begin with a smooth, firm stroking massage (effleurage).

Continue the soft tissue mobilization along the trapezius to the attachments on the clavicles and scapulae (Figure 4-62B).

Repeat the soft tissue mobilization using a kneading technique (petrissage) to the left and right upper trapezius (Figure 4-62C).

With face turned left to put a mild stretch on the left trapezius, repeat the stroking and kneading soft tissue mobilization (Figure 4-62D).

With face turned right to stretch right trapezius, repeat the stroking and kneading soft tissue mobilization (Figure 4-62E).

Sit with left side toward table (Figure 4-62F). With elbow on the table, rest head on hand. With head tilted toward left, soft tissue mobilization can be performed for the right lateral neck muscles. Reverse the above positions to perform soft tissue mobilization of the left lateral neck muscles.

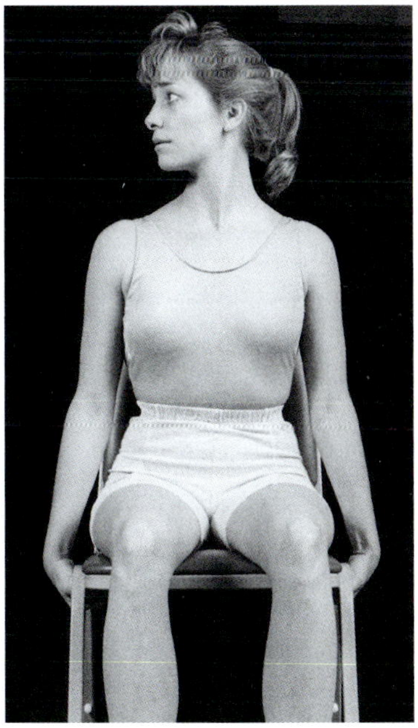

FIGURE 4-63: Exercise to stretch neck rotators.

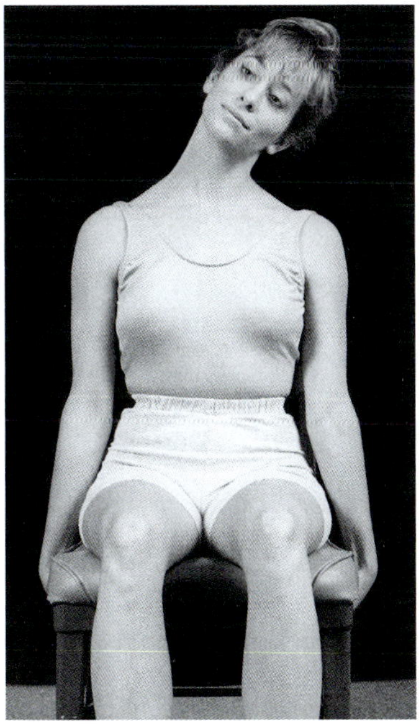

FIGURE 4-64: Exercise to stretch lateral neck flexors.

FIGURE 4-65: Alternate exercise to stretch lateral neck flexors.

Exercises to Stretch Neck Muscles

Stretch Neck Rotators: Sit on a chair with hands grasping seat to keep shoulders down and level (Figure 4-63). Without tilting head, turn toward each side (using opposite neck rotators).

Stretch Lateral Neck Flexors: Sit on a chair with shoulders back and hands grasping seat to hold shoulders down and level (Figure 4-64). Tilt head directly sideways to stretch opposite lateral neck flexors. Exercises for lateral neck stretching may be modified to tilt anterolaterally to stretch opposite posterolateral muscles.

Alternative Stretch Lateral Neck Flexors: Seated or standing, place right hand on left shoulder to hold it down (Figure 4-65). Give assistance with left hand, by grasping right forearm near elbow and pulling it downward. Tilt head directly sideways toward right to stretch left lateral neck flexors. Reverse hands and neck position to stretch right side.

Stretch Neck Extensors: Lie supine (or sit on a stool with back against the wall). With hands up beside head and low back flat (hooklying), press head back with chin down and in, using anterior neck flexors to straighten (i.e., flatten) the neck (Figure 4-66).

Stretch Upper Trapezius by Strengthening Latissimus Dorsi: Sit on a table with padded blocks beside hips (Figure 4-67). Keep body erect, with shoulders in good alignment. Press downward, straightening the elbows, and lift buttocks directly upward from table.

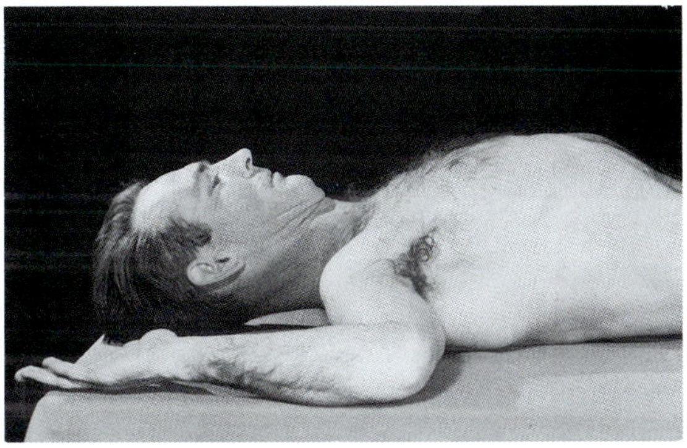

FIGURE 4-66: Exercise to stretch neck extensors.

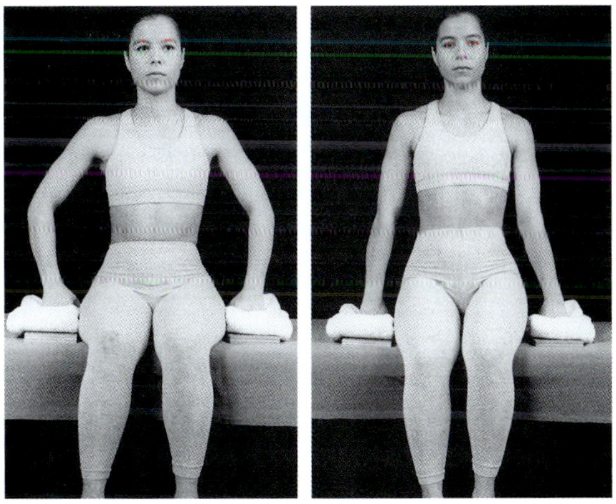

FIGURE 4-67: Exercise to stretch upper trapezius by strengthening latissimus dorsi.

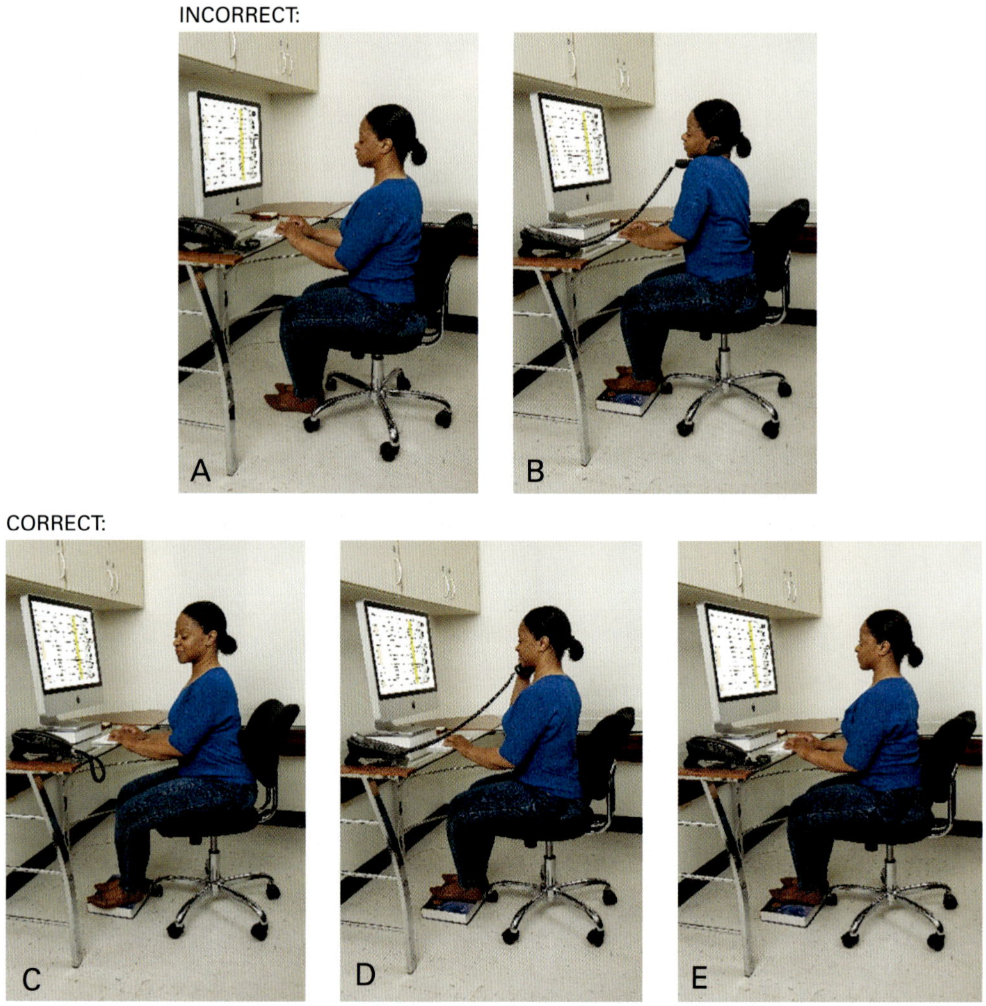

FIGURE 4-68: Computer ergonomics: **A**, Incorrect position while sitting at a computer; **B**, Incorrect position while using the telephone; **C**, Correct position while using speakerphone; **D**, Correct position while using the telephone; **E**, Correct position while using the computer.

Computer Ergonomics

Working at a computer (laptop or desktop) is frequently the cause of neck and upper back discomfort and headaches if basic ergonomic rules are ignored. The office set-up in Figure 4-68 was chosen as an example of how to correct alignment and relieve strain. The key to improving posture is a well-fitting chair that adjusts for proper height, arm support, and back support. Use of a speaker and proper use of the phone handset can relieve neck strain.

Incorrect: In Figure 4-68A, the elbows are positioned lower than the wrist and the knees are lower than the hips. In Part B, the individual's phone is placed on the shoulder, neck tilted left, rotated right and extended; shoulders elevated despite correcting posture for viewing the monitor and typing at the desk.

Correct: Monitor at or below eye level. Use of speaker phone (Part C) and corrected posture for use of the phone handset (Part D) provide a neutral position of the cervical spine to relieve strain. Chair has proper back support and is of appropriate height relative to that of the desk. Legs supported by footrest (book) under desk (Part E).

BACK

Weak Anterior Abdominal Muscles

Sleeping on a firm mattress allows the back to flatten. This change from the lordotic position may give relief and comfort to the patient. The back may be eased in sitting by resting against the back of the chair and avoiding the erect sitting position, which tends to arch the low back. Relief of pain can also come from the use of a proper support to help correct the faulty alignment and relieve the strain on the weak abdominal muscles. (The William's Flexion Brace and the Goldthwait Brace were designed to support the abdomen and correct the lordosis.)

When marked weakness exists, the patient should start an exercise program and continue using the support while working to build muscle strength. This advice is contrary to the often-repeated admonition that the muscles will get weaker if a support is used. Weakness from wearing a support will occur only if the patient does not exercise to build up the muscles. Use of the support helps to maintain alignment and to relieve stretch and strain of the weak muscles until they regain strength through exercise.

With back extensor or hip flexor tightness, it is necessary to treat these muscles to restore normal length before the abdominals can be expected to function optimally. See stretching exercises, under Interventions.

Tight One-Joint Hip Flexors (Chiefly Iliopsoas)

The severity of the lordosis depends directly on the extent of the hip flexor tightness. Stress on the low back in the lordotic position is often relieved by giving in to the tight hip flexors. In standing, this is accomplished by bending the knees slightly. In sitting, the hips are flexed, and the hip flexors are slack. Some people can sit for long periods of time without pain or discomfort but have pain when standing for brief periods. In such cases, these patients should be examined for hip flexor shortness. Lying on the back or on the side with the hips and knees flexed relaxes the pull of the tight hip flexors on the low back. Patients often seek these means to relieve pain in the back—and legitimately so during the acute stage. The problem, however, is that giving in to the tightness by flexing the hips in these various positions aggravates the underlying problem, permitting further adaptive shortening of the very muscles that are causing the problem. After the hip flexors are stretched through appropriate exercises, it is not necessary to flex the hips and knees to be comfortable when supine.

In supine, with the hips flexed enough to allow the back to flatten, the patient may be more comfortable on a firm mattress than on a soft one. On a soft mattress, the pelvis can sink down and tilt anteriorly, causing a hyperlordotic position of the low back.

Prone-lying may not be tolerated because the tight hip flexors hold the back in a lordotic position. The prone position, however, can be made comfortable by placing a firm pillow directly under the abdomen to help flatten the low back and allow slight flexion of the hips.

A back support can provide some relief from a painful back that is held in a lordosis by tight hip flexors, but it cannot help to stretch the tight hip flexors (see hip flexor stretching exercises and exercises to strengthen lower abdominal muscle).

Tight Low Back Muscles

Pain related to tight low back muscles may be relieved or made worse by recumbency. Relief of pain in recumbency results from removing part of the strain caused by the movement or muscle action in maintaining the upright position. Increase of pain in recumbency occurs if the body weight in this position imposes a strain on the tight back muscles. During bed rest in the acute stage, some relief is obtained by giving in to the back by putting a small roll under it. This roll should conform to the contour of and give support to the low back. The pressure against the low back offers some relief as well. When a back support in the form of a corset or a brace is indicated, it sometimes is advisable to use the support when recumbent as well as when weight-bearing.

In addition to the relief that comes from restriction of motion, pain is relieved by pressure from the support against the low back. Steel stays in back support should be bent in to conform to the back, and a pad may be added if it gives additional comfort.

Back Supports

The diminishing pain that may accompany immobilization—and the fear of repeating the movement that brought on an acute onset of pain—may have so relieved the patient that there will be reluctance to cooperate in treatment to restore movement. Recovery depends on cooperation, and this will not be obtained unless the patient understands the procedure.

Giving in to the patient's preferred position and supporting the back in that position for the relief of pain should not be the goal of treatment. Stretching the low back muscles to restore normal flexibility and building up the abdominal muscle strength are the ultimate goals.

Following are several forms of abdominal and back supports (Figures 4-69 and 4-70, respectively).

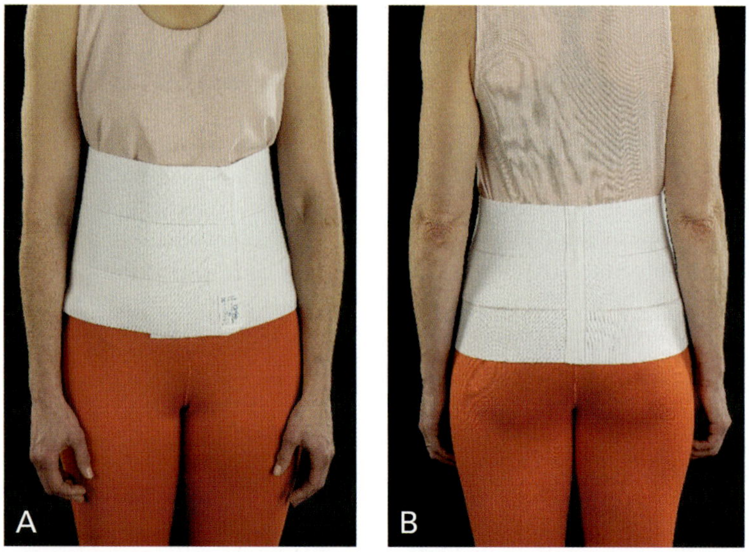

FIGURE 4-69: Abdominal binder. **A**, Front view. **B**, Back view.

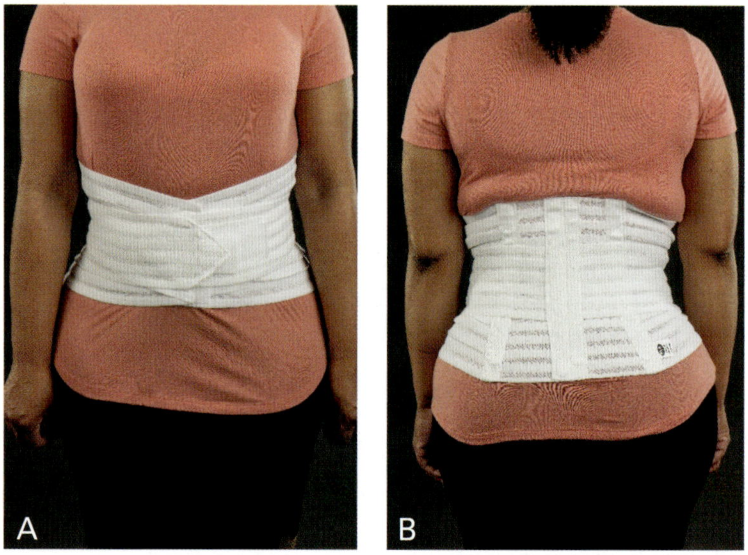

FIGURE 4-70: Back support: **A**, Front view; **B**, Back view.

Excessive Flexion (Hyperflexion)

With a painful low back and hypermobility in flexion, the treatment of choice is a support that prevents excessive range of motion. If the hamstrings are tight and exercises are done to stretch them, one should avoid forward bending, and one should wear the back support while doing passive or active straight-leg raising.

Treatment for Back Weakness

Weakness of the low back is seldom seen in common faulty postures. The low back muscles are an exception to the general rule that muscles that are elongated beyond normal range tend to show weakness. For a striking example, see Chapters 5 and 7 for photographs of a subject who has excessive flexion, but normal back strength.

Marked weakness of erector spinae muscles is not seen except in connection with neuromuscular problems. Even in cases of extensive involvement in some neuromuscular conditions, the back extensor muscles are often spared.

An individual should be able to raise the trunk backward from a prone position to the extent that range of motion of the back permits. If a person does not have the strength to perform this movement, and if there is no contraindication, then the back-extension exercises would be appropriate. Adequate strength in back muscles is important for maintenance of upright posture.

When there is severe weakness, a support is necessary. The type, rigidity, and length of the support depend upon the severity of the weakness. The entire trunk musculature is usually involved if erector spinae are weak. The collapse of the trunk takes place anteroposteriorly and laterally.

Exercise to build up strength in extensors must be gauged according to the patient's tolerance and response. Good alignment must be preserved in recumbent positions and supports must be provided in sitting or standing positions to help maintain any benefits from exercise.

The firmness of a mattress is an important consideration of posture in the lying position. A good sleeping position involves having the various parts of the body in approximately the same horizontal plane. Either sagging springs or too soft a mattress may result in poor body alignment.

Many people who have experienced postural back pain have found that pain has decreased or been eliminated by changing the firmness of a mattress. Others who have been accustomed to sleeping on a firm mattress have found that acute pain may be brought on by sleeping on a soft mattress. A pillow under the waist when sleeping prone or between the knees in a side-lying position can assist in maintaining a more ideal alignment leading to relief of stress on the back.

For some individuals, particularly those who have fixed structural faults of alignment such as exaggerated curves of the spine, a softer mattress may be necessary for sleeping comfort because the mattress will provide more support and comfort as it conforms to the curves.

An adult might be comfortable without a pillow when sleeping supine or prone, but may not be comfortable in a side-lying position. Use of too firm a pillow or use of more than one pillow may contribute to faulty head and shoulder positions. However, a person who is used to sleeping with the head high should not change abruptly to using a low pillow or none at all. A person who has a fixed postural fault of forward head and round upper back should not sleep without a pillow. It is important to have a pillow high enough to compensate for the hyperkyphosis and forward head position. Without a pillow or if the pillow is too low, the head will drop back in hyperextension of the neck.

REFERENCES

1. Gore DR, Sepic SB, Gardner GM. Roentgenographic findings of the cervical spine in asymptomatic people. Spine 1986; 11: 521–524.
2. Harrison DE, Harrison DD, Janik TJ, et al. Comparison of axial and flexural stresses in lordosis and three configurations of the cervical spine. Clin Biomech 2001; 16: 276–284.
3. Soderberg GL. Kinesiology-Application to Pathological Motion. 2nd ed. Baltimore: Williams & Wilkins, 1997.
4. Magee DJ. Orthopedic Physical Assessment. Philadelphia: Saunders, 2002.
5. Norkin C, White DJ. Measurement of Joint Motion: A Guide to Goniometry. Ed. Philadelphia: F.A. Davis, 1985.
6. Sforza C, Grassi G, Fragnito N, et al. Three dimensional analysis of active head and cervical spine range of motion: effect of age in healthy male subjects. Clin Biomech 2002; 17: 611–614.
7. Dvorak J, Antinnes J, Panjabi M, et al. Age and gender-related normal motion of the cervical spine. Spine 1992; 17: 393–398.
8. Palmer ML, Epler ME. Fundamentals of Musculoskeletal Assessment Techniques. 2nd ed. Lippincott: Philadelphia, 1998. pp. 221–224.
9. Clarkson HM. Musculoskeletal Assessment. 2nd ed. Baltimore: Lippincott Williams & Wilkins, 2000, p. 402.
10. Reese NB, Bandy WD. Joint Range of Motion and Muscle Length Testing. Philadelphia: W.B. Saunders, 2002. p. 408.
11. Sobotta-Figge. Atlas of Human Anatomy, Vol 1. Munich: Urban & Schwarzenberg, 1974.
12. Margolis S, Moses S, eds. Johns Hopkins Medical Handbook. New York: Rebus, 1992, pp. 128, 129.
13. Ayub E, Glasheen-Wray M, Kraus S. Head posture: a case study of the effects on the rest position of the mandible. J Orthop Sports Phys Ther 1984; 6: 179–183.
14. DeRosa C, Porterfield JA. A physical therapy model for the treatment of low back pain. Phys Ther 1992; 72(4): 263.
15. Kendall H, Kendall F. Study and Treatment of Muscle Imbalance in Cases of Low Back and Sciatic Pain. Baltimore: Privately Printed, 1936.
16. Kendall H, Kendall F, Boynton D. Posture and Pain. Baltimore: Williams & Wilkins, 1952, pp. 2–73, 156–159.
17. Perolat R, Kastler A, Nicot B, et al. Facet joint syndrome: from diagnosis to interventional management. Insights Imaging 2018; 9(5): 773–789.
18. Williams PC. Lesions of the lumbosacral spine. Part II. Chronic traumatic (postural) destruction of the lumbosacral intervertebral disc. J Bone Joint Surg 1937; 19: 690–703.
19. Ober FR. Relation of the fascia lata to conditions of the lower part of the back. JAMA 1937; 109(8): 554–555.
20. Fahrni WH, Trueman GE. Comparative radiological study of spines of a primitive population with North Americans and North Europeans. J Bone Joint Surg [Br] 1965; 47-B: 552.
21. Pope M, Wilder D, Booth J. The biomechanics of low back pain. In: White AA, Gordon SL, eds. Symposium on Idiopathic Lower Back Pain. St. Louis, MO: C.V. Mosby, 1982.
22. Farfan HF. Mechanical Disorders of the Low Back. Philadelphia: Lea & Febiger, 1973.
23. Chaffin DB. Occupational biomechanics of low back injury. In: White AA, Gordon SL, eds. Symposium on Idiopathic Low Back Pain. St. Louis, MO: C.V. Mosby, 1982.
24. Poulson E, Jorgensen K. Back muscle strength, lifting and stoop working positions. Appl Ergon 1971; 133–137.
25. Tichauer ER, Miller M, Nathan IM. Lordosimetry: a new technique for the measurement of postural response to materials handling. AM Ind Hyg Assoc J 1973; 34: 1–12.
26. Adams MA, Hutton WC. Prolapsed invertebral disc: a hyperflexion injury. In: Industrial Rehabilitation American Therapeutics 1989, pp. 1031–1038. Presented at the 8th annual meeting of the international society for the study of the lumbar spine. Paris, May 18, 1981.

TRUNK AND RESPIRATORY MUSCLES

5

OUTLINE

INTRODUCTION

It is interesting to note that muscles that act in unison for certain movements act in opposition to each other in support of good alignment. In the movement of raising the trunk up sideways, lateral trunk muscles laterally flex the trunk as the hip abductors stabilize the pelvis. In support of good alignment in standing, lateral trunk muscles are assisted by hip abductors on the opposite side.

Photographs and line drawings clearly illustrate the differences between the normal movements that occur during testing and the changes that occur when there is imbalance between muscles that normally act in unison. In many cases, because of the interaction of some abdominal muscles, group tests are more useful than tests of individual muscles.

The section on Muscles of Respiration rightfully belongs in this chapter. The lungs and diaphragm are located in the trunk. Faulty alignment of the skeletal structures and problems of muscle imbalance can adversely affect the respiratory system. The Respiratory Muscle Chart (Figure 5-33) lists the twenty-three muscles (each of which has a right and left component) plus the diaphragm as the muscles of respiration. Most of these muscles also have a function related to posture and muscle balance.

SECTION I
INNERVATION

Innervation to the trunk muscles does not include an intervening plexus between the spinal cord and the peripheral nerves like the cervical, brachial, lumbar, and sacral plexus. Abdominal muscles receive their innervation from the thoracic branches of the ventral divisions of the spinal nerves.

SPINAL NERVE AND MUSCLE CHART: TRUNK

Figure 5-1 shows the spinal nerves and muscles of the trunk.

SPINAL NERVE AND MUSCLE CHART
TRUNK AND LOWER EXTREMITY

Name | Date

	MUSCLE STRENGTH GRADE	MUSCLE	D. T1-12 L1-5, S1-3	V. T1, 2, 3, 4	V. T5, 6	V. T7, 8	V. T9, 10, 11, 12	V. Iliohypogastric T12 L1	V. Ilioinguinal T(12) L1	V. Lumb. Plex. T(12) L1, 2, 3, 4	P. Femoral L(1) 2, 3, 4	A. Obturator L(1) 2, 3, 4	P. Sup. Glut. L4, 5, S1	P. Inf. Glut. L5, S1, 2	V. Sac. Plex. L4, 5, S1, 2, 3	P. Sciatic L4, 5, S1, 2	A. Sciatic L4, 5, S1, 2, 3	P. C. Peroneal L4, 5, S1, 2	A. Tibial L4, 5, S1, 2, 3	L1	L2	L3	L4	L5	S1	S2	S3
		ERECTOR SPINAE	●																	1	2	3	4	5	1	2	3
Thoracic Nerves		SERRATUS POST SUP		●																							
		TRANS THORACIS		●	●	●																					
		INT INTERCOSTALS		●	●	●	●																				
		EXT INTERCOSTALS		●	●	●	●																				
		SUBCOSTALES		●	●	●	●																				
		LEVATOR COSTARUM		●	●	●	●																				
		OBLIQUUS EXT ABD			(●)	●	●																				
		RECTUS ABDOMINIS			●	●	●																				
		OBLIQUUS INT ABD				●	●	●	(●)											1							
		TRANSVERSUS ABD				●	●	●	(●)											1							
		SERRATUS POST INF					●																				
Lumb. Plexus		QUAD LUMBORUM								●										1	2	3					
		PSOAS MINOR								●										1	2						
		PSOAS MAJOR								●										1	2	3	4				

KEY
→
D Dorsal Primary Ramus
V Ventral Primary Ramus
A Anterior Division
P Posterior Division

SENSORY

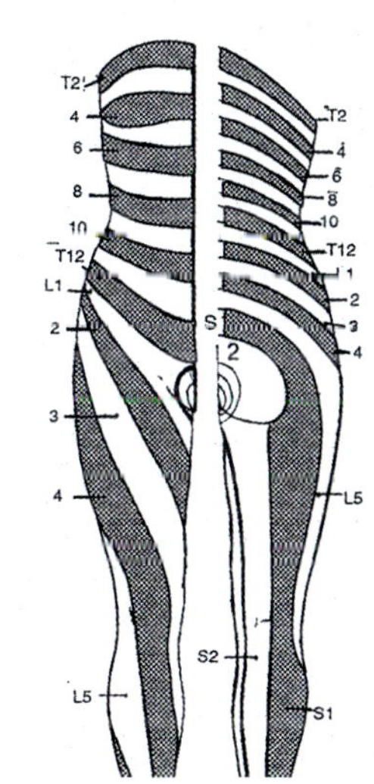

FIGURE 5-1: Spinal nerve and muscles of the trunk. © 2005 Florence P. Kendall.

TRUNK MUSCLES ATTACHED TO PELVIS

With the pelvis pivoting on the femora, the opposing groups of muscles act not only in straight anteroposterior opposition but also combine their pulls to tilt the pelvis forward or backward as well as laterally. There are four main groups of muscles in **anteroposterior opposition**:

- The erector spinae, quadratus lumborum, and other *posterior back muscles* attached to the posterosuperior part of the pelvis exert an upward pull posteriorly.
- The *anterior abdominals*, especially the rectus abdominis with its insertion on the symphysis pubis and the external oblique with its attachment on the anterior iliac crest, exert an upward pull anteriorly.
- The *gluteus maximus and hamstrings*, with attachments on the posterior ilium, sacrum, and ischium, exert a downward pull posteriorly.
- The *hip flexors*, including the rectus femoris, tensor fasciae latae, and sartorius, with attachments on the anterior, superior, and inferior spines of the ilium, and the iliopsoas, with attachment on the lumbar spine and inner surface of the ilium, exert a downward pull anteriorly.

The low back muscles act with the hip flexors (especially the psoas, with its direct pull from the lumbar spine to the femur) to tilt the pelvis downward and forward (i.e., anterior tilt). They are opposed in action by the combined pull of the anterior abdominals, pulling upward anteriorly, and the hamstrings and gluteus maximus, pulling downward posteriorly, to level the pelvis from a position of anterior tilt.

There are two main groups of pelvic muscles in **lateral opposition**:

- The leg abductors (mainly the gluteus minimus and medius), which arise from the lateral surface of the pelvis, pull downward on the pelvis when the leg is fixed as in standing.
- The lateral trunk muscles, attached to the lateral crest of the ilium, pull upward laterally on the pelvis.

Hip abductors on one side and lateral trunk muscles on the other side combine in action to tilt the pelvis laterally: Right abductors pull downward on the right side of the pelvis as left lateral trunk muscles pull upward on the left side, and vice versa. These actions are assisted by hip adductors on the same side as the lateral trunk muscles.

In combination, right hip abductors, left hip adductors, and left lateral trunk muscles oppose left hip abductors, right hip adductors, and right lateral trunk muscles.

SECTION II
ABDOMINAL MUSCLES AND TESTS

ABDOMINAL MUSCLES

Rectus Abdominis

Origin: Pubic crest and symphysis (Figure 5-2).

Insertion: Costal cartilages of the fifth through seventh ribs and xiphoid process of the sternum.

Direction of Fibers: Vertical.

Action: Flexes the vertebral column by approximating the thorax and pelvis anteriorly. With the pelvis fixed, the thorax will move toward the pelvis; with the thorax fixed, the pelvis will move toward the thorax.

Nerve: T5, 6, **T7–11,** T12, ventral rami.

Weakness: Weakness of this muscle results in decreased ability to flex the vertebral column. In the supine position, the ability to tilt the pelvis posteriorly or approximate the thorax toward the pelvis is decreased, making it difficult to raise the head and upper trunk. For anterior neck flexors to raise the head from a supine position, the anterior abdominal muscles (particularly the rectus abdominis) must fix the thorax. With marked weakness of the abdominal muscles, an individual may not be able to raise the head even though the neck flexors are strong. In the erect position, weakness of this muscle permits an anterior pelvic tilt and a hyperlordotic posture (i.e., increased anterior convexity of the lumbar spine).

Cross-Section of Rectus Abdominis and Its Sheath

Above the arcuate line (1), the aponeurosis of the internal oblique (b) divides. Its anterior lamina fuses with the aponeurosis of the external oblique (a) to form the ventral layer of the rectus sheath. Its posterior lamina fuses with the aponeurosis of the transversus abdominis (c) to form the dorsal layer of the rectus sheath.

Below the arcuate line (2), the aponeuroses of all three muscles fuse to form the ventral layer of the rectus sheath, and the transversalis fascia forms the dorsal layer.

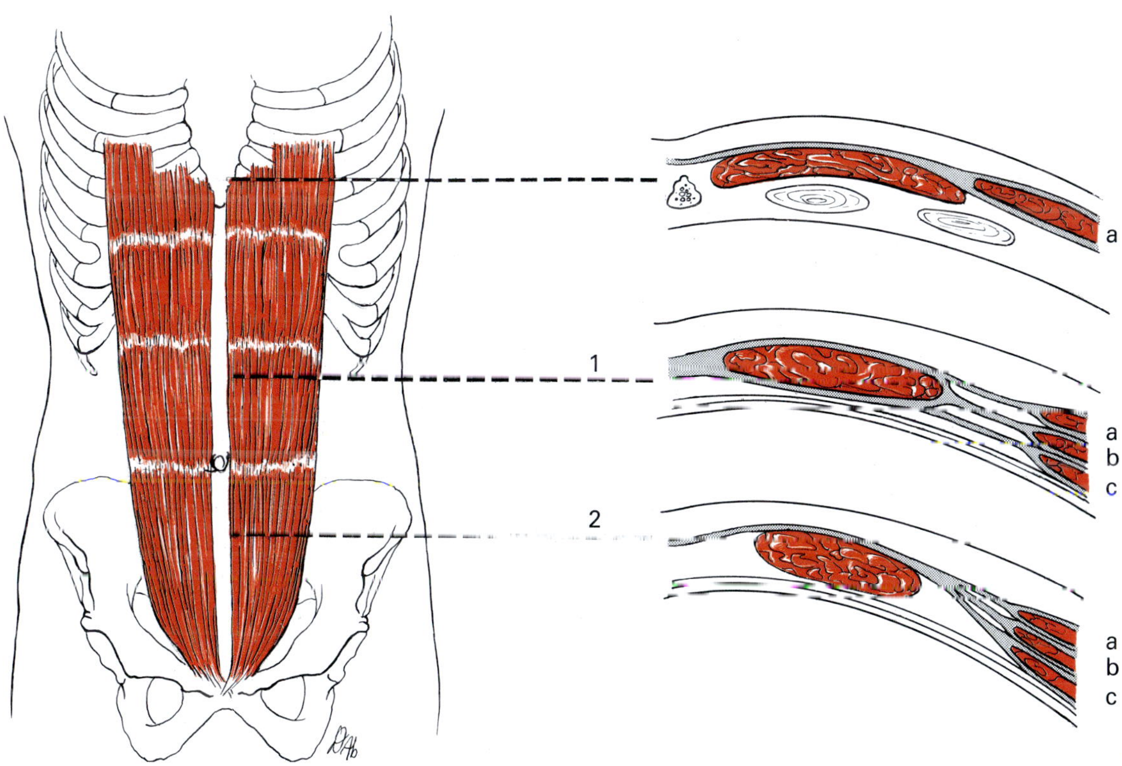

FIGURE 5-2: Anterior and cross-sectional views of the rectus abdominis muscles.

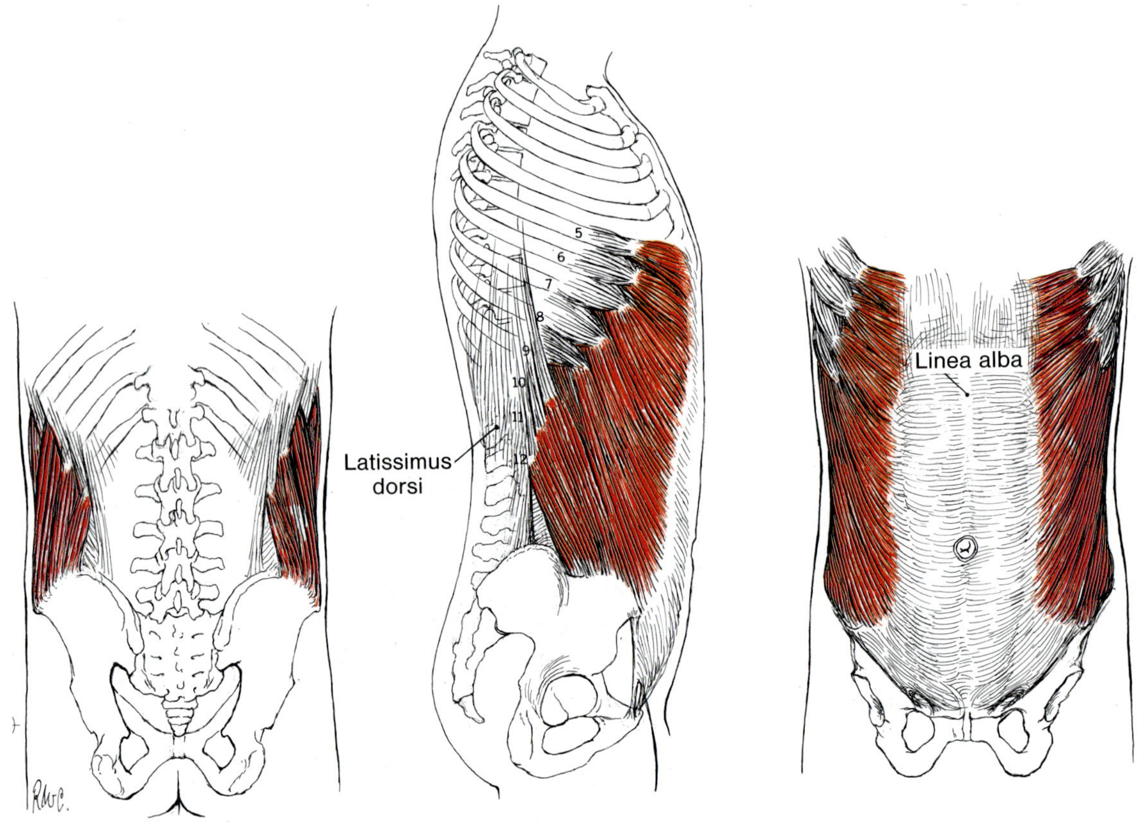

FIGURE 5-3: External oblique muscles.

External Oblique

The external oblique muscles are illustrated in Figure 5-3.

External Oblique, Anterior Fibers

Origin: External surfaces of ribs five through eight interdigitating with the serratus anterior.

Insertion: Into a broad, flat aponeurosis, terminating in the linea alba, which is a tendinous raphe that extends from the xiphoid.

Direction of Fibers: Obliquely downward and medially, with the uppermost fibers more medial.

Action: Acting *bilaterally*, the anterior fibers flex the vertebral column (approximating the thorax and pelvis anteriorly), support and compress the abdominal viscera, depress the thorax, and assist in respiration. Acting *unilaterally* with the anterior fibers of the internal oblique on the opposite side, the anterior fibers of the external oblique rotate the vertebral column, bringing the thorax forward (when the pelvis is fixed), or the pelvis backward (when the thorax is fixed). For example, with the pelvis fixed, the right external oblique rotates the thorax to the left, and the left external oblique rotates the thorax to the right.

Nerves to anterior and lateral fibers: (T5, 6), **T7–11,** T–12

External Oblique, Lateral Fibers

Origin: External surface of the ninth rib, interdigitating with the serratus anterior; external surfaces of the 10th through 12th ribs, interdigitating with the latissimus dorsi.

Insertion: As the inguinal ligament, into the anterosuperior spine and pubic tubercle and into the external lip of the anterior ½ of the iliac crest.

Direction of Fibers: Fibers extend obliquely downward and medially, but more downward than the anterior fibers.

Action: Acting *bilaterally,* the lateral fibers of the external oblique flex the vertebral column with a major influence on the lumbar spine, tilting the pelvis posteriorly (see also action in relation to posture). Acting *unilaterally* with the lateral fibers of the internal oblique on the same side, these fibers of the external oblique laterally flex the vertebral column, approximating the thorax and iliac crest. These external oblique fibers also act with the internal oblique on the opposite side to rotate the vertebral column. The external oblique, in its action on the thorax, is comparable to the sternocleidomastoid in its actions on the head and cervical spine.

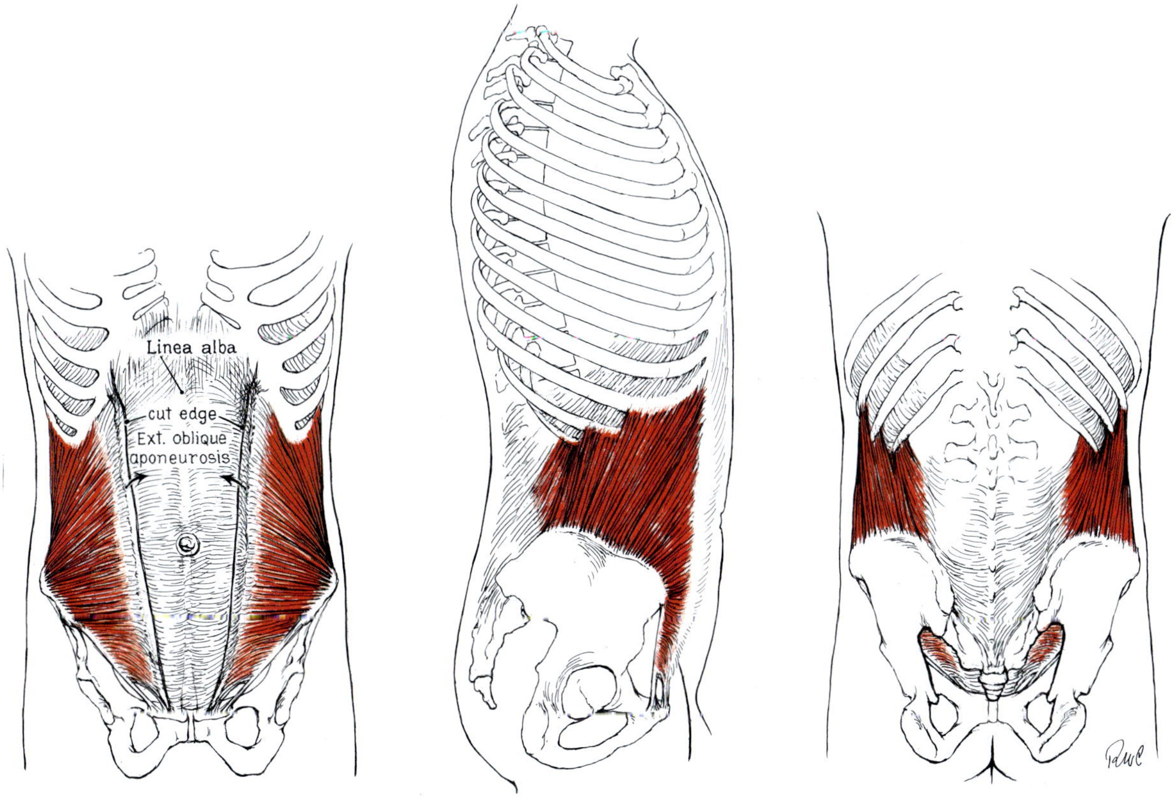

FIGURE 5-4: Internal oblique muscles.

Internal Oblique

The internal oblique muscles are illustrated in Figure 5-4.

Internal Oblique, Lower Anterior Fibers

Origin: Lateral ⅔ of inguinal ligament and short attachment on iliac crest near the anterosuperior iliac spine.

Insertion: With the transverse abdominis into crest of the pubis, medial part of the pectineal line and into the linea alba by means of an aponeurosis.

Direction of Fibers: Transversely across the lower abdomen.

Action: The lower anterior fibers compress and support the lower abdominal viscera in conjunction with the transversus abdominis.

Internal Oblique, Upper Anterior Fibers

Origin: Anterior ⅓ of intermediate line of the iliac crest.

Insertion: Linea alba by means of an aponeurosis.

Direction of Fibers: Obliquely medially and upward.

Action: Acting *bilaterally,* the upper anterior fibers flex the vertebral column (approximating the thorax and pelvis anteriorly), support and compress the abdominal viscera, depress the thorax, and assist in respiration. Acting *unilaterally* in conjunction with the anterior fibers of the external oblique on the opposite side, the upper anterior fibers of the internal oblique rotate the vertebral column, bringing the thorax backward (when the pelvis is fixed), or the pelvis forward (when the thorax is fixed). For example, the right internal oblique rotates the thorax to the right, and the left internal oblique rotates the thorax to the left on a fixed pelvis.

Internal Oblique, Lateral Fibers

Origin: Middle ⅓ of intermediate line of the iliac crest and the thoracolumbar fascia.

Insertion: Inferior borders of the 10th through 12th ribs and the linea alba by means of an aponeurosis.

Direction of Fibers: Obliquely upward and medially, but more upward than the anterior fibers.

Action: Acting *bilaterally,* the lateral fibers flex the vertebral column (approximating the thorax and pelvis anteriorly) and depress the thorax. Acting *unilaterally* with the lateral fibers of the external oblique on the same side, these fibers of the internal oblique laterally flex the vertebral column, approximating the thorax and pelvis. These fibers also act with the external oblique on the opposite side to rotate the vertebral column.

Nerves to Anterior and Lateral Fibers: T7, 8, **T9–12, L1** (iliohypogastric and ilioinguinal), ventral rami.

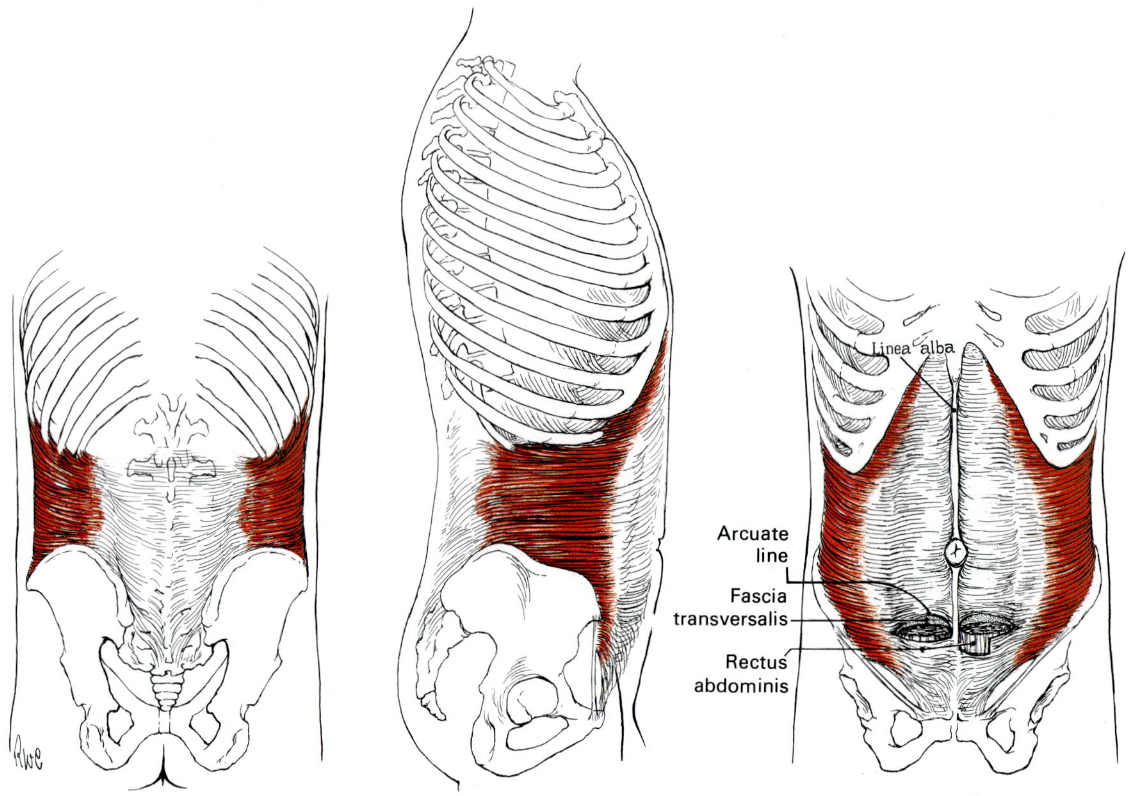

FIGURE 5-5: Transversus abdominis muscles.

Transversus Abdominis

Origin: Inner surfaces of cartilages of the lower six ribs, interdigitating with the diaphragm; thoracolumbar fascia; anterior ¾ of internal lip of the iliac crest; and lateral ⅓ of the inguinal ligament (Figure 5-5).

Insertion: Linea alba by means of a broad aponeurosis, pubic crest, and pecten pubis.

Direction of Fibers: Transverse (horizontal).

Action: Acts likes a girdle to flatten the abdominal wall and compress the abdominal viscera; upper portion helps to decrease the infrasternal angle of the ribs, as in expiration. This muscle has no action in lateral trunk flexion, except that it acts to compress the viscera and to stabilize the linea alba, permitting better action by the anterolateral trunk muscles.

Nerve: T7–12, L1 iliohypogastric and ilioinguinal, ventral divisions.

Weakness: Permits a bulging of the anterior abdominal wall, which indirectly tends to increase hyperextension of the trunk in the prone position, a lateral bulge tends to occur if the transversus abdominis is weak (Figure 5-6).

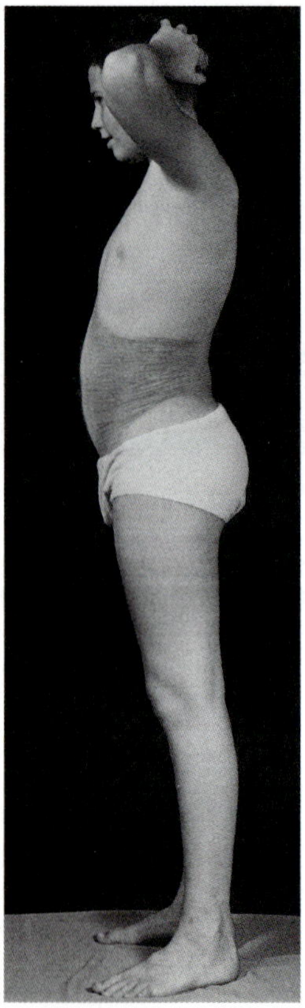

FIGURE 5-6: Bulging of the anterior abdominal wall.

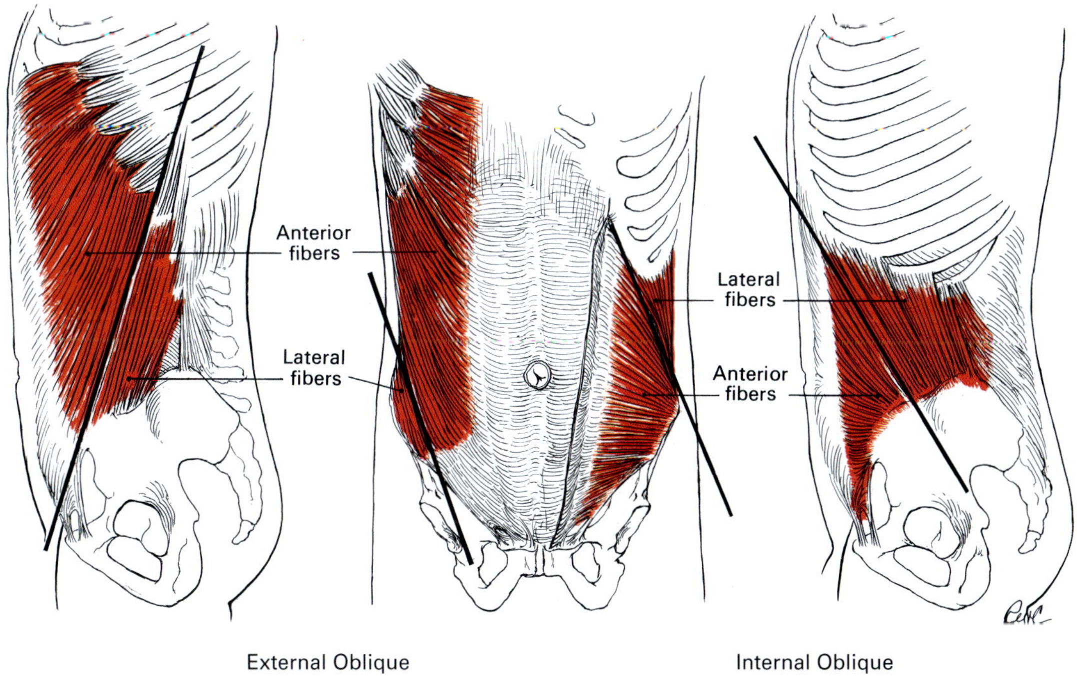

FIGURE 5-7: External and internal oblique muscles.

External and Internal Obliques: Weakness and Shortness

The external and internal oblique muscles are illustrated in Figure 5-7.

Weakness: Moderate or marked weakness of both the external and the internal obliques decreases both respiratory efficiency and support of the abdominal viscera.

Bilateral weakness of external obliques decreases the ability to flex the vertebral column and tilt the pelvis posteriorly. In standing, it results in either anterior pelvic tilt or anterior deviation of the pelvis in relation to the thorax and lower extremities.

Bilateral weakness of the internal obliques decreases the ability to flex the vertebral column.

Cross-sectional weakness of the external oblique on one side and of the internal oblique on the other allows separation of the costal margin from the opposite iliac crest, resulting in rotation and lateral deviation of the vertebral column. With weakness of the right external and left internal obliques (as seen in a right thoracic, left lumbar scoliosis), there is a separation of the right costal margin from the left iliac crest. The thorax deviates toward the right and rotates posteriorly on the right. With weakness of the left external and right internal obliques, the reverse occurs.

Unilateral weakness of lateral fibers of the external and internal obliques on the same side allows separation of the thorax and the iliac crest laterally, resulting in a C-curve that is convex toward the side of weakness. Weakness of the lateral fibers of the left external and internal obliques gives rise to a left C-curve.

Shortness: *Bilateral shortness* of anterior fibers of the external and internal oblique muscles causes the thorax to be depressed anteriorly, contributing to flexion of the vertebral column. In standing, this is seen as a tendency toward kyphosis and depressed chest. In a kyphotic-lordotic posture, the lateral portions of the internal oblique are shortened, and the lateral portions of the external oblique are elongated. These same findings occur in a sway-back posture with anterior deviation of the pelvis and posterior deviation of the thorax.

Cross-sectional shortness of the external oblique on one side and of the internal oblique on the other causes rotation and lateral deviation of the vertebral column. Shortness of the left external and right internal obliques (as seen in advanced cases of right thoracic, left lumbar scoliosis) causes rotation of the thorax forward on the left.

Unilateral shortness of lateral fibers of the external and internal obliques on same side causes approximation of the iliac crest and thorax laterally, resulting in a C-curve that is convex toward the opposite side. Shortness of the lateral fibers of the right internal and external obliques may be seen in a left C-curve.

Divisions of Abdominal Muscles

The divisions of abdominal muscles are illustrated in Figure 5-8.

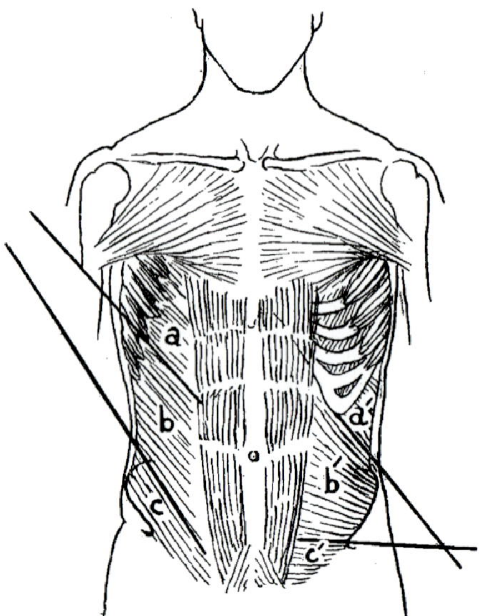

Anterior view of abdomen showing division of right external oblique into a, b, and c portions and left internal oblique into a′, b′, and c′ portions.

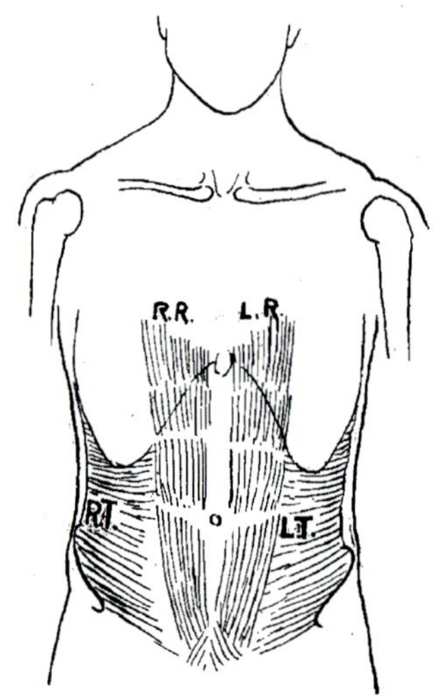

Anterior view showing left and right portions (L. R. and R. R.) of rectus abdominis, and left and right portions (L. T. and R. T.) of transversus abdominis.

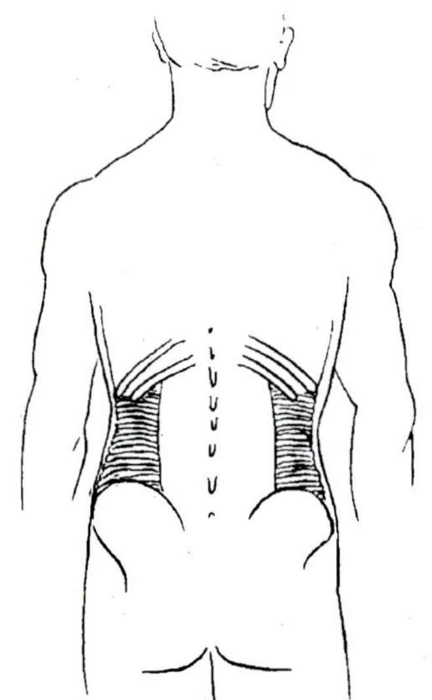

Posterior view showing posterior fibers of transversus abdominis.

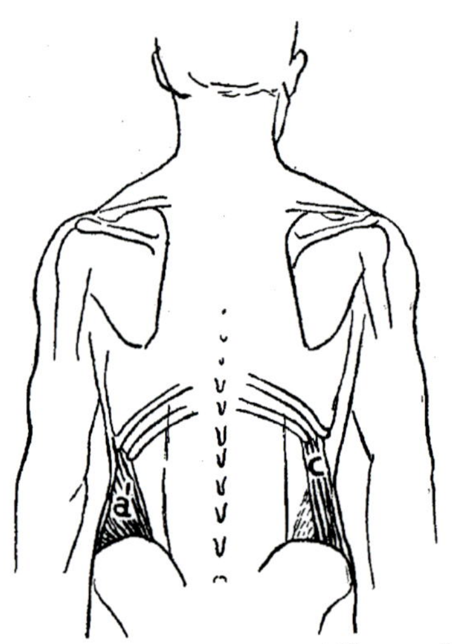

Posterior view showing posterior fibers of left internal oblique, a, and right external oblique c.

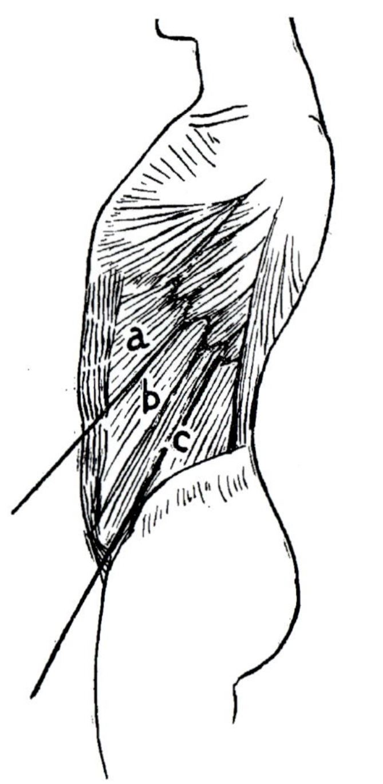

Lateral view of left external oblique showing a, b, and c portions.

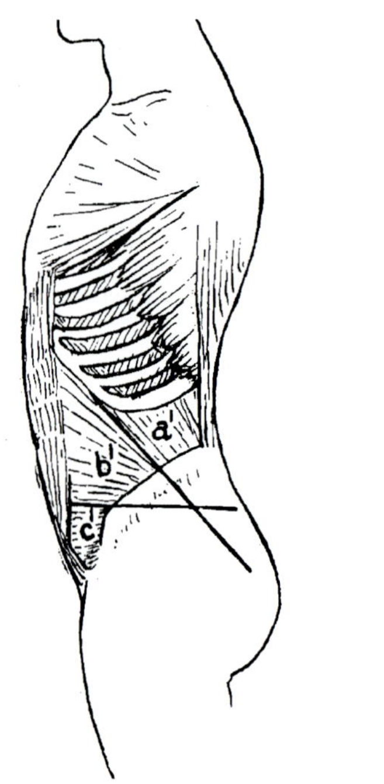

Lateral view of left internal oblique showing a′, b′, and c′ portions.

FIGURE 5-8: Divisions of the abdominal muscles.

ABDOMINAL MUSCLE TESTS

Differentiating Actions of the Upper and Lower Abdominals

The terms *upper* and *lower* differentiate two important strength tests for the abdominal muscles. More often than not, there is a difference between the grades of strength attributed to the upper abdominals compared to those attributed to the lower abdominals.

If the same muscles entered into both tests and the difference in strength resulted from a difference in the difficulty of the tests, there should be a fairly constant ratio between the two measurements.

In order of frequency, the following combinations of strength and weakness are found:

1. Upper strong and lower weak.
2. Upper and lower both weak.
3. Upper and lower both strong.
4. Lower strong and upper weak.

The difference in strength may be remarkable. A subject who can perform as many as 50 or more curled-trunk sit-ups may grade less than fair on the leg-lowering test. This same subject can increase the strength of the lower abdominals to normal by doing exercises specifically localized to the external oblique.

Because the oblique abdominal muscles are essentially fan-shaped, one part of a muscle may function in a somewhat different role than another part of the same muscle. Knowledge of the attachments and the line of pull of the fibers, along with clinical observations of patients with marked weakness and those with good strength, lead to conclusions regarding the action of muscles or segments of abdominal muscles.

The rectus abdominis enters into both tests. There is a distinct difference, however, between action of the internal oblique and that of the external oblique as exhibited by the two tests.

When analyzing which muscles or parts of muscles enter into the various tests, it is necessary to observe the movements that take place and the line of pull of the muscles that enter into the movement.

As trunk flexion is initiated by slowly raising the head and shoulders from a supine position, the chest is depressed and the thorax is pulled toward the pelvis. Simultaneously, the pelvis tilts posteriorly. These movements obviously result from action of the rectus abdominis muscle (Figure 5-9).

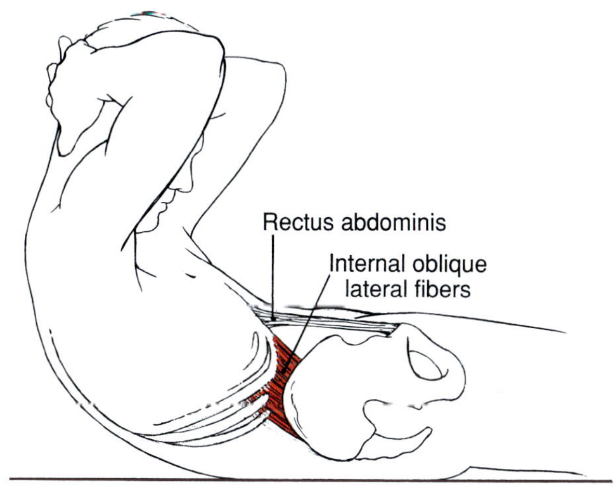

FIGURE 5-9: Movements that result from action of the rectus abdominus muscles.

Along with depression of the chest, the ribs flare outward and the infrasternal angle is increased. These movements are compatible with the action of the internal oblique.

No test movement can cause an approximation of parts to which the lower transverse fibers of the internal oblique are attached, because these fibers extend across the lower abdomen from ilium to ilium like the lower fibers of the transversus abdominis. In posterior pelvic tilt and in trunk-raising movements, however, this part of the internal oblique will act with the transversus to compress the lower abdomen.

As the trunk curl is completed and the movement enters the hip flexion phase, one will observe that the rib cage, which had flared outward, is now being pulled inward and that the infrasternal angle decreases. The anterior fibers of the external oblique now come into play.

If the internal oblique and rectus are strong (as indicated by the ability to perform numerous curled-trunk sit-ups), and if part of the external oblique is also brought into action during this movement, where is the weakness that accounts for the marked difference in the test results of the upper and lower abdominals?

The posterolateral fibers of the external oblique are elongated as the thoracic spine flexes during the trunk curl (Figure 5-10). These fibers of the external oblique help to draw the posterior rib cage toward the anterior iliac crest and in so doing, they tend to extend, not to flex, the thoracic spine.

The action of the external oblique may also be observed in cases of scoliosis with muscle imbalance between the right and left external oblique muscles. It is not uncommon to observe that flexion of the spine may begin with a rather symmetrical pull; however, as the effort is made to raise the trunk in flexion toward the thighs, there will be forward rotation of the thorax with extension of the thoracic spine on the side of the stronger external oblique.

With the trunk held in flexion during the hip flexion phase of trunk raising, the rectus abdominis, anterior fibers of the external oblique, and upper anterior and lateral fibers of the internal oblique shorten. In contrast, the posterolateral fibers of the external oblique elongate. This helps explain why an individual may be able to perform many sits-ups but fail the leg-lowering test.

Figure 5-11 shows a subject with strong external oblique muscles doing a sit-up with the trunk held straight and the lower abdomen pulled upward and inward. This is in sharp contrast to a curled-trunk sit-up, as shown in the illustration on left, or to an arched-back sit-up, as shown later in the abdominal muscle weakness trunk raising photo, Figure 5-35.

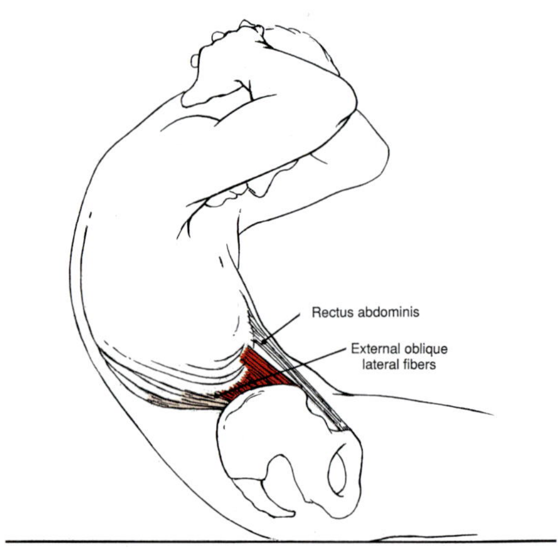

FIGURE 5-10: Posterolateral fibers of the external oblique are elongated as the thoracic spine flexes during the trunk curl.

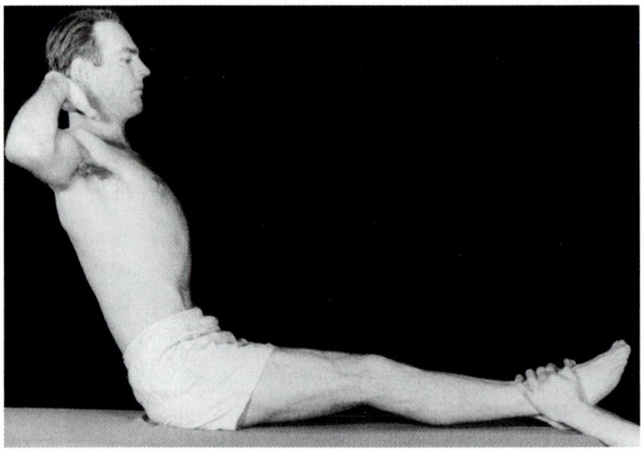

FIGURE 5-11: Subject with strong external oblique muscles doing a sit-up with the trunk held straight and the lower abdomen pulled upward and inward.

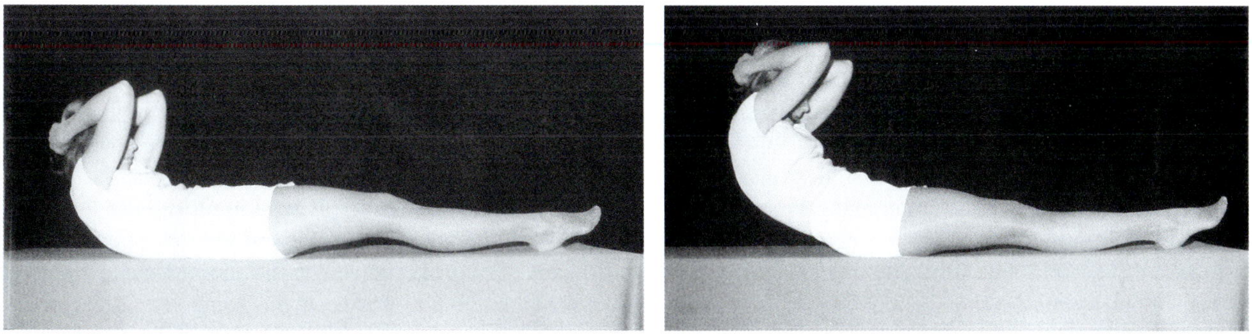

FIGURE 5-12: Curled trunk sit-up with legs extended.

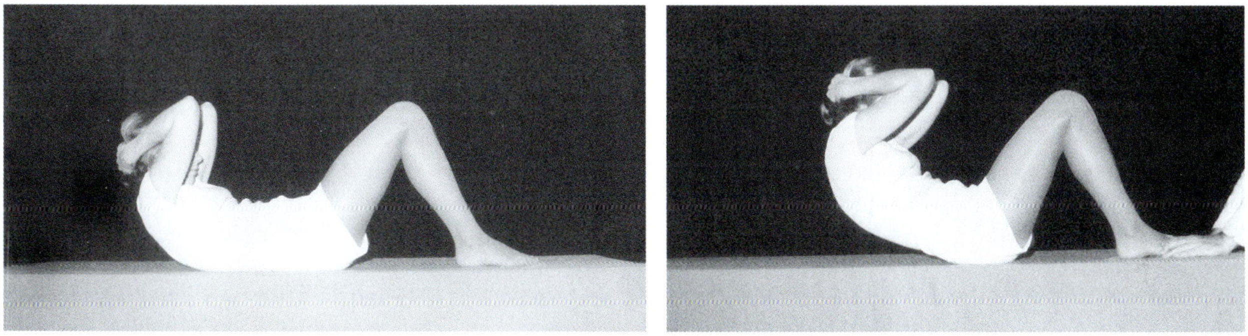

FIGURE 5-13: A curled-trunk sit-up with the hips and knees flexed (i.e., knee-bent sit-up).

Analysis of Movements and Muscle Actions During Curled-Trunk Sit-Ups

Definitions and Descriptions of Trunk Movements

Trunk curl refers to flexion of the spine only (i.e., the upper back curves convexly backward and the low back straightens). When abdominal muscles are strong and the hip flexor muscles very weak, only the trunk curl can be completed when attempting to do a sit-up.

Sitting position is one in which the trunk is upright and the hips are flexed. To **sit down** means to move from an upright to a sitting position by flexing at the hip joints; however, this movement may not require action of the hip flexor muscles. To **sit up** means to move from a reclining to a sitting position by flexing at the hip joints. When done unassisted, this movement can be performed only by hip flexor muscles. Whether alone or in combination, the word *sit* should be used only in connection with movement that involves hip joint flexion.

The **sit-up exercise**, therefore, is the movement of coming from a supine to a sitting position by flexing the hip joints, and it is performed by the hip flexors. It may be combined correctly with trunk and leg positions as illustrated above, or incorrectly as in Figure 5-45 later in this chapter.

A curled-trunk sit-up with the knees extended consists of flexion of the spine (i.e., trunk curl) performed by abdominal muscles followed by flexion of the hip joints (i.e., sit-up) performed by hip flexors (1–3) (Figure 5-12).

A curled-trunk sit-up with the hips and knees flexed (i.e., knee-bent sit-up) starts from a position of hip flexion (i.e., flexion of the thigh toward the pelvis) and consists of flexion of the spine (i.e., trunk curl) performed by the abdominal muscles followed by further flexion of the hip joints (by flexion of the pelvis toward the thigh) performed by the hip flexors (1, 2) (Figure 5-13).

5 TRUNK AND RESPIRATORY MUSCLES

Figures 5-14 and 5-15 show the various stages of movement of the spine and hip joints that occur during a curled-trunk sit-up. Next, the illustrations are repeated with accompanying text that describes the associated muscle actions.

Outlines of the basic features have been made from photographs. Drawings of the femur and pelvis and a dotted line representing part of the vertebral column have been added. The solid line from the anterosuperior spine to the symphysis pubis is the line of reference for the pelvis. A dotted line parallel to the solid line has been drawn through the pelvis to the hip joint, and this line continues as a reference line through the femur to indicate the angle of the hip joint (i.e., the angle of flexion) at the various stages of movement.

Specific degrees, based on the average normal ranges of motion presented here and in Chapter 2, help to explain the movements that occur. Because of individual variations with respect to ranges of motion of the spine and hip joints, the manner in which subjects perform these movements will also vary.

For this particular analysis, the abdominal and erector spinae muscles, as well as the hip flexor and extensor muscles, are assumed to be normal in length and strength. The spine and hip joints are also assumed to permit normal range of motion.

Normal hip joint extension is given as 10°. From the viewpoint of stability in standing, it is desirable to have a few degrees of extension; however, it is not desirable to have more than a few degrees. In the upright or supine position with the hips and knees extended, a posterior pelvic tilt of 10° results in 10° of hip joint extension. This occurs because the pelvis is tilted posteriorly toward the back of the thigh instead of the thigh being moved posteriorly toward the pelvis. Flattening of the lumbar spine accompanies the posterior pelvic tilt. Flexion to the point of straightening or flattening the low back is considered to be normal flexion on the basis that it is an acceptable and desirable range of motion.

With the knee flexed, the hip joint can flex approximately 125° from the zero position to an acute angle of approximately 55° between the femur and the pelvis. With the knee extended (as in the straight-leg-raising test for hamstring length), the lower extremity can be raised approximately 80° from the table. The equivalent of this is a trunk-raising movement, with the knees extended, in which the pelvis is flexed toward the thighs through a range of approximately 80° from the table.

For convenience in measuring joint motion, the trend is to use the anatomical position as zero. Thus, the straight position of the hip joint is considered to be the zero position. However, it is necessary to adhere to geometric terms when describing angles and the number of degrees in angles.

In Figures 5-14 and 5-15, the right column under *Hip Joints* refers to the angle of flexion anteriorly between the reference line through the pelvis and the line through the femur, and degrees are expressed in geometric terms. Changes in the angle of flexion represent corresponding changes in the length of hip flexors. The left column under *Hip Joints* lists the number of degrees from the anatomical position through which the hip joint has moved, first in extension and then in flexion.

MOVEMENTS DURING CURLED-TRUNK SIT-UPS WITH KNEES EXTENDED

Figure 5-14 illustrates movements made during curled-trunk sit-ups with knees extended.

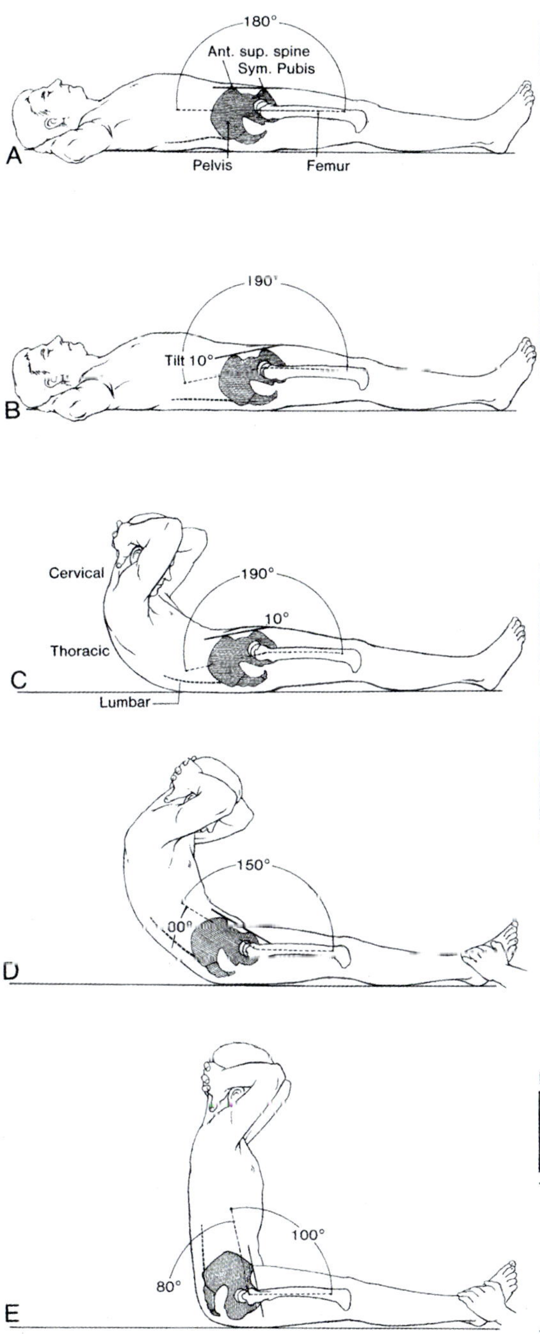

SPINE			PELVIS	HIP JOINTS	
Cervical (Neck)	Thoracic (Upper Back)	Lumbar (Low Back)		Anatomical degrees	Geometric degrees
A					
Position: Supine, hands clasped behind head.					
Zero (Normal anterior curve)	Zero (Normal posterior curve)	Zero (Normal anterior curve)	Neutral position	Zero	180°
B					
Movement: Posterior pelvic tilt, lumbar spine flexion, and hip joint extension.					
Zero	Zero	Flexed (straight)	10° posterior tilt	10° extension	190°
C					
Movement: Cervical and thoracic spine flexion. Figure C represents completion of spine flexion phase and beginning of hip flexion phase.					
Flexed (straight)	Flexed (curled)	Flexed (straight)	10° posterior tilt	10° extension	190°
D					
Movement: Hip joint flexion. The hip joint has moved from a 190° angle of flexion to a 150° angle by pelvis flexing toward femur.					
Flexed (straight)	Flexed (curled)	Flexed (straight)	Posterior tilt in relation to trunk Anterior tilt toward thigh	Flexed 30°	150°
E					
Movement: Hip joint flexion, and return toward zero position of cervical and thoracic spines.					
Toward zero	Toward zero	Flexed (straight)	Posterior tilt in relation to trunk Anterior tilt toward thigh	Flexed 80°	100°

FIGURE 5-14: Movements of the spine and hip joints that occur during curled-trunk sit-ups with legs extended.

MOVEMENTS DURING CURLED-TRUNK SIT-UPS WITH HIPS AND KNEES FLEXED

Figure 5-15 illustrates movements made during curled-trunk sit-ups with hips and knees flexed.

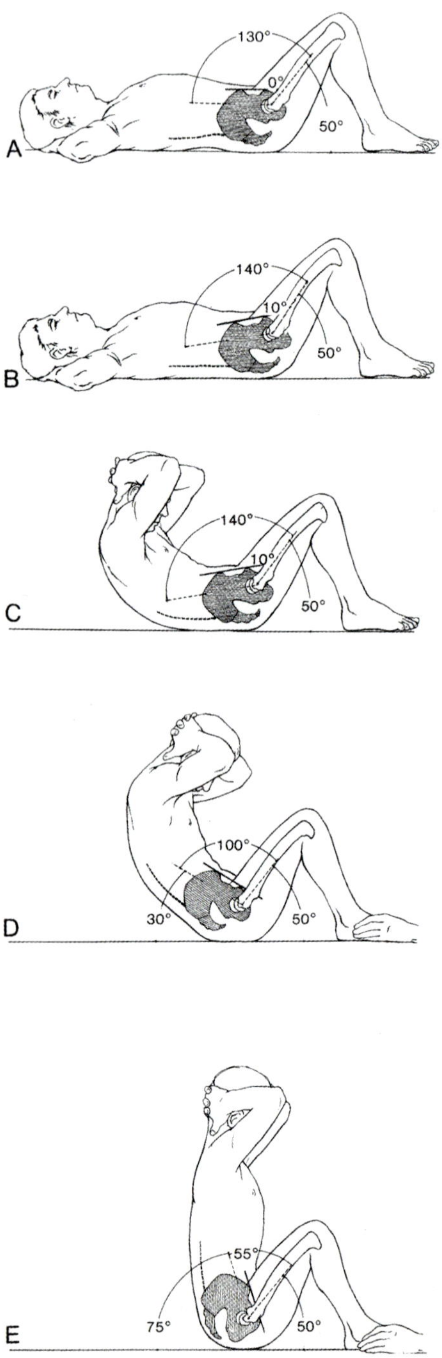

SPINE: Cervical (Neck)	SPINE: Thoracic (Upper Back)	SPINE: Lumbar (Low Back)	PELVIS	HIP JOINTS: Anatomical degrees	HIP JOINTS: Geometric degrees
A — *Position*: Supine, hands clapsed behind head, knees bent.					
Zero (Normal anterior curve)	Zero (Normal posterior curve)	Zero (Normal anterior curve)	Neutral position	50°	130°
B — *Movement*: Lumbar spine flexion and 10° decrease in hip joint flexion by virtue of posterior pelvic tilt.					
Zero	Zero	Flexed (straight)	10° posterior tilt	50° flexion of thigh	140°
C — *Movement*: Cervical and thoracic spine flexion. Figure C represents completion of spine flexion and the beginning of the flexion of the pelvis toward flexed thigh.					
Flexed (straight)	Flexed (curled)	Flexed (straight)	10° posterior tilt	50° flexion of thigh	140°
D — *Movement*: Hip joint flexion. The hip joint has moved from a 140° angle of flexion to a 100° angle by the pelvis flexing toward the femur.					
Flexed (straight)	Flexed (curled)	Flexed (straight)	Posterior tilt in relation to trunk Anterior tilt toward thigh	80° (50° thigh + 30° pel-vis)	100°
E — *Movement*: Hip joint flexion, and a return toward zero position of the cervical and thoracic spines. On the basis of 125° being complete flexion, hip joint has reached the position of complete flexion.					
Toward zero	Toward zero	Flexed (straight)	Posterior tilt in relation to trunk Anterior tilt toward thigh	125° (50° thigh + 75° pel-vis)	55°

FIGURE 5-15: Movements of the spine and hip joints that occur during curled-trunk sit-ups with hips and knees flexed.

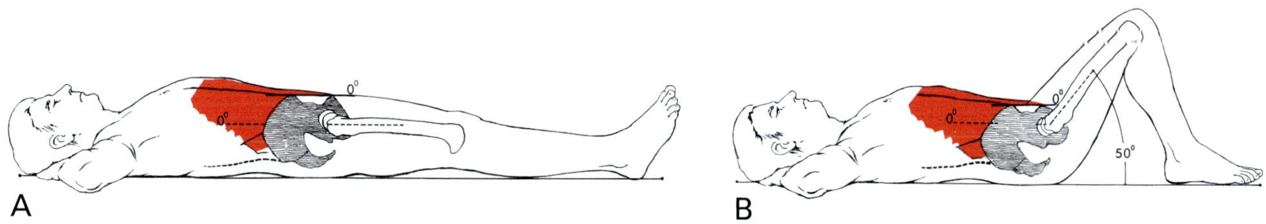

FIGURE 5-16: **A**, Zero position of the spine, pelvis, and hip joints (starting position); **B**, Zero position of the spine and pelvis and flexion of the hip joints.

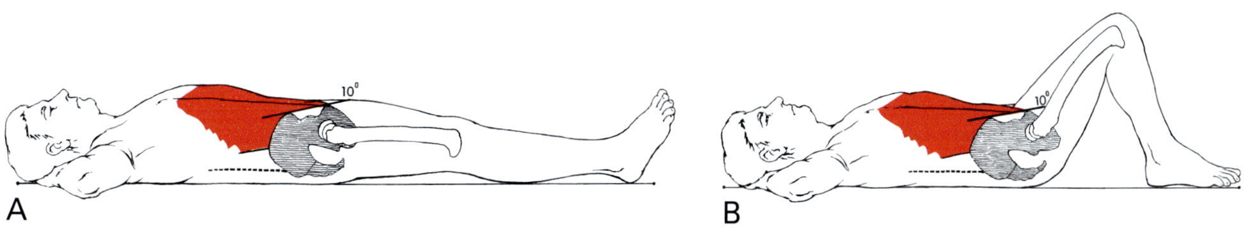

FIGURE 5-17: **A**, Posterior pelvic tilt, lumbar spine flexion, and hip joint extension. **B**, Posterior pelvic tilt, lumbar spine flexion, and hip joint flexion.

ABDOMINAL AND HIP FLEXOR MUSCLES DURING CURLED-TRUNK SIT-UPS

Zero Position of the Spine, Pelvis, and Hip Joints

Figure 5-16A and B may be regarded as hypothetical starting positions. In reality, especially with the knees bent, the low back tends to flatten (i.e., the lumbar spine flexes) when a normally flexible individual assumes the supine position.

In Figure 5-16A, the length of the hip flexors corresponds with the zero position of the hip joints.

Zero Position of the Spine and Pelvis and Flexion of the Hip Joints

In Figure 5-16B, because of the flexed position of the hips, the one-joint hip flexors are shorter in length than those in Figure 5-16A. In relation to its overall length, the iliacus is at approximately 40% of its range of motion, which is within the middle ⅓ of the overall range.

Posterior Pelvic Tilt, Lumbar Spine Flexion, and Hip Joint Extension

Figure 5-17A and B represents a stage of movement in which the pelvis is tilted posteriorly before beginning to raise the trunk. (Note the 10° posterior pelvic tilt.) In testing, this movement often is performed as a separate stage to ensure lumbar spine flexion.

When the posterior tilt is not done as a separate movement, as shown in Figure 5-17A and B, it occurs simultaneously with the beginning phase of trunk raising (i.e., the trunk-curl phase), *unless* the abdominal muscles are extremely weak or the hip flexors are so short that they prevent posterior tilt when the subject is supine with the knees extended.

In Figure 5-17A, the hip flexors have lengthened, and the one-joint hip flexors (chiefly the iliacus) have reached the limit of length permitted by the hip joint extension. At this length, they help to stabilize the pelvis by restraining further posterior pelvic tilt.

Posterior Pelvic Tilt, Lumbar Spine Flexion, and Hip Joint Flexion

In Figure 5-17B, the hip flexor length is slightly more than that in Figure 5-16B, because the pelvis has tilted posteriorly 10° away from the femur. Posterior pelvic tilt exercises are frequently used with the intention of strengthening the abdominal muscles. Too often, however, the tilt is done without any benefit to the abdominals. The subject performs the movement by contracting the buttocks muscles (i.e., the hip extensors) and, in the case of the knee-bent position, by pushing with the feet to help "rock" the pelvis back into posterior tilt.

To ensure that the pelvic tilt is performed by the abdominal muscles, there must be an upward and inward pull by these muscles, with the front and sides of the lower abdomen becoming very firm.

It is necessary to discourage use of the gluteal muscles to force action by the abdominals when performing a posterior pelvic tilt.

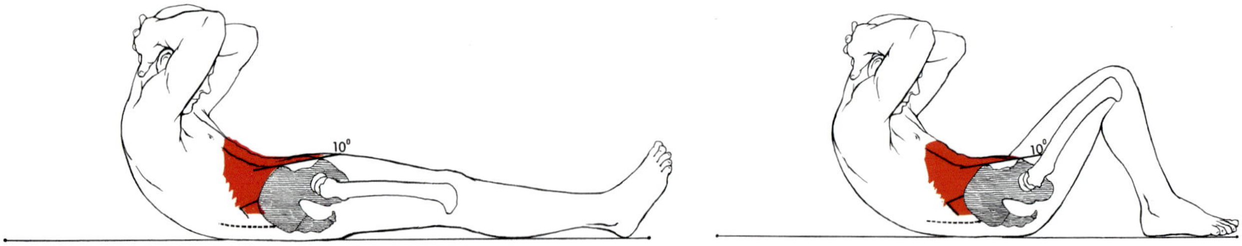

A B

FIGURE 5-18: Spine flexion phase (trunk-curl) completed.

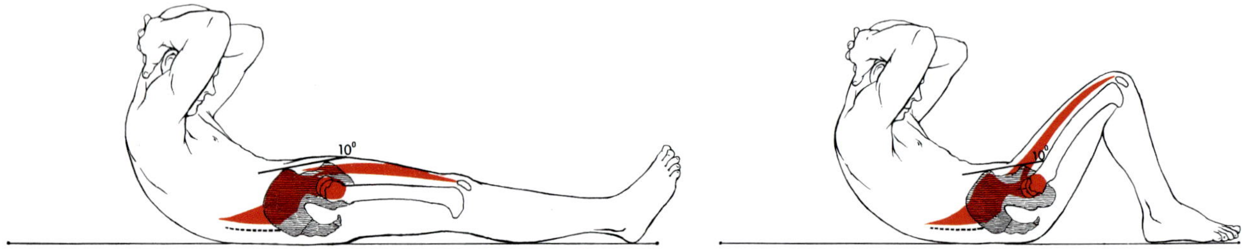

A B

FIGURE 5-19: Hip flexion phase (sit-up) initiated.

Spine Flexion Phase (Trunk-Curl) Completed

In Figure 5-18A and B, the cervical, thoracic, and lumbar spines are flexed. The lumbar spine remained in the same degree of flexion as shown in Figure 5-17A and B, where it reached maximum flexion for this subject.

In Figure 5-18A and B, the abdominal muscles have shortened to their fullest extent with the completion of spine flexion. In Figure 5-18A, the hip flexors have remained lengthened to the same extent as shown in Figure 5-17A.

In Figure 5-18B, the one-joint hip flexors have not reached the limit of their length and, therefore, do not act passively to restrain posterior tilt. The hip flexors contract to stabilize the pelvis, and palpation of the superficial hip flexors provides evidence of firm contraction as the subject begins to lift the head and shoulders from the table.

Hip Flexion Phase (Sit-Up) Initiated

With flexion of the spine complete (as shown in Figure 5-18A and B, and Figure 5-19A and B), no further movement in the direction of coming to a sitting position can occur except by flexion of the hip joints.

Because the abdominal muscles do not cross the hip joint, these muscles cannot flex the hip.

From a supine position, hip flexion can be performed only by the hip flexors acting to bring the pelvis in flexion toward the thighs.

Figure 5-19A and B represents the beginning of the sit-up phase as well as the end of the trunk-curl phase.

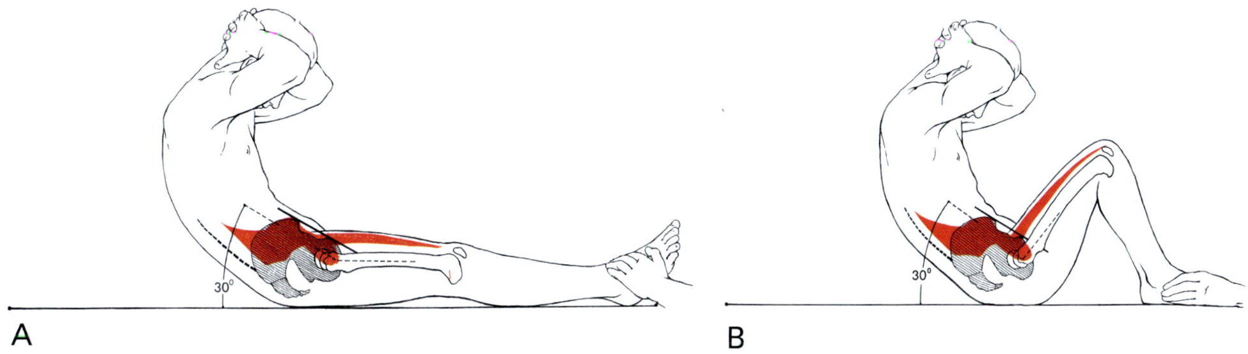

FIGURE 5-20: Hip flexion phase (sit-up) continued.

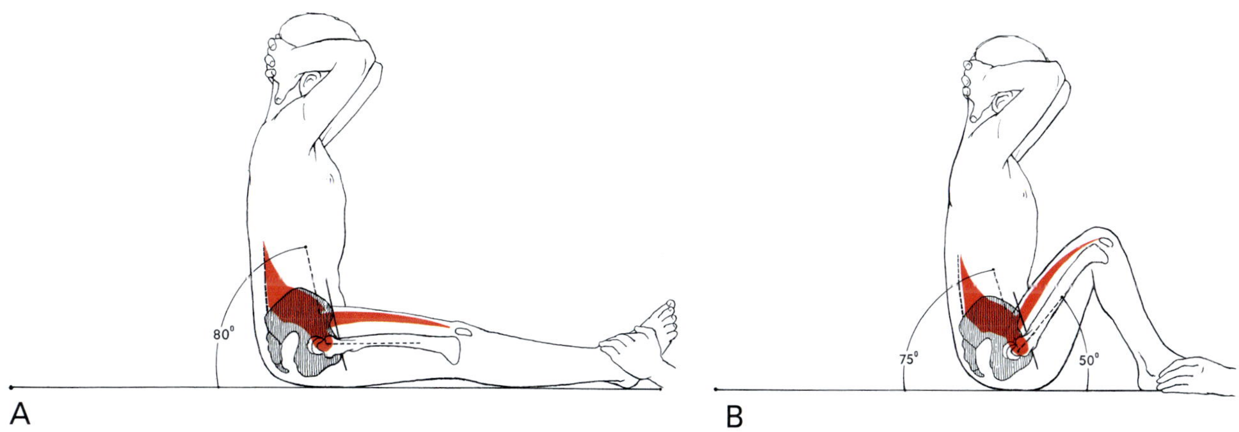

FIGURE 5-21: Hip flexion phase (sit-up) completed.

Hip Flexion Phase (Sit-Up) Continued

Figure 5-20A and B shows a point in the arc of movement between the completed trunk curl (as shown in Figures 5-18A, B, 5-19A, B) and the full sit-up. The abdominal muscles maintain the trunk in flexion, and the hip flexors have lifted the flexed trunk upward toward the sitting position through an arc of approximately 30° from the table.

When necessary, the feet may be held down at the initiation of and during the hip flexion phase. Before the hip flexion phase, the feet must not be held down.

Hip Flexion Phase (Sit-Up) Completed

In Figure 5-21A and B, as the subjects reach the sitting position, the cervical and thoracic spines are no longer fully flexed, and the abdominal muscles relax to some extent.

In Figure 5-21A, the hip flexors have moved the pelvis in flexion toward the thigh, completing an arc of approximately 80° from the table. In this position, with the knees extended and the lumbar spine flexed, the hip joint is as fully flexed as the range of normal hamstring length permits. The lumbar spine remains flexed, because moving from the flexed position of the low back to the zero position (i.e., normal anterior curve) would require that the pelvis tilt 10° more in flexion toward the thigh, which the hamstring length does not permit.

In Figure 5-21B, the hip flexors have moved the pelvis in flexion toward the thigh through an arc of approximately 75° from the table. The lumbar spine remains in flexion, because the hip joint has already reached the 125° of full flexion. Further flexion of the hip joints by tilting the pelvis forward (and bringing the low back into a normal anterior curve) could be done only if the flexion of the thigh were decreased by moving the heels farther from the buttocks in this sitting position.

UPPER ABDOMINAL MUSCLES: TESTING AND GRADING

Analysis of the Trunk-Raising Movement

Before doing this test, examine the flexibility of the back so that any restriction of motion is not interpreted as muscle weakness.

The *trunk-raising movement*, when properly done as a test, consists of two parts: spine flexion (i.e., trunk curl) by the abdominal muscles and hip flexion (i.e., sit-up) by the hip flexors.

During the *trunk-curl phase*, the abdominal muscles contract and shorten, flexing the spine. The upper back rounds, the lower back flattens, and the pelvis tilts posteriorly. On completion of the curl, the spine is fully flexed, with the low back and pelvis still flat on the table. The abdominal muscles act to flex the spine only. During this phase, the heels should remain in contact with the table.

The trunk curl is followed by the *hip flexion phase,* during which the hip flexors contract and shorten, lifting the trunk and pelvis up from the table by flexion at the hip joints and pulling the pelvis in the direction of anterior tilt. Because the abdominal muscles do not cross the hip joints, they cannot assist with the sit-up movement. If the abdominal muscles are strong enough, however, they can continue to hold the trunk curled.

The hip flexion phase is included in this test because it provides resistance against the abdominal muscles. The crucial point in the test is the moment at which the hip flexion phase is initiated. At this point, the feet of some subjects may start to come up from the table. The feet may be held down if the force exerted by the extended lower extremities does not counterbalance that exerted by the flexed trunk. However, if the feet are held down, attention must be focused on whether the trunk maintains the curl because at this point, the strength of the hip flexors can overcome the ability of the abdominals to maintain the curl. If this occurs, the pelvis will quickly tilt anteriorly, the back will arch, and the subject will continue the sit-up movement with the feet stabilized.

The trunk-raising test for the upper abdominal muscles is valuable when performed correctly. However, if the ability to perform a sit-up—regardless of how it is done—is equated with good abdominal strength, this test loses its value.

During a curled-trunk sit-up with the knees extended, the pelvis first tilts posteriorly, accompanied by flattening of the low back and extension of the hip joints. After the trunk-curl phase is completed, the pelvis tilts anteriorly (i.e., forward), toward the thigh, in hip flexion, but it remains in posterior tilt in relation to the trunk, maintaining the flat-back position (see Figures 5-18A and 5-19A).

During a sit-up with the low back arched, the pelvis tilts anteriorly toward the thigh as the sit-up begins, and it remains tilted anteriorly.

Test for Upper Abdominal Muscles

Patient: Supine, with knees extended. If the hip flexor muscles are short and prevent posterior pelvic tilt with flattening of the lumbar spine, place a roll under the knees to passively flex the hips enough to allow the back to flatten. (Arm positions are described below under *Grading*.)

Fixation: None necessary during the initial phase of the test (i.e., trunk curl), in which the spine is flexed and the thorax and pelvis are approximated. *Do not hold the feet down during the trunk-curl phase.* Stabilization of the feet will allow hip flexors to initiate trunk raising by flexion of the pelvis on the thighs.

Test Movement: Have the subject do a trunk curl *slowly*, completing spine flexion and, thereby, the range of motion that can be performed by the abdominal muscles. Without interrupting the movement, have the subject continue into the hip flexion phase (i.e., the sit-up) to obtain strong resistance against the abdominal muscles and, thereby, an adequate strength test.

Resistance: During the trunk-curl phase, resistance is offered by the weight of the head and upper trunk, and by the arms placed in various positions. However, the resistance offered by the weight of the head, shoulders, and arms is not sufficient to provide an adequate test for strength of the abdominal muscles.

The hip flexion phase provides strong resistance against the abdominals. The hip flexors pull strongly downward on the pelvis as the abdominals work to hold the trunk in flexion and the pelvis in the direction of posterior tilt.

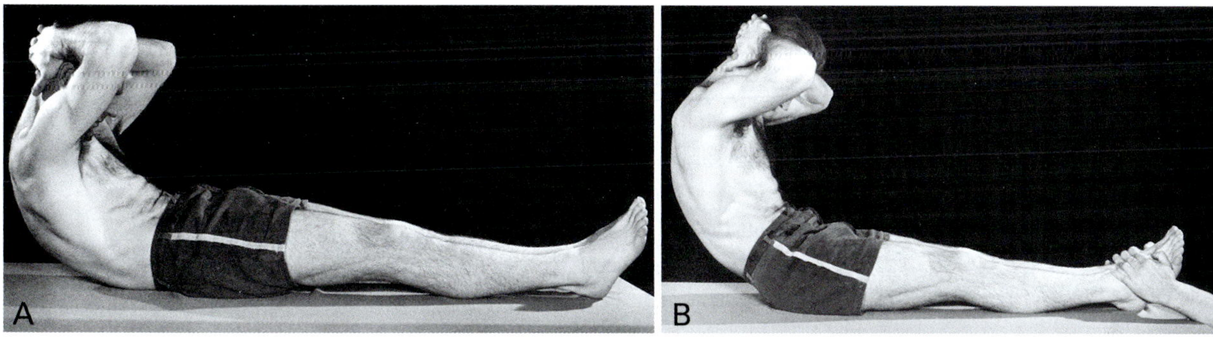

FIGURE 5-22: Testing and grading of upper abdominal muscles; Normal (10) Grade.

Normal (10) Grade[1]: With the hands clasped behind the head, the subject is able to flex the vertebral column (Figure 5-22A) and keep it flexed while entering the hip flexion phase and coming to a sitting position (Figure 5-22B). The feet may be held down during the hip flexion phase, if necessary, but close observation is required to be sure that the subject maintains the flexion of the trunk.

Because many people can do a curled-trunk sit-up with hands clasped behind the head, it is usually permissible to have a subject place the hands in this position (initially) and attempt to perform the test. If the difficulty of this test is a concern, have the subject start with the arms reaching forward, progress to placing arms folded across the chest, and then place the hands behind the head.

Good (8) Grade: With the arms folded across the chest, the subject is able to flex the vertebral column and keep it flexed while entering the hip flexion phase and coming to a sitting position. The strongest force against the abdominals is at the moment the hip flexors start to raise the trunk. Performing only the trunk curl is not sufficient for strength testing (Figure 5-23).

Fair+ (6) Grade: With the arms extended forward, the subject is able to flex the vertebral column and keep it flexed while entering the hip flexion phase and coming to a sitting position (Figure 5-24).

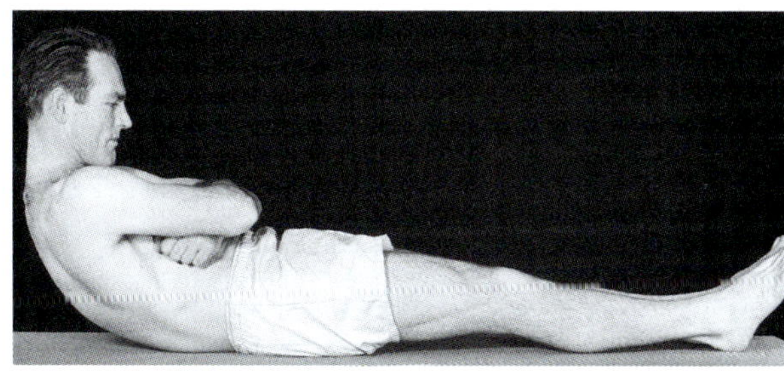

FIGURE 5-23: Testing and grading of upper abdominal muscles; Good (8) Grade.

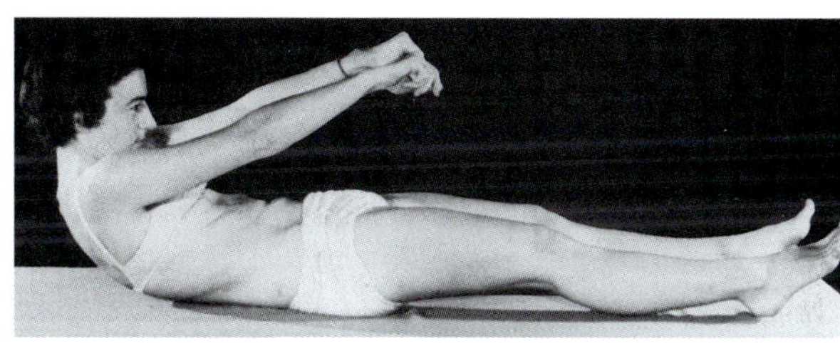

FIGURE 5-24: Testing and grading of upper abdominal muscles; Fair+ (6) Grade.

Fair (5) Grade: With the arms extended forward, the subject is able to flex the vertebral column but is unable to maintain the flexion when attempting to enter the hip flexion phase.

[1] See numerical equivalents for word symbols used in *The Key to Muscle Grading* in Chapter 1.

LOWER ABDOMINAL MUSCLES: TESTING AND GRADING

Anterior Trunk Flexors: Lower Abdominal Muscle Test

Anterior trunk flexion by the lower abdominal muscles focuses on the ability of these muscles to flex the lumbar spine by flattening the low back on the table and then holding it flat against the gradually increasing resistance provided by the leg-lowering movement.

Patient: Supine on a firm surface. A folded blanket may be used, but not a soft pad. Forearms are folded across the chest to ensure that the elbows are not resting on the table for support.

> **NOTE:** ***Avoid*** *extending the arms overhead or clasping the hands behind the head.*

Fixation: No fixation should be applied to the trunk, because this test determines the ability of the abdominal muscles to fix the pelvis in approximation to the thorax against resistance provided by the leg-lowering movement. Giving stabilization to the trunk would be giving assistance. Allowing the patient to hold onto the table, or to rest the hands or elbows on the table, would also be providing assistance.

Test Movement: The examiner assists the patient in raising the legs to a vertical position, or the examiner has the patient raise the legs one at a time to that position, keeping the knees straight. (Hamstring tightness will interfere with obtaining the full starting position.)

Have the subject tilt the pelvis posteriorly to flatten the low back on the table by contracting the abdominal muscles and then hold the low back flat while slowly lowering the legs. Focus attention on the position of the low back and pelvis as the legs are lowered. The subject should not raise the head and shoulders during the test.

Resistance: The force exerted by the hip flexors and the leg-lowering movement tends to tilt the pelvis anteriorly and acts as a strong resistance against the abdominal muscles, which are attempting to hold the pelvis in posterior tilt. As the lower extremities are lowered by the eccentric (i.e., lengthening) contraction of the hip flexors, leverage increases and provides increasing resistance against the abdominal muscles for the purpose of grading the strength of these muscles.

Grading: Strength is graded based on the ability to keep the low back flat on the table while slowly lowering both lower extremities from the vertical position (i.e., 90° angle).

The angle between the extended lower extremities and the table is noted at the moment that the pelvis tilts anteriorly and the low back arches from the table. To help detect the moment when this occurs, the examiner may place one hand at—but not under—the low back and the other hand with the thumb just below the anterosuperior spine of the ilium. When testing patients with weakness or pain, however, place the thumb of one hand just below the anterosuperior spine, and leave the other hand free to support the lower extremities the moment the back starts to arch.

The leg-lowering test for abdominal strength is not applicable to very young children. The weight of their lower extremities is small in relation to the trunk, and the back does not arch as the lower extremities are raised or lowered. Please see Appendix A for further details.

See the numerical equivalents for word symbols used in *The Key to Muscle Grading* in Chapter 1 (Figure 5-25).

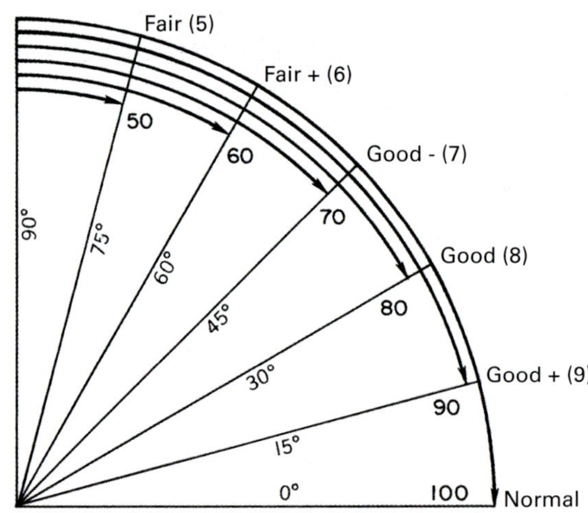

FIGURE 5-25: See the numerical equivalents for word symbols used in The Key to Muscle Grading in Chapter 1.

Fair+ (6) Grade: With arms folded across the chest, the subject is able to keep the low back flat on the table while lowering the lower extremities to an angle of 60° from the table (Figure 5-26).

FIGURE 5-26: Testing and grading of lower abdominal muscles; Fair+ (6) Grade.

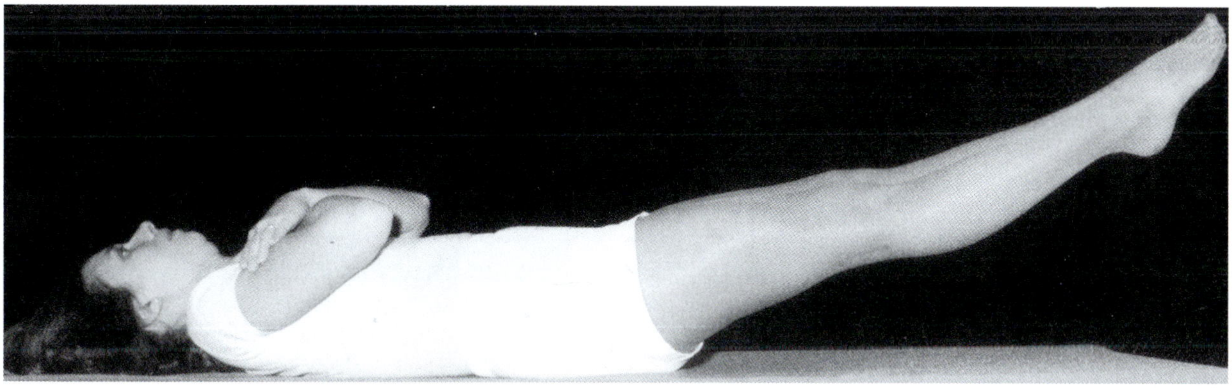

FIGURE 5-27: Testing and grading of lower abdominal muscles; Good (8) Grade.

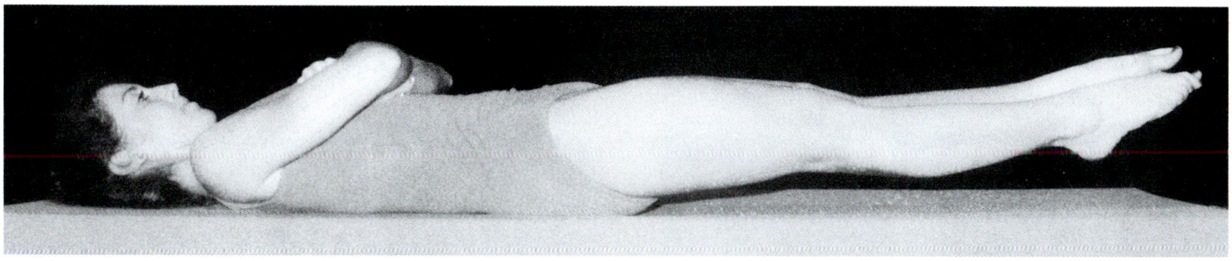

FIGURE 5-28: Testing and grading of lower abdominal muscles; Normal (10) Grade.

Good (8) Grade: With arms folded across the chest, the subject is able to keep the low back flat while lowering the lower extremities to an angle of 30° from the table. (In Figure 5-27, the legs are at a 20° angle.)

Normal (10) Grade: With arms folded across the chest, the subject is able to keep the low back flat on the table while lowering the lower extremities to table level. (In Figure 5-28, the legs are elevated a few degrees.)

Abdominal Muscle Weakness: Lower Extremity Lowering

A subject with marked weakness of abdominal muscles and strong hip flexors can hold the extended extremities in flexion on the pelvis and lower them slowly, but the low back arches increasingly as the lower extremities approach the horizontal. The force exerted by the weight of the extremities, and by the hip flexors holding the extremities in flexion on the pelvis, tilts the pelvis anteriorly, overcoming the force of the weak abdominal muscles that are attempting to pull in the direction of posterior tilt (Figure 5-29).

A

B

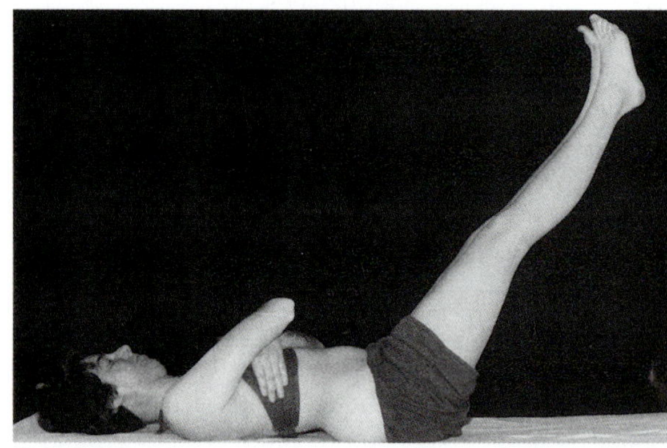

C

D

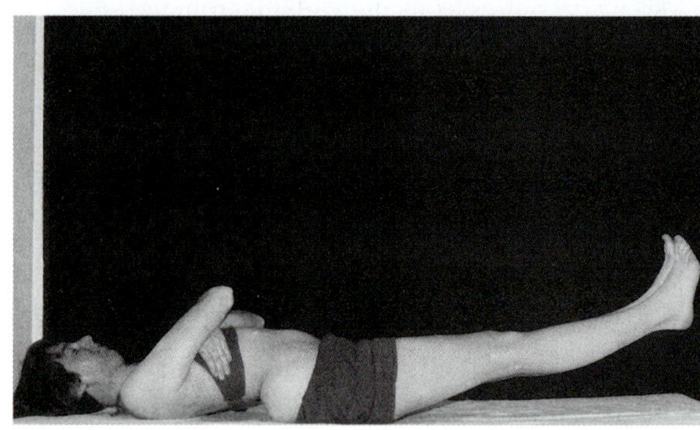

E

FIGURE 5-29A–E: Abdominal weakness, but strong hip flexors during lowering of legs.

MARKED ABDOMINAL MUSCLE WEAKNESS: TESTING AND GRADING

Temporary Use of the Hooklying (Knee-Bent) Position

When the one-joint hip flexors are short, they hold the pelvis in anterior tilt and the low back in hyperextension when standing or when supine with the legs extended. From this position, it is difficult—if not impossible—to do posterior pelvic tilt exercises to strengthen the abdominal muscles. Because the head- and shoulder-raising movement involves a simultaneous posterior pelvic tilt, interference occurs with this exercise as well.

As an effort is made to tilt the pelvis, the short hip flexors become taut and prevent the movement. To release this restraint and make tilting the pelvis easier, the hooklying position can be utilized. This position obviously gives in to the short, tight hip flexors. It also makes it relatively easy to perform the tilt, often merely by pressing the feet against the table to "rock" the pelvis back. With shortness of the hip flexors, the hips and knees should be bent, *only as much as needed,* to allow the pelvis to tilt posteriorly. This position should be maintained passively by using a large-enough roll or pillow under the knees. From this position, the pelvic tilt and trunk-curl exercises may be done to strengthen the abdominal muscles.

Although bending the hips and knees is initially needed and justified, the position should not be continued indefinitely. Therefore, the extent and duration of modifying the exercise become important. Goals should be based on the desired end result, and exercises should be directed toward attaining it. A desired end result in standing is the ability to maintain good alignment of the pelvis with the legs straight (i.e., with the hip joints and knee joints in good alignment). Working toward this goal in exercise is accomplished by minimizing, and then gradually decreasing, the amount of hip flexion that is permitted by the hooklying position.

Tilting the pelvis posteriorly with the lower extremities extended as much as possible moves the pelvis in the direction of elongating the hip flexors while strengthening the abdominals. This movement is not sufficient to stretch the hip flexors, but it helps to establish the necessary pattern of muscle action when attempting to correct a faulty lordotic posture in standing. Concurrent with proper abdominal exercise, the hip flexors should be stretched so that, in time, the individual will be capable of doing the posterior tilt with the lower extremities extended.

Objective grading of the anterolateral abdominal muscles is not difficult when strength is fair (i.e., grade of 5) or above. Below a strength of fair, it is more difficult to grade accurately. The tests and grades described here furnish guidelines for grading weak muscles.

With marked imbalance in the abdominal muscles, one must observe deviations of the umbilicus and rely on palpation for grading.

Before doing the tests that follow, it is necessary to test the strength of the anterior neck muscles.

Anterior Abdominal Muscles (Mainly Rectus Abdominis)

Fair— (4) Grade: In the supine position with knees slightly flexed (i.e., rolled towel under the knees), the patient is able to tilt the pelvis posteriorly and keep the pelvis and thorax approximated as the head is raised from the table.

Poor (2) Grade: In the same position as above, the patient is able to tilt the pelvis posteriorly. As the head is raised, however, the abdominal muscles cannot hold against that resistance anteriorly, and the thorax moves away from the pelvis.

Trace Grade: In the supine position, when the patient attempts to depress the chest or tilt the pelvis posteriorly, a contraction can be felt in the anterior abdominal muscles, but no approximation of the pelvis and thorax is observed.

Oblique Abdominal Muscles

Fair— (4) Grade: In the supine position with the examiner providing moderate resistance against a diagonally downward pull of the arm, the cross-sectional pull of the oblique abdominal muscles will be very firm on palpation and will pull the costal margin toward the opposite iliac crest. If the arm is weak, pushing the shoulder forward in a diagonal direction toward the opposite hip and holding it against pressure may be substituted for the arm movement.

In the supine position with one leg held straight in approximately 60° hip flexion, the examiner applies moderate pressure against the thigh in a downward and outward direction. The oblique muscles should be strong enough to pull the iliac crest toward the opposite costal margin. (This test can be used only if hip flexor strength is good.)

Poor (2) Grade: The patient is able to approximate the iliac crest toward the opposite costal margin.

Trace Grade: A contraction can be felt in the oblique muscle when the patient makes an effort to pull the costal margin toward the opposite iliac crest (i.e., a slight lateral shift of the thorax over the pelvis, but with no approximation of these parts).

Lateral Trunk Muscles

Fair— (4) Grade: In a side-lying position, firm fixation and approximation of the rib cage and iliac crest laterally will be noted during active leg abduction and arm adduction against resistance.

Poor (2) Grade: In the supine position, the patient is able to approximate the iliac crest and rib cage laterally as an effort is made to elevate the pelvis laterally or adduct the arm against resistance.

Trace Grade: In the supine position, a contraction can be felt in the lateral abdominal muscles as an effort is made to elevate the pelvis laterally or adduct the arm against resistance, but no approximation of the thorax and the lateral iliac crest is noted.

Recording Grades of Abdominal Muscle Strength

Abdominal muscle grades are recorded in two different ways. The method chosen depends on the amount of strength.

When strength is fair (i.e., grade of 5) or better in the trunk-raising and leg-lowering tests, it is usually sufficient to grade and record on the basis of these tests (Figure 5-30A). Intrinsic imbalance between parts of the rectus or the obliques seldom necessitates grading parts separately if these tests show a grade of fair or better.

When marked weakness or imbalance exists, it is necessary to indicate the test findings in relation to specific muscles (Figure 5-30B).

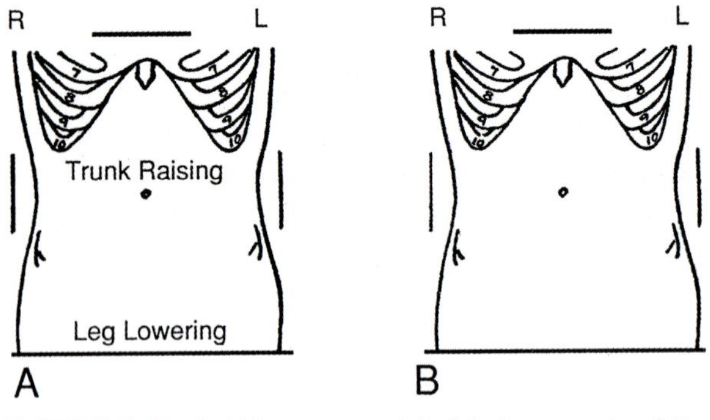

FIGURE 5-30: A, When strength is fair (i.e., grade of 5) or better in the trunk-raising and leg-lowering tests, it is usually sufficient to grade and record on the basis of these tests. **B**, When marked weakness or imbalance exists, it is necessary to indicate the test findings in relation to specific muscles.

Abdominal Muscle Imbalance and Umbilicus Deviations

With marked weakness and imbalance in the abdominal muscles, it is possible, to some degree, to determine the extent of the imbalance by observing the deviations of the umbilicus. The umbilicus will deviate toward a strong segment and away from a weak segment. If, for example, three segments—the left external and the left and right internal obliques—are equally strong and the right external is markedly weak, the umbilicus will deviate decidedly toward the left internal. This happens not because the left internal is the strongest, but because it has no opposition in the right external. This shows deviations away from a weak segment.

On the other hand, deviation may mean that one segment is strong and the other three segments are weak, and the deviation will be toward the strongest segment. In such a case, the relative strengths must be determined by palpation and by the extent to which the umbilicus deviates during the performance of localized test movements.

At times, the umbilicus deviates not because of active muscle contraction but because of a stretching of the muscle. The examiner must be sure that the muscles being tested are actively contracting before deviations of the umbilicus can be used to indicate strength or weakness.

To obtain true deviations, the abdominal muscles should first be in a relaxed position. The knees may be bent sufficiently to relax the back flat on the table. Then, the patient may be asked to attempt raising the head or tilting the pelvis posteriorly (even though the back is already flat). If resistive arm and leg movements are used in testing, they should begin from this relaxed position as well. Movements should be such that they produce actual shortening of the muscle. When weakness is very apparent, the initial test should be a mild, active movement, with resistance gradually applied. Note first to what extent the muscle can approximate its origin and insertion and then how much pressure can be added before the pull "breaks" and the muscle begins to stretch.

An individual unfamiliar with examination of the abdominal muscles may find it very difficult to be sure of the deviations of the umbilicus. If a tape or cord is held transversely and then diagonally over the umbilicus as the test movements are performed, the direction of the deviation can be determined more readily. The umbilicus may deviate upward or downward from the transverse tape, showing an uneven pull of the upper and lower rectus muscles. If it also shows a deviation from the tape held diagonally over the umbilicus, it will exhibit an imbalance between the obliques.

Lines made with ink or a skin pencil on the anterior iliac crests, the costal margins, just above the pubis, and below the sternum also may aid the examiner. As the test movement is done, the tape is held from the umbilicus to the various marks. Actual shortening or stretching of the segments can then be detected as a movement is attempted.

Arm Movements in Testing Abdominal Muscles

Arm movements are performed against resistance or held against pressure during abdominal muscle testing, because unresisted arm movements do not demand appreciable action by the trunk muscles for fixation.

Normally, an upward movement of the arms in the forward plane requires fixation by the back muscles, and a downward movement in the forward plane requires fixation by the abdominal muscles. With abdominal weakness, however, fixation for the downward pull or push of the arm may be provided by the back muscles. For example, if a patient is in a supine position and given resistance to a downward pull of both arms, normal abdominal muscles will contract to fix the thorax firmly toward the pelvis. With extensive abdominal weakness, however, the back will arch from the table, and the thorax will pull away from the pelvis until it is firmly fixed by extension of the thoracic spine. The arching of the back stretches the abdominal muscles, which may become taut and feel firm on palpation. The examiner must be careful not to mistake this tautness for the firmness that accompanies contraction of the muscles.

In cross-sectional or diagonal arm movements if the abdominal muscles are normal, the external oblique on the same side as the arm and the internal oblique on the opposite side contract to fix the thorax to the pelvis. With cross-sectional weakness in that line of pull, however, the opposite oblique muscles may act to give fixation. To perform an accurate examination, the examiner should understand these substitute actions.

SECTION III
MUSCLES OF RESPIRATION AND TESTS

Respiration refers to the exchange of gases between the cells of an organism and the external environment. Numerous neural, chemical, and muscular components are involved. This section, however, relates specifically to the role of the muscles.

Respiration consists of ventilation and circulation. **Ventilation** is the movement of gases into and out of the lungs; **circulation** is the transport of these gases to the tissues via blood. Although the movement of gases in the lungs and tissues is by diffusion, their transport to and from the environment and throughout the body requires work by the respiratory and cardiac pumps.

The respiratory pump is comprised of the muscles of respiration and the thorax, which in turn is made up of the ribs, scapulae, clavicle, sternum, and thoracic spine. This musculoskeletal pump provides the necessary pressure gradients to move gases into and out of the lungs to ensure adequate diffusion of oxygen and carbon dioxide within the lung.

The work of breathing performed by respiratory muscles in overcoming lung, chest wall, and airway resistances normally occurs only during inspiration. Muscular effort is required to enlarge the thoracic cavity and lower the intrathoracic pressure. Expiration results from the elastic recoil of the lungs and relaxation of the inspiratory muscles. The muscles of expiration are active, however, when the demands of breathing are increased. Heavy work, exercise, blowing, coughing, and singing all involve significant work by the expiratory muscles. Also, in conditions like emphysema, in which the elastic recoil is impaired, techniques such as pursed-lip breathing are employed to enhance expiration and minimize effort.

The *Respiratory Muscle Chart* (Figure 5-33) shows the division of muscles according to their major inspiratory or expiratory roles in ventilation. This division, however, does not mean that the listed muscles function only in that singular capacity. For example, abdominal muscles, which are the chief expiratory muscles, also play a role in inspiration. The inspiratory intercostals as well as the diaphragm also perform an important "braking" action during expiration.

The further division on the chart into primary and accessory muscles shows the numerous muscles that can be recruited to assist in the ventilatory process. Exactly which muscles participate as well as the extent of their participation depends not only on the demands of breathing but also on individual differences in breathing habits or needs.

The fact that breathing can be altered by changes in position, emotional state, activity level, disease, and even the wearing of tight garments means that numerous variations in patterns of breathing exist. For example, Duchenne remarked that normal breathing of women in the mid-19th century was "of the costosuperior type" because of compression from corsets on the lower part of the chest (4).

According to Shneerson, "It is better to regard the respiratory muscles as being capable of recruitment according to the pattern of ventilation, posture, wakefulness or stage of sleep, muscle strength, air flow resistance, and compliance of the lungs and chest wall" (5).

Some authorities dispute the accessory role of certain muscles, particularly the upper trapezius and serratus anterior. Other muscles also often are omitted in writings about the accessory respiratory muscles. The rhomboid, for example, which is not included in Figure 5-33, has a role in stabilizing the scapula to assist the serratus in forced inspiration.

All the muscles listed on the chart have the capacity to be recruited, when needed, to facilitate breathing. Many of them perform vital roles in stabilizing parts of the body so that adequate force is provided to move air both into and out of the lungs. As the work of breathing increases, larger volumes of gas must be moved more quickly, and greater pressure generation is required. The ventilatory muscles work harder, and additional muscles are recruited to meet the demands of breathing.

The following quotation emphasizes the importance of *all* respiratory muscles: "The distance runner struggling for air … may use even the platysma for expanding his chest, and the patient in paroxysms of cough probably contracts every muscle of the trunk, thorax and pectoral girdle during forced expiration" (6). Although the numerous muscles of the upper airways, especially the intrinsic and extrinsic muscles of the larynx, are not discussed here, they play an important role in permitting the free flow of air both to and from the lungs (Table 3-2).

In some individuals and under certain circumstances, accessory muscles may be used as the primary muscles. If the diaphragm or intercostals are paralyzed, for example, breathing is still possible through increased use of the accessory muscles. The importance of the accessory muscles was well documented in the case of a patient with a permanent tracheostomy and no movement in his diaphragm or intercostals muscles. He had, surprisingly, a very large vital capacity, breathing with scaleni supplied by the cervical nerves and with the sternocleidomastoid and upper trapezius supplied by the spinal accessory nerve (7).

PRIMARY MUSCLES OF RESPIRATION

Most of the primary and accessory muscles listed in Figure 5-33 perform a postural function. Only the diaphragm and the anterior intercostals may be purely respiratory. Twenty of these muscles have either all or part of their origins or insertions on the ribs or costal cartilages. Any muscle attached to the rib cage is able to influence the mechanics of breathing to some degree. These muscles must be able to help support the skeletal structures of the ventilatory pump and to generate pressures that ensure continued adequate gas exchange at the alveoli.

These pressures can be substantial. To double air flow, a fourfold increase in pressure normally is required. If air flow is to remain constant in the face of a twofold decrease in the radius of an airway, there must be a 16-fold increase in pressure (6). Respiratory complications can arise from a variety of obstructive and restrictive diseases as well as neuromuscular and skeletal disorders. Once a diagnosis is established, treatment is designed to preserve the existing lung function and to eliminate or reduce the problem that is compromising respiration. The goal is to improve a patient's ability to ventilate the lungs.

Of primary importance is the need to lessen the work of breathing and thus reduce the energy expenditure (i.e., oxygen consumption) of the respiratory muscles. Depending on the respiratory disorder, it may be the elastic, resistive, mechanical, or some combination of such work that needs to be alleviated. Respiratory failure can result when the increased work of breathing leads to alveolar hypoventilation and hypoxia.

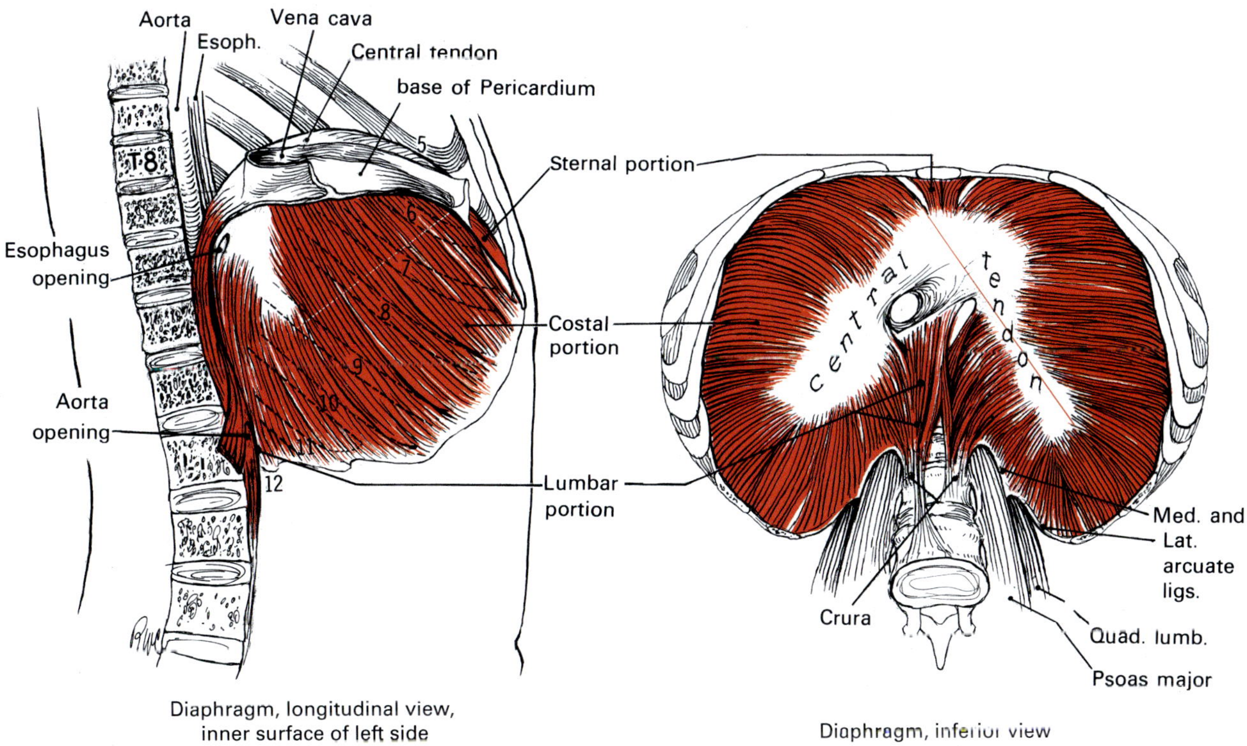

FIGURE 5-31: Diaphragm: Longitudinal and inferior views.

Diaphragm

Origin, Sternal Part: Dorsum of the xiphoid process.

Origin, Costal Part: Inner surfaces of the lower six costal cartilages and lower six ribs on either side, interdigitating with the transversus abdominis.

Origin, Lumbar Part: By two muscular crura from the bodies of the upper lumbar vertebrae and by two fibrous arches on either side, known as the medial and lateral arcuate ligaments, which span from the vertebrae to the transverse processes and from the latter to the 12th rib (Figure 5-31).

Insertion: Into the central tendon, which is a thin, strong aponeurosis with no bony attachment. Because the anterior muscular fibers of the diaphragm are shorter than the posterior muscular fibers, the central tendon is situated closer to the ventral than to the dorsal part of the thorax.

Action: The dome-shaped diaphragm separates the thoracic and abdominal cavities and is the principal muscle of respiration. During inspiration, the muscle contracts, and the dome descends, increasing the volume and decreasing the pressure of the thoracic cavity while decreasing the volume and increasing the pressure of the abdominal cavity. The descent of the dome or central tendon is limited by the abdominal viscera, and when descent occurs, the central tendon becomes the more fixed portion of the muscle. With continued contraction, the vertical fibers that are attached to the ribs elevate and evert the costal margin. The dimensions of the thorax are constantly enlarged craniocaudally, anteroposteriorly, and transversely. During expiration, the diaphragm relaxes, and the dome ascends, decreasing the volume and increasing the pressure of the thoracic cavity while increasing the volume and decreasing pressure of the abdominal cavity.

NOTE: *In cases of pulmonary pathology (e.g., emphysema), the dome of the diaphragm is so depressed that the costal margin or base of the thorax cannot be expanded.*

Nerve: Phrenic, C3, **4**, 5.

Tests: See later in this chapter.

The diaphragm, by virtue of its attachment and actions, serves as a pressure partitioner and force transmitter. Normal length and strength of this muscle are essential for these functions. Limited or excessive excursion of the diaphragm reduces its effectiveness in both inspiration and expiration.

In certain respiratory conditions (e.g., emphysema), the diaphragm is not able to return to a dome-shaped contour on relaxation; rather, it is held in a shortened, flattened position. Both pressure-generating capability and inspiratory capacity are reduced because the lungs remain in a partially inflated state at the resting level. Also, the ability of the diaphragm to act as a force transmitter and to assist in emptying the lungs is reduced.

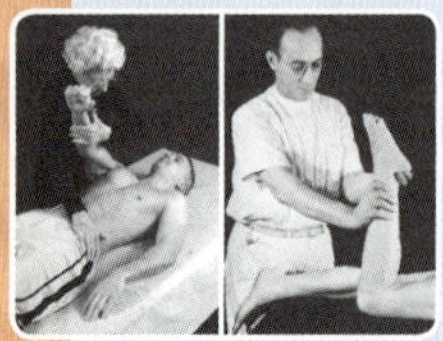

CLASSIC KENDALL

Case Report

A dramatic example of reverse action of the diaphragm was seen in an infant with poliomyelitis who was placed in a respirator. The muscles of the abdomen, which normally are weak in infants, were paralyzed. During the positive-pressure phase, air was forced out of the lungs, and the diaphragm moved upward. During the negative-pressure phase, air was drawn into the lungs with a momentary expansion of the rib cage, followed by excessive descent of the diaphragm into the abdominal cavity. The abdomen ballooned as the viscera moved downward. By virtue of the attachment of the diaphragm to the inner wall of the chest, the ribs were drawn downward and inward, causing the rib cage to cave in as the diaphragm descended into the abdominal cavity—completely defeating the function of this muscle. Within hours, a support in the form of a tiny corset was made and applied to restrict the ballooning of the abdomen and help to prevent the excessive descent of the diaphragm and the devastating effect of this on the rib cage.

The abdominal viscera, supported by the abdominal muscles, normally limit the downward descent of the diaphragm during inspiration and assist with its upward movement during expiration. Under abnormal circumstances, e.g., during respirator therapy, there can even be a reverse action of the diaphragm.

Intercostal Muscles

The **external intercostals** arise from the lower borders of the ribs and attach to the upper borders of the ribs below (Figure 5-32). Similarly, the **internal intercostals** arise on the inner surfaces of the ribs and costal cartilages and insert on the upper borders of the adjacent ribs below. The body has two layers of these rib cage muscles "everywhere except anteriorly in the interchondral region and posteriorly in the areas medial to the costal angle" (8).

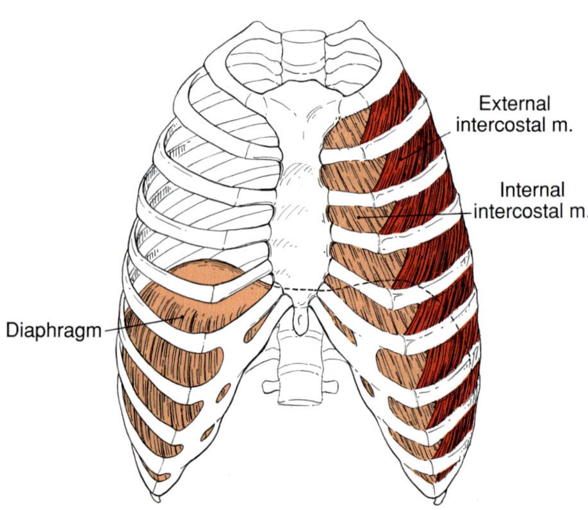

FIGURE 5-32: Primary muscles of respiration.

These muscles play an important postural as well as a respiratory role. They stabilize and maintain the shape and integrity of the rib cage. Anatomically, they appear to be extensions of the external and internal oblique muscles.

Debate persists as to the exact respiratory function of these muscles. It seems that at least the exposed anterior portion of the internal intercostals (i.e., parasternal, intercartilaginous) acts as an inspiratory muscle along with the external intercostals, elevating the ribs and expanding the chest. The posterior portion (i.e., interosseous) of the internal intercostals depresses the ribs and acts in an expiratory capacity.

It has been suggested that the function of these muscles varies with lung volume and depth of respiration as the position and slope of the ribs to which they are attached changes. These muscles are always active during speech. During controlled expiration, they perform an important "braking" action that minimizes the static recoil of the lungs and the chest wall. Singers make much use of this expiratory action of the intercostals.

Breathing is possible when the intercostals are paralyzed, but sucking and blowing capacity is diminished. Movement of the rib cage is also limited, and ability to stabilize the rib cage is decreased.

Abdominal Muscles

The abdominal muscles are the internal obliques, external obliques, rectus abdominis, and transversus abdominis. These muscles are the chief expiratory muscles, but they are also active toward the end of inspiration. The most important muscles at the end of inspiration and the beginning of expiration are those with little or no flexor action. Specifically, the lower fibers of the internal obliques and transversus are most active, along with the lateral fibers of the external obliques.

These muscles must be able to contract sufficiently to raise the intra-abdominal pressure to meet increased demands of breathing—especially sudden, expulsive acts. Pressure generated in this way is transmitted to the thoracic cage by the diaphragm to assist with emptying the lungs.

The transversus arises from the cartilages of the lower six ribs and interdigitates with the diaphragm. The quadratus lumborum, by virtue of its insertion on the 12th rib, anchors the rib cage and so aids diaphragmatic action in inspiration as well as expiration.

The external oblique muscles cover a major portion of the lower thorax because some fibers interdigitate with the lower digitations of the serratus anterior. Increased abdominal activity (particularly of the external oblique) reduces fluctuations of thoracic cage volume and helps to maintain constancy of pressure.

ACCESSORY MUSCLES OF RESPIRATION

Scalenes

The anterior, medial, and posterior scalenes are accessory muscles of inspiration that function as a unit. By elevating and firmly fixing the first and second ribs, they aid in deep inspiration. The scalenes have been observed to be active during quiet breathing and have been classified by some researchers as primary rather than accessory muscles.

The scalenes may become active during expiratory efforts as well. According to Egan, "the expiratory function of the scalene muscles is to fix the ribs against the contraction of the abdominal muscles and to prevent herniation of the apex of the lung during coughing" (9).

Sternocleidomastoid

This muscle is considered by many to be the most important accessory muscle of inspiration. For the sternocleidomastoid to act in this capacity, the head and neck must be held in a stable position by the neck flexors and extensors. This muscle "pulls from its skull insertions and elevates the sternum, increasing the A-P diameter of the chest" (9). It contracts during moderate and deep inspiration. When the lungs are hyperinflated, the sternocleidomastoid is especially active. Electrical activity is sometimes evident during quiet inspiration (5, 10). This muscle is not active during expiration.

Serratus Anterior

This muscle arises from the upper eight or nine ribs, and it inserts on the costal surface of the medial border of the scapula. Its primary action is to abduct and rotate the scapula and hold the medial border firmly against the rib cage.

When the scapula is stabilized in adduction by the rhomboids, thereby fixing the insertion, the serratus can assist in forced inspiration. It helps to expand the rib cage by pulling the origin toward the insertion. Because it takes a stronger serratus to move the rib cage than to move the scapula, a person with only fair strength may be able to move the scapula in abduction but have difficulty expanding the rib cage with the scapula fixed in adduction. Consequently, weakness in this muscle diminishes its ability to be recruited to meet increased inspiratory needs.

Pectoralis Major

The pectoralis major is a large, fan-shaped muscle that is active in deep or forced inspiration but not in expiration. Egan considers this muscle to be the third most important accessory muscle and describes its mechanism of action as follows: "If the arms and shoulders are fixed, as by leaning on the elbows or firmly grasping a table, the pectoralis major can use its insertion as an origin and pull with great force on the anterior chest, lifting up ribs and sternum and increasing thoracic A-P diameter" (9).

Pectoralis Minor

The pectoralis minor assists in forced inspiration by raising the ribs, thereby moving the origin toward the insertion. The insertion must be fixed by stabilizing the scapula in an optimal position that prevents anterior tilt with depression of the coracoid process downward and forward. This stabilization is accomplished by the lower and middle trapezius.

Upper Trapezius

The trapezius muscle is discussed in detail in Chapter 6. The ventilatory role of the upper trapezius is to assist with forced inspiration by helping to elevate the thoracic cage. The insertion of the upper fibers onto the lateral ⅓ of the clavicle ensures the participation of this portion of the muscle whenever clavicular breathing is needed for ventilation.

Latissimus Dorsi

Although the respiratory role of the latissimus dorsi is essentially in forceful expiration, this muscle also has a role in deep inspiration. The anterior fibers, which are active during trunk flexion, assist in expiration; the posterior fibers, which are active during trunk extension, assist in inspiration.

Erector Spinae (Thoracic)

The thoracic erector spinae muscles extend the thoracic spine and aid inspiration by raising the rib cage to permit full expansion of the chest.

Iliocostalis Lumborum

This erector spinae muscle inserts onto the inferior angles of the lower six or seven ribs and can assist as an accessory muscle of expiration.

Quadratus Lumborum

The quadratus lumborum fixes the posterior fibers of the diaphragm by holding down the 12th rib so that it is not elevated along with the others during respiration.

OTHER ACCESSORY MUSCLES

The following muscles cannot be tested manually and are inaccessible to palpation.

Serratus Posterior Superior: This inspiratory muscle is attached to the second through fifth ribs and arises on the spines of the seventh cervical and upper two or three thoracic vertebrae. It lies beneath the fibers of the rhomboids and trapezius, and it expands the chest by raising the ribs to which it is attached.

Serratus Posterior Inferior: This muscle inserts on the lower four ribs and arises on the spines of the lower two thoracic and upper two or three lumbar vertebrae. It acts to draw the ribs backward and downward. Usually, it is considered to be an accessory muscle of expiration, although some research list it as an inspiratory muscle (5, 11).

Levatores Costarum: These 12 strong, fan-shaped muscles are parallel with the posterior borders of the external intercostals. Their action is to elevate and abduct the ribs and to extend and laterally flex the vertebral column. They are considered to be inspiratory muscles. They arise from the transverse processes of the seventh cervical and upper 11 thoracic vertebrae, and they insert onto the rib immediately below each vertebra.

Transversus Thoracis: This muscle (and other muscles of the innermost layer of the thorax) acts in an expiratory capacity to decrease the volume of the thoracic cavity. The transversus thoracis (i.e., triangularis sterni) is an expiratory muscle on the ventral thoracic wall. It narrows the chest by depressing the second through sixth ribs. It arises from the xiphoid cartilage and sternum, and it inserts onto the lower borders of the costal cartilages of these ribs. Its caudal fibers are continuous with the transversus abdominis.

Also in this layer are the intercostales intimi and the subcostales. The latter muscles on the lower dorsal thoracic wall bridge two or three intercostals spaces, and they act to draw the ribs together.

Subclavius: This is a shoulder girdle muscle that arises on the first rib and cartilage and inserts on the undersurface of the clavicle. It draws the clavicle downward and stabilizes it. The action of this muscle suggests it is important in the avoidance of clavicular breathing when this is not appropriate.

Figure 5-33 is the respiratory muscle chart referred to in this chapter.

RESPIRATORY MUSCLE CHART

Patient's Name ______________________________ Clinic # ______________

Left								Right
				Examiner				
				Date				
				Inspiratory Muscles Primary				
				Diaphragm				
. . . . .	. . . . .	. . . . .	. . . . .	Levator costarum (3)	. . . . .	. . . . .	. . . . .	. . . . .
				External intercostals				
				Internal intercostals, anterior (1)				
				Accessory				
				Scaleni				
				Sternocleidomastoid				
				Trapezius				
				Serratus anterior				
. . . . .	. . . . .	. . . . .	. . . . .	Serratus posterior, superior (3)	. . . . .	. . . . .	. . . . .	. . . . .
				Pectoralis major				
				Pectoralis minor				
				Latissimus dorsi				
				Erector spinae, thoracic				
. . . . .	. . . . .	. . . . .	. . . . .	Subclavius (3)	. . . . .	. . . . .	. . . . .	. . . . .
				Expiratory Muscles Primary				
				Abdominal muscles				
				Internal oblique				
				External oblique				
				Rectus abdominis				
				Transversus abdominis				
				Internal intercostals, posterior (2)				
. . . . .	. . . . .	. . . . .	. . . . .	Transversus thoracis (3)	. . . . .	. . . . .	. . . . .	. . . . .
				Accessory				
				Latissimus dorsi				
. . . . .	. . . . .	. . . . .	. . . . .	Serratus posterior inferior (3)	. . . . .	. . . . .	. . . . .	. . . . .
				Quadratus lumborum				
				Iliocostalis lumborum				

Notes: ______________________________

(1) Also called parasternal or intercartilaginous
(2) Also called interosseus
(3) Cannot be tested manually

FIGURE 5-33: Respiratory muscle chart.

The following photographs, Figure 5-34A–D, show the muscles of respiration in inspiration and expiration.

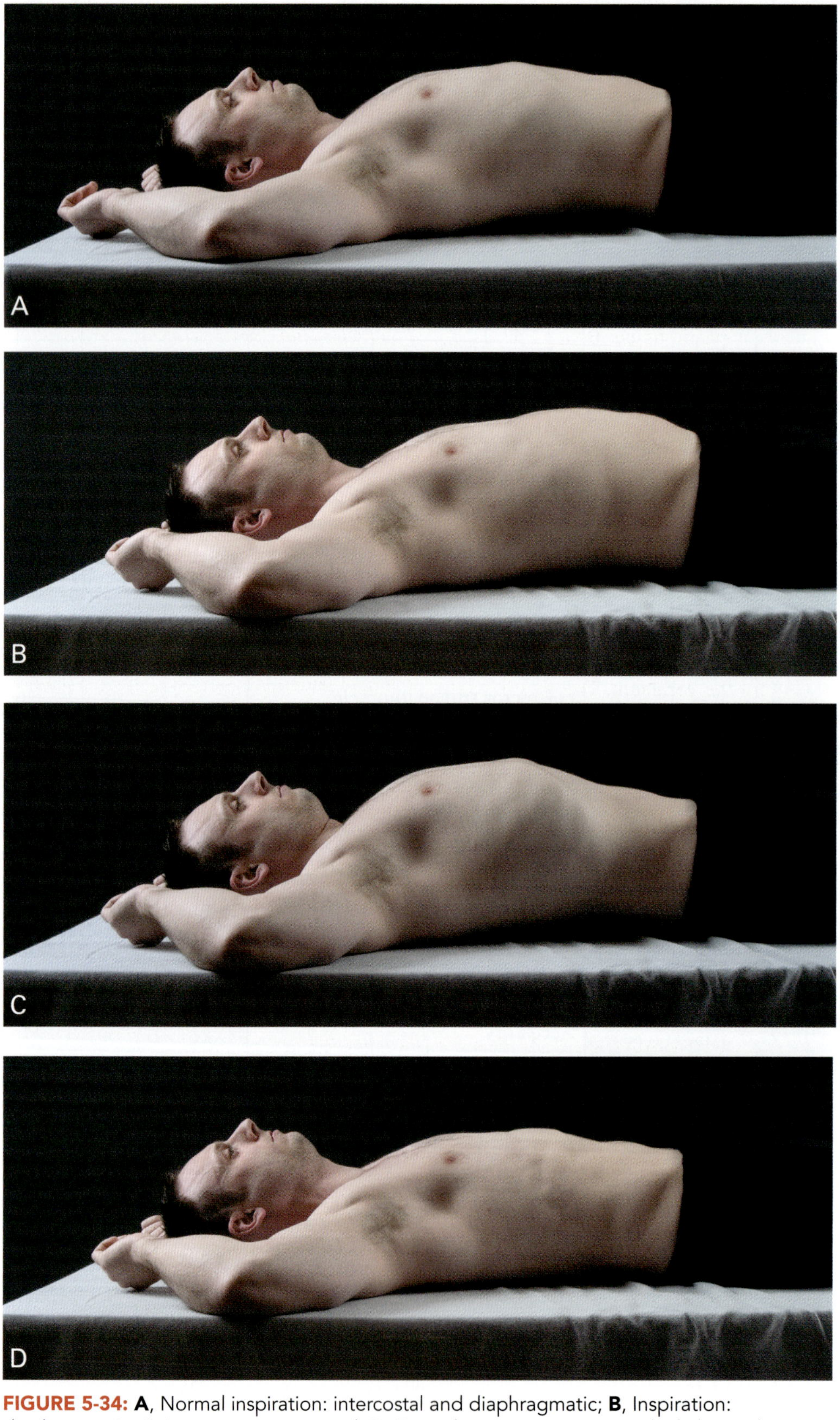

FIGURE 5-34: **A**, Normal inspiration: intercostal and diaphragmatic; **B**, Inspiration: diaphragmatic; **C**, Inspiration: intercostal; **D**, Forced expiration: intercostal, abdominal, and accessory muscles.

SECTION IV
PELVIS

OSTEOLOGY

The pelvic cavity is formed by articulation between the sacrum, coccyx, and left and right innominate bones, ligaments, muscles, and fascial structures that support these articulations. The pelvic cavity contains the terminal organs and structures associated with the digestive and genitourinary systems including their assigned blood vessels, nerves, and lymphatic tissue. Boundaries of the pelvic cavity are the anteroinferior wall, posterior wall, right and left lateral walls, and the pelvic floor. Muscles aligned with the posterior and lateral walls will be discussed in greater detail in Chapter 7.

MUSCLES

The pelvic floor consists of a group of muscles that form the pelvic diaphragm (Table 5-1). These muscles include the levator ani—comprised of the iliococcygeus, puborectalis, and pubococcygeus—and the coccygeus. These muscles and their supporting fasciae separate the pelvic cavity from the perineum. The role of these muscles is to increase intraabdominal pressure, mitigate intrapelvic pressure, maintain continence, support the pelvic contents, and resist migration through the pelvic outlet (12).

TESTING

Observation/Palpation

Visual inspection of the urogenital triangle, perineum, and anal triangle is essential to rule out any gross anatomical deformity. Digital palpation of the pelvic floor muscles is used to assess contraction and relaxation and to evaluate pain. Normal contraction results in an upward lift of the perineum. Excessive straining and impaired contraction of the pelvic floor muscles may result in descent of the perineum (15).

Manual Muscle Testing

Patients may be tested in supine with one pillow under the head, hips, and knees flexed to 60°, and a small towel roll placed under the lumbar spine to place the spine in neutral.

A digital vaginal examination is performed using two fingers while patients are instructed to perform a maximal pelvic floor contraction while breathing normally (12). Patients are instructed to "squeeze the muscle of the bottom of your pelvis as if trying to stop the flow of urine."

The Modified Oxford scale is widely used to evaluate strength of pelvic floor musculature. A 6-point ordinal scale is utilized as follows: 0 = no contraction, 1 = flicker contraction, 2 = weak contraction, 3 = moderate contraction, 4 = good contraction (with lift), and 5 = strong contraction (16).

TABLE 5-1 Muscles, Attachments, and Innervation of the Pelvic Diaphragm

Muscles		Attachments	Innervation
Levator ani	Iliococcygeus	Ischial spine anococcygeal ligament	Pelvic surface—ventral rami; S2–S4 Perineal surface—pudendal nerve
	Pubococcygeus	Body of pubis/coccyx/anococcygeal ligament	
	Puborectalis	Right and left pubic bodies	
Coccygeus		Ischial spine/sacrum/coccyx	Ventral rami; S4–S5
Obturator internus (OI)		Discussed in detail in Chapter 6	Nerve to OI; L5–S2
Piriformis (P)			Nerve to P; S–S2/variable*

*Direct branches of sacral plexus; L5-S2/superior gluteal nerve; S1-S2 (13, 14).

5 | TRUNK AND RESPIRATORY MUSCLES

SECTION V
CLINICAL FINDINGS

ABDOMINAL REGION

Abdominal Muscle Weakness: Trunk Raising

When the abdominal muscles are too weak to curl the trunk, the hip flexors tilt the pelvis forward and hyperextend the lumbar spine as they raise the trunk to a sitting position. Some people cannot perform a sit-up unless the feet are held down from the start. Usually, these subjects have marked weakness of the abdominal muscles. They should practice the trunk curl only and avoid doing the sit-up in the manner illustrated in Figure 5-35.

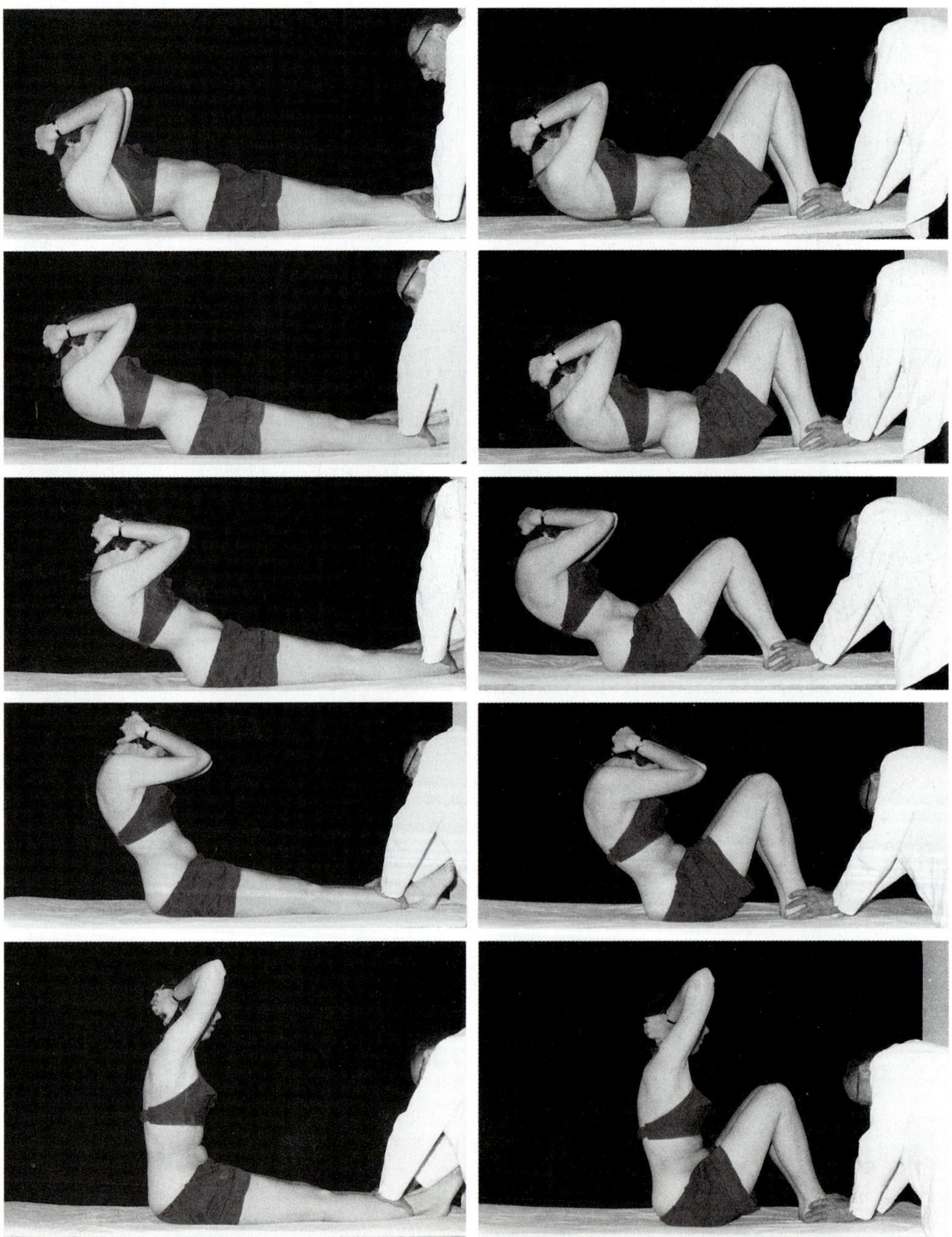

FIGURE 5-35: Subjects who have weak abdominal muscles should practice the trunk curl only and avoid doing the sit-up in the manner illustrated here.

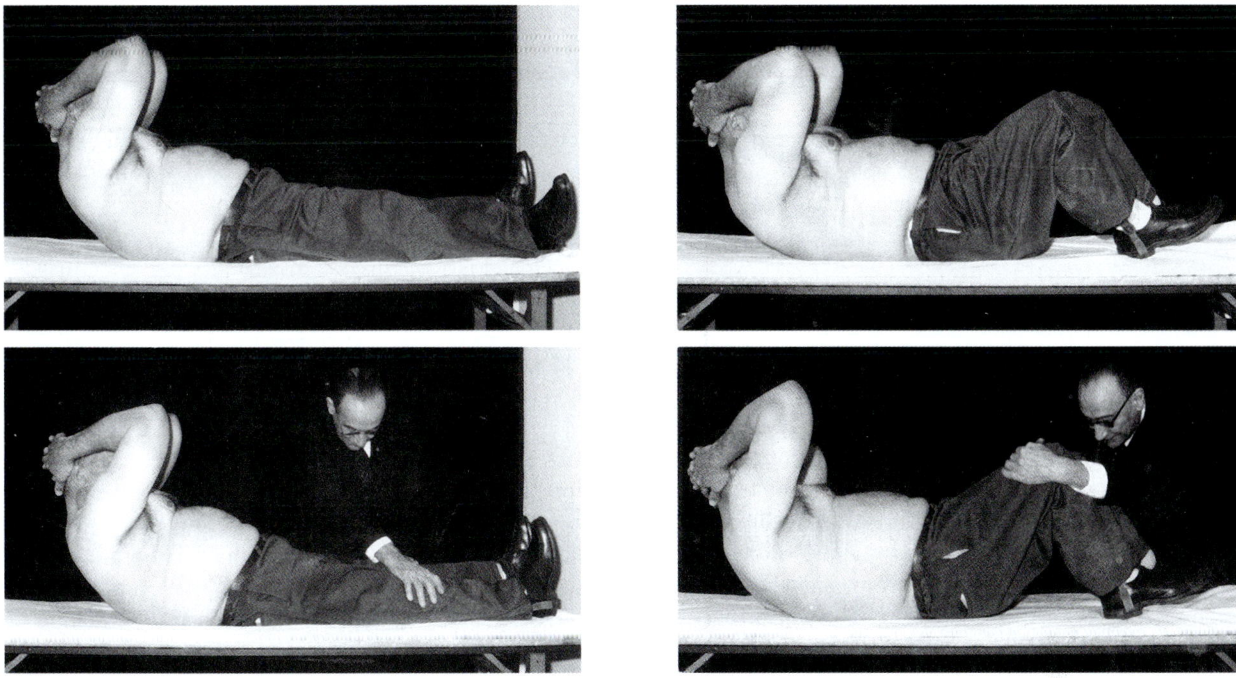

FIGURE 5-36: A subject with strong abdominal muscles and weak hip flexor muscles may only be able to perform the trunk curl.

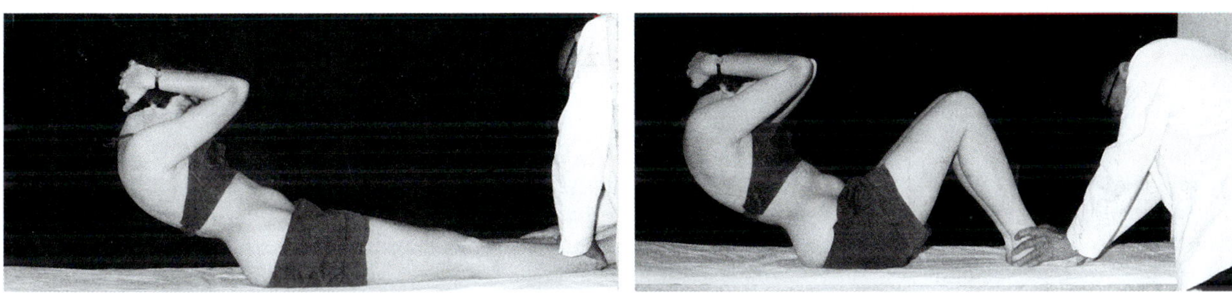

FIGURE 5-37: A subject performing a sit-up with low back arched (with legs extended or flexed), which occurs when the abdominal muscles are very weak but the hip flexors are strong.

Abdominal and Hip Flexor Imbalance

Strong Abdominals, Weak Hip Flexors

A subject with strong abdominal muscles and weak hip flexor muscles may only be able to perform the trunk curl (Figure 5-36). Flexing the trunk toward the thighs (i.e., hip joint flexion) requires action by muscles that cross the hip joint (i.e., the hip flexors). Because the abdominal muscles do not cross the hip joint, they cannot assist in the movement.

It may be noted that the subject does not raise the trunk as high from the table with legs flexed as with legs extended. The pelvis moves more freely in posterior tilt with the legs flexed. As the abdominal muscles shorten, both the pelvis and the thorax move, with the result that the thorax is not raised as high from the table as would occur if the pelvis were stabilized by the legs being in extension. (Leg braces were left on for photos in order to stabilize legs in knee-bent position.)

Strong Hip Flexors, Weak Abdominals

Sit-up with low back arched (with legs extended or flexed) occurs when the abdominal muscles are very weak (Figure 5-37). The movement consists of flexion of the hip joints by action of the hip flexors, accompanied by hyperextension of the low back (i.e., hyperlordosis). With strong hip flexors, the entire trunk-raising movement can be performed. (Compare with the photographs in Figure 5-35, in which no hip joint flexion occurs in the absence of hip flexors.)

PELVIC REGION

Sacroiliac Strain

The type of joint and the amount of movement permitted by the sacroiliac joint are central to any discussion regarding recognition and treatment of sacroiliac strain.

Sacroiliac Joint

Basmajian describes two areas of articulation in the sacroiliac joint. As seen from the side, the wings of the sacrum present anterior and posterior areas. The anterior area is shaped like an ear and is referred to as the **auricular surface**. Its articulation with the ilium is called the **synovial sacroiliac joint**. The posterior area is rough and is referred to as the **tuberosity**. This articulation with the ilium is called the **fibrous sacroiliac joint**, and it "gives attachment to strong interosseous and strong posterior sacroiliac ligaments that bind the bones together and permit only a minimum of movement" (17).

The distinction helps to clarify the confusion that results when the joint is variously described as a **syndesmosis** (i.e., immovable), a **synchondrosis** (i.e., slightly movable), or a **synovial joint** (i.e., freely movable).

Anatomists refer to this joint variously. *Gray's Anatomy* calls it a synchondrosis (18). Sabotta states that it is an almost immovable joint and that the tuberosities are united anteriorly by a joint and posteriorly by a syndesmosis (19).

For those who need to think in terms of inches, 1 mm is approximately 1/25 of an inch. Surely these measurements put this joint clearly in the classification of an almost immovable or, at best, a slightly movable joint. When one also considers that the sacroiliac joints and the symphysis pubis, like the sagittal suture of the skull, hold the two halves of the body together, the concept of an almost immovable joint is very important.

Incontinence

Urinary incontinence (UI) is defined as an involuntary release of urine and affects approximately 25 million people annually. Women are initially affected more than men with incidence being equal in the 9th decade of life. Incidence of UI may be related to pregnancy, multiple or complex vaginal delivery, and physiologic changes related to menopause in women and prostate disease in men (20, 21). Other factors affecting the incidence of UI are age-related changes to the urinary tract, history of smoking, obesity, constipation, pelvic floor weakness, injury, and depression (20). Pizzol et al. found that UI is associated with poor quality of life (22).

Fecal incontinence is defined as an involuntary or uncontrolled release of solid or liquid stool and is found to be prevalent in 7–15% of community dwelling adults. Women are affected more than men. Risk factors include advanced age, smoking, disease comorbidities, decreased physical activity, surgical trauma, and infection (23).

Pelvic organ prolapse is the abnormal herniation of the pelvic organs (uterus, vaginal apex, bladder, or rectum) from their normal position in the pelvis. Weakness of the pelvic floor musculature and elongation of connective tissue attachments to the bony pelvis may lead to this herniation (24).

Coccyalgia

Coccyalgia or coccygodynia refers to pain in the coccyx or neighboring area. Numerous factors, including trauma, are responsible for coccyalgia. Faulty position of the body may have no relation to the onset of symptoms but may result secondarily and become an important factor.

One who has persistent coccyalgia tends to sit in a very erect position, with hyperextension (i.e., hyperlordosis) of the spine in an effort to avoid undue pressure on the painful coccyx. Years of sitting in such a position can result in tightness in the low back and weakness of the gluteus maximus muscles.

SECTION VI
INTERVENTION

SIT-UP EXERCISES

For many years, sit-ups were done most frequently with the legs extended. More recently, emphasis has been placed on doing the exercise in the knee-bent position, which automatically flexes the hips in the supine position. Whether performed with legs straight or bent, the sit-up is a strong hip flexor exercise; the difference between the two leg positions is in the arc of hip joint motion through which the hip flexors act. With the legs extended, the hip flexors act through an arc from zero to approximately 80°. With the hips and knees flexed, the hip flexors act through an arc from approximately 50° (i.e., the starting position) to 125°, a total range of motion of approximately 75°.

Ironically, the knee-bent sit-up has been advocated as a means of minimizing action of the hip flexors. For many years, the idea has persisted, both among professionals and laypeople, that having the hips and knees bent in the hooklying position would put the hip flexors on a slack and eliminate action of the hip flexors while doing a sit-up, and that in this position the sit-up would be performed by the abdominal muscles. The abdominal muscles can only curl the trunk. They cannot perform the hip flexion part (i.e., the major part) of the trunk-raising movement (see Figure 5-35). Furthermore, the iliacus is a one-joint muscle that is expected to complete the movement of hip flexion and, as such, is not put on a slack. The two-joint rectus femoris is also not put on a slack, because it is lengthened over the knee joint while shortened over the hip joint.

If the hip flexors are not short, an individual, when starting the trunk-raising movement with legs extended, will curl the trunk, and the lumbar spine will flatten before the hip flexion phase begins. The danger of hyperextension will occur only if the abdominals are too weak to maintain the curl—a reason not to continue into the sit-up.

The real problem in doing sit-ups with the legs extended compared to the apparent advantage of flexing the hips and knees stems from dealing with many subjects who have short hip flexors. In the supine position, a person with short hip flexors will lie with the lumbar spine hyperextended (i.e., a hyperlordosis). The hazard of doing sit-ups from this position is that the hip flexors will further hyperextend the lumbar spine, causing a stress on that area while doing the exercise, and will increase the tendency toward a hyperlordotic posture in standing. The hooklying position, however, releases the downward pull by the short hip flexors, allowing the pelvis to tilt posteriorly and the lumbar spine to flatten, thereby relieving strain on the lumbar spine.

Instead of recognizing and treating the problem of the short hip flexors, the solution has been to give in to them by flexing the hips and knees. Problems arise from this solution, however. The same hazard of coming up with the lumbar spine hyperextended can occur with the knees bent, and it does occur when the abdominal muscles are too weak to curl the trunk. In trying to come up, the subject requires more pressure than usual to hold the feet down, or more extension of the legs, or is aided by performing the movement quickly with added momentum. Sometimes it is advocated—inadvisably—that the arms be placed overhead and brought quickly forward to help in performing the sit-up. This added momentum enables the subject to do the sit-up, but the lumbar spine is hyperextended, causing strain on the abdominal muscles as well as stress on the lumbar spine.

Indications and Contraindications

This subject, with knees flexed, can flex the vertebral column but cannot raise the trunk any higher from the table than illustrated (Figure 5-38).

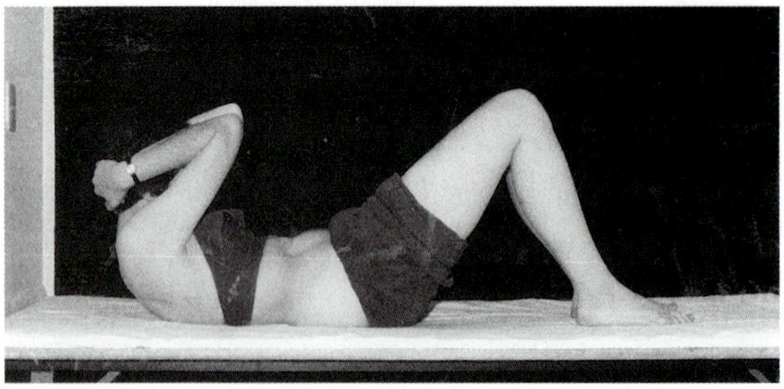

FIGURE 5-38: Subject can flex the vertebral column but cannot raise the trunk any higher than shown.

With the feet held down, the subject immediately begins the hip flexion phase and can continue to a full sitting position (Figure 5-39), as seen in the series of photographs of this same individual in Figure 5-35.

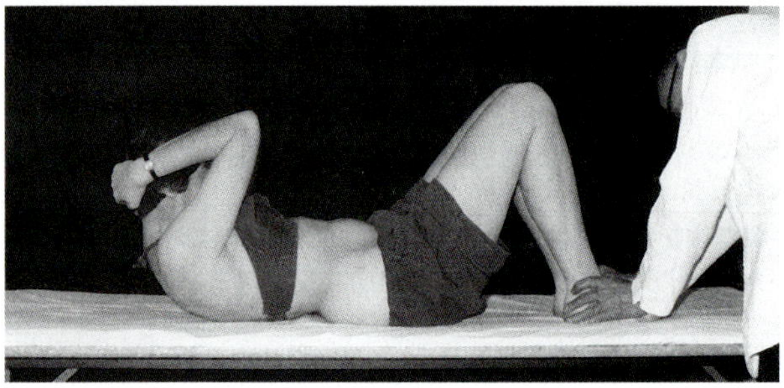

FIGURE 5-39: Subject, with the feet held down, can continue to a full sitting position.

The subject is making an effort to sit up with the arms in an easy test position and the feet not held down (Figure 5-40). It is obvious that the subject goes immediately into the hip flexion phase. Legs tend to extend in an effort to move the center of gravity of the lower extremities more distally and offset the force exerted by the trunk. These same problems exist with respect to stabilization of the feet whether the knees are extended or flexed.

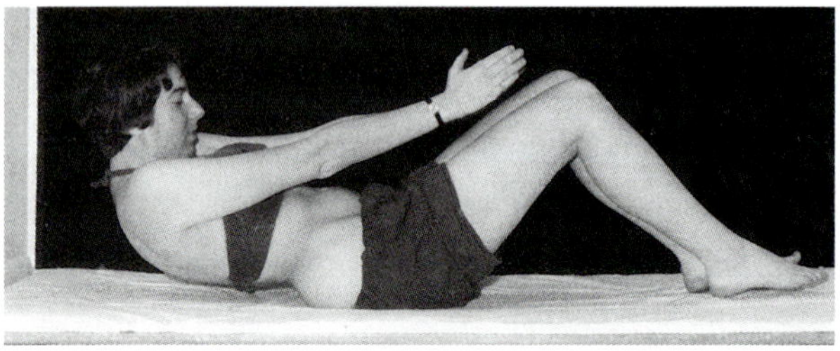

FIGURE 5-40: Subject, without the feet held down, tries to sit up.

The ability to do a curled-trunk sit-up should be considered a normal accomplishment. People should be able to get up easily from a supine position without having to roll over on the side or push themselves up with their arms. When there is weakness in either or both of the muscle groups involved in a curled-trunk sit-up (i.e., abdominal and hip flexor muscles), efforts should be made to correct the weakness and restore the ability to perform the movement correctly. Hip flexors may exhibit some weakness associated with postural problems, but rarely does this occur to the degree that it interferes with performing the sit-up (i.e., hip flexion) movement. The problem in performing the trunk curl results from weakness of the abdominal muscles. Using the sit-up exercise to correct the abdominal weakness is a mistake because, when marked weakness exists, the hip flexors initiate and perform the movement with the lumbar spine hyperextended.

The sit-up is a strong hip flexor exercise whether the knees are bent or the legs are extended. The hip joint moves to completion of hip joint flexion with hips and knees bent, making this type of sit-up more conducive to the development of shortness in the iliopsoas than a sit-up with the knees and hips extended.

Normal flexibility of the back is a desirable feature, but excessive flexibility is not. The hazards of the knee-bent sit-up also relate to the danger of hyperflexion of the trunk (i.e., the spine curving convexly backward). With the body in the anatomical position or supine with the legs extended, the center of gravity is slightly anterior to the first or second sacral segment. With the hips and knees bent, the center of mass moves cranially (i.e., toward the head). The lower extremities exert less force in counterbalancing the trunk during a sit-up with the hips and knees bent than during a sit-up with the legs extended.

Two alternatives exist for accomplishing the sit-up from this knee-bent position: Outside pressure must be exerted to hold the feet down (more than is required for those few who need it with the legs extended) or the trunk must curl excessively to move the center of mass downward. This excessive flexion is portrayed as an exaggerated thoracic curve (i.e., marked rounding of the thoracic spine), as abnormal flexion involving the thoracolumbar area (i.e., roundness extending into the lumbar spine), or both. Abnormal flexion involving the thoracolumbar area is accentuated when the knee-bent sit-up is done without the feet being held down and with the heels placed close to the buttocks.

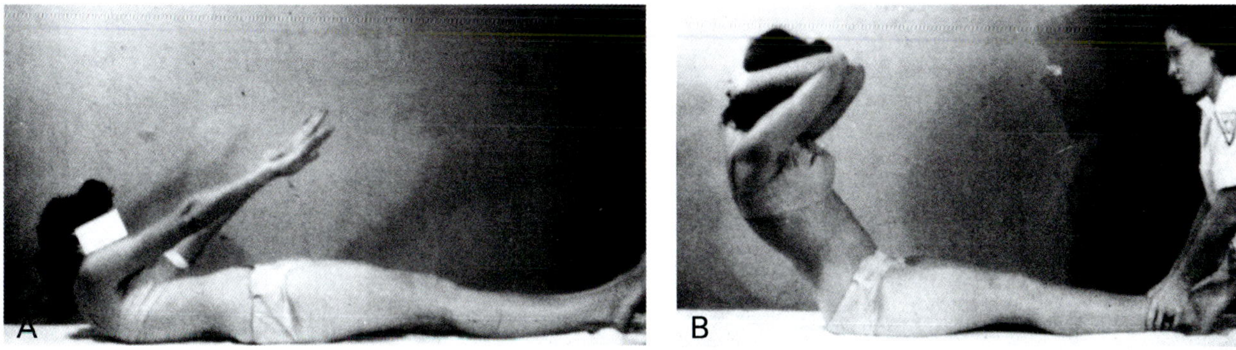

FIGURE 5-41: **A**, Subject with marked abdominal weakness has trouble sitting up when the feet are not held down. **B**, Same subject, with the feet held down.

Effect of Holding Feet Down During Trunk Raising Forward

The center of mass of the body generally is given as approximately the level of the first sacral segment, and this point is above the hip joint. If ½ the body weight is above the center of mass, then more than ½ the body weight is above the hip joint. (Basmajian states that the lower extremities constitute approximately ⅓ of the body weight (25).) For most people, this means that the force exerted by the trunk in the supine position is greater than the force exerted by both lower extremities. Usually, double-leg raising with the knees straight can be initiated without overbalancing the weight of the trunk in the supine position. Seldom, however, can the straight or hyperextended trunk be raised from the supine position toward a sitting position without some outside force being applied (e.g., pressure downward on the feet) in addition to that exerted by the extended extremities.

On the other hand, if the trunk curls sufficiently as the trunk raising is started, the center of mass of the body moves downward, toward, or below the hip joints. As this occurs, the curled trunk can be raised in flexion toward the thighs without the feet needing to be held down. Most adolescents (especially those with long legs in relation to the trunk) and most women can perform a sit-up with legs extended and without the feet being held down. In contrast, many men need to have some added force (usually very little) applied at the point where the trunk curl is completed and the hip flexion phase begins.

For the curled-trunk sit up to be used as a test of abdominal muscle strength, it must be made certain that the ability to curl the trunk is actually being measured. The trunk curl must precede the hip flexion phase in the trunk-raising movement. When the feet are not held down, the pelvis tilts posteriorly as the head and shoulders are raised in initiating the trunk curl. With the feet held down, the hip flexors are given fixation, and the trunk raising can immediately become an arched-back sit-up with flexion at the hip joints. Hence, to help ensure that the test determines the ability to curl the trunk before the hip flexion phase begins, the feet must not be held down during the trunk flexion phase.

If abdominal strength is normal, the feet can be held down if the subject is performing only a few sit-ups, but it should be avoided if many repetitions are being performed. One or two curled-trunk sit-ups, properly done, determine normal strength; they do not determine endurance. An individual may grade normal and perform several sit-ups properly. With repeated sit-ups, however, the abdominal muscles may fatigue, and this same individual may slip into doing an arched-back sit-up. This situation arises frequently, because abdominal muscles do not have the endurance exhibited by the hip flexors.

The transition to an arched-back sit-up could—and would—go undetected if the feet were held down from the beginning of the sit-up. If the feet were not held down during the initial spine flexion phase, however, the inability to curl the trunk would become obvious as fatigue sets in. An individual might be able to do as many as 100 sit-ups with the feet held down, yet no more than 5 sit-ups without the feet held down. This would indicate that the trunk-raising became an arched-back sit-up after the first five.

An individual with marked abdominal muscle weakness who, with arms in a relatively easy test position, is unable to flex the lumbar spine and complete the sit-up when the feet are not held down (Figure 5-41).

The same individual shown in Figure 5-41A who, with arms in a normal-grade test position, is able to perform the sit-up by hip flexor action because the feet are held down. As a test, this only measures hip flexor strength.

THERAPEUTIC EXERCISES: TRUNK CURL

To strengthen the abdominal muscles that show weakness on the trunk-curl test, it is desirable, in most instances, to have the subject perform only the trunk-curl part of the movement. This provides the advantage of abdominal muscle exercise without strong hip flexor exercise. In addition, according to Nachemson and Elfstron, less intradiscal pressure occurs when doing only the trunk curl as compared to completing the sit-up (26).

When the subject can perform the trunk curl to completion of spine flexion, the resistance may be increased by folding the forearms across the chest and completing the curl. Later, more resistance can be added by placing the hands behind the head and completing the curl. At each stage, work to achieve some endurance (i.e., completion of curl, holding it for several seconds and repeating approximately 10 times).

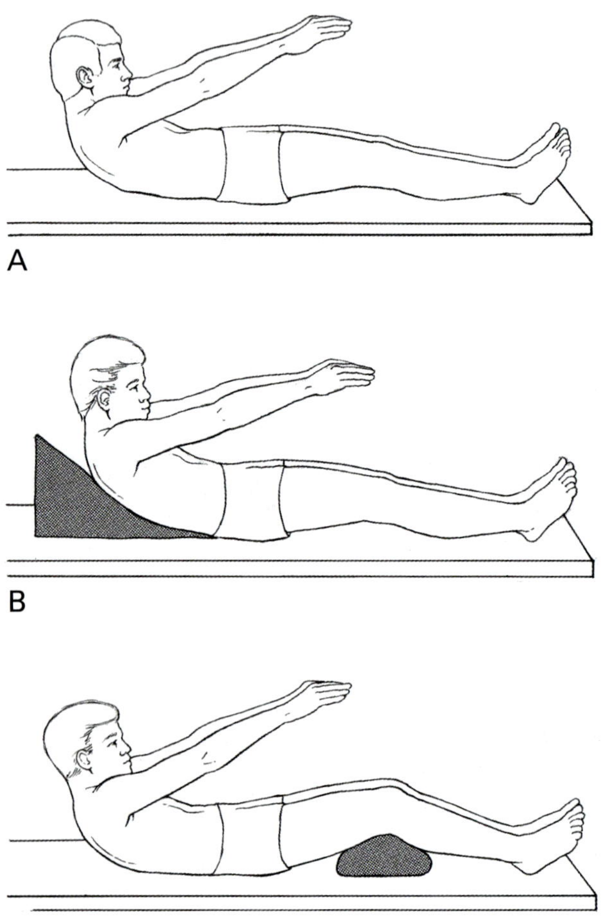

FIGURE 5-42: Abdominal exercises: **A**, Trunk curl; **B**, Assisted trunk curl; **C**, Short hip flexors.

Abdominal Exercise, Trunk Curl: In supine with the lower extremities extended, place a small roll under the knees. Tilt the pelvis to flatten the lower back on the table by pulling upward and inward with the muscles of the lower abdomen. With arms extended forward, raise the head and shoulders from the table. Raise the upper trunk as high as the back will bend, but *do not try to come to a sitting position* (Figure 5-42A).

Abdominal Exercise, Assisted Trunk Curl: If the abdominal muscles are very weak and the subject cannot lift the shoulders from the table, modify the above exercise by placing a wedge-shaped pillow (or the equivalent) under the head and shoulders. This position enables the subject to exercise within a short range of motion. As the ability to hold the completed curl improves, use a smaller pillow and have the subject flex to completion of the curl (Figure 5-42B).

Abdominal Exercise, Short Hip Flexors: When the hip flexor muscles are short and restrict the posterior pelvic tilt, modify the above trunk curl exercise by temporarily placing a pillow under the knees to passively flex the hips, as illustrated in Figure 5-42C.

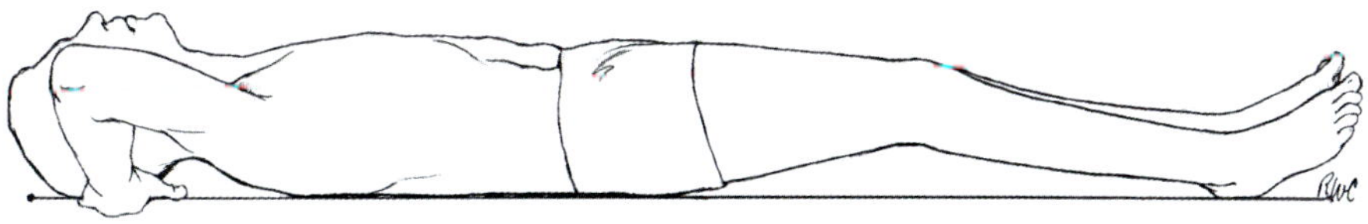

FIGURE 5-43: A subject with strong abdominal muscles and very weak or paralyzed hip flexors cannot lift the legs upward from a supine position.

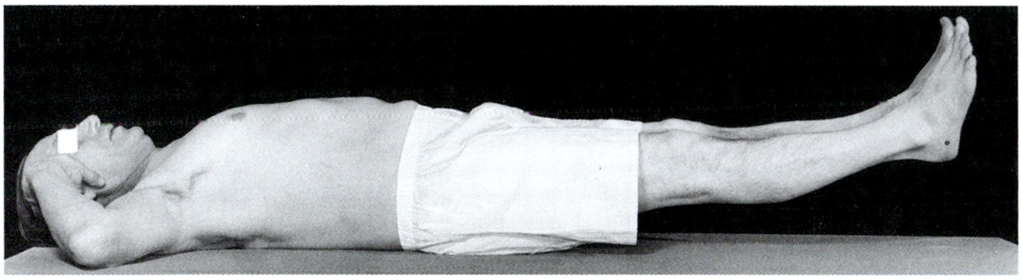

FIGURE 5-44: A subject with strong abdominal muscles can have the back held flat on the table by the abdominals holding the pelvis in posterior tilt during the leg-raising movement.

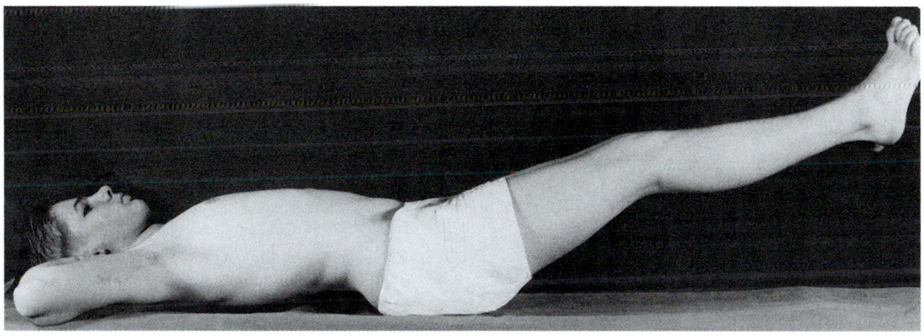

FIGURE 5-45: In a subject with weak abdominal muscles, the pelvis tilts anteriorly as the legs are lifted.

ABDOMINAL MUSCLES DURING LEG LOWERING

Definitions and Descriptions of Abdominal Muscles During Leg Lowering

Double-leg raising from a supine position is flexion of the hips with the knees extended. With the knee extensors holding the knees straight, the hip flexors raise the legs upward. No abdominal muscles cross the hip joints, so these muscles cannot assist directly in the leg-raising movement. The role of the hip flexors is made very clear by observing the loss of function when they are paralyzed, as seen in Figure 5-43.

To perform the double-leg-raising movement from a supine position, the pelvis must be stabilized in some manner. The abdominal muscles cannot enter directly into the leg-raising movement, but the strength or weakness of these muscles directly affects the trunk position and how the pelvis is stabilized. Leg raising through hip flexor action exerts a strong pull downward on the pelvis in the direction of tilting it anteriorly. The abdominal muscles pull upward on the pelvis, in the direction of tilting it posteriorly.

A subject with strong abdominal muscles and very weak or paralyzed hip flexors cannot lift the legs upward from a supine position. In attempting to raise the legs, the only active movement that occurs is that the pelvis is drawn forcefully into posterior tilt. Passively, the thighs may be raised slightly from the table secondary to tilting of the pelvis, as illustrated in Figure 5-43, or they may remain flat on the table if the anterior hip joint structures are relaxed.

If the subject has strong abdominal muscles, the back can be held flat on the table by the abdominals holding the pelvis in posterior tilt during the leg-raising movement (Figure 5-44).

If the abdominal muscles are weak, the pelvis tilts anteriorly as the legs are lifted. As this tilt occurs, the back hyperextends, often causing pain, and the weak abdominal muscles are put on a stretch and are vulnerable to strain (Figure 5-45).

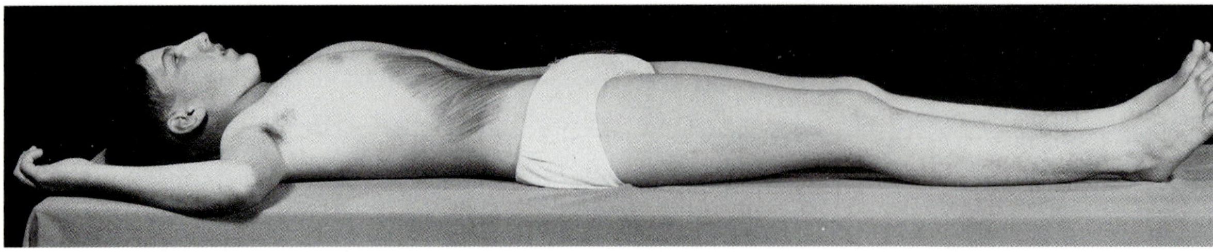

FIGURE 5-46: The lateral fibers of the external oblique act to tilt the pelvis posteriorly.

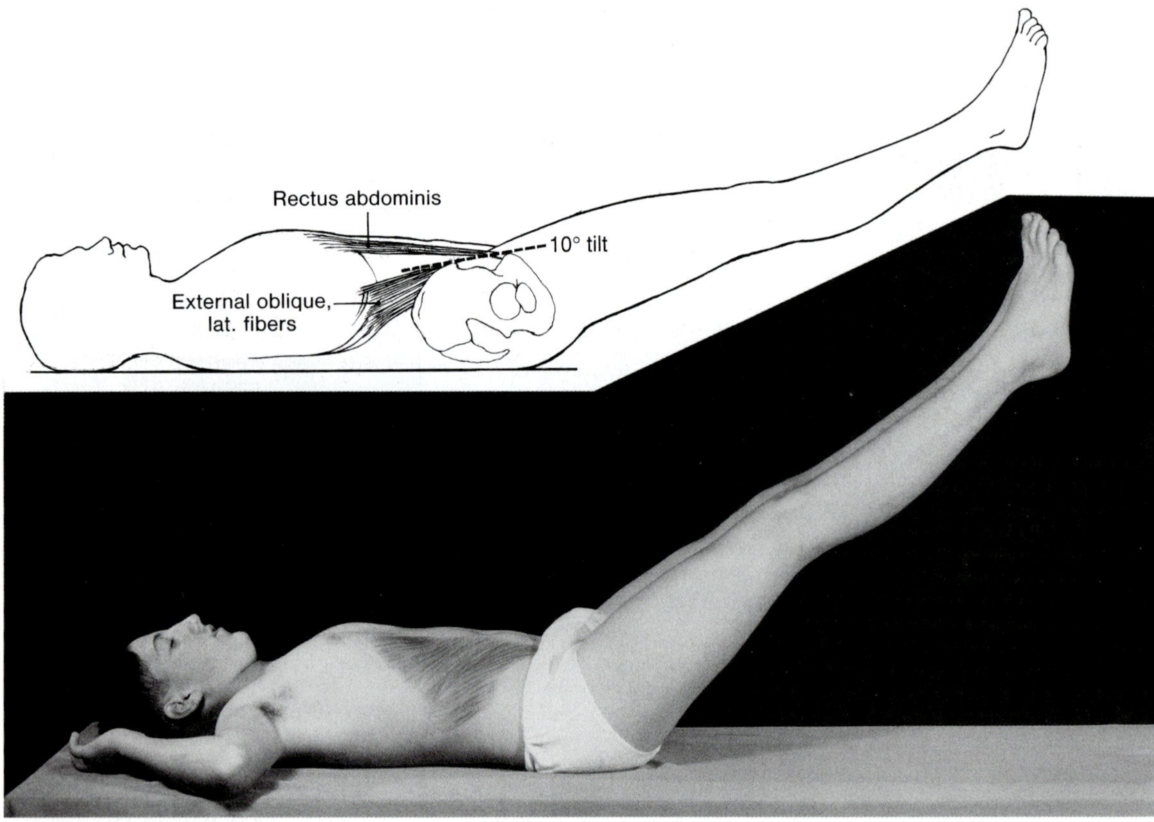

FIGURE 5-47: Action by the rectus abdominis and external oblique is required to maintain the pelvis in posterior tilt and the low back flat on the table as the legs are raised or lowered.

Actions of the Abdominal Muscles

When discussing the actions of the abdominal muscles, it should be recognized that various segments of the abdominal musculature are closely allied and interdependent. The external oblique, however, is essentially fan-shaped, and different segments may have different actions. The pelvis can be tilted posteriorly by an upward pull on the pubis, by an oblique pull in an upward and posterior direction on the anterior iliac crest, or by a downward pull posteriorly on the ischium. The muscles (or parts of muscles) that are aligned in these directions of pull are the rectus abdominis, the lateral fibers of the external oblique, and the hip extensors. These muscles may act to tilt the pelvis posteriorly whether the subject is standing erect or lying supine. In the supine position, however, during double-leg lowering, the hip extensors are not in a position to assist in maintaining flexion of the lumbar spine and posterior pelvic tilt. Consequently, the rectus abdominis and external oblique muscles assume the major role in maintaining the position of the low back and pelvis during the leg-lowering movement.

The lateral fibers of the external oblique act to tilt the pelvis posteriorly, and they may do so with little or no assistance from the rectus abdominis. The subject's arms are placed overhead to expose the drawings on the abdomen (Figure 5-46).

Action by the rectus abdominis and external oblique is required to maintain the pelvis in posterior tilt and the low back flat on the table as the legs are raised or lowered (Figure 5-47).

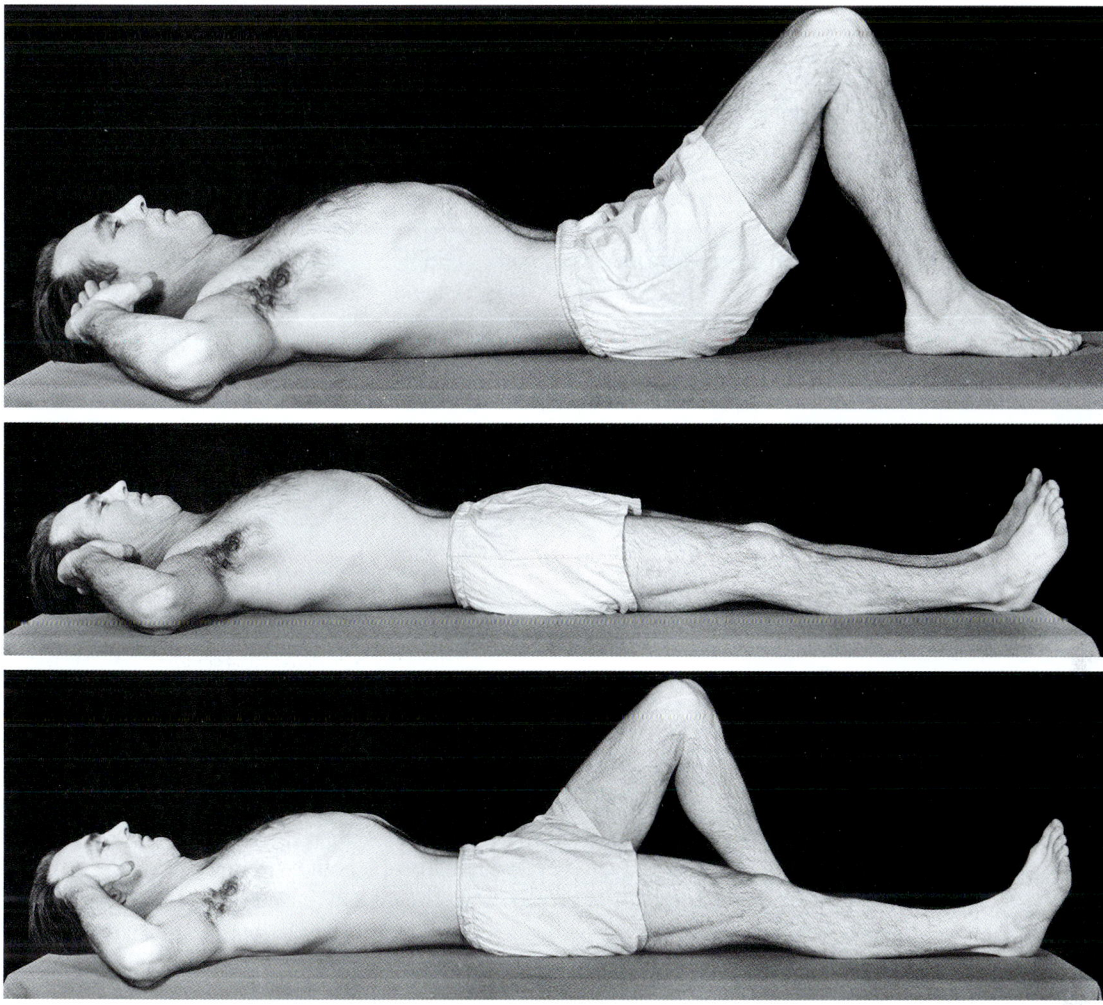

FIGURE 5-48: Posterior pelvic tilt and leg-sliding exercise done correctly to exercise the external oblique.

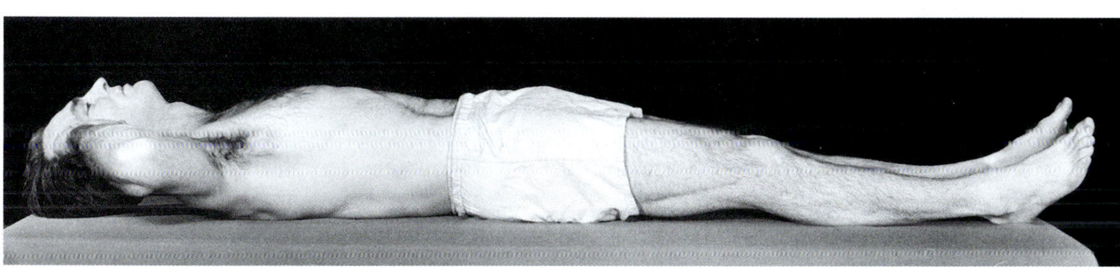

FIGURE 5-49: Posterior pelvic tilt may be done with the rectus abdominis, but it should not be done in this manner when attempting to strengthen the external oblique.

THERAPEUTIC EXERCISES: POSTERIOR PELVIC TILT

Posterior pelvic tilt and leg-sliding exercise done correctly to exercise the external oblique (Figure 5-48).

The lower abdomen is pulled upward and inward, and the pelvis is tilted posteriorly to flatten the low back on the table by action of the external oblique (especially the posterior lateral fibers). The subject should be taught to palpate the lateral fibers of the oblique to ensure their action and to avoid using the gluteus maximus to tilt the pelvis when doing this exercise.

Posterior pelvic tilt may be done with the rectus abdominis, but it should not be done in this manner when attempting to strengthen the external oblique (Figure 5-49).

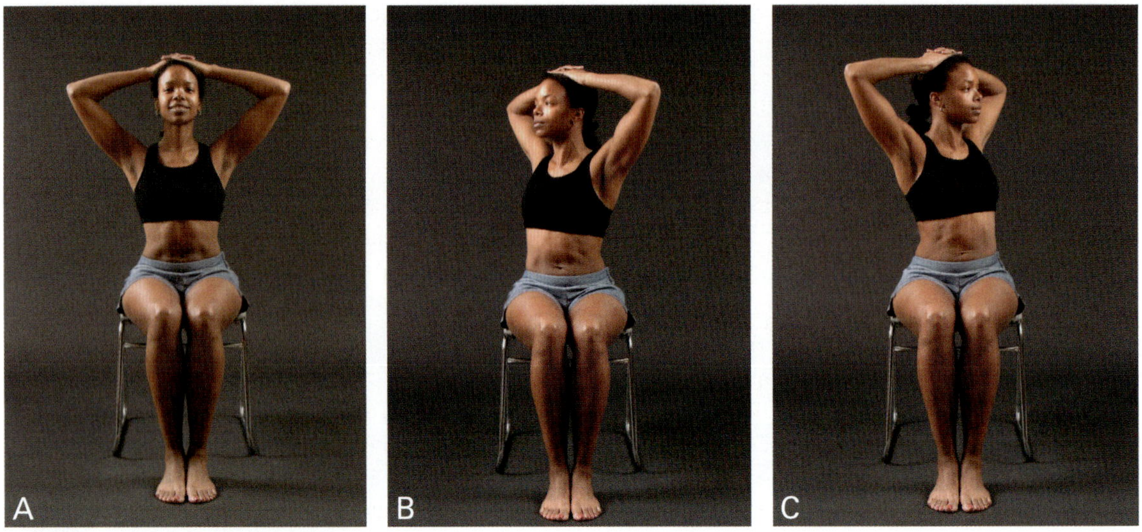

FIGURE 5-50: Trunk rotation through external oblique strength exercise.

THERAPEUTIC EXERCISES: TRUNK ROTATION

External Oblique Strength Exercise

Strong external oblique muscles play an important role in maintaining good postural alignment and in preventing low back pain. Exercise to strengthen these muscles must be specific as illustrated above. Weakness of the external oblique is common in persons performing excessive sit-up exercises because the posterolateral fibers of the external oblique elongate during the trunk curl.

The sitting position offers resistance for the external obliques in holding the lower abdomen up and in and keeping the low back flat. In addition, the rotation of the thorax on the pelvis, as illustrated in Figure 5-50, requires strong unilateral action alternately by the right and left external oblique muscles.

Starting Position: Sit erect in a chair or on a stool, facing forward with feet on the floor and legs together. This position stabilizes the pelvis. Place hands on top of the head to help keep the chest up and the thoracic spine in good alignment.

Exercise: To strengthen the left external oblique, slowly rotate the upper trunk toward the right (clockwise), holding the position for several seconds. Relax and return to midline. To exercise the right external oblique, slowly rotate the upper trunk toward the left (counterclockwise), holding the position for several seconds. Relax and return to midline.

> **NOTE:** *Exercises may be performed in standing but it is more difficult to fix the upper trunk because the pelvis rotates toward the same side as the external oblique.*

TREATMENT FOR COCCYALGIA

Conservative treatment for coccyalgia consists of providing some padding for the coccyx by use of a corset, which is worn low to hold the buttocks close together. Preferably, this corset has back laces that cross over and tighten by lateral straps.

The corset should be tightened with the patient standing. The gluteal muscles thus form a padding for the coccyx in the sitting position. A soft pad may be incorporated into the corset as well. Pain may be alleviated by this simple procedure.

THERAPEUTIC OBJECTIVES FOR MUSCLES OF RESPIRATION

A variety of techniques, procedures, and mechanical devices are used to assist lung function. Treatment must be specific to the ventilatory problem of the patient, but certain principles and practices are basic to all respiratory therapies.

Reduce Fear: The first step in reducing the work of breathing and instituting effective treatment is to reduce the patient's level of fear and anxiety to obtain confidence and compliance. Existing respiratory problems are severely exacerbated by breath-holding, breathlessness, and increased tension in the accessory muscles, all of which frequently accompany a fearful state. When the confidence and cooperation of the patient are obtained, other treatment measures will be far more effective.

Improve Relaxation: Relaxation produces a decrease in oxygen consumption of the skeletal muscles and an increase in compliance of the chest wall. When indicated, diaphragmatic breathing exercises may aid in relaxation and give the patient a better sense of control over respiration. These exercises emphasize abdominal rather than rib cage expansion, and they are helpful in cases with overuse of the accessory muscles of the neck and upper chest. Practicing a pattern of deep breathing and sighing can reduce the work of breathing and help to relax a patient who has attacks of breathlessness or breath-holding.

Improve Posture: Optimal breathing capability derives from a posture of optimal muscle balance. A balanced musculature is most efficient in terms of energy expenditure. Imbalance of the musculature resulting from tightness, weakness, or paralysis may adversely affect the volumes and pressures that can be attained and maintained. Very weak and protruding abdominal muscles are not able to generate maximum expiratory pressures to meet increased demands of breathing brought on by exertion or illness. Weakness of the upper back erector spinae and of the middle and lower trapezius muscles interferes with the ability to straighten the upper back, limiting the ability to raise and expand the chest and thus maximize lung capacity. Postural problems associated with kyphosis, kyphoscoliosis, osteoporosis, and pectus excavum restrict breathing and result in decreased chest wall compliance.

Improve Strength and Endurance of Respiratory Muscles: "Strength is needed for sudden respiratory movements such as coughing and sneezing, and brief spells of extreme exertion, whereas endurance is necessary for more prolonged exercise or to overcome an increase in air flow resistance or a decrease in compliance" (5).

Strong and well-conditioned muscles are more efficient and require less oxygen for a given amount of work than poorly conditioned muscles. Reports are mixed as to the efficacy of muscle strength training of respiratory muscles, but such training may be beneficial if respiratory muscle weakness limits exercise or diminishes inspiratory capacity.

The stronger the abdominal muscles, the greater their ability to compress the abdomen and thus generate additional pressure during expiration. Exercises to strengthen these muscles can help to improve coughing and other expulsive maneuvers that are required to clear the airways and facilitate breathing.

If there is marked weakness of these abdominal muscles, exercises should be supplemented with a support that will reduce the downward pull of the abdomen and help to keep the diaphragm in the most advantageous position for both inspiration and expiration. Such assistance often helps to minimize breathing problems associated with obesity.

Respiratory muscle fatigue may precipitate respiratory failure. Endurance training is intended to increase the capacity of muscles to resist fatigue. Training has been shown to benefit approximately 40% of patients suffering from chronic air-flow obstruction, and slight improvements in endurance have been observed in patients with cystic fibrosis (5).

In disorders of the respiratory muscles, "[r]espiratory failure is usually closely related to the degree of respiratory muscle weakness but occasionally occurs with only mild impairment of muscle function" (5). Because of the high risk of respiratory failure associated with weak respiratory muscles, exercises to strengthen these muscles may be of critical importance, but they also must be very conservative and closely monitored.

Improve Coordination: The oxygen cost of performing a task can be greater than normal in a person who moves in an uncoordinated fashion. When inefficient patterns of breathing and movement are identified, corrective treatment can be instituted, and the work of breathing will be gradually reduced.

Improve Overall Fitness: Cardiovascular fitness can be improved through whole-body exercises (e.g., walking and bicycling) to strengthen ventilatory capability and efficiency. Exercises that involve the legs rather than the arms are preferred initially so that the accessory muscles can be used to aid breathing.

Reduce Weight: Respiratory problems associated with obesity often are very severe. According to Cherniack, the oxygen cost of breathing in an obese person is approximately threefold the normal cost (27). Unlike some skeletal and neuromuscular respiratory disorders, obesity is a condition that can sometimes be reversed, with, in turn, respiration greatly improved.

ABDOMINAL STRENGTHENING EXERCISES

Exercises in the lying position should be done on a firm surface (e.g., a board on the bed, a treatment table, or the floor, with a thin pad or folded blanket placed on the hard surface for comfort).

Stretching exercises should be preceded by gentle heat and massage to help relax tight muscles. (Avoid using heat on weak, overstretched muscles.) Stretching should be done gradually, with a conscious effort to relax. Continue until a firm, but tolerable pull is felt, breathing comfortably while holding the stretch, then return slowly from the stretched position.

Strengthening exercises should also be done slowly, with an effort to feel a strong pull by the muscles being exercised. Hold the completed position for several seconds, then relax and repeat the exercise the number of times indicated by your therapist.

Low Back Stretching

Face-Lying Position: Place a firm pillow under the abdomen (*not* under the hips) and a rolled towel under the ankles. Lying on a firm pillow puts low back muscles on a slight stretch.

Back-Lying Position: Slowly pull both knees toward the chest, gently stretching the low back muscles just enough to flatten low back on table (Figure 5-51).

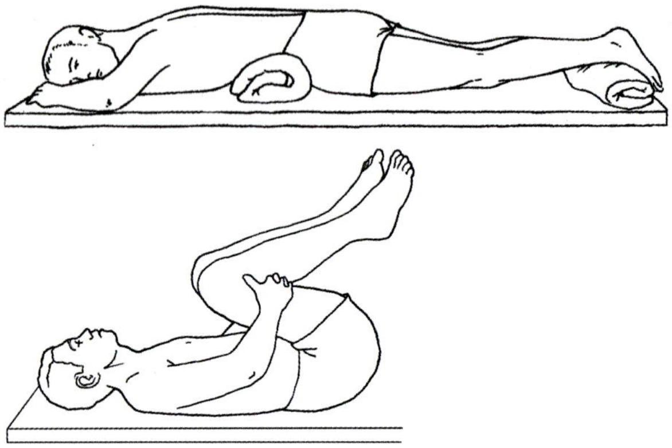

FIGURE 5-51: Low back stretching.

Trunk Rotation in Back-Lying Position

Starting Position: Lying on the floor with knees bent and feet flat (Figure 5-52).

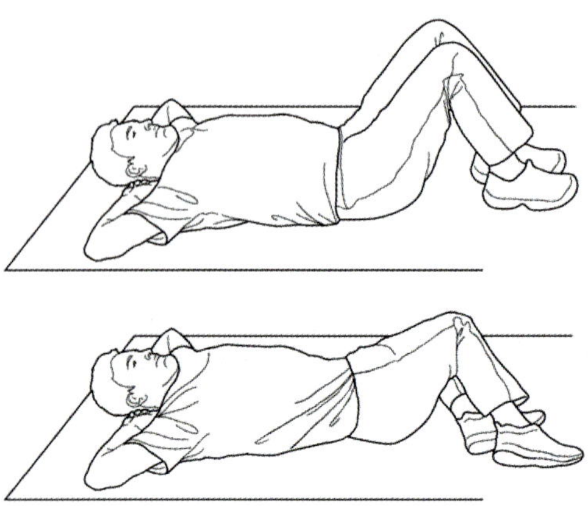

FIGURE 5-52: Trunk rotation in back-lying position.

Slowly move the knees toward the left, rotating the lower trunk. Return to midline and repeat toward the other side. Do NOT move arms from starting position and keep feet on floor during the exercise.

Lower Abdominal Exercise and Low Back Stretching

Back-Lying Position: Bend knees and place feet flat on table (Figure 5-53). With hands up beside head, tilt pelvis to flatten low back on table by pulling up and in with lower abdominal muscles. Keep low back flat and slide heels down along table. Straighten legs as much as possible with back held flat. Keep back flat and return knees to bent position, sliding one leg back at a time. (Do NOT use buttock muscles to tilt pelvis and do NOT lift feet from floor.)

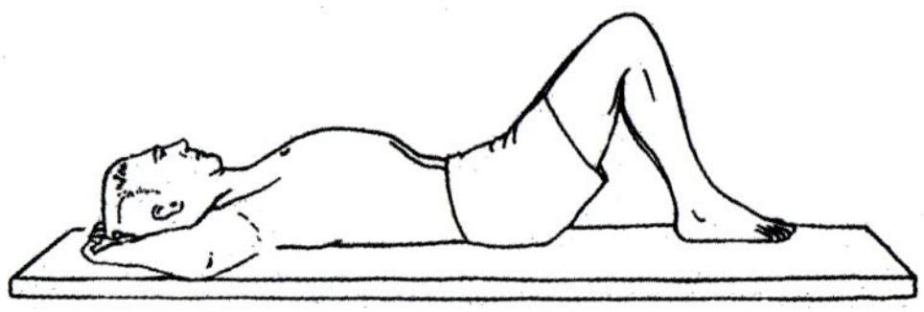

FIGURE 5-53: Lower abdominal exercise and low back stretching.

Lower Abdominal Exercise

Back-Lying Position: Place a rolled towel or small pillow under knees (Figure 5-54). With hands up beside head, tilt pelvis to flatten low back on table by pulling up and in with lower abdominal muscles. Hold back flat and breathe in and out easily, relaxing upper abdominal muscles. There should be good chest expansion during inspiration, but back should not arch. (Do NOT use buttock muscles to tile the pelvis.)

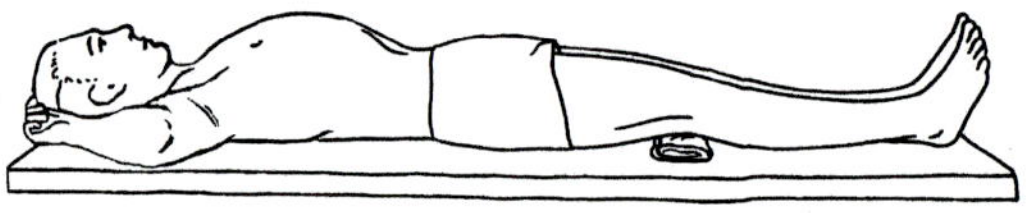

FIGURE 5-54: Lower abdominal exercise.

Wall-Standing Postural Exercises

Stand with back against a wall, heels about 3 inches from wall. Knees should be straight, but *not locked.* Place hands up beside head with elbows touching wall. Tilt pelvis to flatten low back against wall by pulling up and in with the lower abdominal muscles. Keep arms in contact with wall and move arms slowly to a diagonally overhead position (Figure 5-55).

FIGURE 5-55: Wall-standing postural exercises.

Modified Upper Abdominal Strengthening

Modified upper abdominal strengthening is from forearm prop for marked weakness. Maintain trunk curl position without elevation or trunk rotation (Figure 5-56).

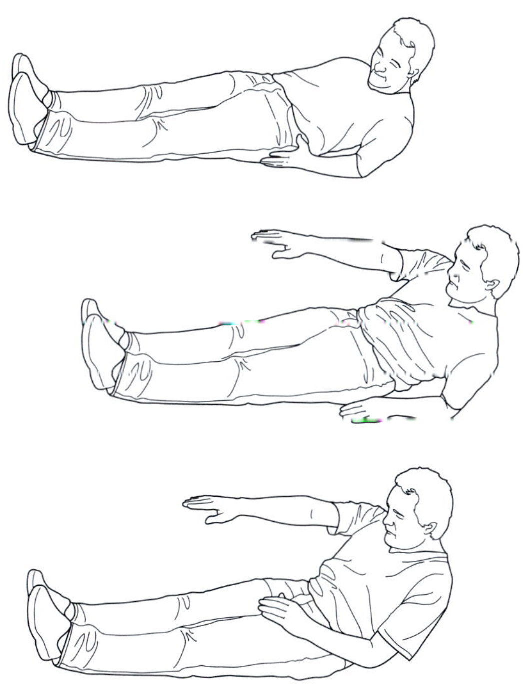

FIGURE 5-56: Modified upper abdominal strengthening.

Starting Position: Forearm prop with trunk curl; head in neutral position.

1. Reach forward with right arm, maintaining trunk curl. Hold. Return to starting position. Repeat with left arm.
2. Reach forward with right arm. Hold. Reach forward with left arm. Return right arm, then left arm to starting position.

External Oblique Abdominal Strengthening

Seated in a chair with feet on floor and knees together and facing forward, slowly rotate trunk toward the left, using diagonal abdominal muscles. Hold. Return to midline and repeat toward the other side (Figure 5-57).

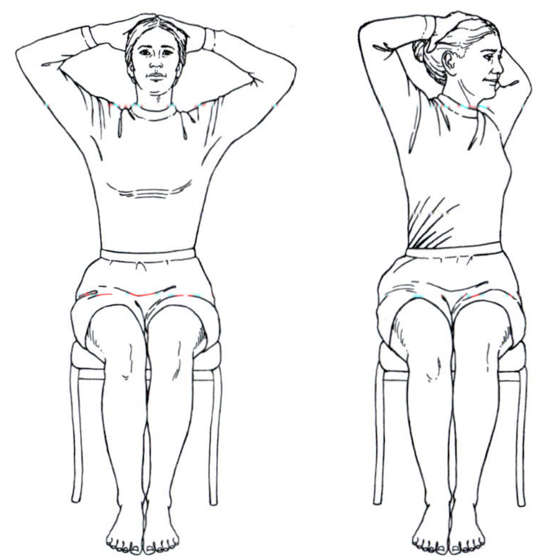

FIGURE 5-57: External oblique abdominal strengthening.

Upper Abdominal Strengthening

In back-lying position, tilt pelvis to flatten low back on table by pulling up and in with lower abdominal muscles. With arms forward, raise head and shoulders up from table. Do NOT attempt to come to sitting position, but raise upper trunk as high as back will bend. As strength progresses, arms may be folded across chest, and later placed behind head to increase resistance during the exercise (Figure 5-58).

FIGURE 5-58: Upper abdominal strengthening.

RATIONALE FOR SACROILIAC JOINT TREATMENT

Sacroiliac strains do exist. As stated by the authors of *Posture and Pain*, "Because the normal range of motion of the joint is small, it takes very little more to be excessive. A tension sufficient to cause ligamentous strain may not appear on x-ray" (28).

Treatment varies from the conservative approach of nothing other than application of support in the form of a belt, corset, or brace to the use of sophisticated mobilization techniques.

In all probability, most sacroiliac strains are the result of undue tension on the ligaments without any displacement. There is no way of knowing how many cases are never brought to professional attention but clear up spontaneously. Very often, application of a belt or some other support gives immediate relief. This response to immobilization is a strong indication of a strain only.

Opinions vary widely with respect to the need for mobilization. In some cases, it may be the treatment of choice and appropriate; in others, it may be unnecessary and unwarranted. If a belt does not offer relief but mobilization does, it is plausible that a minor displacement was corrected by the manipulation. Many individuals will be helped by use of a support after the mobilization treatment. A person who is subject to recurrent attacks is in greater need of a support to protect the joint from becoming too mobile as compared with a person who has had a simple strain.

The sacroiliac joint is supported by strong ligaments. No muscles cross directly over the joint to support it. There would be no useful function for elastic, contractile tissue (e.g., muscle) to act on a joint that has almost no movement. Weakness or tightness of muscles elsewhere, however, can affect the sacroiliac joint. When motion is restricted in an adjacent area (e.g., the back or the hip joints), stress on the sacroiliac joints is increased during any forward-bending movement.

Sacroiliac strain in subjects with flat-back posture and tight hamstrings tends to be more common among men than among women. On the other hand, sacroiliac strain in subjects with a lordosis is found more often among women than among men. Sacroiliac strain may be bilateral but more often is unilateral. There may be more pain in sitting than in standing or walking. The strain can be brought on by sitting in unsupported flexion of the lumbosacral region (e.g., sitting on the floor tailor fashion, squatting, or sitting on a chair or a sofa that is too deep from front to back).

Usually, there is tenderness over the affected sacroiliac area. There also may be diffuse, not easily defined pain through the pelvis, buttock, and into the thigh. Pain may be referred to the lower abdomen and groin area, and at times, there may be associated sciatic symptoms. In some cases, there is pain on hip flexion.

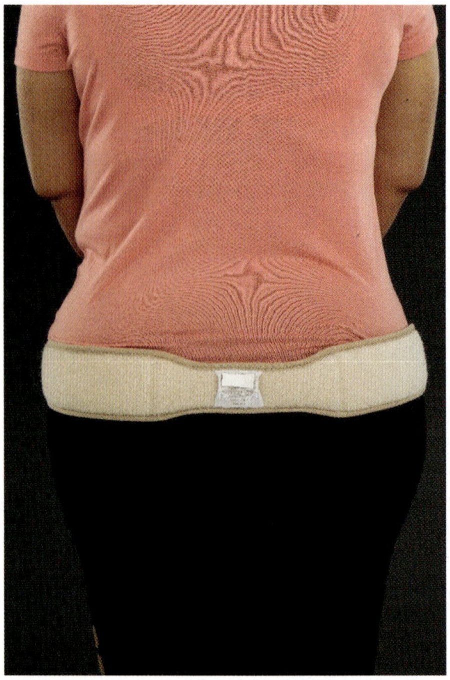

FIGURE 5-59: Support for sacroiliac strain in the form of a belt, corset, or brace.

For immobilization with a belt, commercial belts are usually available and adequate for men. For women, it is more difficult to keep a belt from riding up out of a position of support.

Figure 5-59 shows a panty girdle with a strap approximately 3 inches wide attached to the girdle with Velcro.

PELVIC FLOOR TRAINING

Exercise and Biofeedback

The effectiveness of pelvic floor muscle exercises to improve pelvic floor strength and function has been well documented (29, 30). Patients are instructed to "squeeze the muscle of the bottom of your pelvis as if trying to stop the flow of urine." Feedback for eliciting pelvic floor contraction can be provided via vaginal palpation, the use of biofeedback technology via ultrasound (31).

Biofeedback via anal manometer, using a rectal balloon, has been shown to be effective for addressing fecal incontinence. Patients are instructed to contract the external anal sphincter when they perceive balloon distention. The process is repeated with progressively smaller volumes of distention (32). Pelvic floor muscle training regimen consisting of four 30-minute biofeedback sessions combined with a home exercise regimen were shown to improve continence (33).

REFERENCES

1. Guimaraes ACS, et al. The contribution of the rectus abdominis and rectus femoris in twelve selected abdominal exercises. J Sports Med Phys Fitness 1991; 31: 222–230.
2. Andersson EA, et al. Abdominal and hip flexor muscle activation during various training exercises. Eur J Appl Physiol 1997; 75: 115–123.
3. Wickenden D, Bates S, Maxwell L. An electromyographic evaluation of upper and lower rectus abdominus during various forms of abdominal exercises. N Z J Physiother 1992; August: 17–21.
4. Duchenne GB. Physiology of Motion. Philadelphia: J.B. Lippincott, 1949, p. 480.
5. Shneerson J. Disorders of Ventilation. London: Blackwell Scientific Publications, 1988, pp. 22, 31, 155, 287, 289.
6. Youmans WD, Siebens AA. Respiration. In: Brobeck, ed. Best and Taylors Physiological Basis of Medical Practice. 9th Ed. Baltimore: Williams & Wilkins, 1973, pp. 6–35.
7. Guz A, Noble M, Eisele J, Trenchard D. The role of vagal inflation reflexes. In: Porter R, ed. Breathing: Hering-Breuer Centenary Symposium. A CIBA Foundation Symposium. London: JA Churchill, 1970, pp. 155, 235, 246, 287, 289.
8. Basmajian JV, De Luca DJ. Muscles Alive. 5th Ed. Baltimore: Williams & Wilkins, 1985, pp. 255, 414.
9. Egan DF. Fundamentals of Respiratory Therapy. 3rd Ed. St. Louis, MO: C.V. Mosby, 1977.
10. Basoudan N, Rodrigues A, Gallina A, Garland J, Guenette JA, Shadgan B, Road J, Reid WD. Scalene and sternocleidomastoid activation during normoxic and hypoxic incremental inspiratory loading. Physiol Rep 2020; 8(14): e14522. doi:10.14814/phy2.14522. PMID: 32726513; PMCID: PMC7389984.
11. Moore KL. Clinically Oriented Anatomy. 2nd Ed. Baltimore: Williams & Wilkins, 1985.
12. Thompson J, O'Sullivan P, Briffa K, Neumann P. Assessment of voluntary pelvic floor muscle contraction in continent and incontinent women using transperineal ultrasound, manual muscle testing and vaginal squeeze pressure measurements. Int Urogynecol J Pelvic Floor Dysfunct 2006; 17: 624–630. doi:10.1007/s00192-006-0081-2.
13. Iwanaga J, Eid S, Simonds E, Schumacher M, Loukas M, Tubbs RS. The majority of piriformis muscles are innervated by the superior gluteal nerve. Clin Anat 2019; 32(2): 282–286. doi:10.1002/ca.23311. Epub 2018 Dec 21. PMID: 30408241.
14. Ramirez PT, Frumovitz M, Abu-Rustum NR. Principles of Gynecologic Oncology Surgery E-Book. Elsevier Health Sciences; 2018 July 1, pp. 3–49.
15. Faubion SS, Shuster LT, Bharucha AE. Recognition and management of nonrelaxing pelvic floor dysfunction. Mayo Clin Proc 2012; 87(2): 187–193. doi:10.1016/j.mayocp.2011.09.004.
16. Chevalier F, Fernandez-Lao C, Cuesta-Vargas AI. Normal reference values of strength in pelvic floor muscle of women: a descriptive and inferential study. BMC Womens Health 2014; 14: 143. Published 2014 Nov 25. doi:10.1186/s12905-014-0143-4.
17. Basmajian JV. Primary Anatomy. 5th Ed. Baltimore: Williams & Wilkins, 1964, pp. 29, 61.
18. Goss CM, ed. Gray's Anatomy of the Human Body. 28th Ed. Philadelphia: Lea & Febiger, 1966, pp. 277, 311, 319, 380–381, 968.
19. Sabotta J. Atlas of Human Anatomy. New York: GE Stechert, 1933, p. 142.
20. Demaagd GA, Davenport TC. Management of urinary incontinence. P T 2012; 37(6): 345–361H.
21. Markland AD, Goode PS, Redden DT, Borrud LG, Burgio KL. Prevalence of urinary incontinence in men: results from the national health and nutrition examination survey. J Urol 2010; 184(3): 1022–1027. doi:10.1016/j.juro.2010.05.025.
22. Pizzol D, Demurtas J, Celotto S, et al. Urinary incontinence and quality of life: a systematic review and meta-analysis. Aging Clin Exp Res 2021; 33(1): 25–35. doi:10.1007/s40520-020-01712-y.
23. Bharucha AE, Dunivan G, Goode PS, et al. Epidemiology, pathophysiology, and classification of fecal incontinence: state of the science summary for the National Institute of Diabetes and Digestive and Kidney Diseases (NIDDK) workshop. Am J Gastroenterol 2015; 110(1): 127–136. doi:10.1038/ajg.2014.396.
24. Wallace SL, Miller LD, Mishra K. Pelvic floor physical therapy in the treatment of pelvic floor dysfunction in women. Curr Opin Obstet Gynecol 2019; 31(6): 485–493 doi:10.1097/GCO.0000000000000584.
25. Boileau J, Basmajian JV. Grant's Methods of Anatomy. 7th Ed. Baltimore: Williams & Wilkins, 1965
26. Nachemson A, Elfstron G. Intravital Dynamic Pressure Measurements in Lumbar Discs. Stockholm: Almqvista Wiksell, 1970.
27. Cherniack RM, et al. Respiration in Health and Disease. 2nd Ed. Philadelphia: W.B. Saunders, 1972: 410.
28. Kendall H, Kendall F, Boynton D. Posture and Pain. Baltimore: Williams & Wilkins, 1952, pp. 2–73, 156–159.
29. Kegel AH. Progressive resistance exercise in the functional restoration of the perineal muscles. Am J Obstet Gynecol 1948; 56(2): 238–248, ISSN 0002-9378, doi:10.1016/0002-9378(48)90266-X.
30. Dumoulin C, Cacciari LP, Hay-Smith EJC. Pelvic floor muscle training versus no treatment, or inactive control treatments, for urinary incontinence in women. Cochrane Database Syst Rev 2018; 10(10): CD005654. Published 2018 Oct 4. doi:10.1002/14651858.CD005654.pub4.
31. Lasak AM, Jean-Michel M, Le PU, Durgam R, Harroche J. The role of pelvic floor muscle training in the conservative and surgical management of female stress urinary incontinence: does the strength of the pelvic floor muscles matter? PM R 2018; 10(11): 1198–1210. doi:10.1016/j.pmrj.2018.03.023.
32. Heymen S, Scarlett Y, Jones K, Ringel Y, Drossman D, Whitehead WE. Randomized controlled trial shows biofeedback to be superior to pelvic floor exercises for fecal incontinence. Dis Colon Rectum 2009; 52(10): 1730–1737. doi:10.1007/DCR.0b013e3181b55455.
33. Theofrastous J, Wyman J, Bump R, McClish D, Elser D, Bland D, Fantl J. Effects of pelvic floor muscle training on strength and predictors of response in the treatment of urinary incontinence. Neurourol Urodyn 2002; 21: 486–490. doi:10.1002/nau.10021.

UPPER EXTREMITY 6

OUTLINE

INTRODUCTION

Differential diagnosis of problems of the shoulder girdle requires that special attention be paid to the innervation of the muscles. The shoulder girdle and upper extremity have many muscles that are supplied by nerves that are motor only, meaning they do not have a branch that provides cutaneous innervation to a designated region (i.e., sensation of light touch, pressure, pain). With no sensory innervation, the result can be a loss-of-motor function without symptoms of pain in the same region. An example is the extreme weakness of the serratus anterior muscle, illustrated in Section IV of this chapter. Ordinarily, the terms *joint* and *articulation* are used interchangeably. However, the original authors of this text provided a distinction between the two. Differentiating them served a special purpose. With *joint* referring to a "bone to bone" connection and *articulation* referring to a "bone to muscle to bone" connection, the role of muscle has been made very clear.

This distinction enables a complete description of the shoulder girdle by recognizing such articulations as the vertebroscapular and vertebroclavicular posteriorly, and the costoscapular and costoclavicular anteriorly. Reference to attachments of scapular muscles to the dorsal thorax via the scapulothoracic joint should no longer be necessary. The charts and tables in Section III of this chapter provide information regarding the 10 classifications for 25 articulations of the shoulder girdle.

The glenohumeral joint provides freedom of motion in all directions for the upper extremity as a whole. Stability in certain positions is obtained by the coordinated action of muscles. The elbow joint provides free motion in the direction of flexion and stability in the position of zero extension (180° angle). By virtue of forearm supination and pronation at the proximal and distal radioulnar joints, the extended hand can be moved from the anatomical position facing forward to facing backward. The wrist joints provide for flexion and extension and radial and ulnar deviation, but not for rotation. Text and charts later in this chapter are devoted to range of joint motions and to strength testing of the digits and thumb.

This chapter includes discussions regarding faulty and painful conditions of the upper back and upper extremity. Brief reviews of several cases of nerve injuries show the value of the *Spinal Nerve and Muscle Chart* (Figure 1-8) as an aid in differential diagnosis.

SECTION I
INNERVATION

THE BRACHIAL PLEXUS

The brachial plexus arises between the anterior and middle scalene muscles. The ventral rami of C5, 6, 7, and 8, and the greater part of T1, plus a communicating loop from C4 to C5 and one from T2 (sensory) to T1, form (successively) the roots, trunks, divisions, cords, and terminal branches of the plexus.

Ventral rami containing C5 and C6 fibers unite to form the superior (upper) trunk. The ventral rami of C7 form the middle trunk. Those containing C8 and T1 fibers unite to form the inferior (lower) trunk. Next, the trunks separate into anterior and posterior divisions. The anterior divisions from the superior and middle trunks, composed of C5, 6, and 7 fibers, unite to form the lateral cord. The anterior division from the inferior trunk, composed of C8 and T1 fibers, forms the medial cord, and the posterior divisions from all three trunks, composed of C5 through C8 (but not T1) fibers, unite to form the posterior cord.

The cords then divide and reunite into branches that become peripheral nerves. The posterior cord branches into the axillary and radial nerves. The medial cord, after receiving a branch from the lateral cord, terminates as the ulnar nerve. One branch of the lateral cord becomes the musculocutaneous nerve; the other branch unites with one from the medial cord to form the median nerve. Other peripheral nerves exit directly from various components of the plexus, and some exit directly from the ventral rami (see the left columns and top of Figure 6-1).

The anterior divisions, the lateral and medial cords, and the peripheral nerves arising from them innervate anterior or flexor muscles of the upper extremity. The posterior division, the posterior cord, and the peripheral nerves arising from them innervate the posterior or extensor muscles of the upper extremity.

Figure 6-1 is the brachial plexus portion of the *Spinal Nerve and Muscle Chart* (Figure 1-8). Figure 6-2 is an anatomical illustration of the brachial plexus. Figure 6-3 facilitates interpretation of muscle test findings as recorded in Figure 6-1, and it aids in determining the site or level of lesion.

			PERIPHERAL NERVES																		SPINAL SEGMENT								
	MUSCLE STRENGTH GRADE	MUSCLE	Cervical	Cervical	Cervical	Phrenic	Long Thor.	Dor. Scap	N. to Subcl.	Suprascap	U. Subscap.	Thoracodor	L. Subscap.	Lat. Pect.	Med. Pect.	Axillary	Musculocu.	Radial	Median	Ulnar	C1	C2	C3	C4	C5	C6	C7	C8	T1
		Cervical	1-8	1-8	1-4	3, 4, 5	5, 6, 7, (8)	4, 5	5, 6	4, 5, 6	(4), 5, 6, (7)	(5), 6, 7, 8	5, 6, (7)	5, 6, 7	(6), 7, 8	5, 6	(4), 5, 6, 7	5, 6, 7, 8	5, 6, 7, 8	7, 8									
		T.	1												1			1	1	1									
			D.	V.	V.	V.	P.R.	P.R.	S.T.	S.T.	P.	P.	P.	L.	M.	P.	L.	P.	L.M.	M.									
Brachial Plexus: Root		Serratus Anterior					•																		5	6	7	8	
Brachial Plexus: Root		Rhomboids Maj & Min						•																4	5				
Brachial Plexus: Trunk		Subclavius							•																5	6			
Brachial Plexus: Trunk		Supraspinatus								•														4	5	6			
Brachial Plexus: Trunk		Infraspinatus								•														(4)	5	6			
Brachial Plexus: P Cord		Subscapularis									•		•												5	6	7		
Brachial Plexus: P Cord		Latissimus Dorsi										•														6	7	8	
Brachial Plexus: P Cord		Teres Major											•												5	6	7		
Brachial Plexus: L		Pectoralis Maj (Upper)												•											5	6	7		
Brachial Plexus: M&L		Pectoralis Maj (Lower)												•	•											6	7	8	1
Brachial Plexus: M&L		Pectoralis Minor													•											(6)	7	8	1

KEY →
- D. = Dorsal Prim. Ramus
- V. = Vent. Prim. Ramus
- P.R. = Plexus Root
- S.T. = Superior Trunk
- P. = Posterior Cord
- L. = Lateral Cord
- M. = Medial Cord

FIGURE 6-1: Spinal nerve and muscle chart: brachial plexus. © 2005 Florence P. Kendall.

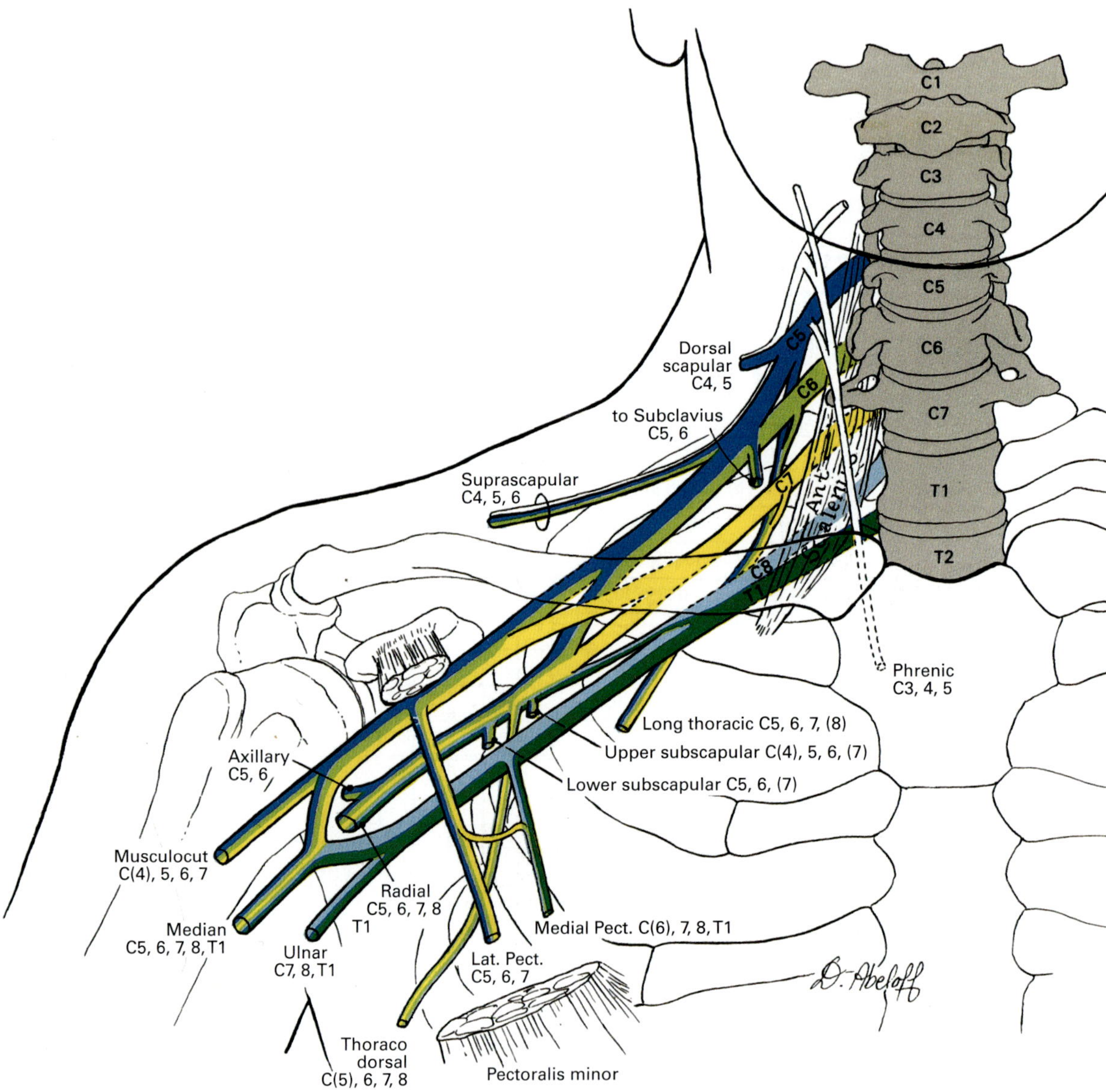

FIGURE 6-2: Brachial plexus.

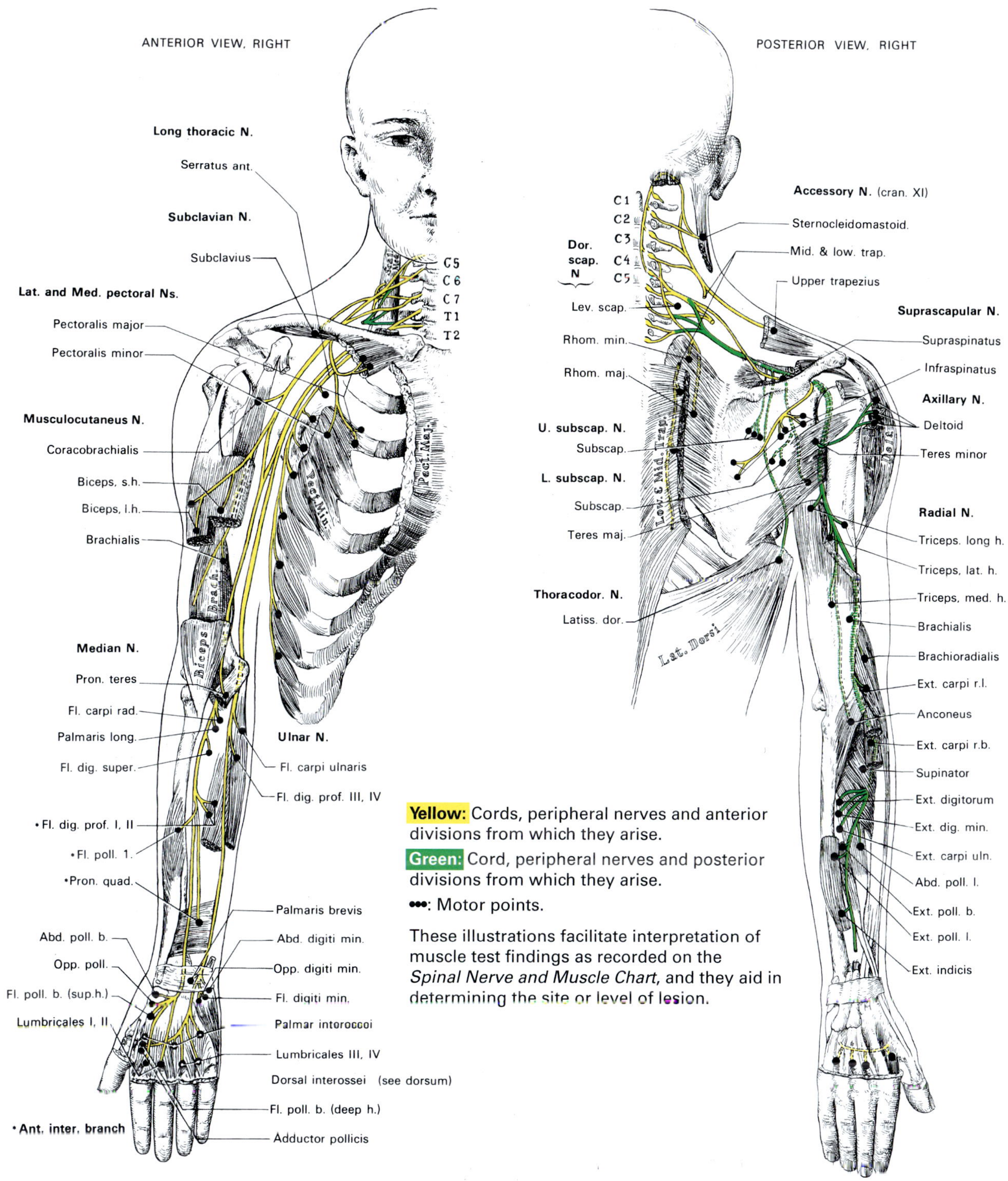

FIGURE 6-3: Spinal nerve and motor point chart. © 1993 Florence P. Kendall.

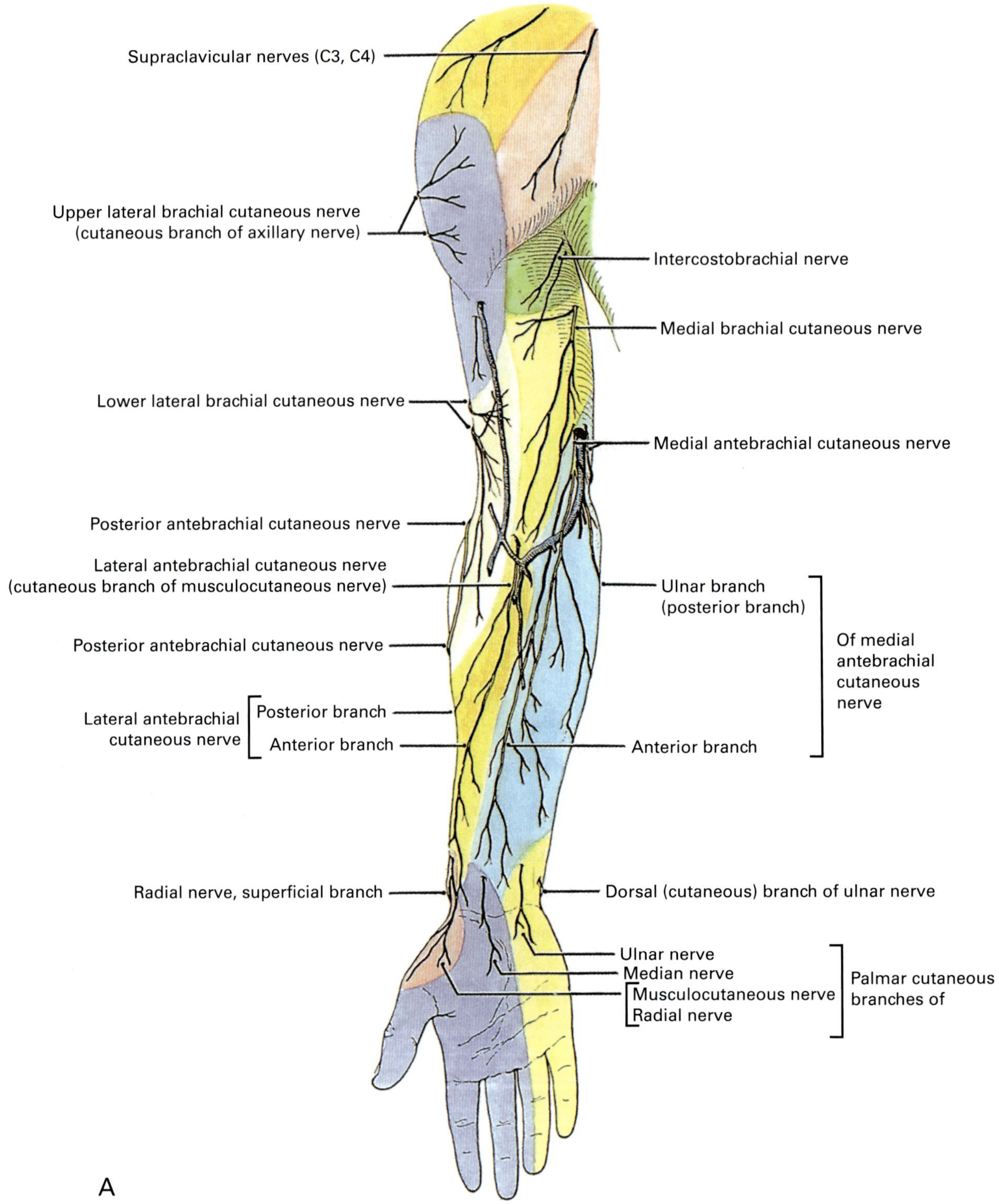

FIGURE 6-4: A, Cutaneous nerves of the upper limb: anterior view.

CUTANEOUS NERVES OF THE UPPER LIMB

Anterior View

Of the five terminal branches of the brachial plexus—musculocutaneous, median, ulnar, radial, and axillary nerves—the first four contribute cutaneous branches to the hand (Figure 6-4A). The posterior cord of the plexus is represented by five cutaneous nerves. One of these, the upper lateral brachial cutaneous nerve is a branch of the axillary nerve.

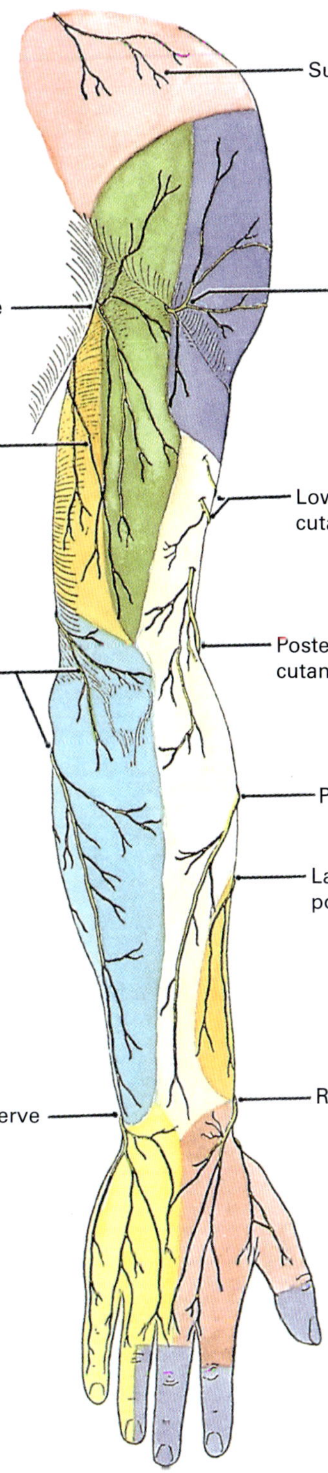

FIGURE 6-4: *(continued)* **B,** Cutaneous nerves of the upper limb: posterior view. From *Grant's Atlas of Anatomy* (1).

Posterior View

The other branches of the posterior cord are the posterior brachial cutaneous nerve, the lower lateral brachial cutaneous nerve, the posterior antebrachial cutaneous nerve, and the superficial branch of the radial nerve (Figure 6-4B).

DERMATOMES AND CUTANEOUS NERVE DISTRIBUTIONS

Table 6-1 lists and Figure 6-5 illustrates dermatomes and cutaneous nerve distributions.

TABLE 6-1 Scapular Muscle Chart

	Spinal Segment														
	Cervical							Th			Downward or Medial	Upward or Lateral			
Scapular Muscles	2	3	4	5	6	7	8	1	Elevation	Adduction	Rotation	Rotation	Depression	Abduction	Anterior Tilt
Trapezius	2	3	4						Upper Trapezius	Trapezius		Trapezius	Lower Trapezius		
Levator scapulae		3	4	5					Levator scapulae		Levator scapulae				
Rhomboids, major and minor			4	5					Rhomboids	Rhomboids	Rhomboids				
Serratus anterior				5	6	7	8		Upper Serratus anterior			Serratus anterior	Lower Serratus anterior	Serratus anterior	
Pectoralis minor					(6)	7	8	1							Pectoralis minor

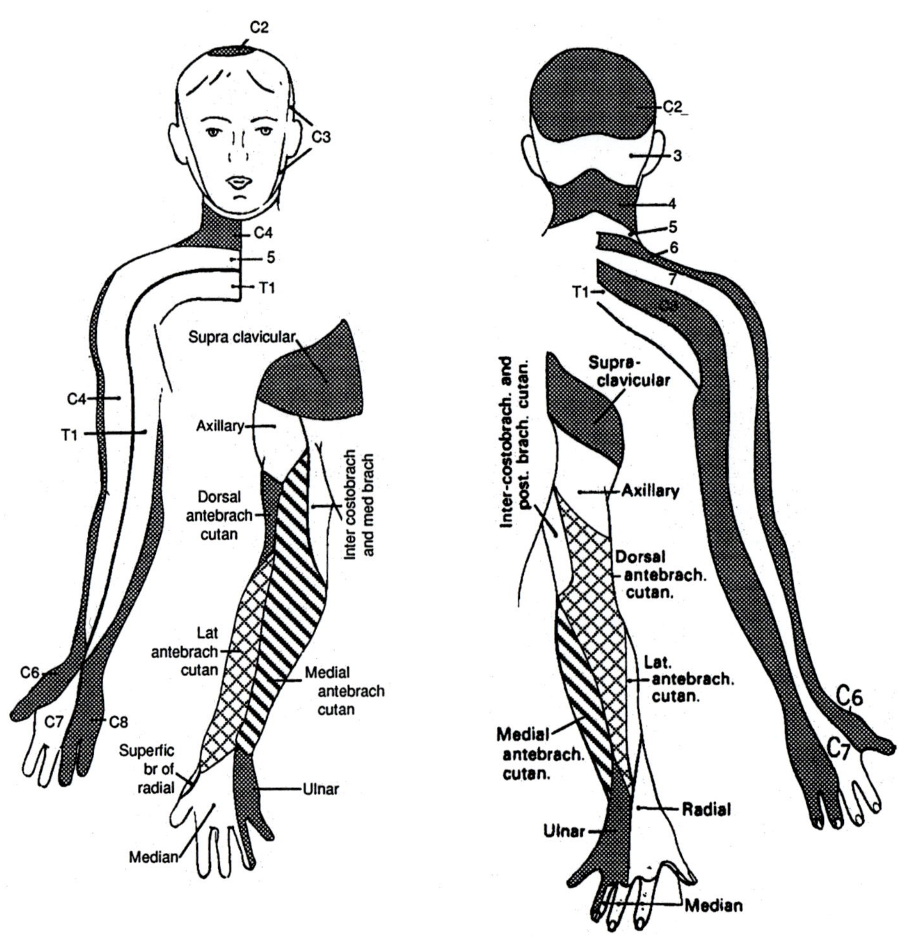

FIGURE 6-5: Dermatomes and cutaneous nerve distribution.

NERVES

Motor and Sensory

The following is a brief description of the relationship of the nerves and muscles. This material is chiefly from *Gray's Anatomy* (2).

Axillary: Leaves the axilla through the quadrangular space bounded by the surgical neck of the humerus, teres major, teres minor, and long head of the triceps, and innervates the deltoid and teres minor.

Musculocutaneous: Pierces the coracobrachialis and innervates this muscle as well as the biceps brachii and brachialis.

Radial: The *posterior interosseous* branch divides into a muscular and an articular branch. The muscular branch supplies the extensor carpi radialis brevis and the supinator before passing between the superficial and deep layers of the supinator muscle. After passing through the supinator, it supplies the remaining muscles that are innervated by the radial nerve.

Median: The median nerve is formed from contributions of the medial and lateral cords of the brachial plexus and descends along the medial aspect of the arm to the cubital fossa. After passing through the cubital fossa, the median nerve passes between the two heads of the pronator teres and under the flexor retinaculum. It is distributed to the forearm and hand. (See Figure 6-3 for a list of muscles innervated.)

Spinal Accessory: The spinal accessory nerve is a cranial nerve that exits the skull at the jugular foramen and descends deep to the sternocleidomastoid. It then passes through the posterior triangle of the neck and terminates in the subtrapezial plexus. Anatomy texts describe the spinal accessory nerve as being purely motor. A 1999 study by Bremner-Smith and Unwin, however, "shows that the spinal accessory nerve does not contain solely motor fibers, it also contains small unmyelinated C fibers associated with pain, temperature and mechanoreceptive reflex responses" (3).

Ulnar: The medial cord of the plexus terminates as the ulnar nerve. It supplies branches to the elbow joint, to the flexor carpi ulnaris and to the flexor digitorum profundus. (See Figure 6-3 for a complete list of muscles supplied by the ulnar nerve.)

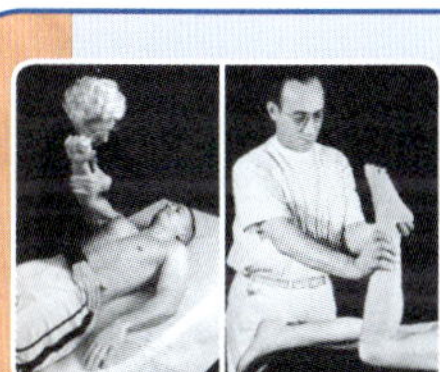

CLASSIC KENDALL

Motor Only

For years, Florence P. Kendall, the first author of the last three editions, gathered information about which muscles are supplied by purely motor nerves. Some typewritten pages listing the peripheral nerves and whether they were sensory, motor, or both date back to the late 1930s, but these pages had no reference source noted. A 1932 *Dorland's Medical Dictionary* had a table of nerves that included this information (4). An article on serratus anterior paralysis stated, "The long thoracic nerve or external respiratory nerve of Bell is almost unique in that it arises directly from the spinal nerve roots, carries no known sensory fibers, and goes to a single muscle of which it is the sole innervation of consequence" (5). Later, a table was found in Taber's dictionary (6). The 1988 edition of *Dorland's* did not have the tables included in an earlier edition, but the information was found in conjunction with the description of each nerve (7). Finally, scattered bits of information have come from some of the many books and articles on nerve injuries, compression, and entrapments (8–14).

Surprisingly, as the information was compiled, a very interesting pattern developed. Figure 6-6 shows that the nerves from the roots, trunks, and cords of the brachial plexus to muscles are motor nerves. In addition, the anterior and posterior interosseous nerves, which are branches of the median and radial nerves, respectively, are purely motor to the muscles they supply (6, 12–14). Several of the nerves have sensory branches to joints. Of the suprascapular nerve, Hadley et al. state that it "gives off motor branches to the muscles and sensory branches to the shoulder and acromioclavicular joints" (9). In addition, Dawson et al. state, "Since there is no cutaneous territory for this nerve, there are no characteristic sensory symptoms or findings in any lesion of this nerve" (15). Conway et al. state that "entrapment of the posterior interosseous nerve is purely motor and has no associated sensory loss of dysesthetic pain" (11).

The lack of sensory fibers provides the explanation for the lack of sensory symptoms in the muscles supplied by nerves that are only motor. (See discussion and examples later in this chapter.) There may be a sensory branch to a joint or joints but not to the muscle.

Spinal Segment

Cervical					Th		SHOULDER						ELBOW		FOREARM	
4	**5**	**6**	**7**	**8**	**1**	MUSCLE	Abduction	Lat. Rotat.	Flexion	Med. Rotat.	Extension	Adduction	Flexion	Extension	Supination	Pronation
4	**5**	6				Supraspinatus	Supraspin.									
(4)	**5**	**6**				Infraspinatus		Infraspin.								
	5	**6**				Teres minor		Teres mi.								
	5	**6**				Deltoid	Deltoid	Delt., post.	Delt., ant.	Delt., ant.	Delt., post.					
	5	**6**				Biceps	Biceps, l.h.		Biceps			Biceps, s.h.	Biceps		Biceps	
	5	**6**				Brachialis							Brachialis			
	5	**6**				Brachioradialis							Brachiorad.		Brachiorad.	Brachiorad.
	5	**6**	**7**			Pectoralis maj., upp.			Pect. mj., u.	Pect. mj., u.		Pect. mj., u.				
	5	**6**	7			Subscapularis				Subscap.						
	5	**6**	(7)			Supinator									Supinator	
	5	**6**	7			Teres major				Teres mj.	Teres mj.	Teres mj.				
	5	**6**	**7**	8		Ext. carpi rad. l. & b.							Ext. c. r. l.			
		6	**7**			Coracobrachialis			Coracobr.			Coracobr.				
		6	**7**			Pronator teres							Pron. teres			Pron. teres
		6	**7**	8		Flex. carpi rad.							Fl. c. rad.			Fl. c. rad.
		6	**7**	**8**		Latissimus dorsi				Lat. dorsi	Lat. dorsi	Lat. dorsi				
		6	**7**	**8**		Ext. digitorum										
		6	**7**	**8**		Ext. digit. min.										
		6	**7**	**8**		Ext. carpi ulnaris										
		6	**7**	**8**		Abd. poll. long.										
		6	**7**	**8**		Ext. poll. brev.										
		6	**7**	**8**		Ext. poll. long.										
		6	**7**	**8**		Ext. indicis										
		6	**7**	**8**	**1**	Pect. maj., lower						Pect. mj., l.				
		6	**7**	**8**	1	Triceps					Tri., l.h.	Tri., l.h.		Triceps		
		(6)	**7**	**8**	1	Palmaris long.							Palm. l.			
		(6)	7	**8**	**1**	Flex. poll. long.										
		(6)	7	**8**	**1**	Lumb. I & II										
		6	7	8	1	Abd. poll. brev.										
		6	7	8	1	Opponens poll.										
		6	7	8	1	Flex. poll br. (s. h.)										
			7	**8**		Anconeus								Anconeus		
			7	**8**	1	Flex. carpi ulnaris							Fl. c. ul.			
			7	**8**	**1**	Flex. digit. super.										
			7	**8**	**1**	Flex. digit. prof.										
			7	**8**	**1**	Pronator quad.										Pron. quad.
			(7)	**8**	**1**	Abd. digiti min.										
			(7)	**8**	**1**	Opp. digiti min.										
			(7)	**8**	**1**	Flex. digiti min.										
			(7)	**8**	**1**	Lumb. III & IV										
				8	**1**	Dor. interossei										
				8	**1**	Palm. interossei										
				8	**1**	Flex. poll. br. (d.h.)										
				8	**1**	Add. pollicis										

FIGURE 6-6: Upper extremity muscle chart. © 1993 Florence P. Kendall.

WRIST				CARPOMETACARPAL OF THUMB & LITTLE FINGER AND METACARPOPHALANGEAL JOINTS					DIG. 2-5 PROX INTERPHAL. JTS.		DIG. 1-5 DISTAL INTERPHAL. JTS.	
Extension	Flexion	Abduction	Adduction	Extension	Abduction	Flexion	Opposition	Adduction	Extension	Flexion	Extension	Flexion
Ext. c. r. l & b		Ext. c. r. l & b										
	Fl. c. rad.	Fl. c. rad.										
Ext. dig.		Ext. dig.		Ext. dig.	Ext. dig.				Ext. dig.		Ext. dig.	
				Ext. dig. min.	Ext. dig. min.				Ext. dig. min.		Ext. dig. min.	
Ext. c. ul.			Ext. c. ul.									
	Abd. poll. l.	Abd. poll. l.		Abd. poll. l.	Abd. poll. l.							
		Ext. poll. b.		Ext. poll. b.	Ext. poll. b.							
Ext. poll. l.		Ext. poll. l		Ext. poll. l.							Ext. poll. l.	
				Ext. ind.				Ext. Ind.	Ext. ind.		Ext. ind.	
	Palm. l.											
	Fl. poll. l.					Fl. poll. l.						Fl. poll. l.
						Lumb. I, II			Lumb. I, II		Lumb. I, II	
				Abd. poll. b.	Abd. poll. b.	Abd. poll. b.	Abd. poll. b.				Abd. poll. b.	
							Opp. poll.					
						Fl. poll. b. (s)	Fl. poll. br. (s)				Fl. poll. br. (s)	
	Fl. c. ul.		Fl. c. ul.									
	Fl. dig. sup.					Fl. dig. sup.				Fl. dig. sup.		
	Fl. dig. pro.					Fl. dig. pro.				Fl. dig. pro.		Fl. dig. pro.
					Abd. d. min.	Abd. d. min.	Abd. d. min.		Abd. d. min.		Abd. d. min.	
							Opp. d. min.					
						Fl. d. min.	Fl. d. min.					
						Lumb. II, III			Lumb III, IV		Lumb. III, IV	
					Dor. int.	Dor. int.			Dor. int.		Dor. int.	
						Palm. int.		Palm. int.	Palm. int.		Palm. int.	
						Fl. poll. b. (d)	Fl. poll. b. (d)					
				Add. poll.		Add. poll.	Add. poll.	Add. poll.				

FIGURE 6-6: *(continued)*

SECTION II

HAND, WRIST, FOREARM, AND ELBOW STRENGTH AND EXTENSIBILITY TESTS

Table 6-2 lists the joints, movements, starting positions, and available ROMs of the hand, wrist, forearm, and elbow.

TABLE 6-2 Joints, Movements, and Range of Motion

Joint	Joint Type	Movements	Starting Position	Available ROMs
1st IP joint	Ginglymus (hinge) joint	Thumb *flexion* and *extension* in an ulnar and a radial direction	MCP and IP joints in extension	0°–80° flexion; 10° of hyperextension can be achieved
1st MCP joint	Condyloid	Flexion/extension, abduction/adduction, and rotation	MCP and IP joints in extension	0°–50°
1st CM joint	Reciprocal reception or saddle	Flexion/extension, abduction/adduction, medial/lateral rotation, and circumduction	1st and 2nd metacarpals adjacent and parallel to each other	
Digital IP joints	Ginglymus or hinge joint	Flexion/extension	Wrist and digits in anatomical position	PIP: 0°–100° DIP: 0°–80°
MCP joints digits 2–5	Condyloid joints	Flexion/extension, abduction/adduction, and circumduction	Wrist and digits in anatomical position	0°–90° flexion Hyperextension is possible. The line of reference for abduction and adduction of the digits is the axial line through the third digit. Abduction is movement in the plane of the palm away from the axial line, spreading the digits widely apart. The third digit may move in abduction both ulnarly and radially from the axial line. Adduction is movement in the plane of the palm toward the axial line (i.e., closing the extended digits together sideways). *Circumduction* is the combination of flexion, abduction, extension, and adduction movements performed consecutively, in either direction, at the metacarpophalangeal joints of the digits. Extension in these condyloid joints is somewhat limited; therefore, the base of the cone described by the digit tip is relatively small.
Digital carpometacarpal joints				
Joint	**Joint Type**	**Movements**	**Zero Position**	**Noteworthy ROMs?**
Wrist	Condyloid	Flexion/extension, abduction/adduction (radial/ulnar deviation), and circumduction		Starting with the wrist straight (as in the anatomical position) as the zero position, the range of flexion is approximately 80°–85° and that of extension approximately 70°–85°. With the anatomical position as the zero position, the range of adduction is approximately 35°–45° and that of abduction approximately 20°.
Radioulnar	Trochoid or pivot	Supination/pronation	From the anatomical position with the elbow extended, midway between supination and pronation (thumb will be directed forward)	The normal range of motion is 90° from the zero position in either direction.
Elbow	Ginglymus or hinge	Flexion/extension	Straight elbow	From the zero positions to a fully bent position is approximately 145°.

IP = interphalangeal, MCP = metacarpophalangeal, CM = carpometacarpal. PIP = proximal interphalangeal, DIP = distal interphalangeal

Figure 6-7 is a chart to document the analysis of muscle imbalance.

Name:.. *Date: 1st. Ex.-*.............................. *2nd. Ex.-*..............................

Diagnosis:.. *Onset:*.............................. *Exam. of*..............................*extremity*

		2nd. EX.	1st. EX.	1st. EX.	2nd. EX.		
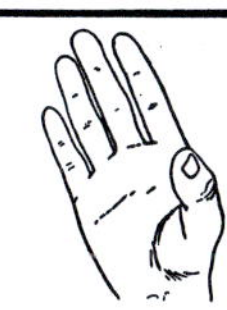	FLEXOR POLLICIS BREVIS					EXTENSOR POLLICIS BREVIS	
	FLEXOR POLLICIS LONGUS					EXTENSOR POLLICIS LONGUS	
	OPPONENS POLLICIS					ADDUCTOR POLLICIS	
	ABDUCTOR POLLICIS LONGUS					1 PALMAR INTEROSSEUS	
	ABDUCTOR POLLICIS BREVIS					1 DORSAL INTER. (THUMB ADD.)	
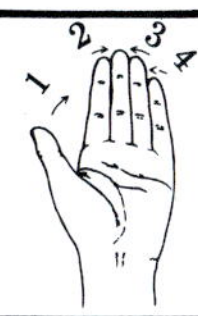	PALMAR INTEROSSEUS 2					1 DORSAL INTER. (INDEX ABD.)	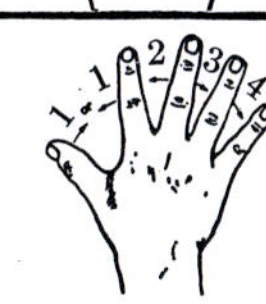
	(DORSAL INTEROSSEUS 3)					2 DORSAL INTEROSSEUS	
	(DORSAL INTEROSSEUS 2)					3 DORSAL INTEROSSEUS	
	PALMAR INTEROSSEUS 3					4 DORSAL INTEROSSEUS	
	PALMAR INTEROSSEUS 4					ABDUCTOR DIGITI MINIMI	
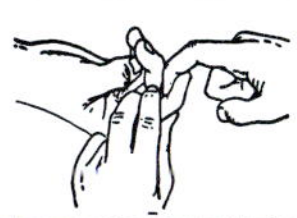	1					1	
	FLEXOR DIGITORUM 2					2 DISTAL INTER-PHALANGEAL	
	PROFUNDUS 3					3 JOINT EXTENSORS	
	4					4	
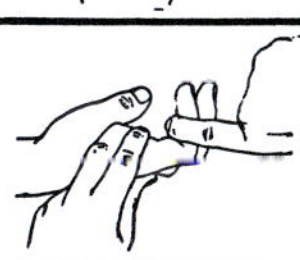	1					1	
	FLEXOR DIGITORUM 2					2 PROXIMAL INTER-PHALANGEAL	
	SUPERFICIALIS 3					3 JOINT EXTENSORS	
	4					4	
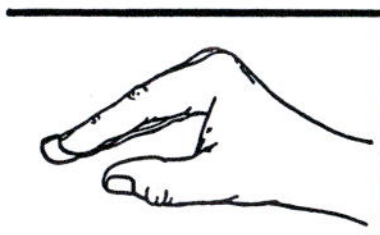	1					1 EXT. DIGIT. & INDICIS	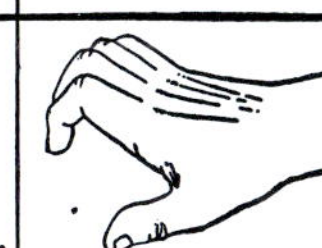
	LUMBRICALES & INTEROSSEI 2					2 EXT. DIGIT.	
	3					3 EXT. DIGIT.	
	& FLEXOR DIGITI MINIMI 4					4 EXT. DIGIT. COM. & DIG. MIN.	
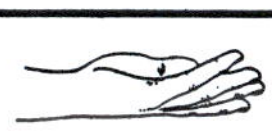	OPPONENS DIGITI MINIMI						
	PALMARIS BREVIS						
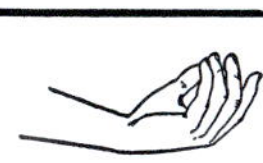	PALMARIS LONGUS					EXTENSOR CARPI RADIALIS LONGUS & BREVIS	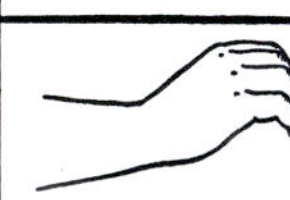
	FLEXOR CARPI ULNARIS						
	FLEXOR CARPI RADIALIS					EXTENSOR CARPI ULNARIS	
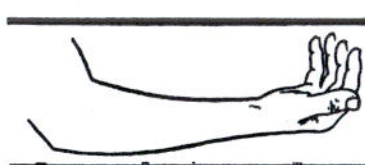	BICEPS .... } SUPINATORS					PRONATORS { .... QUADRATUS	
	SUPINATOR ...... }					{ .... TERES	
	BRACHIORADIALIS ..... } ELBOW FLEXORS						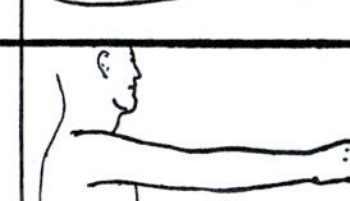
	BRACHIALIS .... }					ELBOW EXTENSORS { .... TRICEPS	
	BICEPS .... }					{ .... ANCONEUS	
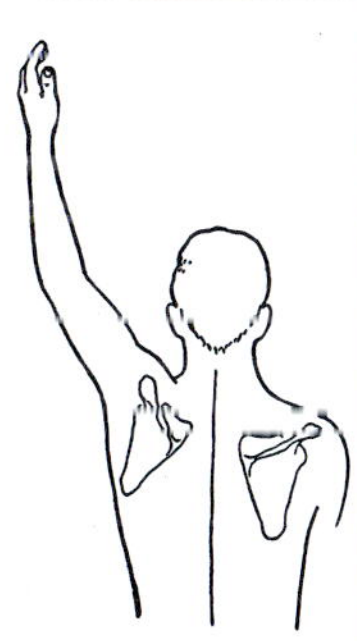	CORACOBRACHIALIS						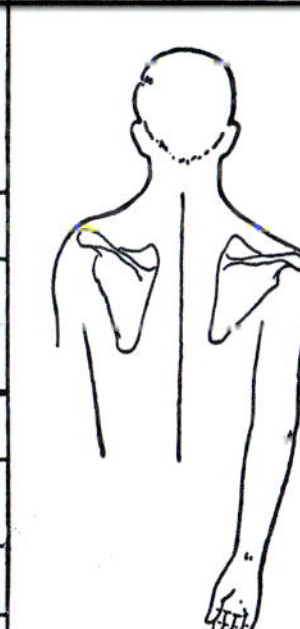
	ANTERIOR DELTOID						
	MIDDLE DELTOID					LATISSIMUS DORSI	
	POSTERIOR DELTOID					CLAV. PECTORALIS MAJOR	
	SUPRASPINATUS					STER. PECTORALIS MAJOR	
	TERES MINOR & INFRASPINATUS					TERES MAJOR & SUBSCAPULARIS	
	SERRATUS ANTERIOR					RHOMBOIDS & LEV. SCAP.	
	UPPER TRAPEZIUS					LATISSIMUS DORSI	
	MIDDLE TRAPEZIUS					PECTORALIS MAJOR	
	LOWER TRAPEZIUS					PECTORALIS MINOR	

FIGURE 6-7: Analysis of muscle imbalance. © 1993 Florence P. Kendall.

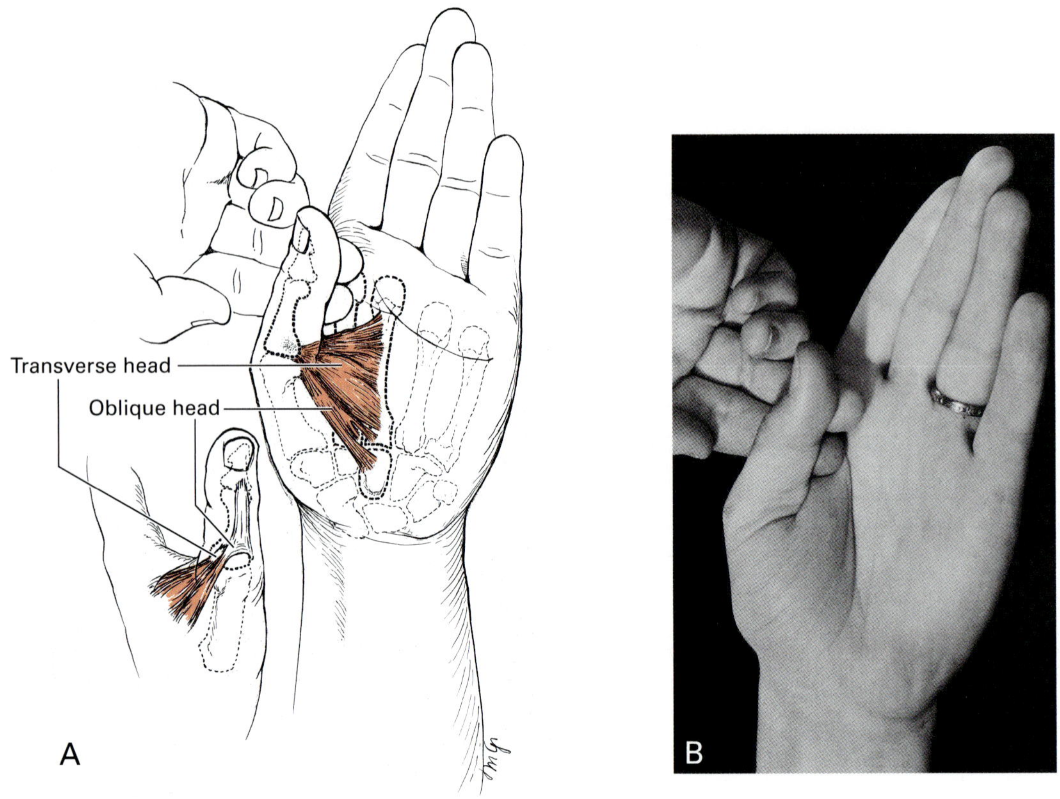

FIGURE 6-8: A, Adductor pollicis; **B,** Muscle testing of adductor pollicis.

Strength Tests

Adductor Pollicis

Origin of Oblique Fibers: Capitate bone and bases of the second and third metacarpal bones (Figure 6-8A).

Origin of Transverse Fibers: Palmar surface of the third metacarpal bone.

Insertion: Transverse head into ulnar side of base of the proximal phalanx of the thumb and oblique head into extensor expansion.

Action: Adducts the carpometacarpal joint, and both adducts and assists in flexion of the metacarpophalangeal joint, so that the thumb moves toward the plane of the palm. Aids in opposition of the thumb toward the fifth digit. May assist in extending the interphalangeal joint.

Nerve: Ulnar, **C8, T1**.

Patient: Sitting or supine.

Fixation: The hand may be stabilized by the examiner or rest on the table for support (as illustrated).

Test: Adduction of the thumb toward the palm (Figure 6-8B).

Pressure: Against the medial surface of the thumb, in the direction of abduction away from the palm.

Weakness: Results in inability to clench the thumb firmly over the closed fist.

Shortness: Adduction deformity of the carpometacarpal joint.

NOTE: *A test that is frequently used to determine the strength of the adductor pollicis is the ability to hold a piece of paper between the thumb and second metacarpal. However, an individual with a well-developed adductor, the bulk of the muscle itself prevents close approximation of these parts.*

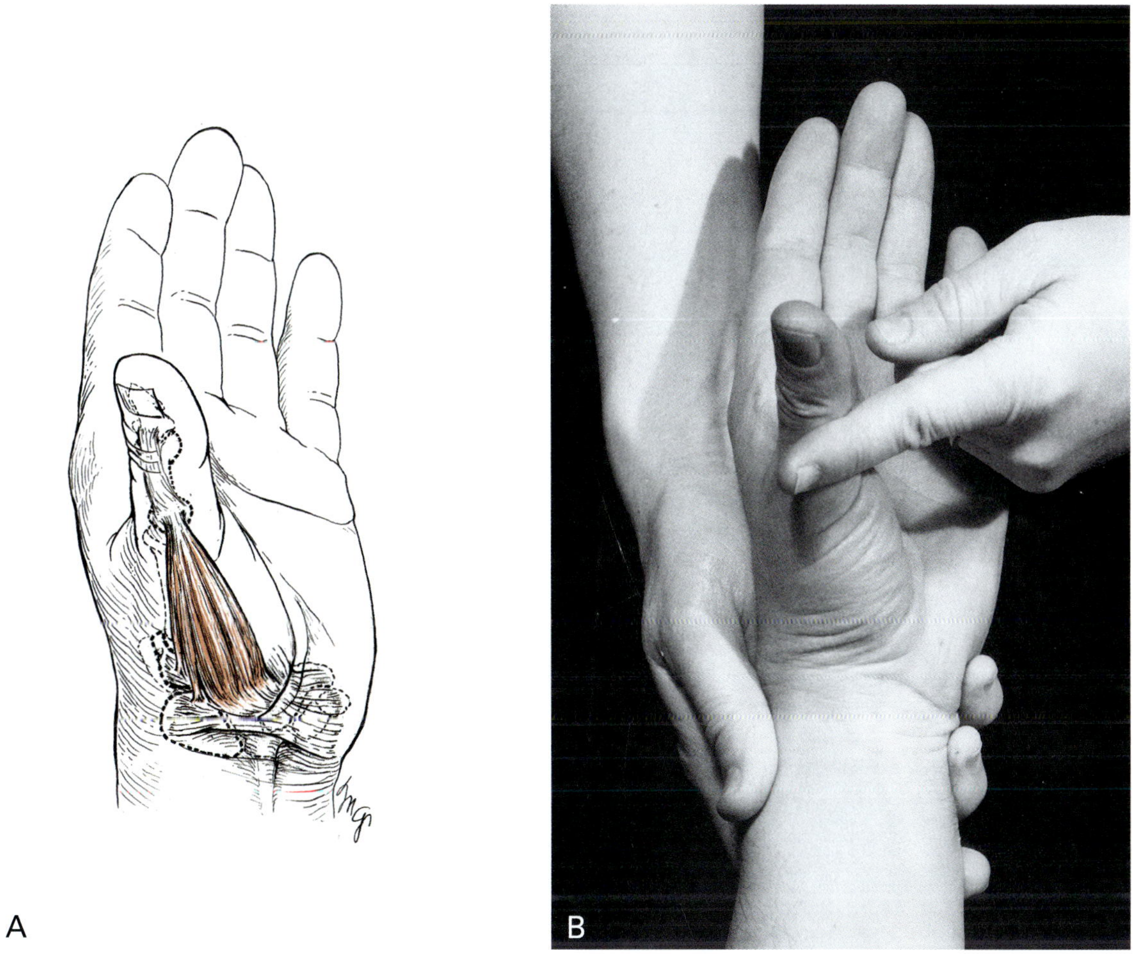

FIGURE 6-9: A, Abductor pollicis brevis; **B,** Muscle testing of abductor pollicis brevis.

Abductor Pollicis Brevis

Origin: Flexor retinaculum, tubercle of the trapezium bone, and tubercle of the scaphoid bone (Figure 6-9A).

Insertion: Base of the proximal phalanx of the thumb, radial side.

Action: Abducts the carpometacarpal and metacarpophalangeal joints of the thumb in a ventral direction perpendicular to the plane of the palm. Assists in opposition and may assist in flexion and medial rotation of the metacarpophalangeal joint.

Nerve: Median, C6, 7, 8, T1.

Patient: Sitting or supine.

Fixation: The hand may rest on the table for support (as illustrated) or the examiner may stabilize the hand.

Test: Abduction of the thumb ventrally from the palm (Figure 6-9B).

Pressure: Against the proximal phalanx, in the direction of adduction toward the palm.

Weakness: Decreases the ability to abduct the carpometacarpal joint, making it difficult to grasp a large object. An adduction deformity of the thumb may result from marked weakness.

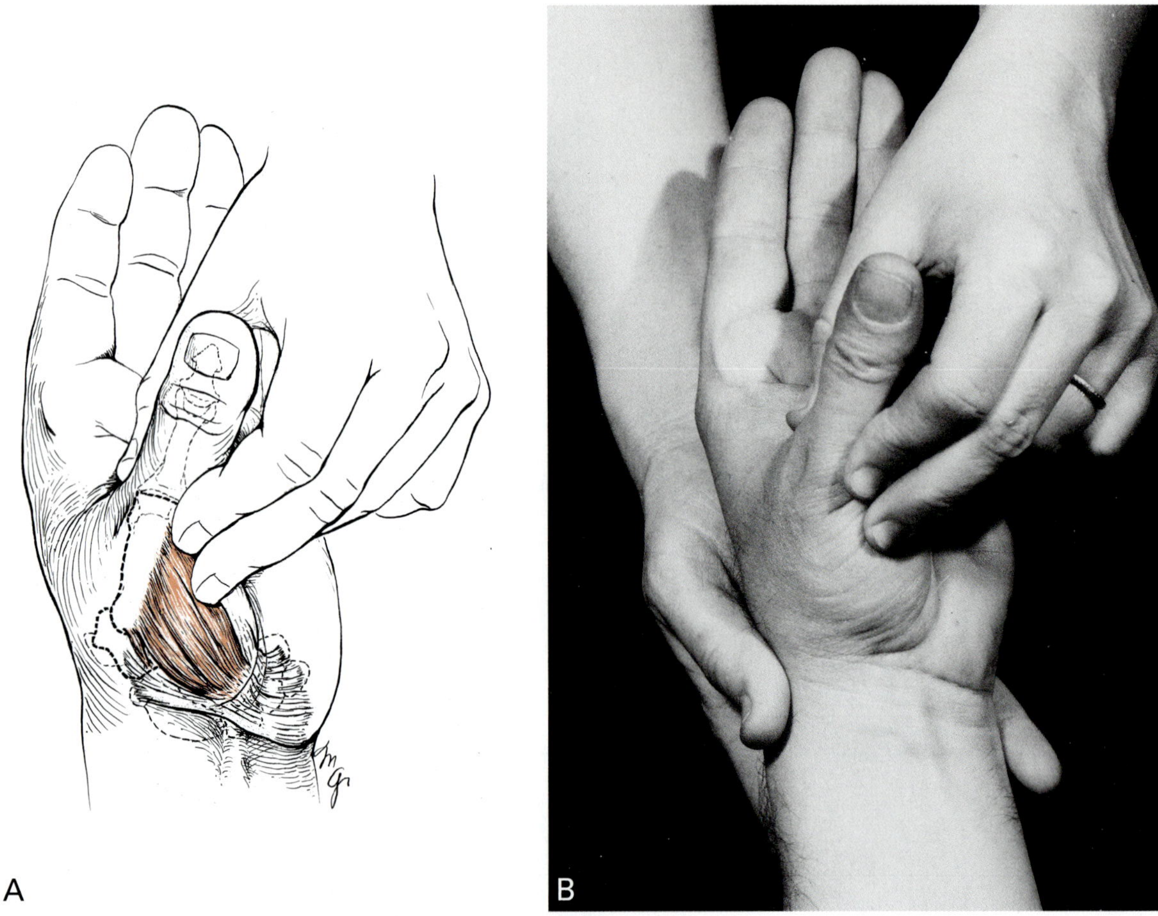

A B

FIGURE 6-10: A, Opponens pollicis; **B,** Muscle testing of opponens pollicis.

Opponens Pollicis

Origin: Flexor retinaculum and tubercle of the trapezium bone (Figure 6-10A).

Insertion: Entire length of the first metacarpal bone, radial side.

Action: Opposes (i.e., flexes and abducts with slight medial rotation) the carpometacarpal joint of the thumb, placing the thumb in a position so that, by flexion of the metacarpophalangeal joint, it can oppose the digits. For true opposition of the thumb and fifth digit, the pads of these digits come in contact. Bringing the tips of these digits together can be done without action of the opponens by excessive flexion at the metacarpophalangeal and interphalangeal joints.

Nerve: Median, C6, 7, 8, T1.

Patient: Sitting or supine.

Fixation: The examiner stabilizes the hand.

Test: Flexion, abduction, and slight medial rotation of the metacarpal bone so that the thumbnail shows in the palmar view (Figure 6-10B).

Pressure: Against the metacarpal bone, in the direction of extension and adduction with lateral rotation.

Weakness: Results in a flattening of the thenar eminence, extension and adduction of the first metacarpal, and difficulty in holding a pencil for writing or in grasping objects firmly between the thumb and other digits.

> **NOTE:** *The attachment of the palmaris longus and the opponens pollicis to the flexor retinaculum accounts for contraction of the palmaris longus during the opponens test.*

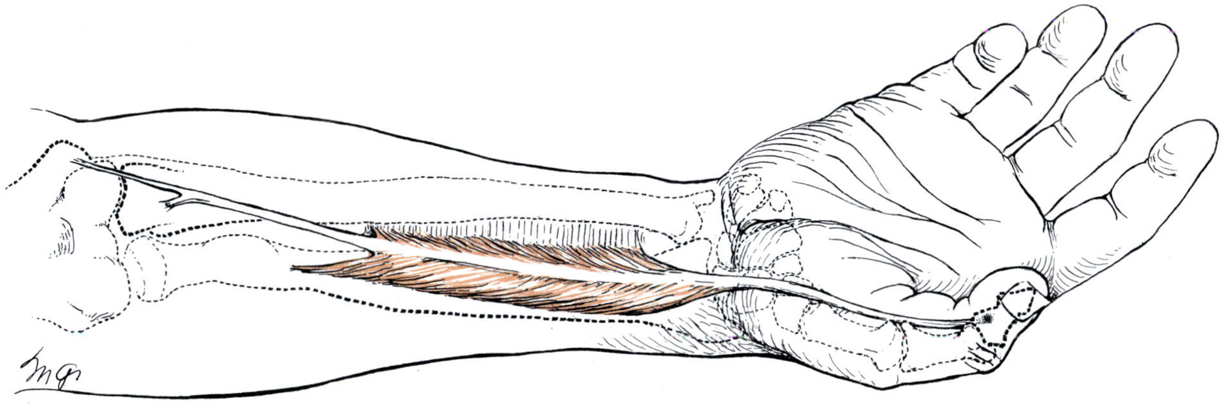

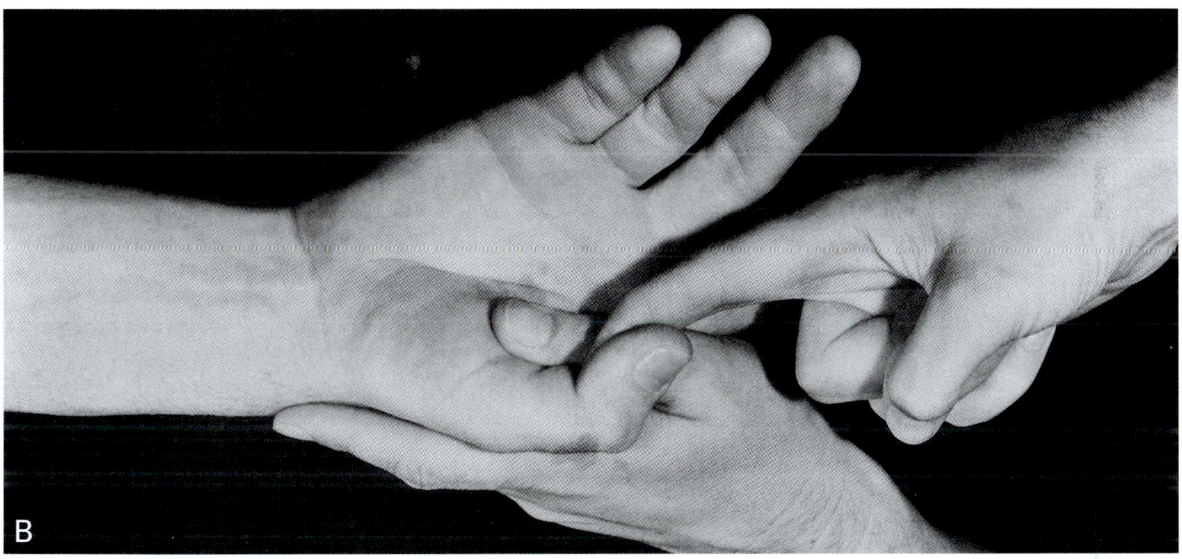

FIGURE 6-11: A, Flexor pollicis longus; **B,** Muscle testing of flexor pollicis longus.

Flexor Pollicis Longus

Origin: Anterior surface of the body of the radius below the tuberosity, interosseous membrane, medial border of the coronoid process of the ulna, and/or the medial epicondyle of the humerus (Figure 6-11A).

Insertion: Base of the distal phalanx of the thumb, palmar surface.

Action: Flexes the interphalangeal joint of the thumb. Assists in flexion of the metacarpophalangeal and carpometacarpal joints and may assist in flexion of the wrist.

Nerve: Median, C(6), 7, **8, T1**.

Patient: Sitting or supine.

Fixation: The hand may rest on the table for support (as illustrated), with the examiner stabilizing the metacarpal bone and proximal phalanx of the thumb in extension. Alternatively, the hand may rest on its ulnar side, with the wrist in slight extension and the examiner stabilizing the proximal phalanx of the thumb in extension.

Test: Flexion of the interphalangeal joint of the thumb (Figure 6-11B).

Pressure: Against the palmar surface of the distal phalanx, in the direction of extension.

Weakness: Decreases the ability to flex the distal phalanx, making it difficult to hold a pencil for writing or to pick up minute objects between the thumb and other digits. Marked weakness may result in a hyperextension deformity of the interphalangeal joint.

Contracture: Flexion deformity of the interphalangeal joint.

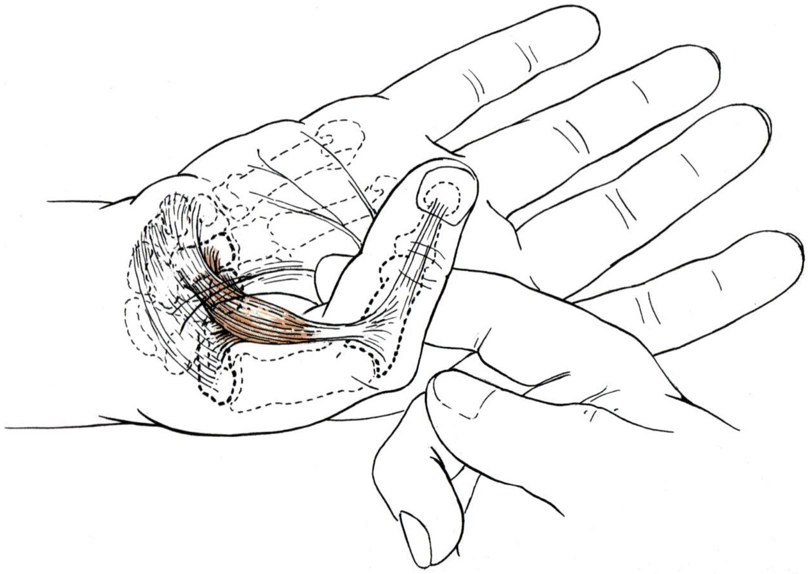

A

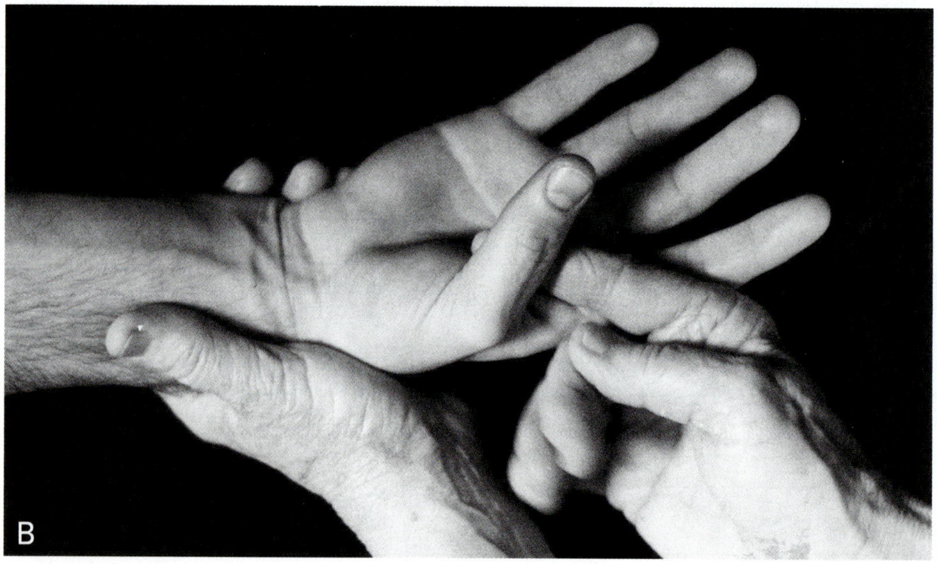

B

FIGURE 6-12: **A,** Flexor pollicis brevis; **B,** Muscle testing of flexor pollicis brevis.

Flexor Pollicis Brevis

Origin of Superficial Head: Flexor retinaculum and trapezium bone (Figure 6-12A).

Origin of Deep Head: Trapezoid and capitate bones.

Insertion: Base of the proximal phalanx of the thumb, radial side, and extensor expansion.

Action: Flexes the metacarpophalangeal and carpometacarpal joints of the thumb and assists in opposition of the thumb toward the fifth digit.

Nerve to Superficial Head: Median, C6, 7, 8, T1.

Nerve to Deep Head: Ulnar, **C8, T1**.

Patient: Sitting or supine.

Fixation: The examiner stabilizes the hand.

Test: Flexion of the metacarpophalangeal joint of the thumb without flexion of the interphalangeal joint (Figure 6-12B).

Pressure: Against the palmar surface of the proximal phalanx, in the direction of extension.

Weakness: Decreases the ability to flex the metacarpophalangeal joint, making it difficult to grip objects firmly between the thumb and other digits. Marked weakness may result in a hyperextension deformity of the metacarpophalangeal joint.

Contracture: Flexion deformity of the metacarpophalangeal joint.

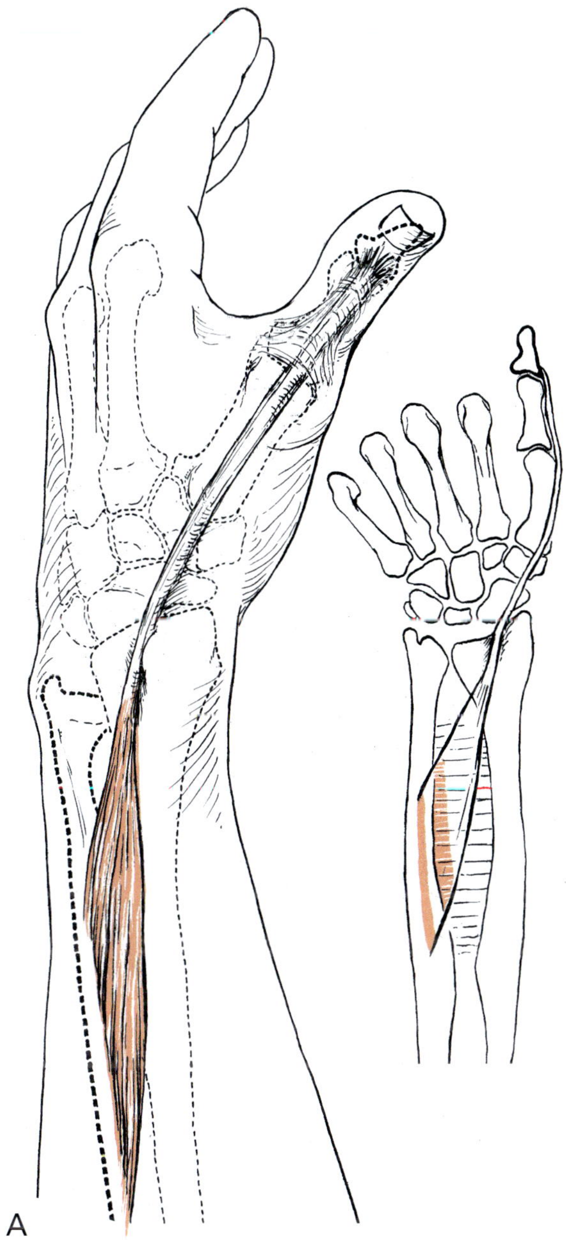

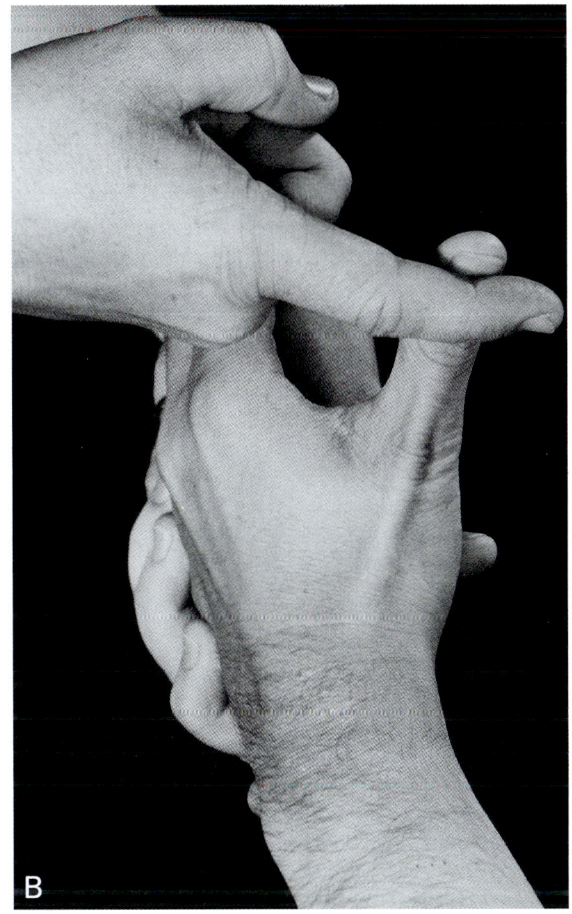

FIGURE 6-13: **A,** Extensor pollicis longus; **B,** Muscle testing of extensor pollicis longus.

Extensor Pollicis Longus

Origin: Middle 1/3 of the posterior surface of the ulna distal to origin of the abductor pollicis longus and to the interosseous membrane (Figure 6-13A).

Insertion: Base of the distal phalanx of the thumb, dorsal surface.

Action: Extends the interphalangeal joint and assists in extension of the metacarpophalangeal and carpometacarpal joints of the thumb. Assists in abduction and extension of the wrist.

Nerve: Radial, C6, **7**, **8**.

Patient: Sitting or supine.

Fixation: The examiner stabilizes the hand and gives counterpressure against the palmar surface of the first metacarpal and proximal phalanx.

Test: Extension of the interphalangeal joint of the thumb (Figure 6-13B).

Pressure: Against the dorsal surface of the interphalangeal joint of the thumb in the direction of flexion.

Weakness: Decreases the ability to extend the interphalangeal joint and may result in a flexion deformity of that joint.

> NOTE: *In a radial nerve lesion, the interphalangeal joint of the thumb may be extended by action of the abductor pollicis brevis, the flexor pollicis brevis, the oblique fibers of the adductor pollicis, or the first palmar interosseous by virtue of their insertions into the extensor expansion of the thumb. Interphalangeal joint extension in an otherwise complete radial nerve lesion should not be interpreted as regeneration or partial involvement if only this one action is observed.*

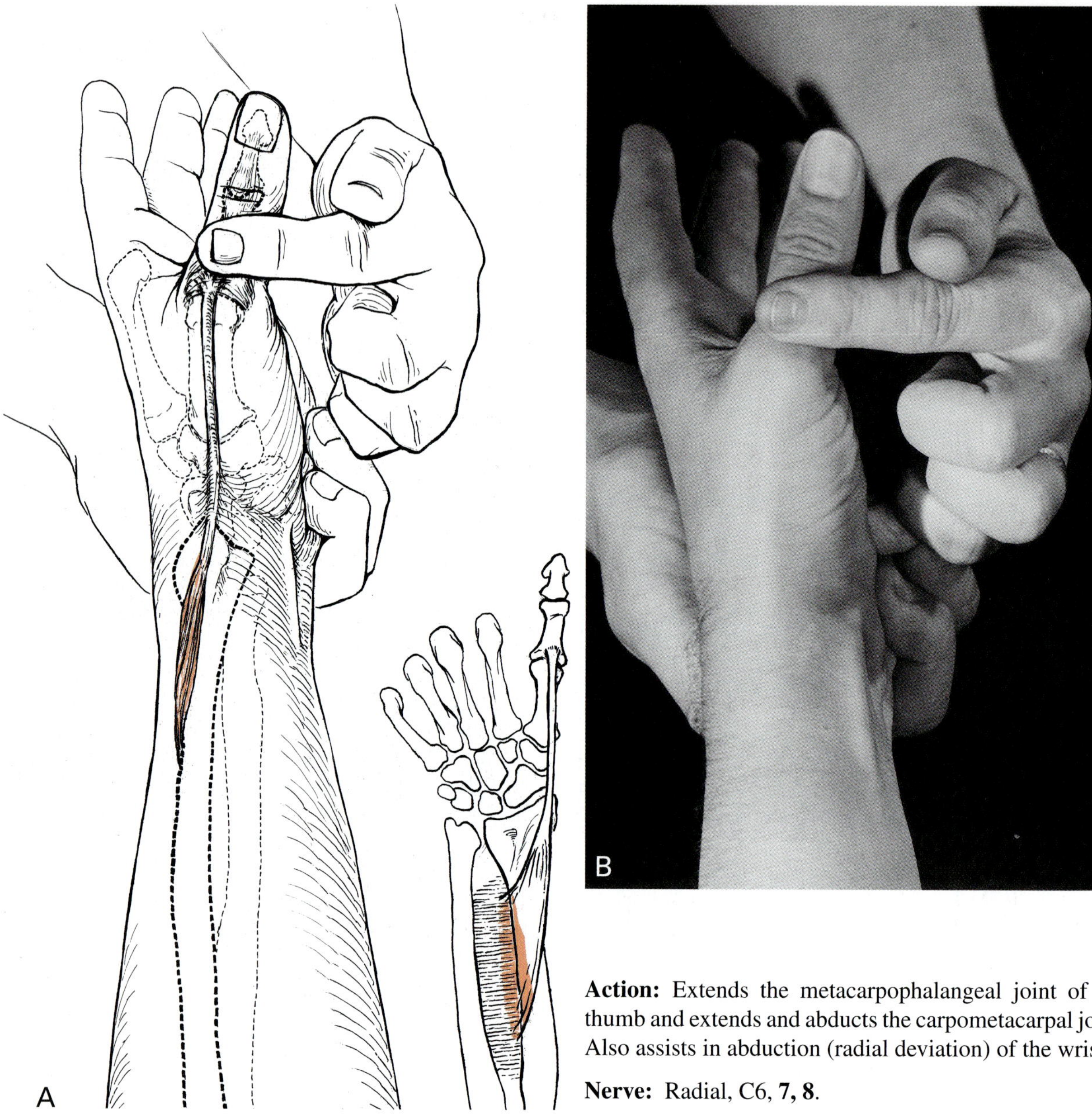

FIGURE 6-14: **A,** Extensor pollicis brevis; **B,** Muscle testing of extensor pollicis brevis.

Extensor Pollicis Brevis

Origin: Posterior surface of the body of radius distal to origin of abductor pollicis longus and to the interosseous membrane (Figure 6-14A).

Insertion: Base of the proximal phalanx of the thumb, dorsal surface.

Action: Extends the metacarpophalangeal joint of the thumb and extends and abducts the carpometacarpal joint. Also assists in abduction (radial deviation) of the wrist.

Nerve: Radial, C6, **7, 8**.

Patient: Sitting or supine.

Fixation: The examiner stabilizes the wrist.

Test: Extension of the metacarpophalangeal joint of the thumb (Figure 6-14B).

Pressure: Against the dorsal surface of the proximal phalanx, in the direction of flexion.

Weakness: Decreases the ability to extend the metacarpophalangeal joint and may result in a position of flexion of that joint.

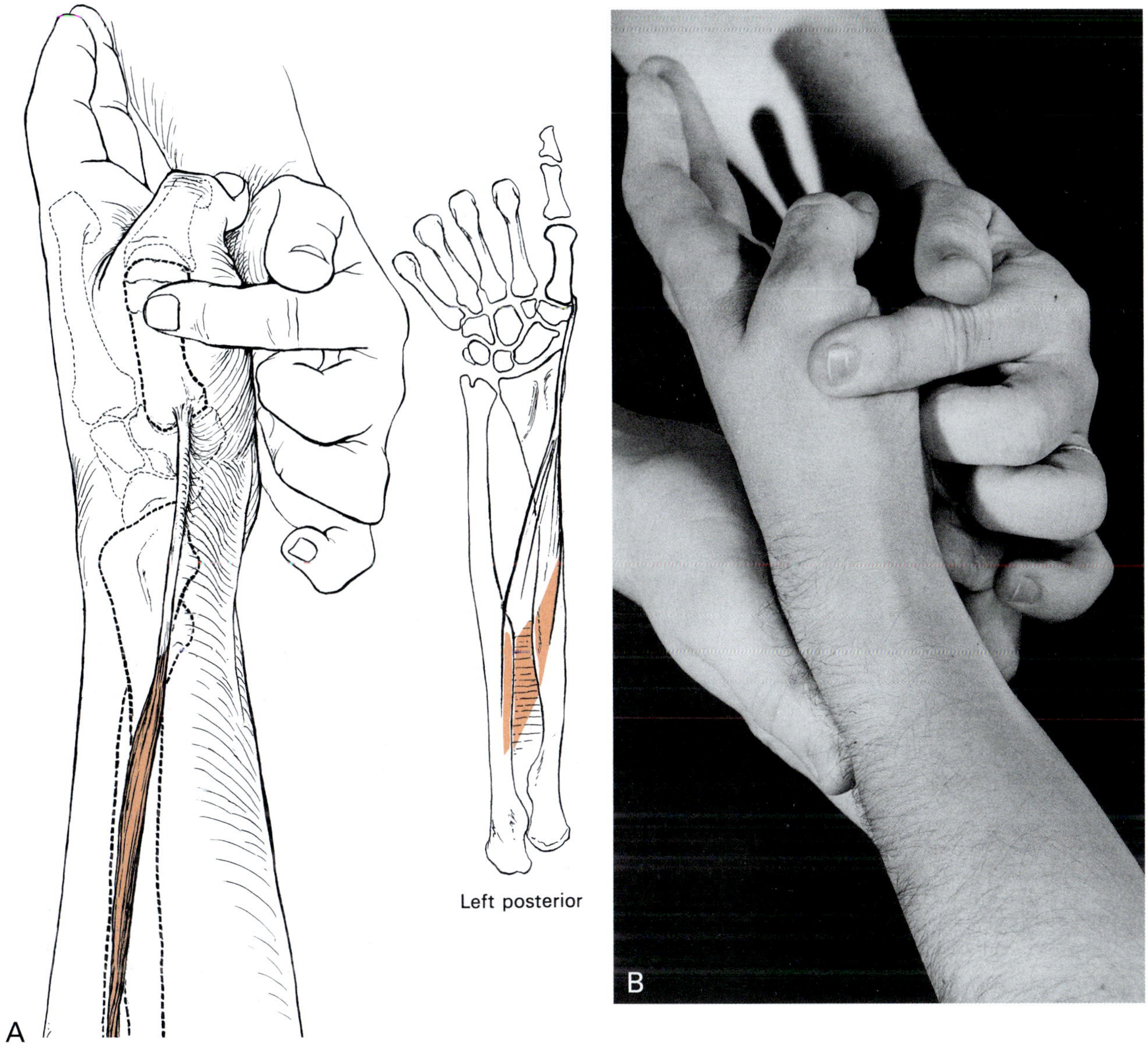

FIGURE 6-15: **A,** Abductor pollicis longus; **B,** Muscle testing of abductor pollicis longus.

Abductor Pollicis Longus

Origin: Posterior surface of the body of the ulna distal to origin of the supinator, interosseous membrane, and posterior surface of middle 1/3 of the body of the radius (Figure 6-15A).

Insertion: Base of the first metacarpal bone, radial side.

Action: Abducts and extends the carpometacarpal joint of the thumb and abducts (radial deviation) and assists in flexing the wrist.

Nerve: Radial, C6, **7**, **8**.

Patient: Sitting or supine.

Fixation: The examiner stabilizes the wrist.

Test: Abduction and slight extension of the first carpometacarpal joint (Figure 6-15B).

Pressure: Against the lateral surface of the distal end of the first metacarpal, in the direction of adduction and flexion.

Weakness: Decreases the ability to abduct the first metacarpal and the wrist.

Contracture: Abducted and slightly extended position of the first metacarpal, with slight radial deviation of the hand.

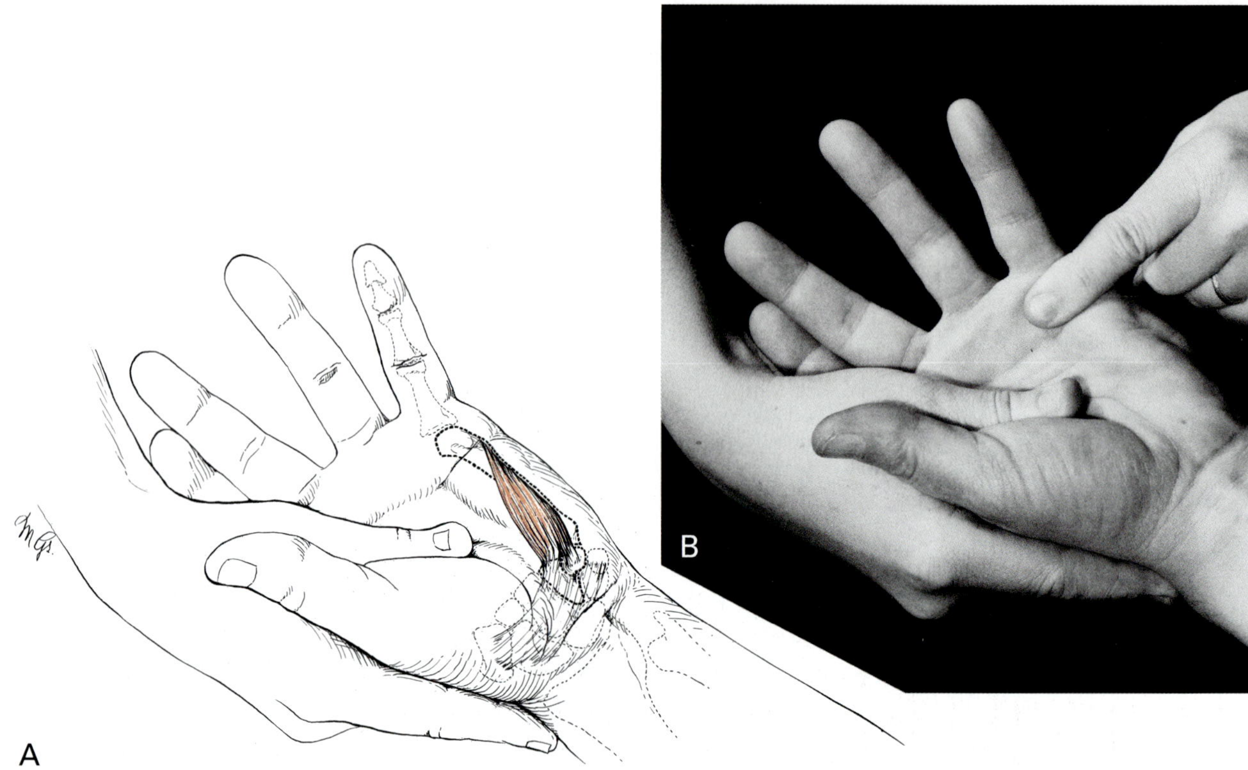

FIGURE 6-16: A, Opponens digiti minimi; **B,** Muscle testing of opponens digiti minimi.

Opponens Digiti Minimi

Origin: Hook of the hamate bone and the flexor retinaculum (Figure 6-16A).

Insertion: Entire length of the fifth metacarpal bone, ulnar side.

Action: Opposes (i.e., flexes with slight rotation) the carpometacarpal joint of the fifth digit, lifting the ulnar border of the hand to a position in which the metacarpophalangeal flexors can oppose the fifth digit to the thumb. Helps to cup the palm of the hand.

Nerve: Ulnar, C(7), **8, T1**.

Patient: Sitting or supine.

Fixation: The hand may be stabilized by the examiner or rest on the table for support. The first metacarpal is held firmly by the examiner.

Test: Opposition of the fifth metacarpal toward the first (Figure 6-16B).

Pressure: Against the palmar surface, along the fifth metacarpal, in the direction of flattening the palm of the hand. In the illustration, one-digit pressure was used to avoid obscuring the belly of the muscle; usually, the thumb is used to apply pressure along the fifth metacarpal.

Weakness: Results in a flattening of the palm, and makes it difficult, if not impossible, to oppose the fifth digit to the thumb.

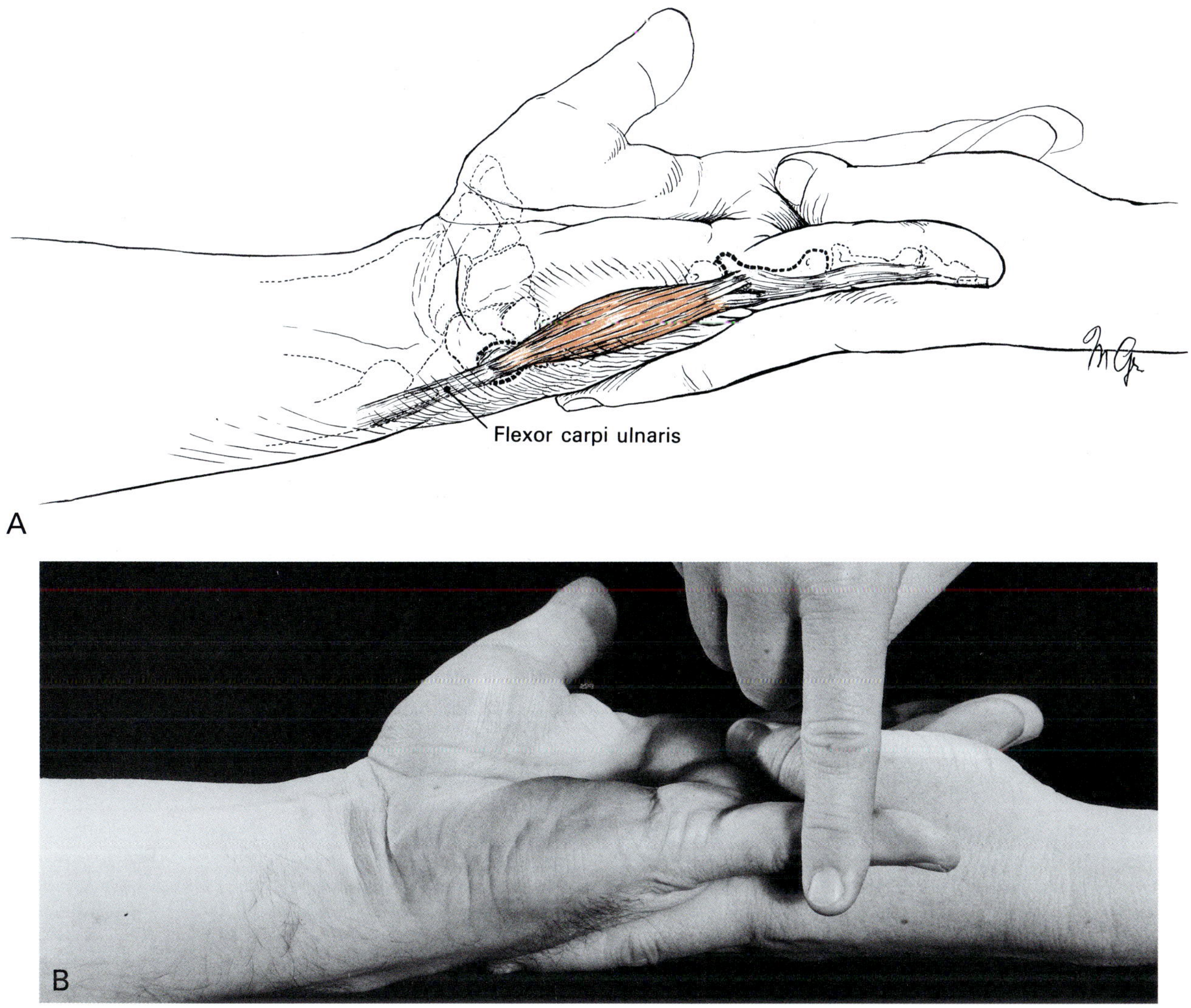

FIGURE 6-17: A, Abductor digiti minimi; **B,** Muscle testing of abductor digiti minimi.

Abductor Digiti Minimi

Origin: Tendon of the flexor carpi ulnaris and the pisiform bone (Figure 6-17A).

Insertion: By two slips: one into base of proximal phalanx of fifth digit, ulnar side and one into the ulnar border of the extensor expansion.

Action: Abducts, assists in opposition, and may assist in flexion of the metacarpophalangeal joint of the fifth digit. By virtue of its insertion into the extensor expansion, may assist in extension of the interphalangeal joints.

Nerve: Ulnar, C(7), **8, T1**.

Patient: Sitting or supine.

Fixation: The hand may be stabilized by the examiner or rest on the table for support.

Test: Abduction of the fifth digit (Figure 6-17B).

Pressure: Against the ulnar side of the fifth digit, in the direction of adduction toward the midline of the hand.

Weakness: Decreases the ability to abduct the fifth digit and results in adduction of this digit.

> **NOTE:** *One should be consistent regarding the placement of pressure during all digit abduction and adduction tests. Pressure against the sides of the middle phalanges seems to be most appropriate for all these tests.*

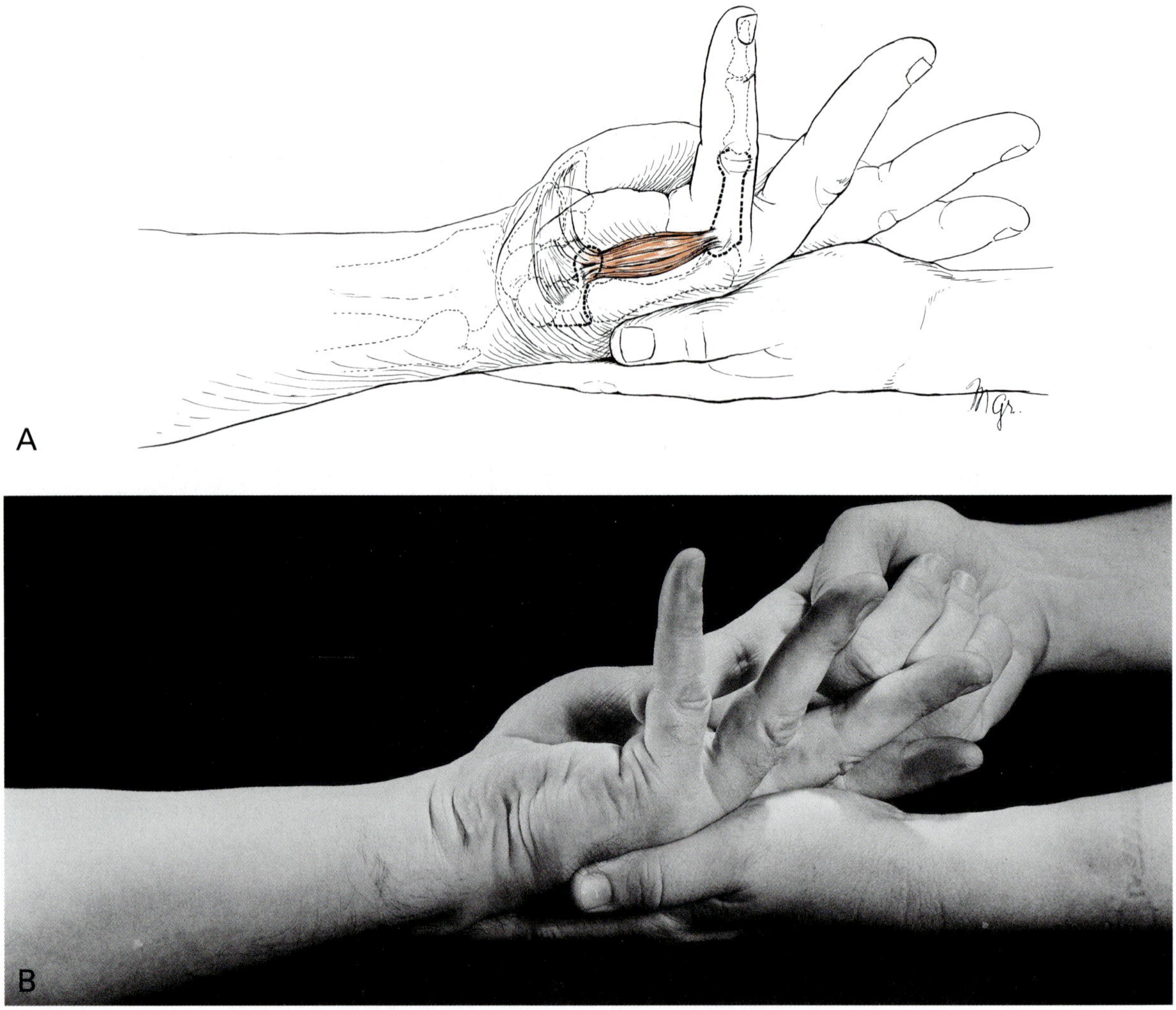

FIGURE 6-18: **A,** Flexor digiti minimi; **B,** Muscle testing of flexor digiti minimi.

Flexor Digiti Minimi

Origin: Hook of the hamate bone and the flexor retinaculum (Figure 6-18A).

Insertion: Base of the proximal phalanx of the fifth digit, ulnar side.

Action: Flexes the metacarpophalangeal joint of the fifth digit and assists in opposition of the fifth digit toward the thumb.

Nerve: Ulnar, C(7), **8, T1**.

Patient: Sitting or supine.

Fixation: The hand may rest on the table for support or be stabilized by the examiner.

Test: Flexion of the metacarpophalangeal joint, with the interphalangeal joints extended (Figure 6-18B).

Pressure: Against the palmar surface of the proximal phalanx, in the direction of extension.

Weakness: Decreases the ability to flex the fifth digit and oppose it toward the thumb.

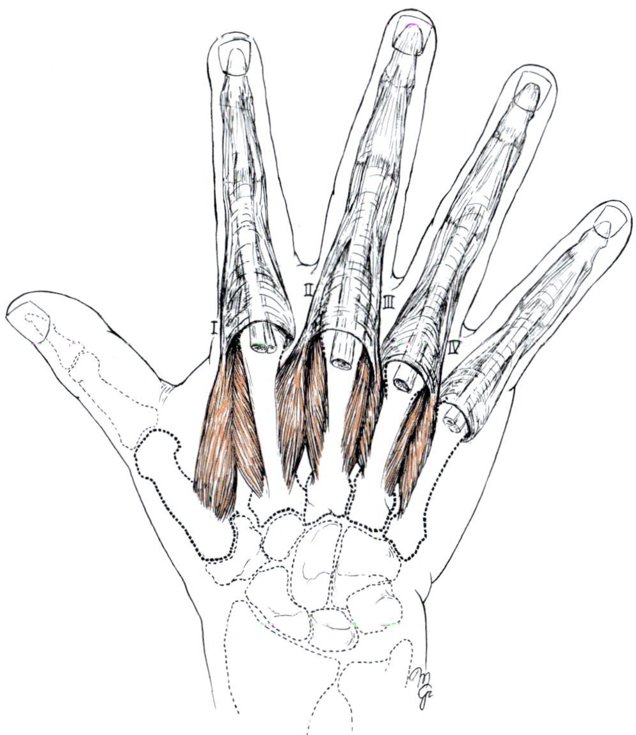

FIGURE 6-19: Dorsal interossei.

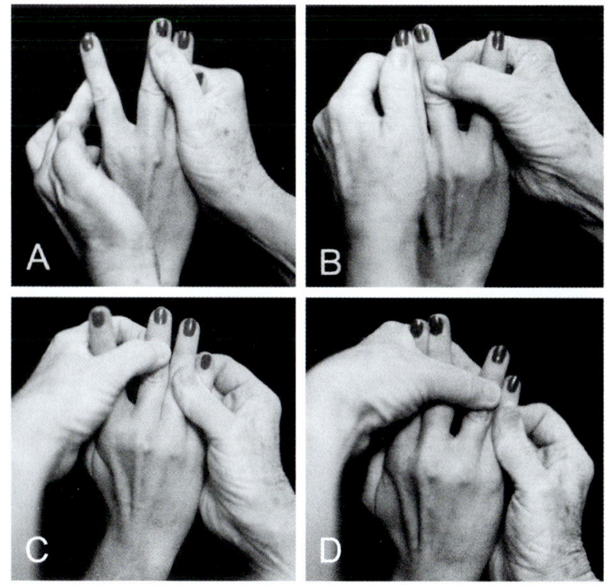

FIGURE 6-20: **A–D,** Testing of dorsal interossei.

Dorsal Interossei

Origins:

First, lateral head: Proximal 1/2 of the ulnar border of the first metacarpal bone.

First, medial head: Radial border of the second metacarpal bone.

Second, third, and fourth: Adjacent sides of the metacarpal bones in each interspace.

Insertions: Into extensor expansion and to the base of the proximal phalanx as follows:

First: Radial side of the second digit, chiefly to the base of the proximal phalanx.

Second: Radial side of the third digit.

Third: Ulnar side of the third digit, chiefly into extensor expansion.

Fourth: Ulnar side of the fourth digit.

Action: Metacarpophalangeal abduction of the second, third, and fourth digits from the midline through the third digit. Assists in flexion of the metacarpophalangeal joints and extension of the interphalangeal joints of the same digits. The first assists in adduction of the thumb (Figure 6-19).

Nerve: Ulnar, **C8, T1**.

Patient: Sitting or supine.

Fixation: In general, stabilization of the adjacent digits to give fixation of the digit toward which the digit being tested is moved and to prevent assistance from the digit on the other side.

Test and Pressure or Traction: Against the middle phalanx (Figure 6-20):

First: (Part A) Abduction of the second digit toward the thumb. Apply pressure against the radial side of the second digit in the direction of the third digit.

Second: (Part B) Abduction of the third digit toward the second digit. Hold the third digit and pull in the direction of the fourth digit.

Third: (Part C) Abduction of the third digit toward the fourth digit. Hold the third digit and pull in the direction of the second digit.

Fourth: (Part D) Abduction of the fourth digit toward the fifth digit. Hold the fourth digit and pull in the direction of the third digit.

Weakness: Decreases the ability to abduct the second, third, and fourth digits. Decreases the strength of extension of the interphalangeal joints and flexion of the metacarpophalangeal joints of the second, third, and fourth digits.

Shortness: Abducted second and fourth digits. Limited adduction ulnarly or radially of the third digit.

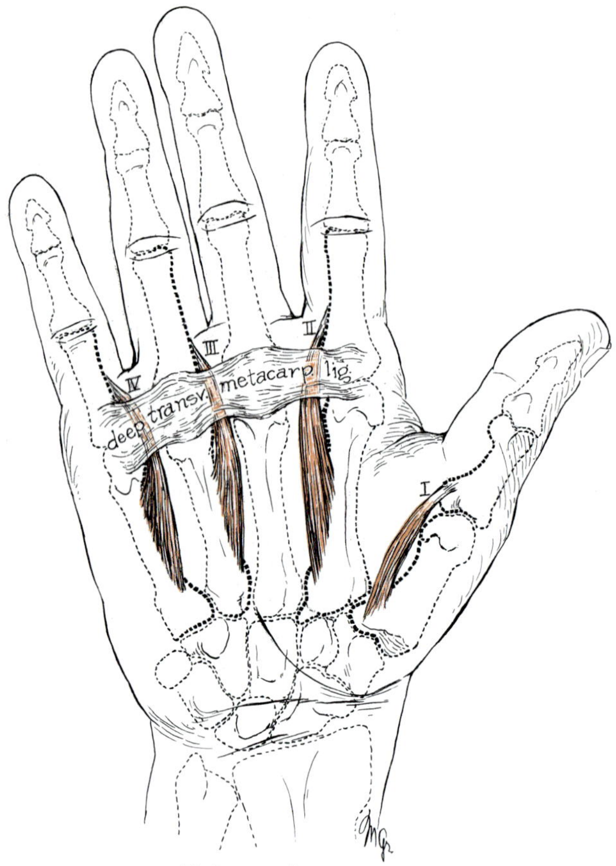

Palmar view

FIGURE 6-21: Palmar interossei.

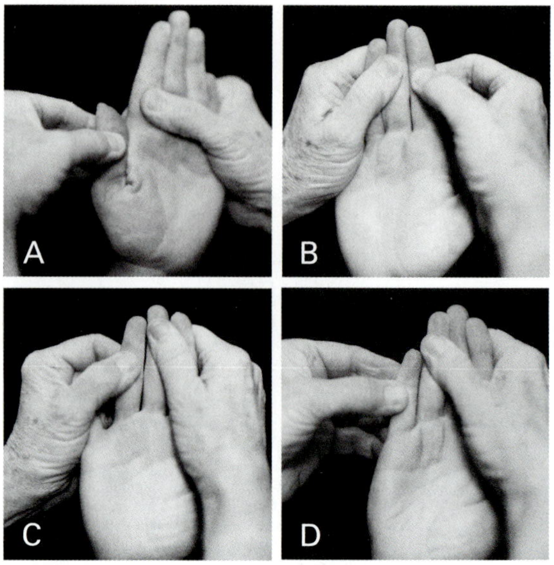

FIGURE 6-22: **A–D,** Testing of palmar interossei.

Palmar Interossei

Origins:

First: Base of the first metacarpal bone, ulnar side.

Second: Length of the second metacarpal bone, ulnar side.

Third: Length of the fourth metacarpal bone, radial side.

Fourth: Length of the fifth metacarpal bone, radial side.

Insertions: Chiefly into extensor expansion of the respective digit, with possible attachment to the base of the proximal phalanx as follows:

First: Ulnar side of the thumb.

Second: Ulnar side of the second digit.

Third: Radial side of the fourth digit.

Fourth: Radial side of the fifth digit.

Action: Metacarpophalangeal adduction of the first, second, third, and fourth digits toward the midline through the third digit. Assist in flexion of the metacarpophalangeal joints and extension of the interphalangeal joints of the three digits (Figure 6-21).

Nerve: Ulnar, **C8, T1**.

Patient: Sitting or supine.

Fixation: In general, stabilization of adjacent digits to give fixation of the digit toward the digit that is moved and to prevent assistance from the digit on the other side.

Test and Traction: Against the middle phalanx (Figure 6-22):

First: (A) Adduction of the thumb toward the second digit (acting with the adductor pollicis and first dorsal interosseous). Hold the thumb and pull in the radial direction.

Second: (B) Adduction of the second digit toward the third digit. Hold the second digit and pull in the direction of the thumb.

Third: (C) Adduction of the fourth digit toward the third digit. Hold the fourth digit and pull in the direction of the fifth digit.

Fourth: (D) Adduction of the fifth digit toward the fourth digit. Hold the fifth digit and pull in the ulnar direction.

Weakness: Decreases the ability to adduct the thumb and second, third, and fourth digits. Decreases strength in flexion of the metacarpophalangeal joints and extension of the interphalangeal joints of the second, fourth, and fifth digits.

Shortness: Digits held in adduction. May result from prolonged immobilization with the digits in adduction.

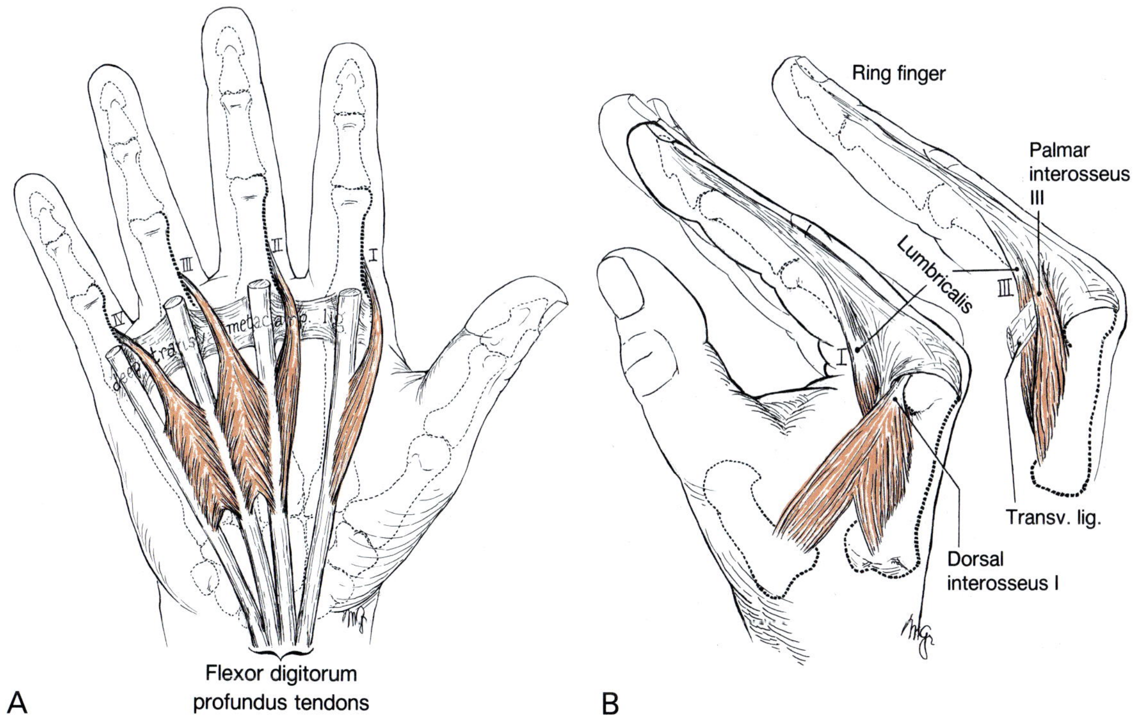

FIGURE 6-23: A, B, Lumbricales.

Lumbricales

Origins:

First and second: Radial surface of the flexor profundus tendons of the second and third digits, respectively.

Third: Adjacent sides of the flexor profundus tendons of the third and fourth digits.

Fourth: Adjacent sides of the flexor profundus tendons of the fourth and fifth digits.

Insertion: Into the radial border of the extensor expansion on the dorsum of the respective digits.

Action: Extend the interphalangeal joints and simultaneously flex the metacarpophalangeal joints of the second through fifth digits. The lumbricals also extend the interphalangeal joints when the metacarpophalangeal joints are extended. As the digits are extended at all joints, the flexor digitorum profundus tendons offer a form of passive resistance to this movement. Because the lumbricals are attached to the flexor profundus tendons, they can diminish this resistive tension by contracting and pulling these tendons distally, and this release of tension decreases the contractile force needed by the muscles that extend the digit joints (Figure 6-23).

Nerves:

First and second: Median (Anterior Interosseous Branch), C(6), 7, **8, T1**.

Third and fourth: Ulnar, C(7), **8, T1**.

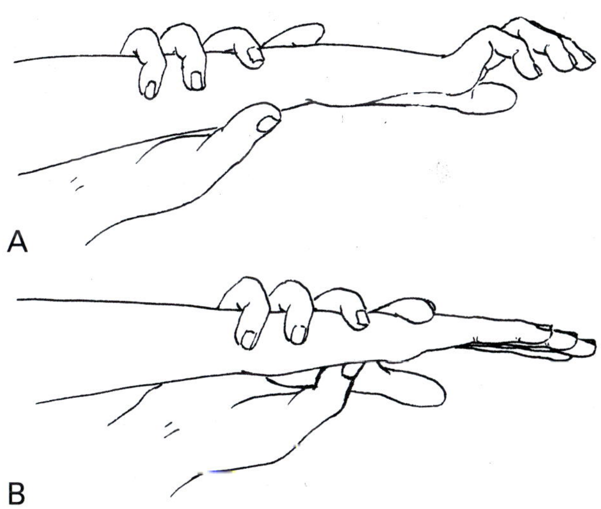

FIGURE 6-24: A, B, Testing of lumbricales.

Hyperextension of the metacarpophalangeal joints, resulting from weakness of the lumbricals and interossei, prevents normal function of the extensor digitorum in extending the interphalangeal joints (Figure 6-24A).

When the examiner offers fixation that normally is afforded by the lumbricals and interossei, a strong extensor digitorum will extend the digits (Figure 6-24B).

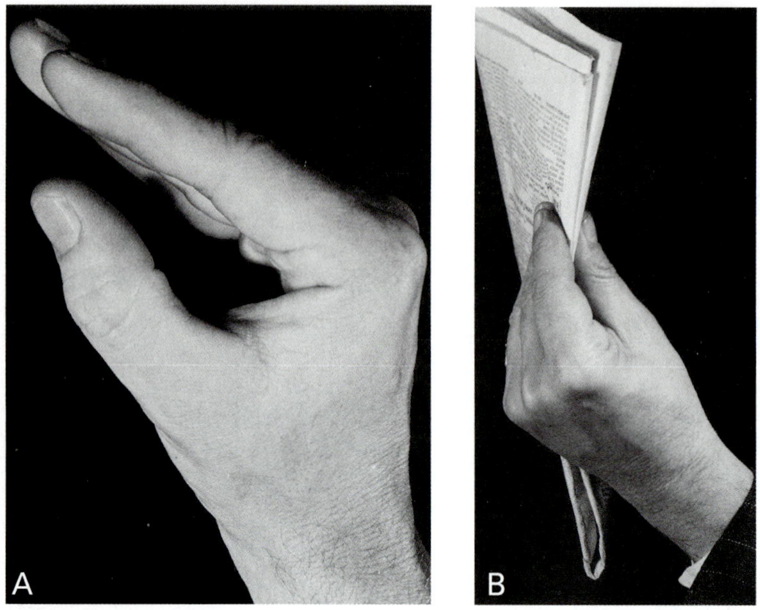

FIGURE 6-25: **A, B,** Lumbricales and interossei.

Lumbricals and Interossei

Patient: Sitting or supine.

Fixation: The examiner stabilizes the wrist in slight extension if there is any weakness of the wrist muscles.

Test: Extension of the interphalangeal joints, with simultaneous flexion of the metacarpophalangeal joints.

Pressure: First, against the dorsal surface of the middle and distal phalanges, in the direction of flexion, and second, against the palmar surface of the proximal phalanges, in the direction of extension. Pressure is not illustrated in the photograph, because it is *applied in two stages,* not simultaneously.

Weakness: Results in claw-hand deformity.

Shortness: Metacarpophalangeal joint flexion with interphalangeal joint extension.

NOTE: *An important function of the lumbricals and interossei is illustrated in Figure 6-25. With marked weakness or paralysis of these muscles, an individual cannot hold a newspaper or a book upright in one hand. The complaint by the patient that a newspaper could not be held in one hand was a clue to this type of weakness.*

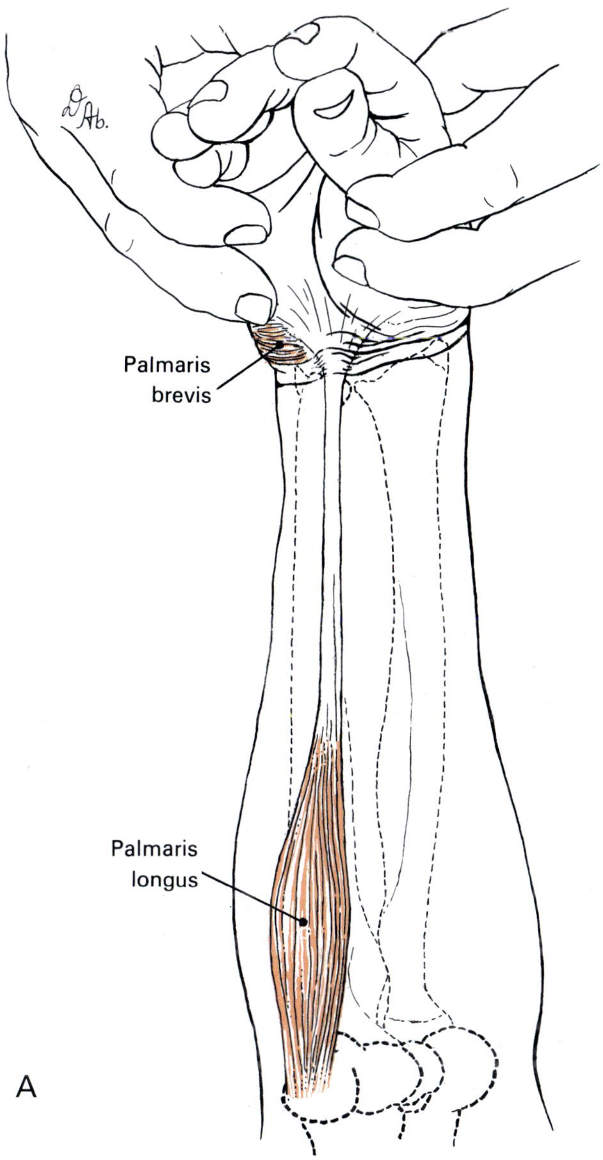

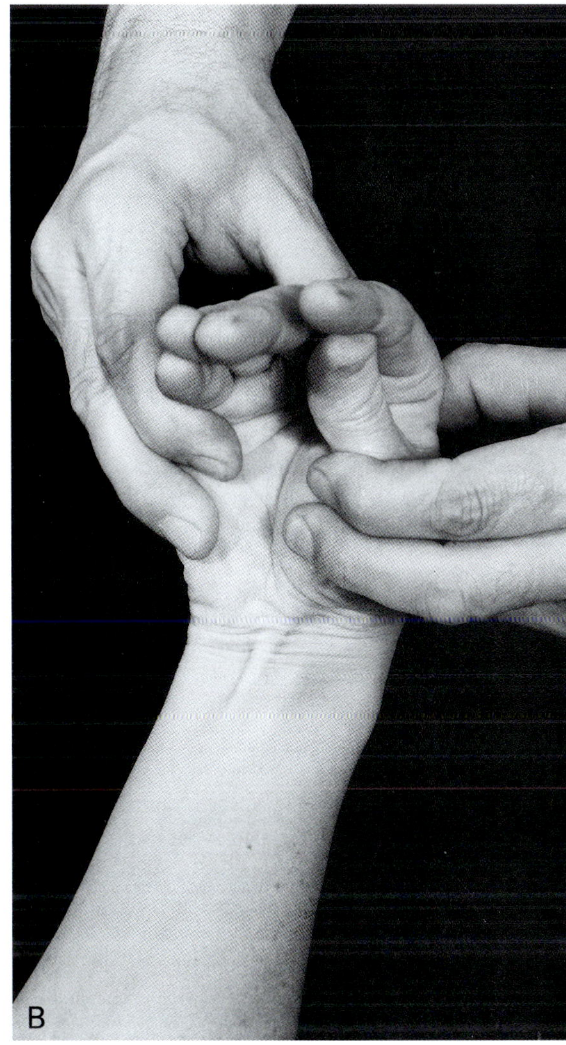

FIGURE 6-26: A, Palmaris longus and brevis; **B,** Muscle testing of palmaris longus.

Palmaris Longus

Origin: Common flexor tendon from the medial epicondyle of the humerus and the deep antebrachial fascia (Figure 6-26A).

Insertion: Flexor retinaculum and palmar aponeurosis.

Action: Tenses the palmar fascia, flexes the wrist, and may assist in flexion of the elbow.

Nerve: Median, C(6), **7, 8,** T1.

Patient: Sitting or supine.

Fixation: The forearm rests on the table for support, in a position of supination.

Test: Tensing of the palmar fascia by strongly cupping the palm of the hand, and flexion of the wrist (Figure 6-26B).

Pressure: Against the thenar and hypothenar eminences in the direction of the flattening of the palm of the hand, and against the hand in the direction of extending the wrist.

Weakness: Decreases the ability to cup the palm of the hand. Strength of wrist flexion is also diminished.

Palmaris Brevis

Origin: Ulnar border of the palmar aponeurosis and palmar surface of the flexor retinaculum.

Insertion: Skin on the ulnar border of the hand.

Action: Corrugates the skin on the ulnar side of the hand.

Nerve: Ulnar, C(7), **8, T1**.

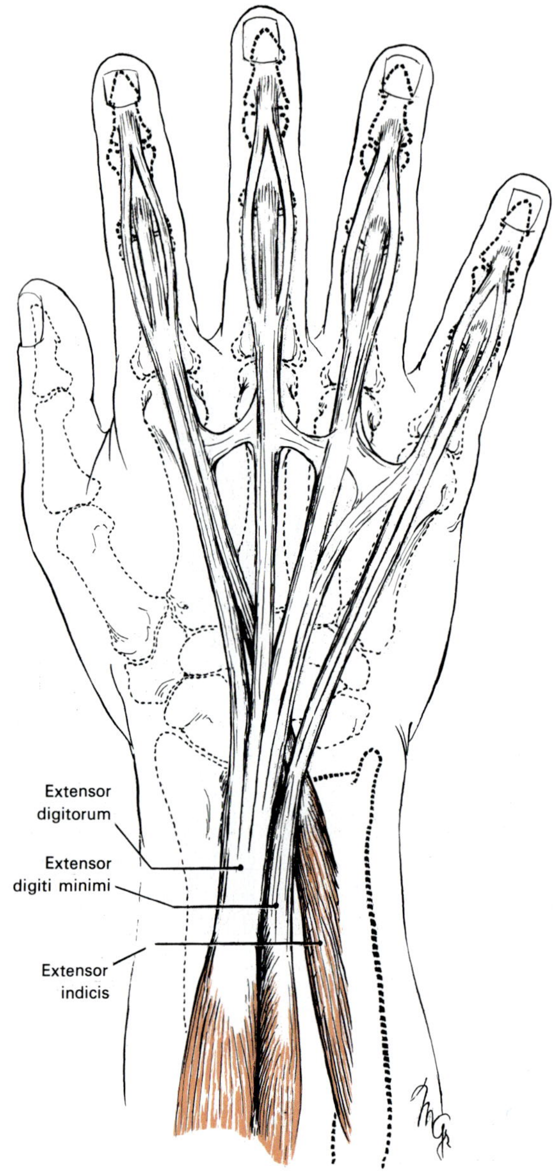

FIGURE 6-27: Extensor indicis and digiti minimi.

EXTENSOR INDICIS AND DIGITI MINIMI

Extensor Indicis

Origin: Posterior surface of the body of the ulna distal to the origin of the extensor pollicis longus, and interosseous membrane (Figure 6-27).

Insertion: Into extensor expansion of the second digit with the extensor digitorum longus tendon.

Action: Extends the metacarpophalangeal joint and, in conjunction with the lumbrical and interossei, extends the interphalangeal joints of the second digit. May assist in adduction of the second digit.

Nerve: Radial, C6, **7, 8**.

Extensor Digiti Minimi

Origin: Common extensor tendon from the lateral epicondyle of the humerus and deep antebrachial fascia.

Insertion: Into extensor expansion of the fifth digit with the extensor digitorum tendon.

Action: Extends the metacarpophalangeal joint and, in conjunction with the lumbrical and interosseous, extends the interphalangeal joints of the fifth digit. Assists in abduction of the fifth digit.

Nerve: Radial, C6, **7, 8**.

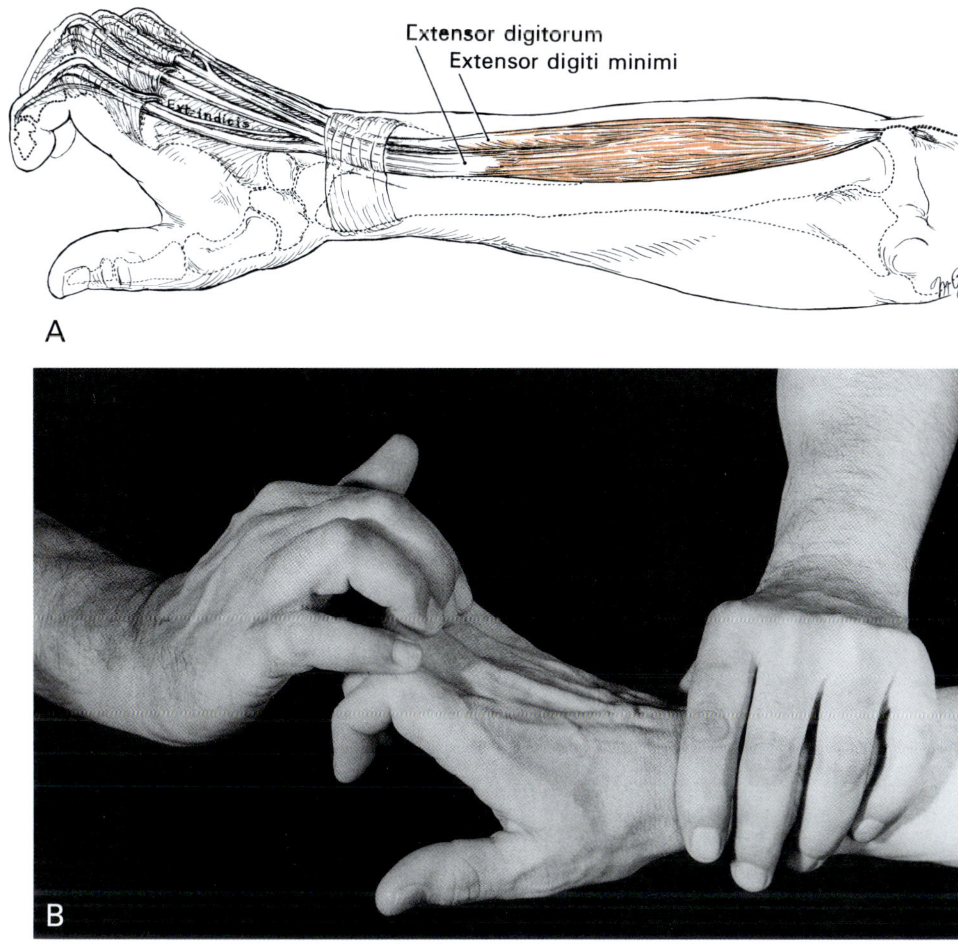

FIGURE 6-28: A, Extensor digitorum; **B,** Muscle testing of extensor digitorum.

Extensor Digitorum

Origin: Common extensor tendon from the lateral epicondyle of the humerus and deep antebrachial fascia (Figure 6-28A).

Insertion: By four tendons, each penetrating a membranous expansion on the dorsum of the second through fifth digits and dividing over the proximal phalanx into a medial and two lateral bands. The medial band inserts into the base of the middle phalanx; the lateral bands reunite over the middle phalanx and insert into the base of the distal phalanx.

Action: Extends the metacarpophalangeal joints and, in conjunction with the lumbricals and interossei, extends the interphalangeal joints of the second through fifth digits. Assists in abduction of the second, fourth, and fifth digits and in extension and abduction of the wrist.

Nerve: Radial, **C6, 7, 8.**

Patient: Sitting or supine.

Fixation: The examiner stabilizes the wrist, avoiding full extension.

Test: Extension of the metacarpophalangeal joints of the second through fifth digits with the interphalangeal joints relaxed (Figure 6-28B).

Pressure: Against the dorsal surfaces of the proximal phalanges, in the direction of flexion.

Weakness: Decreases the ability to extend the metacarpophalangeal joints of the second through fifth digits and may result in a position of flexion of these joints. The strength of wrist extension is also diminished.

Contracture: Hyperextension deformity of the metacarpophalangeal joints and limitations of wrist flexion.

Shortness: Hyperextension of the metacarpophalangeal joints if the wrist is flexed, or extension of the wrist if the metacarpophalangeal joints are flexed.

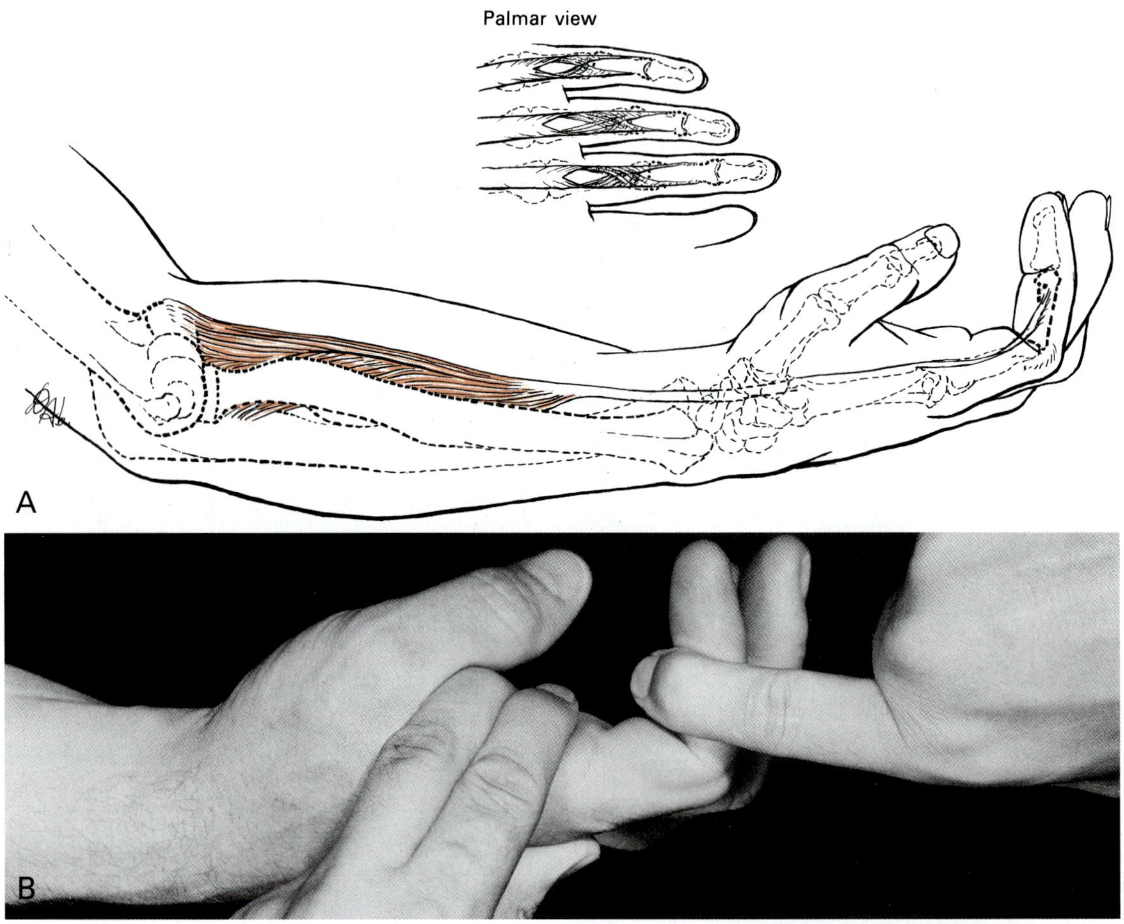

FIGURE 6-29: A, Flexor digitorum superficialis; **B,** Muscle testing of flexor digitorum superficialis.

Flexor Digitorum Superficialis

Origin of Humeral Head: Common flexor tendon from the medial epicondyle of the humerus, ulnar collateral ligament of the elbow joint, and deep antebrachial fascia.

Origin of Ulnar Head: Medial side of the coronoid process.

Origin of Radial Head: Oblique line of the radius.

Insertion: By four tendons into the sides of the middle phalanges of the second through fifth digits.

Action: Flexes the proximal interphalangeal joints of the second through fifth digits and assists in flexion of the metacarpophalangeal joints and in flexion of the wrist (Figure 6-29A).

Nerve: Medial, **C7, 8, T1**.

Patient: Sitting or supine.

Fixation: The examiner stabilizes the metacarpophalangeal joint, with the wrist in neutral position or slight extension.

Test: Flexion of the proximal interphalangeal joint, with the distal interphalangeal joint extended, of the second, third, fourth, and fifth digits. (See *Note.*) Each digit is tested as illustrated for the second digit (Figure 6-29B).

Pressure: Against the palmar surface of the middle phalanx, in the direction of extension.

Weakness: Decreases the strength of grip and wrist flexion. Interferes with digit function in activities in which the proximal interphalangeal joint is flexed, while the distal joint is extended, such as typing, playing the piano, and playing some stringed instruments. Weakness causes loss of joint stability at the proximal interphalangeal joints of the digits so that during digit extension, these joints hyperextend.

Contracture: Flexion deformity of the proximal interphalangeal joint of the second through fifth digits. A flexion deformity of the wrist may also be observed.

Shortness: Flexion of the proximal interphalangeal joint of the second through fifth digits if the wrist is extended, or flexion of the wrist if the digits are extended.

NOTE: *It appears to be the exception rather than the rule to obtain isolated flexor superficialis action in the fifth digit.*

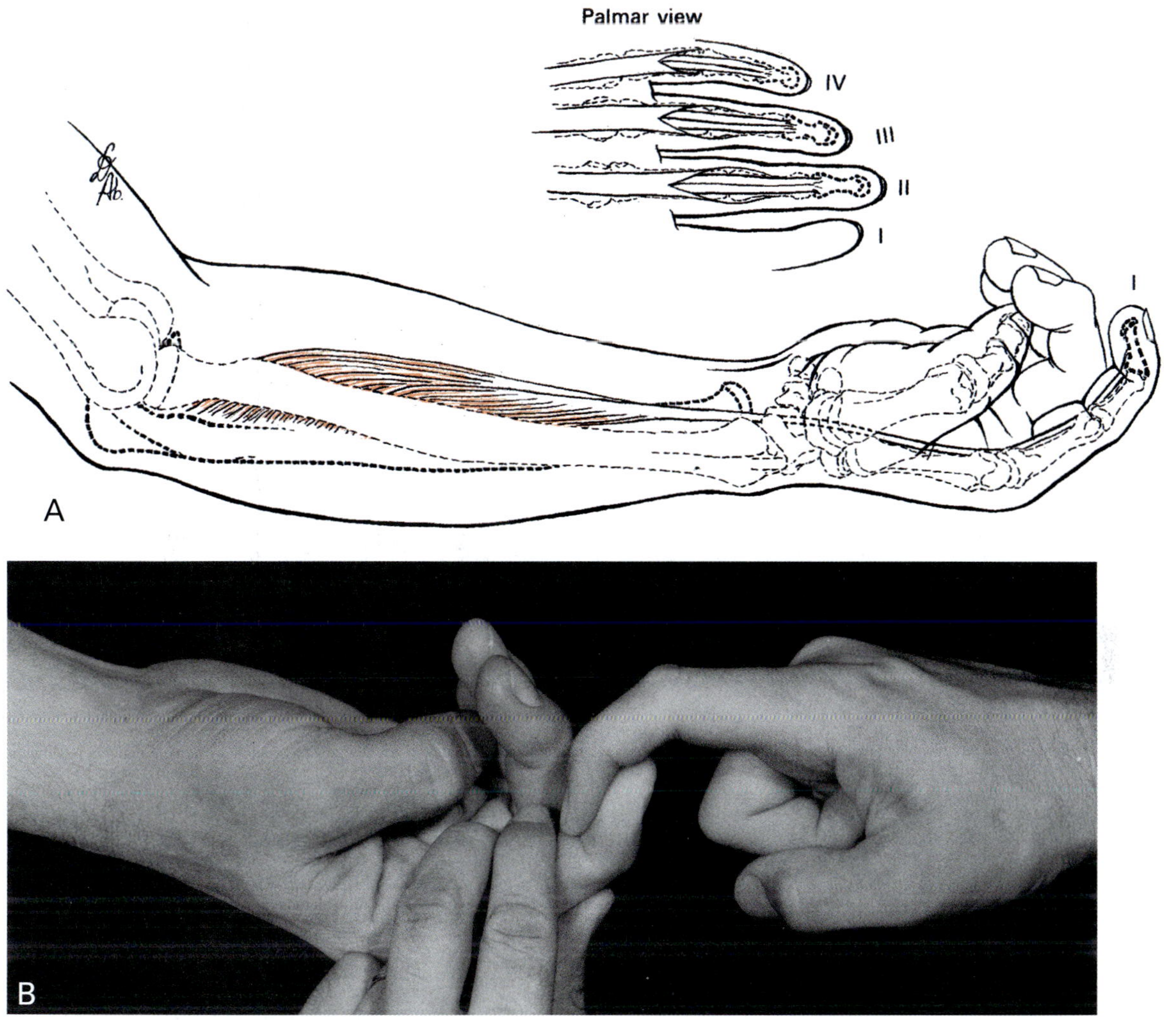

FIGURE 6-30: **A,** Flexor digitorum profundus; **B,** Muscle testing of flexor digitorum profundus.

Flexor Digitorum Profundus

Origin: Anterior and medial surfaces of the proximal 3/4 of the ulna, interosseous membrane, and deep antebrachial fascia (Figure 6-30A).

Insertion: By four tendons into the bases of the distal phalanges, anterior surface.

Action: Flexes the distal interphalangeal joints of the second through fifth digits and assists in flexion of the proximal interphalangeal and metacarpophalangeal joints. May assist in flexion of the wrist.

Nerves:

First and second: Median, C7, **8, T1**.

Third and fourth: Ulnar, C7, **8, T1**.

Patient: Sitting or supine.

Fixation: With the wrist in slight extension, the examiner stabilizes the proximal and middle phalanges.

Test: Flexion of the distal interphalangeal joint of the second, third, fourth, and fifth digits. Each digit is tested as illustrated above for the second digit (Figure 6-30B).

Pressure: Against the palmar surface of the distal phalanx, in the direction of extension.

Weakness: Decreases the ability to flex the distal joints of the digits in direct proportion to the extent of weakness because this is the only muscle that flexes the distal interphalangeal joints. Flexion strength of the proximal interphalangeal, metacarpophalangeal, and wrist joints may be diminished.

Contracture: Flexion deformity of the distal interphalangeal joint of digits two through five and limited wrist extension.

Shortness: Flexion of the distal interphalangeal joint of digits two through five if the wrist is extended, or flexion of the wrist if the digits are extended.

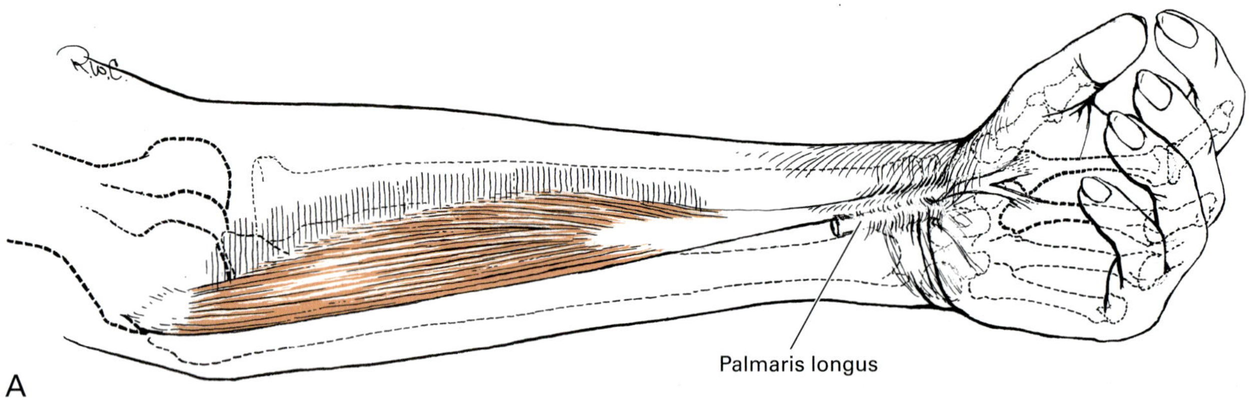

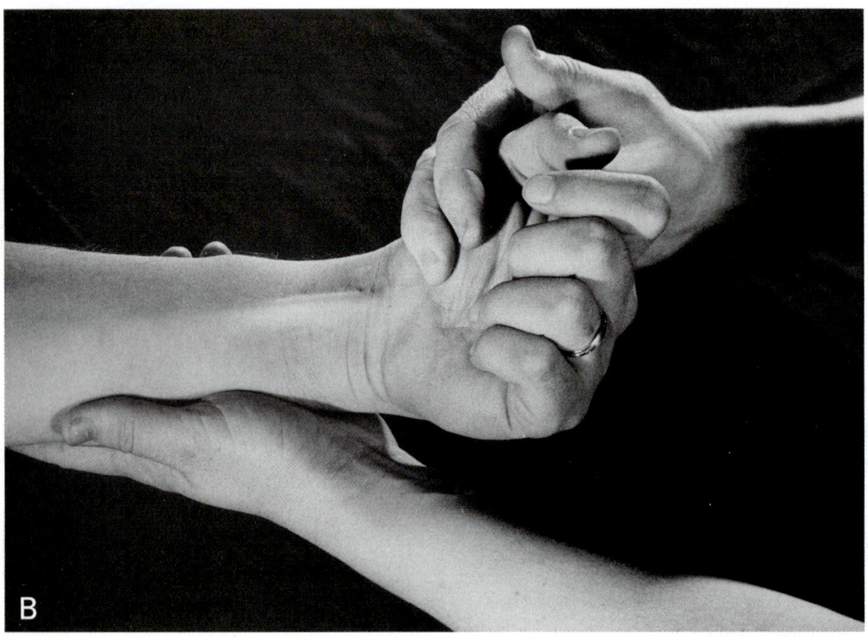

FIGURE 6-31: A, Flexor carpi radialis (palmaris longus reflected); **B,** Muscle testing of flexor carpi radialis.

Flexor Carpi Radialis

Origin: Common flexor tendon from the medial epicondyle of the humerus and deep antebrachial fascia (Figure 6-31A). (Fascia are indicated by parallel lines.)

Insertion: Base of the second metacarpal bone and a slip to the base of the third metacarpal bone.

Action: Flexes and abducts the wrist and may assist in pronation of the forearm and in flexion of the elbow.

Nerve: Median, **C6, 7,** 8.

Patient: Sitting or supine.

Fixation: The forearm is in slightly less than full supination and either rests on the table for support or is supported by the examiner.

Test: Flexion of the wrist with radial deviation (Figure 6-31B). (See *Note* on the facing page.)

Pressure: Against the thenar eminence, in the direction of extension toward the ulnar side.

Weakness: Decreases the strength of wrist flexion, and pronation strength may be diminished. Allows an ulnar deviation of the hand.

Shortness: Wrist flexion with radial deviation.

NOTE: *The palmaris longus cannot be ruled out in this test.*

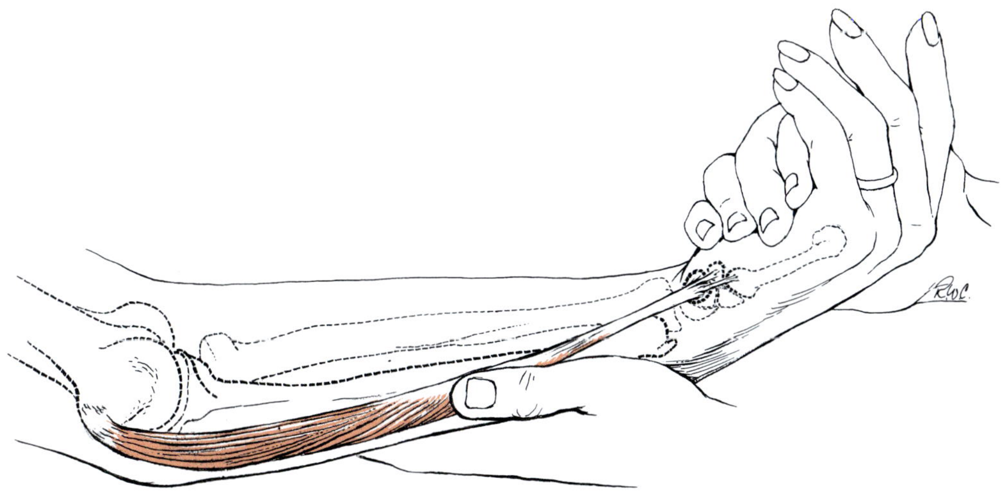
A

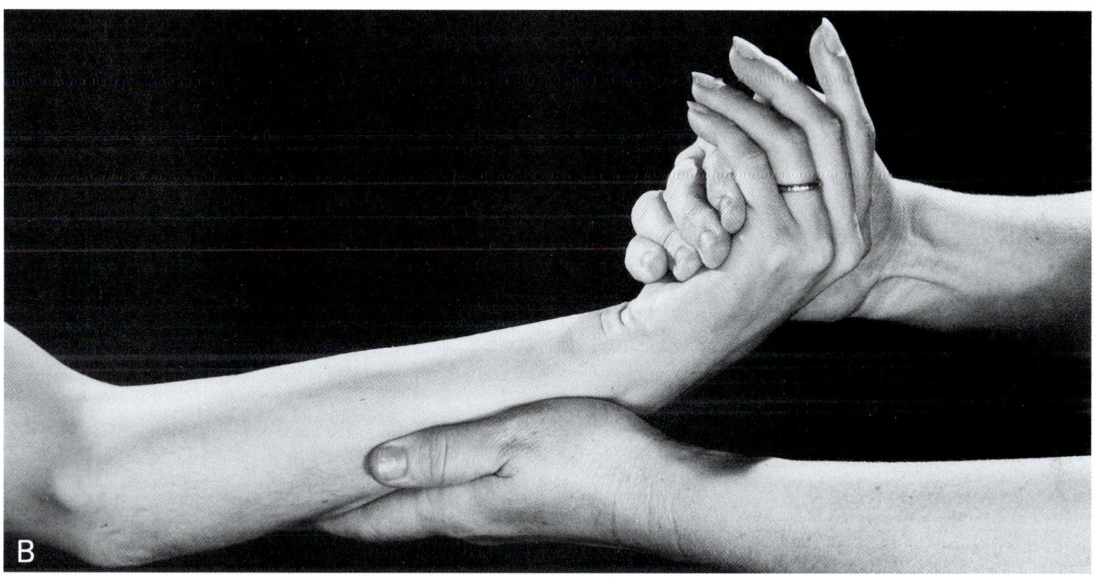

FIGURE 6-32: A, Flexor carpi ulnaris; **B,** Muscle testing of flexor carpi ulnaris.

Flexor Carpi Ulnaris

Origin of Humeral Head: Common flexor tendon from the medial epicondyle of the humerus.

Origin of Ulnar Head: By aponeurosis from the medial margin of the olecranon, proximal 2/3 of the posterior border of the ulna, and the deep antebrachial fascia.

Insertion: Pisiform bone and, by ligaments, to the hamate and fifth metacarpal bones.

Action: Flexes and adducts the wrist and may assist in flexion of the elbow (Figure 6-32A).

Nerve: Ulnar, C7, **8,** T1.

Patient: Sitting or supine.

Fixation: The forearm is in full supination and either rests on the table for support or is supported by the examiner.

Test: Flexion of the wrist with ulnar deviation (Figure 6-32B).

Pressure: Against the hypothenar eminence, in the direction of extension toward the radial side.

Weakness: Decreases the strength of wrist flexion and may result in a radial deviation of the hand.

Shortness: Wrist flexion with ulnar deviation.

NOTE: *Normally, the digits will be relaxed when the wrist is flexed. If the digits actively flex as wrist flexion is initiated, however, the digit flexors (profundus and superficialis) are attempting to substitute for the wrist flexors.*

6 | UPPER EXTREMITY

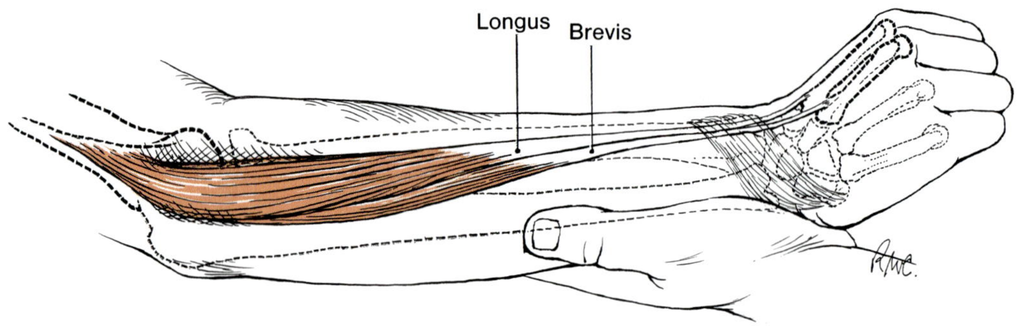

FIGURE 6-33: Extensor carpi radialis longus and brevis.

EXTENSOR CARPI RADIALIS LONGUS AND BREVIS

Extensor Carpi Radialis Longus

Origin: Distal 1/3 of the lateral supracondylar ridge of the humerus and lateral intermuscular septum (Figure 6-33).

Insertion: Dorsal surface of the base of second metacarpal bone, radial side.

Action: Extends and abducts the wrist and assists in flexion of the elbow.

Nerve: Radial, C5, **6, 7,** 8.

Extensor Carpi Radialis Brevis

Origin: Common extensor tendon from the lateral epicondyle of the humerus, radial collateral ligament of elbow joint, and deep antebrachial fascia.

Insertion: Dorsal surface of the base of the third metacarpal bone.

Action: Extends and assists in radial deviation of the wrist.

Nerve: Radial, **C6, 7,** 8.

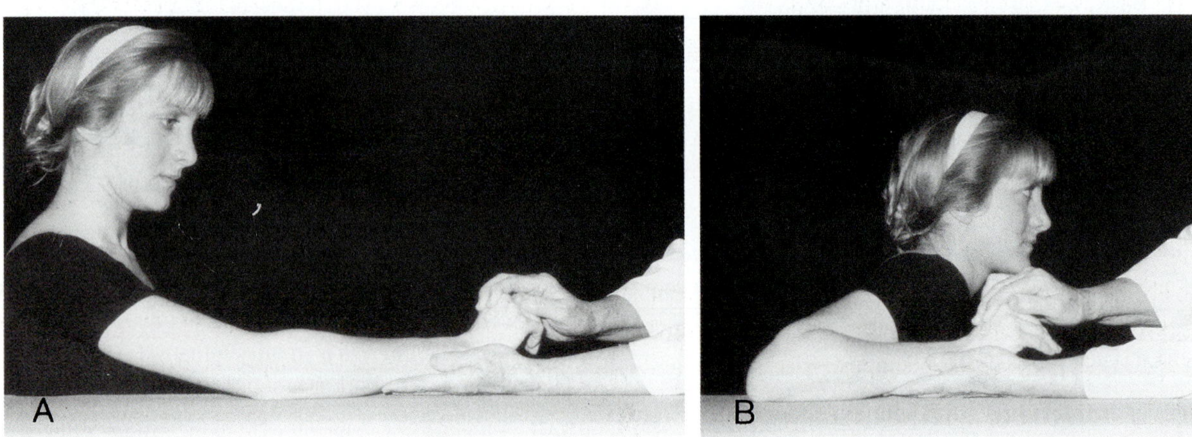

FIGURE 6-34: **A,** Muscle testing of extensor carpi radialis longus and brevis; **B,** Muscle testing of extensor carpi radialis brevis.

Test of Extensor Carpi Radialis Longus and Brevis

Patient: Sitting with the elbow approximately 30° from zero extension (Figure 6-34A).

Fixation: The forearm is in slightly less than full pronation and rests on the table for support.

Test: Extension of the wrist with radial deviation. (Interphalangeal joints should be allowed to flex as the wrist is extended.)

Pressure: Against the dorsum of the hand, along the second and third metacarpal bones, in the direction of flexion toward the ulnar side.

Weakness: Decreases the strength of wrist extension and allows an ulnar deviation of the hand.

Shortness: Wrist extension with radial deviation.

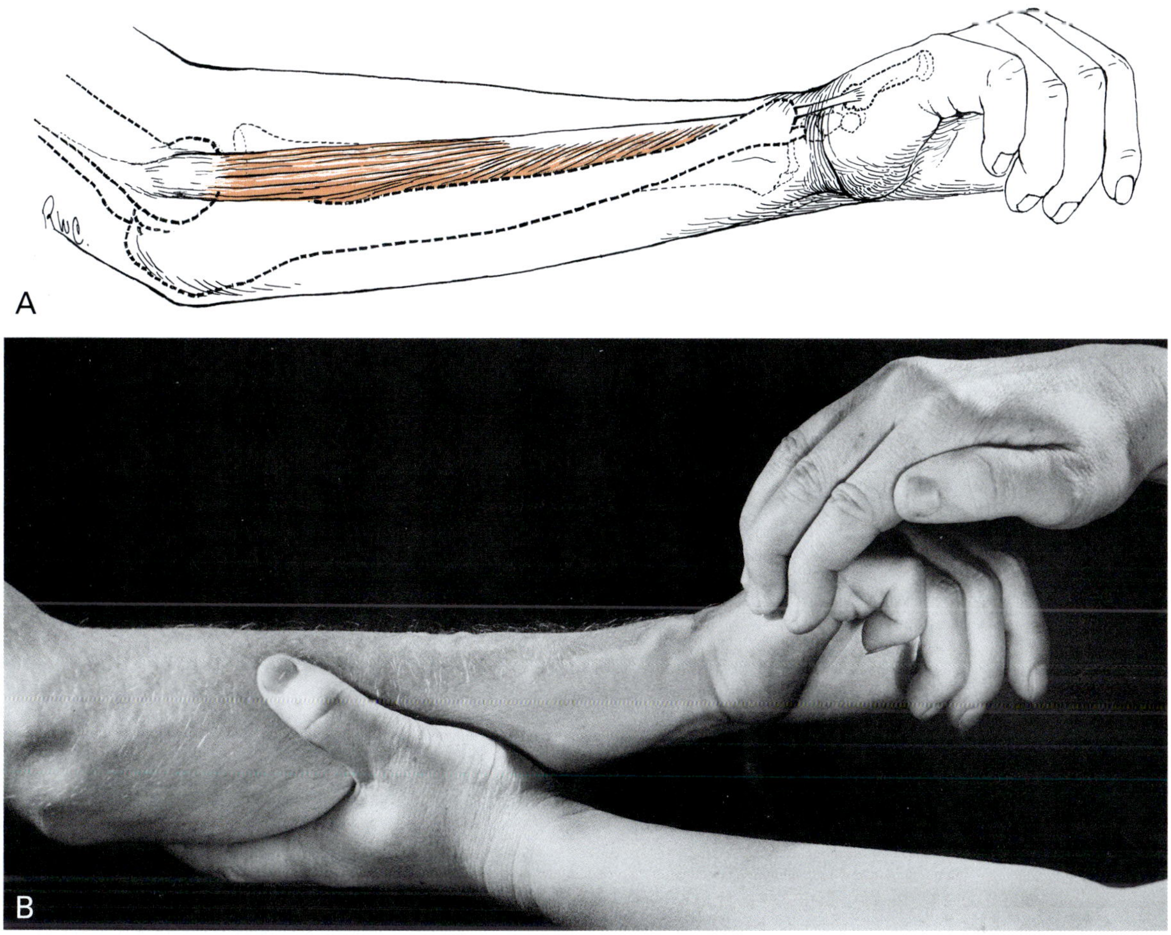

FIGURE 6-35: **A,** Extensor carpi ulnaris; **B,** Muscle testing of extensor carpi ulnaris.

Test of Extensor Carpi Radialis Brevis

Patient: Sitting with the elbow fully flexed (Figure 6-34B). (Have the subject lean forward to flex the elbow.)

Fixation: The forearm is in slightly less than full pronation and rests on the table for support.

Test: Extension of the wrist with radial deviation.

Pressure: Against the dorsum of the hand, along the second and third metacarpal bones, in the direction of flexion toward the ulnar side.

Extensor Carpi Ulnaris

Origin: Common extensor tendon from the lateral epicondyle of the humerus, by the aponeurosis from the posterior border of the ulna and deep antebrachial fascia (Figure 6-35A).

Insertion: Base of the fifth metacarpal bone, ulnar side.

Action: Extends and ulnarly deviates the wrist.

Nerve: Radial, C6, **7**, **8**.

Patient: Sitting or supine.

Fixation: The forearm is in full pronation and either rests on the table for support or is supported by the examiner.

Test: Extension of the wrist with ulnar deviation (Figure 6-35B).

Pressure: Against the dorsum of the hand, along the fifth metacarpal bone, in the direction of flexion toward the radial side.

Weakness: Decreases the strength of wrist extension and may result in a radial deviation of the hand.

Shortness: Wrist ulnar deviation with slight extension.

> **NOTE:** *Normally, the digits will be in a position of passive flexion when the wrist is extended. If the digits actively extend as wrist extension is initiated, however, the digit extensors (digitorum, indicis, and digiti minimi) are attempting to compensate for weak wrist extensors.*

6 UPPER EXTREMITY

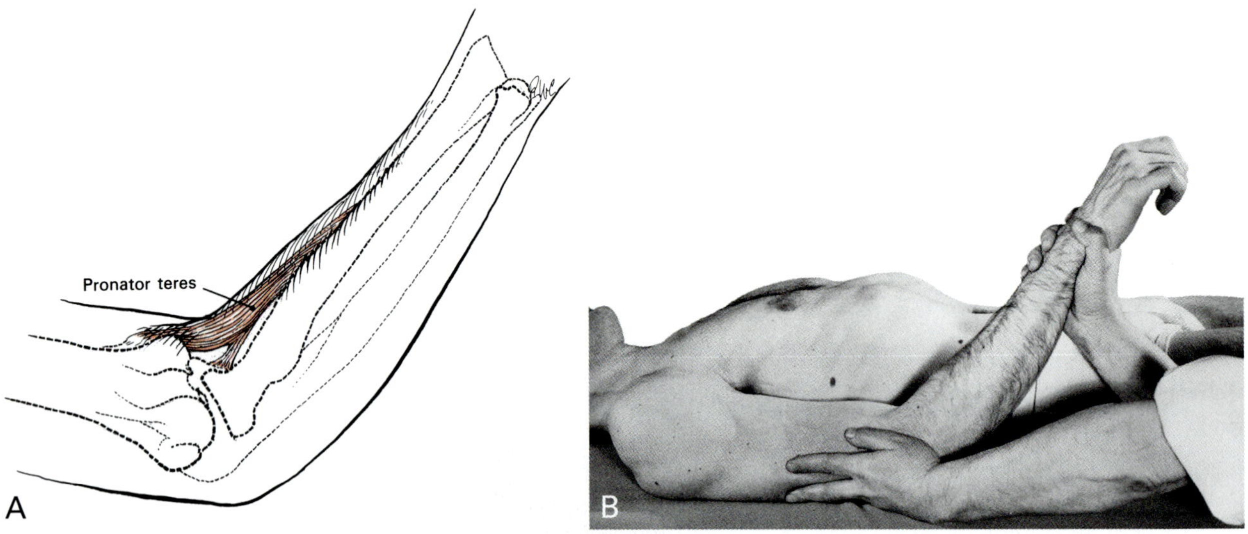

FIGURE 6-36: **A,** Pronator teres; **B,** Muscle testing of pronator teres and pronator quadratus.

PRONATOR TERES AND QUADRATUS

Pronator Teres

Origin of Humeral Head: Immediately proximal to the medial epicondyle of the humerus, common flexor tendon, and deep antebrachial fascia (Figure 6-36A).

Origin of Ulnar Head: Medial side of the coronoid process of the ulna.

Insertion: Middle of the lateral surface of the radius.

Action: Pronates the forearm and assists in flexion of the elbow joint.

Nerve: Median, **C6, 7**.

Pronator Quadratus

Origin: Medial side, anterior surface of the distal 1/4 of the ulna (Figure 6-37A).

Insertion: Lateral side, anterior surface of the distal 1/4 of the radius.

Action: Pronates the forearm.

Nerve: Median, C7, **8, T1**.

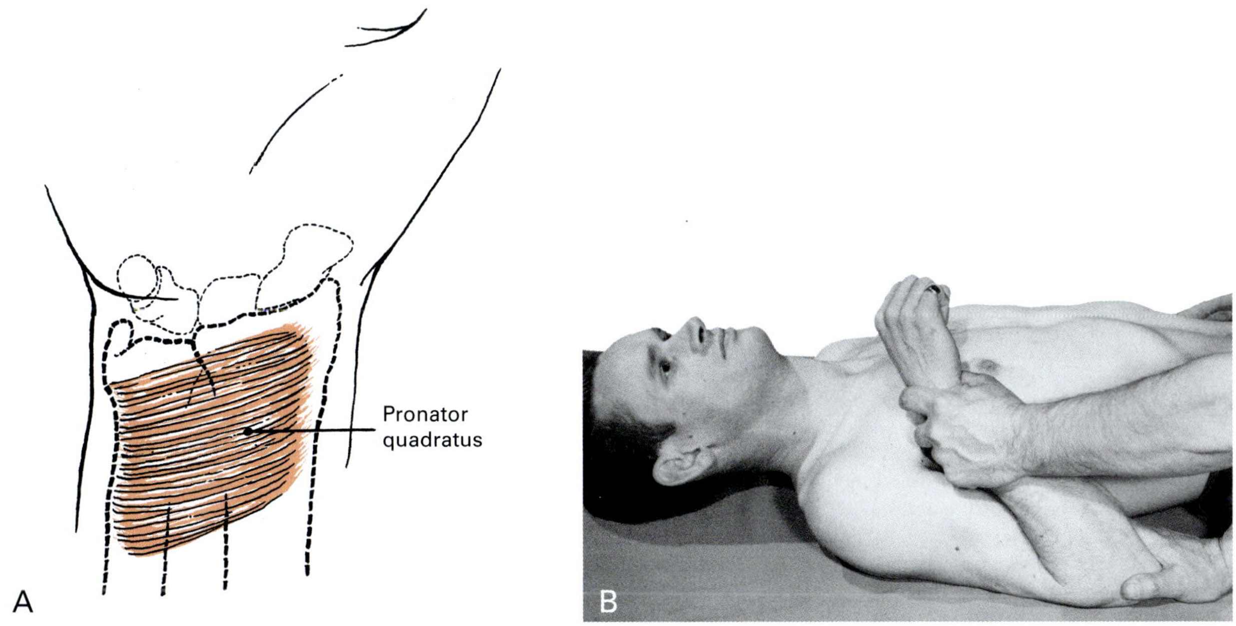

FIGURE 6-37: A, Pronator quadratus; **B,** Muscle testing of pronator quadratus.

Test for Pronators Teres and Quadratus

Patient: Supine or sitting.

Fixation: The elbow should be held against the patient's side or be stabilized by the examiner to avoid any shoulder abduction movement.

Test: Pronation of the forearm, with the elbow partially flexed (Figure 6-36B).

Pressure: At the lower forearm, above the wrist (to avoid twisting the wrist), in the direction of supinating the forearm.

Weakness: Allows a supinated position of the forearm and interferes with many everyday functions, such as turning a doorknob, using a knife to cut meat, and turning the hand downward in picking up a cup or other object.

Contracture: With the forearm held in a position of pronation, interferes markedly with many normal functions of the hand and forearm that require moving from pronation to supination.

NOTE: *Avoid squeezing the radius and ulna together because this may be painful.*

Test for Pronators Quadratus

Patient: Supine or sitting.

Fixation: The elbow should be held against the patient's side (either by the patient or by the examiner) to avoid shoulder abduction.

Test: Pronation of the radioulnar joints, with the elbow completely flexed to make the humeral head of the pronator teres less effective by being in a shortened position (Figure 6-37B).

Pressure: At the lower forearm, above the wrist (to avoid twisting the wrist), in the direction of supinating the forearm.

NOTE: *Avoid squeezing the radius and ulna together because this may be painful.*

6 UPPER EXTREMITY

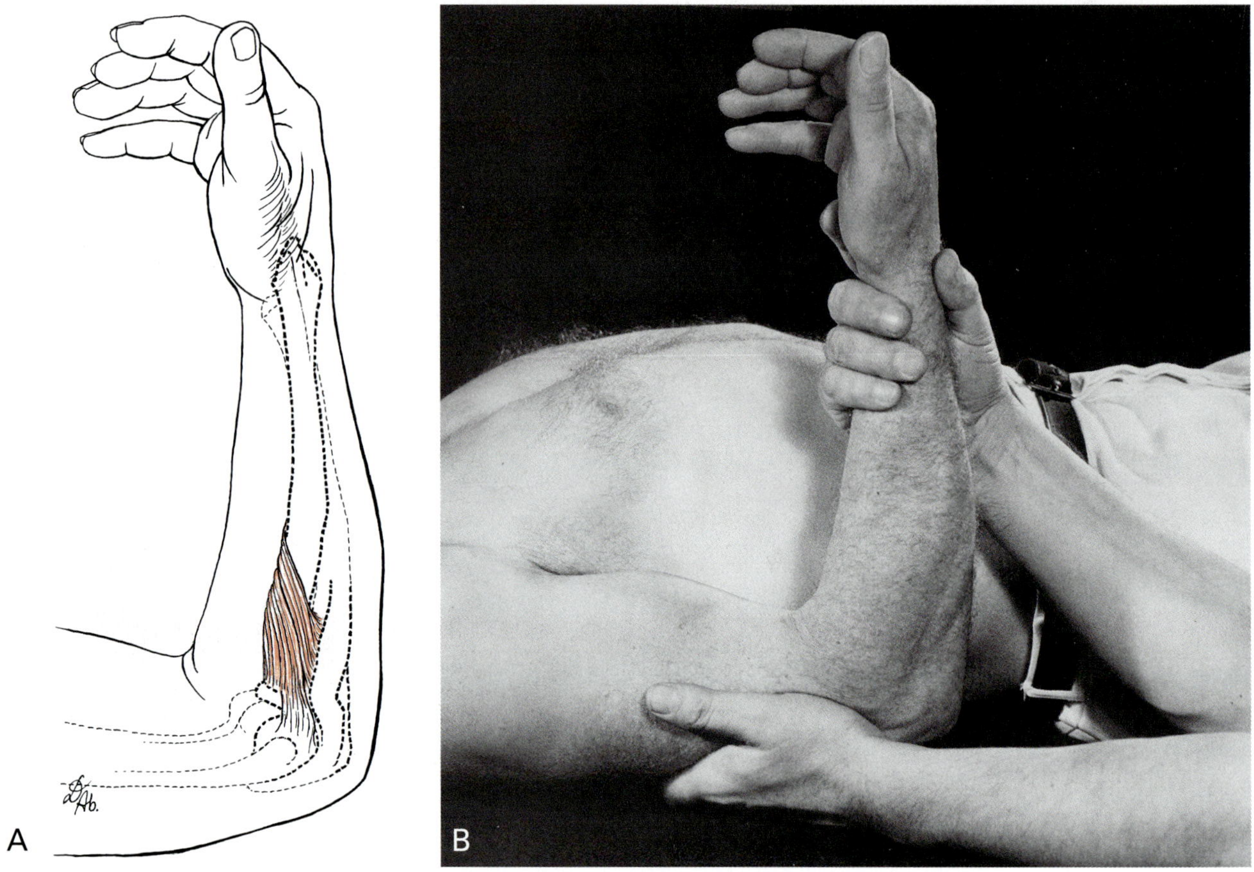

FIGURE 6-38: **A,** Supinator and biceps; **B,** Position for muscle testing for supinator and biceps.

SUPINATOR AND BICEPS

Supinator

Origin: Lateral epicondyle of the humerus, radial collateral ligament of the elbow joint, annular ligament of the radius, and supinator crest of the ulna (Figure 6-38A).

Insertion: Lateral surface of the upper 1/3 of the body of the radius, covering part of the anterior and posterior surfaces.

Action: Supinates the forearm at the radioulnar joint.

Nerve: Deep Radial, C5, **6**, (7).

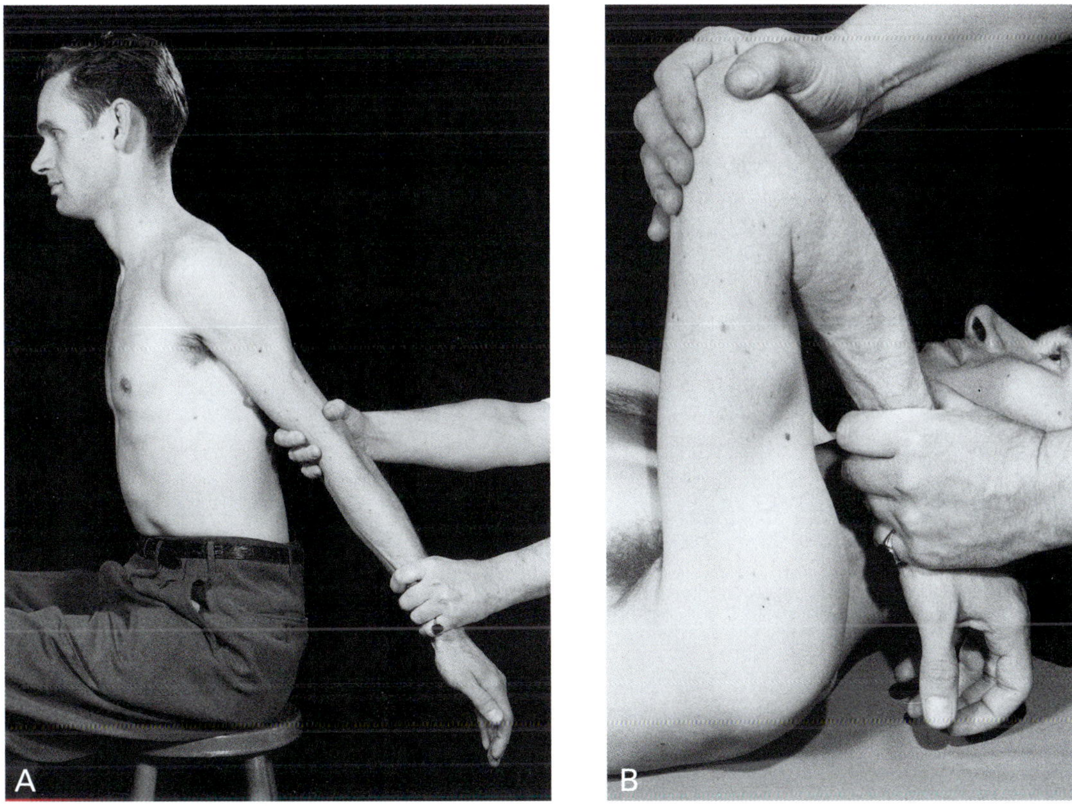

FIGURE 6-39: Muscle testing of the supinator with (**A**) the biceps elongated, and (**B**) the biceps in a shortened position.

Tested with the biceps elongated:

Patient: Sitting or standing.

Fixation: The examiner holds the shoulder and elbow in extension with the biceps elongated.

Test: Supination of the forearm (Figure 6-39A).

Pressure: At the distal end of the forearm, above the wrist, in the direction of pronation. The subject may attempt to rotate the humerus laterally to make it appear that the forearm remains in supination as pressure is applied and the forearm starts to pronate.

Tested with the biceps in a shorted position:

Patient: Supine.

Fixation: The examiner holds the shoulder in flexion, with the elbow completely flexed and the biceps in a shortened position. It is usually advisable to have the subject close the digits to keep them from touching the table, which may be done to brace the forearm in the test position.

Test: Supination of the forearm (Figure 6-39B).

Pressure: At the distal end of the forearm, above the wrist, in the direction of pronation. Take care to avoid maximum pressure because as strong pressure is applied, the biceps contracts and may develop a cramp in this shortened position. A severe cramp may leave the muscle sore for several days. This test should be used merely as a differential diagnostic aid.

Test for Supinator and Biceps

Patient: Supine.

Fixation: The elbow should be held against the patient's side to avoid shoulder movement.

Test: Supination of the forearm, with the elbow at a right angle or slightly below (Figure 6-38B).

Pressure: At the distal end of the forearm, above the wrist (to avoid twisting the wrist), in the direction of pronating the forearm.

Weakness: Allows the forearm to remain in a pronated position. Interferes with many functions of the extremity, particularly those involved with feeding oneself.

Contracture: Elbow flexion with forearm supination. Interferes markedly with functions of the extremity that involve the change from a supinated to a pronated position of the forearm.

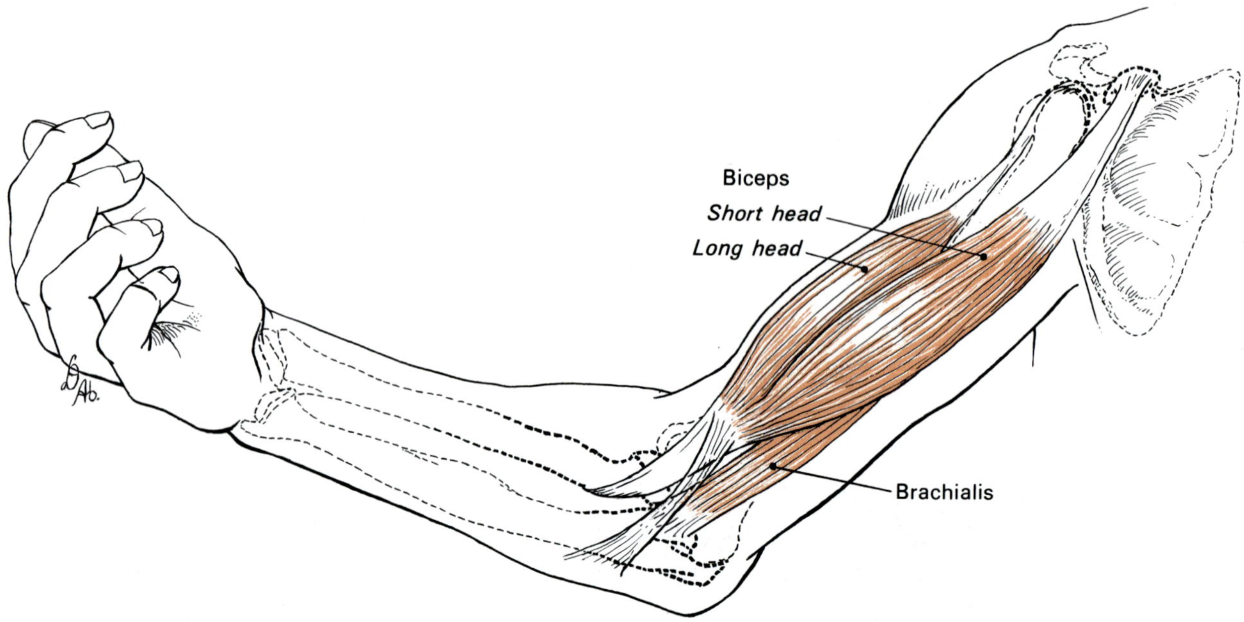

FIGURE 6-40: Biceps brachii and brachialis.

BICEPS BRACHII AND BRACHIALIS

Biceps Brachii

Origin of Short Head: Apex of the coracoid process of the scapula.

Origin of Long Head: Supraglenoid tubercle of the scapula.

Insertion: Tuberosity of the radius and aponeurosis of the biceps brachii (lacertus fibrosus).

Action: Flexes the shoulder joint. The short head assists with shoulder adduction. The long head may assist with abduction if the humerus is externally rotated. With the origin fixed, flexes the elbow joint, moving the forearm toward the humerus and supinates the forearm. With the insertion fixed, flexes the elbow joint, moving the humerus toward the forearm, as in a pull-up exercise (Figure 6-40).

Nerve: Musculocutaneous **C5, 6**.

Brachialis

Origin: Distal 1/2 of the anterior surface of the humerus and both medial and lateral intermuscular septa.

Insertion: Tuberosity and coronoid process of the ulna.

Action: With the origin fixed, flexes the elbow joint, moving the forearm toward the humerus. With the insertion fixed, flexes the elbow joint, moving the humerus toward the forearm, as in a pull-up exercise.

Nerve: Musculocutaneous, small branch from radial, **C5, 6**.

Test for Biceps Brachii and Brachialis

Patient: Supine or sitting.

Fixation: The examiner places one hand under the elbow to cushion it from table pressure.

Test: Elbow flexion slightly less than or at a right angle, with the forearm in supination.

Pressure: Against the distal forearm, in the direction of extension.

Weakness: Decreases the ability to flex the elbow against gravity. Interferes markedly with daily activities such as feeding oneself or combing the hair.

Shortness: Flexion deformity of the elbow.

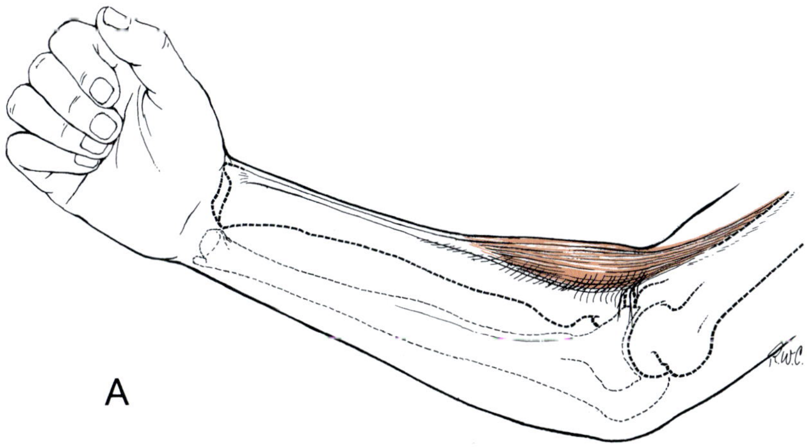

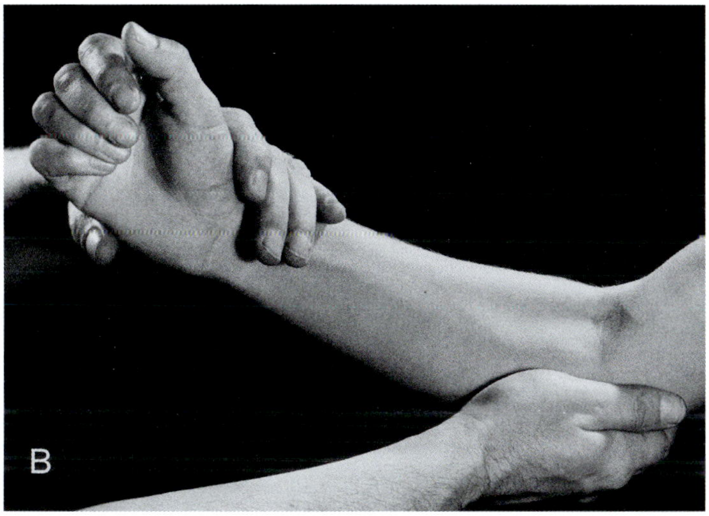

FIGURE 6-41: A, Brachioradialis; **B,** Muscle testing off brachioradialis.

Brachioradialis

Origin: Proximal 2/3 of the lateral supracondylar ridge of the humerus, and lateral intermuscular septum (Figure 6-41A).

Insertion: Lateral side of the base of the styloid process of the radius.

Action: Flexes the elbow joint. Also has a secondary action of forearm pronation (16).

Nerve: Radial, **C5, 6**

Patient: Supine or sitting.

Fixation: The examiner places one hand under the elbow to cushion it from table pressure.

Test: Flexion of the elbow, with the forearm neutral between pronation and supination. The belly of the brachioradialis (Figure 6-41B) must be seen and felt during this test because the movement can also be produced by other muscles that flex the elbow.

Pressure: Against the distal forearm, in the direction of extension.

Weakness: Decreases the strength of elbow flexion and of resisted pronation to midline.

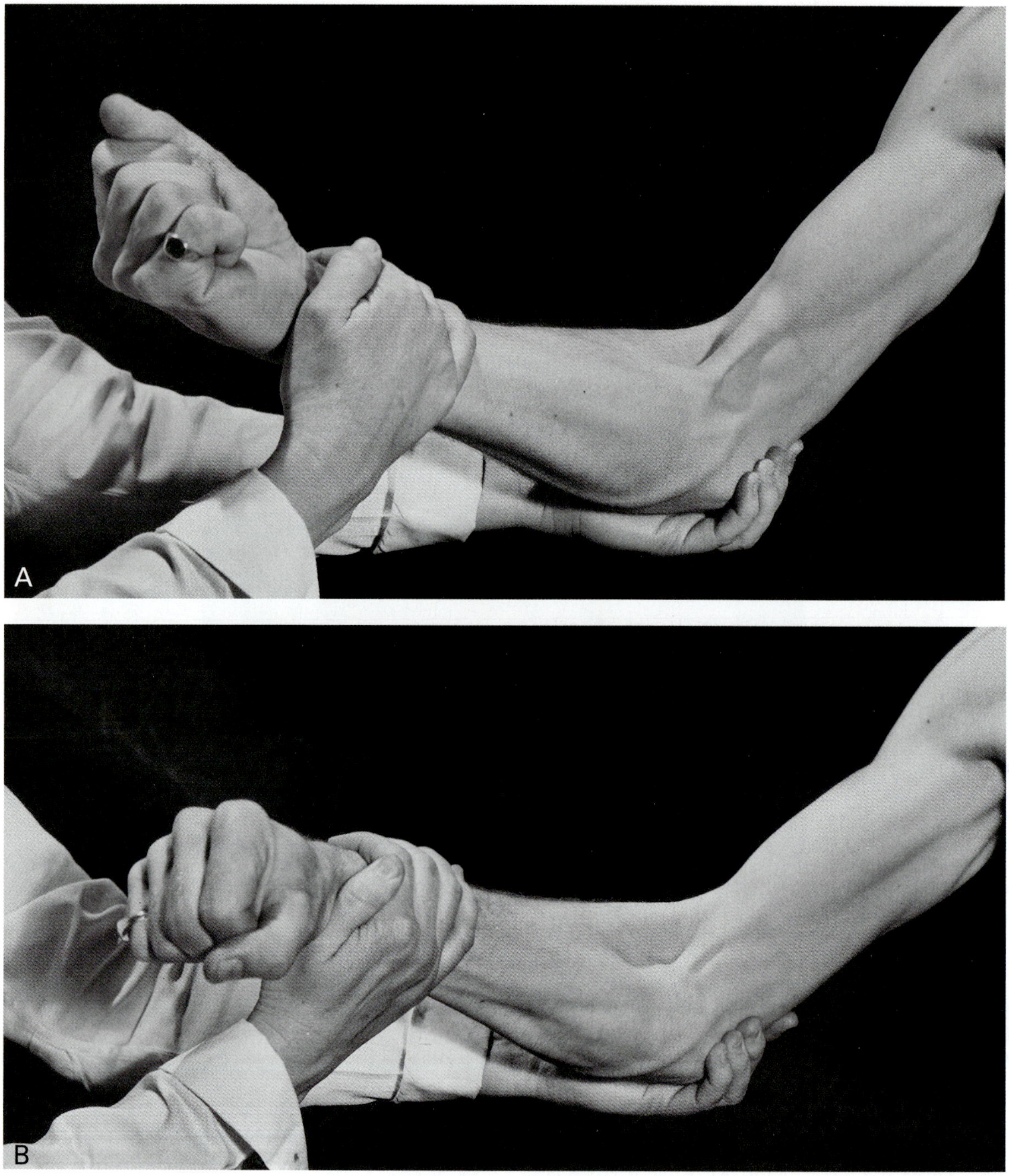

FIGURE 6-42: **A,** Elbow flexion with the forearm supinated; **B,** Elbow flexion with the forearm pronated.

Elbow Flexors

Figure 6-42 illustrates elbow flexion. Part B shows that, against resistance, the biceps acts in flexion even though the forearm is in pronation. Because the brachialis is inserted on the ulna, the position of the forearm, whether in supination or in pronation, does not affect the action of this muscle in elbow flexion. The brachioradialis appears to have a slightly stronger action in the pronated position of the forearm during the elbow flexion test than in the supinated position, although its strongest action in flexion is with the forearm in midposition (17).

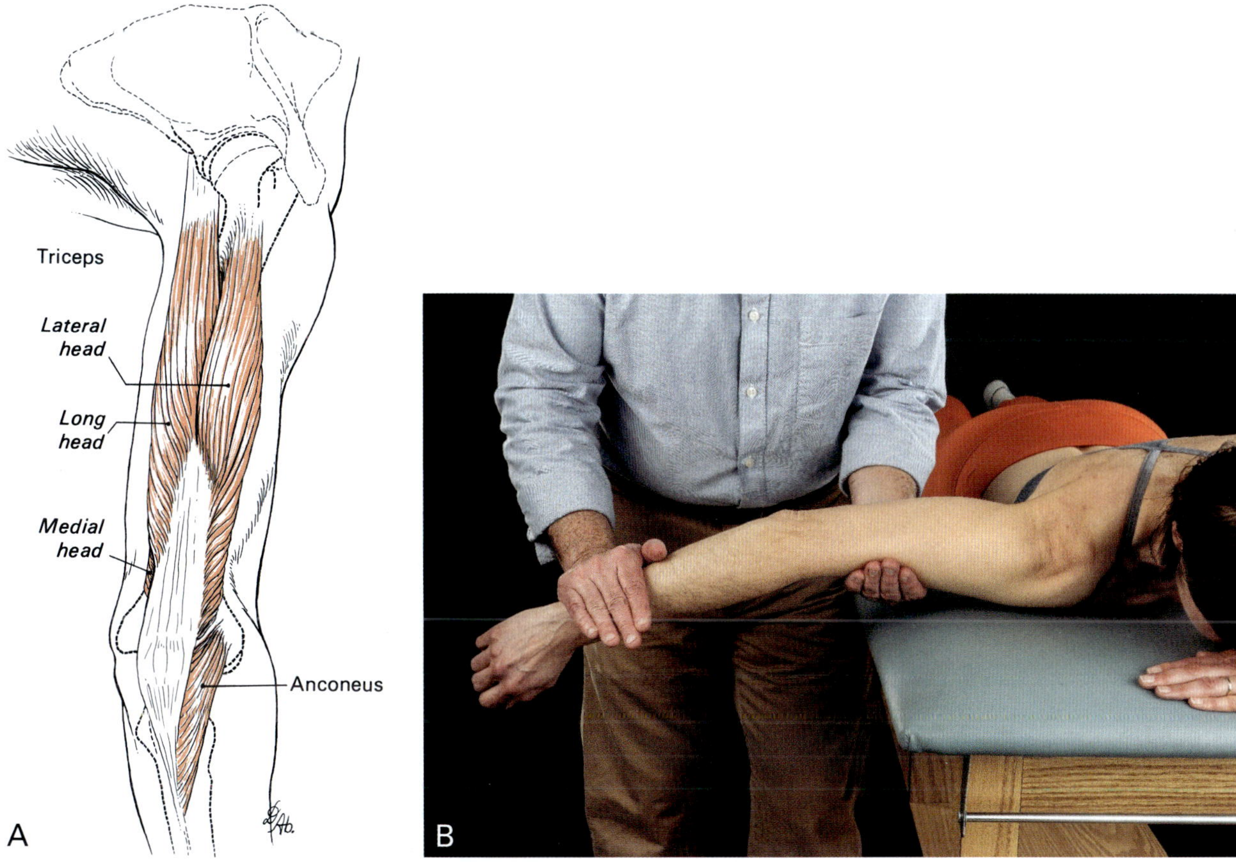

FIGURE 6-43: **A,** Triceps brachii and anconeus; **B,** Muscle testing for triceps brachii.

TRICEPS BRACHII AND ANCONEUS

Triceps Brachii

Origin of Long Head: Infraglenoid tubercle of the scapula.

Origin of Lateral Head: Lateral and posterior surfaces of the proximal 1/2 of the body of the humerus and lateral intermuscular septum.

Origin of Medial Head: Distal 2/3 of the medial and posterior surfaces of the humerus below the radial groove and from the medial intermuscular septum.

Insertion: Posterior surface of the olecranon process of the ulna and antebrachial fascia.

Action: Extends the elbow joint. The long head also assists in adduction and extension of the shoulder joint (Figure 6-43A).

Nerve: Radial, C6, **7, 8,** T1.

Anconeus

Origin: Lateral epicondyle of the humerus, posterior surface.

Insertion: Lateral side of the olecranon process and upper 1/4 of the posterior surface of the body of the ulna.

Action: Extends the elbow joint and may stabilize the ulna during pronation and supination.

Nerve: Radial, **C7, 8**.

Triceps Brachii and Anconeus

Patient: Prone.

Fixation: The shoulder is at 90° abduction, neutral with regard to rotation, and with the arm supported between the shoulder and the elbow by the table. The examiner places one hand under the arm near the elbow to cushion the arm from table pressure (Figure 6-43B).

Test: Extension of the elbow joint (to slightly less than full extension).

Pressure: Against the forearm, in the direction of flexion.

Patient: Supine.

Fixation: The shoulder is at approximately 90° flexion, with the arm supported in a position perpendicular to the table (Figure 6-44).

Test: Extension of the elbow (to slightly less than full extension).

Pressure: Against the forearm in the direction of flexion.

Weakness: Results in the inability to extend the forearm against gravity. Interferes with everyday functions that involve elbow extension, such as reaching upward toward a high shelf. Results in loss of ability to throw objects or to push them with the extended elbow. Also handicaps the individual in using crutches or a cane because of inability to extend the elbow and transfer weight to the hand.

Contracture: Extension deformity of the elbow. Interferes markedly with everyday functions that involve elbow flexion.

NOTE: *When the shoulder is horizontally abducted with the elbow extended, the long head of the triceps is shortened over both the shoulder and elbow joints. When the shoulder is horizontally adducted with the elbow extended, the long head of the triceps is shortened over the elbow joint but elongated over the shoulder joint. Because of this two-joint action, the long head is made less effective in the prone position by being shortened over both joints, with the result that the triceps withstands less pressure when tested in the prone position than when tested in the supine position.*

The triceps and anconeus act together in extending the elbow joint, but it may be useful to differentiate between these two muscles. Because the belly of the anconeus muscle is distal to the elbow joint, it can be distinguished from the triceps by palpation. The branch of the radial nerve to the anconeus arises near the midhumeral level and is quite long. It is possible for a lesion to involve only this branch, leaving the triceps unaffected. Impaired action of the anconeus reduces the strength of elbow extension. One may find that a grade of good elbow extension strength is actually the result of a normal triceps and an impaired anconeus.

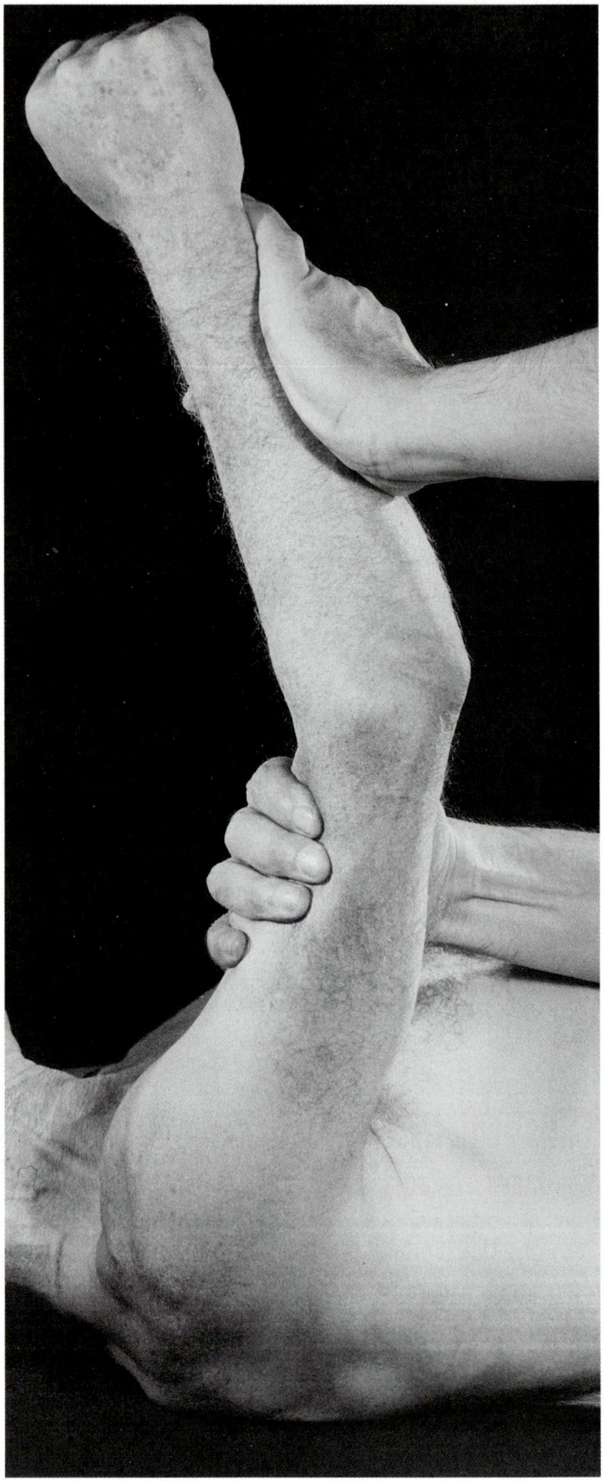

FIGURE 6-44: Muscle testing for anconeus.

Range of Motion: Thumb and Digits

The references in Figure 6-45 demonstrate the lack of consensus regarding normative values for thumb and digit range of motion. Previous authors chose ranges that are representative of both established sources and clinical practice. When mobility is limited, the measurement should be documented in parenthesis, and when it is excessive, hypermobility should be indicated by a circle around the measured number.

Strength Testing of Thumb and Digits

Box 6-1 lists muscle testing grades for thumb and digits.

Joint		Kendall	Palmer (17)	Reese (18)	Clarkson (19)	AAOS (20)	AMA (21)
Joint	Thumb						
CMC	Flexion	15	0–15	0–15	0–15	0–15	
	Extension	20	0–70	0–20	0–20	0–20	
	Abduction	60	0–60	0–70	0–70	0–70	0–50
	Opposition	Pad of thumb to					
		pad of 5th digit					
MCP	Flexion	50	0–50	0–50	0–50	0–50	0–60
	Extension	0	50–0	0		0	0
IP	Flexion	80	0–80	0–65	0–80	0–80	0–80
	Extension	0	80–0	0–10–20		0–20	0–10
	2nd–5th Digits						
MCP	Flexion	90	0–90	0–90	0–90	0–90	0–90
	Extension	0	90–0	0–20	0–45	0–45	
	Abduction	20	0–20				
PIP	Flexion	100	0–120	0–100	0–100	0–100	0–100
	Extension	0	120–0	0		0	
DIP	Flexion	70	0–80	0–70	0–90	0–90	0–70
	Extension	0	80–0	0	0		

FIGURE 6-45: Range of motion: thumb and digits.

BOX 6-1

Muscle Testing Grades for Thumbs and Digits

Grade	Description
0	No contraction is felt in the muscle
1	Feeble contraction in muscle belly or tendon is prominent
2	Muscle moves part through a small arc of motion
3	Muscle moves part through a moderate arc of motion
4	Muscle moves part through an almost complete arc of motion
5	Muscle moves part through complete arc of motion
6–7	Moves part through complete arc of motion, holds against slight pressure
8–9	Same as above, but holds against moderate pressure
10	Same as above against maximum pressure

SECTION III
SHOULDER AND SCAPULA STRENGTH TESTS

MOVEMENTS OF THE SHOULDER GIRDLE (COMPLEX)

Joints

The **shoulder girdle** is a complex structure, efficient in the performance of many movements yet vulnerable to injury because of the many and varied stresses it encounters. The joints of the shoulder girdle are named according to the skeletal structures that are held together. Joints of the shoulder girdle are listed below.

1. *Sternocostal:* connects the sternum with sternal ends of 10 ribs (seven directly and three indirectly).
2. *Sternoclavicular:* connects the manubrium of the sternum with the medial end of the clavicle.
3. *Acromioclavicular:* connects the acromial process of the scapula with the lateral end of the clavicle.
4. *Glenohumeral:* connects the head of the humerus and the socket of the glenoid (hence, a ball-and-socket joint).
5. *Costovertebral:* includes the connections of the head of each rib with two adjacent vertebral bodies and the connection of the tubercle of each rib with the transverse process of the vertebra.

Glenohumeral Joint

The shoulder joint, also called the **glenohumeral joint**, is a spheroid or ball-and-socket joint formed by the head of the humerus and the glenoid cavity of the scapula. It is the most mobile and least stable joint in the body (18), very vulnerable to injury, and dependent upon neighboring musculoskeletal articulations for stability and positioning. Because of the mobility of this joint and the many movements performed by the shoulder and scapular muscles, maintaining a balanced musculature is vital to the stability of the joint. Actions of the muscles of the neck and shoulder are closely related and compensatory patterns may be seen in response to weakness or muscle shortness. In addition to six basic joint movements, it is necessary to define circumduction and two movements in the horizontal plane.

Flexion and *extension* are movements about a coronal axis. Flexion is movement in the anterior direction and may begin from a position of extension 45° beyond the midaxillary line. It describes an arc forward through the zero anatomical position and on to the 180° overhead position. However, the 180° overhead position is attained only by the combined movement of the joints of the shoulder girdle. The glenohumeral joint can be flexed only to approximately 120°. The remaining 60° is attained from the abduction and upward rotation of the scapula, which allows the glenoid cavity to face more anteriorly and superiorly and the humerus to flex to a fully vertical position. The scapular motion can be quite variable at first, but after 60° of flexion or 30° of abduction, a relatively constant relationship exists between the movement of the humerus and the movement of the scapula. Inman et al. found that between the 30° and 170° range of flexion, the glenohumeral joint provided 10°, and scapular rotation 5°, for every 15° of motion (19)

Extension is movement in the posterior direction and technically refers to the arc of motion from 180° of flexion to 45° of extension. If the elbow joint is flexed, the range of shoulder joint extension will be increased, because the tension of the biceps will be released.

Abduction and *adduction* are movements about a sagittal axis. *Abduction* is movement in a lateral direction through a range of 180° to a vertical overhead position. This end position is the same as that attained in flexion, and it coordinates the movements of the shoulder girdle and of the glenohumeral joint. *Adduction* is movement toward the midsagittal plane in a medial direction and technically refers to the arc of motion from full elevation overhead through the zero anatomical position to a position obliquely upward and across the front of the body.

Horizontal abduction and *horizontal adduction* are combination movements of abduction and adduction that occur at a degree of shoulder flexion. These movements occur in the transverse plane about a longitudinal axis. *Horizontal abduction* is movement in a lateral and posterior direction, and *horizontal adduction* is movement in an anterior and medial direction. The end position of complete horizontal adduction is the same as that for adduction obliquely upward across the body. In one instance, the arm moves horizontally to that position; in the other, it moves obliquely upward to that position.

The range of horizontal abduction, being determined to a great extent by the length of the pectoralis major, is extremely variable. With the shoulder in 90° of abduction as the zero position for measurement, the normal range should be approximately 45° in horizontal abduction and approximately 135° in horizontal adduction, most readily judged by the ability to place the palm of the hand on top of the opposite shoulder (20).

Medial (internal) rotation and *lateral (external) rotation* are movements about a longitudinal axis through the humerus. *Medial rotation* is movement in which the anterior surface of the humerus turns toward the midsagittal plane. *Lateral rotation* is movement in which the anterior surface of the humerus turns away from the midsagittal plane.

The extent of medial or lateral rotation varies with the degree of elevation in abduction or flexion. For purposes of joint measurement, the zero position is one in which the shoulder is at 90° abduction, the elbow is at 90° flexion, and the forearm is at a right angle to the coronal plane. From this position, lateral rotation of the shoulder describes an arc of 90° to a position in which the forearm is parallel with the head. With the scapula stabilized, medial rotation describes an arc of approximately 70°. If the scapula is allowed to tilt anteriorly, an arc of 90° of medial rotation can be achieved.

As the shoulder is abducted or flexed from the anatomical position, lateral rotation continues to be free, but medial rotation is limited. As the shoulder is adducted or extended, the range of medial rotation remains free, and that of lateral rotation decreases. In treatment to restore the motion of a restricted shoulder joint, one must be concerned with obtaining lateral rotation as a prerequisite to full flexion or full abduction (21).

Circumduction combines, consecutively, the movements of flexion, abduction, extension, and adduction as the upper limb circumscribes a cone with its apex at the glenohumeral joint (Figure 6-46). This succession of movements can be performed in either direction, and it is used to increase the overall range of motion of the shoulder joint, as in Codman's or shoulder-wheel exercises.

STERNOCLAVICULAR JOINT AND SCAPULA

Sternoclavicular Joint

Because of its joint structure (saddle or sellar), the **sternoclavicular joint** permits motion in the anterior and posterior directions about a longitudinal axis, in the cranial and caudal directions about a sagittal axis, and in rotation about a coronal axis. These movements are slightly enhanced and transmitted by the acromioclavicular joint to the scapula, allowing scapular motion. Additional motions of the shoulder girdle described here are those of the scapula.

Scapula

The **scapula** connects with the humerus at the glenohumeral joint and with the clavicle at the acromioclavicular joint. With the upper back in good alignment, the scapulae lie against the thorax approximately between the levels of the second and seventh ribs. In addition, the medial borders are essentially parallel and approximately 4 inches apart.

Muscles that attach the scapula to the thorax anteriorly and to the vertebral column posteriorly provide support and motion (Table 6-3). They are obliquely oriented so that their directions of pull can produce rotatory as well as linear motions of the bone. As a result, the movements ascribed to the scapula do not occur individually as pure movements (Figure 6-47). Because the contour of the thorax is rounded, some degree of rotation or tilt of the scapula accompanies abduction and adduction and, to a lesser extent, elevation and depression.

Although no pure linear movements occur, eight basic movements of the scapula are described:

1. *Adduction:* Gliding movement in which the scapula moves toward the vertebral column.
2. *Abduction:* Gliding movement in which the scapula moves away from the vertebral column and, following the contour of the thorax, assumes a posterolateral position in full abduction.
3. *Lateral or upward rotation:* Movement about a sagittal axis in which the inferior angle moves laterally and the glenoid cavity moves cranially.
4. *Medial or downward rotation:* Movement about a sagittal axis in which the inferior angle moves medially and the glenoid cavity moves caudally.
5. *Anterior tilt:* Movement about a coronal axis in which the coracoid process moves in an anterior and caudal direction while the inferior angle moves in a posterior and cranial direction. The coracoid process may be said to be depressed anteriorly. This movement is associated with elevation.
6. *Posterior tilt*: Movement about a coronal axis in which the coracoid process moves in a posterior and cephalad direction while the inferior angle moves in an anterior and cephalad direction.
7. *Elevation:* Gliding movement in which the scapula moves cranially, as in "shrugging" the shoulder.
8. *Depression:* A gliding movement in which the scapula moves caudally. This movement is the reverse of both elevation and anterior tilt.

NOTE: *It is suggested to avoid use of the terms "protraction" and "retraction" to describe movements of the scapula because they lack the precision and detail needed to explain scapular position and movement. The scapula must abduct for "protraction" of the arm and shoulder to occur, but lateral rotation of the inferior angle, anterior tilt, and elevation may also be present. "Retraction" of the arm and shoulder requires adduction and (usually) medial rotation of the scapula, with the possibility of elevation or depression.*

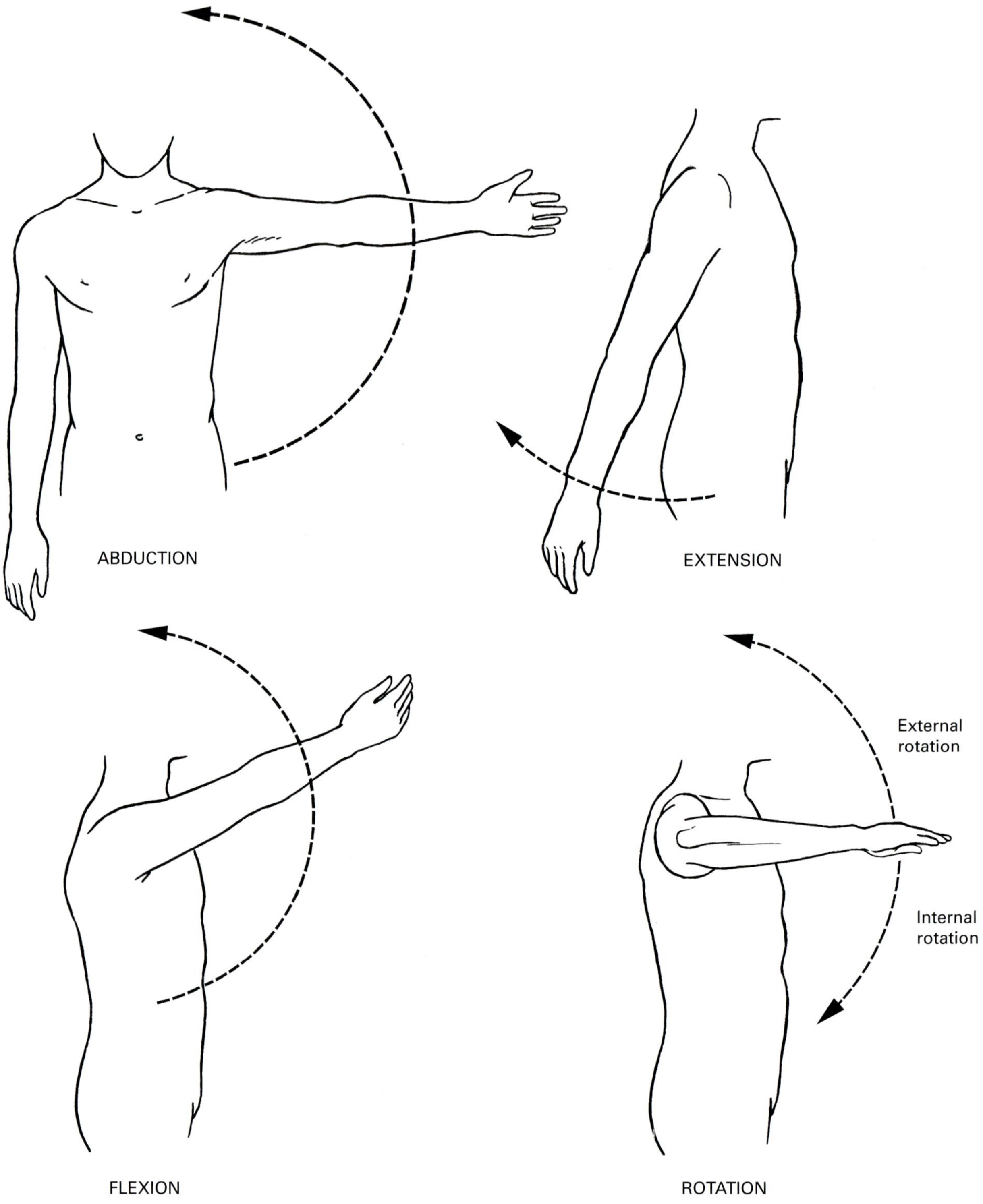

FIGURE 6-46: Movements of the glenohumeral joint.
Note: horizontal adduction and abduction are not illustrated.

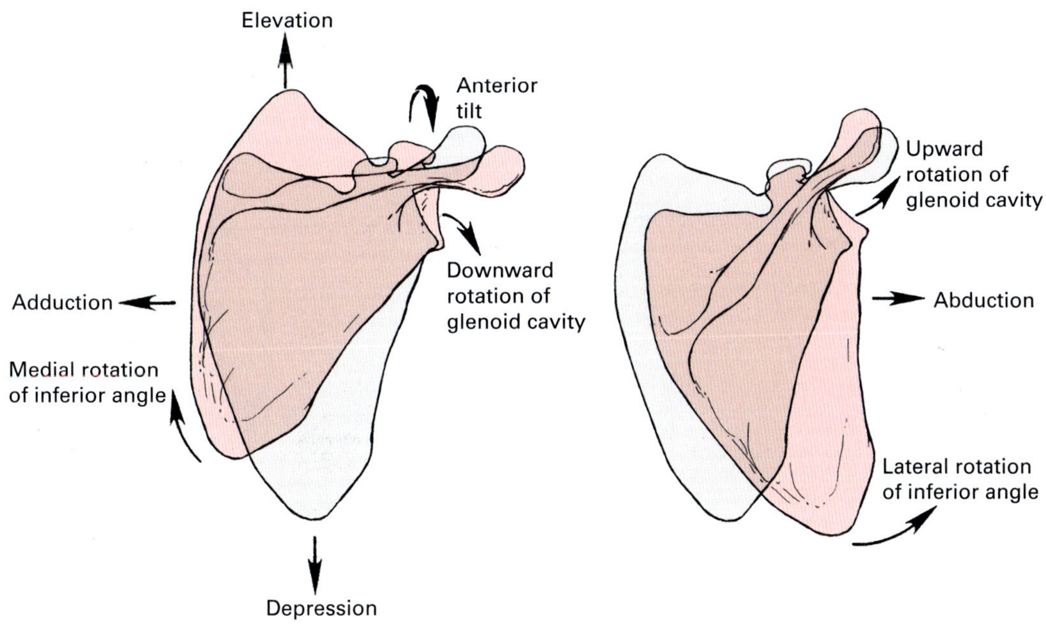

FIGURE 6-47: Movements of the scapula.

TABLE 6-3 Combined Shoulder and Scapular Muscles

Movement	Shoulder Muscles	Scapular Muscles
Full flexion (to 180°)	*Flexors:* Anterior deltoid Biceps Pectoralis major, upper Coracobrachialis *Lateral rotators:* Infraspinatus Teres minor Posterior deltoid	*Abductor:* Serratus anterior *Lateral rotators:* Serratus anterior Trapezius
Full abduction (to 180°)	*Abductors:* Deltoid Supraspinatus Biceps, long head *Lateral rotators:* Infraspinatus Teres minor Posterior deltoid	*Adductor:* Trapezius, acting to stabilize scapula in adduction *Lateral rotators:* Trapezius Serratus anterior
Full extension (to 45°)	*Extensors:* Posterior deltoid Teres major Latissimus dorsi Triceps, long head	*Adductors, medical rotators, and elevators:* Rhomboids Levator scapulae *Anterior tilt of scapula by:* Pectoralis minor
Full adduction to side against resistance	*Adductors:* Pectoralis major Teres major Latissimus dorsi Triceps, long head Biceps, short head	*Adductors:* Rhomboids Trapezius

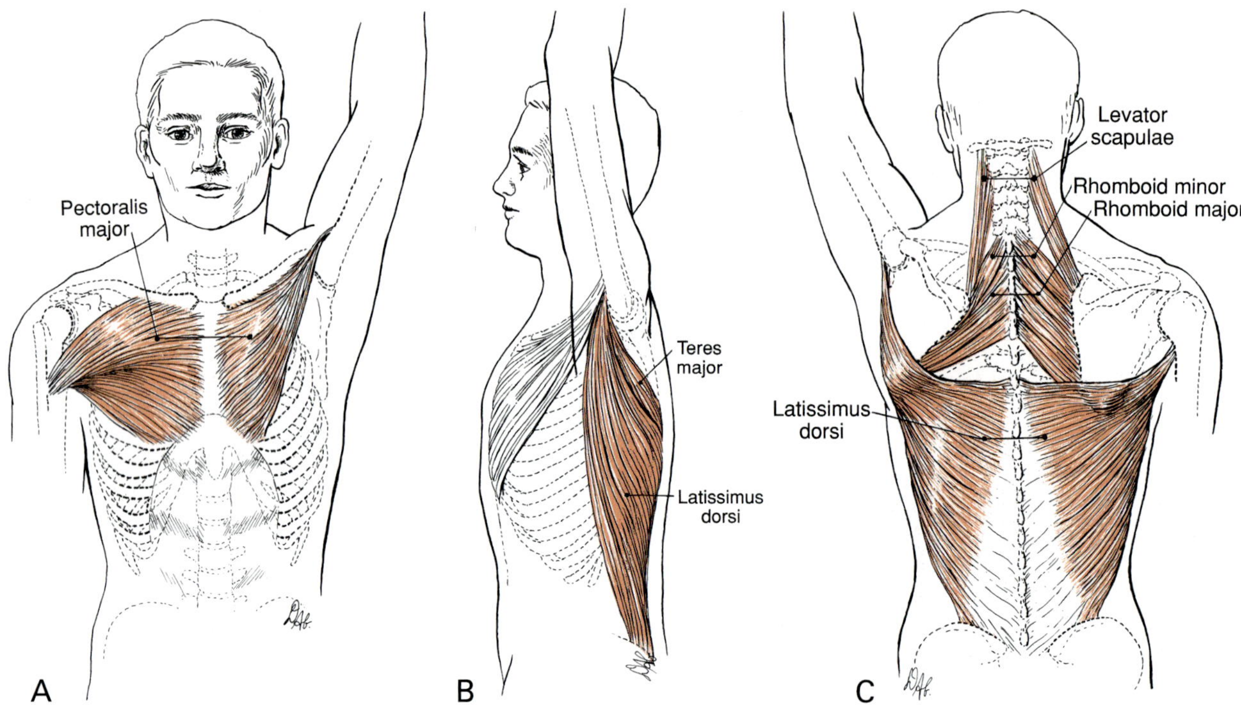

FIGURE 6-48: A–C, Length of humeral and scapular muscles.

MUSCLE LENGTH TESTS

Length of Humeral and Scapular Muscles

Full range of scapulohumeral and scapular motion for normal overhead elevation of the arm in flexion or in abduction requires adequate length in the pectoralis major, pectoralis minor, latissimus dorsi, teres major, subscapularis, rhomboids, biceps, and triceps (Figure 6-48).

Full range of shoulder motion into lateral rotation requires normal length of the medial rotators—namely the anterior deltoid, pectoralis major, latissimus dorsi, teres major, and subscapularis. Full range of motion into medial rotation requires normal length of the lateral rotators—namely the teres minor, infraspinatus, and posterior deltoid.

To test accurately for the various movements, there must be no substitution by movements of the trunk. The trunk position must be standardized, with the subject supine, the knees bent and the low back flat on a flat surface (i.e., hooklying) (Figure 6-49). The table should not have a soft pad; however, a folded blanket may be used for the comfort of the subject.

If the low back arches up from the table, the amount of shoulder flexion or lateral rotation will appear to be greater and the amount of medial rotation will appear to be less than the actual range of shoulder and scapular motion. If the chest is depressed (as in a thoracic hyperkyphosis or tight upper abdominals), the amount of shoulder flexion and lateral rotation will appear to be less and the amount of medial rotation will appear to be greater than the actual range of motion.

If the trunk bends laterally with convexity toward the tested side, the amount of abduction will appear to be greater than the actual range of shoulder and scapular motion.

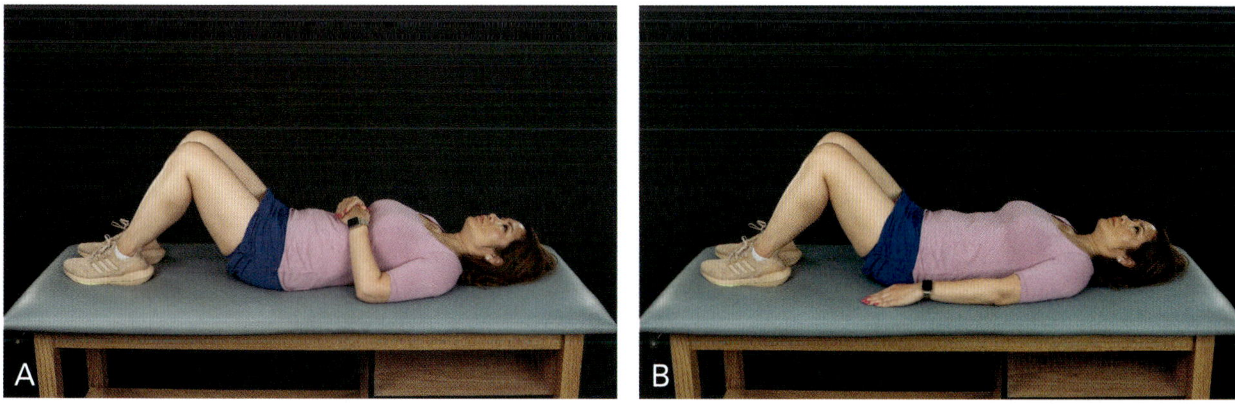

FIGURE 6-49: Hooklying position: **A,** Hands on abdomen; **B,** Arms extended.

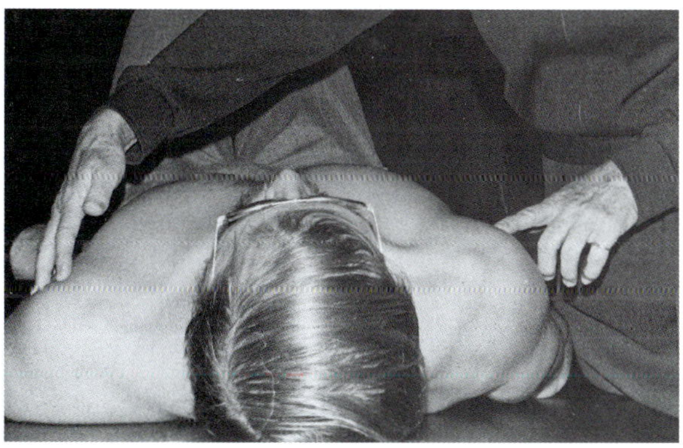

FIGURE 6-50: Hooklying position, arms extended. Left, normal length; right, short length, holding the shoulder forward.

Test for Length of Pectoralis Minor

Equipment: Firm table; unpadded.

Starting Position: Hooklying with the arms at the sides, elbows extended, and palms upward.

Test: The examiner stands at the head of the table and observes the position of the shoulder girdle. Figure 6-50 shows normal length of the left pectoralis minor and shortness of the right. The amount of tightness is measured by the extent to which the shoulder is raised from the table and by the amount of resistance to downward pressure on the shoulder. Tightness may be recorded as slight, moderate, or marked.

Test for Tightness of Muscles That Depress the Coracoid Process Anteriorly

The subject assumes a hooklying position on a firm table with arms at the sides, elbows extended, and palms upward. There appears to be some anterior tilt of the shoulder suggesting tightness of the pectoralis minor.

The examiner stands on the left side of the subject, with the palm of the examiner's left hand on the subject's anterior shoulder region, and presses firmly down toward the table in the manner of rolling the shoulder region back to correct anterior tilting of the shoulder (as opposed to anterior tilting of the scapula). The amount of resistance indicates the extent of tightness in the group of muscles attached to the coracoid process (Figure 6-51).

To focus on the pectoralis minor, it is necessary to slacken each of the other muscles. While maintaining continuous pressure with the left hand on the anterior shoulder region, the examiner's right hand bends the subject's elbow to full flexion to slacken the biceps. If the shoulder region can be displaced somewhat downward, this provides evidence that the biceps are contributing to the clinical presentation.

Maintaining continuous pressure on the anterior shoulder region, the examiner then lifts the subject's elbow approximately 6 or 8 inches up from the table to slacken the coracobrachialis. If the shoulder can be further depressed, this provides evidence the coracobrachialis is contributing to the clinical presentation. Any tightness that remains should be attributed to the pectoralis minor.

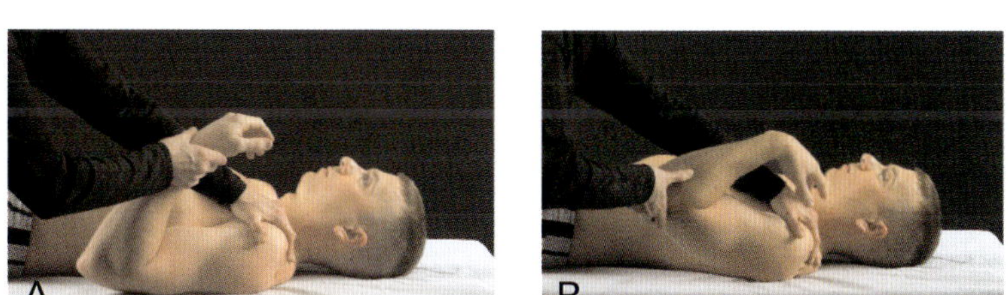

FIGURE 6-51: **A, B,** Test for tightness of muscles that depress the coracoid process anteriorly.

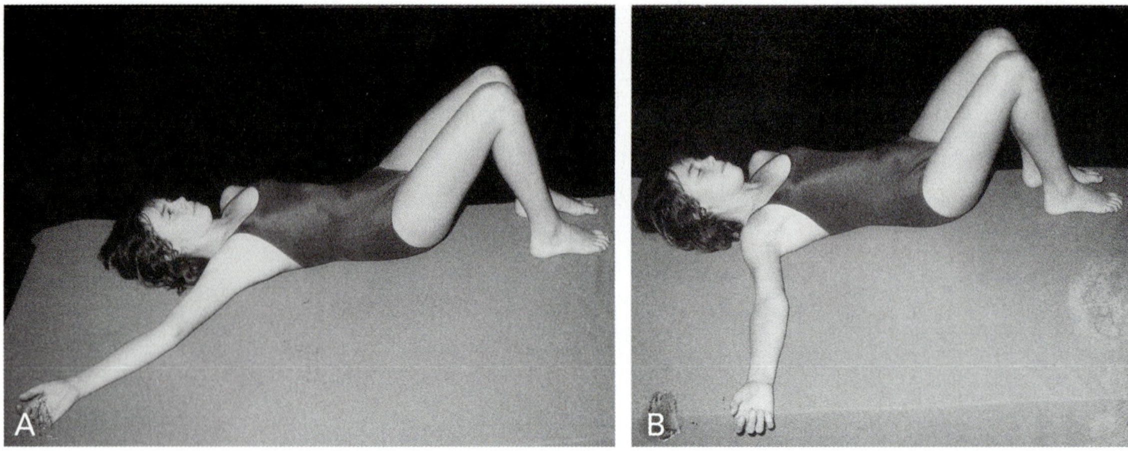

FIGURE 6-52: Test for length of pectoralis major: **A,** Normal length of lower fibers; **B,** Normal length of upper fibers.

Test for Length of Pectoralis Major

Equipment: Firm table; no soft padding.

Starting Position: Hooklying.

Test Movement for Lower (Sternal) Part: The examiner places the subject's upper extremity in a position of approximately 135° of shoulder abduction (in line with the lower fibers), with the elbow extended. The shoulder will be in external rotation.

Normal Length: Upper extremity drops to table level, with the low back remaining flat on the table (Figure 6-52A).

Shortness: The extended upper extremity does not drop down to table level. Limitation may be recorded as slight, moderate, or marked, measured in degrees, using a goniometer or inches, using a ruler, to record the number of inches from the lateral epicondyle to the table.

Test Movement for Upper (Clavicular) Part: The examiner places the subject's shoulder in 90° of abduction with the elbow extended and the shoulder in external rotation (palm toward the ceiling).

Normal Length: 90° of shoulder abduction, with external rotation, the upper extremity flat on the table, and without trunk rotation (Figure 6-52B).

Shortness: The upper extremity does not drop down to table level. Limitation may be recorded as slight, moderate, or marked, measured in degrees using a goniometer, or inches using a ruler, to record the number of inches between the table and the lateral epicondyle. Marked limitation is seldom found in this test.

NOTE: *Tightness of the clavipectoral fascia can interfere with length testing of the clavicular portion.*

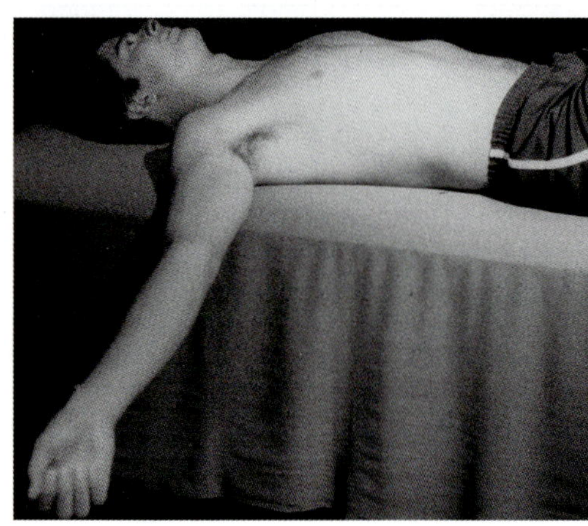

FIGURE 6-53: Test for excessive length in upper (clavicular) part of pectoralis major.

Test for Excessive Length: Position the subject with the shoulder joint at the edge of the table so that the arm can abduct horizontally below table level. Record excessive range as slight, moderate, or marked, or measure in degrees using a goniometer. Excessive range of motion is not uncommon (Figure 6-53).

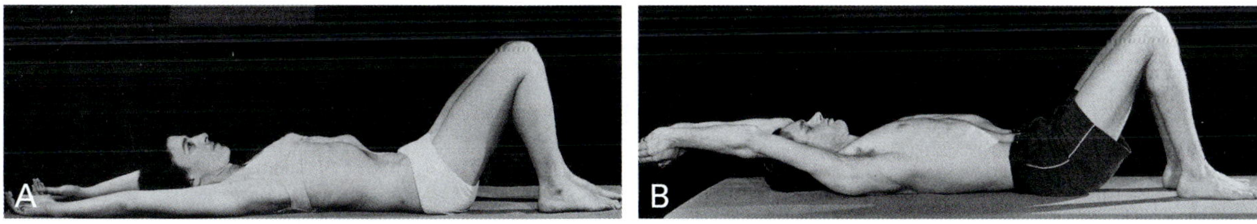

FIGURE 6-54: Tests for length of teres major, latissimus dorsi, and rhomboids: **A,** Normal length; **B,** Short.

Tests for Length of Teres Major, Latissimus Dorsi, and Rhomboids

Equipment: Firm table; unpadded.

Starting Position: Hooklying, with the upper extremities at the sides, elbows extended.

Test Movement: Subject raises both upper extremities in flexion overhead, keeping them close to the head and bringing them down toward the table (maintaining a flat low back) (Figure 6-54).

Normal Length: The ability to bring the upper extremities down to table level, keeping them close to the head.

Shortness: Indicated by the inability to get the upper extremities to table level. Record measurements as slight, moderate, or marked; measure the angle between the table and humerus to determine the number of degrees of limitation, or measure the number of inches between the table and the lateral epicondyle.

NOTE: *Tightness of the upper abdominals will depress the chest and tend to pull the shoulder forward, interfering with the test. Likewise, a hyperkyphosis of the thoracic spine will make it impossible to get the shoulder down on the table.*

A contracted pectoralis minor tilts the scapula anteriorly, pulling the shoulder girdle downward and forward. With the change in alignment of the shoulder girdle, flexion of the glenohumeral joint will appear to be limited even if the range is actually normal, because the arm cannot be brought down to touch the table.

Tightness of the pectoralis minor is an important factor in many cases of arm pain. With attachment of the pectoralis minor on the coracoid process, tightness of this muscle depresses the coracoid anteriorly and can cause pressure and impingement on the cords of the brachial plexus and the axillary blood vessels that lie between the coracoid and the rib cage.

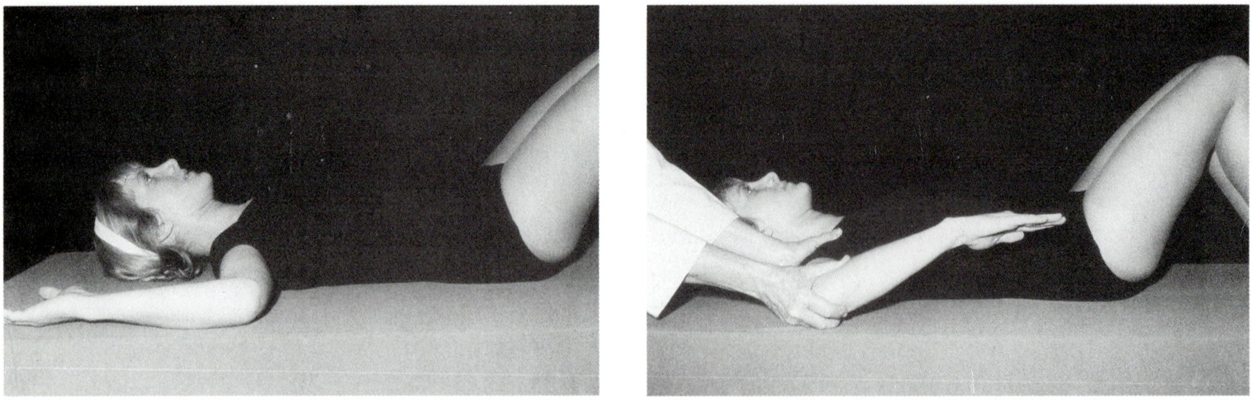

FIGURE 6-55: Tests for length of shoulder rotators: **A,** Medial rotators; **B,** Lateral rotators.

Tests for Length of Shoulder Medial Rotators

Equipment: Firm table; no soft padding.

Starting Position: Supine, with the low back flat on the table, shoulder at 90° abduction, elbow at the edge of the table and flexed to 90° (forearm perpendicular to the table).

Test for Length of Medial Rotators: External rotation of the shoulder, bringing the forearm down toward table level (Figure 6-55A). (Do not allow the back to arch up from the table.)

Normal Range of Motion: 90° of external rotation (forearm flat on the table while maintaining the low back flat on the table).

> **NOTE:** *If the test for tightness of the teres major and latissimus dorsi shows limitation, but external rotation (as in Figure 6-56A) shows normal range, then tightness is in the latissimus dorsi but not in the teres major.*

Test of Length of Shoulder Lateral Rotators

Equipment: Firm table; no soft padding.

Starting Position: Supine, with the low back flat on the table, shoulder at 90° abduction, elbow at the edge of the table and flexed to 90° (forearm perpendicular to the table).

Test for Length of Lateral Rotators: Internal rotation of the shoulder, bringing the forearm downward toward table, while the examiner stabilizes the shoulder to prevent substitution by the shoulder girdle (Figure 6-55B). (Do not allow forward thrust of the shoulder girdle.)

Normal Range of Motion: 70° of internal rotation.

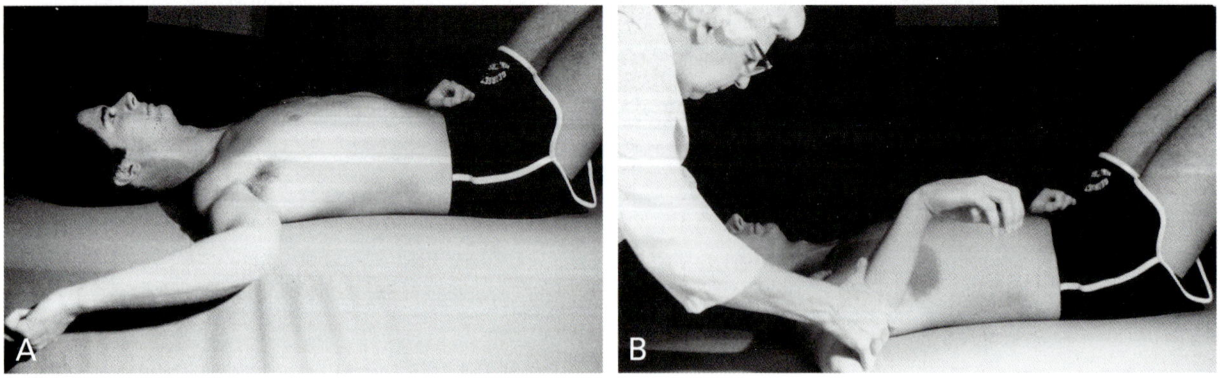

FIGURE 6-56: Tests for length of shoulder rotators: **A,** Muscle test for excessive range of motion in lateral rotation; **B,** Imbalance—limitation of medial rotation, and excessive lateral rotation.

To test for excessive range of motion in external rotation, it is necessary to have the elbow slightly beyond the edge of the table to allow the forearm to drop below table level (Figure 6-56A).

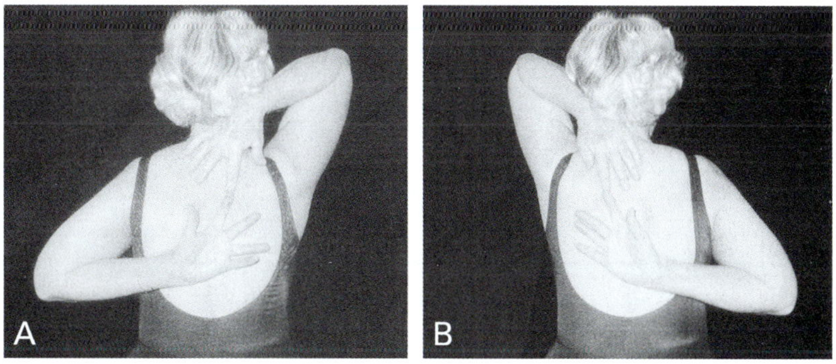

FIGURE 6-57: A, B, Placing the hands behind the back to test range of shoulder joint rotation.

This subject demonstrated marked limitation of internal rotation and excessive external rotation—an imbalance often seen in baseball players (22).

Placing the hands behind the back, as illustrated in Figure 6-57, requires normal range of shoulder joint rotation without abnormal shoulder-girdle movement.

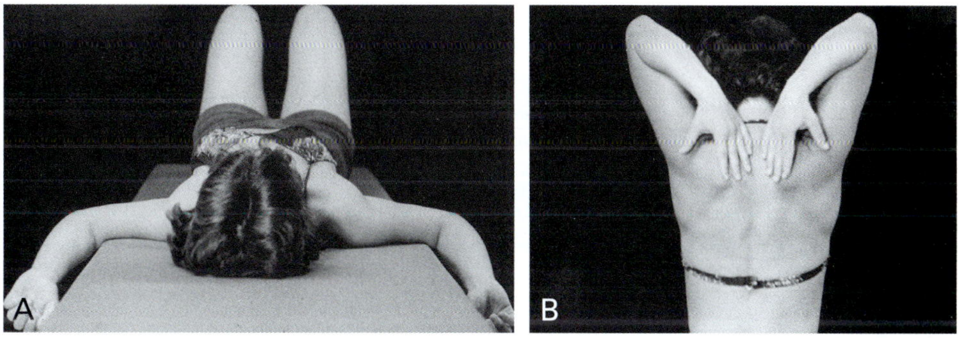

FIGURE 6-58: A, Slightly excessive shoulder joint lateral rotation; **B,** Placing the hands easily on the upper back.

Slightly excessive shoulder joint external rotation. The hands can easily be placed on the upper back (Figure 6-58).

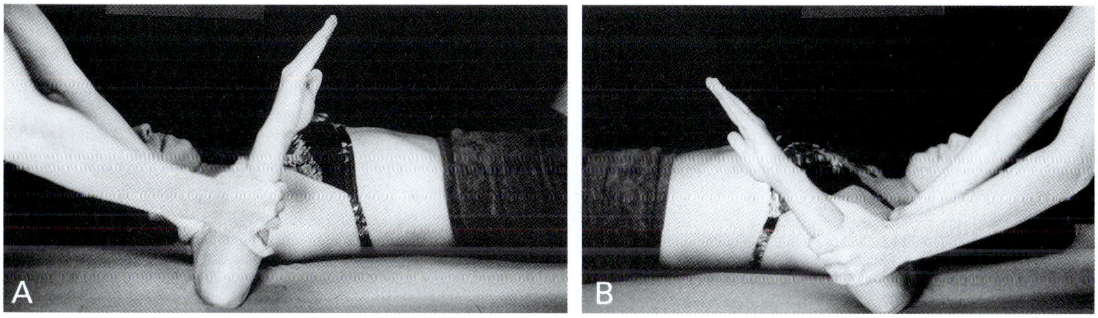

FIGURE 6-59: A, B, Limited shoulder joint medial rotation, more in the right than in the left.

Limited shoulder joint internal rotation, more in the right than in the left (Figure 6-59). The shoulder girdle is held down to prevent substitution of shoulder-girdle motion for shoulder joint motion.

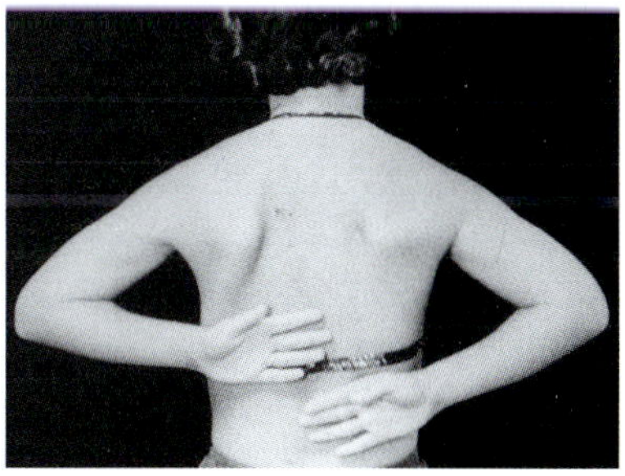

FIGURE 6-60: Substitution by shoulder girdle motion.

Substitution by shoulder-girdle motion permits the subject to place the hands behind the back (Figure 6-60). However, encouraging or permitting such substitution can have adverse effects by contributing to overdevelopment of the pectoralis minor. (See pectoralis minor.)

HISTORICAL NOTE

Upper Extremity Chart

PATIENT'S NAME CLINIC No.

LEFT														RIGHT
							Examiner Date							
							Trapezius, upper							
							Trapezius, middle							
							Trapezius, lower							
							Serratus anterior							
							Rhomboids							
							Pectoralis minor							
							Pectoralis major							
							Latissimus dorsi							
							Shoulder medial rotators							
							Shoulder lateral rotators							
							Deltoid, anterior							
							Deltoid, middle							
							Deltoid, posterior							
							Biceps							
							Triceps							
							Brachioradialis							
							Supinators							
							Pronators							
							Flexor carpi radialis							
							Flexor carpi ulnaris							
							Extensor carpi radialis							
							Extensor carpi ulnaris							
						1	Flexor digitorum profundus	1						
						2	Flexor digitorum profundus	2						
						3	Flexor digitorum profundus	3						
						4	Flexor digitorum profundus	4						
						1	Flexor digit. superficialis	1						
						2	Flexor digit. superficialis	2						
						3	Flexor digit. superficialis	3						
						4	Flexor digit. superficialis	4						
						1	Extensor digitorum	1						
						2	Extensor digitorum	2						
						3	Extensor digitorum	3						
						4	Extensor digitorum	4						
						1	Lumbricalis	1						
						2	Lumbricalis	2						
						3	Lumbricalis	3						
						4	Lumbricalis	4						
						1	Dorsal interosseus	1						
						2	Dorsal interosseus	2						
						3	Dorsal interosseus	3						
						4	Dorsal interosseus	4						
						1	Palmar interosseus	1						
						2	Palmar interosseus	2						
						3	Palmar interosseus	3						
						4	Palmar interosseus	4						
							Flexor pollicis longus							
							Flexor pollicis brevis							
							Extensor pollicis longus							
							Extensor pollicis brevis							
							Abductor pollicis longus							
							Abductor pollicis brevis							
							Adductor pollicis							
							Opponens pollicis							
							Flexor digiti minimi							
							Abductor digiti minimi							
							Opponens digiti minimi							

NOTES:

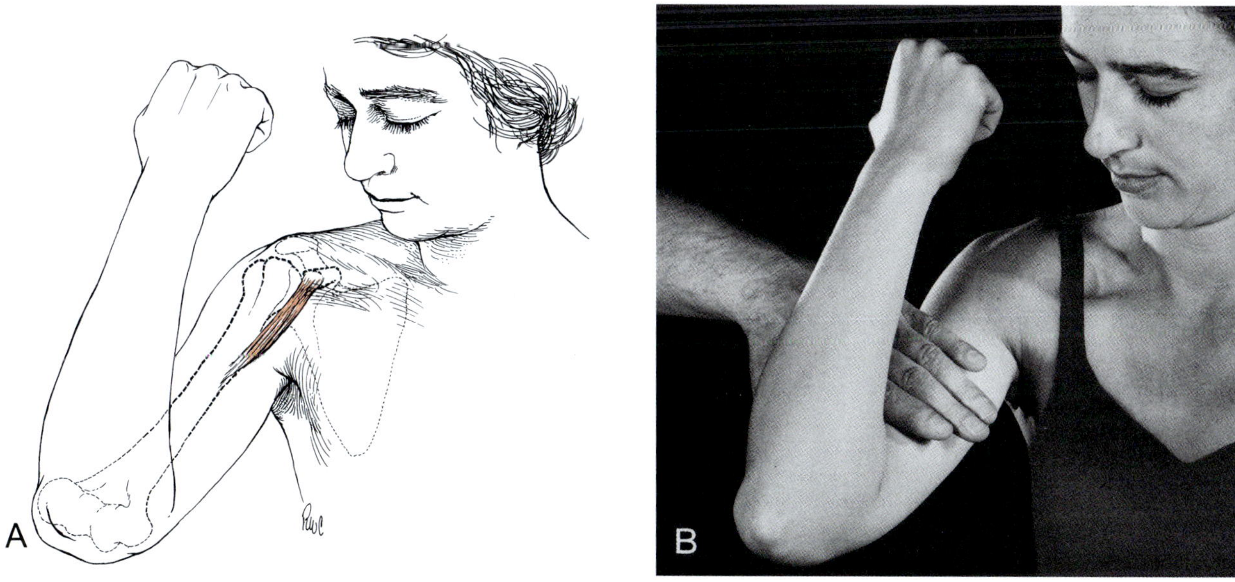

FIGURE 6-61: **A,** Coracobrachialis; **B,** Muscle testing of coracobrachialis.

MUSCLE STRENGTH TESTS

Coracobrachialis

Origin: Apex of the coracoid process of the scapula (Figure 6-61A).

Insertion: Medial surface of the middle of the shaft of the humerus, opposite to the deltoid tuberosity.

Action: Flexes and adducts the shoulder joint.

Nerve: Musculocutaneous, C5, **6, 7**.

Patient: Sitting or supine.

Weakness: Decreases the strength of shoulder flexion, particularly in movements that involve complete elbow flexion and supination, such as combing the hair.

Shortness: The coracoid process is depressed anteriorly when the upper extremity is down at the side.

Fixation: If trunk is stable, no fixation by the examiner should be necessary.

Test: Shoulder flexion in external rotation, with the elbow completely flexed and the forearm supinated. Assistance from the biceps in shoulder flexion is decreased in this test position because the complete elbow flexion and forearm supination place the muscle in too short a position to be effective in shoulder flexion.

Pressure: Against the anteromedial surface of the lower 1/3 of the humerus, in the direction of extension and slight abduction (Figure 6-61B).

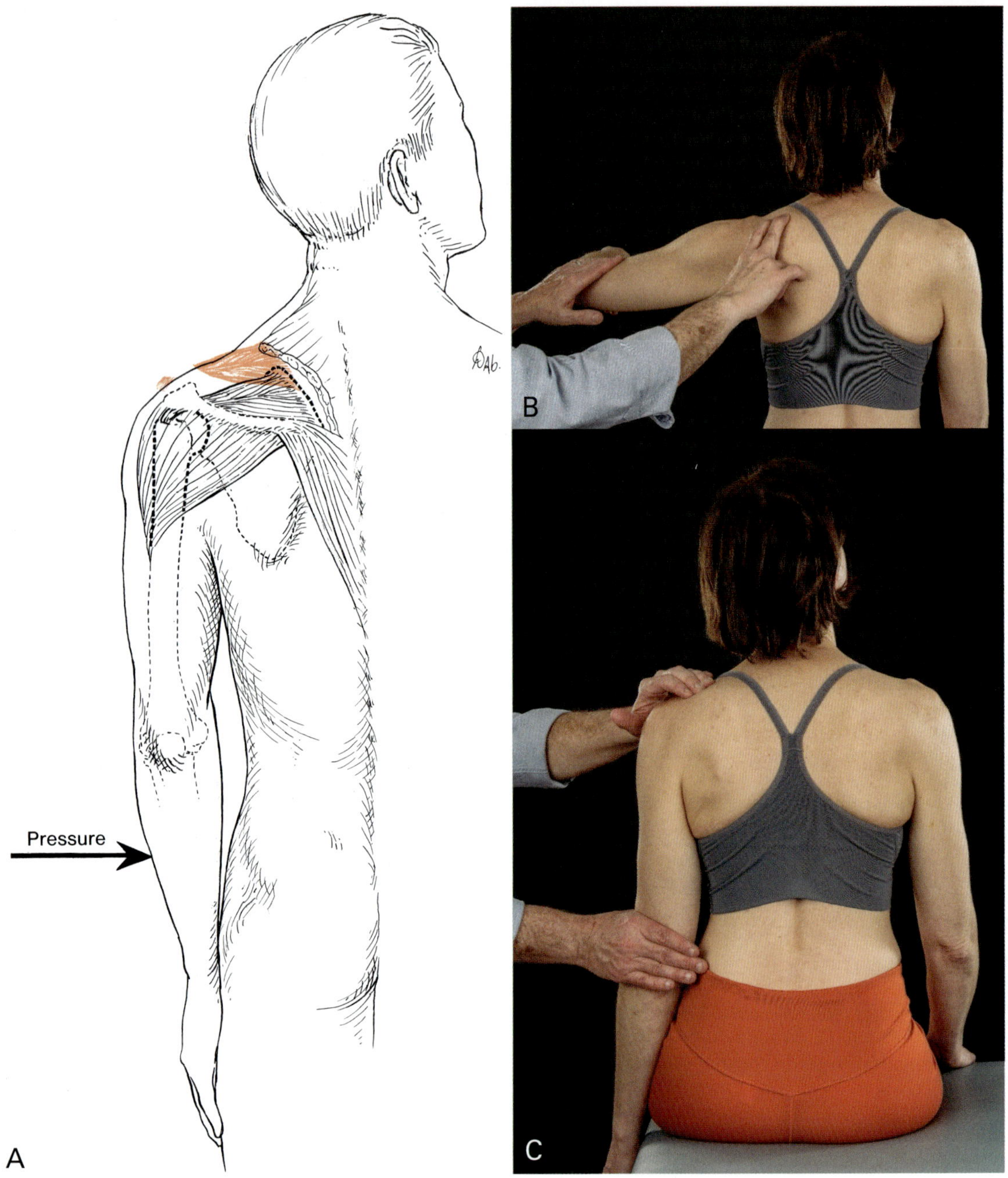

FIGURE 6-62: A, Supraspinatus; **B, C,** Muscle testing of supraspinatus.

Supraspinatus

Origin: Medial 2/3 of the supraspinous fossa of the scapula (Figure 6-62A).

Insertion: Superior facet of the greater tubercle of the humerus and shoulder joint capsule.

Action: Abducts and laterally rotates the shoulder joint. Stabilizes the head of the humerus in the glenoid cavity during movements of this joint.

Nerve: Suprascapular, C4, **5,** 6.

Patient: Sitting or standing, with shoulder abducted 90°, the head and neck extended and laterally flexed toward the same side, and the face rotated toward the opposite side.

Fixation: The sitting position provides greater stabilization of the trunk than a standing position.

NOTE: *No effort is made to distinguish the supraspinatus from the deltoid in the strength test for the purpose of grading because these muscles act simultaneously in abducting the shoulder. However, the supraspinatus can be palpated to determine whether it is active. Because the supraspinatus is completely covered by the upper and middle fibers of the superior trapezius, the trapezius should be as relaxed as possible to palpate the supraspinatus. This is accomplished by extending and laterally flexing the head and neck so that the face is rotated toward the opposite side, as illustrated in Figure 6-62B, and by palpating the supraspinatus at the beginning of the abduction movement, when the activity of the trapezius is at a low level. The deltoid and supraspinatus act together in initiating abduction, and this test is not to be construed to mean that the supraspinatus is responsible for the first few degrees of abduction.*

Test movement: Initiation of abduction of the humerus (23).

Resistance: Against the forearm, in the direction of adduction.

NOTE: *The supraspinatus needs to be tested in its most shortened position because it is a one-joint muscle and tests strongest in its most shortened position.*

The "empty can" test does not meet the requirements for testing the strength of this muscle. Rowlands et al. concluded "our study found that the empty can test does not allow selective activation of the supraspinatus muscle" (24)

Weakness: The tendon of the supraspinatus is firmly attached to the superior surface of the capsule of the shoulder joint. Weakness of the muscle or rupture of the tendon decreases shoulder joint stability, allowing the head of the humerus to alter its relationship with the glenoid cavity.

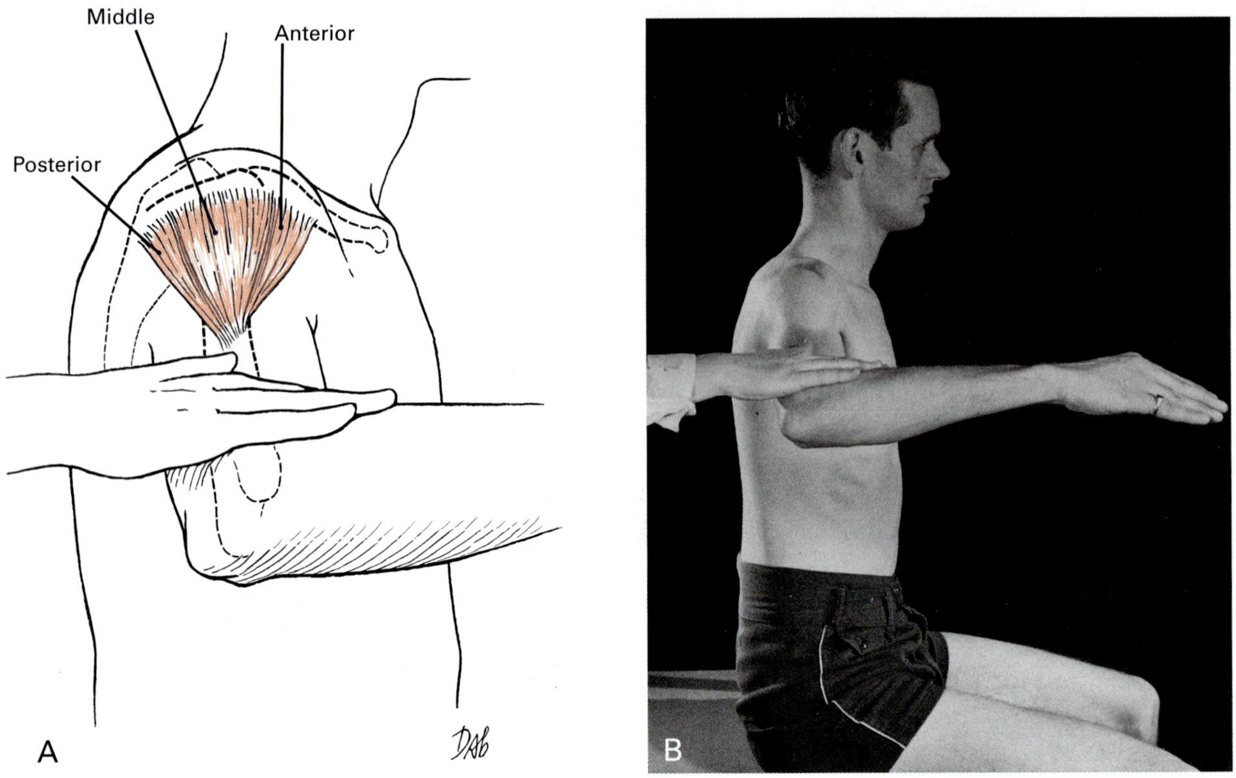

FIGURE 6-63: A, Deltoid; **B,** Muscle testing of deltoid.

Deltoid

Origin of Anterior Fibers: Anterior border, superior surface, and lateral 1/3 of the clavicle (Figure 6-63A).

Origin of Middle Fibers: Lateral margin and superior surface of the acromion.

Origin of Posterior Fibers: Inferior lip of the posterior border of the spine of the scapula.

Insertion: Deltoid tuberosity of the humerus.

Action: Abduction of the shoulder joint, performed chiefly by the middle fibers, with stabilization by the anterior and posterior fibers. In addition, the anterior fibers flex and internally rotate the shoulder joint. The posterior fibers extend and externally rotate the shoulder joint.

Nerve: Axillary, **C5, 6**.

Patient: Sitting.

Fixation: The position of the trunk in relation to the arm in this test is such that a stable trunk will need no further stabilization by the examiner. If the scapular fixation muscles are weak, then the examiner must stabilize the scapula.

Test: Shoulder abduction without rotation (Figure 6-63B). When placing the shoulder in the test position, the elbow should be flexed to indicate the neutral position of rotation. The elbow may be extended, however, after the shoulder position is established so that the extended extremity can be used for a longer lever. The examiner should be consistent in the technique during subsequent tests.

Pressure: Against the dorsal surface of the distal aspect of the arm if the elbow is flexed, or against the distal forearm if the elbow is extended.

Weakness: Results in inability to lift the arm in abduction against gravity. With paralysis of the entire deltoid and supraspinatus, the humerus tends to sublux inferiorly if the arm remains unsupported against gravity. The capsule of the shoulder joint permits almost an inch of separation between the head of the humerus and the glenoid cavity. In cases of axillary nerve involvement in which the deltoid is weak but the supraspinatus is not affected, relaxation of the joint is not as marked, but it tends to progress if the deltoid strength does not return.

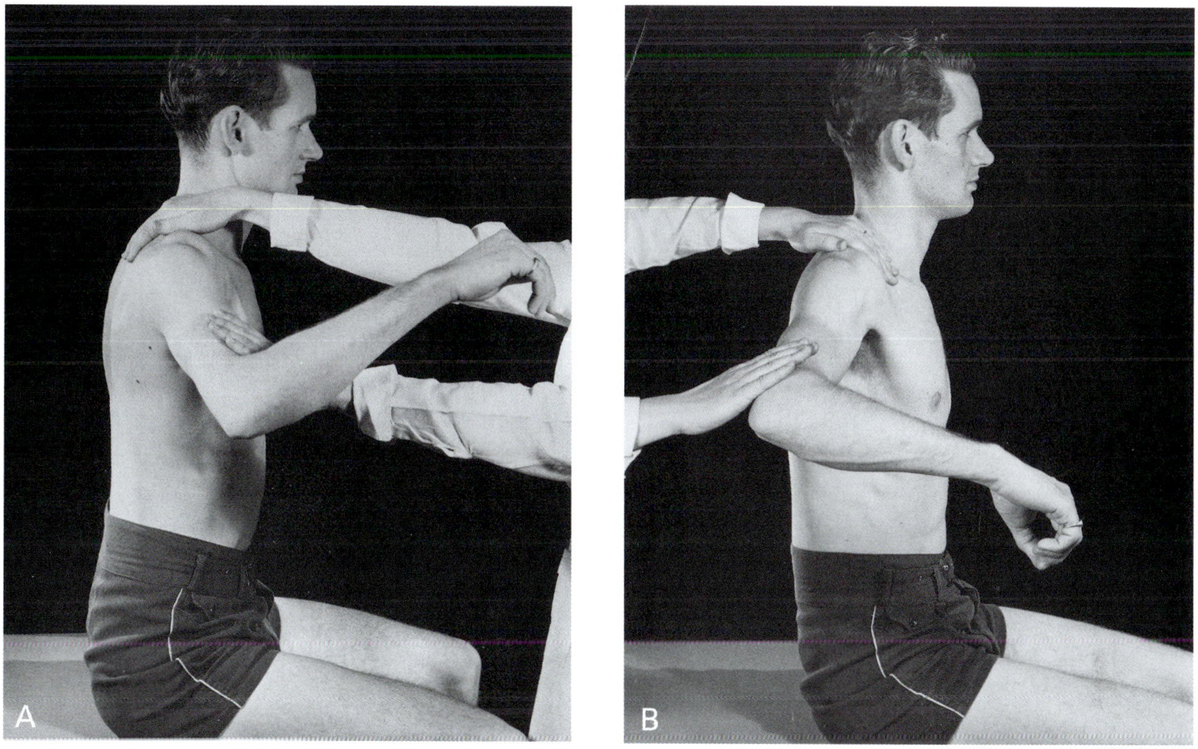

FIGURE 6-64: **A,** Muscle testing of the anterior deltoid, sitting; **B,** Muscle testing of the posterior deltoid, sitting.

Anterior Deltoid

Patient: Sitting.

Fixation: If scapular fixation muscles are weak, the scapula must be stabilized by the examiner. As pressure is applied on the arm, counterpressure is applied posteriorly to the shoulder girdle.

Test: Shoulder abduction in slight flexion and slight external rotation (Figure 6-64A). In the erect sitting position, it is necessary to place the shoulder in slight external rotation to increase the effect of gravity on the anterior fibers.

Pressure: Against the anteromedial surface of the arm, in the direction of adduction and slight extension.

Posterior Deltoid

Patient: Sitting.

Fixation: If scapular fixation muscles are weak, the scapula must be stabilized by the examiner. As pressure is applied on the arm, counterpressure is applied anteriorly on the shoulder girdle.

Test: Shoulder abduction in slight extension and internal rotation (Figure 6-64B). In the erect sitting position, it is necessary to place the shoulder in slight internal rotation to have the posterior fibers in an antigravity position.

Pressure: Against the posterolateral surface of the distal arm, in the direction of adduction and slight flexion.

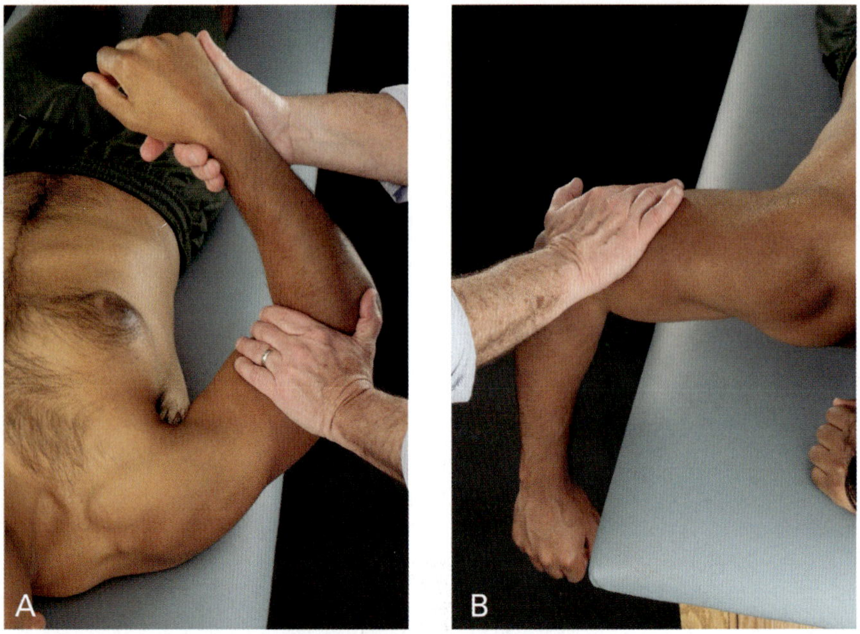

FIGURE 6-65: **A,** Muscle testing of the anterior deltoid, supine; **B,** Muscle testing of the posterior deltoid, prone.

Anterior Deltoid

Patient: Supine.

Fixation: The trapezius and serratus anterior should stabilize the scapula *in all the deltoid tests*, but if these muscles are weak, the examiner should stabilize the scapula.

Test: Shoulder abduction in the position of slight flexion and internal rotation (Figure 6-65A). One hand of the examiner is placed under the patient's wrist to make sure the elbow is not lifted by reverse action of the wrist extensors, which may occur if the patient is allowed to press their hand down on the chest.

Pressure: Against the anterior surface of the distal arm, in the direction of adduction toward the side of the body.

Posterior Deltoid

Patient: Prone.

Fixation: The scapula must be held stable, either by the scapular muscles or by the examiner.

Test: Shoulder abduction to 90° with slight extension and external rotation (Figure 6-65B).

Pressure: Against the posterolateral surface of the arm, in a direction obliquely downward and midway between adduction and flexion.

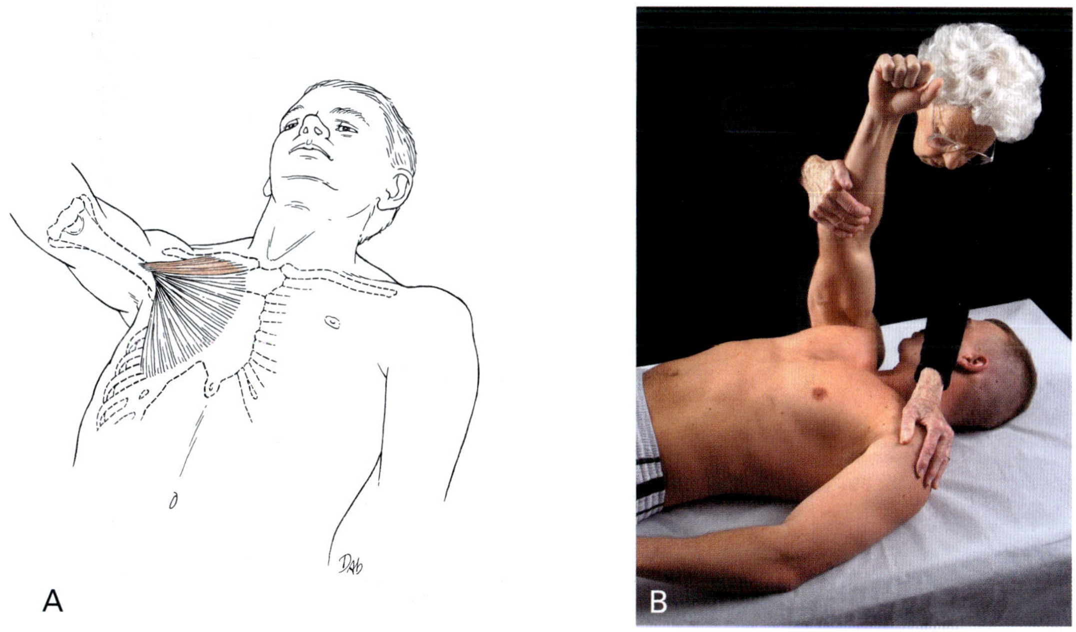

FIGURE 6-66: **A,** Pectoralis major, clavicular portion; **B,** Testing of pectoralis major, clavicular portion.

Pectoralis Major, Clavicular Portion

Origin, Clavicular Portion: Anterior surface of the sternal 1/2 of the clavicle (Figure 6-66A).

Insertion: Crest of greater tubercle of the humerus. Fibers are more anterior and caudal on the crest than the lower fibers.

Action: Flex and medially rotate the shoulder joint, and horizontally adduct the humerus toward the opposite shoulder.

Nerve: Lateral pectoral, **C5, 6, 7,** 8, T1.

Patient: Supine.

Fixation: The examiner holds the opposite shoulder firmly on the table. The triceps maintains the elbow in extension.

Test: Starting with the elbow extended and with the shoulder in 90° flexion and slight internal rotation, the upper extremity is adducted toward the sternal end of the clavicle (Figure 6-66B).

Pressure: Against the proximal medial forearm, in the direction of abduction.

Weakness: Decreases the ability to draw the arm in adduction across the chest, making it difficult to touch the hand to the opposite shoulder. Decreases the strength of shoulder flexion and internal rotation.

Shortness: The range of motion in flexion, abduction, and external rotation of the shoulder is decreased. Shortness of the pectoralis major holds the arm in internal rotation and adduction and, secondarily, results in abduction of the scapula from the spine.

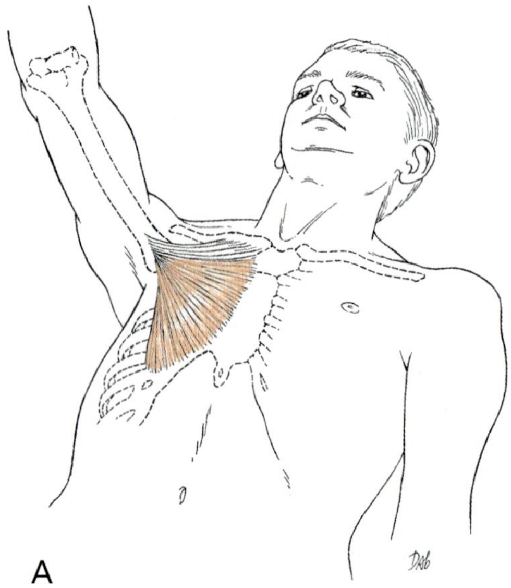
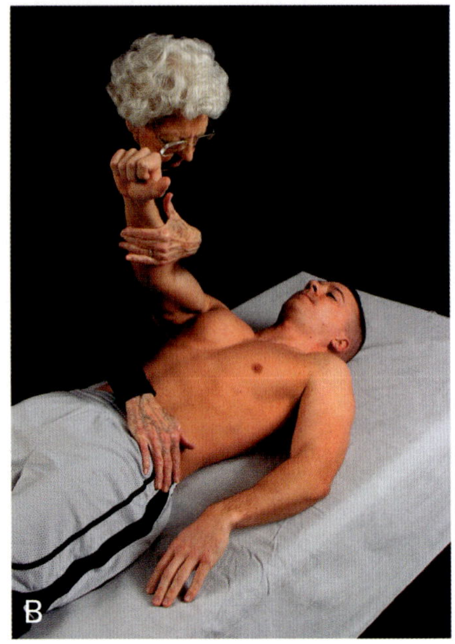

FIGURE 6-67: A, Pectoralis major, sternocostal portion; **B,** Testing of pectoralis major, sternocostal portion.

Pectoralis Major, Sternocostal Portion

Origin, Sternocostal Portion: Anterior surface of the sternum, cartilages of first six or seven ribs, and aponeurosis of the external oblique (Figure 6-67A).

Insertion: Crest of the greater tubercle of the humerus. The fibers twist on themselves and are more posterior and cranial than the upper fibers.

Action: Depress the shoulder girdle by virtue of the attachment on the humerus and obliquely adduct the humerus toward the opposite iliac crest.

Nerves: Lateral and medial pectoral, C5, **6, 7, 8, T1**.

Action of Muscle as a Whole: With the origin fixed, the pectoralis major adducts and medially rotates the humerus. With the insertion fixed, it may assist in elevating the thorax, as in forced inspiration. In crutch-walking or parallel-bar work, it will assist in supporting the weight of the body.

Patient: Supine.

Fixation: The examiner places one hand on opposite iliac crest to hold the pelvis firmly on the table. The anterior parts of the external and internal oblique muscles stabilize the thorax on the pelvis. In cases of abdominal weakness, the thorax, instead of the pelvis, must be stabilized. The triceps maintains the elbow in extension.

Test: Starting with the elbow extended and with the shoulder in flexion and slight internal rotation, adduction of the shoulder and upper extremity obliquely toward the opposite iliac crest (Figure 6-67B).

Pressure: Against the anteromedial forearm obliquely, in shoulder abduction and flexion.

Weakness: Decreases the strength of adduction obliquely toward the opposite hip. Continuity of muscle action is on the same side also lost from the pectoralis major to the external oblique on the same side and internal oblique on the opposite side, with the result that chopping or striking movements are difficult. From a supine position, if the subject's arm is placed diagonally overhead, it will be difficult to lift the arm from the table. The subject will also have difficulty holding any large or heavy object in both hands either at or near waist level.

Shortness: A forward depression of the shoulder girdle from the pull of the pectoralis major on the humerus often accompanies the pull of a tight pectoralis minor on the scapula. Flexion and abduction ranges of motion overhead are limited.

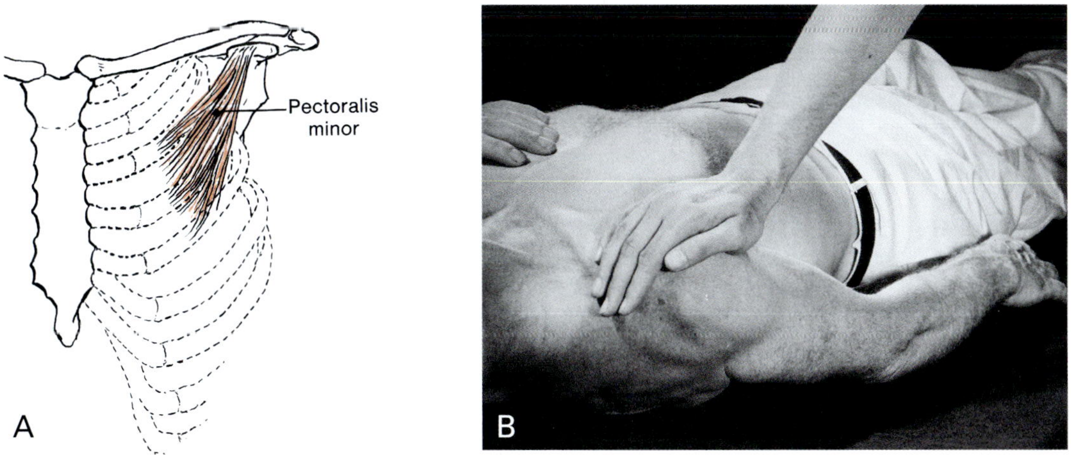

FIGURE 6-68: **A,** Pectoralis minor; **B,** Testing of pectoralis minor.

Pectoralis Minor

Origin: Superior margins; outer surfaces of the second, third, fourth, and fifth ribs near the costal cartilages; fascia over corresponding intercostal muscles (Figure 6-68A).

Insertion: Medial border, superior surface of the coracoid process of the scapula.

Action: With the origin fixed, depresses, downwardly rotates, and adducts the scapula. With the scapula stabilized, to fix the insertion, the pectoralis minor assists in forced inspiration.

Nerve: Medial pectoral, with fibers from a communicating branch of the lateral pectoral; C8, T1.

Patient: Supine.

Fixation: None by the examiner unless the abdominal muscles are weak, in which case the rib cage on the same side should be held down firmly.

Test: Forward thrust of the shoulder, with the arm at the side. The subject must exert no downward pressure through the subject's hand to force the shoulder forward (Figure 6-68B). (If necessary, raise the subject's hand and elbow off the table.)

Pressure: Against the anterior aspect of the shoulder, downward toward the table.

Weakness: Strong extension of the shoulder is dependent on fixation of the scapula by the rhomboids and levator scapulae posteriorly and by the pectoralis minor anteriorly. With weakness of the pectoralis minor, the strength of shoulder extension is diminished.

With the scapula stabilized in a position of good alignment, the pectoralis minor acts as an accessory muscle of inspiration. Weakness of this muscle will increase respiratory difficulty in patients already suffering involvement of the respiratory muscles.

Shortness: With the origin of this muscle on the ribs and the insertion on the coracoid process of the scapula, a contracture tends to depress the coracoid process of the scapula both forward and downward. Such muscle contracture is an important contributing factor in many cases of upper extremity pain. With the cords of the brachial plexus and the axillary blood vessels lying between the coracoid process, the tendon of the pectoralis minor, and the rib cage, contracture of the pectoralis minor may produce an impingement on these large vessels and nerves.

A contracted pectoralis minor restricts flexion of the shoulder joint by limiting scapular rotation and preventing the glenoid cavity from attaining the cranial orientation necessary for complete flexion of the joint.

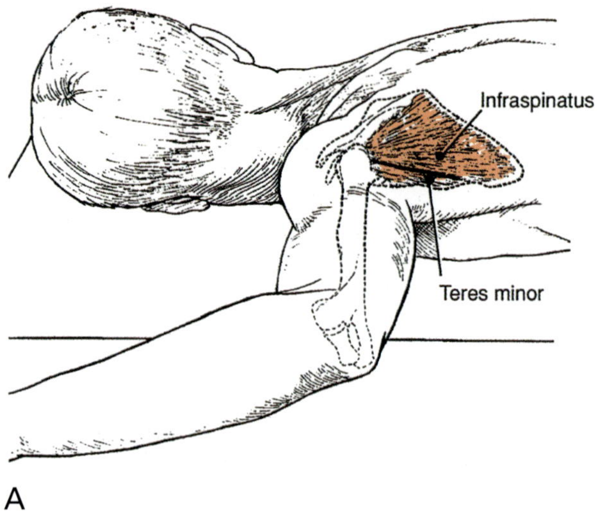

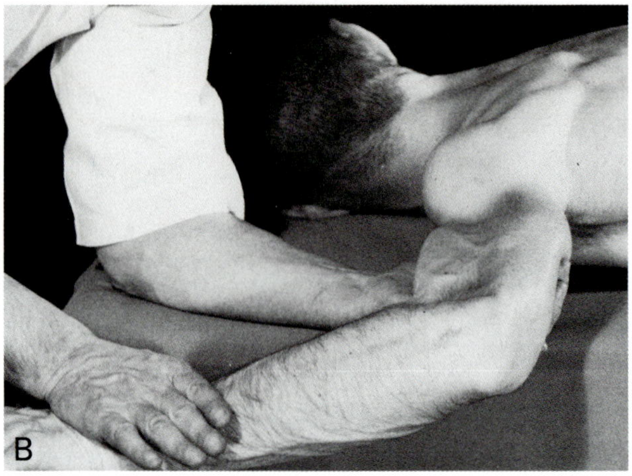

FIGURE 6-69: **A,** Infraspinatus; **B,** Muscle testing of infraspinatus.

Shoulder External Rotators

Infraspinatus

Origin: Medial 2/3 of the infraspinous fossa of the scapula (Figure 6-69A).

Insertion: Middle facet of the greater tubercle of the humerus and shoulder joint capsule.

Action: Externally rotates the shoulder joint and stabilizes the head of the humerus in the glenoid cavity during movements of this joint.

Nerve: Suprascapular, **C5, 6**.

Patient: Prone.

Fixation: The arm rests on the table. The examiner places one hand under the distal arm and stabilizes the humerus to ensure rotation by preventing shoulder adduction or abduction. The examiner's hand cushions against the table pressure. This test requires strong fixation by the scapular muscles, particularly the middle and lower trapezius, and in performing this test, one must observe whether the lateral rotators of the scapula or the external rotators of the shoulder break when pressure is applied.

Test: External rotation of the shoulder, with the elbow at 90° of flexion (Figure 6-69B).

Pressure: Using the forearm as a lever, pressure is applied in the direction of internal rotation of the shoulder.

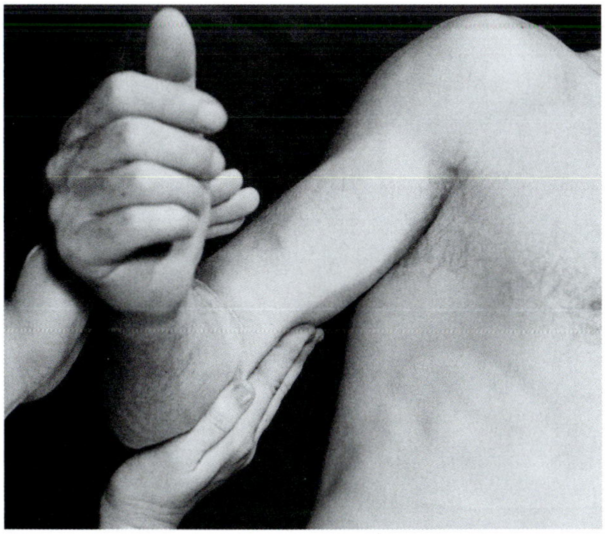

FIGURE 6-70: Muscle testing of teres minor.

Teres Minor

Origin: Upper 2/3, dorsal surfaces of the lateral border of the scapula (Figure 6-69A).

Insertion: Inferior facet of the greater tubercle of the humerus and shoulder joint capsule.

Action: Externally rotates the shoulder joint and stabilizes the head of the humerus in the glenoid cavity during movements of this joint.

Nerve: Axillary, **C5, 6**.

Patient: Supine.

Fixation: Counterpressure is applied by the examiner against the inner aspect of the distal end of the arm to ensure rotation.

Test: External rotation of the shoulder, with the elbow held at 90° of flexion (Figure 6-70).

Pressure: Using the forearm as a lever, pressure is applied, in the direction of shoulder internal rotation.

Weakness: The shoulder assumes a position of internal rotation. External rotation in antigravity positions is difficult or impossible.

For the purpose of objectively grading a weak external rotator group against gravity and for palpation of the rotator muscles, the test in the prone position is preferred over the teres minor and infraspinatus test in the supine position. For action of these two rotators without much assistance from the posterior deltoid and without the necessity of maximal trapezius fixation, the test in the supine position is preferred.

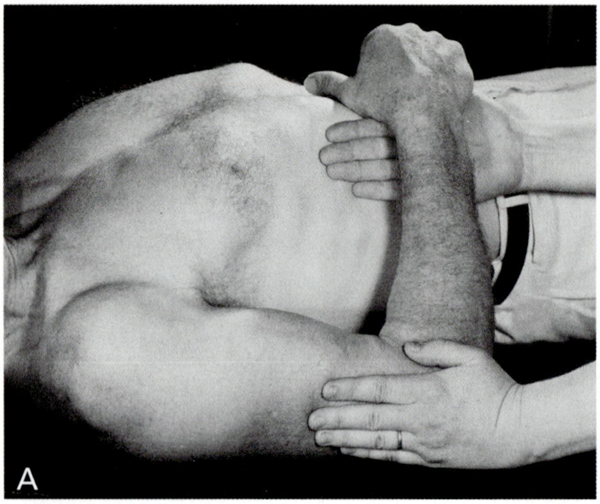

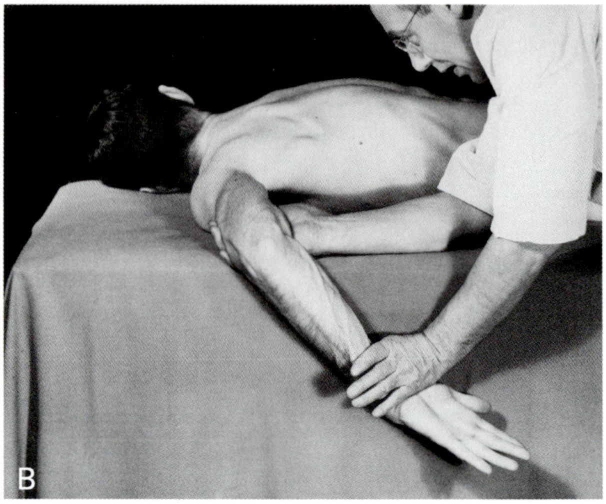

FIGURE 6-71: A, B, Internal rotators (group test).

Shoulder Internal Rotators (Group Test)

The chief muscles acting in this shoulder internal rotation test are the latissimus dorsi, pectoralis major, subscapularis, and teres major.

Patient: Supine.

Fixation: The examiner applies counterpressure against the outer aspect of the distal end of the humerus to ensure a rotation motion.

Test: Internal rotation of the humerus, with the arm at the side and the elbow held at 90° of flexion.

Pressure: Pressure is applied at the distal, anterior forearm in the direction of external rotation of the shoulder.

> **NOTE:** *For the purpose of objectively grading a weak internal rotator group against gravity, the test in the prone position (see Figure 6-71B) is preferred over the test in the supine position. For a maximum strength test, greater scapular fixation is required. For this reason, the test in supine position is preferred because it provides the necessary scapular fixation.*

Patient: Prone.

Fixation: The arm rests on the table. The examiner's hand, proximal to the elbow, cushions against table pressure and stabilizes the humerus to ensure rotation by preventing any shoulder adduction or abduction. The rhomboids provide fixation of the scapula.

Test: Internal rotation of the shoulder, with the elbow held at 90° of flexion (Figure 6-71).

Pressure: Pressure is applied at the distal, anterior forearm in the direction of external rotation of the shoulder.

Weakness: Because the internal rotators are also strong adductors, the ability to perform both shoulder internal rotation and adduction is decreased.

Shortness: Range of both shoulder flexion overhead and external rotation are limited.

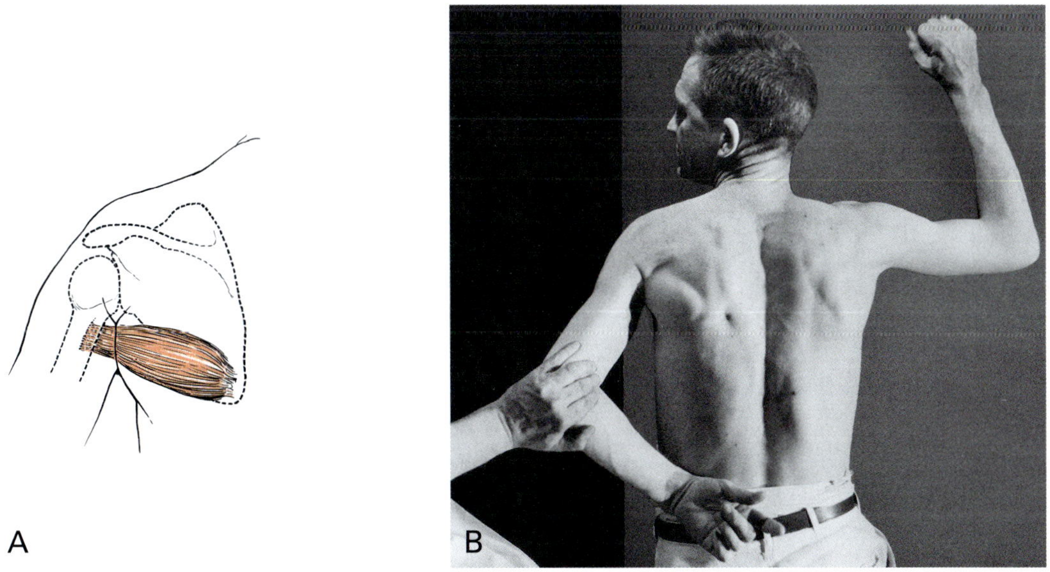

FIGURE 6-72: A, Teres major; **B,** Muscle testing of teres major.

Teres Major

Origin: Dorsal surfaces of the inferior angle and lower 1/3 of the lateral border of the scapula (Figure 6-72A).

Insertion: Crest of the lesser tubercle of the humerus.

Action: Internally rotates, adducts, and extends the shoulder joint.

Nerve: Lower subscapular, C5, **6**.

Patient: Prone.

Fixation: None usually is necessary because the weight of the trunk is sufficient fixation. If additional fixation is necessary, however, the opposite shoulder may be held down on the table.

Test: Extension and adduction of the shoulder in the internally rotated position, with the hand resting on the ipsilateral posterior iliac crest (Figure 6-72B).

Pressure: Against the arm, proximal to the elbow, in the direction of shoulder abduction and flexion.

Weakness: Diminishes the strength of internal rotation as well as shoulder adduction and extension.

Shortness: Prevents the full range of shoulder external rotation and abduction. With tightness of the teres major, the scapula will begin to upwardly rotate almost simultaneously with shoulder flexion or abduction. Scapular movements that accompany shoulder flexion and abduction may be influenced by the degree of muscle shortness of the teres major and subscapularis.

Subscapularis

Figure 6-73 illustrates the subscapularis.

Origin: Subscapular fossa of the scapula.

Insertion: Lesser tubercle of the humerus and shoulder joint capsule.

Action: Internally rotates the shoulder joint and stabilizes the head of the humerus in the glenoid cavity during movements of this joint.

Nerve: Upper and lower subscapular, **C5, 6,** 7.

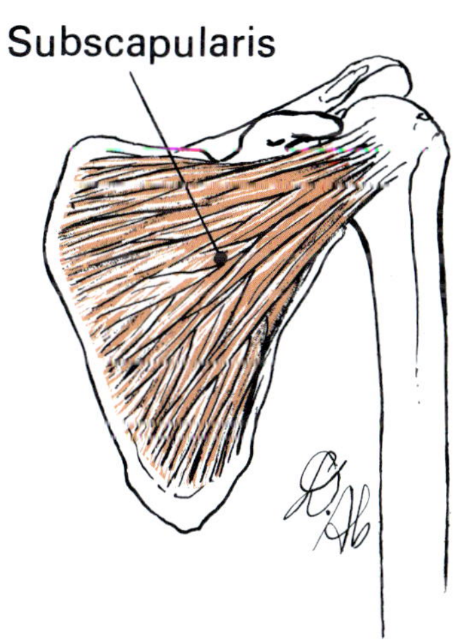

FIGURE 6-73: Subscapularis (anterior view).

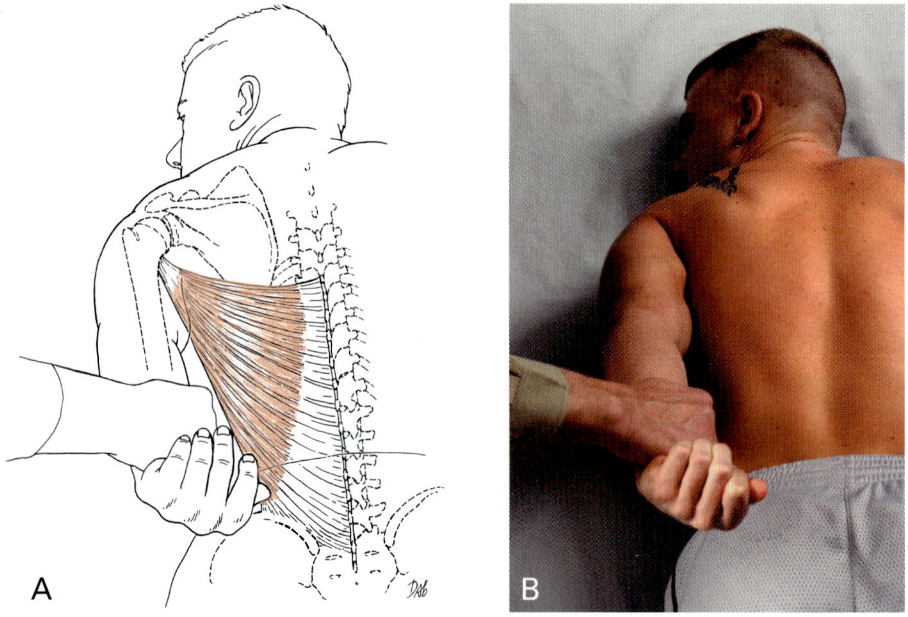

FIGURE 6-74: **A,** Latissimus dorsi; **B,** Muscle testing of latissimus dorsi.

Latissimus Dorsi

Figure 6-74A shows the attachments of the latissimus dorsi to the spine and pelvis, emphasizing the importance of this muscle with regard to its many functions. See preferred position of forearm in Figure 6-75.

Origin: The spinous processes of last seven (T6–T12) thoracic vertebrae, the last three or four ribs, through the thoracolumbar fascia from the lumbar and sacral vertebrae and posterior 1/3 of external lip of iliac crest, and a slip from the inferior angle of the scapula.

Insertion: The floor of the intertubercular groove of humerus.

Action: With the origin fixed, internally rotates, adducts, and extends the shoulder joint. By continued action, depresses the shoulder girdle and assists in lateral flexion of the trunk. With the insertion fixed, assists in tilting the pelvis both anteriorly and laterally. Acting bilaterally, this muscle can assist in hyperextending the spine and anteriorly tilting the pelvis. Depending on the degree of upper extremity elevation, the latissimus dorsi can also flex the spine. The latissimus dorsi also may act as an accessory muscle of respiration (25).

Nerve: Thoracodorsal, **C6, 7, 8**.

Patient: Prone.

Fixation: One hand of the examiner may apply counterpressure laterally over the posterior iliac crest.

Test: Shoulder adduction with extension, in the internally rotated position (Figure 6-74B).

Pressure: Against the distal forearm, in the direction of shoulder abduction and slight flexion.

Weakness: Weakness interferes with activities that involve shoulder adduction toward the body or of the body toward the arm. The strength of lateral trunk flexion is diminished.

> **NOTE:** *See Figure 6-75 regarding shortness of latissimus dorsi.*

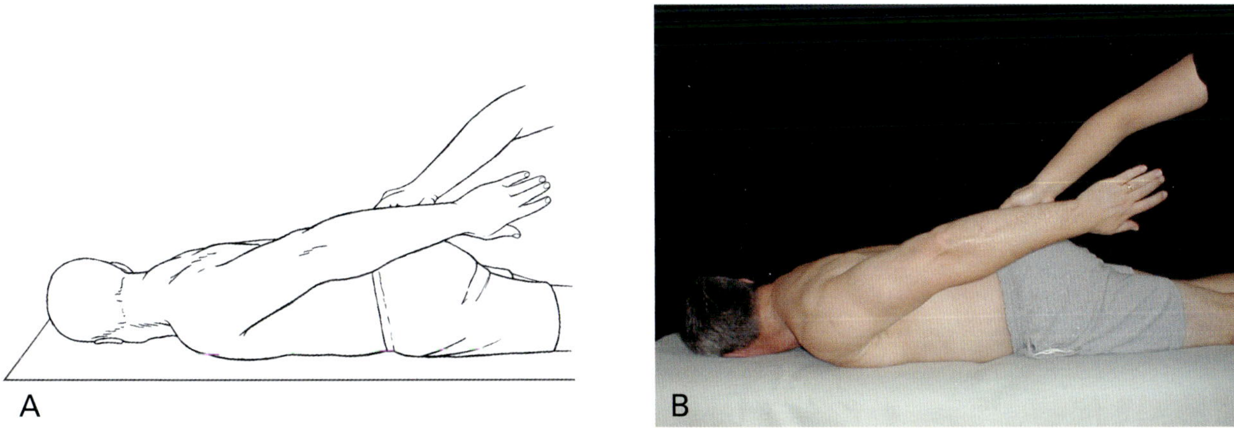

FIGURE 6-75: Latissimus dorsi: **A,** Side view of test position; **B,** Preferred position for forearm.

Shortness: Shortness of the latissimus dorsi results in a limitation of shoulder elevation into flexion and abduction. It tends to depress the shoulder girdle downward and anteriorly. In a right C-curve of the spine (named by the convexity), the lateral fibers of the left latissimus dorsi are usually shortened. In a marked kyphosis, the anterior fibers are shortened bilaterally. Shortness of this muscle may be found in individuals who have walked with crutches for a prolonged period of time.

This muscle is important in relation to movements involved in activities such as climbing, walking with crutches, and hoisting the body on parallel bars, in which the muscles act to lift the body toward the fixed arms. The strength of the latissimus dorsi is a factor in forceful arm movements such as swimming, rowing, and chopping. All adductors and internal rotators act in these strong movements, but the latissimus dorsi may be of major importance.

NOTE: *In the coronal plane, the latissimus dorsi (shoulder girdle depressor) is the most direct opponent of the upper trapezius (shoulder girdle elevator). Test the strength of the latissimus when the shoulder is elevated. Restoration of the muscle balance may require stretching the trapezius and strengthening the latissimus.*

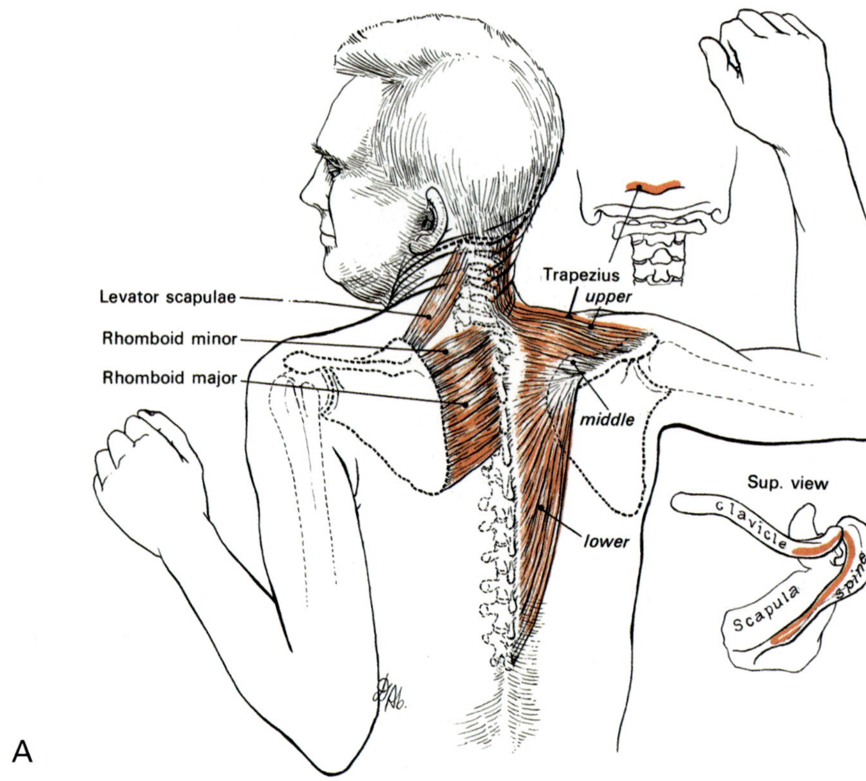

FIGURE 6-76: A, Rhomboids, levator scapulae, and trapezius.

Rhomboids, Levator Scapulae, and Trapezius

Rhomboid Major

Origin: Spinous processes of second through fifth thoracic vertebrae (Figure: 6-76A).

Insertion: By fibrous attachment to medial border of the scapula between spine and inferior angle.

Action: Along with rhomboid minor, adduction, elevation, and downward rotation of the scapula.

Nerve: Dorsal scapular, **C5**.

Rhomboid Minor

Origin: Ligamentum nuchae, spinous processes of seventh cervical and first thoracic vertebrae.

Insertion: Medial border at root of spine of the scapula.

Action: Along with rhomboid major, adduction, elevation, and downward rotation of the scapula.

Nerve: Dorsal scapular, **C5**.

Test for Rhomboids and Levator Scapulae

Patient: Prone.

Fixation: None on the part of the examiner is necessary, but it is assumed that the adductors of the shoulder joint have been tested and found to be strong enough to support the upper extremity for use as a lever in this test.

Test: Adduction, elevation, and downward rotation of the scapula. To obtain this position of the scapula and leverage for pressure in the test, the upper extremity is placed in the position as in Figure 6-76B. With the elbow flexed, the shoulder is adducted toward the side of the body in slight extension and slight external rotation. The test is to determine the ability of the rhomboids to hold the scapula in the test position as pressure is applied against the upper extremity. (See Alternate Rhomboid Test.)

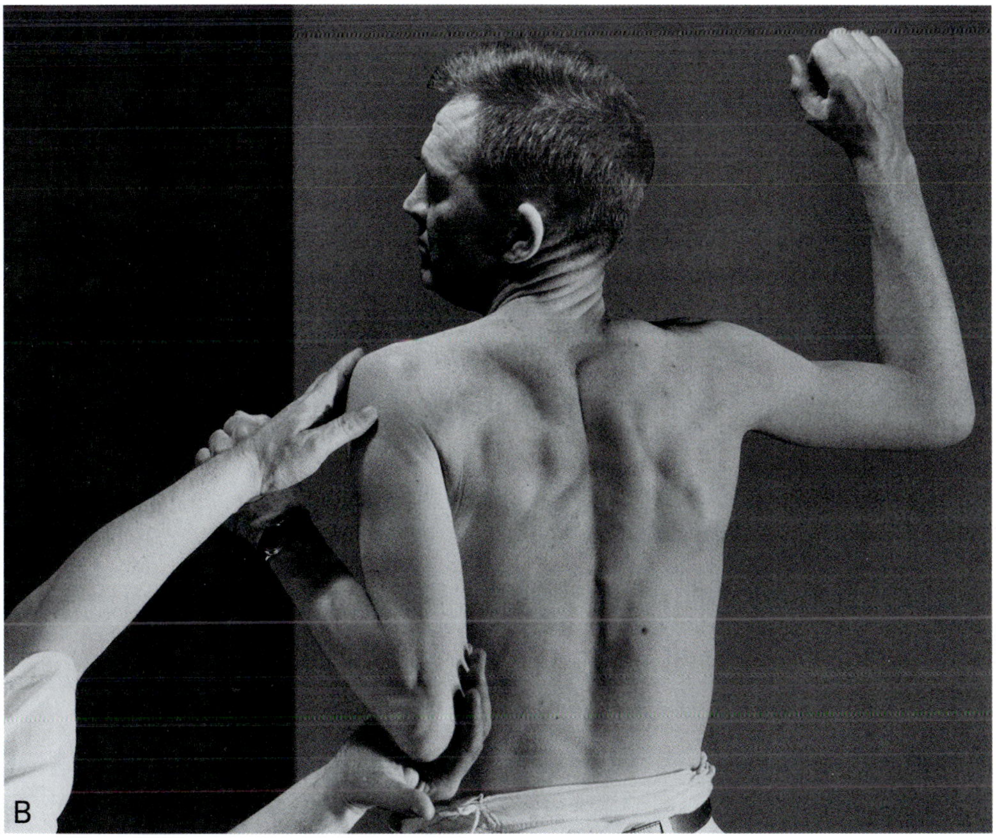

FIGURE 6-76: *(continued)* **B,** Muscle testing of rhomboids and levator scapulae.

Pressure: The examiner applies pressure with one hand on the posterior distal elbow in the direction of scapular abduction and upward rotation. The other hand applies pressure against the patient's shoulder in the direction of depression.

Weakness: The scapula abducts and upwardly rotates. The strength of adduction and extension of the shoulder is diminished by loss of rhomboid fixation of the scapula. Ordinary function of the upper extremity is less affected by loss of the rhomboid strength than by loss of either trapezius or serratus anterior strength.

Shortness: The scapula is drawn into a position of adduction and elevation. Shortness tends to accompany paralysis or weakness of the serratus anterior, because the rhomboids are direct opponents of the serratus.

Modified Test: If the shoulder muscles are weak, the examiner places the scapula in the test position and attempts to abduct, depress, and downwardly rotate the scapula.

NOTE: *Figure 6-76B shows the rhomboids in a state of contraction. (See right rhomboids in neutral position and left rhomboids in elongated position).*

Levator Scapulae

Origin: Transverse processes of the first four cervical vertebrae.

Insertion: Medial border of scapula, between superior angle and root of spine.

Action: With the origin fixed, elevates the scapula, and assists in downward rotation. With the insertion fixed and acting unilaterally, ipsilaterally flexes the cervical vertebrae and rotates them toward the same side. Acting bilaterally, the levator may assist in extension of the cervical spine.

Nerve: Direct branches of cervical ventral rami **C3, 4, and dorsal scapular, C5**.

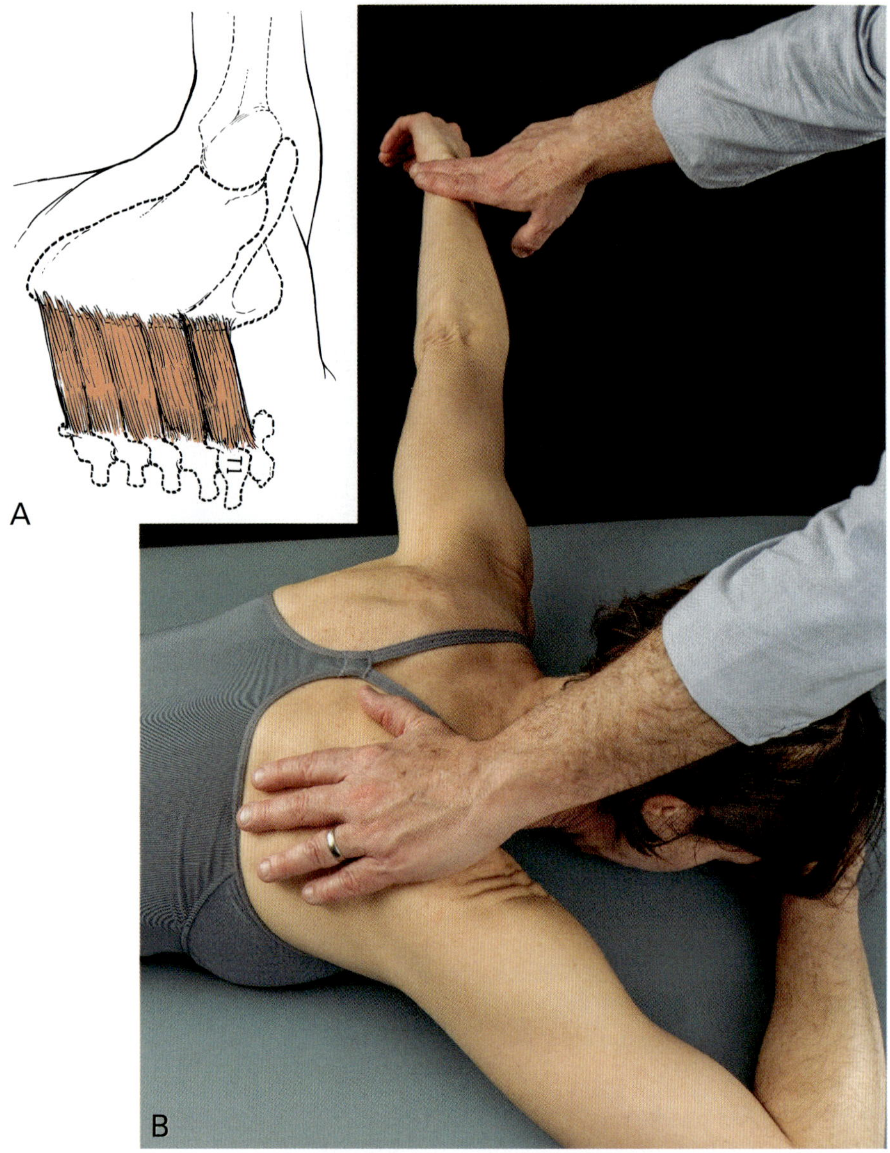

FIGURE 6-77: **A,** Rhomboids; **B,** With shoulder in medial rotation.

Alternate Rhomboid

If a position of internal rotation of the shoulder and elevation of the scapula is permitted during testing of the middle trapezius, it ceases to be a trapezius test. As seen in Figure 6-77, the shoulder is internally rotated, the scapula is elevated, depressed anteriorly, and the scapula is adducted by rhomboid action rather than by middle trapezius action. A comparison of this photograph with Figure 6-78 gives an example of what is meant by obtaining the specific action in which a muscle is the prime mover.

The marked difference that often exists between strength of the rhomboids and of the trapezius is dramatically demonstrated by careful testing.

Patient: Prone.

Fixation: Same as for middle trapezius, except the middle deltoid does not assist as an intervening muscle and the elbow extensors are necessary intervening muscles.

Test: Adduction, elevation, and downward rotation of the scapula. The position of the scapula is obtained by placing the shoulder in 90° abduction and in sufficient internal rotation to move the scapula into the test position. The palm of the hand faces in a caudal direction.

Pressure: Against the distal forearm, in a downward direction toward the table.

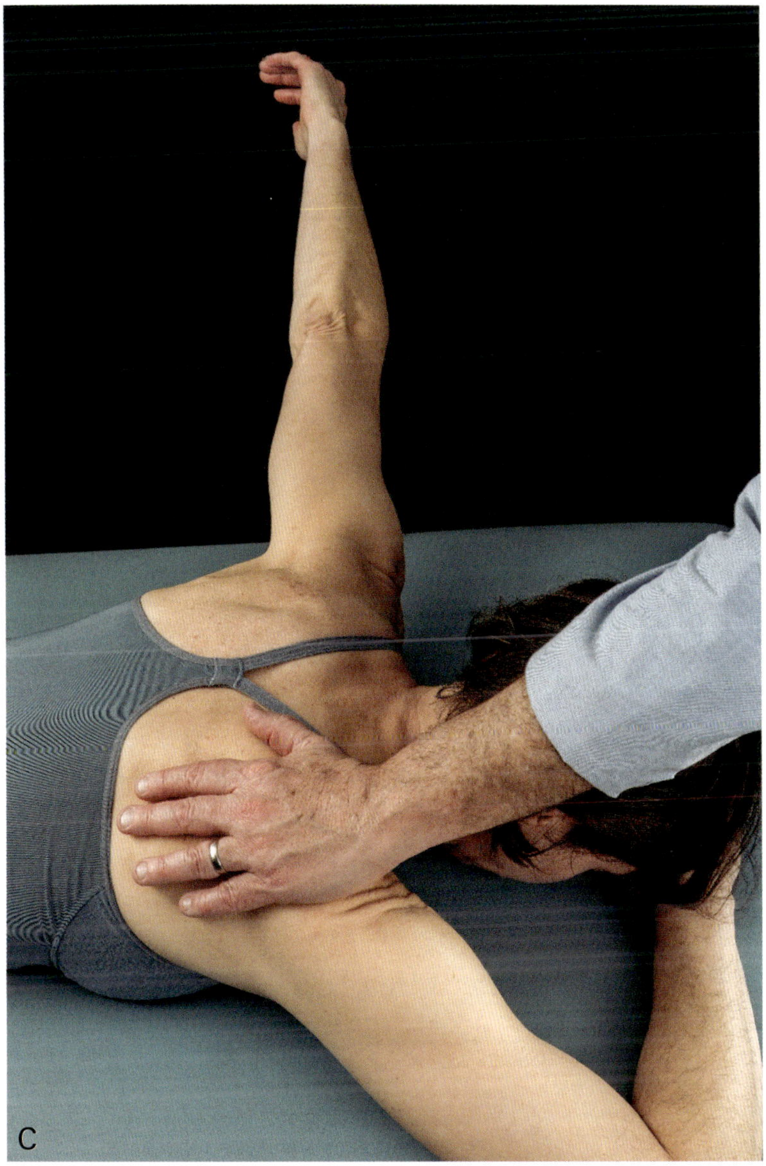

FIGURE 6-77: **C,** Rhomboids with shoulder in medial rotation (unobstructed view).

Trapezius

Origin of Upper Fibers: External occipital protuberance, medial 1/3 of superior nuchal line, ligamentum nuchae, and spinous process of the seventh cervical vertebra.

Origin of Middle Fibers: Spinous processes of the first through fifth thoracic vertebrae.

Origin of Lower Fibers: Spinous processes of the sixth through 12th thoracic vertebrae.

Insertion of Upper Fibers: Lateral 1/3 of the clavicle and acromion process of the scapula.

Insertion of Middle Fibers: Medial margin of the acromion and superior lip of the spine of the scapula.

Insertion of Lower Fibers: Tubercle at the apex of the spine of the scapula.

Action: With origin fixed, adduction of scapula, performed chiefly by the middle fibers, with stabilization by upper and lower fibers. Upward rotation of scapula, performed chiefly by upper and lower fibers, with stabilization by middle fibers. In addition, upper fibers elevate, and lower fibers depress, the scapula. With the insertion fixed and acting unilaterally, upper fibers extend, laterally flex, and rotate the cervical intervertebral joints to the opposite side. With insertion fixed and acting bilaterally, the upper trapezius extends the cervical intervertebral joints. The trapezius also acts as an accessory muscle of respiration.

Nerve: Spinal portion of cranial nerve XI (accessory) and ventral ramus, **C3, 4**.

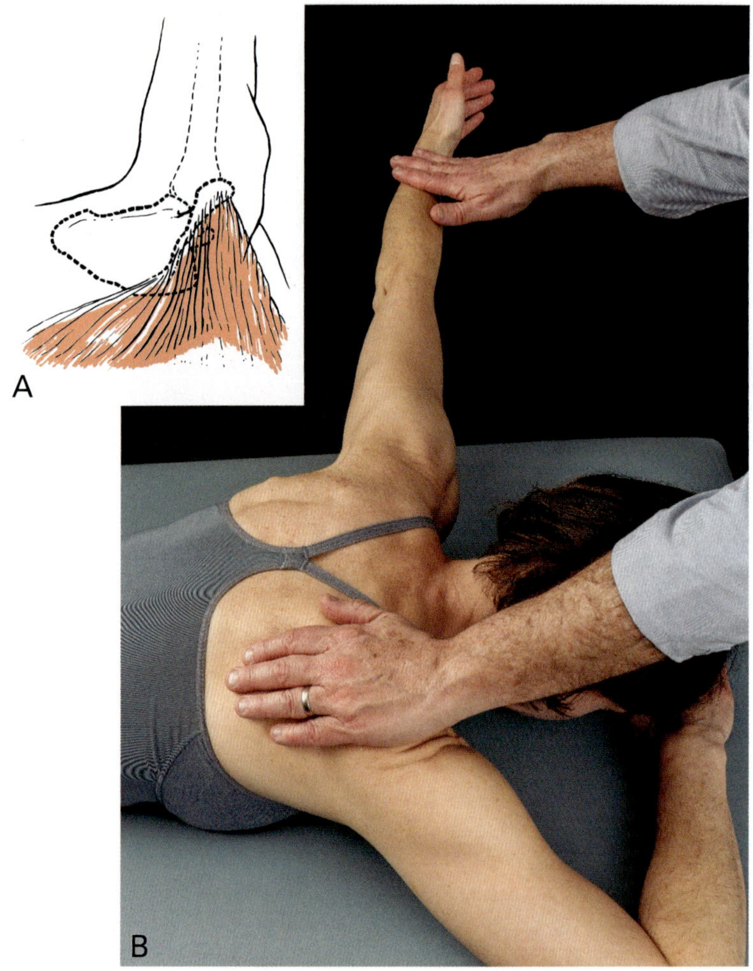

FIGURE 6-78: **A,** Middle trapezius; **B,** Muscle testing of middle trapezius.

Middle Trapezius

Patient: Prone.

Fixation: The intervening shoulder joint extensors (posterior deltoid, teres minor, and infraspinatus, with assistance from the middle deltoid) must give necessary stabilization of the shoulder to use the upper extremity as a lever. To a lesser extent, the elbow extensors may need to give some stabilization of the elbow. However, with the shoulder externally rotated, the elbow is also rotated into a position so that downward pressure on the forearm is exerted against the elbow laterally rather than in the direction of elbow flexion.

The examiner provides fixation by placing one hand on the opposite scapular area to prevent trunk rotation, as illustrated in Figure 6-78. (The examiner's hand in the photograph merely indicates the downward direction of pressure.)

Test: Adduction and upward rotation of the scapula, without elevation of the shoulder girdle. The test position is obtained by placing the shoulder in 90° abduction and in external rotation sufficient to bring the scapula into upward rotation.

The teres major is an internal rotator attached along the axillary border of the scapula. Traction on this muscle as the shoulder is externally rotated draws the scapula into upward rotation. The degree of shoulder external rotation necessary to produce the effect on the scapula will vary according to the tightness or laxity of the internal rotators. Usually, shoulder external rotation positions the hand so that the palm faces cranially and facilitates good positioning of the scapula.

Both the trapezius and the rhomboids adduct the scapula, but they differ in their action of rotation. Differentiating these muscles in testing is based on their rotation actions.

In addition to placing the parts in precise test position, it is necessary to observe the scapula during the testing to make sure that rotation is maintained as pressure is applied.

Pressure: Against the distal forearm, in a downward direction toward the table.

Weakness: Results in abduction of the scapula and a forward position of the shoulder. The middle and lower trapezius reinforce the thoracic spine extensors. Weakness of these fibers of the trapezius increases the tendency toward excessive kyphosis.

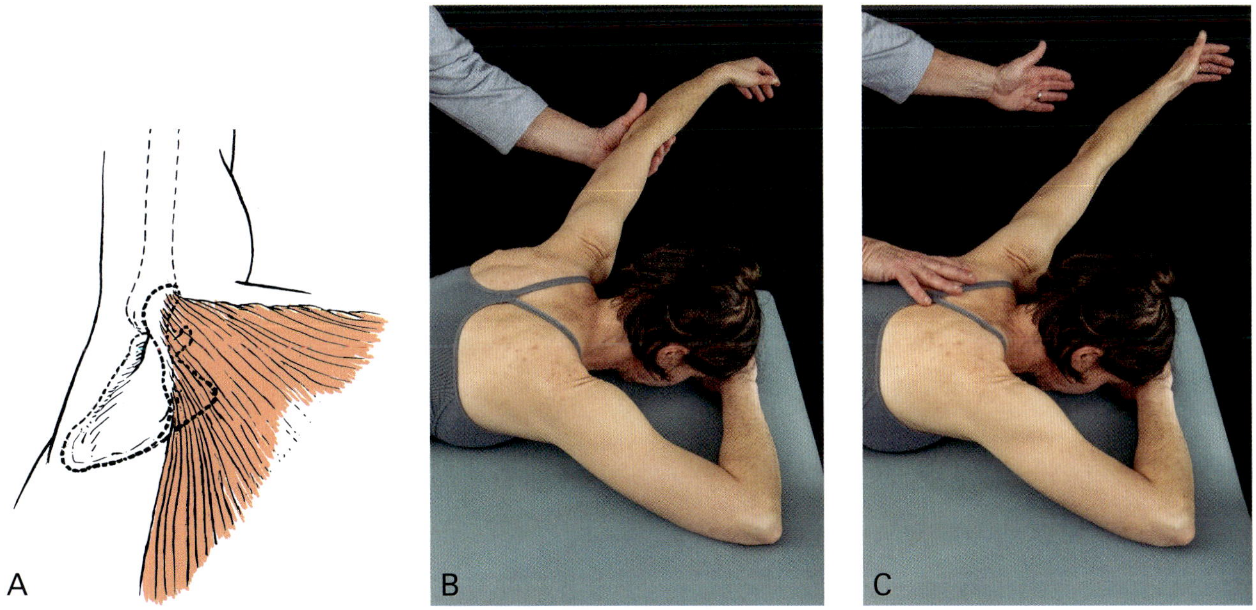

FIGURE 6-79: **A,** Lower trapezius; **B, C,** Muscle testing of lower trapezius.

Lower Trapezius

Patient: Prone.

Fixation: The intervening shoulder extensors, particularly the posterior deltoid, must give the necessary stabilization of the shoulder, and to a lesser extent, the elbow extensors need to stabilize the elbow in extension. (See explanation under Fixation in the section on the Middle Trapezius.)

The examiner provides fixation by placing one hand below the scapula on the opposite side (not shown).

Test: Adduction, depression, and upward rotation of the scapula. The upper extremity is placed diagonally overhead, in line with the lower fibers of the trapezius. External rotation of the shoulder joint occurs along with elevation, so it usually is not necessary to further rotate the shoulder to bring the scapula into upward rotation (Figure 6-79). (See explanation under Middle Trapezius.)

Pressure: Against the distal forearm, in a downward direction toward the table.

Weakness: Allows the scapula to ride upward and tilt forward, with depression of the coracoid process. If the upper trapezius is tight, it helps to pull the scapula upward and acts to oppose a weak lower trapezius.

> **NOTE:** *Tests for the lower and middle trapezius are especially important during examination of cases with faulty shoulder position or with upper back or arm pain (26).*

Modified Trapezius

For use when the posterior shoulder joint muscles are weak.

Patient: Prone, with the shoulder at the edge of the table and the upper extremity hanging down over the side of the table.

Fixation: None.

Test: Supporting the weight of the upper extremity, the examiner places the scapula in a position of adduction, some upward rotation, and without elevation of the shoulder girdle.

Pressure: As the support of the upper extremity is removed, the weight of the suspended upper extremity will exert a force that tends to abduct the scapula. A very weak trapezius will not hold the scapula adducted against this force. If the trapezius can hold the scapula in adduction against the weight of the suspended upper extremity, then resist against the middle portion of the scapula by pressure in the direction of abduction and against the lower portion of the scapula by pressure in a diagonal direction toward abduction and elevation. When recording the grade of strength, note that pressure was applied on the scapula, because the upper extremity could not be used as a lever.

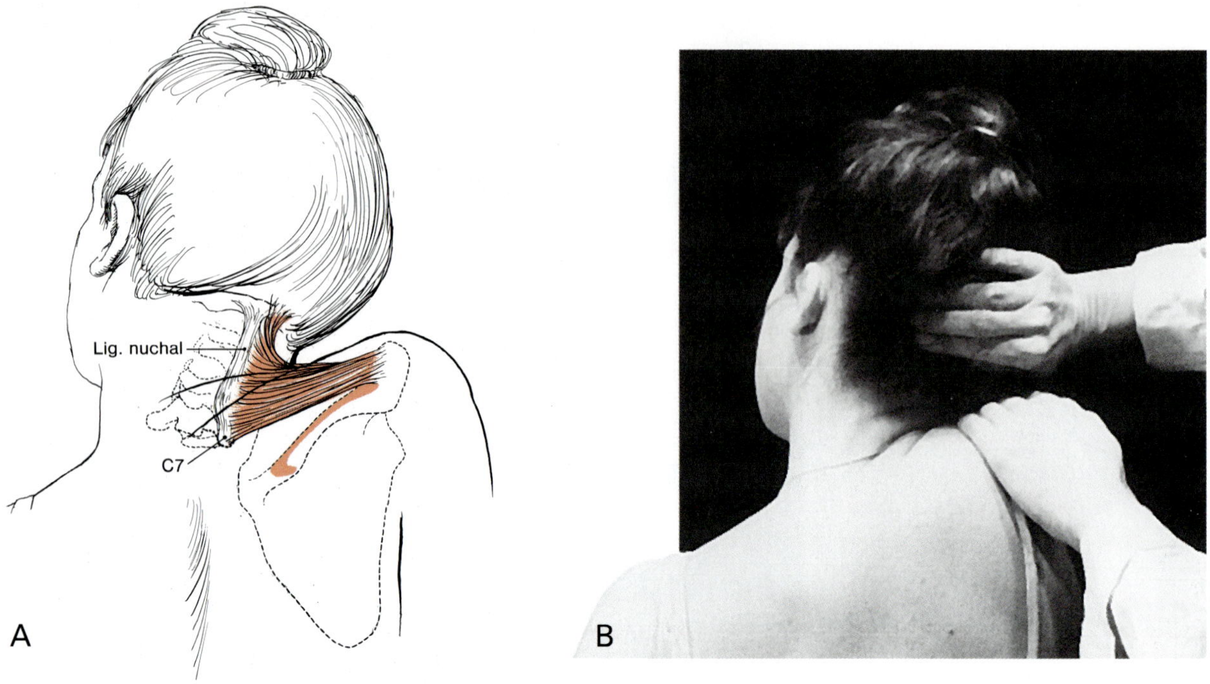

FIGURE 6-80: **A,** Upper trapezius; **B,** Muscle testing of upper trapezius.

Upper Trapezius

Patient: Sitting.

Fixation: None necessary.

Test: Elevation of the acromial end of the clavicle and scapula, with cervical extension and opposite side rotation (bringing the occiput toward the elevated shoulder with the face turned in the opposite direction).

The upper trapezius can be differentiated from other elevators of the scapula because it is the only one that elevates the acromial end of the clavicle and the scapula. It also upwardly rotates the scapula as it elevates it, in contrast to the straight elevation that occurs when all elevators contract, as in shrugging the shoulders (Figure 6-80).

Pressure: Against the shoulder, in the direction of depression, and against the head, in the direction of flexion anterolaterally.

Weakness: *Unilaterally*, weakness decreases the ability to approximate the acromion and the occiput. *Bilaterally*, weakness decreases the ability to extend the cervical spine (e.g., to raise the head from a prone position).

Shortness: Results in a position of elevation of the shoulder girdle. In a faulty posture with a forward head position and kyphosis, the cervical spine is in extension, and the upper trapezius muscles are in a shortened position.

Contracture: Unilateral contracture is frequently seen in cases of torticollis. For example, the right upper trapezius is usually contracted along with a contracture of the right sternocleidomastoid and scaleni.

Weakness of Whole Trapezius: Results in abduction and downward rotation of the scapula, with depression of the acromion, and interferes with ability to raise the upper extremity in abduction overhead. (See Figure 6-85, which shows the posture of shoulder when the entire trapezius is paralyzed.)

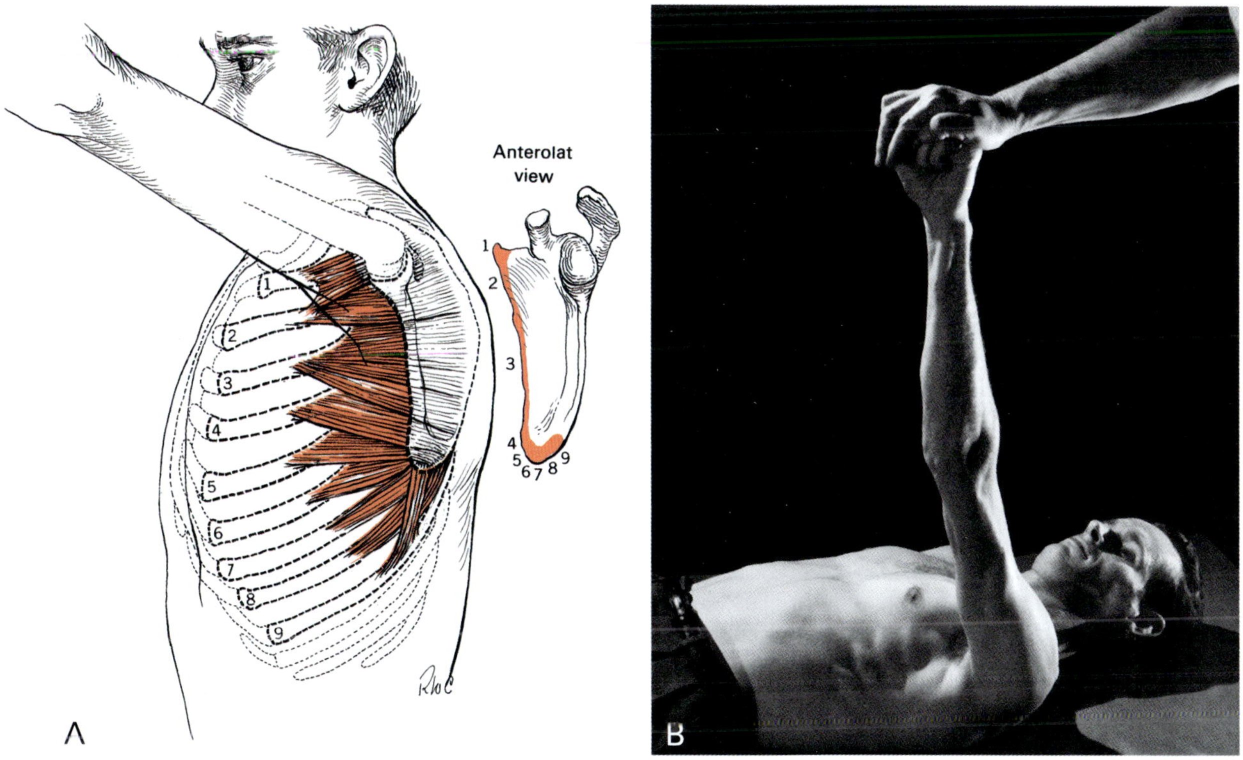

FIGURE 6-81: **A,** Serratus anterior; **B,** Muscle testing of serratus anterior.

Serratus Anterior

Origin: Outer surfaces and superior borders of the upper eight or nine ribs (Figure 6-81A).

Insertion: Costal surface of the medial border of the scapula.

Action: With the origin fixed, abducts and upwardly rotates the scapula, and holds the medial border of the scapula firmly against the rib cage. In addition, the lower fibers may depress the scapula, and the upper fibers may elevate it slightly.

Starting from a position with the shoulder fixed in flexion and the hands against a wall (see the standing serratus test), the serratus acts to displace the thorax posteriorly as the effort is made to push the body away from the wall. Another example of this type of action is a properly executed push-up.

With the scapula stabilized in adduction by the rhomboids, thereby fixing the insertion, the serratus may act in forced inspiration.

Nerve: Long thoracic, **C5, 6, 7**.

Patient: Supine.

Fixation: None necessary, unless the shoulder or elbow muscles are weak, in which case the examiner supports the upper extremity in the perpendicular position during the test.

Test: Abduction of the scapula, projecting the upper extremity anteriorly (upward from the table) (Figure 6-81B). *Movement of the scapula must be observed and the inferior angle palpated to ensure that the scapula is abducting.* Projection of the extremity can be accomplished by action of the pectoralis minor (aided by the levator and rhomboids) when the serratus is weak, in which case the scapula tilts forward at the coracoid process and the inferior angle moves posteriorly and in the direction of downward rotation. The firm surface of the table supports the scapula. Therefore, there will be no winging, and the pressure against the hand may elicit what appears to be normal strength. Because this type of substitution can occur during this test, the test in the sitting position (as described next) is preferred (27).

Pressure: Against the subject's fist, transmitting the pressure downward through the extremity to the scapula in the direction of adducting the scapula. *Slight* pressure may be applied against the lateral border of the scapula as well as against the fist.

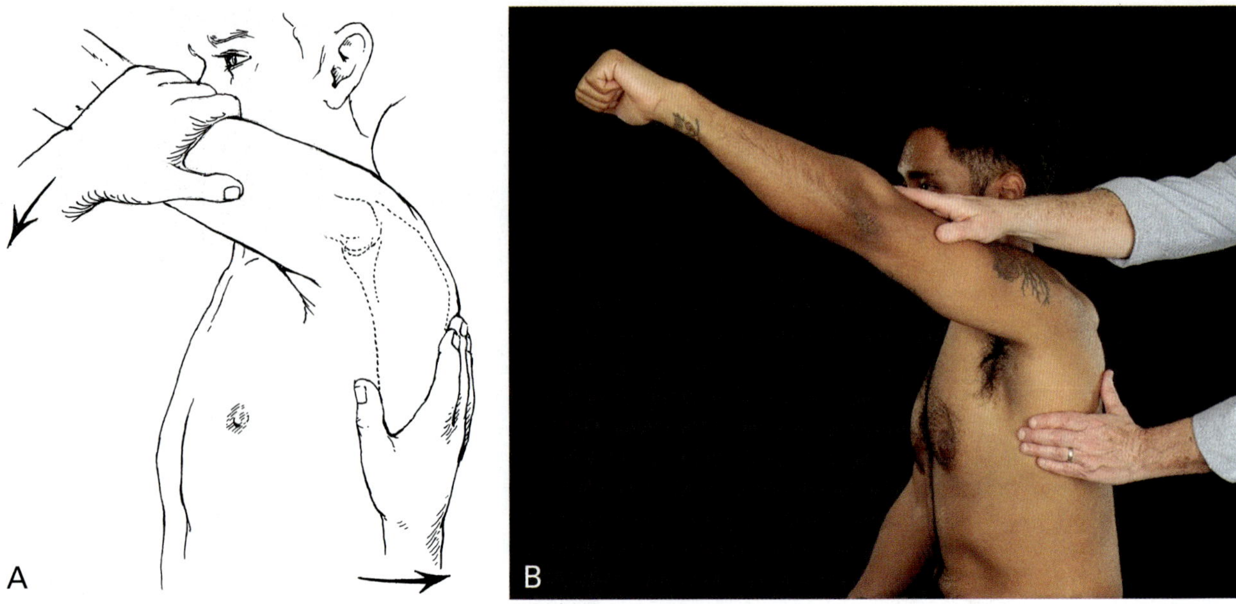

FIGURE 6-82: A, Serratus anterior; **B,** Preferred test of serratus anterior.

Preferred Test for Serratus Anterior

Patient: Sitting.

Fixation: If the trunk is stable, none by the examiner should be necessary. However, the shoulder flexors must be strong to use the arm as a lever in this test. Allow the subject to hold on to the table with one hand.

Test: The ability of the serratus to stabilize the scapula in a position of abduction and upward rotation, with the shoulder in a position of approximately 120° to 130° of flexion with slight internal rotation. This test emphasizes the upward rotation action of the serratus in the abducted position, as compared with the emphasis on the abduction action shown during the supine and standing tests.

Pressure: Against the dorsal surface of the arm, between the shoulder and elbow, downward in the direction of extension, and *slight* pressure against the lateral border of the scapula, in the direction of downward rotation. The thumb against the lateral border (as shown in Figure 6-82A) acts more to track the movement of the scapula than to offer pressure.

In practice, it is preferable to stand beside the subject and apply pressure as illustrated in Figure 6-82A. It is not advisable to use a long lever by applying pressure on the forearm or at the wrist, because the intervening shoulder flexors will often break before the serratus.

To help the reader clearly visualize how the pressure is applied, the examiner in Part B is standing behind the subject and applying pressure with the fingertips on the scapula.

Weakness: Makes it difficult to raise the arm in flexion. Results in winging of the scapula. With marked weakness, the test position cannot be held. With moderate or slight weakness, the scapula cannot hold the position when pressure is applied on the arm. Because the rhomboids are direct opponents of the serratus, the rhomboids become shortened in some cases of serratus weakness.

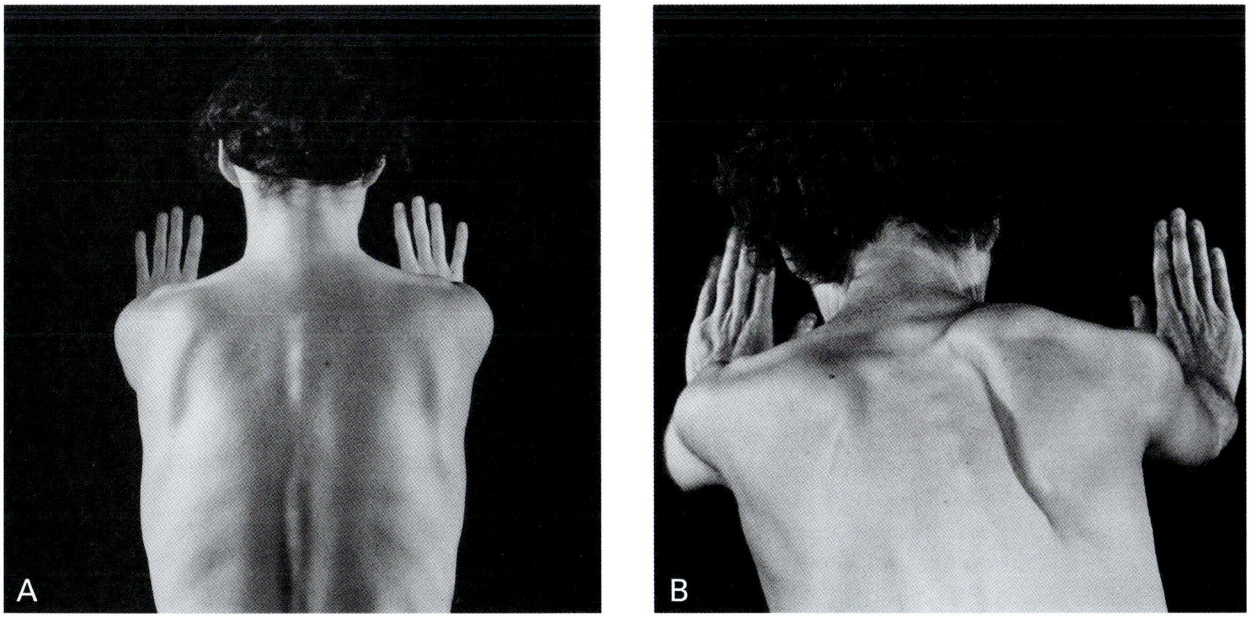

FIGURE 6-83: A, B, Standing test for serratus anterior.

Standing Test for Serratus Anterior

Patient: Standing.

Fixation: None necessary.

Test movement: Facing a wall and with the elbows straight, the subject places both hands against the wall, either at shoulder level or slightly above (Figure 6-83A). To begin, the thorax is allowed to sag forward so that the scapulae are in a position of some adduction. The subject then pushes hard against the wall, displacing the thorax backward, until the scapulae are in a position of abduction.

Resistance: The thorax acts as resistance in this test. By fixation of the hands and extended elbows, the scapulae become relatively fixed, and the anterolateral rib cage is drawn backward toward the scapulae. (By contrast, the scapula is pulled forward, toward the fixed rib cage, during the forward thrust of the arm in the supine test shown in Figure 6-81.) Because the resistance of displacing the weight of the thorax makes this a strenuous test, it will differentiate only between strong and weak for purposes of grading.

Weakness: Winging of scapula (Figure 6-83B shows weakness of the R serratus as evidenced by winging on that side.)

SECTION IV
CLINICAL FINDINGS

SHORTNESS OF INTRINSIC MUSCLES OF THE HAND

In Figure 6-84, a middle-aged woman presented with a complaint that her middle digit occasionally pained rather severely and that a constant, tight, "drawing" feeling was felt along the sides of this digit. She did not feel that the pain was actually in the joints of the digit. A medical checkup had revealed no arthritis. This person was an avid card player, and the condition was present in the left hand, which was the hand in which she held her cards.

Part A shows the position of the subject's hand in holding a hand of cards. This position is one of strong lumbrical and interosseous action. Just as in holding a newspaper, the middle digit is the one that strongly opposes the thumb.

On testing for length of the intrinsic muscles, evidence of shortness was found, chiefly in the muscles in the middle digit.

The patient could close the digits to make a fist, as shown in **Part B**. This was possible even though some shortness existed in the lumbricals and interossei, because the muscles were being elongated over the interphalangeal joints only, not over the metacarpophalangeal joints.

When attempting to close the hand into a claw-hand position, as shown in **Part C**, the shortness became apparent. In closing the digits into this position, the lumbricals and interossei must elongate over all three joints at the same time. The middle digit shows the greatest limitation. The fourth digit shows slight limitation, which is demonstrated by the lack of distal joint flexion as well by decreased hyperextension of the metacarpophalangeal joint.

The patient could extend the digits, as shown in **Part D**. This was possible because the muscles were being elongated over the metacarpophalangeal joints only, not over the interphalangeal joints. In **Part D**, the distal phalanx of the middle digit, which opposes the thumb in holding the cards, is in slight hyperextension.

The fact that the digits could be spread apart, as shown in **Part E**, and closed sideways, as shown in **Part F**, suggests that the shortness may have been in the lumbricals more than in the interossei.

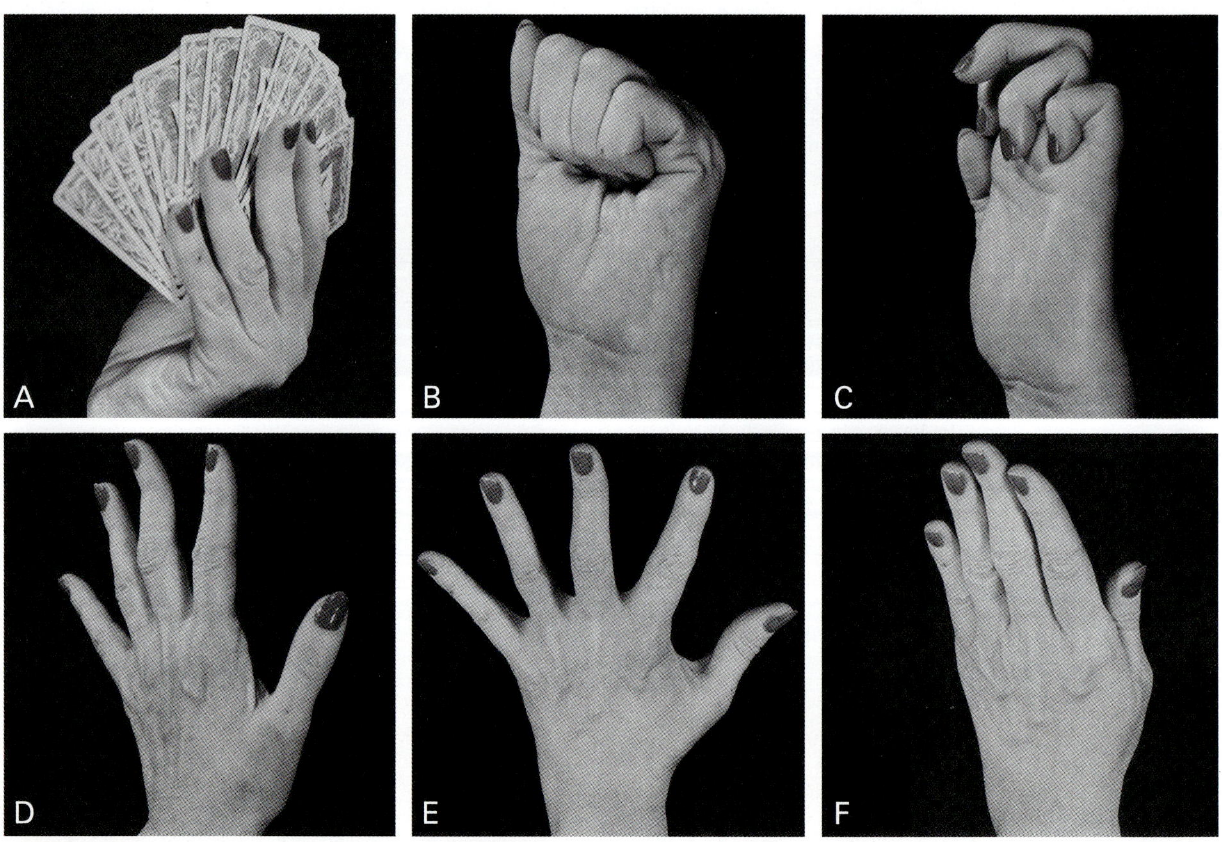

FIGURE 6-84: A–F, Shortness of intrinsic muscles of the hand.

NERVE ENTRAPMENTS

Upper extremity peripheral nerve entrapments can occur at any point along the nerve's path. Clinical presentations can vary dependent on the site of entrapment (28, 29). Table 6-4 shows common areas of entrapment as well as details regarding some clinical presentations and differential diagnoses. It is not intended to be an all-compassing list of upper extremity diagnoses. Table 6-5 lists radial nerve syndromes.

TABLE 6-4 Common Areas of Nerve Entrapment

Nerve	Entrapment Regions	Differential Diagnoses
Radial (30)	• Post. humerus • Elbow • Forearm • Wrist	• Median neuropathy • Multifocal motor neuropathy • Radial neuropathy • Ulnar neuropathy • High radial N. palsy • Radial tunnel syndrome
Median (31)	• L. of Struthers • Between the two heads of the pronator teres • Between the pronator teres and the flexor digitorum sublimus • The lacertus fibrosus extension of the biceps tendon • From the pronator teres and the two potential entrapment sites at the location of the pronator teres (pronator syndrome) (incl. ant. interosseous n. branching?) • Flexor retinaculum of the wrist (carpal tunnel)	
Ulnar	• Cubital tunnel (32) • Guyon's canal	• Thoracic outlet syndrome • Medial epicondylopathy • Carpal tunnel syndrome • Hamate fracture • Ulnar collateral ligament sprain • Brachial plexopathy • Guyon's tunnel syndrome
Musculocutaneous (33)	• Coracobrachialis • Distal arm	• Lateral (antebrachial) cutaneous nerve syndrome • Biceps tendinopathy (distal) • Brachial plexopathy • Thoracic outlet syndrome

TABLE 6-5 Radial Nerve Syndromes

Syndrome	Etiologies	Symptoms	Signs
High radial nerve palsy	• Trauma to the radial nerve as it winds around the posterior aspect of the humerus • Tourniquet use • Humeral fractures • "Saturday night palsy" • Repetitive activities that involve elbow extension	• P!/paresthesia in radial nerve distribution (may radiate proximal or distal to elbow)	• Triceps, supinator, wrist extensor, thumb extensor, thumb abductor, and MP or motor involvement • Nonsegmental sensation loss in radial N. distribution (post. forearm, dorsal hand, dorsal aspect of the radial 3 and ½ of 4th digit up to DIP)

Radial tunnel syndrome (RTS) (34-39)	• Compression of the radial nerve as it travels through the radial tunnel.
	• Compression sites include:
	  • Fibrous bands anterior to the radiohumeral joint (anterior to the radial head)
	  • Leash of Henry
	  • Fibrous edge of the extensor carpi radialis brevis
	  • The arcade of Frohse
	  • The distal edge of the radial tunnel
		• Lateral elbow and dorsal forearm pain that may radiate to the wrist and dorsum of the fingers
		• Night pain that may interfere with sleeping
		• Symptom aggravation with repeated forearm supination/pronation
		• Arm fatigue
			• Localized tenderness over the radial nerve 5 cm distal to the lateral epicondyle
			• P! With adverse neural tension (active elbow, forearm pronation, and wrist flexion)
			• Exacerbation of the pain with resisted supination
			• Exacerbation of the pain with wrist hyperextension against resistance
			• The Rule of Nine? (39)

(continues on the next page)

TABLE 6-5 (*continued*)

Syndrome	Etiologies	Symptoms	Signs
Posterior interosseous syndrome (PIS)	• Entrapment of the post. interosseous nerve (PIN), which is a branch of the radial nerve, that supplies motor innervation to the extensor compartment of the forearm. PIS, like RTS, is often found to be insidious in onset and may involve the same sites of compression as RTS.	• Dorsal forearm pain	• Weakness in finger and thumb • Preserved wrist • Negative sensory changes • When asked to make a fist, the wrist may deviate radially due to extensor carpi ulnaris weakness (40-41).
Ulnar nerve syndromes			
Guyon's tunnel syndrome	• May result from direct trauma (i.e., hook of hamate fractures) • Repetitive activities • Compression from a sustained posture • Space-occupying cysts/tumors • Inflammation • More anatomically proximal structures	• Pain and/or paresthesia in the ulnar nerve distribution of the hand and digits • Motor loss and atrophy in the hypothenar muscles • Clawing of the fourth and fifth digits may be observed	
Cubital tunnel syndrome	• Compression of the ulnar nerve in the cubital tunnel • Frequently seen in patients who participate in overhead sports that require repetitive elbow flexion/extension activities like throwing	• Pain and paresthesia along the medial forearm, ulnar side of hand, 5th digit and the ulnar half of 4th digit • Can be brought on with sustained elbow flexion (1–3 min.)	• Sensory loss over the ulnar aspect of the hand and 5th and the ulnar half of fourth digits and occasionally a snapping/popping of the nerve with flexion/extension of the elbow. • With more advanced involvement, motor involvement may be found in the intrinsic hand muscles. • + ulnar nerve neurodynamic testing
Median nerve syndromes			
Carpal tunnel syndrome	• Past injury • Previous neurodynamic impairments • Any acute trauma, injury, or condition (e.g., tendinopathy, inflammatory disorder) that decreases the space of the carpal tunnel • Cumulative trauma (overuse) • Frequency use of vibratory tools • Sustained postures (e.g., wrist flexion)	• May present with pain and/or paresthesia of the radial hand and digits • Waking at night with paresthesia is not uncommon.	• Motor loss and atrophy in the thenar muscles • Impaired grip • + Phalen's Test, carpal compression, Tinel's test, and adverse neural tension test.
Pronator syndrome	• Trauma • Repetitive forearm pronation and elbow flexion movements • Repetitive wrist flexion movements	• Pain in the anterior forearm with radicular paresthesia along the median nerve distribution of the hand	• Symptom onset with resisted wrist flexion, forearm pronation, and elbow flexion • + Pronator compression tests, Tinel's test, and ANTTs

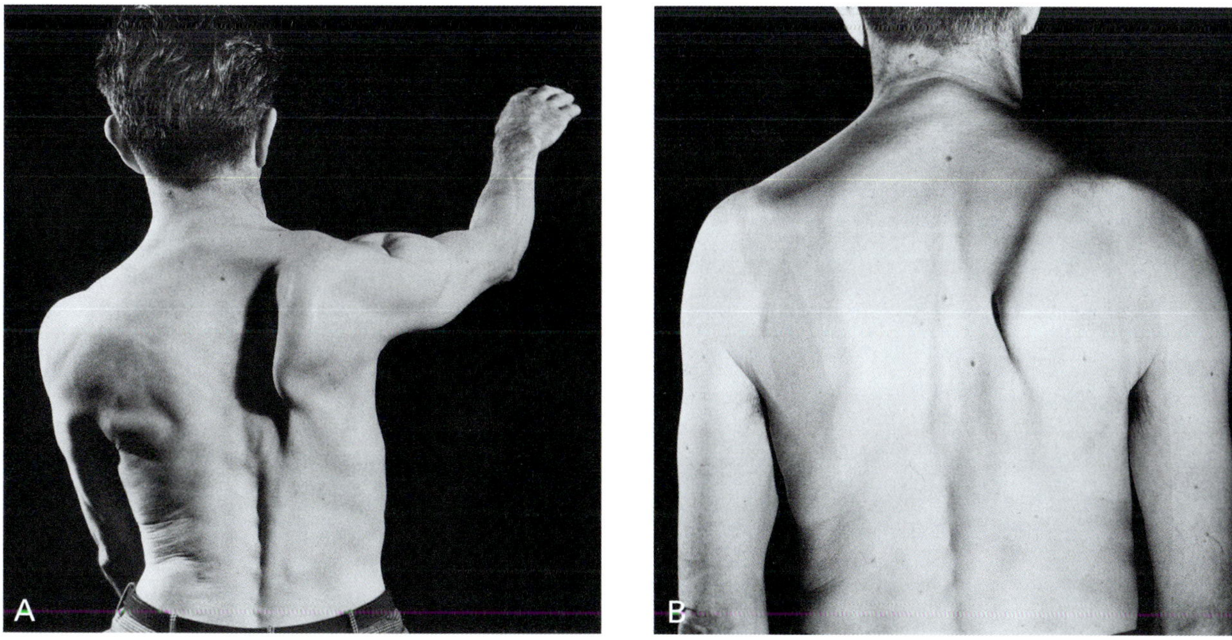

FIGURE 6-85: Impairment of right trapezius and serratus anterior: **A,** Inability to raise the arm overhead; **B,** Abnormal position of the right scapula.

IMPAIRMENTS INVOLVING TRAPEZIUS AND SERRATUS ANTERIOR

Impairment of Right Trapezius and Serratus Anterior

Figure 6-85A shows the subject's inability to raise the arm overhead when both the serratus and the trapezius are paralyzed. The winging of the medial border of the scapula makes it appear that the rhomboids were weak even though, in fact, they were not.

Figure 6-85B shows the abnormal position of the right scapula that results from a paralysis of both the trapezius and the serratus anterior. The scapula is downwardly rotated and elevated. The rhomboids were found to be strong.

Impairment of Right Trapezius with Normal Serratus

Abducting the shoulder joint is accompanied by isolated upward rotation of the scapula. With paralysis of the trapezius, the isolated upward rotation of the scapula cannot occur. Hence, shoulder abduction is limited, as seen in Figure 6-86A.

Flexing the shoulder joint requires that the scapula upwardly rotate in the abducted position. With an intact serratus, the arm could be raised higher in flexion than in abduction, as seen in Figure 6-86B.

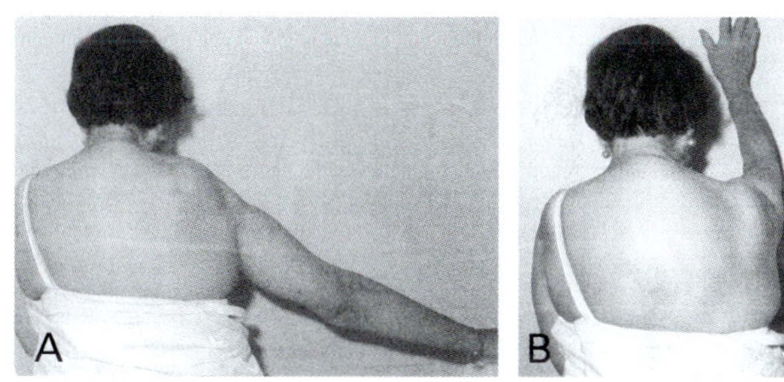

FIGURE 6-86: **A, B,** Impairment of right trapezius with normal serratus.

With a weak serratus and strong trapezius, the arm could be raised higher in abduction than in flexion.

Figure 6-87 illustrates the posture of the shoulders and scapulae as seen in some cases of mild serratus weakness.

FIGURE 6-87: Posture of the shoulders and scapulae as seen in some cases of mild serratus weakness.

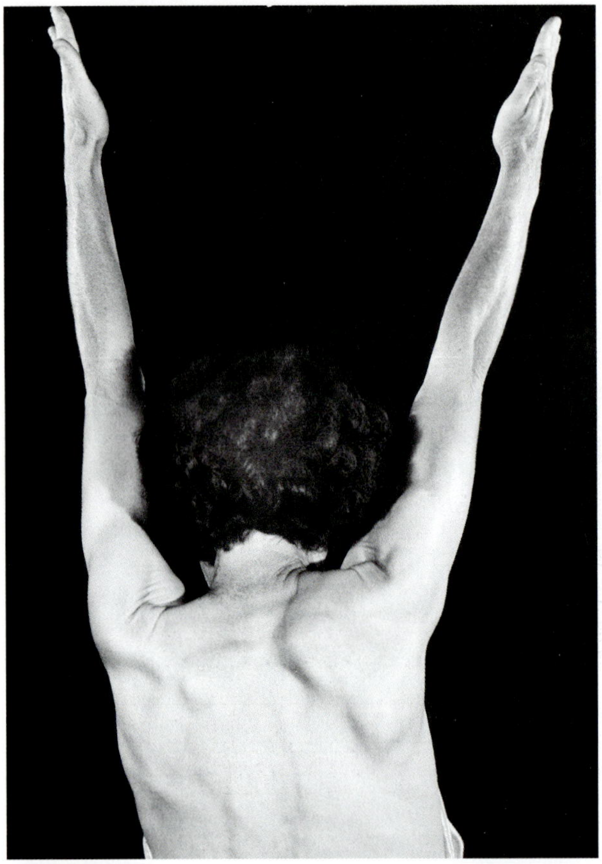

FIGURE 6-88: Upper extremity elevation with right serratus anterior impairment.

Slight winging of the scapulae is readily visible on the right because the thoracic spine is straight (lacking a typical kyphosis). However, one must not assume the presence of serratus weakness only on the basis of appearance. When the thoracic spine is straight, the scapulae may be prominent even if the serratus is normal in strength because of the curved shape of the scapula on the flattened costal cage (42).

Mild weakness of the serratus anterior is more prevalent than generally realized, and weakness tends to be more on the left than on the right, regardless of handedness (42, 43). When weakness exists, it can be aggravated by attempting strenuous exercises, such as push-ups.

Figure 6-88 shows the extent to which the right upper extremity could be elevated overhead with the subject in a standing position with impairment of the right serratus anterior. In this situation, the upper extremity could not be raised anteriorly, and the right scapula could neither be abducted nor fully rotated as on the normal (left) side. The upper and lower fibers of the trapezius are clearly seen because they are compensating to upwardly rotate the scapula. In repeating the movement five or six times, however, the muscle fatigued, and the ability to raise the arm above shoulder level decreased.

Subjects without any serratus anterior weakness show a wide range of strength in the lower and middle trapezius. This variation in strength is associated with postural or occupational stress on these muscles. The grade of strength will range from fair to normal. Because of these wide differences, variations are also found in the ability to raise an arm overhead among those who develop marked weakness or isolated paralysis of the serratus. If an individual already has marked trapezius weakness of a postural or an occupational nature and, subsequently, incurs paralysis of the serratus, that person will not be able to raise the arm overhead as in Figure 6-88.

Aside from the usual distribution, it has been necessary, at times, to have a separate category for persons who exhibit good strength through part of the range of motion of abduction while attempting to support the weight of the arm in flexion. The scapula can be passively brought forward into the test position by pulling the arm diagonally upward and forward, but it immediately slips back as the subject attempts to hold the arm in test position. This weakness can best be described as a stretch weakness of the serratus. Stretching that has taken place is graphically illustrated in Figure 6-89A–F. Invariably, those who fall into this special category are persons who have engaged in many push-ups, bench presses, or activities involving strong rhomboid action. A person may start doing push-ups properly, but when the serratus fatigues, the scapulae remain adducted and the push-up is continued by the action of the pectoralis major and the triceps—to the detriment of the serratus.

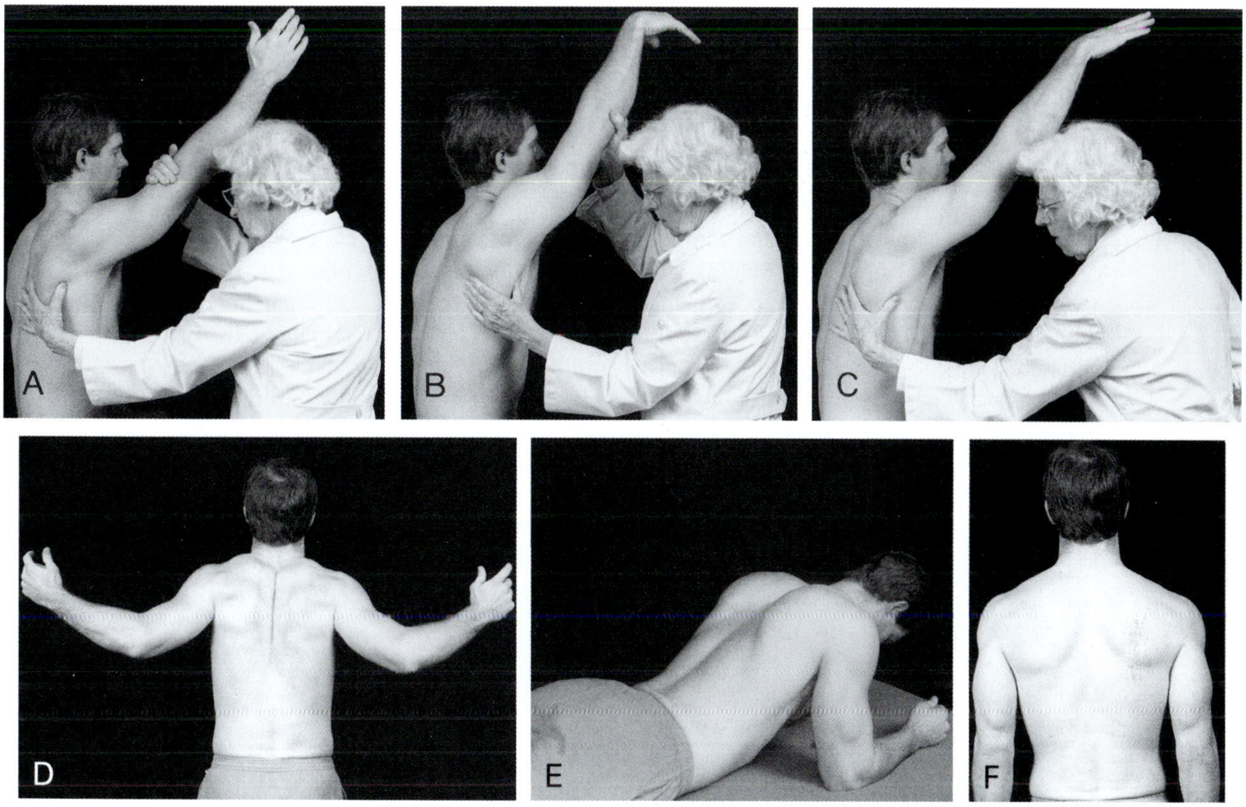

FIGURE 6-89: A–F, Stretch weakness of the serratus.

Part A: When the arm is raised into shoulder flexion, to position the scapula for the serratus test, the scapula does not move to the normal position of abduction. However, the serratus appears to test strong in that position (potential compensation due to over development of shoulder flexors). Part F below shows the same subject. The winging of the scapula clearly indicates weakness of the serratus anterior

Part B: The scapula can be brought forward to almost normal abduction if the subject relaxes the weight of the arm and allows the examiner to draw the arm diagonally forward into the test position.

Part C: The scapula cannot hold the abducted and upwardly rotated position when the examiner releases the arm, and the subject attempts to hold it in position.

Part D: This subject has routinely performed both bench presses and shoulder adduction exercises, including seated rowing and bent-over rowing with heavy weights. As seen in the photographs (parts D–F), the rhomboids have become overdeveloped. The rhomboids are antagonists to the serratus, and this type of exercise is not recommended in the presence of serratus weakness.

Part E: In a prone position, resting on the forearms, winging of the scapulae is observed. This patient's serratus is unable to hold the abducted position against resistance offered by the weight of the trunk in this position.

Part F: This photograph shows the abnormal position that the scapulae assume at rest.

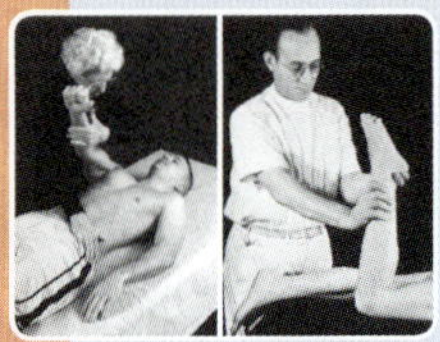

CLASSIC KENDALL

Cases of Serratus Anterior Paralysis

During a time of hospital affiliation, the Kendalls examined and treated numerous cases of serratus anterior paralysis. Depending on the etiology, some patients had pain associated with the paralysis, but not in the area of the muscle itself. Additionally, some patients did not complain of pain before, during, or for a while after the onset of paralysis. Early complaints were about the inability to use the arm normally. In some cases, when the onset was gradual, patients made no complaints until weakness became more and more pronounced. When the effects of serratus weakness created secondary problems involving other structures, patients complained of pain or discomfort in areas other than that of the serratus muscle, such as the neck or shoulder. Significant to such history is the fact that *the long thoracic nerve to the serratus is purely motor*. (See also Appendix D.)

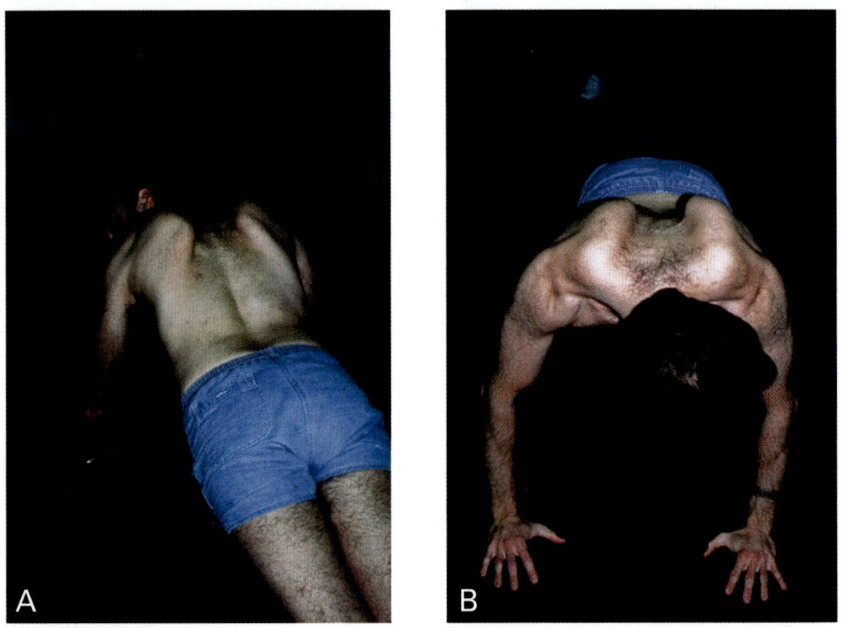

FIGURE 6-90: A, B, Subject performed, without complaint, a push-up despite extreme weakness of the serratus anterior.

Figure 6-90A and B shows two views of the same subject. He performed a push-up in spite of extreme weakness of the serratus anterior and without complaint of pain.

> **NOTE:** *See the earlier section in the chapter regarding muscles that are supplied by nerves that are motor only.*

PAINFUL CONDITIONS OF THE UPPER BACK

Management of painful conditions of the upper extremity requires careful evaluation, including a detailed history and objective observation and testing. Although it is necessary to test for both range of motion and strength before establishing a diagnosis, pain management, through support and protection of the injured or painful part, is a priority. Understanding that the onset of pain may be delayed in conditions involving nerves whose primary function is motor but may have a sensory cutaneous role, as well, is an important consideration relating to the duration of the problem.

Pain may occur both in and around joints as a result of changes in alignment of the scapula and shoulder girdle. Alternatively, pain may be most pronounced in the area of muscle attachments to bone.

The loss of normal movement in one area may result in excessive movement in another. Whatever the cause of related pain, the treatment of choice is restoration of muscle balance to facilitate normal movement, both through stretching tight muscles and strengthening weak muscles and through use of supports when indicated.

Weakness of Upper Back

Weakness of the upper back erector spinae can develop as the shoulders slump forward and the thoracic kyphosis increases. If the back has not become fixed in this faulty position, exercises are indicated to help strengthen the upper back extensors and to stretch the opposing anterior trunk muscles if they have begun to shorten. Proper shoulder supports are indicated when the muscles are very weak (44, 45).

The middle and lower portions of the trapezius muscles reinforce the upper back extensors and help to hold the shoulders back. The manner in which these muscles are exercised is very important. (The wall-sitting and wall-standing exercises are illustrated in Chapter 2, Figures 2-60 and 2-51, and later in this chapter.)

It is necessary to check whether opposing tightness limits the range of motion before attempting the exercises. Tests for length of the latissimus dorsi, teres major, pectoralis major, and pectoralis minor should be performed. Tightness in the upper anterior abdominal muscles and restriction of chest expansion will also interfere with efforts to straighten the upper back.

As a general rule, exercises for the rhomboid muscles are not indicated. Although these muscles pull the shoulders back, they do so in a manner that elevates the shoulder girdle and tends to tip the scapula forward in a faulty postural position. Additionally, the rhomboids are usually strong.

Short Rhomboids

Rhomboids may shorten as a result of frequent repetitive forceful exercises in the direction of adduction, elevation, and downward rotation of the scapula. They also may shorten as a result of weakness or paralysis of the serratus anterior (42), which is a direct opponent of the rhomboids. Treatment by massage and stretching of the rhomboids would be indicated (42). Placing the shoulder in flexion normally produces scapular abduction and upward rotation. When the rhomboids are shortened, it is difficult to obtain this scapular position merely by positioning the upper extremity. To lengthen the rhomboids, it is necessary to apply pressure against the vertebral border of the scapula in the direction of scapular abduction and upward rotation.

Middle and Lower Trapezius Strain

Middle and lower trapezius strain is a prevalent, painful condition that may be acute in nature but is often chronic, resulting from tension on the middle and lower trapezius muscles. It does not have an acute onset unless associated with injury, but chronic symptoms may become very painful. Symptoms of pain do not appear early. Weakness may be present for some time without many complaints. It appears, however, that complaints of pain are associated with traction by the muscle on its bony attachments along the spine. Patients may complain of a focal area of pain, or palpation may elicit pain or acute tenderness, in the areas of vertebral or scapular attachments of the middle and lower trapezius.

The stretch weakness of the muscles that precedes the chronic muscle strain may result from a habitual position of forward shoulders, thoracic hyperkyphosis, or the combination of these two faults. It also may result from the shoulders being pulled forward by overdeveloped, short anterior shoulder-girdle muscles. Repetitive movements associated with some sports, such as baseball, may contribute to overdevelopment of shoulder adductor muscles. Occupations that require continuous movement with the arms in a forward position, as in using a computer or piano playing, contribute to tension of the trapezius muscles.

Some occupations require sustained positions for extended periods of time. An example is the dentist bending forward over the patient, putting tension on the upper back muscles and stress on the anterior surfaces of the bodies of the thoracic vertebrae (Figure 6-91).

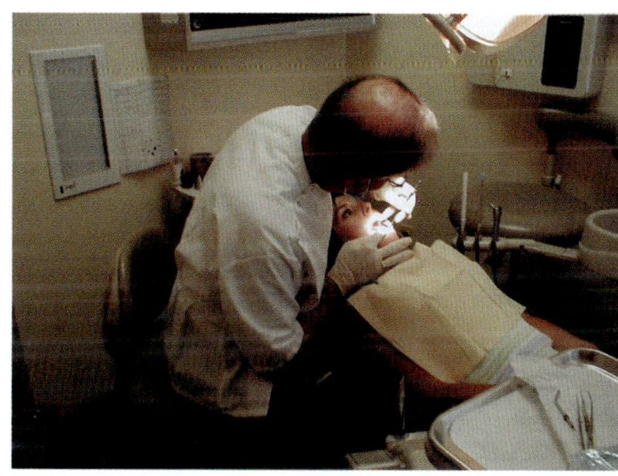

FIGURE 6-91: This dentist stands for long periods of time, putting tension on the upper back muscles and stress on the anterior surfaces of the bodies of the thoracic vertebrae.

For some individuals, recumbency or change of sitting posture may remove the element of continuous tension on the trapezius, but in individuals with tight shoulder adductors and coracoclavicular fascia, tension is continuously present. Change of position does not change the alignment of the part when such tightness exists. Pain is relieved very little—if at all—by recumbency.

Tests for length of the shoulder adductors and internal rotators should be done to determine whether tightness exists. If tightness is present, gradual stretching of the tight muscles and fascia is indicated. Some effective relief of pain should be achieved in a short time if gentle treatment is given daily.

With marked weakness of the middle and lower trapezius, regardless of whether opposing tightness exists, a shoulder support is often indicated. Such a support can effectively assist in the effort to hold the shoulders back in a position that relieves tension on the muscles.

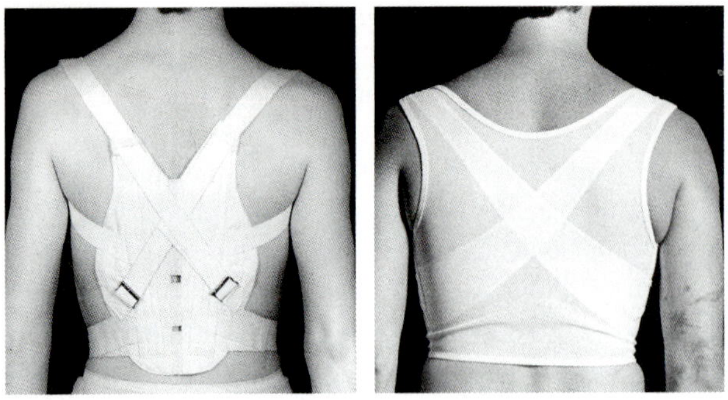

FIGURE 6-92: Shoulder supports: **A,** Stays in back to support upper back and hold shoulders back; **B,** Elastic, vest-type support to hold shoulders back.

Middle and Upper Back Pain due to Osteoporosis

Thoracic hyperkyphosis is a primary deformity in osteoporosis, usually accompanied by compensatory extension of the cervical spine. Complaints of back pain are common and can best be treated by gentle efforts to reduce the postural deformity and prevent further progression before it becomes a fixed structural fault. If a support can be tolerated, a patient could use one to help maintain the best possible alignment (Figure 6-92). Exercises should be done as tolerated to help maintain functional range of motion and develop strength (46).

For patients (usually older adults) with a fixed hyperkyphosis of the thoracic spine, little correction can be obtained. Some correction of the forward shoulders may be possible, but the basic faults cannot be altered. A Taylor-type brace may be used to prevent progression of the deformity and to give some relief from painful symptoms (47).

Subjects with a thoracic hyperkyphosis often develop symptoms in the posterior neck. As the thoracic spine flexes into a hyperkyphosis, the head is carried forward, the eyes seek eye level to preserve the erect position of the head, and the upper cervical spine is hyperextended. Symptoms associated with this problem are described under *Tightness of Posterior Neck Muscles* in Chapter 4.

The subject in Figure 6-93A–E exhibits a posture typical of osteoporosis—thoracic hyperkyphosis, posterior pelvic tilt with protruding abdomen, and compensatory cervical hyperextension. Because the deformity was still somewhat flexible, correction was achieved by connecting an upper back support with posterior stays to a panty brief by using soft strapping and Velcro. This provides support to the thoracic spine in standing and sitting, with improved head and neck alignment.

A vest-type upper back posture support can also be effective in improving alignment (Figure 6-93F, G).

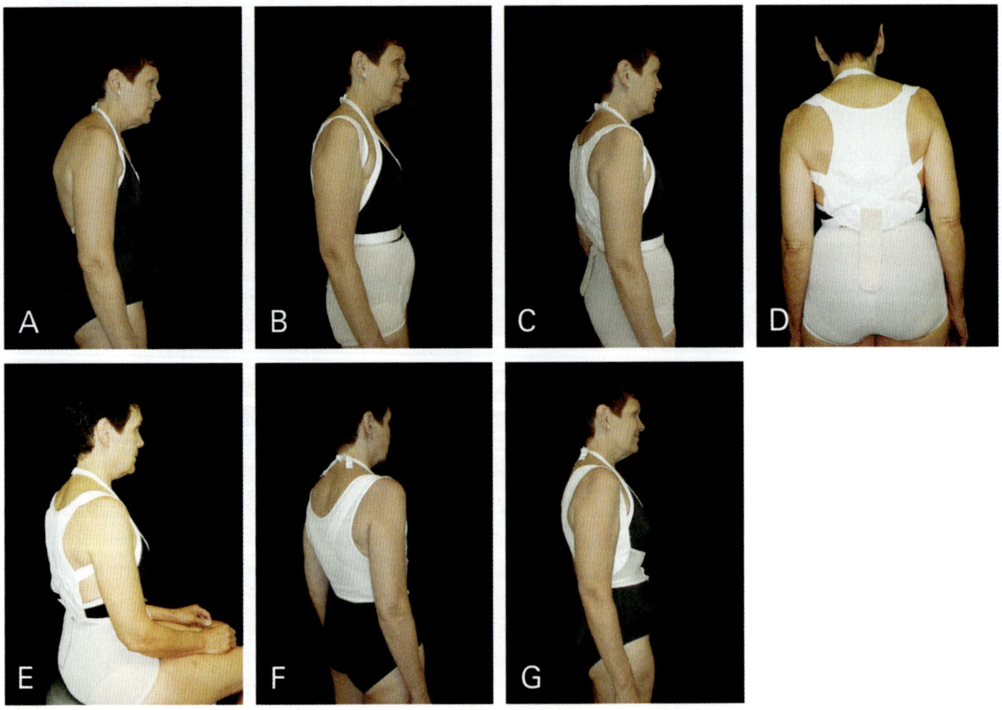

FIGURE 6-93: **A,** Typical posture seen with osteoporosis; **B–G,** Posture correction of osteoporosis.

Scheuermann's Disease

Scheuermann's disease (Scheuermann's kyphosis, juvenile kyphosis) is a degenerative process of the vertebra at the interface of the disc and vertebral body. Defects in the endplate, due to impaired blood flow to the cartilage that does not allow it to mature to bone, disrupt the growth plate anteriorly, which leads to vertebral collapse anteriorly, making the vertebra wedge-shaped. Pain usually starts a few years after deformity becomes visible. Patients complain of slowly increasing pain brought on by heavy labor or simply as the day progresses. Patients frequently complain of fatigue in the mid-back. Thoracolumbar problems can develop later in life as a result of overcompensation (48).

PAINFUL CONDITIONS OF THE UPPER EXTREMITY

Localized or radiating pain in the upper extremity is often the result of faulty alignment that causes compression or tension on nerves, blood vessels, or supporting soft tissues. Faulty alignment may be primarily in the cervical region, thoracic region, or the shoulder girdle. All three areas may be involved, as well, and treatment must be directed toward overall correction.

Under normal conditions and through normal range of motion, it may be presumed that a muscle will not irritate a nerve that lies in close proximity to or pierces the muscle. A muscle that is drawn taut, however, becomes firm and has the potential for exerting a force of compression or friction. A muscle that has developed adaptive shortness moves through less range and becomes taut before reaching normal length; a lengthened muscle moves through more than normal range before becoming taut. A taut muscle, especially a weight-bearing muscle, can cause friction on a nerve during repetitive movements.

In mild cases, the symptoms may be discomfort and dull ache rather than sharp pain when the muscles contract or elongate. Sharp pain may be elicited by vigorous movements. More often, however, it tends to be intermittent, because the subject finds ways to avoid the painful movements.

Recognizing this phenomenon during the early stages can increase the likelihood of finding ways to counteract or prevent the more painful or disabling problems that develop later. Physical therapists who implement stretching and strengthening exercises have the opportunity to observe early signs of impingement among their patients. Examples of such impingement include:

- Teres major with the axillary nerve
- Supinator with the radial nerve (15), (49)
- Pronator with the median nerve (12), (15), (49)
- Flexor carpi ulnaris with the ulnar nerve (50)
- Lateral head of the triceps with the radial nerve (15), (49)
- Trapezius with the greater occipital nerve (50)
- Scalenus medius with C5 and C6 root of the plexus and the long thoracic nerve (50)
- Coracobrachialis with the musculocutaneous (12), (15)

Guyon's Tunnel Syndrome

Guyon's tunnel syndrome (GTS) results from injury to the ulnar nerve at Guyon's canal. The canal is bordered by the palmar carpal ligament, transverse carpal ligament, hook of hamate, and pisiform. Injury to the ulnar nerve at Guyon's canal may result from direct trauma (i.e., hook of hamate fractures) or repetitive activity, compression from a sustained posture (i.e., use of dropdown handlebars when cycling), space-occupying cysts/tumors, inflammation, or more anatomically proximal factors (51, 52).

The ulnar nerve is a mixed sensory and motor nerve as it enters the canal. For this reason, patients may present with pain and/or paresthesia in the ulnar nerve distribution of the hand and digits as well as motor loss and atrophy in the hypothenar muscles. Clawing of the fourth and fifth digits may be observed. Motor/sensory loss presentation is dependent upon the site of compression. If the clinical presentation and history suggest GTS as a differential diagnosis, further testing of the muscles innervated by the ulnar nerve proximal and distal to Guyon's canal as well as assessment of dermatomal involvement would be indicated to confirm (51, 52).

Carpal Tunnel Syndrome

Carpal tunnel syndrome (CTS) results from compression of the median nerve under the flexor retinaculum of the wrist. Patients may be predisposed to onset due to past medical history, which may include injury or previous neurodynamic impairments, but any acute trauma, injury, or condition (e.g., tendinopathy, inflammatory disorder) that decreases space of the carpal tunnel can be the root cause as well as joint instability. Cumulative trauma, frequent use of vibratory tools, and sustained postures (i.e., wrist flexion) are possible etiologies, as well.

Patients may present with pain and/or paresthesia of the radial hand and digits as well as motor loss and atrophy in the thenar muscles, impacting grip ability and strength. Symptoms can radiate to the wrist and on occasion as far distal as the arm and shoulder. Waking at night with paresthesia is not uncommon. Symptoms can be produced with Phalen's test, carpal compression, Tinel's test, and adverse neural tension test (ANTT). If the clinical presentation and history suggest possible CTS as a differential diagnosis, further assessment of proximal myotomal and dermatomal presentation would be indicated to rule in or out potential proximal causes (53).

Pronator Syndrome

Pronator syndrome (PS) presents as pain in the anterior forearm with radicular paresthesia along the median nerve distribution of the hand (54). Potential points of compression distal to the antecubital fossa include the bicipital aponeurosis where it connects to the pronator teres muscle or between the two heads of the pronator teres (55). Anatomical variations such as the ligament of Struthers can also play a role in compression when present. Aggravating factors are trauma, repetitive forearm pronation, and elbow flexion movements, as well as wrist flexion (56). Symptoms can be brought on with resisted wrist flexion, forearm pronation, elbow flexion, pronator compression tests, Tinel's test, and ANTTs (54).

Cubital Tunnel Syndrome

Cubital tunnel syndrome (CBTS or CUTS) results from compression of the ulnar nerve in the cubital tunnel. The tunnel is bordered by the medial edge of the olecranon and trochlea, medial epicondyle, ulnar collateral ligament, and the cubital tunnel retinaculum (arcuate ligament). CBTS is frequently seen in patients who participate in overhead sports that require repetitive elbow flexion/extension activities like throwing. Symptoms include pain and paresthesia along the medial forearm, ulnar side of hand, fifth digit, and the ulnar half of the fourth digit. Signs of CBTS include a sensory loss over the ulnar aspect of the hand, the fifth digit, and the ulnar half of the fourth digit, and occasionally a snapping/popping of the nerve with flexion and extension of the elbow. With more advanced involvement, motor involvement may be found in the intrinsic hand muscles. Symptoms can be brought on with sustained elbow flexion (1–3 min) and ulnar never neurodynamic testing (57, 58).

Thoracic Outlet Syndrome

Thoracic outlet syndrome (TOS) results from compression of either the subclavian artery or the brachial plexus within the channel bordered by the anterior and middle scalenes. The diagnosis is often puzzling and controversial as etiology can stem from postural faults, cervical foraminal stenoses, and other syndromes such as anterior scalene, hyperabduction, costoclavicular, costodorsal outlet, pectoralis minor, and cervical disc and rib (59). In rare instances, a tumor should be considered when patients do not respond to indicated treatment as expected. Postural presentations may include forward head, thoracic hyperkyphosis, 1st rib elevation, and protracted, depressed, or downwardly rotated scapulae.

Symptoms are varied and may be neurogenic or vascular in origin. Paresthesia and diffuse aching pain over the whole arm are common. The condition is aggravated by carrying, lifting, or engaging in activities that require a sustained upper extremity position, such as playing a musical instrument.

When present, muscle atrophy usually affects all the intrinsic muscles of the hand. Tendon reflexes are not altered. Arterial compression is a less common cause than was once thought, but symptoms such as coldness, aching in the muscles, and loss of strength with continued use can reflect vascular compromise. As Dawson et al. state, "The proper diagnostic test should be the production of the neurologic symptoms of arm abduction, whether or not there is a change in the pulse or the appearance of a bruit" (15).

Coracoid Pressure Syndrome

The pectoralis minor is often cited as a potential cause of TOS symptoms, as well. However, coracoid pressure syndrome (CPS) is a condition of arm pain that involves compression of the brachial plexus that was first reported by the Kendalls in 1942. It was presented at a Joint Meeting of the Baltimore and Philadelphia Orthopedic Society, March 17, 1947, by E. David Weinberg, M.D., and was later referred to in an article by Dr. Irvin Stein (60–62).

CPS is associated with muscle imbalance and faulty postural alignment (63). At the level of attachment of the pectoralis minor to the coracoid process of the scapula, the three cords of the plexus and the axillary artery and vein pass between these structures and the rib cage (Figure 6-94). In normal alignment of the shoulder girdle, there should be no compression on the nerves or blood vessels. Forward depression of the coracoid process, which occurs in some types of faulty postural alignment, tends to narrow this space.

The coracoid process may be tilted downward and forward either because of tightness in certain muscles or because weakness of other muscles allows it to ride into that position. The painful upper extremity conditions are more often found when tightness is present.

The muscle that acts on the coracoid process anteriorly to depress the scapula is chiefly the pectoralis minor. The rhomboids and levator scapulae aid in the elevation of the scapula that goes along with the anterior tilt. Tightness of the latissimus dorsi and sternal part of the pectoralis major affects the position indirectly through its action to depress the head of the humerus. In some instances, tightness of the biceps and coracobrachialis can be a factor as well. Muscle tightness may be ascertained by the shoulder adductor and internal rotator length tests.

Weakness of the lower trapezius enables the scapula to elevate and tilt down anteriorly, creating a faulty shoulder position. This faulty shoulder position can cause a stretch weakness of the lower trapezius, which allows the scapula to then ride upward and tilt down anteriorly, as well, favoring an adaptive shortening of the pectoralis minor.

In the acute stage, moderate or even slight pressure over the coracoid process usually elicits pain down the upper extremity. Soreness is acute in the area of the coracoid process and in the area described by the pectoralis minor muscle along the chest wall.

The upper extremity pain may be generalized or predominantly of lateral or medial cord distribution. There may be tingling, numbness, or weakness in the upper extremity. The patient often complains of loss of grip strength. Evidence of circulatory congestion, with puffiness of the hand and engorgement of the blood vessels, may be present (effort vein thrombosis). In cases of marked disturbance, the hand may be somewhat cyanotic in appearance (subclavian artery). The patient will complain of increased pain when wearing a heavy overcoat, trying to lift a heavy weight, or carrying a suitcase with that arm. Pressure can also be caused by a backpack or a shoulder bag.

Frequently, the area from the occiput to the acromion process, which corresponds to the upper trapezius muscle, is sensitive and painful. This muscle may be in a state of protective spasm in an effort to lift the weight of the shoulder girdle and thus relieve pressure on the plexus. The muscle tends to remain in a state of contraction unless effective treatment is instituted.

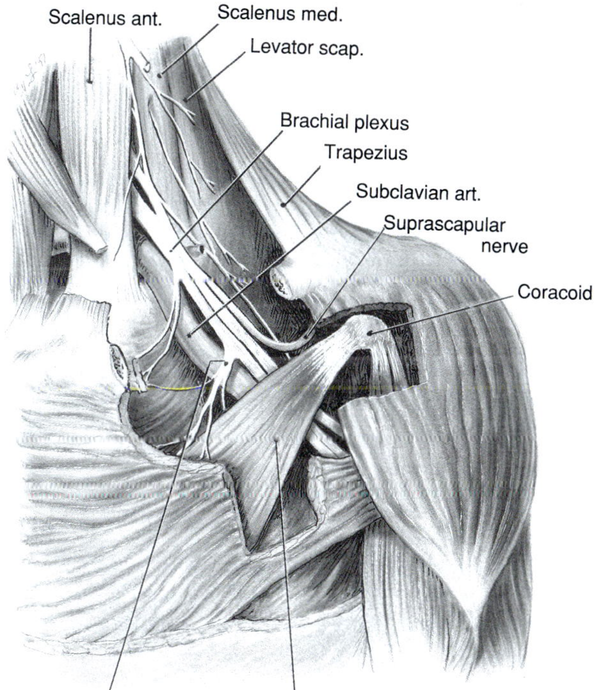

FIGURE 6-94: Regions affected by coracoid pressure syndrome.

Teres Syndrome (Quadrilateral Space Syndrome)

Teres syndrome was described in *Posture and Pain* in 1952 (63). A book published in 1980 contains a very interesting discussion of this syndrome in which it is called quadrilateral space syndrome (64).

The quadrilateral (or quadrangular) space in the axilla is bounded by the teres major, teres minor, long head of the triceps, and humerus (Figure 6-95). The axillary nerve and posterior circumflex artery emerge through this space to supply the deltoid and teres minor. The area of sensory distribution of the cutaneous branch of the axillary nerve is shown in Figure 6-4.

This syndrome is characterized by shoulder pain and limitation of shoulder joint motion, particularly rotation and abduction (65). Pain extends into the area of cutaneous distribution of the sensory branch of the axillary nerve. Tenderness may be elicited by palpation of the quadrilateral space between the teres major and teres minor. A slight or moderate pressure over the space may elicit sharp pain radiating into the area of the deltoid muscle.

Pain that is more marked during active motion indicates friction on the axillary nerve by the teres muscle. The teres major, when tight, holds the humerus in internal rotation, which creates tension on the posterior cord and axillary nerve. Internal or external rotation, whether done actively or passively, is painful. With limitation of external rotation, abduction movements are also painful because the humerus does not rotate outward as it normally should during abduction.

When stretching a tight teres major, a patient may complain of a shooting pain in the area of cutaneous

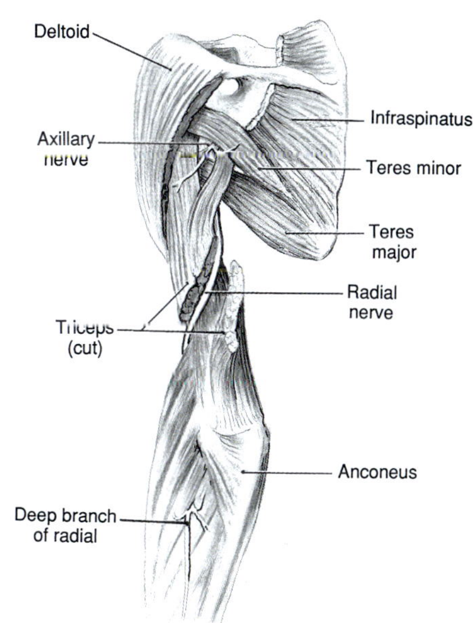

FIGURE 6-95: Regions affected by teres syndrome (quadrilateral space syndrome).

sensory distribution of the axillary nerve. The assumption is that the axillary nerve is being compressed or stretched against the tight teres major. The pain that results from direct irritation to the nerve is in contrast with the discomfort that is often associated with the usual stretching of tight muscles, and is not unlike the pain encountered in cases of subdeltoid bursitis.

Tight Shoulder External Rotators

There may be significant differences in range of motion depending on a person's occupation. According to one source, "major league pitchers have different ranges of motion for each shoulder. In the pitching arm, with the shoulder in abduction, there are 11 degrees less extension, 15 degrees less internal rotation, and 9 degrees more external rotation" (66).

Cervical Rib

A cervical rib is a rare, congenital, bony abnormality that may—or may not—give rise to symptoms of nerve irritation (67).

A painful arm condition appearing in a young or middle-aged adult can be related to the presence of a cervical rib. The posture of the individual with a cervical rib often determines whether painful symptoms will occur. The appearance of symptoms only after the person has reached adulthood may be explained by the fact that the posture of the individual has gradually become more faulty in alignment, thus causing the relationship of the rib and the adjacent nerve trunks to change unfavorably.

The faulty alignment most likely to cause irritation is the type characterized by a hyperkyphosis and forward head. Care of a patient with painful symptoms resulting from a cervical rib requires postural correction of the upper back and neck. This treatment may relieve the symptoms completely and obviate the need for a surgical procedure.

USE OF CHARTS FOR DIFFERENTIAL DIAGNOSIS

The authors of previous editions of this text included case studies using charts to assist with assessment of muscle and neural involvement based upon distribution of muscles presenting as weak (Figures 6-96 to 6-101). Samples of these charts can be found in Appendix D for review.

The following pages include actual case studies completed by the previous authors. The forms were and still could be used to summarize what is described in the following paragraphs. The forms provide a visual summation of the clinical information intake and reasoning any clinician could and should be doing at the time of an assessment.

Muscle strength grades are recorded in the column to the left of the muscle names. The grades may be in numeral or letter symbols. Either system may be used and grades can be translated as indicated in Table 1-4, the Key to Grading Symbols.

After the grades have been recorded, the nerve involvement is plotted, when applicable, by circling the dot(s) under peripheral supply and the number(s) under the spinal segment distribution that corresponds with each involved muscle.

The involvement of peripheral nerves and/or parts of the plexus is ascertained from the encircled dots by following the vertical lines upward to the top of the chart, or the horizontal lines to the left margin. Where there is evidence of involvement at spinal segment level, the level of lesion may be indicated by a heavy black line drawn vertically to separate the involved from the uninvolved spinal segments.

As a rule, muscles graded good (9) and above may be considered as not being involved from a neurological standpoint. This degree of weakness may be the result of such factors as inactivity, stretch weakness, or lack of fixation by other muscles. It should be borne in mind, however, that a grade of good might indicate a deficit of a spinal segment that minimally innervates the muscle.

Weakness with grades of fair or less may occur as a result of inactivity, disuse atrophy, immobilization, or from neurological problems. Faulty posture of the thoracic spine and shoulders may cause weakness of middle and lower trapezius.

It is not uncommon to find bilateral weakness of these muscles with grades as low as fair−. It is unlikely that there is a neurological problem with involvement of the spinal accessory nerve in cases of isolated weakness of these muscles unless there is involvement of the upper trapezius also.

The use of the *Spinal Nerve and Muscle Chart* (Chapter 1) is illustrated by the six case studies that follow, which are examples of different neuromuscular problems. The subjects were referred to the Kendalls for manual muscle testing to aid in establishing a diagnosis. They were not seen for follow-up treatment. The results of the manual muscle testing, recorded on the Spinal Nerve and Muscle Chart, became an important aid at that time in determining the extent and level of lesion.

CASE 1: RADIAL NERVE LESION

Neck, Diaphragm, and Upper Extremity

Name | Date

RIGHT

PERIPHERAL NERVES — KEY → D. = Dorsal Prim. Ramus; V. = Vent. Prim. Ramus; P.R. = Plexus Root; S.T. = Superior Trunk; P. = Posterior Cord; L. = Lateral Cord; M. = Medial Cord

Peripheral nerve columns (key; cervical segments; T.; nerve): D., 1-8, 1, Cervical | V., 1-8, Cervical | V., 1-4, Cervical | V., 3, 4, 5, Phrenic | P.R., 5, 6, 7, (8), Long Thor. | P.R., 4, 5, Dor. Scap | S.T., 5, 6, N. to Subcl. | S.T., 4, 5, 6, Suprascap | P., (4), 5, 6, (7), U. Subscap. | P., (5), 6, 7, 8, Thoracodor | P., 5, 6, (7), L. Subscap. | L., 5, 6, 7, Lat. Pect. | M., (6), 7, 8, 1, Med. Pect. | P., 5, 6, Axillary | L., (4), 5, 6, 7, Musculocu. | P., 5, 6, 7, 8, 1, Radial | L.M., 5, 6, 7, 8, 1, Median | M., 7, 8, 1, Ulnar

	MUSCLE STRENGTH GRADE	MUSCLE	Peripheral nerve (●)	C1	C2	C3	C4	C5	C6	C7	C8	T1
		HEAD & NECK EXTENSORS	Cervical (D.)	1	2	3	4	5	6	7	8	1
Cervical nerves		INFRAHYOID MUSCLES	Cervical (V. 1-4)	1	2	3						
Cervical nerves		RECTUS CAP ANT. & LAT.	Cervical (V. 1-4)	1	2							
Cervical nerves		LONGUS CAPITIS	Cervical (V. 1-4)	1	2	3	(4)					
Cervical nerves		LONGUS COLLI	Cervical (V. 1-8)		2	3	4	5	6	(7)		
Cervical nerves		LEVATOR SCAPULAE	Cervical (V. 1-4); Dor. Scap			3	4	5				
Cervical nerves		SCALENI (A. M. P.)	Cervical (V. 1-8)			3	4	5	6	7	8	
Cervical nerves		STERNOCLEIDOMASTOID	Cervical (V. 1-4)	(1)	2	3						
Cervical nerves		TRAPEZIUS (U. M. L.)	Cervical (V. 1-4)		2	3	4					
Cervical nerves		DIAPHRAGM	Phrenic			3	4	5				
Brachial Plexus – Root		SERRATUS ANTERIOR	Long Thor.					5	6	7	8	
Brachial Plexus – Root		RHOMBOIDS MAJ & MIN	Dor. Scap				4	5				
Brachial Plexus – Trunk		SUBCLAVIUS	N. to Subcl.					5	6			
Brachial Plexus – Trunk		SUPRASPINATUS	Suprascap				4	5	6			
Brachial Plexus – Trunk		INFRASPINATUS	Suprascap				(4)	5	6			
Brachial Plexus – P Cord		SUBSCAPULARIS										
Brachial Plexus – P Cord		LATISSIMUS DORSI										
Brachial Plexus – P Cord		TERES MAJOR										
Brachial Plexus – L		PECTORALIS MAJ (UPPER)										
Brachial Plexus – M&L		PECTORALIS MAJ (LOWER)										
Brachial Plexus – M&L		PECTORALIS MINOR										
Axil.		TERES MINOR										
Axil.		DELTOID										
Musculocutan		CORACOBRACHIALIS	Musculocu.						6	7		
Musculocutan		BICEPS	Musculocu.					5	6			
Musculocutan		BRACHIALIS	Musculocu.					5	6			
Radial	10	TRICEPS	Radial						6	7	8	1
Radial	10	ANCONEUS	Radial							7	8	
Radial – Lat.M	—	BRACHIALIS (SMALL PART)	Radial					5	6			
Radial – Lat.M	0	BRACHIORADIALIS	Radial (circled)					5	6			
Radial – Lat.M	0	EXT CARPI RAD L	Radial (circled)					5	6	7	8	
Radial – Post Inter	0	EXT CARPI RAD B	Radial (circled)						6	7	(8)	
Radial – Post Inter	0	SUPINATOR	Radial (circled)					5	6	(7)		
Radial – Post Inter	0	EXT DIGITORUM	Radial (circled)						6	7	8	
Radial – Post Inter	0	EXT DIGITI MINIMI	Radial (circled)						6	7	8	
Radial – Post Inter	0	EXT CARPI ULNARIS	Radial (circled)						6	7	8	
Radial – Post Inter	0	ABD POLLICIS LONGUS	Radial (circled)						6	7	8	
Radial – Post Inter	0	EXT POLLICIS BREVIS	Radial (circled)						6	7	8	
Radial – Post Inter	0	EXT POLLICIS LONGUS	Radial (circled)						6	7	8	
Radial – Post Inter	0	EXT INDICIS	Radial (circled)						6	7	8	
Median	10	PRONATOR TERES	Median						6	7		
Median	10	FLEX CARPI RADIALIS	Median						6	7	8	
Median	10	PALMARIS LONGUS	Median						(6)	7	8	1
Median	10	FLEX DIGIT SUPERFICIALIS	Median							7	8	1
Median – A Inter	10	FLEX DIGIT PROF I & II	Median							7	8	1
Median – A Inter	10	FLEX POLLICIS LONGUS	Median						(6)	7	8	1
Median – A Inter	10	PRONATOR QUADRATUS	Median							7	8	1
Median	10	ABD POLLICIS BREVIS	Median						6	7	8	1
Median	10	OPPONENS POLLICIS	Median						6	7	8	1
Median	10	FLEX POLL BREV (SUP. H)	Median						[illegible]	7	[illegible]	1
Median	*	LUMBRICALES I & II	Median						(6)	7	8	1
Ulnar	10	FLEX CARPI ULNARIS	Ulnar							7	8	1
Ulnar	10	FLEX DIGIT. PROF. III & IV	Ulnar							7	8	1
Ulnar	—	PALMARIS BREVIS	Ulnar							(7)	8	1
Ulnar	10	ABD DIGITI MINIMI	Ulnar							(7)	8	1
Ulnar	10	OPPONENS DIGITI MINIMI	Ulnar							(7)	8	1
Ulnar	10	FLEX DIGITI MINIMI	Ulnar							(7)	8	1
Ulnar	10	PALMAR INTEROSSEI	Ulnar								8	1
Ulnar	10	DORSAL INTEROSSEI	Ulnar								8	1
Ulnar	*	LUMBRICALES III & IV	Ulnar							(7)	8	1
Ulnar	10	ADDUCTOR POLLICIS	Ulnar								8	1
Ulnar	10	FLEX POLL BREV. (DEEP H.)	Ulnar								8	1

Case 1: Radial nerve lesion below the level of the branches to the triceps following a fracture of the humerus. Initially, the triceps was weak, but recozvery was complete.

SENSORY

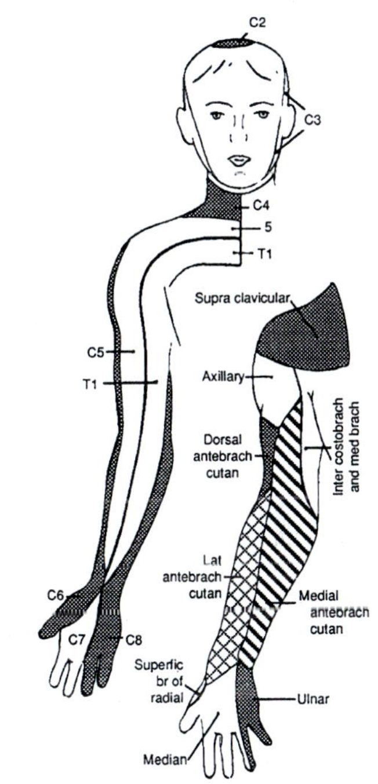

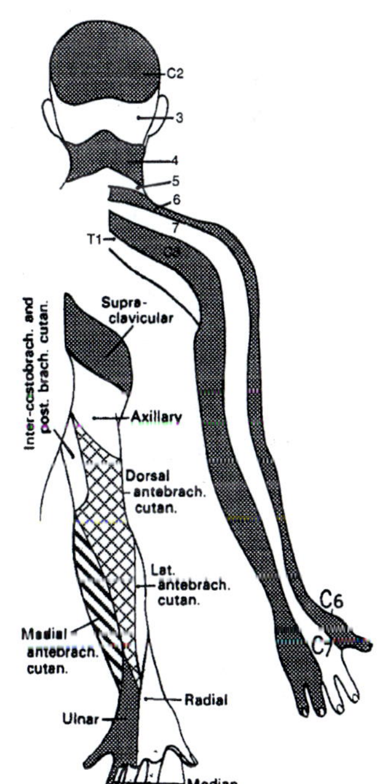

Dermatomes redrawn from Keegan and Garrett Anat Rec 102. 409. 437. 1948
Cutaneous Distribution of peripheral nerves redrawn from *Gray's Anatomy of the Human Body*. 28th ed

FIGURE 6-96: Case 1: Radial nerve lesion. © 2005 Florence P. Kendall.

CASE 2: RADIAL, MEDIAN, AND ULNAR NERVE LESION

Neck, Diaphragm, and Upper Extremity

Name Date

PERIPHERAL NERVES

KEY →
D. = Dorsal Prim. Ramus
V. = Vent. Prim. Ramus
P.R. = Plexus Root
S.T. = Superior Trunk
P. = Posterior Cord
L. = Lateral Cord
M. = Medial Cord

Nerve	Key	Roots
Cervical	D.	Cervical 1-8, T. 1
Cervical	V.	1-8
Cervical	V.	1-4
Phrenic	V.	3, 4, 5
Long Thor.	P.R.	5, 6, 7, (8)
Dor. Scap	P.R.	4, 5
N. to Subcl.	S.T.	5, 6
Suprascap	S.T.	4, 5, 6
U. Subscap.	P.	(4), 5, 6, (7)
Thoracodor	P.	(5), 6, 7, 8
L. Subscap.	P.	5, 6, (7)
Lat. Pect.	L.	5, 6, 7
Med. Pect.	M.	(6), 7, 8, 1
Axillary	P.	5, 6
Musculocu.	L.	(4), 5, 6, 7
Radial	P.	5, 6, 7, 8, 1
Median	L.M.	5, 6, 7, 8, 1
Ulnar	M.	7, 8, 1

LEFT

	MUSCLE STRENGTH GRADE	MUSCLE	PERIPHERAL NERVES	C1	C2	C3	C4	C5	C6	C7	C8	T1
		HEAD & NECK EXTENSORS	Cervical (D.) •	1	2	3	4	5	6	7	8	1
Cervical nerves		INFRAHYOID MUSCLES	Cervical (V. 1-4) •	1	2	3						
Cervical nerves		RECTUS CAP ANT. & LAT.	Cervical (V. 1-4) •	1	2							
Cervical nerves		LONGUS CAPITIS	Cervical (V. 1-4) •	1	2	3	(4)					
Cervical nerves		LONGUS COLLI	Cervical (V. 1-8) •		2	3	4	5	6	(7)		
Cervical nerves		LEVATOR SCAPULAE	Cervical (V. 1-4) •, Dor. Scap •			3	4	5				
Cervical nerves		SCALENI (A. M. P.)	Cervical (V. 1-8) •			3	4	5	6	7	8	
Cervical nerves		STERNOCLEIDOMASTOID	Cervical (V. 1-4) •	(1)	2	3						
Cervical nerves		TRAPEZIUS (U. M. L.)	Cervical (V. 1-4) •		2	3	4					
Cervical nerves		DIAPHRAGM	Phrenic •			3	4	5				
Brachial Plexus – Root		SERRATUS ANTERIOR	Long Thor. •					5	6	7	8	
Brachial Plexus – Root		RHOMBOIDS MAJ & MIN	Dor. Scap •				4	5				
Brachial Plexus – Trunk		SUBCLAVIUS	N. to Subcl. •									
Brachial Plexus – Trunk		SUPRASPINATUS	Suprascap •									
Brachial Plexus – Trunk		INFRASPINATUS	Suprascap •									
Brachial Plexus – P Cord		SUBSCAPULARIS	U. Subscap. •, L. Subscap. •									
Brachial Plexus – P Cord		LATISSIMUS DORSI	Thoracodor •									
Brachial Plexus – P Cord		TERES MAJOR	L. Subscap. •									
Brachial Plexus – L		PECTORALIS MAJ (UPPER)	Lat. Pect. •									
Brachial Plexus – M&L		PECTORALIS MAJ (LOWER)	Lat. Pect. •, Med. Pect. •									
Brachial Plexus – M&L		PECTORALIS MINOR	Med. Pect. •									
Axil.		TERES MINOR	Axillary •									
Axil.		DELTOID	Axillary •									
Musculo-cutan	—	CORACOBRACHIALIS	Musculocu. •									
Musculo-cutan	N	BICEPS	Musculocu. •					5	6			
Musculo-cutan	N	BRACHIALIS	Musculocu. •					5	6			
Radial	N	TRICEPS	Radial •						6	7	8	1
Radial	N	ANCONEUS	Radial •							7	8	
Radial – Lat.M	—	BRACHIALIS (SMALL PART)	Radial •					5	6			
Radial – Lat.M	N	BRACHIORADIALIS	Radial •					5	6			
Radial – Lat.M	N	EXT CARPI RAD L	Radial •					5	6	7	8	
Radial – Post Inter	N	EXT CARPI RAD B	Radial •					5	6	7	8	
Radial – Post Inter	N	SUPINATOR	Radial •					5	6	(7)		
Radial – Post Inter	F	EXT DIGITORUM	Radial ⊙						6	7	8	
Radial – Post Inter	O	EXT DIGITI MINIMI	Radial ⊙						6	7	8	
Radial – Post Inter	O	EXT CARPI ULNARIS	Radial ⊙						6	7	8	
Radial – Post Inter	O	ABD POLLICIS LONGUS	Radial ⊙						6	7	8	
Radial – Post Inter	O	EXT POLLICIS BREVIS	Radial ⊙						6	7	8	
Radial – Post Inter	O	EXT POLLICIS LONGUS	Radial ⊙						6	7	8	
Radial – Post Inter	O	EXT INDICIS	Radial ⊙						6	7	8	
Median	G	PRONATOR TERES	Median •						6	7		
Median	N	FLEX CARPI RADIALIS	Median •						6	7	8	
Median	G	PALMARIS LONGUS	Median •						(6)	7	8	1
Median	O	FLEX DIGIT SUPERFICIALIS	Median ⊙							7	8	1
Median – A Inter	O	FLEX DIGIT PROF I & II	Median ⊙							7	8	1
Median – A Inter	O	FLEX POLLICIS LONGUS	Median ⊙						(6)	7	8	1
Median – A Inter	O	PRONATOR QUADRATUS	Median ⊙							7	8	1
Median	O	ABD POLLICIS BREVIS	Median ⊙						6	7	8	1
Median	O	OPPONENS POLLICIS	Median ⊙						6	7	8	1
Median	O	FLEX POLL BREV (SUP. H)	Median ⊙						6	7	8	1
Median	O	LUMBRICALES I & II	Median ⊙						(6)	7	8	1
Ulnar	G	FLEX CARPI ULNARIS	Ulnar •							7	8	1
Ulnar	O	FLEX DIGIT. PROF. III & IV	Ulnar ⊙							7	8	1
Ulnar	—	PALMARIS BREVIS	Ulnar ⊙							(7)	8	1
Ulnar	O	ABD DIGITI MINIMI	Ulnar ⊙							(7)	8	1
Ulnar	O	OPPONENS DIGITI MINIMI	Ulnar ⊙							(7)	8	1
Ulnar	O	FLEX DIGITI MINIMI	Ulnar ⊙							(7)	8	1
Ulnar	O	PALMAR INTEROSSEI	Ulnar ⊙								8	1
Ulnar	O	DORSAL INTEROSSEI	Ulnar ⊙								8	1
Ulnar	O	LUMBRICALES III & IV	Ulnar ⊙							(7)	8	1
Ulnar	O	ADDUCTOR POLLICIS	Ulnar ⊙								8	1
Ulnar	O	FLEX POLL BREV. (DEEP H.)	Ulnar ⊙								8	1

SENSORY

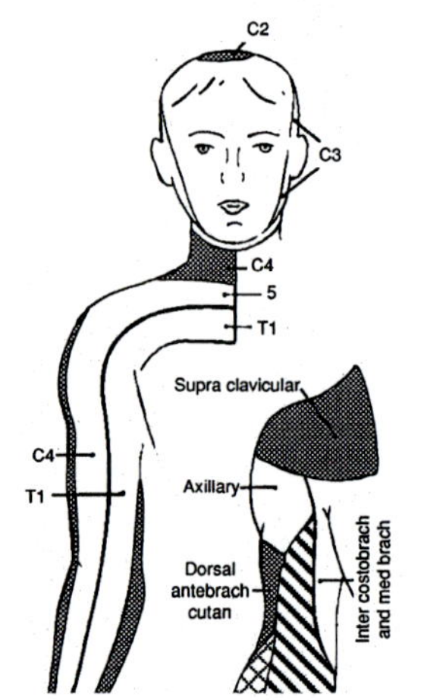

Case 2: The radial, median, and ulnar nerves are all involved at approximately the same level of the forearm, just below the elbow. (Refer to Figure 6-100.) This type of involvement may be caused by pressure from a tourniquet, bandage or a cast. The etiology is not clear-cut in this particular instance, but the history indicates that bandaging may have been a factor.

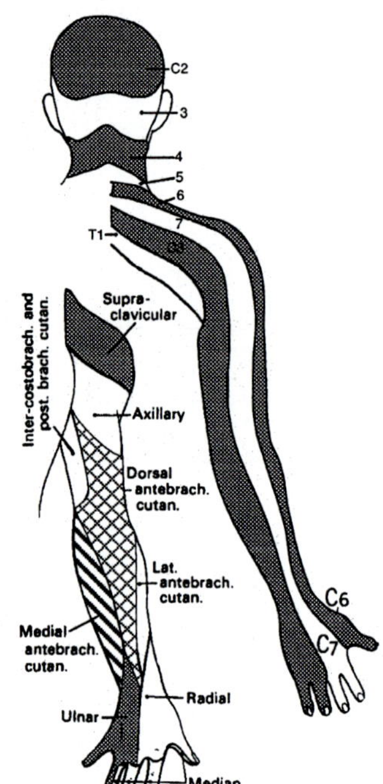

Dermatomes redrawn from Keegan and Garrett Anat Rec 102. 409. 437. 1948
Cutaneous Distribution of peripheral nerves redrawn from *Gray's Anatomy of the Human Body*. 28th ed

FIGURE 6-97: Case 2: Radial, median, and ulnar nerve lesion. © 2005 Florence P. Kendall.

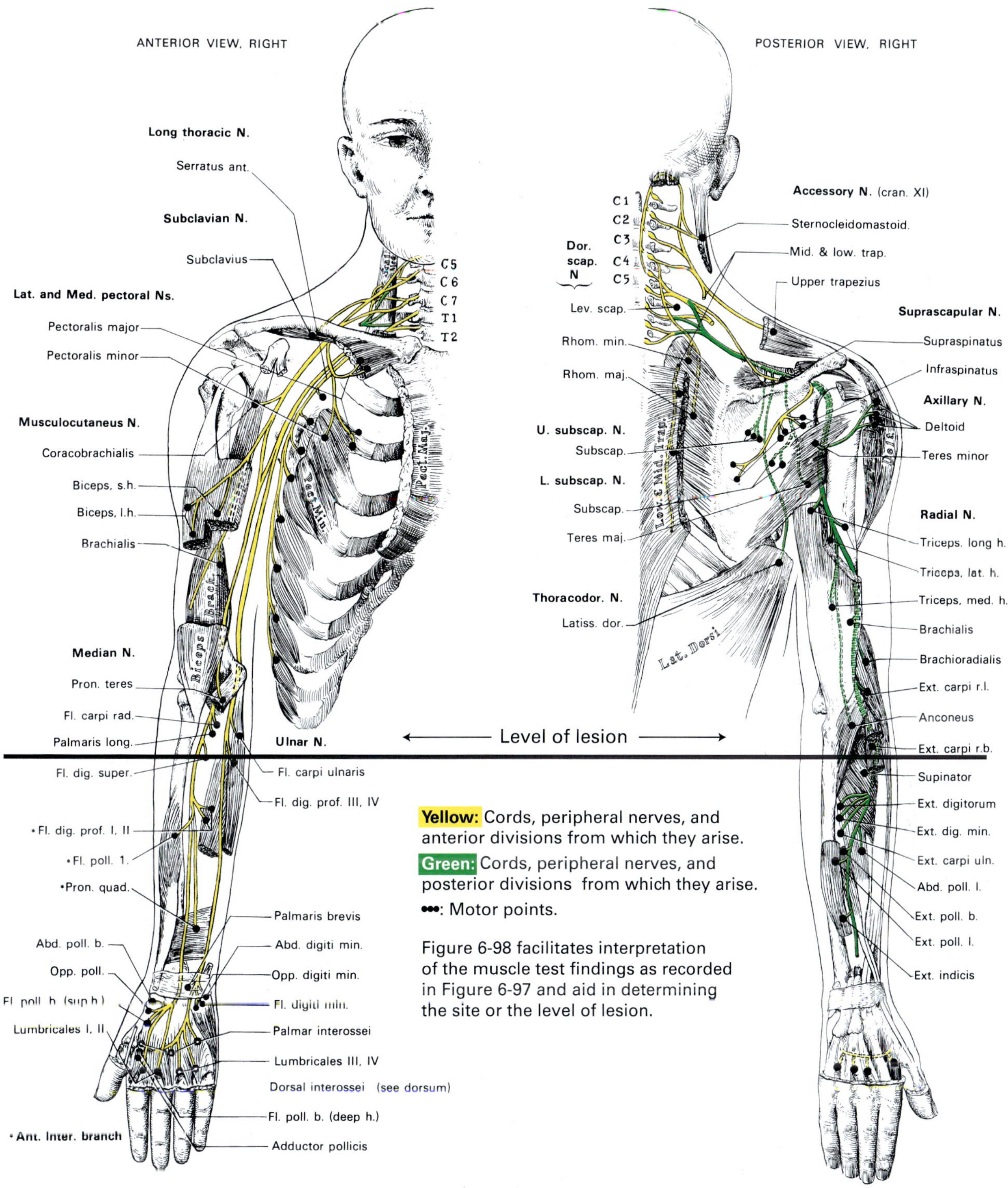

FIGURE 6-98: Determining the site or level of lesion. © 2005 Florence P. Kendall.

CASE 3: PROBABLE C5 LESION

Neck, Diaphragm, and Upper Extremity

Name Date

LEFT

KEY → D. = Dorsal Prim. Ramus; V. = Vent. Prim. Ramus; P.R. = Plexus Root; S.T. = Superior Trunk; P. = Posterior Cord; L. = Lateral Cord; M. = Medial Cord

PERIPHERAL NERVES: Cervical (D., Cervical 1-8, T. 1); Cervical (V., 1-8); Cervical (V., 1-4); Phrenic (V., 3, 4, 5); Long Thor. (P.R., 5, 6, 7, (8)); Dor. Scap (P.R., 4, 5); N. to Subcl. (S.T., 5, 6); Suprascap (S.T., 4, 5, 6); U. Subscap. (P., (4), 5, 6, (7)); Thoracodor (P., (5), 6, 7, 8); L. Subscap. (P., 5, 6, (7)); Lat. Pect. (L., 5, 6, 7); Med. Pect. (M., (6), 7, 8, 1); Axillary (P., 5, 6); Musculocu. (L., (4), 5, 6, 7); Radial (P., 5, 6, 7, 8, 1); Median (L.M., 5, 6, 7, 8, 1); Ulnar (M., 7, 8, 1)

	Muscle Strength Grade	Muscle	Peripheral nerves (•)	Spinal Segment (C1 C2 C3 C4 C5 C6 C7 C8 T1)
		HEAD & NECK EXTENSORS	Cervical (D. 1-8)	1 2 3 4 5 6 7 8 1
Cervical nerves		INFRAHYOID MUSCLES	Cervical (V. 1-4)	1 2 3
Cervical nerves		RECTUS CAP ANT. & LAT.	Cervical (V. 1-4)	1 2
Cervical nerves		LONGUS CAPITIS	Cervical (V. 1-4)	1 2 3 (4)
Cervical nerves		LONGUS COLLI	Cervical (V. 1-8)	2 3 4 5 6 (7)
Cervical nerves		LEVATOR SCAPULAE	Cervical (V. 1-4), Dor. Scap	3 4 5
Cervical nerves		SCALENI (A. M. P.)	Cervical (V. 1-8)	3 4 5 6 7 8
Cervical nerves		STERNOCLEIDOMASTOID	Cervical (V. 1-4)	(1) 2 3
Cervical nerves	N?	TRAPEZIUS (U. M. L.)	Cervical (V. 1-4)	2 3 4
Cervical nerves	*	DIAPHRAGM	Phrenic	3 4 5
Brachial Plexus – Root	G	SERRATUS ANTERIOR	Long Thor.	5 6 7 8
Brachial Plexus – Root	P	RHOMBOIDS MAJ & MIN	Dor. Scap	4 5
Brachial Plexus – Trunk	—	SUBCLAVIUS	N. to Subcl.	5 6
Brachial Plexus – Trunk	?	SUPRASPINATUS	Suprascap	4 5 6
Brachial Plexus – Trunk	T	INFRASPINATUS	Suprascap	(4) 5 6
Brachial Plexus – P Cord	—	SUBSCAPULARIS	U. Subscap., L. Subscap.	5 6 7
Brachial Plexus – P Cord	G	LATISSIMUS DORSI	Thoracodor	6 7 8
Brachial Plexus – P Cord	G	TERES MAJOR	L. Subscap.	5 6 7
Brachial Plexus – L	G-	PECTORALIS MAJ (UPPER)	Lat. Pect.	5 6 7
Brachial Plexus – M&L	G	PECTORALIS MAJ (LOWER)	Lat. Pect., Med. Pect.	6 7 8 1
Brachial Plexus – M&L	G	PECTORALIS MINOR	Med. Pect.	(6) 7 8 1
Axil.	T	TERES MINOR	Axillary	5 6
Axil.	T	DELTOID	Axillary	5 6
Musculo-cutan	—	CORACOBRACHIALIS	Musculocu.	6 7
Musculo-cutan	P+	BICEPS	Musculocu.	5 6
Musculo-cutan	P+	BRACHIALIS	Musculocu.	5 6
Radial	G+	TRICEPS	Radial	6 7 8 1
Radial	G+	ANCONEUS	Radial	7 8
Radial – Lat.M	P	BRACHIALIS (SMALL PART)	Radial	5 6
Radial – Lat.M	F-	BRACHIORADIALIS	Radial	5 6
Radial – Lat.M	G	EXT CARPI RAD L	Radial	5 6 7 8
Radial – Lat.M	G	EXT CARPI RAD B	Radial	5 6 7 8
Radial – Post Inter	F+	SUPINATOR	Radial	5 6 (7)
Radial – Post Inter	N	EXT DIGITORUM	Radial	6 7 8
Radial – Post Inter	N	EXT DIGITI MINIMI		
Radial – Post Inter	N	EXT CARPI ULNARIS		
Radial – Post Inter	N	ABD POLLICIS LONGUS		
Radial – Post Inter	N	EXT POLLICIS BREVIS		
Radial – Post Inter	N	EXT POLLICIS LONGUS		
Radial – Post Inter	N	EXT INDICIS		
Median	N	PRONATOR TERES		
Median	N	FLEX CARPI RADIALIS		
Median	N	PALMARIS LONGUS		
Median	N	FLEX DIGIT SUPERFICIALIS		
Median – A Inter	N	FLEX DIGIT PROF I & II		
Median – A Inter	N	FLEX POLLICIS LONGUS		
Median – A Inter	N	PRONATOR QUADRATUS	Median	7 8 1
Median	N	ABD POLLICIS BREVIS	Median	6 7 8 1
Median	N	OPPONENS POLLICIS	Median	6 7 8 1
Median	N	FLEX POLL BREV (SUP. H)	Median	6 7 8 1
Median	N	LUMBRICALES I & II	Median	(6) 7 8 1
Ulnar	N	FLEX CARPI ULNARIS	Ulnar	7 8 1
Ulnar	N	FLEX DIGIT. PROF. III & IV	Ulnar	7 8 1
Ulnar	—	PALMARIS BREVIS	Ulnar	(7) 8 1
Ulnar	N	ABD DIGITI MINIMI	Ulnar	(7) 8 1
Ulnar	N	OPPONENS DIGITI MINIMI	Ulnar	(7) 8 1
Ulnar	N	FLEX DIGITI MINIMI	Ulnar	(7) 8 1
Ulnar	N	PALMAR INTEROSSEI	Ulnar	8 1
Ulnar	N	DORSAL INTEROSSEI	Ulnar	8 1
Ulnar	N	LUMBRICALES III & IV	Ulnar	(7) 8 1
Ulnar	N	ADDUCTOR POLLICIS	Ulnar	8 1
Ulnar	N	FLEX POLL BREV. (DEEP H.)	Ulnar	8 1

Case 3: Muscle test findings indicate a probably C5 lesion. The test findings in this case compare very closely with those in a known C5 lesion.

SENSORY

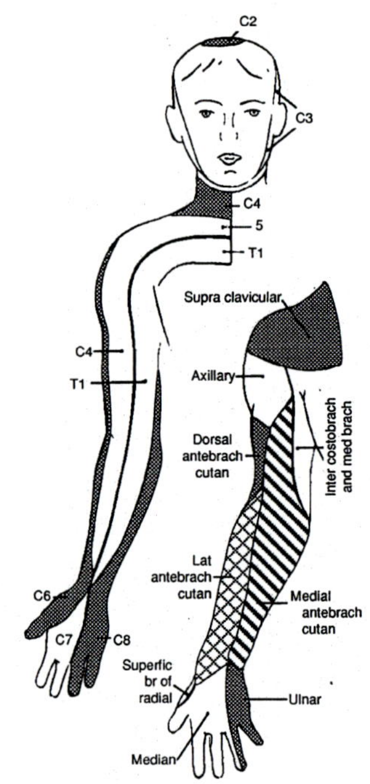

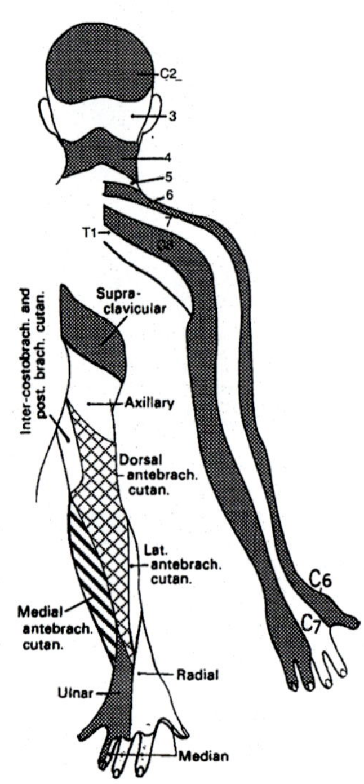

Dermatomes redrawn from Keegan and Garrett Anat Rec 102. 409. 437. 1948
Cutaneous Distribution of peripheral nerves redrawn from Gray's *Anatomy of the Human Body*. 28th ed

FIGURE 6-99: Case 3: Probable C5 lesion. This patient's breathing seemed to be slightly labored. The patient stated that breathing was difficult for about a week after onset.

CASE 4: LATERAL AND MEDIAL CORD LESION

Neck, Diaphragm, and Upper Extremity

Name Date

LEFT

PERIPHERAL NERVES — KEY → D. = Dorsal Prim. Ramus; V. = Ventral Prim. Ramus; P.R. = Plexus Root; S.T. = Sup. Trunk; P. = Post. Cord; L. = Lat. Cord; M. = Med. Cord

		MUSCLE STRENGTH GRADE	MUSCLE	Cervical D. 1-8 T. 1	Cervical V. 1-8	Cervical V. 1-4	Phrenic V. 3, 4, 5	Long Thor. P.R. 5, 6, 7, (8)	Dor. Scap P.R. 4, 5	N. to Subcl. S.T. 5, 6	Suprascap S.T. 4, 5, 6	U. Subscap. P. (4), 5, 6, (7)	Thoracodor P. (5), 6, 7, 8	L. Subscap. P. 5, 6, (7)	Lat. Pect L. 5, 6, 7	Med. Pect. M. (6), 7, 8, 1	Axillary P. 5, 6	Musculocu. L. (4), 5, 6, 7	Radial P. 5, 6, 7, 8, 1	Median L.M. 5, 6, 7, 8, 1	Ulnar M. 7, 8, 1	SENSORY					
			HEAD & NECK EXTENSORS	•																							
Cervical nerves			INFRAHYOID MUSCLES			•																					
Cervical nerves			RECTUS CAP ANT. & LAT.			•																					
Cervical nerves			LONGUS CAPITIS			•																					
Cervical nerves			LONGUS COLLI		•																						
Cervical nerves			LEVATOR SCAPULAE			•			•																		
Cervical nerves			SCALENI (A. M. P.)		•																						
Cervical nerves			STERNOCLEIDOMASTOID			•																					
Cervical nerves			TRAPEZIUS (U. M. L.)			•																					
Cervical nerves			DIAPHRAGM				•																				
Brachial Plexus	Root	N	SERRATUS ANTERIOR					•																			
Brachial Plexus	Root	N	RHOMBOIDS MAJ & MIN						•																		
Brachial Plexus	Trunk	—	SUBCLAVIUS							•																	
Brachial Plexus	Trunk	N	SUPRASPINATUS								•																
Brachial Plexus	Trunk	N	INFRASPINATUS								•																
Brachial Plexus	P Cord	N	SUBSCAPULARIS									•		•													
Brachial Plexus	P Cord	N	LATISSIMUS DORSI										•														
Brachial Plexus	P Cord	N	TERES MAJOR											•													
Brachial Plexus	L	N	PECTORALIS MAJ (UPPER)												•												
Brachial Plexus	M&L	G	PECTORALIS MAJ (LOWER)												•	•											
Brachial Plexus	M&L	P	PECTORALIS MINOR													⊙								(6)	7	8	1
Axil.		N	TERES MINOR														•						5	6			
Axil.		N	DELTOID														•						5	6			
Musculo-cutan		G−	CORACOBRACHIALIS															⊙						6	7		
Musculo-cutan		F+	BICEPS															⊙					5	6			
Musculo-cutan		F+	BRACHIALIS															⊙					5	6			
Radial		N	TRICEPS																•			Post. cord		6	7	8	1
Radial		N	ANCONEUS																•						7	8	
Radial	Lat.M	—	BRACHIALIS (SMALL PART)																—				5	6			
Radial	Lat.M	G−	BRACHIORADIALIS																⊙				5	6			
Radial	Lat.M	N	EXT CARPI RAD L																•				5	6	7	8	
Radial	Post Inter	N	EXT CARPI RAD B																•				5	6	7	8	
Radial	Post Inter	N	SUPINATOR																•				5	6	(7)		
Radial	Post Inter	G−	EXT DIGITORUM																⊙					6	7	8	
Radial	Post Inter	—	EXT DIGITI MINIMI																—					6	7	8	
Radial	Post Inter	F+	EXT CARPI ULNARIS																⊙					6	7	8	
Radial	Post Inter	G−	ABD POLLICIS LONGUS																⊙					6	7	8	
Radial	Post Inter	G−	EXT POLLICIS BREVIS																⊙					6	7	8	
Radial	Post Inter	G−	EXT POLLICIS LONGUS																⊙					6	7	8	
Radial	Post Inter	—	EXT INDICIS																—					6	7	8	
Median		P	PRONATOR TERES																	⊙		Med. + Lat. cord		6	7		
Median		F+	FLEX CARPI RADIALIS																	⊙				6	7	8	
Median		—	PALMARIS LONGUS																	⊙				(6)	7	8	1
Median		G−	FLEX DIGIT SUPERFICIALIS																	⊙					7	8	1
Median	A Inter	F+	FLEX DIGIT PROF I & II																	⊙					7	8	1
Median	A Inter	F+	FLEX POLLICIS LONGUS																	⊙				(6)	7	8	1
Median	A Inter	P	PRONATOR QUADRATUS																	⊙					7	8	1
Median		F+	ABD POLLICIS BREVIS																	⊙				6	7	8	1
Median		F−	OPPONENS POLLICIS																	⊙				6	7	8	1
Median		P	FLEX POLL BREV (SUP. H)																	⊙				6	7	8	1
Median		O	LUMBRICALES I & II																	⊙				(6)	7	8	1
Ulnar		F+	FLEX CARPI ULNARIS																		⊙	Med. cord			7	8	1
Ulnar		P	FLEX DIGIT. PROF. III & IV																		⊙				7	8	1
Ulnar		—	PALMARIS BREVIS																		⊙				(7)	8	1
Ulnar		O	ABD DIGITI MINIMI																		⊙				(7)	8	1
Ulnar		O	OPPONENS DIGITI MINIMI																		⊙				(7)	8	1
Ulnar		O	FLEX DIGITI MINIMI																		⊙				(7)	8	1
Ulnar		O	PALMAR INTEROSSEI																		⊙					8	1
Ulnar		O	DORSAL INTEROSSEI																		⊙					8	1
Ulnar		O	LUMBRICALES III & IV																		⊙				(7)	8	1
Ulnar		O	ADDUCTOR POLLICIS																		⊙					8	1
Ulnar		O	FLEX POLL BREV. (DEEP H.)																		⊙					8	1

Case 4: A manual muscle test was done before surgery, and the findings indicated the following:

Slight involvement of the muscles supplied by the radial nerve below the level of innervation to the triceps.

Moderate involvement of the lateral cord below the level of the lateral pectoral nerve.

Probably complete involvement of the medial cord above the level of the medical pectoral nerve, interrupting the C8 and T1 supply (i.e., inferior trunk).

That the pectoralis minor, flexor carpi ulnaris, and flexor digitorum profundus III and IV show some strength can mislead one to assume that C8 and T1 are intact. These muscles, along with some of the intrinsic muscles of the hand, also receive C7 innervation, and there may be slight evidence of power in these muscles from C7 without the medial cord being intact.

At surgery, it was found that the medial cord had been interrupted by a bullet above the level of the medial pectoral nerve, as had been indicated by the muscle testing

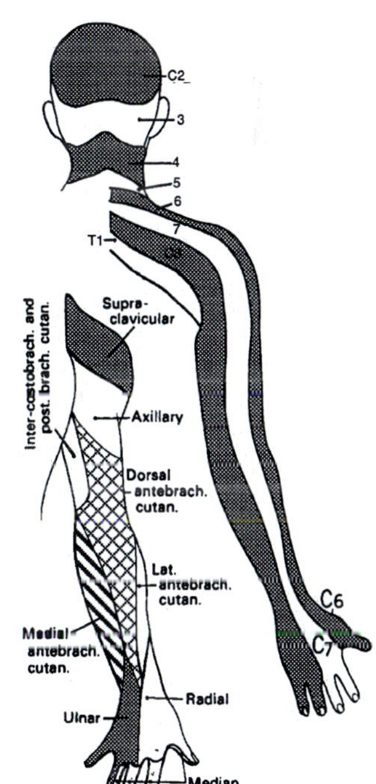

Dermatomes redrawn from Keegan and Garrett Anat Rec 102. 409. 437. 1948
Cutaneous Distribution of peripheral nerves redrawn from *Gray's Anatomy of the Human Body*. 28th ed

FIGURE 6-100: Case 4: Lateral and medial cord lesion. © 2005 Florence P. Kendall.

CASE 5: PARTIAL BRACHIAL PLEXUS LESION

Neck, Diaphragm, and Upper Extremity

Name Date

RIGHT

PERIPHERAL NERVES — KEY:
D. = Dorsal Prim. Ramus
V. = Vent. Prim. Ramus
P.R. = Plexus Root
S.T. = Superior Trunk
P. = Posterior Cord
L. = Lateral Cord
M. = Medial Cord

Region	Muscle Strength Grade	Muscle	Cervical (D., 1-8)	Cervical (V., 1-8)	Cervical (V., 1-4)	Phrenic (V., 3, 4, 5)	Long Thor. (P.R., 5, 6, 7, (8))	Dor. Scap (P.R., 4, 5)	N. to Subcl. (S.T., 5, 6)	Suprascap (S.T., 4, 5, 6)	U. Subscap. (P., (4), 5, 6, (7))	Thoracodor (P., (5), 6, 7, 8)	L. Subscap. (P., 5, 6, (7))	Lat. Pect. (L., 5, 6, 7)	Med. Pect. (M., (6), 7, 8, T1)	Axillary (P., 5, 6)	Musculocu. (L., (4), 5, 6, 7)	Radial (P., 5, 6, 7, 8, T1)	Median (L.M., 5, 6, 7, 8, T1)	Ulnar (M., 7, 8, T1)	C1	C2	C3	C4	C5	C6	C7	C8	T1
	—	HEAD & NECK EXTENSORS	•																		1	2	3	4	5	6	7	8	1
Cervical nerves	—	INFRAHYOID MUSCLES			•																1	2	3						
Cervical nerves	—	RECTUS CAP ANT. & LAT.			•																1	2							
Cervical nerves	—	LONGUS CAPITIS			•																1	2	3	(4)					
Cervical nerves	—	LONGUS COLLI		•																		2	3	4	5	6	(7)		
Cervical nerves	N	LEVATOR SCAPULAE			•			•															3	4	5				
Cervical nerves	—	SCALENI (A. M. P.)		•																			3	4	5	6	7	8	
Cervical nerves	N	STERNOCLEIDOMASTOID			•																(1)	2	3						
Cervical nerves	N	TRAPEZIUS (U. M. L.)			•																	2	3	4					
Cervical nerves	—	DIAPHRAGM				•																	3	4	5				
Brachial Plexus – Root	N	SERRATUS ANTERIOR					•																		5	6	7	8	
Brachial Plexus – Root	N	RHOMBOIDS MAJ & MIN						•																4	5				
Brachial Plexus – Trunk	—	SUBCLAVIUS							•																5	6			
Brachial Plexus – Trunk	F	SUPRASPINATUS								⊙														4	5	6			
Brachial Plexus – Trunk	F+	INFRASPINATUS								⊙														(4)	5	6			
Brachial Plexus – P Cord	O?	SUBSCAPULARIS									⊙		⊙												5	6	7		
Brachial Plexus – P Cord	P?	LATISSIMUS DORSI										⊙														6	7	8	
Brachial Plexus – P Cord	O?	TERES MAJOR											⊙												5	6	7		
Brachial Plexus – L	O	PECTORALIS MAJ (UPPER)												⊙											5	6	7		
Brachial Plexus – M&L	G	PECTORALIS MAJ (LOWER)												•	•											6	7	8	1
Brachial Plexus – M&L	G+	PECTORALIS MINOR													•											(6)	7	8	1
Axil.	P	TERES MINOR														⊙									5	6			
Axil.	T	DELTOID														⊙									5	6			
Musculo-cutan	P	CORACOBRACHIALIS															⊙									6	7		
Musculo-cutan	O	BICEPS															⊙								5	6			
Musculo-cutan	O	BRACHIALIS															⊙								5	6			
Radial	P	TRICEPS																⊙								6	7	8	1
Radial	—	ANCONEUS																⊙									7	8	
Radial – Lat.M	—	BRACHIALIS (SMALL PART)																⊙							5	6			
Radial – Lat.M	O	BRACHIORADIALIS																⊙							5	6			
Radial – Lat.M	O	EXT CARPI RAD L																⊙							5	6	7	8	
Radial – Post Inter	O	EXT CARPI RAD B																⊙							5	6	7	8	
Radial – Post Inter	O	SUPINATOR																⊙							5	6	(7)		
Radial – Post Inter	O	EXT DIGITORUM																⊙								6	7	8	
Radial – Post Inter	O	EXT DIGITI MINIMI																⊙								6	7	8	
Radial – Post Inter	O	EXT CARPI ULNARIS																⊙								6	7	8	
Radial – Post Inter	O	ABD POLLICIS LONGUS																⊙								6	7	8	
Radial – Post Inter	O	EXT POLLICIS BREVIS																⊙								6	7	8	
Radial – Post Inter	O	EXT POLLICIS LONGUS																⊙								6	7	8	
Radial – Post Inter	O	EXT INDICIS																⊙								6	7	8	
Median	G	PRONATOR TERES																	•							6	7		
Median	G	FLEX CARPI RADIALIS																	•							6	7	8	
Median	G	PALMARIS LONGUS																	•							(6)	7	8	1
Median	N	FLEX DIGIT SUPERFICIALIS																	•								7	8	1
Median – A Inter	N	FLEX DIGIT PROF I & II																	•								7	8	1
Median – A Inter	N	FLEX POLLICIS LONGUS																	•							(6)	7	8	1
Median – A Inter	G	PRONATOR QUADRATUS																	•								7	8	1
Median	G	ABD POLLICIS BREVIS																	•							6	7	8	1
Median	N	OPPONENS POLLICIS																	•							6	7	8	1
Median	G	FLEX POLL BREV (SUP. H)																	•							6	7	8	1
Median	N	LUMBRICALES I & II																	•							(6)	7	8	1
Ulnar	N	FLEX CARPI ULNARIS																		•							7	8	1
Ulnar	N	FLEX DIGIT. PROF. III & IV																		•							7	8	1
Ulnar	—	PALMARIS BREVIS																		•							(7)	8	1
Ulnar	N	ABD DIGITI MINIMI																		•							(7)	8	1
Ulnar	N	OPPONENS DIGITI MINIMI																		•							(7)	8	1
Ulnar	N	FLEX DIGITI MINIMI																		•							(7)	8	1
Ulnar	N	PALMAR INTEROSSEI																		•								8	1
Ulnar	N	DORSAL INTEROSSEI																		•								8	1
Ulnar	N	LUMBRICALES III & IV																		•							(7)	8	1
Ulnar	N	ADDUCTOR POLLICIS																		•								8	1
Ulnar	N	FLEX POLL BREV. (DEEP H.)																		•								8	1

SENSORY

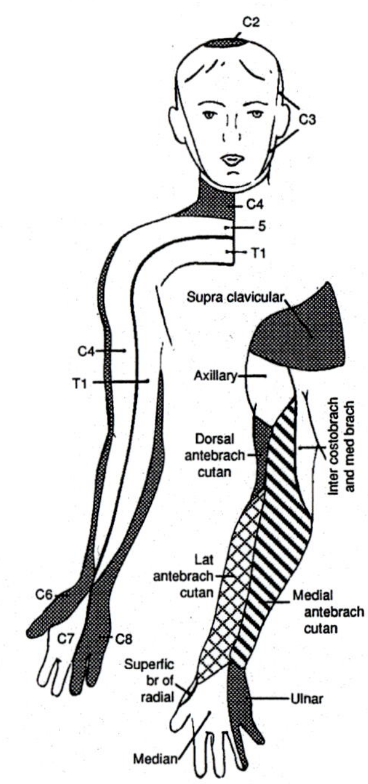

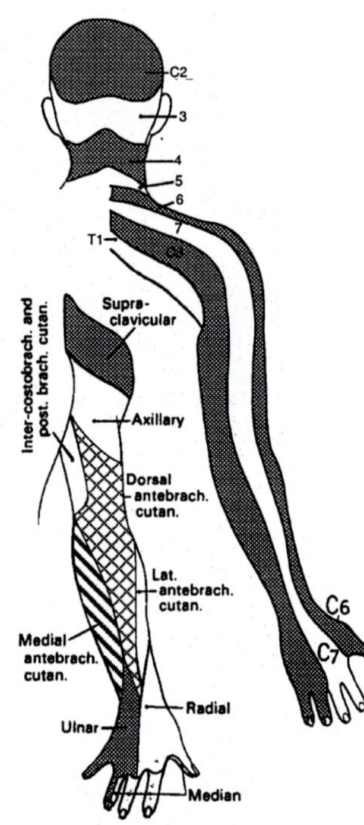

Dermatomes redrawn from Keegan and Garrett Anat Rec 102. 409. 437. 1948
Cutaneous Distribution of peripheral nerves redrawn from Gray's *Anatomy of the Human Body*. 28th ed

FIGURE 6-101: Case 5: Partial brachial plexus lesion. © 2005 Florence P. Kendall.

Case 5: A 30-year-old male fell from a moving automobile and was unconscious for approximately 20 minutes. He was treated in the emergency room of a local hospital for minor abrasions and then released. During the next three weeks, he was seen and treated by several physicians because of paralysis and edema of the right arm and pains in the chest and neck.

Twenty-two days after the accident, he was admitted to the University of Maryland Hospital. A neuromuscular evaluation, including a manual muscle test and an electromyographic study, was performed at that time and showed extensive involvement of the right upper extremity.

The decision was made to defer surgical exploration and treat the patient conservatively with an airplane splint and follow-up therapy in the outpatient clinic. Unfortunately, the patient did not report to the outpatient clinic until five months later. Subsequently, a detailed manual test as well as electrodiagnostic and further electromyographic studies were made.

Sensory and Reflex Tests

Sensation to pinprick was absent over the area of sensory distribution of the axillary, musculocutaneous, and radial nerves. No deep tendon reflexes of the biceps or triceps muscles were observed.

Manual Muscle Test

Figure 6-101 indicates, at a glance, that the muscles supplied by the ulnar nerve were graded as normal, those by the median as either normal or good, and those by the radial, musculocutaneous, and axillary as either poor or zero. At the level of the brachial plexus, the involvement was more complicated, as noted by the grades ranging all the way from normal to zero. Concurrent charting of the involved peripheral nerves and spinal segments, however, furnished additional information and provided the basis for determination of the sites of lesions as follows:

1. *A lesion of the posterior cord of the brachial plexus:* The muscles supplied by the upper and lower subscapular, thoracodorsal, axillary, and radial nerves, which arise from the posterior cord, show complete paralysis or major weakness. Involvement of the subscapularis muscle places the site of lesion proximal to the point where the upper subscapular nerve arises ("c" in Figure 6-102).
2. *No involvement of the medial cord of the plexus:* The muscles supplied by the ulnar nerve, which is the terminal branch of the medial cord, graded normal. The sternal part of the pectoralis major and the pectoralis minor (C5–T1) and some muscles receiving median nerve supply (C6–T1) graded good. It is logical to assume that the slight weakness is attributable to the C5 and C6 deficit and not to any involvement of the medial cord.
3. *A lesion of either the upper trunk (formed by C5 and C6 roots of the plexus) or the anterior division of the upper trunk before it joins with the anterior division of the middle trunk (C7) to form the lateral cord:* Confirmation of this statement requires an explanation of how it is ascertained that the lesion is in this area and that it is no more proximal than "a" or more distal than "b" in Figure 6-102.

The complete paralysis of the biceps and brachialis (from C5 and C6) raises the question of the level of involvement of these muscles—musculocutaneous nerve

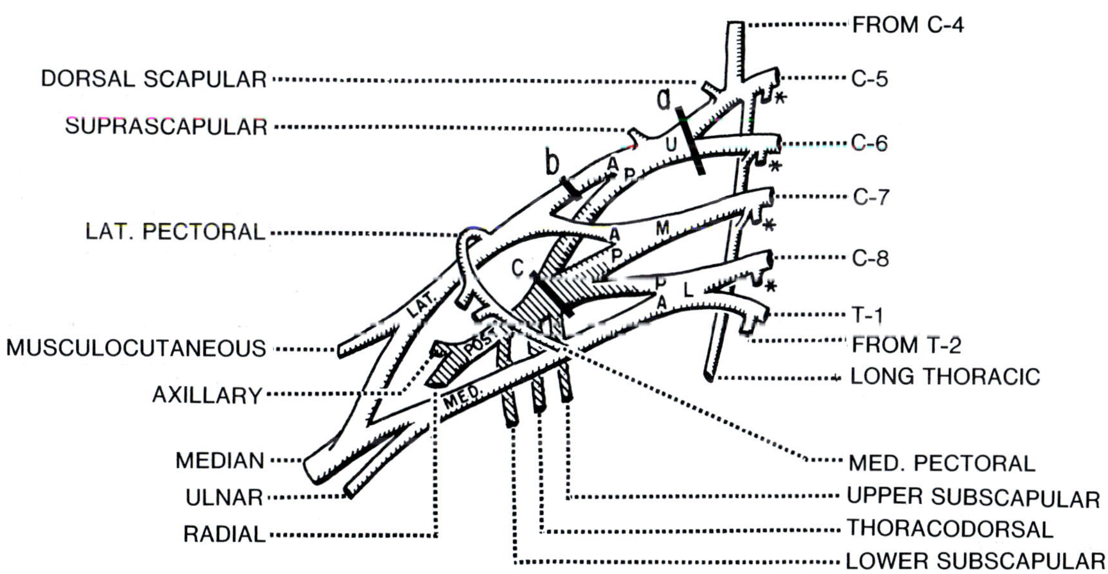

FIGURE 6-102: Brachial plexus with possible sites of lesions (a, b, and c). U, upper; M, middle; L, lateral trunks; A, anterior divisions; P, posterior divisions; *, to longus coli and scaleni; LAT, lateral cord; MED, medial cord; POST, posterior cord. Reprinted from (23).

(C5, C6, and C7), lateral cord (C5, C6, and C7), trunk, or spinal nerve root.

That the coracobrachialis showed some strength rules out complete involvement at the musculocutaneous level. A complete lesion at the level of the lateral cord (C5, C6, and C7) is refuted by several findings that indicate the C7 component is not involved.

The flexor digitorum superficialis, flexor digitorum profundus I and II, and lumbricals I and II, which have C7, C8, and T1 supply through the median nerve, graded normal. Other muscles supplied by the median nerve, which have C6, C7, C8, and T1 supply, graded good and, undoubtedly, would have exhibited more weakness had C7 been involved.

The sternal part of the pectoralis major and pectoralis minor, which are supplied chiefly by the medial pectoral (C8 and T1) and, to some extent, by the lateral pectoral (C5, C6, and C7), graded good and good+. Had C7 been involved, the weakness would undoubtedly have been greater.

The presence of some strength in the coracobrachialis is thus explained on the basis of the C7 components being intact, and it further confirms that such is the case. The stretch weakness, superimposed on this muscle by the shoulder joint subluxation and the weakness of the deltoid and biceps, could account for the coracobrachialis grading as no more than poor. Thus, with C7 not involved, the most distal point of lesion may be considered as "b" in Figure 6-102.

The possibility of C5 and C6 being involved more proximal than "a" (see Figure 6-102) at the level of the roots of the plexus is ruled out, because the rhomboids and serratus anterior muscles graded normal. Whether the lesion is proximal or distal to the point where the suprascapular nerve arises depends on whether involvement of the supraspinatus and infraspinatus muscles is on a neurogenic or a stretch-weakness basis.

The supraspinatus and infraspinatus (C4, C5, and C6) graded fair, and if this partial weakness resulted from a neurological deficit, the lesion must be proximal to the point where the suprascapular nerve arises. Most logically, the presence of fair strength would then be interpreted as a result of regeneration during the 7 months since onset.

On the other hand, the weakness in these muscles may be of a secondary stretch-weakness type and not neurogenic. The patient had not worn the airplane splint that was applied 23 days after injury, and subluxation of the joint and stretching of the capsule were found. Additionally, the weakness was not as pronounced as in the other muscles supplied by C5 and C6, a fullness of contraction could be felt on palpation, and these muscles had been subjected to undue stretch. If the weakness had resulted from stretch, the initial site of lesion would have been distal to the point where the suprascapular nerve arises.

CASE 6: STRETCH WEAKNESS SUPERIMPOSED ON A PERIPHERAL NERVE INJURY

The following is an example of stretch weakness superimposed on a peripheral nerve injury:

A woman was lifting a heavy rock while she was gardening. Her hands were in supination. The rock suddenly fell, turning her forearms into pronation. She felt a sharp pain in her right upper forearm. Weakness developed in the muscles supplied by the radial nerve below the level of the supinator. She was examined by several doctors, including a neurosurgeon who said that he had seen some cases, and knew of others reported in the literature, in which the radial nerve had been similarly involved at the level where it passes through the supinator.

The patient was first seen by a physical therapist 18 months after onset. The wrist extensors and the extensor digitorum showed marked weakness, but not complete paralysis, grading poor and poor+. A splint was applied, and in 2 weeks, the strength had improved to grades of poor+ and fair+. Then, the condition reached a stalemate. The patient had started doing more work with the hand, and she left the splint off most of the time. Three months went by, but rather than give up, it was decided by the patient, the doctor, and the physical therapist that a period of more complete immobilization be tried. A plaster cock-up splint, including extension of the metacarpophalangeal joints, was applied. This protected the wrist extensors and the extensor digitorum but allowed use of the interphalangeal joints in flexion and extension. The splint was removable, but the patient was cautioned to keep it on as much of the 24 hours in a day as possible and not to move the wrist and digits into full flexion whenever the splint was off. After 2 weeks, the wrist and digit muscles were much improved. The patient played the piano and typed for the first time in 2 years.

CUMULATIVE TRAUMA (OVERUSE) INJURIES

A cumulative trauma, or overuse, injury may be defined as damage caused by repetitive movements performed for a length of time that is beyond the tolerance of the tissues involved. The time involved may be short if the load lifted or the force required is excessive in relation to the ability of the subject. Overuse injuries often extend over a prolonged time with the activity causing an irritation or breakdown of muscle, tendon, or capsule and subsequent pain and inflammation.

Cumulative trauma injuries of the upper extremity are common (Box 6-2). These injuries cause numerous problems for over 2.3 million individuals in the United States who have disabilities requiring the use of a manual wheelchair (34). Repetitive hand and arm movements associated with a person's occupational or recreational activities can give rise to a variety of strains, inflammatory processes, or nerve involvements that result in mild to debilitating conditions (68). The most commonly occurring pathology is shoulder impingement syndrome involving the rotator cuff, biceps tendon, and/or the subacromial bursa.

For cumulative trauma injuries such as tennis elbow (i.e., lateral epicondylitis), golfer's elbow (i.e., medial epicondylitis), swimmer's shoulder (i.e., impingement syndrome), repetitive strain injury from excessive keyboard or computer use, or push-ups done to excess, appropriate treatment depends, in part, on the specificity provided by manual muscle testing.

For example, accurate testing may help to avoid diagnoses such as CTS when the problem is, in fact, a pronator teres syndrome. A Mayo Clinic study showed that 7 of 35 patients who were operated on for carpal tunnel were later found to have pronator teres syndrome (56).

BOX 6-2

Common Overuse Areas in the Upper Extremity

Following are areas of most concern for the upper extremity:

- Wrist joint extension, extensor muscles, radial nerve (C5, 6, 7, 8)
- Wrist joint flexion, flexor muscles, ulnar nerve (C7, 8, T1)
- Wrist joint flexion, flexor muscles, median nerve (C6, 7, 8)

*Carpal tunnel syndrome

- Radioulnar joint (forearm), pronator teres, median nerve (C6, 7)

*Pronator teres syndrome

- Elbow joint, flexor muscles, musculocutaneous nerve (C4, 5, 6)

*Lateral epicondylitis (tennis elbow)

*Medial epicondylitis (golfer's elbow)

- Shoulder joint, abductor: supraspinatus, musculocutaneous nerve (C4, 5, 6)
- Shoulder joint, lateral rotation: supraspinatus, infraspinatus (C4, 5, 6)

*Teres minor (C5, 6)

- Shoulder joint, medial rotation: subscapularis, teres major (C5, 6, 7)

*Latissimus dorsi (C6, 7, 8)

SECTION V
INTERVENTION

The treatment suggestions in this section focus on important basics in protection, support, alignment, and restoration of both length and strength as well as suggestions specific to the clinical findings described already. As with most plans of care, an emphasis must be placed on a home program to be performed regularly by the patient. This approach is often all that is needed to achieve a positive outcome. It is beyond the scope of this text to include other, more specific treatment options such as electrical stimulation, isokinetics, weight/fitness training, and manual techniques.

General Corrective Exercise Principles

Exercises in the lying position should be done on a firm surface (e.g., a board on the bed, a treatment table, or the floor, with a thin pad or folded blanket placed on the hard surface for comfort).

Stretching exercises may be performed actively or passively. Active stretching occurs when an individual actively moves to stretch a targeted muscle. Passive stretching occurs when an external force is applied to stretch a specific muscle such as when patients use equipment, gravity, or the assistance of another to achieve a stretch. Stretching should be preceded by a gentle warm-up to prepare targeted tissues. Stretching should be done gradually, with a conscious effort to relax. Best practice is to continue the stretch until a firm but tolerable pull is felt. The patient should be able to breathe comfortably while holding the stretch before returning slowly from the stretched position.

Strengthening exercises should also be done slowly, with the patient focusing attention on the muscles targeted by the exercise (69). The clinician should prescribe a specific number of repetitions as well as duration of muscle contraction and relaxation depending upon the goal of the exercise. The Holten Diagram, which was developed as a means to determine exercise intensity, may assist the prescription of resistance and repetitions for a wide variety of patient populations (70).

Nerve Entrapments

Treatment/management of nerve entrapments will depend on the degree of neural injury. Conservative therapy is often appropriate with a focus on rest, medications as indicated for symptom relief, and/or modifying or removing external barriers or repetitive activities that may be compressing the involved nerve. More significant injuries or those unresponsive to conservative care may benefit from splinting or surgical intervention.

Stretch of Pectoralis Minor

To stretch the pectoralis minor, place the subject in a supine position and press the shoulder backward and downward (Figure 6-103). One hand should be cupped just medial to the glenoids, avoiding direct pressure on the shoulder joints using firm, uniform pressure that helps to rotate the shoulder girdle back.

After strain has been relieved by support and by stretching of the tight opposing muscles, specific exercises are indicated for the middle and lower trapezius. If the overall posture is faulty, general postural correction is needed.

Certain exercises to stretch the pectoralis minor may be contraindicated and so should be approached with caution because they will increase compression in the anterior shoulder region. Head and shoulder raising from a supine position, as in trunk curls, would do the same and so should be avoided, because this movement rounds the thoracic spine and depresses the coracoid anteriorly.

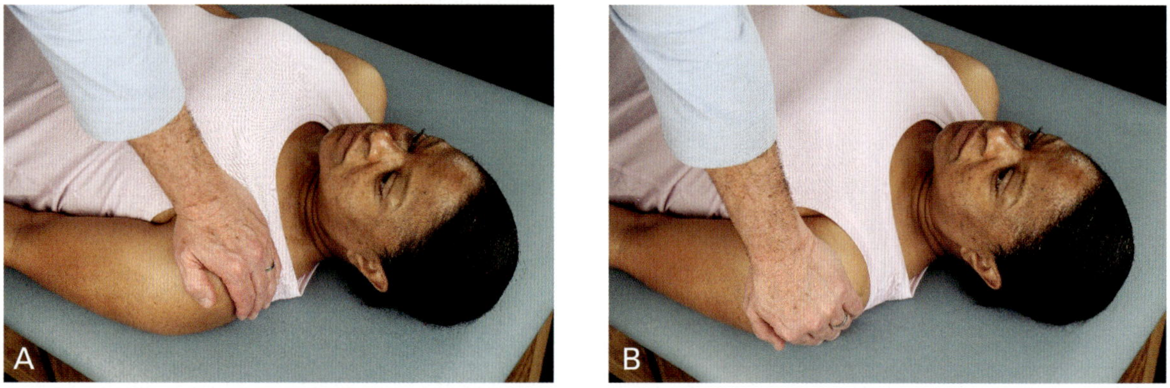

FIGURE 6-103: A, Stretch of pectoralis minor with correct hand placement; **B,** Incorrect hand placement.

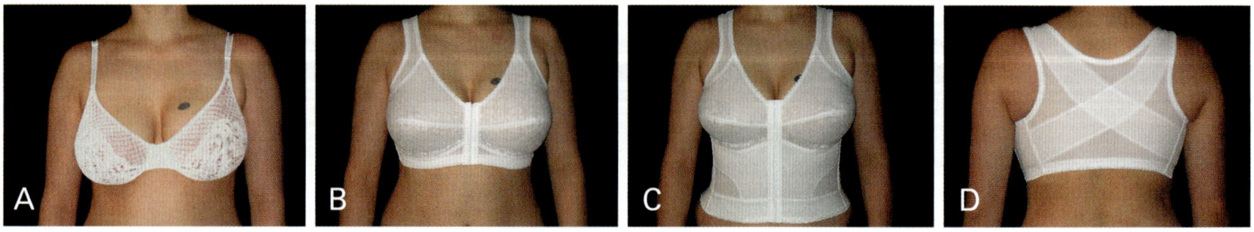

FIGURE 6-104: A, Example of a bra with inadequate support; **B,** Regular posture bra (front view); **C,** Long-line posture bra; **D,** Regular posture bra (back view).

NOTE: *Among women with very large breasts, the faulty alignment may be accentuated by pressure from brassiere straps. In addition, the weight of the breasts pulling forward and downward can contribute to upper and middle back discomfort. A "posture bra" that is readily available in stores can provide effective support for the breasts and relieve bra strap pressure (Figure 6-104).*

Teres Syndrome

Treatment consists of heat and massage to the areas of muscle tightness and active, assisted exercises to stretch the internal rotators and adductors of the shoulder. Stretching of the upper extremity overhead into flexion or abduction and external rotation is done very gradually.

With tightness in the teres major, the scapula is pulled into abduction as the upper extremity is raised into flexion or abduction and externally rotated. To ensure that stretching is localized to the teres, it is necessary to press against the axillary border of the scapula when raising the upper extremity to restrict excessive abduction of the scapula. If the scapula moves excessively in the direction of abduction, the teres, which is a scapulohumeral muscle, will not be stretched, and the rhomboids, which attach the scapula to the vertebral column, will stretch too much.

Assisted Stretch of Teres Major and Latissimus Dorsi

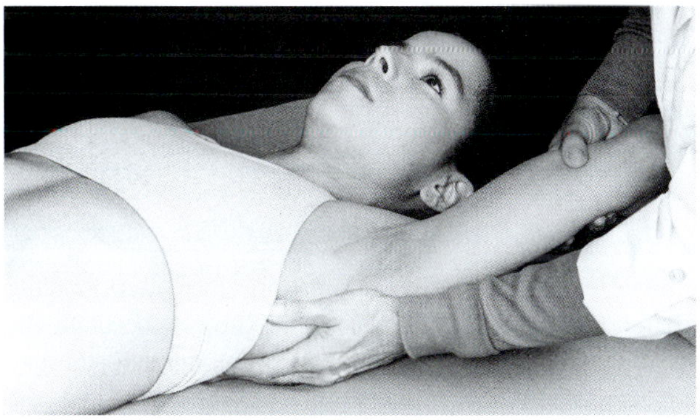

FIGURE 6-105: Assisted stretch of teres major and latissimus dorsi.

Assisted stretch of the teres major and latissimus dorsi is performed with the patient in a hooklying position (supine, with the hips and knees flexed, feet flat on table, and the low back flat) (Figure 6-105). Hold the scapula to prevent excessive abduction, localize stretch to the shoulder joint adductors and prevent excessive stretch of the rhomboids. The therapist provides slight traction on the upper extremity while stretching it overhead.

Pain from Shoulder Subluxation

Shoulder pain resulting from traction on the shoulder joint because of loss of tone and malalignment of the joint requires special treatment considerations. The cause may be paresis secondary to stroke, trauma to the brachial plexus, or a lesion of the axillary nerve. Effective management requires maintaining joint approximation during rest as well as during treatment to restore motion and to improve motor control.

A special sling, called a shoulder-arm support, provides joint approximation and support to protect the subluxed shoulder when the patient is sitting or standing (Figure 6-106A, B) (71, 72). When used to hold the humerus in the glenoid, the shoulder girdle carries the weight of the arm, and the sling does not hang on the neck (Figure 6-106B) The clinician needs to understand careful measurements are taken for the sling to provide the best approximation of the joint and to prevent further stretch, instability, and pain in the weakened upper extremity. Measurements are taken with the elbow flexed to 90°. A tape measure is held at the top of the shoulder, looped down around the forearm, and then back up to the shoulder. The number of inches determines the size of the sling. Clinicians will need to evaluate the sling post application to ensure that the appropriate biomechanical alignment is achieved.

The patient should be taught how to protect the shoulder when not wearing the sling. Proper alignment and approximation can be maintained when sitting in an armchair by having the affected arm supported on the arm rest. In this position, the patient can use the opposite hand to press downward on top of the shoulder, making the humerus feel snug in the glenoid cavity. Shoulder joint approximation must be maintained during active, assisted exercises to restore joint motion and function (73). In other words, do not let the joint be subluxed at any time. Attention needs to be paid to eliminate the tendency to produce excessive shoulder girdle elevation.

The weight of the arm is carried by the neck and opposite shoulder. The patient should be taught how to protect the shoulder so that it does not sublux when not wearing the sling. The ideal position would avoid excessive shoulder girdle elevation so as to maintain joint approximation. In sitting, this can be achieved by having

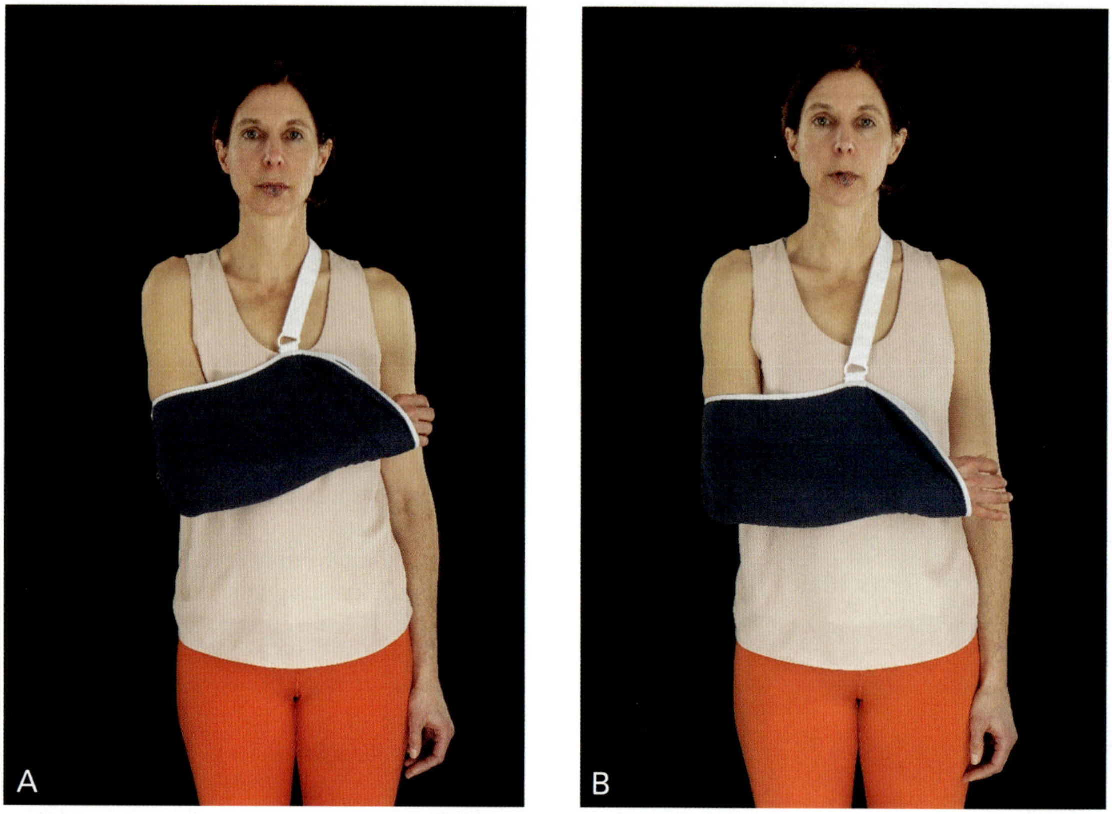

FIGURE 6-106: Shoulder-arm support; **A,** with forearm elevated (increased pressure on neck); **B,** Aligned for addressing subluxed shoulder.

the affected arm supported on an arm rest. In this position, the patient could then also use the opposite hand to press downward on top of the shoulder, making the humerus feel snug in the glenoid cavity.

Assisted Stretch of Shoulder External Rotators

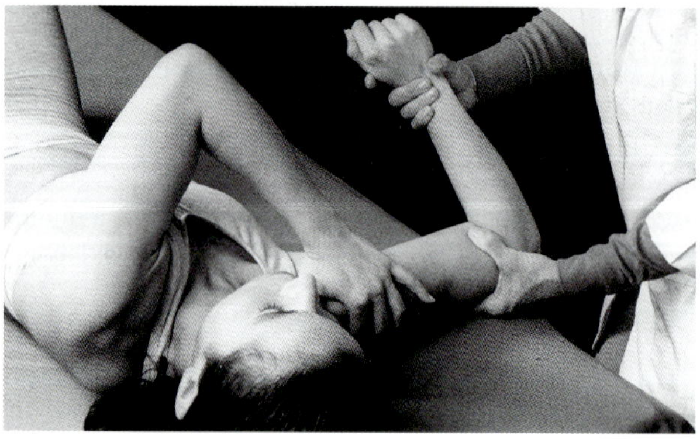

FIGURE 6-107: Assisted stretch of shoulder external rotators.

Assisted stretch of the shoulder external rotators is performed with the patient in a hooklying position (Figure 6-107). The shoulder is abducted 90° with the elbow flexed to 90°. The therapist provides traction at the glenohumeral joint and assists the subject to rotate the shoulder into medial rotation. The clinician can stabilize the shoulder if the positioning listed above affects the scapular stabilization on the side being treated.

Postural Correction of Cervical Rib Presentation

The faulty alignment most likely to cause irritation is the type characterized by a round upper back and a forward head. Care of a patient with painful symptoms resulting from a cervical rib requires postural correction of the upper back and neck (67). This treatment may relieve the symptoms completely and obviate the need for a surgical procedure.

Thoracic Outlet Syndrome

Unless symptoms are severe and clearly defined, conservative treatment of TOS should emphasize increasing the space of the thoracic outlet by improving posture, correcting muscle imbalance, and modifying occupational, recreational, and sleeping habits that adversely affect the posture of the head, neck, and upper back. Cooperation by the patient is essential to success. The patient should be taught self-stretching exercises to relieve tightness in the scalenes, sternocleidomastoid, pectoral muscles, and neck extensors. Learning to do diaphragmatic breathing will lessen involvement of the accessory respiratory muscles, some of which are in need of stretching. Sleeping in a prone position should be avoided, and activities that involve raising the arms overhead should be kept to a minimum. Research has shown that "with conservative therapy [and] … exercises designed to correct slumping shoulder posture, … at least two out of three patients improve to a satisfactory degree" (15)

CORACOID PRESSURE SYNDROME

Treatment in the acute stage of CPS consists first of applying a sling (see Figure 6-106B) that supports the weight of the upper extremity and shoulder girdle, relieving pressure on the plexus and taking the workload off the upper trapezius. Heat and massage may be applied to the upper trapezius and other muscles that exhibit tightness. Massage should be gentle and relaxing, progressing after a few treatments to gentle kneading and stretching. Slow, passive stretching of the pectoralis minor can be initiated (Figure 6-108). If tightness is also present in the pectoralis major, latissimus, or both, the involved upper extremity should be placed carefully overhead, if tolerated, to place the muscles on a slight stretch. Gentle traction is applied with one hand while massage is applied with the other. A shoulder support usually is needed to help maintain the correction of alignment and to relieve strain on the lower trapezius muscle during the recovery period.

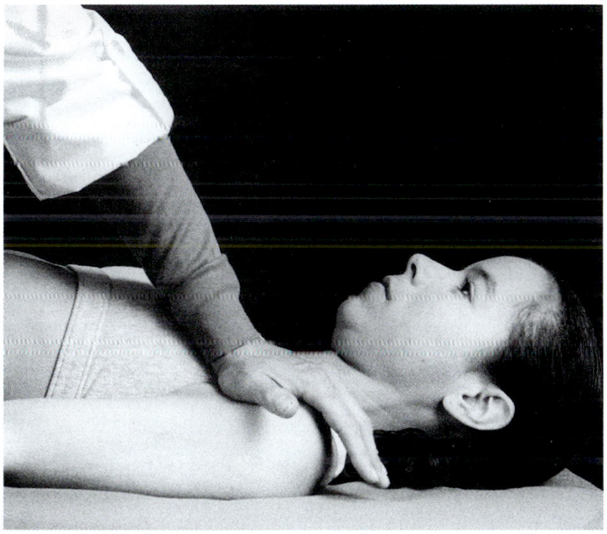

FIGURE 6-108: Slow, passive stretching of the pectoralis minor can be initiated as treatment for coracoid pressure syndrome.

Cumulative Trauma (Overuse) Injuries

The aim of conservative treatment is to relieve pain, reduce excessive use, and alleviate further strain. Periodic use of appropriate wrist, arm, shoulder, or upper back supports can help minimize the debilitating effects of overuse injuries and restore more optimal functioning of the muscles involved.

REFERENCES

1. Agur AMR. Grant's Atlas of Anatomy. 9th ed. Baltimore: Williams & Wilkins, 1991.
2. Goss CM, ed. Gray's Anatomy of the Human Body. 28th ed. Philadelphia: Lea & Febiger, 1966.
3. Bremner-Smith AT, Unwin AJ, Williams WW. Sensory pathways in the spinal accessory nerve. J Bone Joint Surg [BR] 1999; 81-B:226–228.
4. Dorland WA. The American Illustrated Medical Dictionary. Philadelphia: W.B. Saunders, 1932.
5. Johnson JYH, Kendall HO. Isolated paralysis of the serratus anterior muscle. J Bone Joint Surg[Am] 1955;37-A:567; Ortho Appl J 1964;18:201.
6. Taber CW. Taber's Cyclopedic Medical Dictionary. Philadelphia: F.A. Davis, 1969, pp. 1–25, Appendix 45–50.
7. Dorland's Illustrated Medical Dictionary. 27th ed. Philadelphia: W.B. Saunders, 1988, pp. 1118–1125.
8. O'Neill DB, Zarins B, Gelbermaen RH, Keating TM, Louis D. Compression of the anterior interosseous nerve after use of a sling for dislocation of the acromioclavicular joint. J Bone Joint Surg [Am] 1990;72-A(7):1100.
9. Hadley MN, Sonntag VKH, Pittman HW. Suprascapular nerve entrapment. J Neurosurg 1986; 64:843–848.
10. Post M, Mayer J. Suprascapular nerve entrapment. Clin Orthop Realt Res 1987;223:126–135.
11. Conway SR, Jones HR. Entrapment and compression neuropathies. In: Tollison CD, ed. Handbook of Chronic Pain Management. Baltimore: Williams & Wilkins, 1989.
12. Sunderland S. Nerve Injuries and Their Repair: A Critical Appraisal. London: Churchill Livingstone, 1991, p. 161.
13. Nakano KK. Neurology of Musculoskeletal and Rheumatic Disorders. Boston: Houghton Mifflin, 1978, pp. 191, 200.
14. Geiringer SR, Leonard JA. Posterior interosseus palsy after dental treatment: Case report. Arch Phys Med Rehabil 1985:66.
15. Dawson DM, Hallett M, Millender LH. Entrapment Neuropathies. 2nd ed. Boston: Little, Brown and Company, 1990.
16. Boland MR, Spigelman T, Uhl TL. The Function of Brachioradialis. J Hand Surg 2008;33(10):1853–1859.
17. Jamison JC, Caldwell GE. Muscle synergies and isometric torque production: influence of supination and pronation level on elbow flexion. J Neurophysiol 1993;70(3):947–960.
18. Culham E, Peat M. Functional anatomy of the shoulder complex. J Orthop Sports Phys Ther 1993 Jul;18(1):342–350. doi: 10.2519/jospt.1993.18.1.342. PMID: 8348135.
19. Inman VT, Saunders JB, de CM, Abbott LC. Observations on the function of the shoulder joint. J Bone Joint Surg 1944;26:1.
20. Clarkson HM. Musculoskeletal Assessment. 2nd ed. Baltimore: Lippincott Williams & Wilkins, 1989, p. 403.
21. Schenkman M, Rugo de Cartaya V. Kinesiology of the shoulder complex. J Orthop Sports Phys Ther 1987;8(9):438–450. doi: 10.2519/jospt.1987.8.9.438. PMID: 18797034.
22. Burkhart SS, Morgan CD, Kibler WB. The disabled throwing shoulder: Spectrum of pathology part I: Pathoanatomy and biomechanics. Arthrosc: J Arthrosc Relat Surg 2003;19(4):404–420, ISSN 0749-8063.

23. Forbush, Steven W et al. The comparison of the empty can and full can techniques and a new diagonal horizontal adduction test for supraspinatus muscle testing using cross-sectional analysis through ultrasonography. Int J Sports Phys Ther 2003;8(3): 237–247.
24. Rowlands LK, Wertsch JJ, Primack SJ, Spreitzer AM, Roberts MM. Kinesiology of the empty can test. Am J Phys Med Rehabil 1995;74(4):302–304. doi: 10.1097/00002060-199507000-00007
25. Eur Respir J, 1995, 8: 441–445. doi: 10.1183/09031936.95.08030441. Printed in UK – all rights reserved.
26. Park SH, Lee MM. Effects of lower trapezius strengthening exercises on pain, dysfunction, posture alignment, muscle thickness and contraction rate in patients with neck pain. Randomized Controlled Trial. Med Sci Monit 2020;26:e920208. Published March 23, 2020. doi: 10.12659/MSM.920208
27. Jspeert J, Kerstens HCJW, Janssen RMJ et al. Validity and reliability of serratus anterior hand held dynamometry. BMC Musculoskelet Disord 2019;20:360. doi: 10.1186/s12891-019-2741-7
28. Dididze M, Tafti D, Sherman AI. StatPearls [Internet]. StatPearls Publishing; Treasure Island (FL): July 8, 2020. Pronator Teres Syndrome.
29. Latef TJ, Bilal M, Vetter M, Iwanaga J, Oskouian RJ, Tubbs RS. Injury of the radial nerve in the arm: A review. Cureus. 2018 Feb 16;10(2):e2199.
30. Buchanan BK, Maini K, Varacallo M. Radial nerve entrapment. February 12, 2022. In: StatPearls [Internet]. Treasure Island (FL): StatPearls Publishing; January 2022. PMID: 28613749.
31. Wertsch JJ, Melvin J. Median nerve anatomy and entrapment syndromes: A review. Arch Phys Med Rehabil. 1982 Dec;63(12):623–627. PMID: 6756339.
32. Andrews K, Rowland A, Pranjal A, Ebraheim N. Cubital tunnel syndrome: Anatomy, clinical presentation, and management [published correction appears in J Orthop. December 14, 2022;23:275]. J Orthop 2018;15(3):832–836. Published August 16, 2018. doi: 10.1016/j.jor.2018.08.010
33. Swain R. Musculocutaneous nerve entrapment: A case report. Clin J Sport Med 1995 Jul;5(3):196–198. doi: 10.1097/00042752-199507000-00010. PMID: 7670976.
34. Moradi A, Ebrahimzadeh MH, Jupiter JB. Radial tunnel syndrome, diagnostic and treatment dilemma. Arch Bone Jt Surg 2015;3(3):156–162.
35. Wheeler R, DeCastro A. Posterior interosseous nerve syndrome. October 27, 2020. In: StatPearls [Internet]. Treasure Island (FL): StatPearls Publishing; January 2021. PMID: 31082090.
36. Shamrock AG, Das MJ. Radial tunnel syndrome. March 17, 2021. In: StatPearls [Internet]. Treasure Island (FL): StatPearls Publishing; January 2021. PMID: 32310397.
37. Bolster MA, Bakker XR. Radial tunnel syndrome: Emphasis on the superficial branch of the radial nerve. J Hand Surg Eur 2009;34(3):343–347.
38. Hagert CG, Lundborg G, Hansen T. Entrapment of the posterior interosseous nerve. Scand J Plast Reconstr Surg 1977;11(3): 205–212.
39. Loh YC, Lam WL, Stanley JK, Soames RW. A new clinical test for radial tunnel syndrome-the Rule-of-Nine test: A cadaveric study. J Orthop Surg (Hong Kong) 2004;12:83–86.
40. Wheeler R, DeCastro A. Posterior interosseous nerve syndrome. January 19, 2022. In: StatPearls [Internet]. Treasure Island (FL): StatPearls Publishing; January 2022. PMID: 31082090.
41. Kim DH, Murovic JA, Kim YY, Kline DG. Surgical treatment and outcomes in 45 cases of posterior interosseous nerve entrapments and injuries. J Neurosurg 2006 May;104(5):766–777.
42. Martin RM, Fish DE. Scapular winging: Anatomical review, diagnosis, and treatments. Curr Rev Musculoskelet Med 2008;1(1):1–11. doi: 10.1007/s12178-007-9000-5
43. Shih YF, Kao YH. Influence of pain location and hand dominance on scapular kinematics and EMG activities: An exploratory study. BMC Musculoskelet Disord 2011;12:267. Published November 24, 2011. doi: 10.1186/1471-2474-12-267
44. de Mauroy JC, Vallèse P, Fender P, Lecante C. Historical Lyonaise brace treatment for adolescent hyperkyphosis. Results of 272 cases reviewed 2 years minimum after removal of the brace. Scoliosis. 2010;5(Suppl 1):O69. Published September 10, 2010. doi: 10.1186/1748-7161-5-S1-O69
45. Lowe TG, Line BG. Evidence based medicine: Analysis of Scheuermann kyphosis. Spine 2007;32(19 Suppl):S115–S119.
46. Seidi F, Rajabi R, Ebrahimi I, Alizadeh MH, Minoonejad H. The efficiency of corrective exercise interventions on thoracic hyperkyphosis angle. J Back Musculoskelet Rehabil 2014;27(1):7–16. doi: 10.3233/BMR-130411. PMID: 23948845.
47. Etemadifar MR, Jamalaldini MH, Layeghi R. Successful brace treatment of Scheuermann's kyphosis with different angles. J Craniovertebr Junction Spine 2017;8(2):136–143. doi: 10.4103/jcvjs.JCVJS_38_16
48. Gurd DP. Back pain in the young athlete. Sports Med Arthrosc Rev 2011;19(1):7–16
49. Spinner M. Management of nerve compression lesions of the upper extremity. In: Omer GE, Spinner M, eds. Management of Peripheral Nerve Problems. Philadelphia: W.B. Saunders, 1980.
50. Sunderland S. Nerves and Nerve Injuries. 2nd ed. New York: Churchill Livingstone, 1978.
51. Aleksenko D, Varacallo M. Guyon canal syndrome. February 12, 2022. In: StatPearls [Internet]. Treasure Island (FL): StatPearls Publishing; January 2022. PMID: 28613717.
52. Depukat P, Mizia E, Kuniewicz M, Bonczar T, Mazur M, Pełka P, Mróz I, Lipski M, Tomaszewski K. Syndrome of canal of Guyon: Definition, diagnosis, treatment and complication. Folia Med Cracov 2015;55(1):17–23.
53. Huisstede BM, Hoogvliet P, Franke TP, Randsdorp MS, Koes BW. Carpal tunnel syndrome: Effectiveness of physical therapy and electrophysical modalities. An updated systematic review of randomized controlled trials. Arch Phys Med Rehabil 2018 Aug;99(8):1623–1634.e23. doi: 10.1016/j.apmr.2017.08.482. Epub September 20, 2017. PMID: 28942118.
54. Rodner CM, Tinsley BA, O'Malley MP. Pronator syndrome and anterior interosseous nerve syndrome. J Am Acad Orthop Surg 2013 May;21(5):268–275. doi: 10.5435/JAAOS-21-05-268. PMID: 23637145.
55. Dididze M, Tafti D, Sherman AL. Pronator teres syndrome. February 5, 2022. In: StatPearls [Internet]. Treasure Island (FL): StatPearls Publishing; January 2022. PMID: 30252346.
56. Hartz CR, Linscheid RL, Gramse RR, Daube JR. The pronator teres syndrome: Compressive neuropathy of the median nerve. J Bone Joint Surg Am 1981 Jul;63(6):885–890.
57. Cutts, S. Cubital tunnel syndrome, Postgrad Med J 2007;83: 28–31. doi: 10.1136/pgmj.2006.047456
58. Kooner S, Cinats D, Kwong C, Matthewson G, Dhaliwal G. Conservative treatment of cubital tunnel syndrome: A systematic review. Orthop Rev (Pavia) 2019;11(2):7955. Published June 12, 2019. doi: 10.4081/or.2019.7955
59. Pratt NE. Neurovascular entrapment in the regions of the shoulder and posterior triangle of the neck. Phys Ther 1986;66(12): 1894–1900.
60. Okoro T, Reddy VR, Pimpelnarkar A. Coracoid impingement syndrome: A literature review. Curr Rev Musculoskelet Med 2009;2(1):51–55. doi: 10.1007/s12178-009-9044-9
61. Gigante A, Bottegoni C, Barbadoro P. Coracoid syndrome: A neglected cause of anterior shoulder pain. Joints 2016;4(1):31–38. Published June 13, 2016. doi: 10.11138/jts/2016.4.1.031

62. Stein, I. Painful conditions of the shoulder joint. Phys Ther Rev 1948:28(6): 275–279.
63. Kendall HO, Kendall FP, Boyton D. Posture and Pain. Baltimore: Williams & Wilkins, 1952.
64. Cahill BR. Quadrilateral space syndrome. In: Omer GE, Spinner M, eds. Management of Peripheral Nerve Problems. Philadelphia: W.B. Saunders, 1980, pp. 602–606.
65. Hangge PT, Breen I, Albadawi H, Knuttinen MG, Naidu SG, Oklu R. Quadrilateral space syndrome: Diagnosis and clinical management. J Clin Med 2018;7(4):86. Published April 21, 2018. doi: 10.3390/jcm7040086
66. Brown LP, Niehues SL, Harrah A, et al. Upper extremity range of motion and isokinetic strength of internal and external rotators in major league baseball players. In: McMahon PJ, Sallis RE, eds. The Painful Shoulder, Postgraduate Medicine, 1999;106(7): 36–38, 41–43, 47–49.
67. Spadliński Ł, Cecot T, Majos A, et al. The epidemiological, morphological, and clinical aspects of the cervical ribs in humans. Biomed Res Int 2016;2016:8034613. doi: 10.1155/2016/8034613
68. Banks KP, Ly JQ, Beall DP, Grayson DE, Bancroft LW, Tall MA. Overuse injuries of the upper extremity in the competitive athlete: Magnetic resonance imaging findings associated with repetitive trauma. Curr Probl Diagn Radiol. 2005 July–August;34(4):127–142. doi: 10.1067/j.cpradiol.2005.04.001. PMID: 16012484.
69. Moseley JB Jr, Jobe FW, Pink M, Perry J, Tibone J. EMG analysis of the scapular muscles during a shoulder rehabilitation program. Am J Sports Med 1992 March–April;20(2):128–134. doi: 10.1177/036354659202000206. PMID: 1558238.
70. Lorenz DS, Reiman MP, Walker JC. Periodization: Current review and suggested implementation for athletic rehabilitation. Sports Health 2010 Nov;2(6):509–518. doi: 10.1177/1941738110375910. PMID: 23015982; PMCID: PMC3438871.
71. Ada L, Foongchomcheay A, Canning C. Supportive devices for preventing and treating subluxation of the shoulder after stroke. Cochrane Database Syst Rev 2005;2005(1):CD003863. Published January 25, 2005. doi: 10.1002/14651858.CD003863.pub
72. Nadler M, Pauls M. Shoulder orthoses for the prevention and reduction of hemiplegic shoulder pain and subluxation: Systematic review. Clin Rehabil 2017 Apr;31(4):444–453. doi: 10.1177/0269215516648753. Epub July 10, 2016. PMID: 27184582.
73. Burstein D. Joint compression for treatment of shoulder pain. Clin Man 1985;5(2):9.

LOWER EXTREMITY

7

OUTLINE

INTRODUCTION

The lower extremities provide both support and mobility for the body as a whole. Fulfilling these roles requires that good muscle balance of the lower extremity muscles be established and maintained (1).

Unlike the upper extremity where one plexus supplies the arm muscles, the lower extremity is supplied by both the lumbar and sacral plexuses. Differential diagnosis of joint movement impairments in the hip region requires particular attention because of the different origins of the nerves and the multitude of muscles that can be involved. Many of the muscles cross the hip and the knee, and distinguishing impairments among the muscles can be challenging (2). Different impairments can give rise to similar symptoms.

Effective treatment of musculoskeletal impairments depends upon an accurate assessment of the length and strength of the muscles. Serious mistakes can result from failure to pay attention to details. Because actions of the muscles of the hip are closely related, there can be compensations with cases of muscle weakness or muscle length variation. Failure to detect such compensations, or enabling them through incorrect test positions or movements, will render test results invalid.

In order to begin the process of establishing a differential diagnosis and developing a successful treatment plan, it is necessary to have a comprehensive knowledge of the innervation, joint movements, alignment of body segments, and precise testing procedures for length and strength of these muscles. In addition, unique case studies with charts showing test findings are included in this chapter to demonstrate approaches to address clinical diagnoses related to lower extremity dysfunction.

SECTION I
INNERVATION

LUMBAR PLEXUS

The **lumbar plexus** is formed by the ventral primary rami of L1, L2, and L3, a part of L4, and frequently, with a small contribution from T12. Within the substance of the psoas major muscle, the rami branch into anterior and posterior divisions. Peripheral nerves from the anterior divisions innervate adductor muscles on the medial side of the thigh; those from the posterior divisions innervate hip flexors and knee extensors on the anterior aspect of the thigh (Figure 7-1).

SACRAL PLEXUS

The sacral plexus receives contributions from ventral primary rami of L4, L5, S1, S2, and S3. The L4 and L5 ventral rami unite to form the lumbosacral trunk, which enters the pelvic cavity. There, it is joined by the ventral rami of S1, S2, and S3, forming the plexus, which then branches into anterior and posterior divisions. The anterior divisions and the peripheral nerves arising from them include cutaneous nerves and nerves that innervate muscles of the posterior thigh, posterior leg, and intrinsic muscles of the plantar surface of the foot. The posterior divisions, and the peripheral nerves arising from them, innervate the gluteal region on the lateral side of the thigh, the short head of the biceps femoris, and the extensor (dorsiflexor) muscles of the ankle and toes (Figure 7-2).

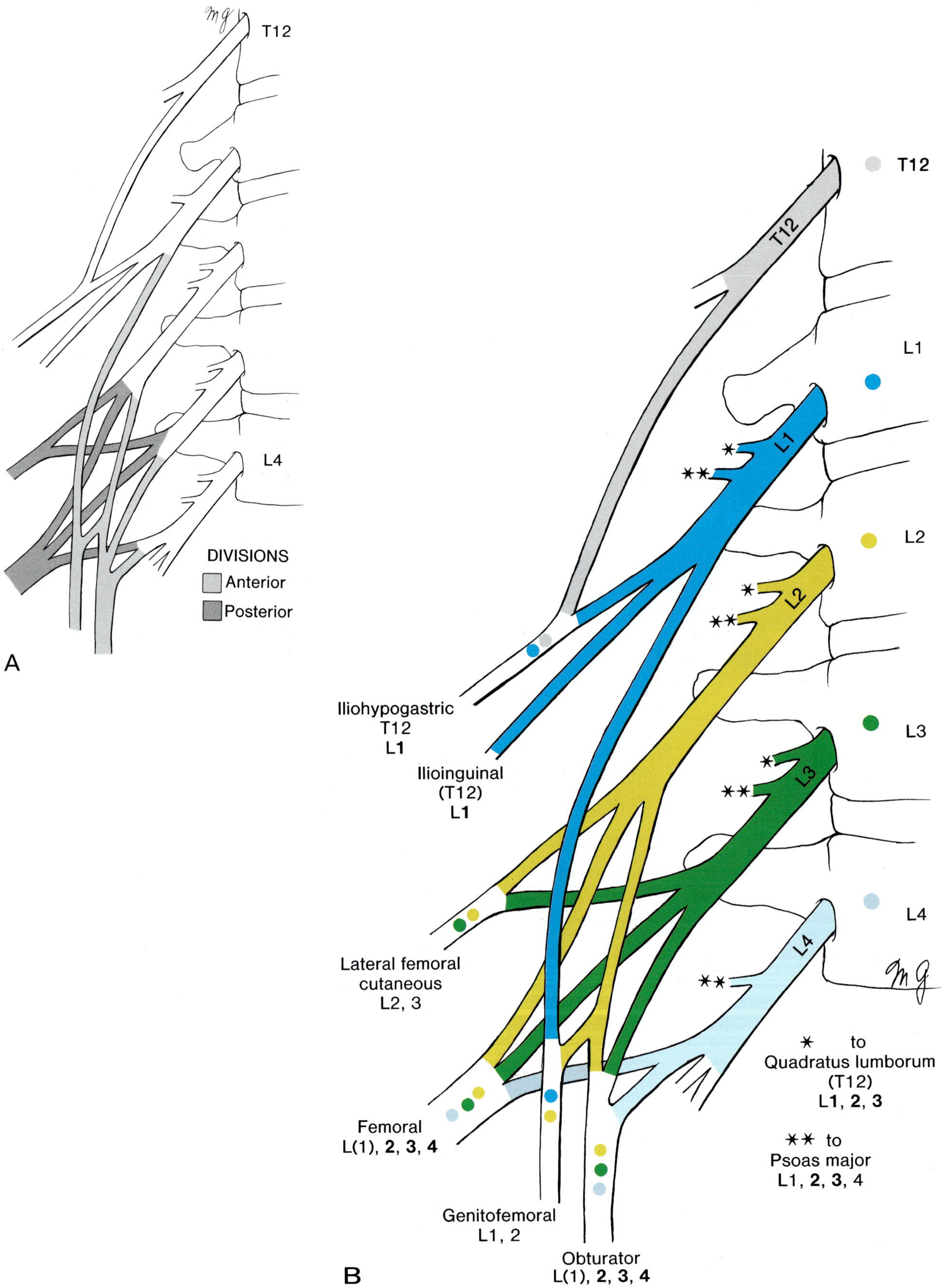

FIGURE 7-1: The lumbar plexus.

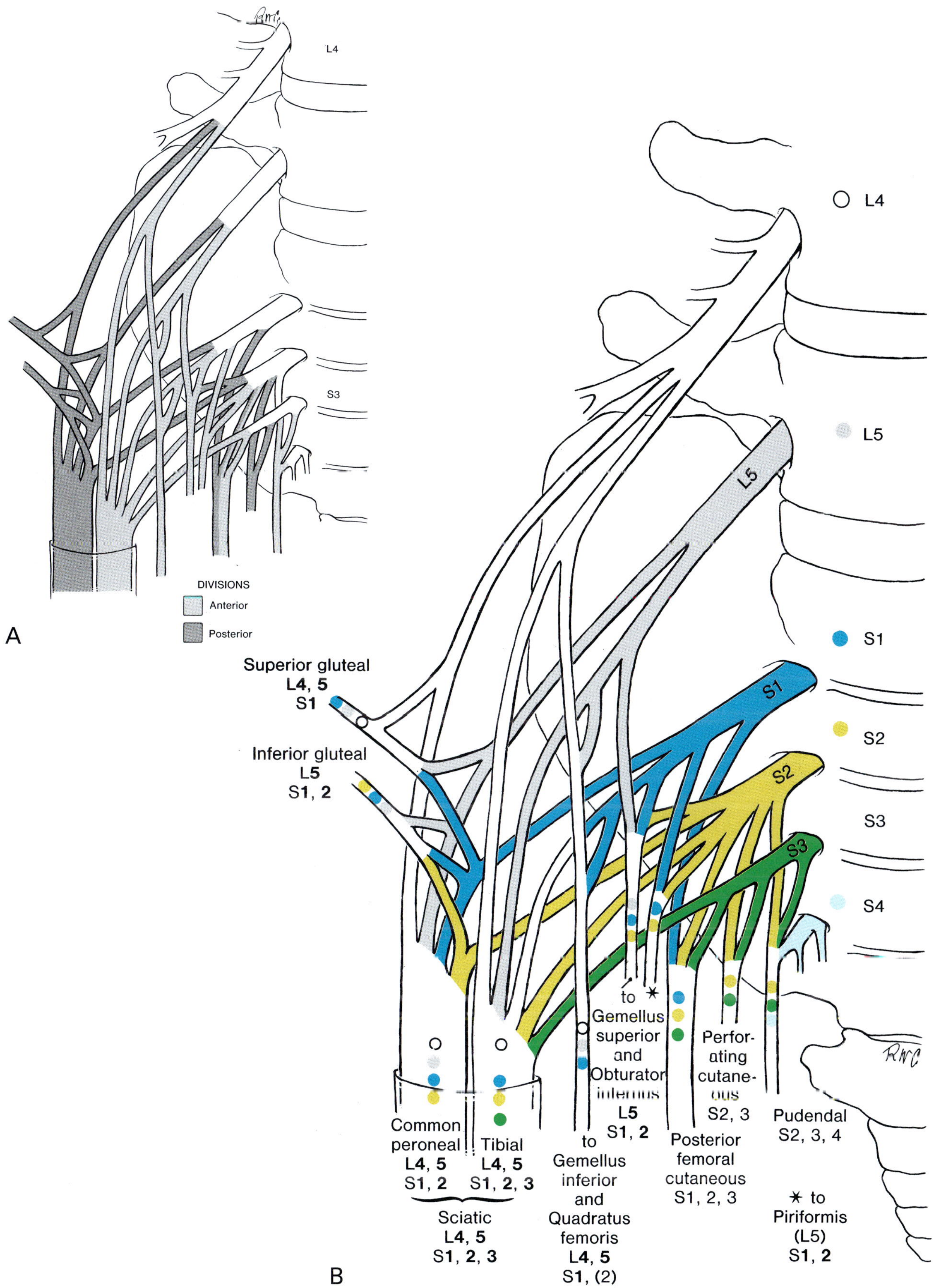

FIGURE 7-2: The sacral plexus.

SPINAL NERVE AND MUSCLE CHART

Lower Extremity

Figure 7-3 shows the lower extremity portion of the spinal nerve and muscle chart.

Name Date

KEY
D Dorsal Primary Ramus
V Ventral Primary Ramus
A Anterior Division
P Posterior Division

		MUSCLE	MUSCLE STRENGTH GRADE	D. T1-12, L1-5, S1-3	V. T1, 2, 3, 4	V. T5, 6	V. T7, 8	V. T9, 10, 11, 12	V. Iliohypogastric T12 L1	V. Ilioinguinal T(12) L1	V. Lumb. Plex. T(12) L1, 2, 3, 4	P. Femoral L(1) 2, 3, 4	A. Obturator L(1) 2, 3, 4	P. Sup. Glut. L4, 5, S1	P. Inf. Glut. L5, S1, 2	V. Sac. Plex. L4, 5, S1, 2, 3	P. Sciatic L4, 5, S1, 2	A. Sciatic L4, 5, S1, 2, 3	P. C. Peroneal L4, 5, S1, 2	A. Tibial L4, 5, S1, 2, 3	L1	L2	L3	L4	L5	S1	S2	S3
Lumb.	Plexus	QUAD LUMBORUM									•										1	2	3					
Lumb.	Plexus	PSOAS MINOR									•										1	2						
Lumb.	Plexus	PSOAS MAJOR									•										1	2	3	4				
Femoral		ILIACUS										•									(1)	2	3	4				
Femoral		PECTINEUS										•	(•)									2	3	4				
Femoral		SARTORIUS										•										2	3	(4)				
Femoral		QUADRICEPS										•										2	3	4				
Obturator	Ant.	ADDUCTOR BREVIS											•									2	3	4				
Obturator	Ant.	ADDUCTOR LONGUS											•									2	3	4				
Obturator	Ant.	GRACILIS											•									2	3	4				
Obturator	Post	OBTURATOR EXT											•										3	4				
Obturator	Post	ADDUCTOR MAGNUS											•					•				2	3	4	5	1		
Gluteal	Sup	GLUTEUS MEDIUS												•										4	5	1		
Gluteal	Sup	GLUTEUS MINIMUS												•										4	5	1		
Gluteal	Sup	TENSOR FAS LAT												•										4	5	1		
Gluteal	In.	GLUTEUS MAXIMUS													•										5	1	2	
Sacral	Plexus	PIRIFORMIS														•									(5)	1	2	
Sacral	Plexus	GEMELLUS SUP														•									5	1	2	
Sacral	Plexus	OBTURATOR INT														•									5	1	2	
Sacral	Plexus	GEMELLUS INF														•								4	5	1	(2)	
Sacral	Plexus	QUADRATUS FEM														•								4	5	1	(2)	
Sciatic	P	BICEPS (SHORT H)															•								5	1	2	
Sciatic	Tibial	BICEPS (LONG H)																•							5	1	2	3
Sciatic	Tibial	SEMITENDINOSUS																•						4	5	1	2	
Sciatic	Tibial	SEMIMEMBRANOSUS																•						4	5	1	2	
Common Peroneal	Deep	TIBIALIS ANTERIOR																	•					4	5	1		
Common Peroneal	Deep	EXT HALL LONG																	•					4	5	1		
Common Peroneal	Deep	EXT DIGIT LONG																	•					4	5	1		
Common Peroneal	Deep	PERONEUS TERTIUS																	•					4	5	1		
Common Peroneal	Deep	EXT DIGIT BREVIS																	•					4	5	1		
Common Peroneal	Sup	PERONEUS LONGUS																	•					4	5	1		
Common Peroneal	Sup	PERONEUS BREVIS																	•					4	5	1		
Tibial	Tibial	PLANTARIS																		•				4	5	1	(2)	
Tibial	Tibial	GASTROCNEMIUS																		•						1	2	
Tibial	Tibial	POPLITEUS																		•				4	5	1		
Tibial	Tibial	SOLEUS																		•					5	1	2	
Tibial	Tibial	TIBIALIS POSTERIOR																		•				(4)	5	1		
Tibial	Tibial	FLEX DIGIT LONG																		•					5	1	(2)	
Tibial	Tibial	FLEX HALL LONG																		•					5	1	2	
Tibial	Med Pl	FLEX DIGIT BREVIS																		•				4	5	1		
Tibial	Med Pl	ABDUCTOR HALL																		•				4	5	1		
Tibial	Med Pl	FLEX HALL BREVIS																		•				4	5	1		
Tibial	Med Pl	LUMBRICALIS I																		•				4	5	1		
Tibial	Lat Plant	ABD DIGITI MIN																		•						1	2	
Tibial	Lat Plant	QUAD PLANTAE																		•						1	2	
Tibial	Lat Plant	FLEX DIGITI MIN																		•						1	2	
Tibial	Lat Plant	OPP. DIGITI MIN																		•						1	2	
Tibial	Lat Plant	ADDUCTORS HALL																		•						1	2	
Tibial	Lat Plant	PLANT INTEROSSEI																		•						1	2	
Tibial	Lat Plant	DORSAL INTEROSSEI																		•						1	2	
Tibial	Lat Plant	LUMB II,III,IV																		•				(4)	(5)	1	2	

SENSORY

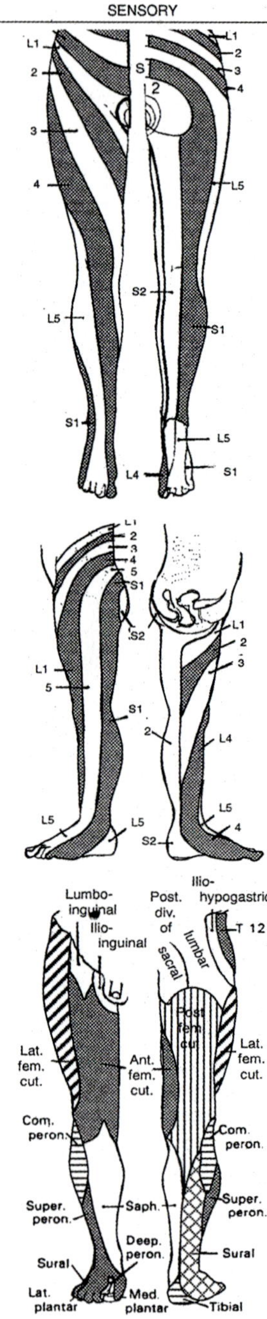

FIGURE 7-3: Spinal nerve and muscle chart: Lower extremity. © 1993 Florence P. Kendall.

SPINAL NERVE AND MOTOR POINT CHART

Figure 7-4 illustrates the spinal nerve and motor points.

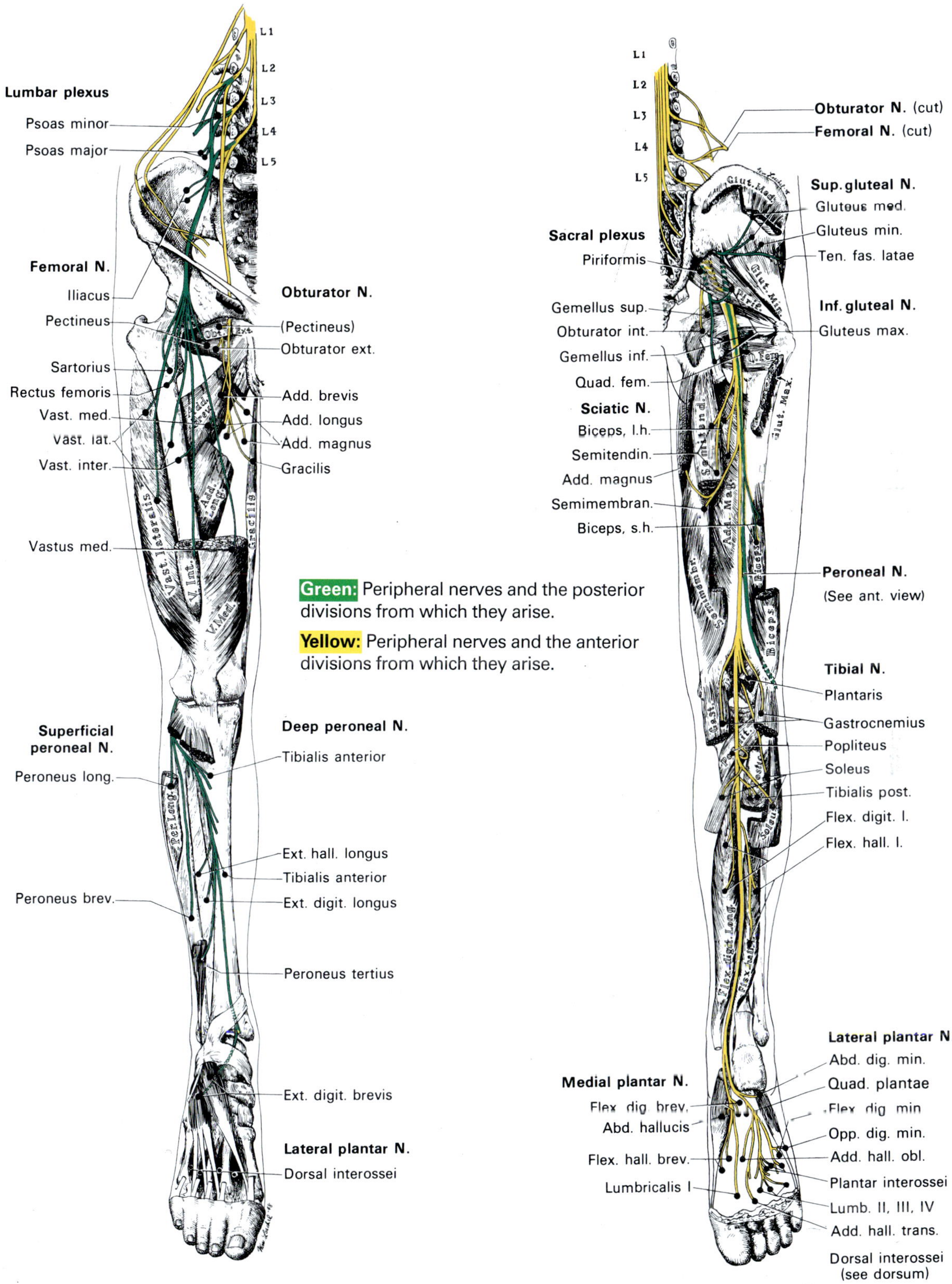

FIGURE 7-4: Spinal nerve and motor point chart. © 1993 Florence P. Kendall.

LOWER EXTREMITY MUSCLES CHART

Figure 7-5 lists the lower extremity muscles according to spinal segment innervation and grouped according to joint action.

Spinal Segment																	
Lumb.					Sac.				HIP						KNEE		
																In flexion	
1	2	3	4	5	1	2	3	Muscle	Flexion	Adduction	Med. Rotat.	Abduction	Lat. Rotat.	Extension	Extension	Lat. Rotat.	Med. Rotat.
1	2	3	4					Psoas major	Psoas maj.			Psoas maj.	Psoas maj.				
(1)	2	3	4					Iliacus	Iliacus			Iliacus	Iliacus				
	2	3	(4)					Sartorius	Sartorius			Sartorius	Sartorius				Sartorius
	2	3	4					Pectineus	Pectineus	Pectineus							
	2	3	4					Adductor long.	Add. long.	Add. long.	Add. long.						
	2	3	4					Adductor brev.	Add. brev.	Add. brev.	Add. brev.						
	2	3	4					Gracilis		Gracilis							Gracilis
	2	3	4					Quadriceps	Rect. fem.						Quadriceps		
	2	3	4					Add. mag. (ant.)	Add. m. (ant.)	Add. mag.							
		3	4					Obturator ext.		Obt. ext.			Obt. ext.				
			4	5	1			Add. mag. (post.)		Add. mag.				Ad. m. post.			
			4	5	1			Tibialis ant.									
			4	5	1			Ten. fas. lat.	Tensor f.l.		Tensor f.l.	Tensor f.l.			Tensor f.l.		
			4	5	1			Gluteus minimus	Glut. min.		Glut. min.	Glut. min.					
			4	5	1			Gluteus medius	G. med., ant.		G. med., ant.	Glut. med.	G. med., post.	G. med., post.			
			4	5	1			Popliteus									Popliteus
			4	5	1			Ext. dig. long.									
			4	5	1			Peroneus tertius									
			4	5	1			Ext. hall. long.									
			4	5	1			Ext. dig. brev.									
			4	5	1			Flex. dig. brev.									
			4	5	1			Flex. hall. brev.									
			4	5	1			Lumbricalis I									
			4	5	1			Abductor hall.									
			4	5	1			Peroneus longus									
			4	5	1			Peroneus brevis									
			(4)	5	1			Tibialis post.									
			4	5	1	(2)		Gemelli inferior				Gem. inf.	Gem. inf.				
			4	5	1	(2)		Quadratus fem.					Quadratus f.				
			4	5	1	(2)		Plantaris									
			4	5	1	2		Semimembranosus			Semimemb.			Semimemb.			Semimemb.
			4	5	1	2		Semitendinosus			Semitend.			Semitend.			Semitend.
				5	1	(2)		Flex. dig. long.									
				5	1	2		Gluteus maximus		G. max., low.		G. max., upp.	Glut. max.	Glut. max.			
				5	1	2		Biceps, short h.								Bic., s.h.	
				5	1	2		Flex. hall. long.									
				5	1	2		Soleus									
				(5)	1	2		Piriformis				Piriformis	Piriformis	Piriformis			
				5	1	2		Gemelli superior				Gem. sup.	Gem. sup.				
				5	1	2		Obturator int.				Obt. int.	Obt. int.				
				5	1	2	3	Biceps, long h.					Biceps l.h.	Biceps l.h.		Bic., l.h.	
			(4)	(5)	1	2		Lumb. II, III, IV									
					1	2		Gastrocnemius									
					1	2		Dorsal inteross.									
					1	2		Plantar inteross.									
					1	2		Abd. dig. min.									
					1	2		Adductor hall.									

FIGURE 7-5: Lower extremity muscles chart. © 1993 Florence P. Kendall.

KNEE	ANKLE		FOOT		METATARSOPHALANGEAL JOINT				Digs. 2–5 Prox. Interphal. Jts.		Digs. 1–5 Distal Interphal. Jts.	
Flexion	Dorsiflex.	Plant flex.	Eversion	Inversion	Extension	Flexion	Abduction	Adduction	Extension	Flexion	Extension	Flexion
Sartorius												
Gracilis												
	Tib. ant.			Tib. ant.								
Popliteus												
	Ext. d. long.		Ext. d. long.		(2-5 dig.) Ext. d. long.				(2-5 dig.) Ext. d. long.		(2-5 dig.) Ext. d. long.	
	Peroneus t.		Peroneus t.									
	Ext. hall. l.			Ext. hall. l.	Ext. hall. l.						Ext. hall l.	
					(1-4 dig.) Ext. dig. br.				(1-4 dig.) Ext. dig. br.		(1-4 dig.) Ext. dig. br.	
						(2-5 dig.) Flex. dig. br.				(2-5 dig.) Flex. dig. br.		
						Flex. hall. br.						
						2nd dig. Lumb. I			(2nd dig.) Lumb. I		(2nd dig.) Lumb. I	
						Abd. hall.	Abd. hall.					
		Peroneus l.	Peroneus l.									
		Peroneus b.	Peroneus b.									
		Tib. post.		Tib. post.								
Plantaris		Plantaris										
Semimemb.												
Semitend.												
		Flex. dig. l.		Flex. dig. l.		(2-5 dig.) Flex. dig. l.				(2-5 dig.) Flex. dig. l.		(2-5 dig.) Flex. dig. l.
Bic., s.h.												
		Flex. hall. l.		Flex. hall. l.		Flex. hall. l.						Flex. hall. l.
		Soleus										
Bic., l.h.												
						(3-5 dig.) Lumb. II-IV			(3-5 dig.) Lumb. II-IV		(3-5 dig.) Lumb. II-IV	
Gastroc.		Gastroc.										
						(2-4 dig.) Dor. int.	(2-4 dig.) Dor. int.		(2-4 dig.) Dor. int.		(2-4 dig.) Dor. int.	
						(3-5 dig.) Plant. int.		(3-5 dig.) Plant. int.	(3-5 dig.) Plant. int.		(3-5 dig.) Plant. int.	
							Abd. d. min.					
						Add. hall.		Add. hall.				

FIGURE 7-5: (*Continued*).

NERVES TO MUSCLES: MOTOR AND SENSORY OR MOTOR

Table 7-1 lists the innervation of nerves to muscles.

Femoral: Pierces the psoas major at the distal part of the lateral border and supplies the iliacus, pectineus, sartorius, and quadriceps. The largest and longest branch of the femoral nerve is the *saphenous nerve,* which innervates the skin over the medial side of the leg.

Obturator: From L2 through L4. Through its *muscular* branch, it supplies the obturator externus, adductor magnus (horizontal and middle portions), adductor longus, and adductor brevis. Through its *articular branch,* it is distributed to the synovial membrane of the knee joint.

Sciatic: From L4, 5, and S1, 2, 3. In most instances, the sciatic nerve exits the pelvis by passing through the greater sciatic foramen inferior to the piriformis muscle and lies superficial to the obturator internus, gemelli, and quadratus femoris (see Figure 7-112 later in this chapter). Variations exist, however, in which the piriformis muscle is split and either one (usually the fibular) or both parts of the sciatic nerve pass through the muscle belly. The sciatic nerve is comprised of anterior (tibial) and posterior (common fibular) divisions that eventually divide within the proximal aspect of the popliteal fossa.

Common Fibular: Passes between the biceps femoris and the lateral head of the gastrocnemius to the head of the fibula and deep to the fibularis longus (see Figure 7-92 later in this chapter). It supplies muscles of the anterior (deep fibular) and lateral compartments (superficial fibular) of the leg, then continues to innervate the extensor hallucis and digitorum brevis.

Tibial: Innervates the posterior thigh muscles and posterior leg muscles before dividing into its terminal branches, the medial and lateral plantar nerves.

TABLE 7-1 Nerves to Muscles: Motor and Sensory or Motor

Source		Spinal Segment	Nerve	Motor/Sensory to Muscle	Muscle
Lumbar plexus	Ventral primary ramus	T12, L1	Iliohypogastric	Motor and sensory	Internal oblique, transverses abdominis
		L1, 2, 3, 4	Lumbar plexus	Motor and sensory	Quadratus lumborum, psoas major, psoas minor
	Posterior division	L2, 3, 4	Femoral	Motor and sensory	Iliacus, pectineus, sartorius, quadriceps
	Anterior division	L2, 3, 4	Obturator	Motor and sensory	Hip adductors
Lumbosacral plexus	Posterior division	L4, 5, S1	Gluteal, superior	Motor[a]	Gluteus medius, gluteus minimus, tensor fasciae latae
	Posterior division	L5, S1, 2	Gluteal, inferior	Motor and sensory	Gluteus maximus
Sciatic nerve	Posterior division	L4, 5, S1, 2	Peroneal	Motor and sensory	Short head of biceps, tibialis anterior, toe extensors, peroneals
	Anterior division	L4, 5, S1, 2, 3	Tibial	Motor and sensory	Semimembranosus, semitendinosus, long head of biceps, 19 ankle and foot muscles
Sacral plexus	Ventral primary ramus	L4, 5, S1, 2, 3	Sacral plexus	Motor and sensory	Piriformis, gemelli superior and inferior, obturator internus, and quadrates femoris

[a]Sensory to hip joint

CUTANEOUS NERVES OF THE LOWER LIMB

Note in Figure 7-6B, *sural* is Latin for the calf. In this illustration, the medial sural cutaneous nerve is joined just proximal to the ankle by a communicating branch (not labeled) of the lateral sural cutaneous nerve to form the sural nerve. The level of the junction is variable, however, being very low in Figure 7-6B.

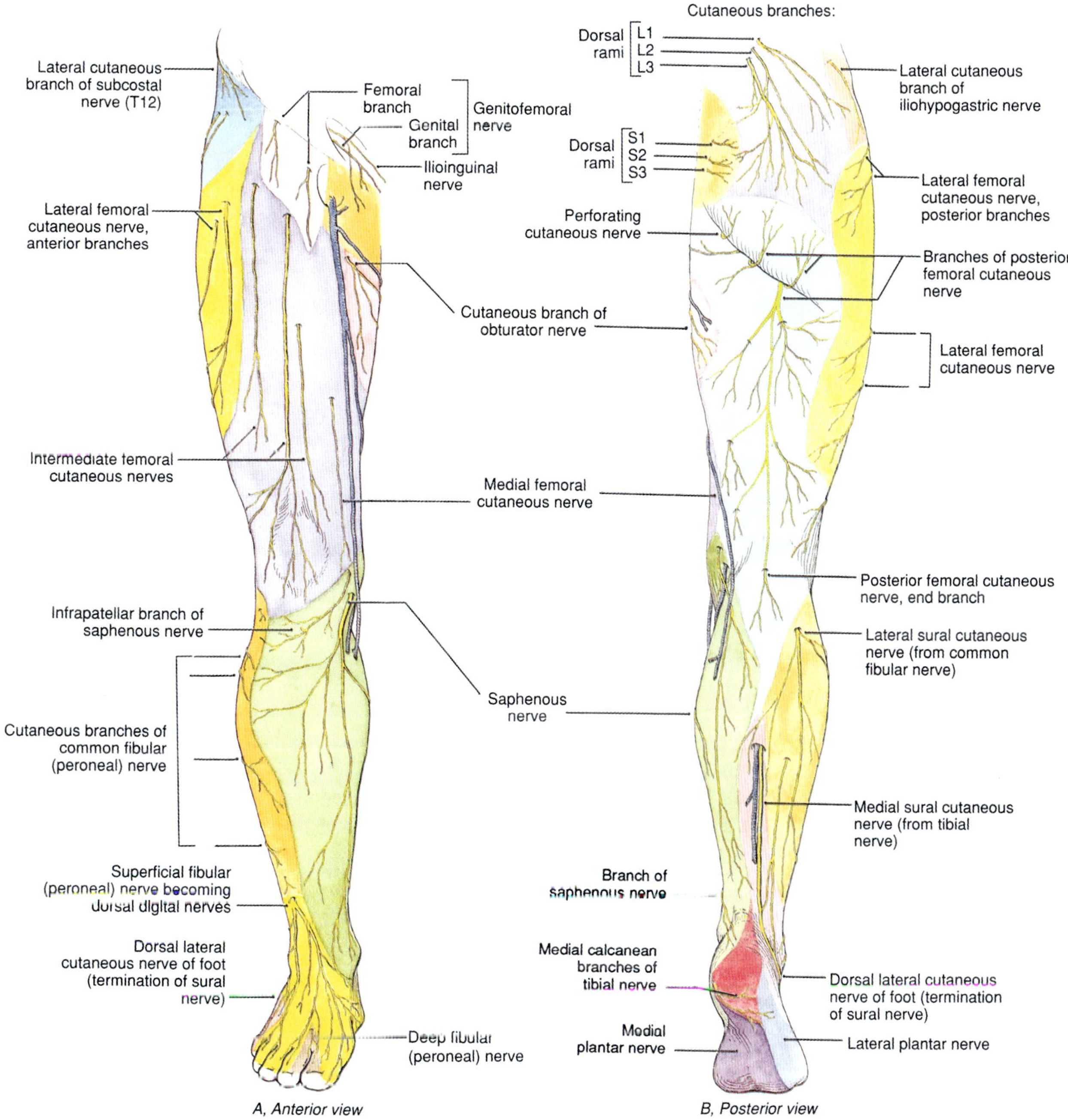

FIGURE 7-6: Cutaneous nerves of the lower limb: **A**, Anterior view; **B**, Posterior view. From Agur AMR, *Grant's atlas of anatomy*. 9th ed. Baltimore: Williams and Wilkins; 1991: 263 (3).

SECTION II
FOOT AND ANKLE STRENGTH AND EXTENSIBILITY TESTS

JOINT MOVEMENTS

Interphalangeal Joints of Toes

The interphalangeal joints are ginglymus or hinge joints that connect the adjacent surfaces of phalanges. Movements observed at the interphalangeal joints include flexion and extension about a coronal axis, with flexion directed in a caudal direction and extension directed in a cranial direction.

Metatarsophalangeal Joints

The metatarsophalangeal joints are condyloid, formed by joining the distal ends of the metatarsals with the adjacent ends of the proximal phalanges.

Flexion and extension occur at the metatarsal phalangeal joints and are similar to the description provided above for the interphalangeal joints. The range of motion in adults is variable, but 40° of flexion (45° for great toe) and 40° of extension (70° for great toe) may be considered an average range for good function of the toes.

Adduction and *abduction* are movements about a sagittal axis. The line of reference for both adduction and abduction of the toes is the axial line projected distally in line with the second metatarsal and extending through the second digit. Adduction is movement toward the axial line, and abduction is movement away from it, as in spreading the toes apart. Because abduction of the toes is restricted by the wearing of shoes, this movement is markedly limited in most adults, and little attention is paid to the lack of ability to abduct.

Subtalar Joint and Transverse Tarsal Joints

The subtalar joint is a multiaxial, modified plane, or gliding joint connecting the talus and calcaneus. The talus also connects with the navicular, and the talonavicular joint is involved in movements of the transverse tarsal joint.

Supination and pronation are movements permitted by the subtalar and talocalcaneonavicular joint. Supination is rotation of the foot in which the sole of the foot moves in a medial direction; pronation is rotation in which the sole of the foot moves in a lateral direction.

The transverse tarsal joint is a compilation of two joints. They are formed by articulation between the talus, calcaneus, and navicular bones (talocalcaneonavicular), and the calcaneus with the cuboid bones (calcaneocuboid).

Adduction and abduction of the forefoot are movements permitted by the transverse tarsal joints. Adduction is movement of the forefoot in a medial direction, and abduction is movement of the forefoot in a lateral direction.

Inversion is a combination of supination and forefoot adduction. More inversion range of motion is available in plantar flexion compared with dorsiflexion.

Eversion is a combination of pronation and forefoot abduction. More eversion range of motion is available in dorsiflexion compared to plantar flexion.

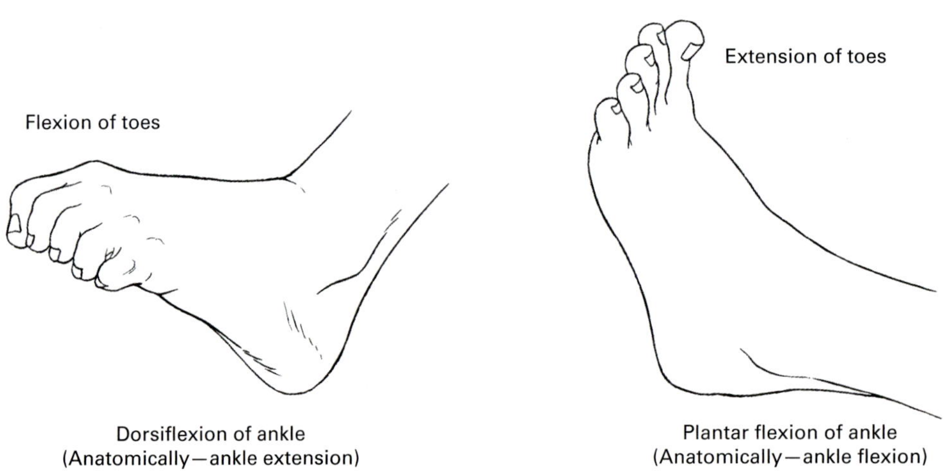

FIGURE 7-7: Movements of the ankle (talocrural) joint.

Movements of Ankle (Talocrural) Joint

The ankle joint is a ginglymus or hinge joint uniting the tibia and fibula with the talus. The axis about which motion takes place extends obliquely from the postero-lateral aspect of the fibular malleolus to the anteromedial aspect of the tibial malleolus.

Plantar flexion (flexion) and dorsiflexion (extension) are the two observable movements that occur about the oblique axis (Figure 7-7). Plantar flexion is movement of the foot in which the plantar surface moves in a caudal and posterior direction. Dorsiflexion is movement of the foot in which the dorsal surface moves in an anterior and cranial direction.

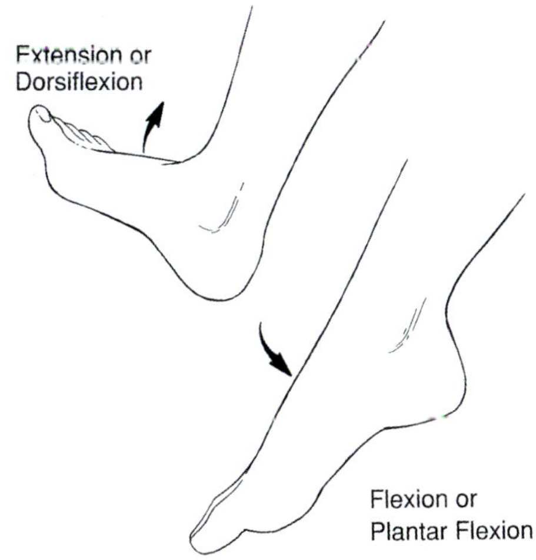

FIGURE 7-8: Ankle movements about the oblique axis.

TESTS FOR LENGTH OF ANKLE PLANTAR FLEXORS

One-Joint Plantar Flexors

Soleus and Popliteus

Action: Ankle plantar flexion

Length Test: Ankle dorsiflexion, *with the knee in flexion.*

Starting Position: Sitting or supine, with the hip and knee flexed.

Test Movement: With the knee flexed 90° or more to make the two-joint gastrocnemius and plantaris slack over the knee joint, dorsiflex the foot (Figure 7-8).

Normal Range: The foot can be dorsiflexed approximately 20°.

Sit forward in a chair with knees bent and feet pulled back toward chair enough to raise the heels slightly from the floor. Press down on thigh to help force heel to the floor (Figure 7-9).

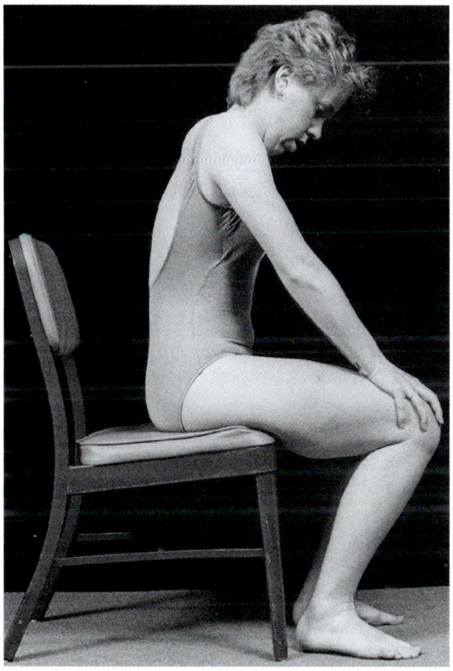

FIGURE 7-9: Testing one-joint plantar flexors.

FIGURE 7-10: Testing two-joint plantar flexors.

Two-Joint Plantar Flexors

Gastrocnemius and Plantaris

Action: Ankle plantar flexion and knee flexion.

Length Test: Ankle dorsiflexion and knee extension.

Starting Position: Supine or sitting, with the knees extended unless hamstring tightness causes the knee to flex.

Test Movement: With the knee in extension to elongate the gastrocnemius and plantaris over the knee joint, dorsiflex the foot.

Normal Range: With the knee fully extended, the foot can be dorsiflexed approximately 10°.

Stand erect on board inclined at a 10° angle, with feet in approximately 8°–10° of out-toeing (Figure 7-10).

Muscle Strength Tests

ABDUCTOR HALLUCIS AND ADDUCTOR HALLUCIS

Abductor Hallucis

Origin: Medial process of tuberosity of the calcaneus, flexor retinaculum, plantar aponeurosis, and adjacent intermuscular septum (Figure 7-11A).

Insertion: Medial side of the base of the proximal phalanx of the great toe. Some fibers are attached to the medial sesamoid bone, and a tendinous slip may extend to the base of the proximal phalanx of the great toe.

Action: Abducts and assists in flexion of the metatarsophalangeal joint of the great toe, and assists with adduction of the forefoot.

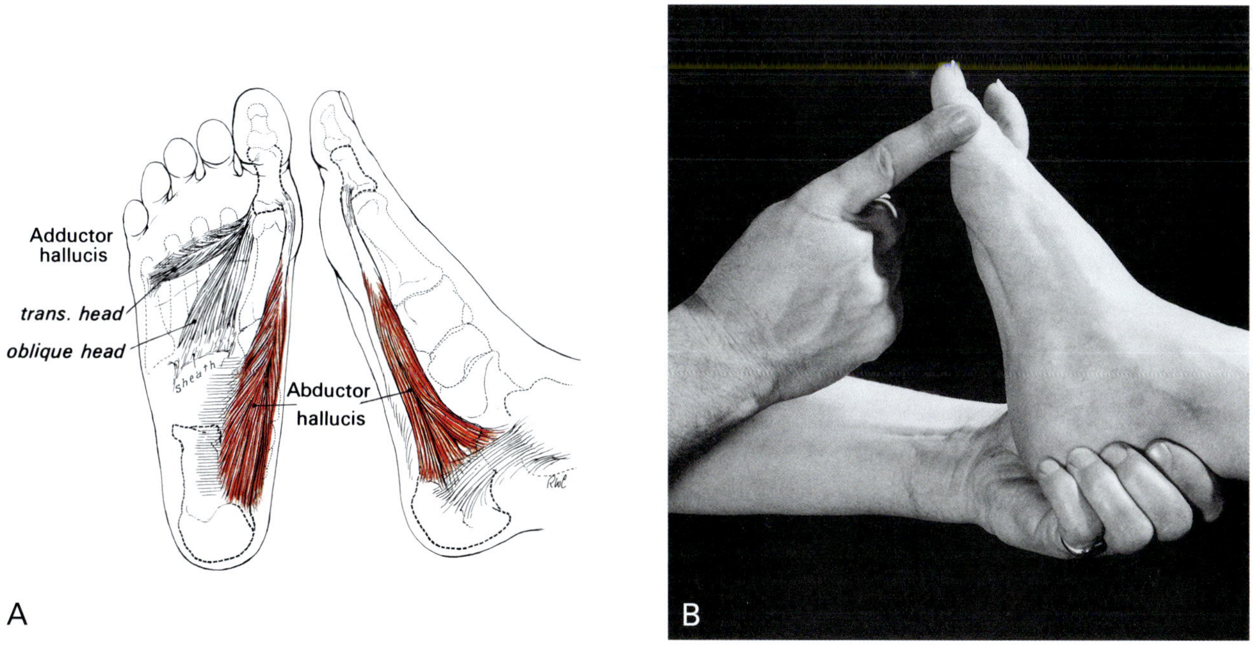

FIGURE 7-11: **A**, Abductor hallucis and adductor hallucis; **B**, Testing abductor hallucis.

Nerve: Tibial, L4, **5**, S**1**.

Patient: Supine or sitting.

Fixation: The examiner grips the heel firmly.

Test: If possible, abduction of the big toe from the axial line of the foot. This is difficult for the average individual, and the action may be demonstrated by having the patient pull the forefoot in adduction against pressure by the examiner (Figure 7-11B).

Pressure: Against the medial side of the first metatarsal and proximal phalanx. The muscle can be palpated and often seen along the medial border of the foot.

Weakness: Allows forefoot valgus, hallux valgus, and medial displacement of the navicular.

Contracture: Pulls the foot into forefoot varus, with the big toe abducted.

Adductor Hallucis

Origin:

Oblique head: From the bases of the second through fourth metatarsal bones and the sheath of the tendon of the fibularis longus.

Transverse head: From the plantar metatarsophalangeal ligaments of the third through fifth digits and the deep transverse metatarsal ligament (Figure 7-11A).

Insertion: Lateral side of the base of the proximal phalanx of the great toe.

Action: Adducts and assists in flexing the metatarsophalangeal joint of the great toe.

Nerve: Tibial, S**1**, **2**.

Contracture: Adduction deformity of the great toe (i.e., hallux valgus).

NOTE: *No test illustrated for adductor hallucis.*

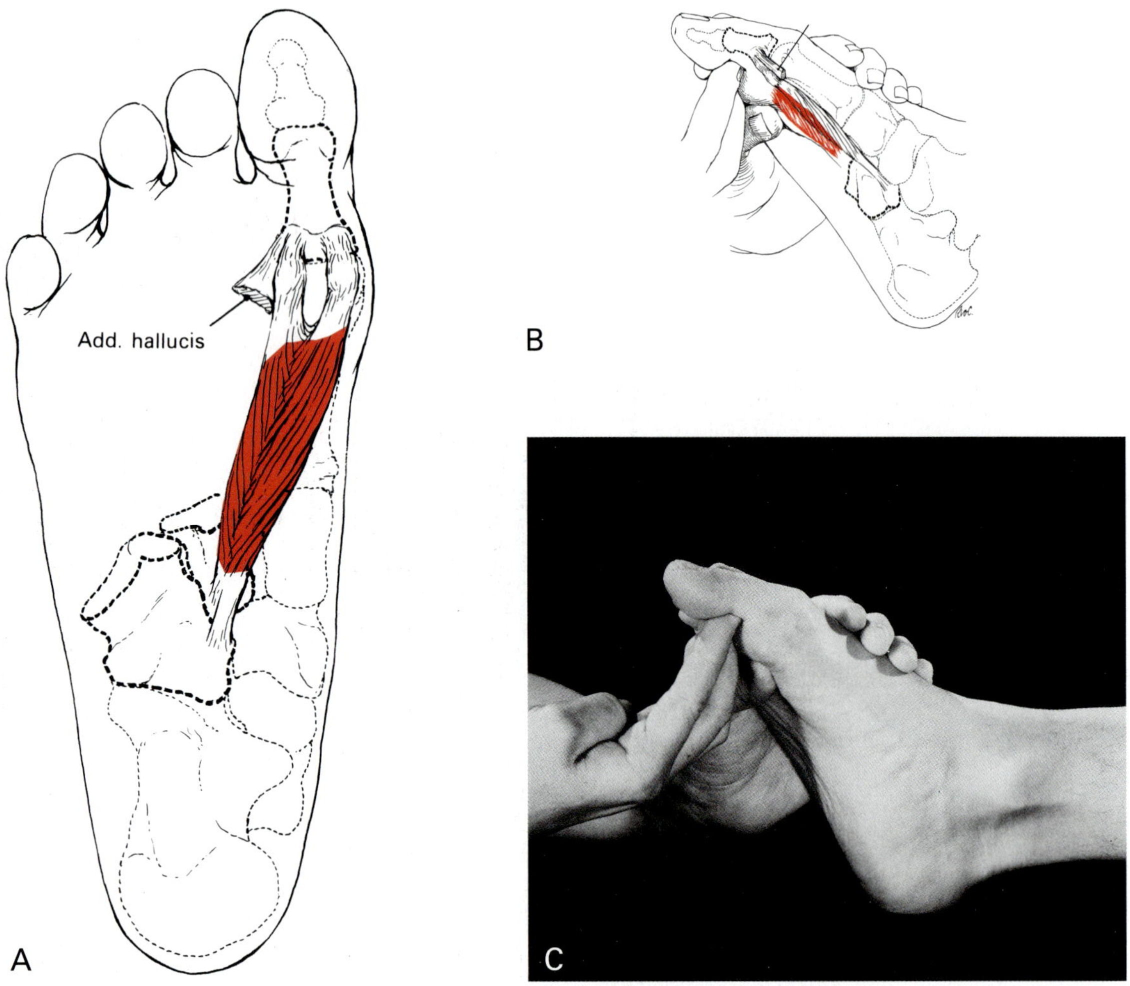

FIGURE 7-12: **A**, Flexor hallucis brevis (adductor hallucis reflected); **B**, **C**, Testing flexor hallucis brevis.

Flexor Hallucis Brevis

Origin: Medial part of the plantar surface of the cuboid bone, adjacent part of the lateral cuneiform bone, and from prolongation of the tendon of the tibialis posterior (Figure 7-12, Part A and B).

Insertion: Medial and lateral sides of the base of the proximal phalanx of the great toe.

Action: Flexes the metatarsophalangeal joint of the great toe.

Nerve: Tibial, L4, **5**, **S1**.

Patient: Supine or sitting.

Fixation: The examiner stabilizes the foot proximal to the metatarsophalangeal joint and maintains a neutral position of the foot and ankle (Figure 7-12C). (A plantar flexed position may cause restriction of the test movement by tension of the opposing long toe extensor muscles.)

Test: Flexion of the metatarsophalangeal joint of the great toe.

Pressure: Against the plantar surface of the proximal phalanx, in the direction of extension.

NOTE: *When the flexor hallucis longus is paralyzed and the brevis is active, the action of the brevis is clear, because the toe flexes at the metatarsophalangeal joint without any flexion of the interphalangeal joint. When the flexor hallucis brevis is paralyzed and the longus is active, the metatarsophalangeal joint hyperextends, and the interphalangeal joint flexes.*

Weakness: Allows a hammer-toe position of the great toe, and lessens stability of the longitudinal arch.

Contracture: The proximal phalanx is held in flexion.

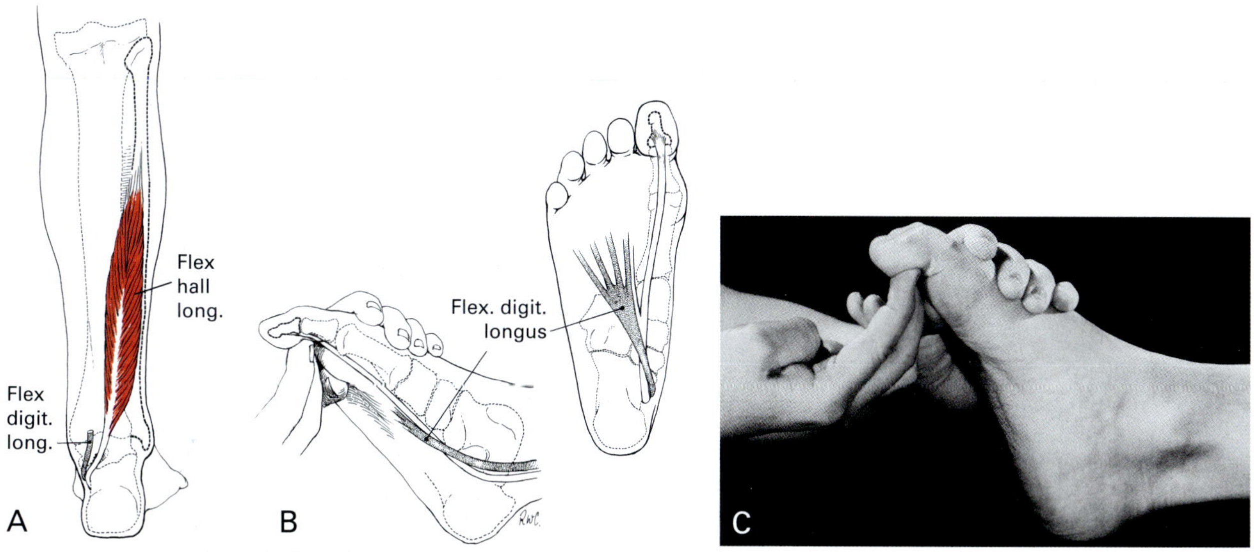

FIGURE 7-13: **A**, Flexor hallucis longus; **B**, **C**, Testing flexor hallucis longus.

Flexor Hallucis Longus

Origin: Posterior surface of the distal ⅔ of the fibula, interosseous membrane, and adjacent intermuscular septa and fascia (Figure 7-13A).

Insertion: Base of the distal phalanx of the great toe, plantar surface (Figure 7-13B).

> NOTE: *The flexor hallucis longus is connected to the flexor digitorum longus by a strong tendinous slip.*

Action: Flexes the interphalangeal joint of the great toe, and assists in flexion of the metatarsophalangeal joint, plantar flexion of the ankle joint, and inversion of the foot.

Nerve: Tibial, L5, S1, 2.

Patient: Supine or sitting.

Fixation: The examiner stabilizes the metatarsophalangeal joint in a neutral position and maintains the ankle joint approximately midway between dorsal and plantar flexion (Figure 7-13C). (Full dorsiflexion may produce passive flexion of the interphalangeal joint, and full plantar flexion would allow the muscle to shorten too much to exert its maximum force.) If the flexor hallucis brevis is very strong and the flexor hallucis longus weak, it is necessary to restrict the tendency for the metatarsophalangeal joint to flex by holding the proximal phalanx in slight extension.

Test: Flexion of the interphalangeal joint of the great toe.

Pressure: Against the plantar surface of the distal phalanx, in the direction of extension.

Weakness: Results in a tendency toward hyperextension of the interphalangeal joint and hammer-toe deformity of the great toe. Decreases the strength of inversion of the foot and plantar flexion of the ankle. In weight-bearing, permits a tendency toward pronation of the foot.

Contracture: Claw-toe deformity of the great toe.

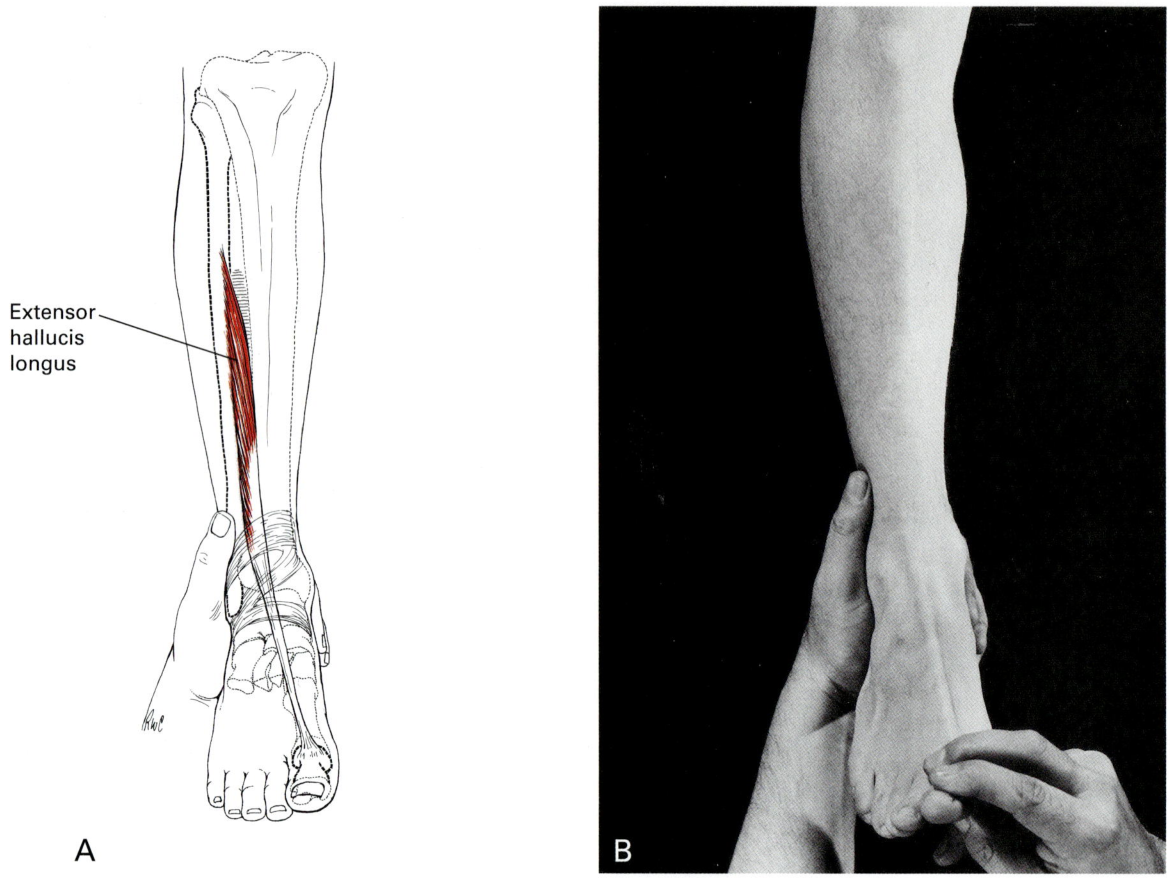

FIGURE 7-14: **A**, Extensor hallucis longus; **B**, Testing extensor hallucis longus (brevis test not shown).

EXTENSOR HALLUCIS LONGUS AND BREVIS

Extensor Hallucis Longus

Origin: Middle two quarters of the anterior surface of the fibula and adjacent interosseous membrane (Figure 7-14A).

Insertion: Base of the distal phalanx of the great toe.

Action: Extends the metatarsophalangeal and interphalangeal joints of the great toe, and assists in inversion of the foot and dorsiflexion of the ankle joint.

Nerve: Deep fibular, L4, **5**, S**1**.

Extensor Hallucis Brevis (Medial Slip of Extensor Digitorum Brevis)

Origin: Distal part of the superior and lateral surfaces of the calcaneus, lateral talocalcaneal ligament, and apex of inferior extensor retinaculum.

Insertion: Dorsal surface of the base of the proximal phalanx of the great toe.

Action: Extends the metatarsophalangeal joint of the great toe.

Nerve: Deep fibular, L4, **5**, S**1**.

Patient: Supine or sitting.

Fixation: The examiner stabilizes the foot in slight plantar flexion (Figure 7-14B).

Test: Extension of the metatarsophalangeal and interphalangeal joints of the great toe.

Pressure: Against the dorsal surface of the distal and proximal phalanges of the great toe in the direction of flexion.

Weakness: Decreases the ability to extend the great toe, and allows a position of flexion. The ability to dorsiflex the ankle joint is decreased.

Contracture: Extension of the great toe, with the head of the first metatarsal driven downward.

> **NOTE:** *Paralysis of the extensor hallucis brevis (first slip of the extensor digitorum brevis) cannot be determined accurately in the presence of a strong extensor hallucis longus. In paralysis of the extensor hallucis longus, however, the action of the extensor hallucis brevis is clear. The distal phalanx does not extend, and the proximal phalanx extends in the direction of adduction (i.e., toward the axial line of the foot).*

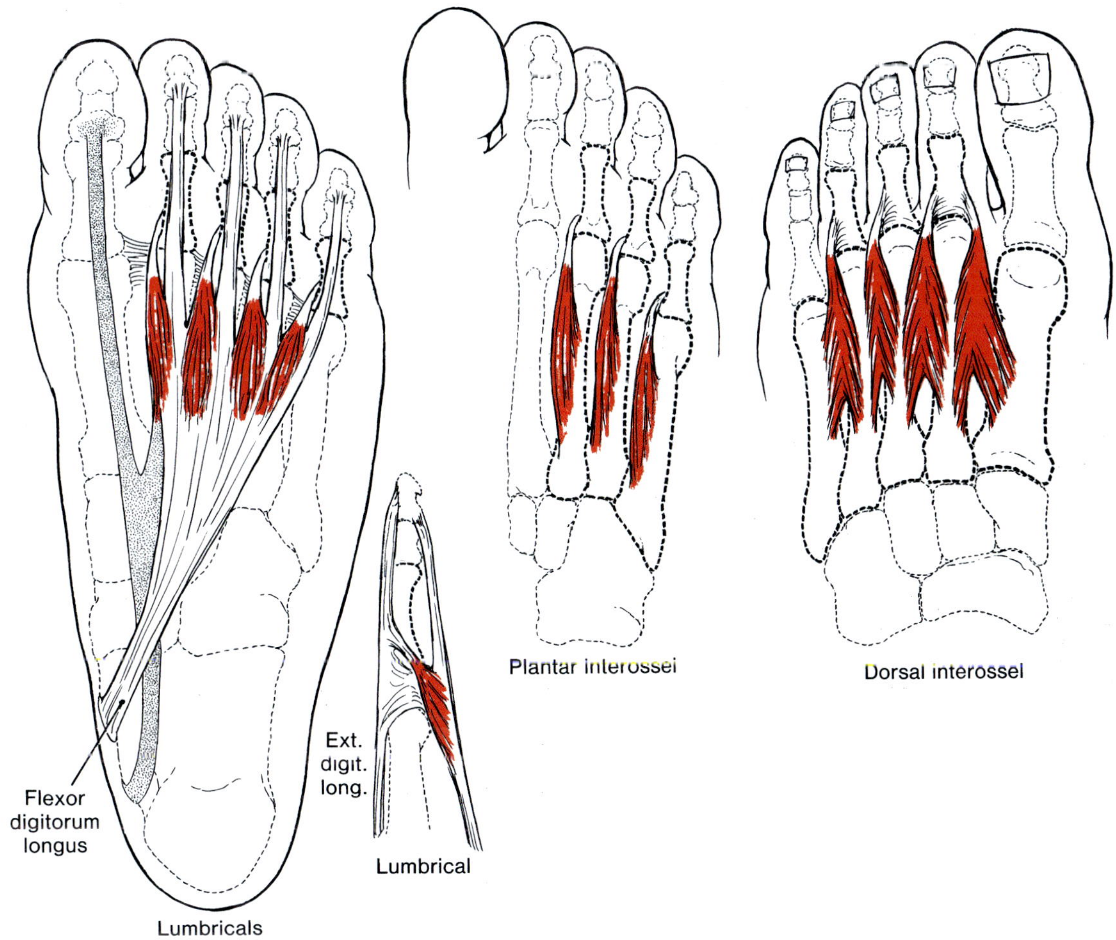

FIGURE 7-15: Lumbricals and interossei.

LUMBRICALS AND INTEROSSEI

Lumbricals (Four)

Origin:

First: From the medial side of the first flexor digitorum longus tendon (Figure 7-15).

Second: From the adjacent sides of the first and second flexor digitorum longus tendons.

Third: From the adjacent sides of the second and third flexor digitorum longus tendons.

Fourth: From the adjacent sides of the third and fourth flexor digitorum longus tendons.

Insertions: Medial side of the proximal phalanx and dorsal expansion of the extensor digitorum longus tendon of the second through fifth digits.

Actions: Flexes the metatarsophalangeal joints, and assists in extending the interphalangeal joints of the second through fifth digits.

Nerve to Lumbricals I: Medial plantar nerve, L4, 5, S1.

Nerve to Lumbricals II, II, and IV: Lateral plantar nerve L(4), (5), S1, 2.

Plantar Interossei (Three)

Origin: Bases and medial sides of the bodies of the third through fifth metatarsal bones (Figure 7-15).

Insertion: Medial sides of the bases of the proximal phalanges of the same digit.

Action: Adduct the third, fourth, and fifth digits toward the axial line through the second digits. Assist in flexion of the metatarsophalangeal joints and may assist in extension of interphalangeal joints of the third, fourth, and fifth digit.

Nerve: Lateral plantar, **S1**, **2**.

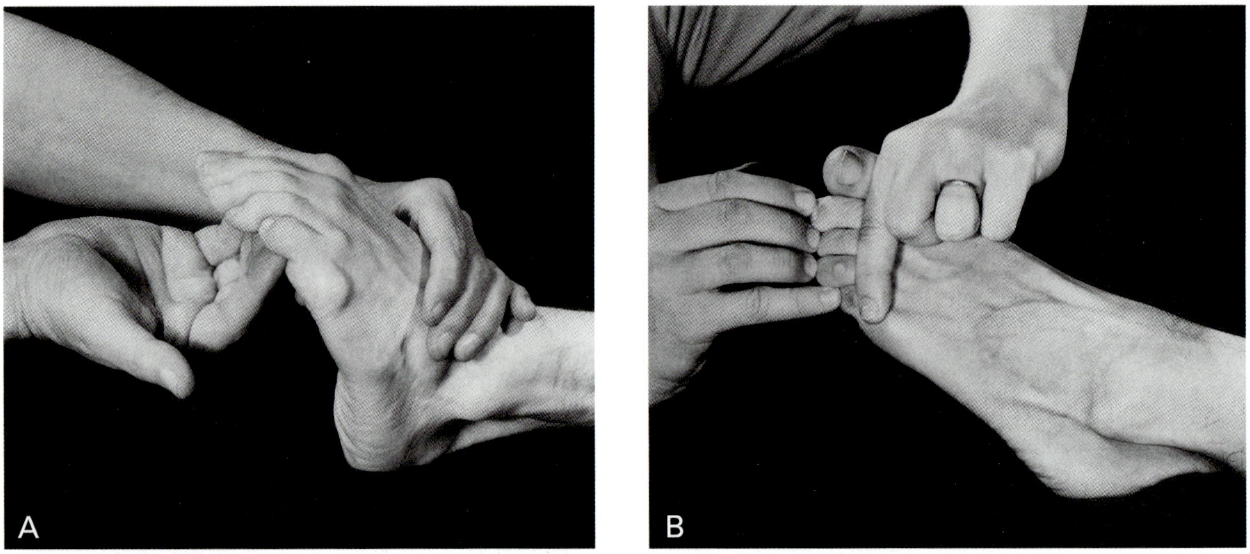

FIGURE 7-16: Testing strength of lumbricals and interossei: **A**, Flexion; **B**, Extension.

Dorsal Interossei (Four)

Origin: Each by two heads from the adjacent sides of the metatarsal bones (Figure 7-15).

Insertions: Side of the proximal phalanx and capsule of the metatarsophalangeal joint.

First: To the medial side of the second digit.

Second through fourth: To the lateral sides of the second through fourth digits.

Action: Abducts the second through fourth digits from the axial line through the second digit. Assists in flexion of the metatarsophalangeal joints and may assist in extension of interphalangeal joints of the second through fourth digits.

Nerve: Lateral plantar, S1, 2.

Patient: Supine or sitting.

Fixation: The examiner stabilizes the midtarsal region and maintains a neutral position of the foot and ankle (Figure 7-16A).

Test: Flexion of the metatarsophalangeal joints of the second through fifth digits, with an effort to avoid flexion of the interphalangeal joints.

Pressure: Against the plantar surface of the proximal phalanges of the four lateral toes.

Weakness: When these muscles are weak and the flexor digitorum longus is active, hyperextension occurs at the metatarsophalangeal joints. The distal joints flex, causing a hammer-toe position of the four lateral toes. Muscular support of the transverse arch is decreased.

Patient: Supine or sitting.

Fixation: The examiner stabilizes the metatarsophalangeal joints and maintains the foot and ankle in approximately 20°– 30° of plantar flexion (Figure 7-16B).

Test: Extension of the interphalangeal joints of the four lateral toes. (A separate test for adduction and abduction of the interossei is not practical, because most individuals cannot perform these movements of the toes.)

Pressure: Against the dorsal surface of the distal phalanges, in the direction of flexion.

NOTE: *Testing for strength of the lumbricals is important in cases of hammer toe and of metatarsal arch strain.*

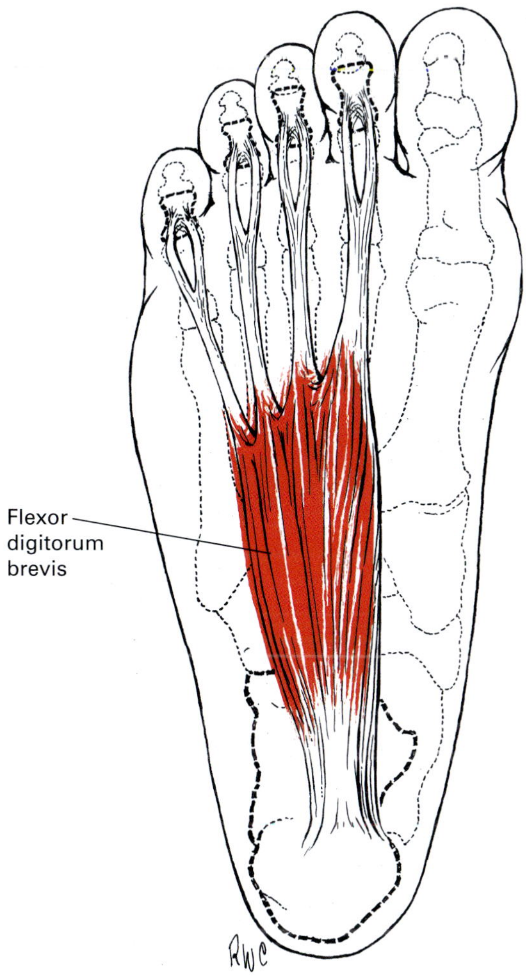

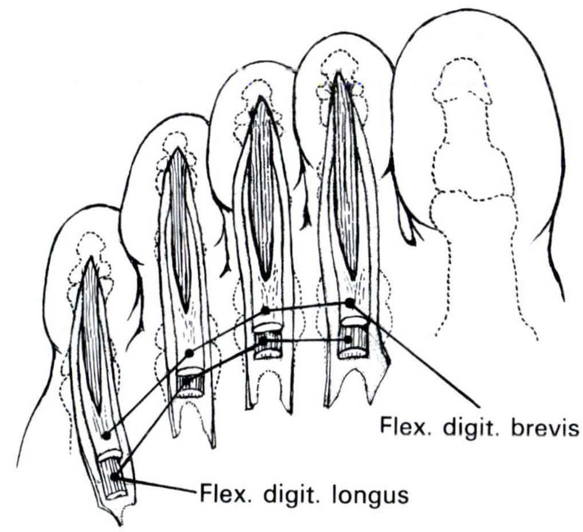

FIGURE 7-17: Flexor digitorum brevis.

Flexor Digitorum Brevis

Origin: Medial process of the tuberosity of the calcaneus, central part of the plantar aponeurosis, and adjacent intermuscular septa.

Insertion: Middle phalanx of the second through fifth digits.

Action: Flexes the proximal interphalangeal joints, and assists in flexion of the metatarsophalangeal joints of the second through fifth digits.

Nerve: Tibial, L4, 5, S1.

Patient: Supine or sitting.

Fixation: The examiner stabilizes the proximal phalanges and maintains a neutral position of the foot and ankle (Figure 7-17). If the gastrocnemius and soleus are paralyzed, the examiner must stabilize the calcaneus, which is the bone of origin, during the toe flexor test.

Test: Flexion of the proximal interphalangeal joints of the second through fifth digits.

Pressure: Against the plantar surface of the middle phalanx of the four toes, in the direction of extension.

> **NOTE:** *When the flexor digitorum longus is paralyzed and the brevis is active, the toes flex at the middle phalanx while the distal phalanx remains extended.*

Weakness: The ability to flex the proximal interphalangeal joints of the four lateral toes is decreased, and the muscular support of the longitudinal and transverse arches is diminished.

Contracture: Restriction of extension of the toes. The middle phalanges flex, and there is a tendency toward a cavus if the gastrocnemius and soleus are weak.

> **NOTE:** *Testing for strength of the flexor digitorum brevis is important in cases of longitudinal arch strain. Often, a point of acute tenderness is found at the origin of this muscle on the calcaneus.*

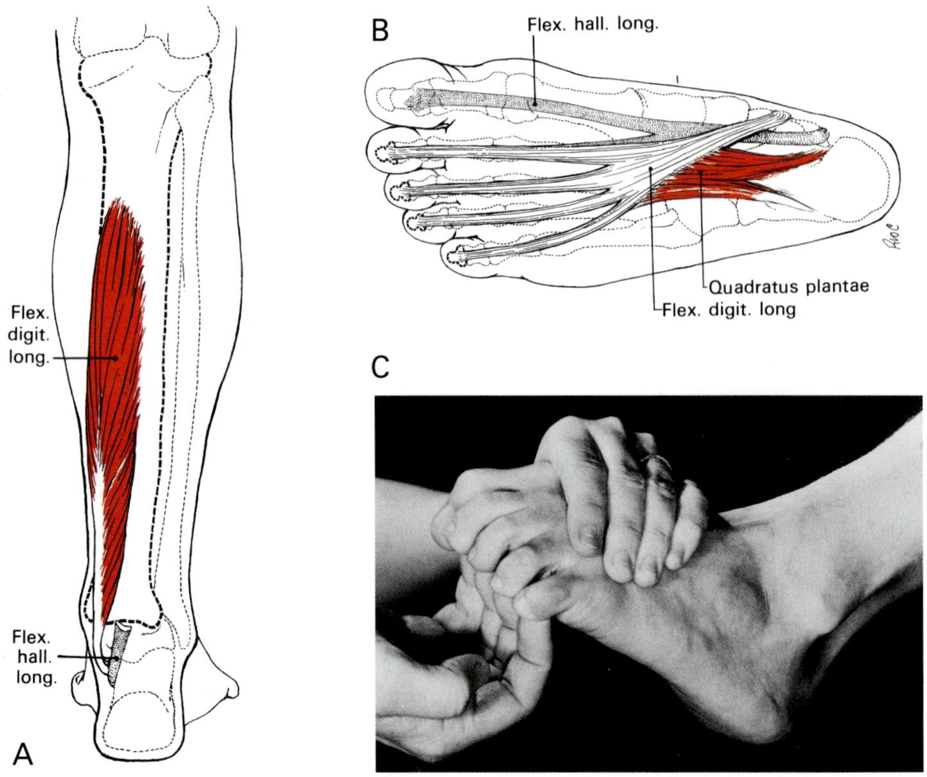

FIGURE 7-18: **A**, Flexor digitorum longus (Flexor hallucis longus refected); **B**, Quadratus Plantae; **C**, Testing for flexor digitorum longus.

FLEXOR DIGITORUM LONGUS AND QUADRATUS PLANTAE

Flexor Digitorum Longus

Origin: Middle 3/5 of the posterior surface of the body of the tibia and from fascia covering the tibialis posterior (Figure 7-18A).

Insertion: Bases of the distal phalanges of the second through fifth digits.

Action: Flexes proximal and distal interphalangeal and metatarsophalangeal joints of the second through fifth digits. Assists in plantar flexion of the ankle joint and inversion of the foot.

Nerve: Tibial, L5, S1, (2).

Patient: Supine or sitting. With gastrocnemius tightness, the knee should be flexed to permit a neutral position of the ankle.

Fixation: The examiner stabilizes the metatarsals and maintains a neutral position of the foot and ankle (Figure 7-18C).

Test: Flexion of the distal interphalangeal joints of the second through fifth digits. The flexor digitorum is assisted by the quadratus plantae.

Pressure: Against the plantar surface of the distal phalanges of the four toes in the direction of extension.

Weakness: Results in a tendency toward hyperextension of the distal interphalangeal joints of the four toes. Decreases the ability to invert the foot and plantar flex the ankle. In weight-bearing, weakness permits a tendency toward eversion of the subtalar and transverse tarsal joints.

Contracture: Flexion deformity of distal phalanges of the four lateral toes, with restriction of dorsiflexion of the ankle and eversion of the subtalar and transverse tarsal joints.

Quadratus Plantae (Flexor Accessorius)

Origin of Medial Head: Medial surface of the calcaneus and medial border of the long plantar ligament (Figure 7-18B).

Origin of Lateral Head: Lateral border of the plantar surface of the calcaneus and lateral border of the long plantar ligament.

Insertion: Lateral margin and the dorsal and plantar surfaces of the tendon of the flexor digitorum longus.

Action: Modifies the line of pull of the flexor digitorum longus tendons, and assists in flexing the second through fifth digits.

Nerve: Tibial, S1, 2.

> **NOTE:** *No test illustrated for Quadratus Plantae (Flexor Accessorius).*

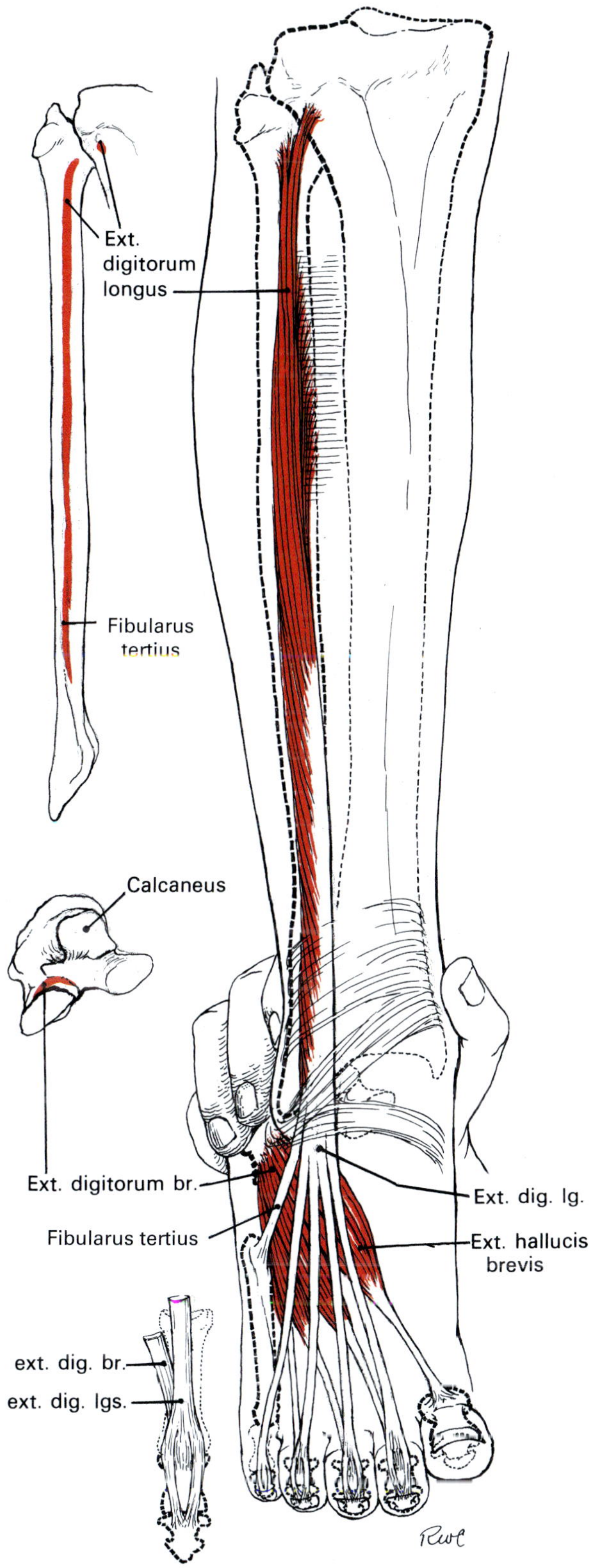

FIGURE 7-19: Extensor digitorum longus and brevis, and fibularis tertius.

EXTENSOR DIGITORUM LONGUS AND BREVIS AND FIBULARIS TERTIUS

Extensor Digitorum Longus

Origin: Lateral condyle of the tibia, proximal 3/4 of the anterior surface of the body of the fibula, proximal part of the interosseous membrane, adjacent intermuscular septa, and deep fascia (Figure 7-19).

Insertion: By four tendons to the second through fifth digits. Each tendon forms an expansion on the dorsal surface of the toe, dividing into an intermediate slip attached to the base of the middle phalanx and two lateral slips attached to the base of the distal phalanx.

Action: Extends the metatarsophalangeal joints and assists in extending the interphalangeal joints of the second through fifth digits. Assists in dorsiflexion of the ankle joint and eversion of the foot.

Nerve: Fibular, L4, 5, S1.

Extensor Digitorum Brevis/Extensor Hallucis Brevis

Origin: Distal part of the superior and lateral surfaces of the calcaneus, lateral talocalcaneal ligament and apex of the inferior extensor retinaculum (Figure 7-19).

Insertion: By four tendons to the first through fourth digits. The extensor hallucis brevis inserts into the dorsal surface of the base of the proximal phalanx of the great toe. The extensor digitorum brevis joins the lateral sides of the tendons of the extensor digitorum longus to the second, third, and fourth digits and contributes to the extensor expansion.

Action: Extends the metatarsophalangeal joints of the first through fourth digits and assists in extending the interphalangeal joints of the second through fourth digits.

Nerve: Deep Fibular, L4, 5, S1.

NOTE: *Because the extensor digitorum brevis tendons fuse with the tendons of the extensor longus to the second through fourth digits, the brevis as well as the longus will extend all joints of these toes. Without an extensor longus, however, no extension of the fifth digit will occur at the metatarsophalangeal joint. To differentiate, palpate the tendon of the longus and the belly of the brevis, and try to detect any difference in movement of the toes.*

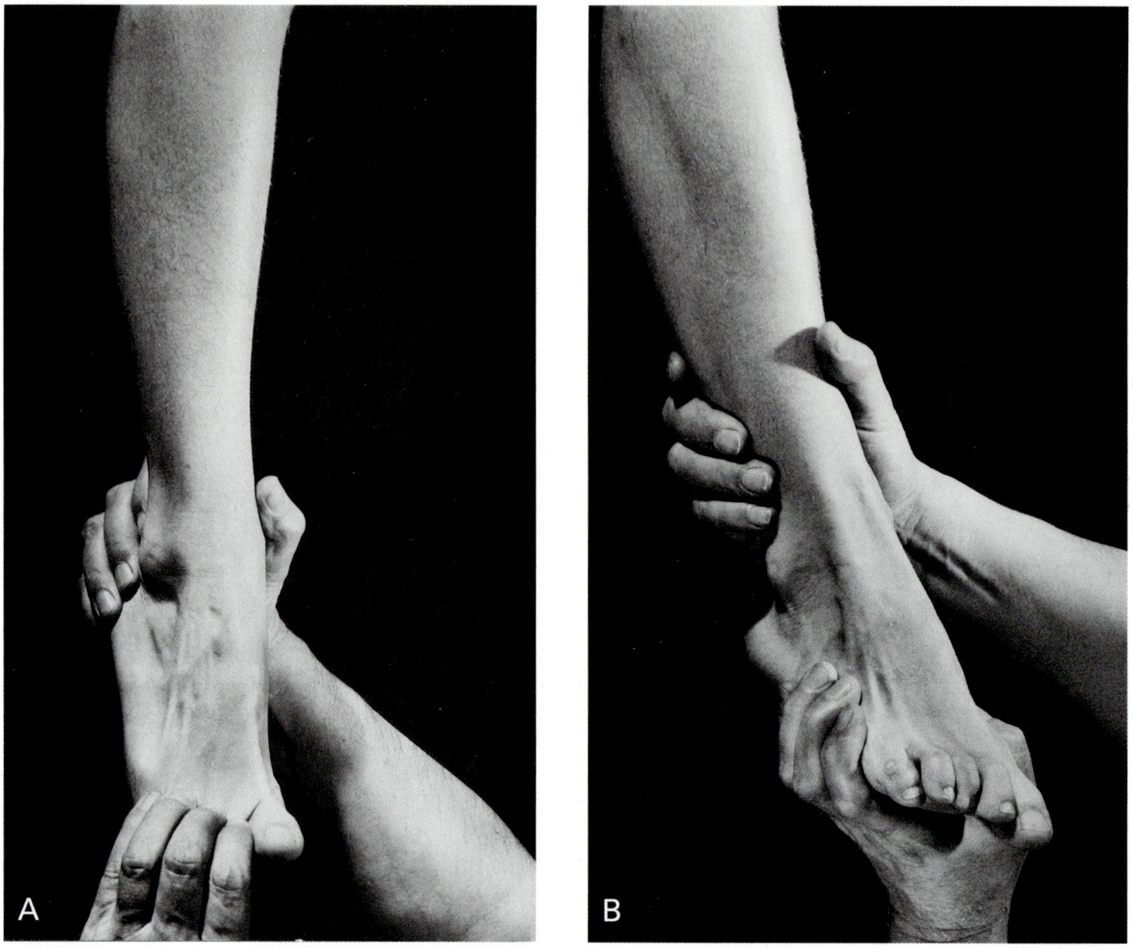

FIGURE 7-20: **A**, Testing extensor digitorum longus and brevis; **B**, Testing fibularis tertius.

Fibularis Tertius

Origin: Distal ⅓ of the anterior surface of the fibula, interosseous membrane, and adjacent intermuscular septum.

Insertion: Dorsal surface, base of the fifth metatarsal.

Action: Dorsiflexes ankle joint, and everts foot.

Nerve: Deep fibular, L4, L5, S1.

Patient: Supine or sitting.

Fixation: The examiner supports the leg above the ankle joint (Figure 7-20B).

Test: Dorsiflexion of the ankle joint, with eversion of the foot.

> **NOTE:** *The fibularis tertius is assisted in this test by the extensor digitorum longus, of which it is a part.*

Pressure: Against the lateral side, dorsal surface of the foot, in the direction of plantar flexion and inversion.

Weakness: Decreases the ability to evert the foot and dorsiflex the ankle joint.

Contracture: Dorsiflexion of the ankle joint and eversion of the foot.

Extensor Digitorum Longus and Brevis

Patient: Supine or sitting.

Fixation: The examiner stabilizes the foot in slight plantar flexion (Figure 7-20A).

Test: Extension of all joints of the second through fifth digits.

Pressure: Against the dorsal surface of the toes, in the direction of flexion.

Weakness: Allows a tendency toward dropfoot and forefoot varus. Diminishes the ability to dorsiflex the ankle joint and evert the foot. Many cases of flat feet (i.e., collapse of the long arch) also have accompanying weakness of the toe extensors.

Contracture: Hyperextension of the metatarsophalangeal joints.

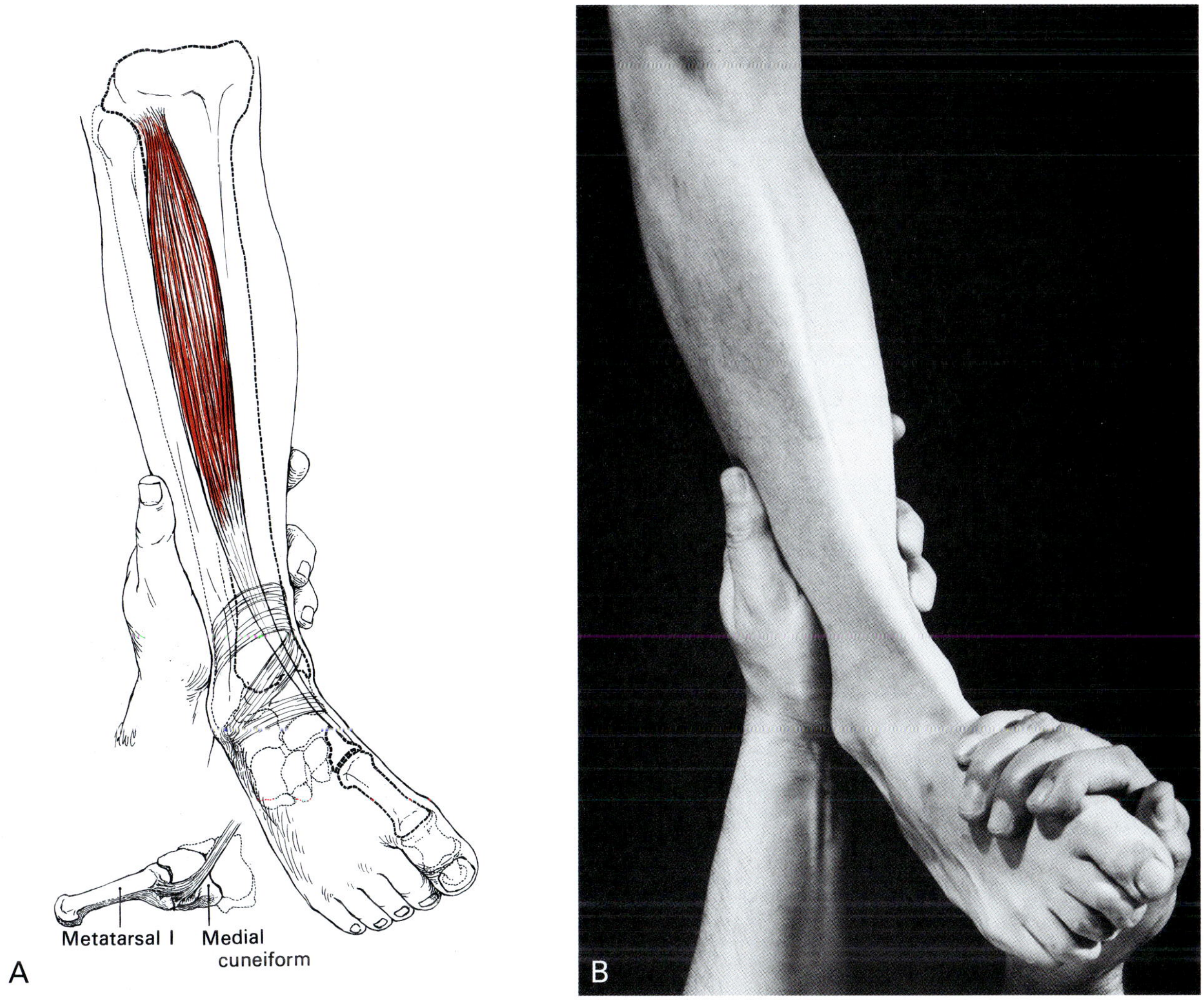

FIGURE 7-21: **A**, Tibialis anterior; **B**, Testing tibialis anterior.

Tibialis Anterior

Origin: Lateral condyle and proximal half of the lateral surface of the tibia, interosseous membrane, deep fascia, and lateral intermuscular septum (Figure 7-21A).

Insertion: Medial and plantar surface of medial cuneiform bone, base of the first metatarsal bone.

Action: Dorsiflexes the ankle joint, and assists in inversion of the foot.

Nerve: Deep fibular, L4, 5, S1.

Patient: Supine or sitting (with knee flexed if any gastrocnemius tightness is present).

Fixation: The examiner supports the leg, just above the ankle joint (Figure 7-21B).

Test: Dorsiflexion of the ankle joint and inversion of the foot, without extension of the great toe.

Pressure: Against the medial side, dorsal surface of the foot, in the direction of plantar flexion of the ankle joint and eversion of the foot.

Weakness: Decreases the ability to dorsiflex the ankle joint, and allows a tendency toward eversion of the foot. This may be seen as a partial dropfoot and tendency toward pronation.

Contracture: Dorsiflexion of ankle joint, with inversion of the foot (i.e., calcaneovarus position of the foot).

> **NOTE:** *Although tibialis anterior weakness may be found in conjunction with a pronated foot, such weakness is seldom found in cases of congenital flatfoot.*

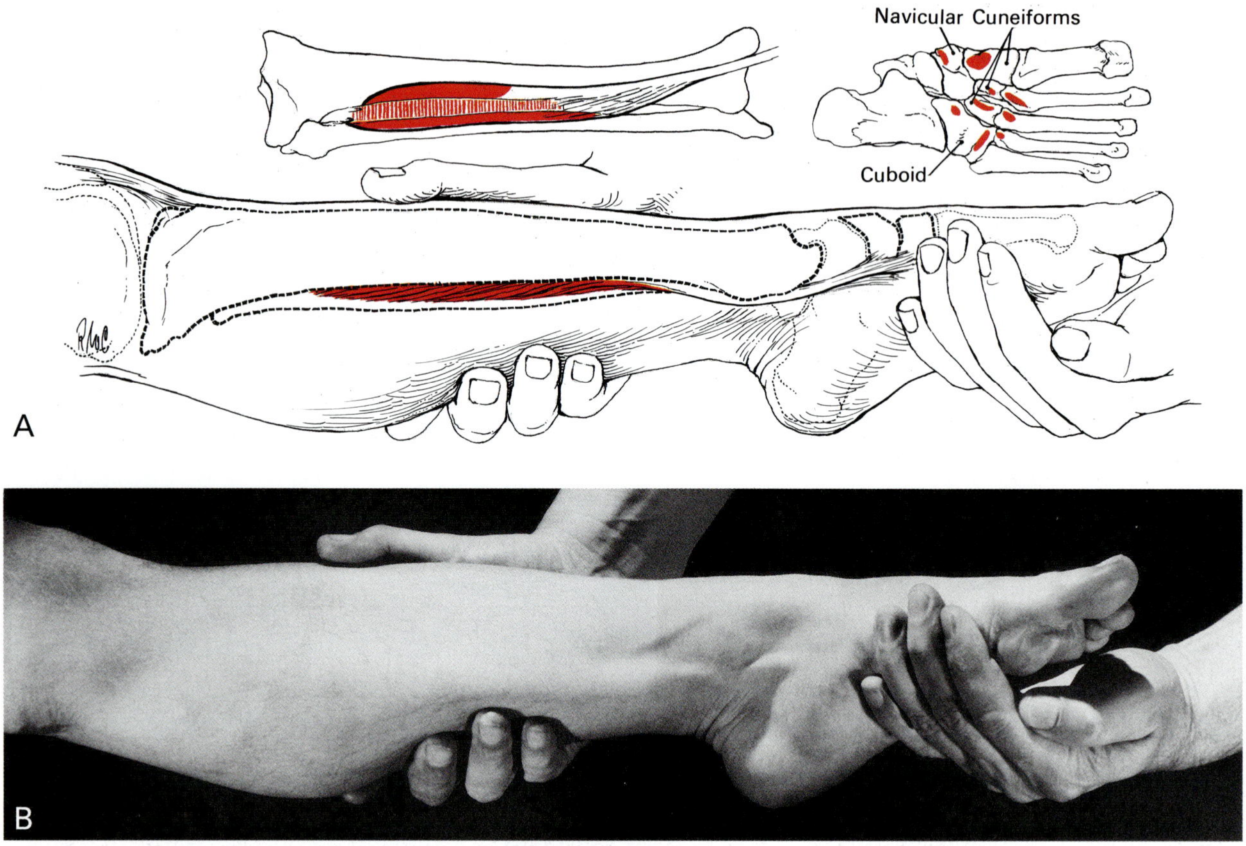

FIGURE 7-22: **A**, Tibialis posterior; **B**, Testing tibialis posterior.

Tibialis Posterior

Origin: Most of the interosseous membrane, lateral portion of the posterior surface of the tibia, proximal ⅔ of the medial surface of the fibula, adjacent intermuscular septa, and deep fascia (Figure 7-22A).

Nerve: Tibial, L(4), **5**, **S1**.

Insertion: Tuberosity of the navicular bone and by fibrous expansions to the sustentaculum tali, three cuneiforms, cuboid, and bases of the second through fourth metatarsal bones.

Action: Inverts the foot, and assists in plantar flexion of the ankle joint.

Patient: Supine, with the extremity in lateral rotation.

Fixation: The examiner supports the leg, above the ankle joint (Figure 7-22B).

Test: Inversion of the foot, with plantar flexion of the ankle joint.

Pressure: Against the medial side and plantar surface of the foot, in the direction of dorsiflexion of the ankle joint and eversion of the foot.

> **NOTE:** *If the flexor hallucis longus and flexor digitorum longus are being substituted for the tibialis posterior, the toes will be strongly flexed as pressure is applied.*

Weakness: Decreases the ability to invert the foot and plantar flex the ankle joint. Results in pronation of the foot and decreased support of the longitudinal arch. Interferes with the ability to rise on the toes, and inclines toward what is commonly called a gastrocnemius limp.

Contracture: In non-weight-bearing, equinovarus position; in weight-bearing, a supinated position of the heel with forefoot varus.

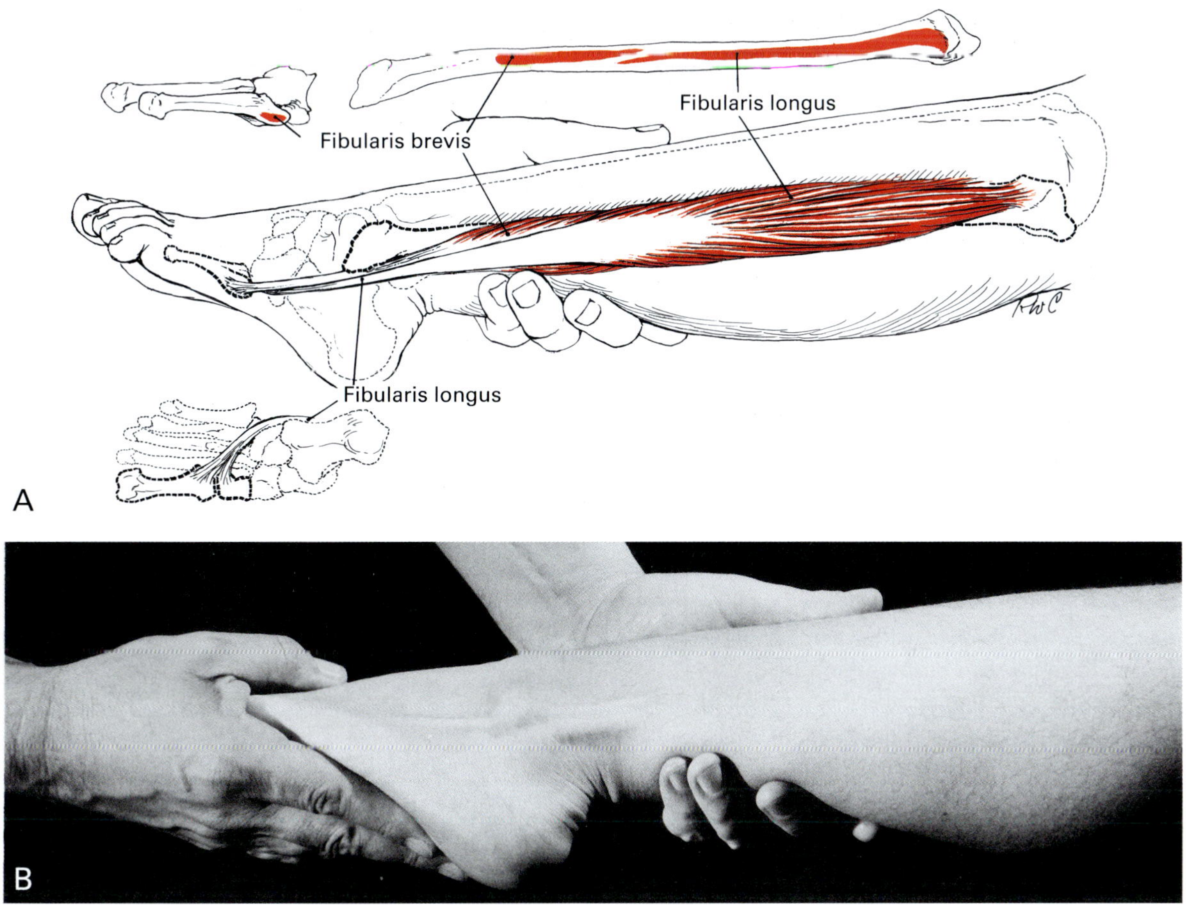

FIGURE 7-23: **A**, Fibularis longus and brevis; **B**, Testing fibularis longus and brevis.

FIBULARIS LONGUS AND BREVIS

Fibularis Longus

Origin: Lateral condyle of the tibia, head and proximal ⅔ of the lateral surface of the fibula, intermuscular septa, and adjacent deep fascia (Figure 7-23A).

Insertion: Lateral side of the base of the first metatarsal and of the medial cuneiform bone.

Action: Everts the foot, assists in plantar flexion of the ankle joint, and depresses the head of the first metatarsal.

Nerve: Superficial fibular, L4, **5**, **S1**.

Patient: Supine, with the extremity medially rotated, or side-lying (on the opposite side).

Fixation: The examiner supports the leg, above the ankle joint (Figure 7-23B).

Test: Eversion of the foot, with plantar flexion of the ankle joint (test for both longus and brevis).

Pressure: Against the lateral border and sole of the foot, in the direction of inversion of the foot and dorsiflexion of the ankle joint.

Fibularis Brevis

Origin: Distal ⅔ of the lateral surface of the fibula and adjacent intermuscular septa.

Insertion: Tuberosity at the base of the fifth metatarsal bone, lateral side.

Action: Everts the foot and assists in plantar flexion of the ankle.

Nerve: Superficial fibular, L4, **5**, **S1**.

Weakness: Decreases the strength of eversion of the foot and plantar flexion of the ankle joint. Allows a varus position of the foot, and lessens the ability to rise on the toes. Lateral stability of the ankle is decreased.

Contracture: Results in an everted or valgus position of the foot.

> **NOTE:** *In weight-bearing, with a strong pull on its insertion at the base of the first metatarsal, the fibularis longus causes the head of the first metatarsal to be pressed downward, into the supporting surface.*

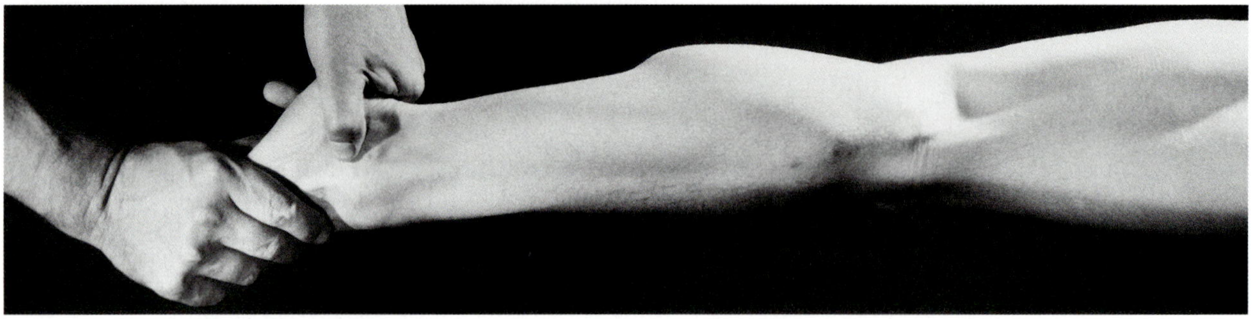

FIGURE 7-24: Test of ankle plantar flexors.

ANKLE PLANTAR FLEXORS

Patient: Prone, with the knee extended and the foot projecting over the end of the table.

Fixation: The weight of the extremity, resting on a firm table, should be sufficient fixation of the part.

Test: Plantar flexion of the foot, with emphasis on pulling the heel upward more than pushing the forefoot downward. This test does not attempt to isolate action of the gastrocnemius from that of the other plantar flexors, but the presence or absence of gastrocnemius contraction can be determined by careful observation during the test (4, 5) (Figure 7-24).

Pressure: For maximum pressure in this position, apply pressure against the forefoot as well as against the calcaneus. If the muscle is very weak, pressure against the calcaneus is sufficient.

The gastrocnemius usually can be seen and always can be palpated if it is contracting during the plantar flexion test. Movements of the toes and forefoot should be observed carefully during the test to detect substitutions. The patient may be able to flex the anterior part of the foot by the toe flexors, tibialis posterior, and fibularis longus without a direct upward pull on the heel by the tendo calcaneus. If the gastrocnemius and soleus are weak, the heel will be *pushed* up secondary to flexion of the anterior part of the foot rather than *pulled* up simultaneously with the flexion of the forepart of the foot. If pressure is applied to the heel rather than to the ball of the foot, it is possible to isolate, at least partially, the combined action of the gastrocnemius and soleus from that of the other plantar flexors. Movement of the foot toward eversion or inversion will show imbalance in the opposing lateral and medial muscles and, if pronounced, will show an attempt to substitute the fibularis muscles or tibialis posterior for the gastrocnemius and soleus.

Action of the gastrocnemius often can be demonstrated in the knee flexion test when the hamstrings are weak. In the prone position, with the knees fully extended, the patient is asked to bend the knee against resistance. If the gastrocnemius is strong, plantar flexion at the ankle will occur as the gastrocnemius acts to initiate knee flexion, followed by ankle dorsiflexion as the knee flexes.

Weakness: Results in hyperextension of the knee and inability to rise on the toes. The inability to transfer weight may result in gait deviations.

Contracture: Equinus position of the foot, and flexion of the knee.

Shortness: Restriction of dorsiflexion of the ankle when the knee is extended, and restriction of knee extension when the ankle is dorsiflexed. During the stance phase of walking, shortness limits normal dorsiflexion of the ankle joint, and the subject toes out during the transfer of weight from heel to forefoot.

Soleus

Origin: Posterior surfaces of the head of the fibula and proximal ⅓ of its body, soleal line and middle ⅓ of the medial border of the tibia, and tendinous arch between the tibia and fibula (Figure 7-25A).

Insertion: With the tendon of the gastrocnemius, into the posterior surface of the calcaneus.

Action: Plantar flexes the ankle joint.

Nerve: Tibial, L5, **S1**, **2**.

Patient: Prone, with the knee flexed at least 90°.

Fixation: The examiner supports the leg, proximal to the ankle.

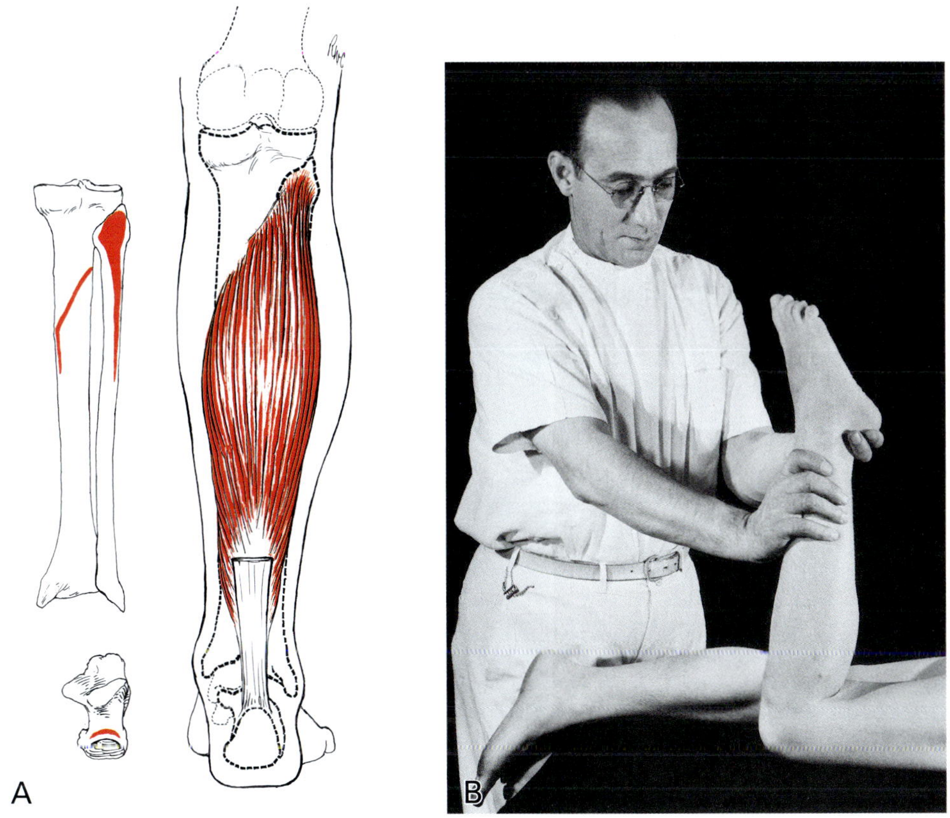

FIGURE 7-25: **A**, Soleus; **B**, Testing soleus.

Test: Plantar flexion of the ankle joint, without inversion or eversion of the foot.

Pressure: Against the calcaneus (as illustrated in Figure 7-25B), pulling the heel in a caudal direction (i.e., in the direction of dorsiflexing the ankle). When weakness is marked, the patient may not be able to hold against pressure at the heel. When weakness is not marked, more leverage is necessary and is obtained by applying pressure simultaneously against the sole of the foot.

> **NOTE:** *Inversion of the foot indicates substitution by the tibialis posterior and toe flexors. Eversion indicates substitution by the fibularis muscles. Extension of the knee is evidence of attempting to assist with the gastrocnemius. That is, the gastrocnemius is at a disadvantage with the knee flexed 90° or more, and to bring it into a stronger action, the patient will attempt to extend the knee.*

Weakness: Predisposes to a cavus. Results in inability to rise on the toes. In standing, the insertion of the soleus muscle on the calcaneus becomes the fixed point for action of this muscle in maintaining normal alignment of the leg in relation to the foot. The deviation that results from weakness of the soleus may appear as a slight knee flexion fault in posture, but it more often results in an anterior displacement of the body weight from the normal plumb line distribution, as seen when the plumb line is hung slightly anterior to the outer malleolus.

A nonparalytic type of weakness may result from sudden trauma to the muscle, as in landing from a jump in a position of ankle dorsiflexion and knee flexion, or from gradual trauma, as in repeated deep knee bending in which the ankle is fully dorsiflexed. The gastrocnemius escapes the stretch because of the knee flexion.

Contracture: Equinus position of the foot, both in weight-bearing and in non-weight-bearing.

Shortness: A tendency toward hyperextension of the knee in the standing position. When walking barefoot, the shortness is compensated for by toeing out, thereby transferring the weight from the posterolateral heel to the anteromedial forefoot. In shoes with heels, the shortness may go unnoticed.

> **NOTE:** *This test is important in the examination of cases with forward deviation of the body from the plumb line. It is also advisable to test this muscle in cases with a change in height of the longitudinal arch.*

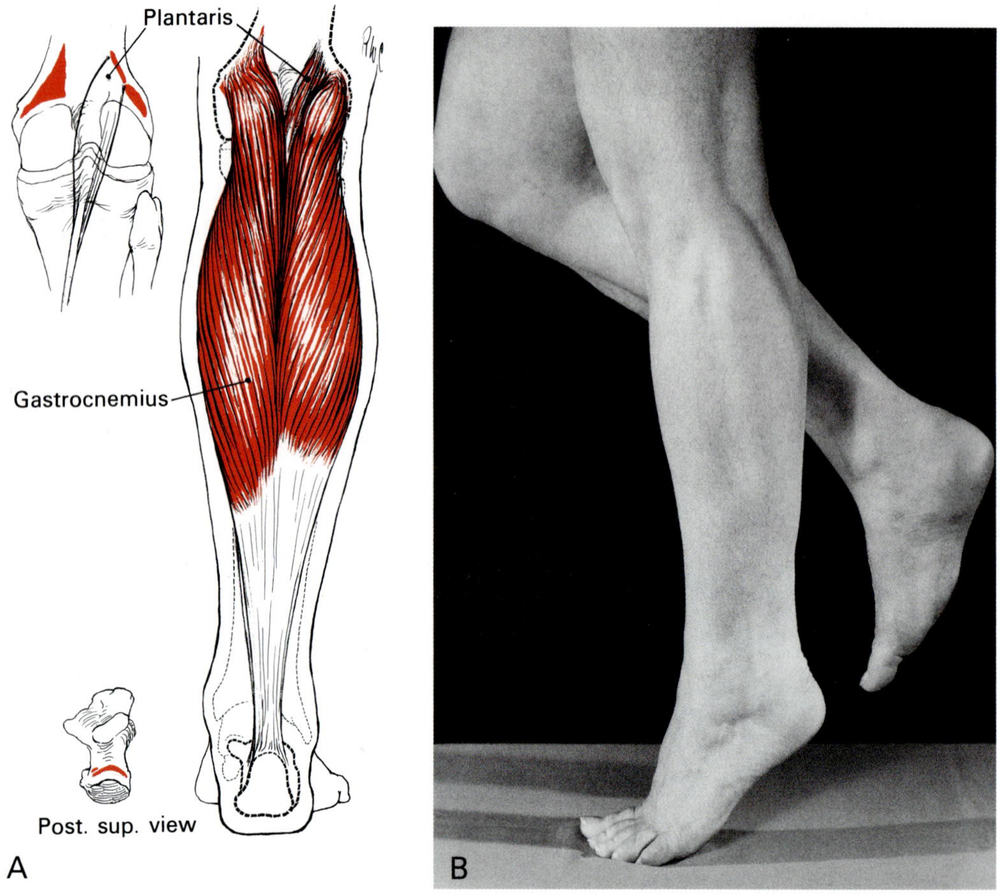

FIGURE 7-26: **A**, Gastrocnemius and plantaris; **B**, Testing ankle plantar flexors.

GASTROCNEMIUS AND PLANTARIS

Gastrocnemius

Origin of Medial Head: Proximal and posterior part of the medial condyle and adjacent part of the femur, capsule of the knee joint (Figure 7-26A).

Origin of Lateral Head: Lateral condyle and posterior surface of the femur, capsule of the knee joint.

Insertion: Middle part of the posterior surface of the calcaneus.

Nerve: Tibial, S**1**, **2**.

Plantaris

Origin: Distal part of the lateral supracondylar line of the femur, adjacent part of its popliteal surface and oblique popliteal ligament of knee joint (Figure 7-26A).

Insertion: Posterior part of the calcaneus.

Nerve: Tibial, L4, **5**, S**1**, (2).

Action: The gastrocnemius and the plantaris plantar flex the ankle joint and assist in flexion of the knee joint.

Ankle Plantar Flexors

Patient: Standing. (Patients may steady themselves with a hand on the table, but they should not take any weight on the hand.)

Test Movement: Rising on toes, pushing the body weight directly upward (Figure 7-26B).

Resistance: Body weight.

> **NOTE:** *Inclining the body forward and flexing the knee is evidence of weakness. The patient dorsiflexes the ankle joint, attempting to clear the heel from the floor by tension of the plantar flexors, as the body weight is thrown forward.*

Shortness: Shortness of the gastrocnemius and soleus muscles tends to develop among women who frequently wear high-heeled shoes.

See Box 7-1 for muscles that act in plantar flexion.

BOX 7-1

Muscles That Act in Plantar Flexion

Soleus Gastrocnemius Plantaris	Ankle joint plantar flexors (tendo calcaneus group)
Tibialis posterior Peroneus longus Peroneus brevis	Forefoot and ankle joint plantar flexors
Flexor hallucis longus Flexor digitorum longus	Toe, forefoot, and ankle joint plantar flexors

SECTION III
KNEE AND HIP STRENGTH AND FLEXIBILITY TESTS

JOINT MOVEMENTS

Movements of Knee Joint

The knee should be flexed when measuring dorsiflexion. With the knee flexed, the ankle joint can be dorsiflexed approximately 20°. If the knee is extended, the gastrocnemius will limit the range of motion to approximately 10° of dorsiflexion. The range of motion in plantar flexion is approximately 45°.

The knee joint is a modified ginglymus or hinge joint formed by the articulation of the condyles of the femur with the condyles of the tibia, and by the patella articulating with the patellar surface of the femur.

Flexion and *extension* are movements about a coronal axis. Flexion is movement in a posterior direction, approximating the posterior surfaces of the lower leg and thigh. Extension is movement in an anterior direction to a position of straight alignment of the thigh and lower leg (0°). From the position of zero extension, the range of flexion is approximately 140°. The hip joint should be flexed when measuring full knee joint flexion to avoid restriction of motion by the rectus femoris, but the joint should not be fully flexed when measuring knee joint extension to avoid restriction by the hamstring muscles.

Hyperextension is an abnormal or unnatural movement beyond the zero position of extension. For the sake of stability in standing, the knee normally is expected to be in a position of only a very few degrees of extension beyond zero. If extended beyond these few degrees, the knee is said to be hyperextended (Figure 7-27).

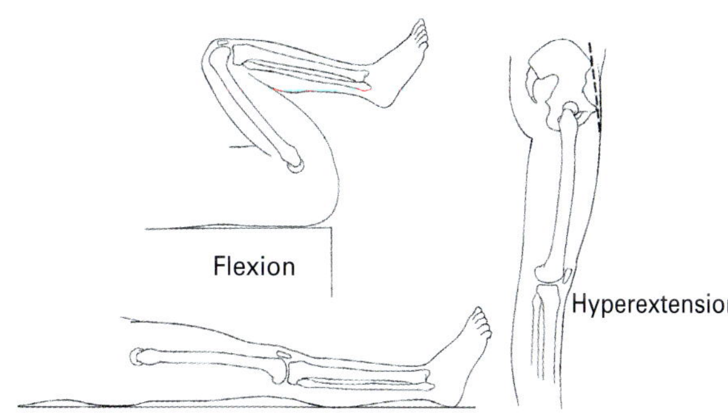

Extension

FIGURE 7-27: Knee joint flexion, extension, and hyperextension.

Lateral rotation and *medial rotation* are movements about a longitudinal axis. Medial rotation is rotation of the anterior surface of the leg toward the midsagittal plane. Lateral rotation is rotation away from the midsagittal plane.

The extended knee (in zero position) is essentially locked, preventing any rotation. Rotation occurs with flexion, combining movement between the tibia and the menisci as well as between the tibia and the femur.

With the thigh fixed, the movement that accompanies flexion is medial rotation of the tibia on the femur. With the leg fixed, the movement that accompanies flexion is lateral rotation of the femur on the tibia.

With the thigh fixed, the movement that accompanies extension is lateral rotation of the tibia on the femur. With the leg fixed, the movement that accompanies extension is medial rotation of the femur on the tibia.

MOVEMENTS OF HIP JOINT

The hip joint is a spheroid or ball-and-socket joint formed by the articulation of the acetabulum of the innominate bone with the head of the femur.

Ordinarily, descriptions of joint movement refer to movement of the distal part on a fixed proximal part. In the upright weight-bearing position, movement of the proximal part on the more fixed distal part becomes of equal—if not primary—importance. For this reason, movements of the pelvis on the femur are mentioned as well as movements of the femur on the pelvis.

Flexion and *extension* are movements about a coronal axis. Flexion is movement in an anterior direction. This movement may be one of bringing the thigh toward the fixed pelvis, as in supine straight-leg raising. Or it may be bringing the pelvis toward the fixed thighs, as in coming up from a supine to a sitting position, bending forward from a standing position, or tilting the pelvis anteriorly in standing. Extension is movement in a posterior direction. This movement may be one of bringing the thigh posteriorly, as in leg-raising backward, or one of bringing the trunk posteriorly, as in returning from a standing forward-bent position, or as in tilting the pelvis posteriorly when standing or lying prone (Figure 7-28).

The range of hip joint flexion from zero is approximately 125°, and the range of extension is approximately 10°, making a total range of approximately 135°. The knee joint should be flexed when measuring hip joint flexion to avoid restriction of motion by the hamstring muscles, and the joint should be extended when measuring hip joint extension to avoid restriction of motion by the rectus femoris.

Abduction and *adduction* are movements about a sagittal axis. Abduction is movement away from a midsagittal plane in a lateral direction. In a supine position, the movement may be one of moving the thigh laterally on a fixed trunk or of moving the trunk so that the pelvis tilts laterally (i.e., downward) toward a fixed thigh. Adduction is movement of the thigh toward the midsagittal plane in a medial direction. In a supine position, the movement may be one of moving the thigh medially on a fixed trunk or of moving the trunk so that the pelvis tilts laterally (i.e., upward) away from a fixed thigh. (For abduction and adduction of the hip joints accompanying lateral pelvic tilt, see below.)

From zero, the range of abduction is approximately 45°, and the range of adduction is approximately 30°, making the total range approximately 75°.

Lateral rotation and *medial rotation* are movements about a longitudinal axis. Medial rotation is movement in which the anterior surface of the thigh turns toward the midsagittal plane. Lateral rotation is movement in which the anterior surface of the thigh moves away from the midsagittal plane. Rotation also may result from movement of the trunk on the femur, for example, when standing with the legs fixed, a counterclockwise rotation of the pelvis will result in a lateral rotation of the right hip joint and a medial rotation of the left.

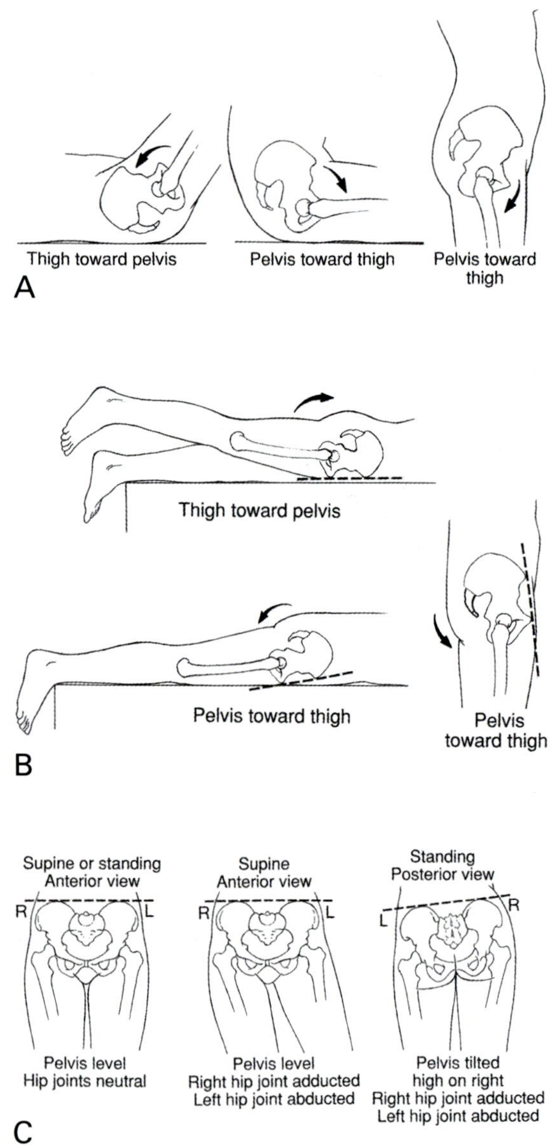

FIGURE 7-28: Movements of the hip joint: **A**, Hip joint flexion; **B**, Hip joint extension, thigh toward pelvis, and pelvis toward thigh; **C**, Hip joint abduction and adduction.

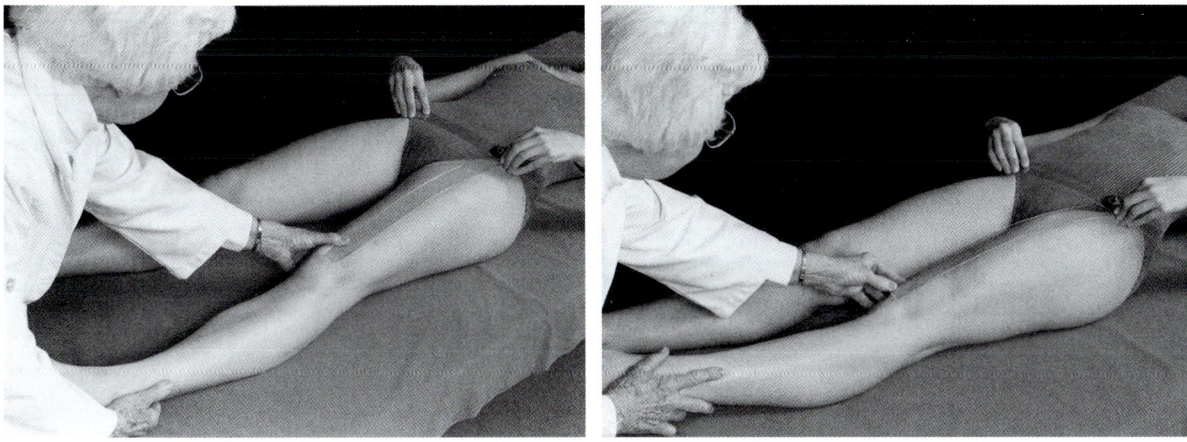

FIGURE 7-29: Hip adduction: Testing range of motion.

HIP ADDUCTION: RANGE OF MOTION

Equipment: Goniometer.

Starting Position: Supine, with the pelvis in a neutral position, like the anatomical position in standing. Place the left leg in a neutral position and the right leg in enough abduction to allow adduction of the left leg. The stationary arm is held firmly against the inferior surface of the anterior and superior iliac spines by the subject, as illustrated. The movable arm is set at a 90° angle (as the zero position) and placed in line with midline of the extremity. Alternatively, the movable arm may be set at an angle that coincides with the axis of the femur (i.e., some adduction), in which case a reading is taken before moving the leg into adduction and the number of degrees are then subtracted from the number measured at the completion of adduction(Figure 7-29).

Test: The movable arm of the goniometer is held in line with the thigh as the left leg is passively and *slowly* moved into adduction without any rotation. At the moment the pelvis starts to move downward on the side of the adducted leg, the movement of the leg in adduction is stopped, and the measurement is recorded (Figure 7-30).

Normal Range of Motion: Random testing has shown that adduction is often less than 10° and seldom more than 10° in the supine position unless the hip joint is in flexion by virtue of anterior pelvic tilt. (With the hip joint flexed, as in sitting, the range of adduction is about 20°.) With the thigh maintained in the coronal plane, as in the modified Ober test, 10° of adduction should be considered as normal.

JOINT MEASUREMENT CHART

JOINT MEASUREMENT CHART

Name...Identification #..................................

Diagnosis...Age..

Onset...Doctor..

LOWER EXTREMITY

					Date Examiner	Motion*	Average Range	Date Examiner					
					Left Hip	Extension	10	Right Hip					
						Flexion	125						
						Range	135						
						Abduction	45						
						Adduction	10						
						Range	55						
						Lateral Rotation	45						
						Medial Rotation	45						
						Range	90						
					Left Knee	Extension	0	Right Knee					
						Flexion	140						
						Range	140						
					Left Ankle	Plantar Flexion	45	Right Ankle					
						Dorsiflexion	20						
						Range	65						
					Left Foot	Inversion	40	Right Foot					
						Eversion	20						
						Range	60						

*Use either anatomical or geometric basis for measurement. Cross off one not used. 180° is the plane of reference for the geometric basis of measurement. The zero position is the plane of reference for all the others. When a part moves in the direction of zero but fails to reach the zero position, the degrees designating the joint motion obtained are recorded with a minus sign and subtracted in computing the range of motion.

Notes:

FIGURE 7-30: Joint measurement chart.

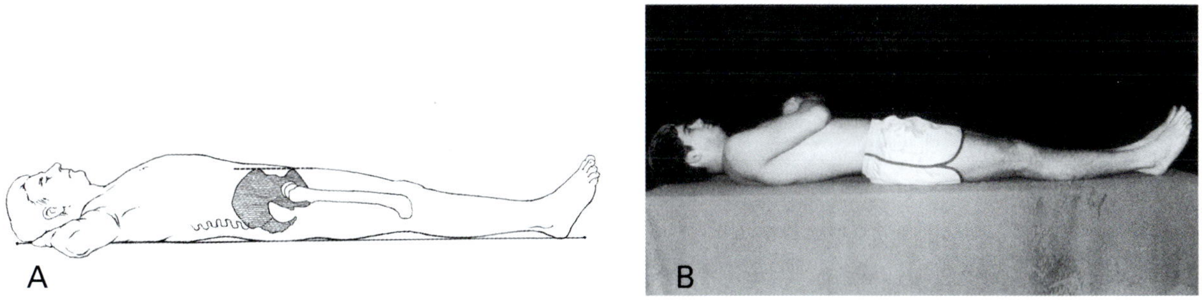

FIGURE 7-31: Testing hip flexor muscles: **A**, Pelvis in neutral position; **B**, Low back in lordotic position.

TESTS FOR LENGTH OF HIP FLEXOR MUSCLES

The psoas major, iliacus, pectineus, adductors longus and brevis, rectus femoris, tensor fasciae latae, and sartorius compose the hip flexor group of muscles. The iliacus, pectineus, and adductor longus and brevis are one-joint muscles. The psoas major and the iliacus (as the iliopsoas) act essentially as a one-joint muscle. The rectus femoris, tensor fasciae latae, and sartorius are two-joint muscles, crossing the knee joint as well as the hip joint. All three muscles flex the hip. However, the rectus femoris and, to some extent, the tensor fasciae latae extend the knee, whereas the sartorius flexes the knee.

The test for hip flexor length is often referred to as the **Thomas test**. Tests to distinguish between one-joint and two-joint hip flexor tightness were first described in *Posture and Pain* in 1952 (6).

Iliopsoas

Action: Hip flexion.

Length test: Hip extension, with the knee in extension.

Rectus Femoris

Action: Hip flexion and knee extension

Length test: Hip extension and knee flexion.

Tensor Fasciae Latae

Action: Hip abduction, flexion, and internal rotation as well as knee extension.

Length test: See Ober Test and Modified Ober Test.

Sartorius

Action: Hip flexion, abduction, and external rotation as well as knee flexion.

Length test: Hip extension, adduction, and internal rotation as well as knee extension.

Equipment:

A stable, firm table that will not tilt with the subject seated at one end
Goniometer and ruler
Chart for recording findings

Starting Position: Seated at the end of the table, with the thighs half off the table because the body position shifts as the subject lies down and brings one knee toward the chest. The end position for the start of testing is with the other knee just at the edge of the table so that the knee is free to flex and the thigh is full length on the table. The examiner places one hand behind the subject's back and the other hand under one knee, flexing the thigh toward the chest and giving assistance as the subject lies down. The subject then holds the thigh, pulling the knee toward the chest *only enough* to flatten the low back and sacrum on the table. (*Do not* bring both knees toward the chest, because that allows excessive posterior tilt, which results in apparent [not actual] hip flexor shortness.)

> **NOTE:** *If testing for excessive hip flexor length, the hip joint should be at the edge of the table, with the thigh off the table.*

Test Movement: If the right knee is flexed toward the chest, the left thigh is allowed to drop toward the table, with the left knee flexed over the end of the table. With four muscles involved in the length test, variations occur that require interpretations as described on the following pages.

In Figure 7-31, Part A, the pelvis is shown in neutral position, the lower back in a normal anterior curve, and the hip joint in zero position. Normal hip joint extension is considered to be approximately 10°. Normal length of hip flexors permits this range of motion in extension. The length may be demonstrated by moving the thigh in a posterior direction with the pelvis in a neutral position; or by moving the pelvis in the direction of posterior tilt with the thigh in zero position.

In a subject with hip flexors of normal length, the low back will tend to flatten in the supine position. If the low back remains in a lordotic position, as in Figure 7-31, Part B, some hip flexor shortness usually is present.

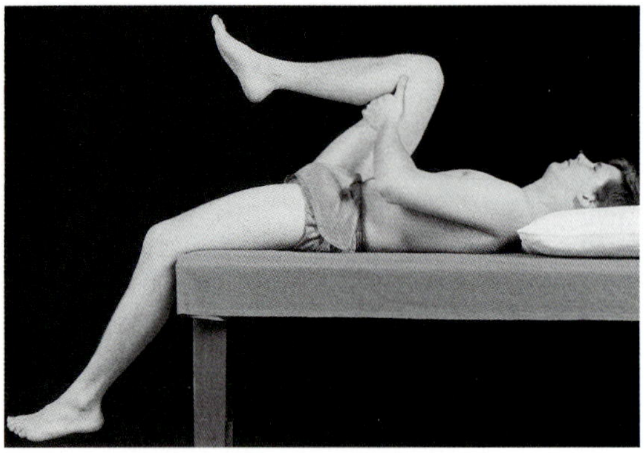

A

FIGURE 7-32: A, Correct test for length of hip flexor muscles.

Correct Test for Length of Hip Flexor Muscles

The low back and sacrum are flat on the table. The thigh touches the table indicating normal length of the one-joint hip flexors (Figure 7-32A). The angle of knee flexion indicates little or no tightness in the two-joint hip flexors. Figure 7-33A shows an error in testing the same subject.

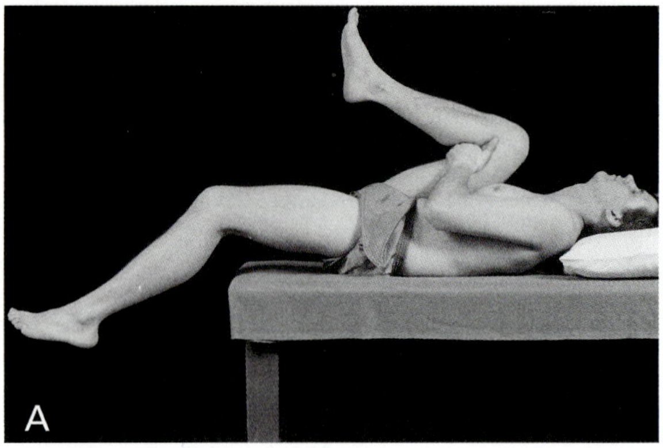

A

FIGURE 7-33: A, Error in testing for length of hip flexor muscles.

Error in Testing

This subject has excessive back flexibility (Figure 7-33A). When he pulls the knee too far toward the chest, the thigh comes up from the table, and the sacrum is no longer flat on the table. The result is that the one-joint hip flexors, which are normal in length, appear to be tight.

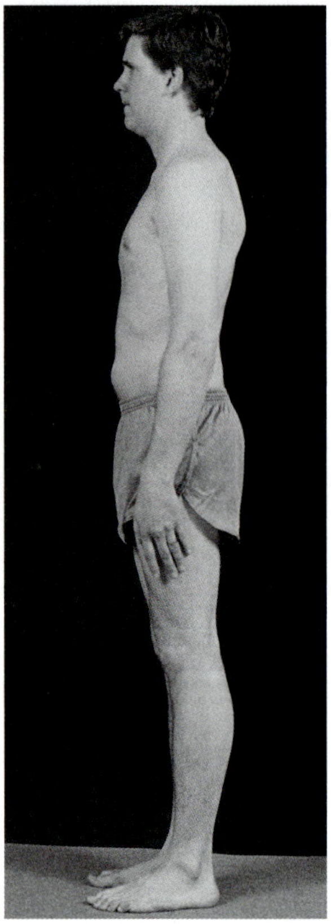

B

FIGURE 7-32: B, Good posture in standing.

This subject has good postural alignment in standing (Figure 7-32B). Examination of posture in standing does not, however, provide a clue regarding the extent of back flexibility in this subject.

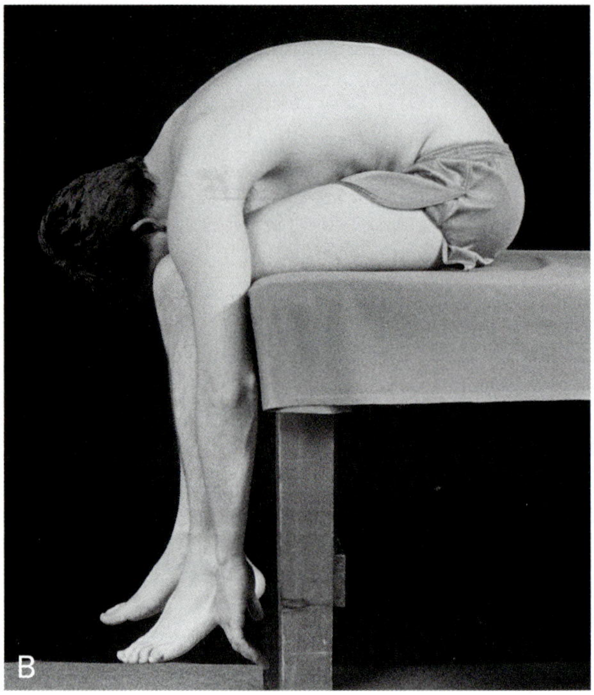

B

FIGURE 7-33: B, Forward bending test; excessive flexion of low back.

Excessive flexion of the low back is clearly demonstrated by the forward bending test, as illustrated in Figure 7-33B.

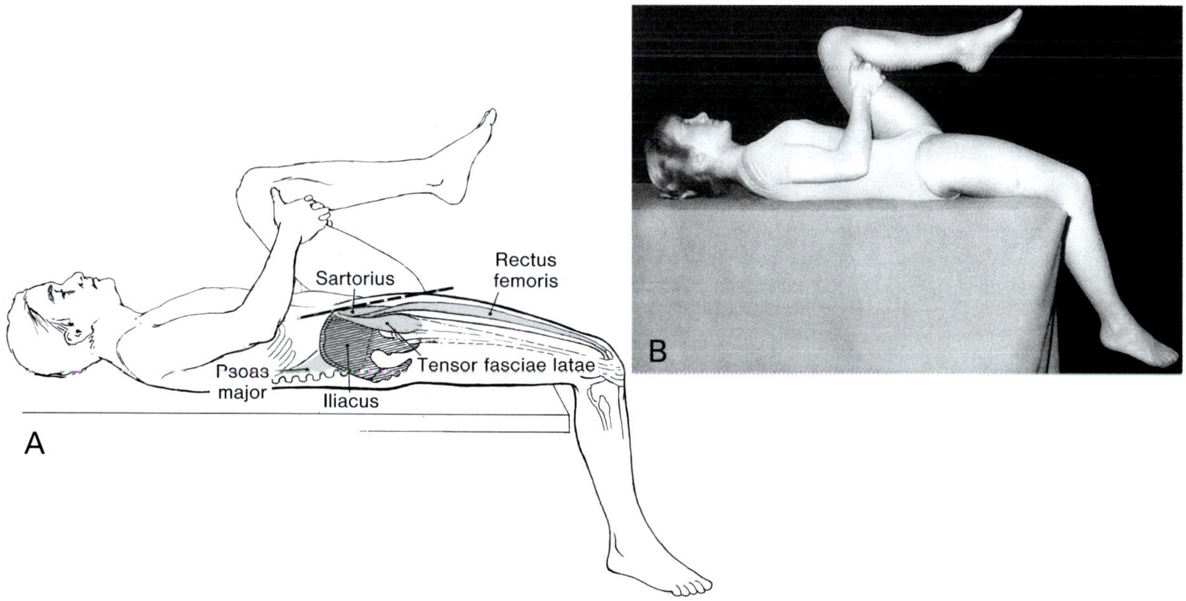

FIGURE 7-34: A, B, Normal length of hip flexors.

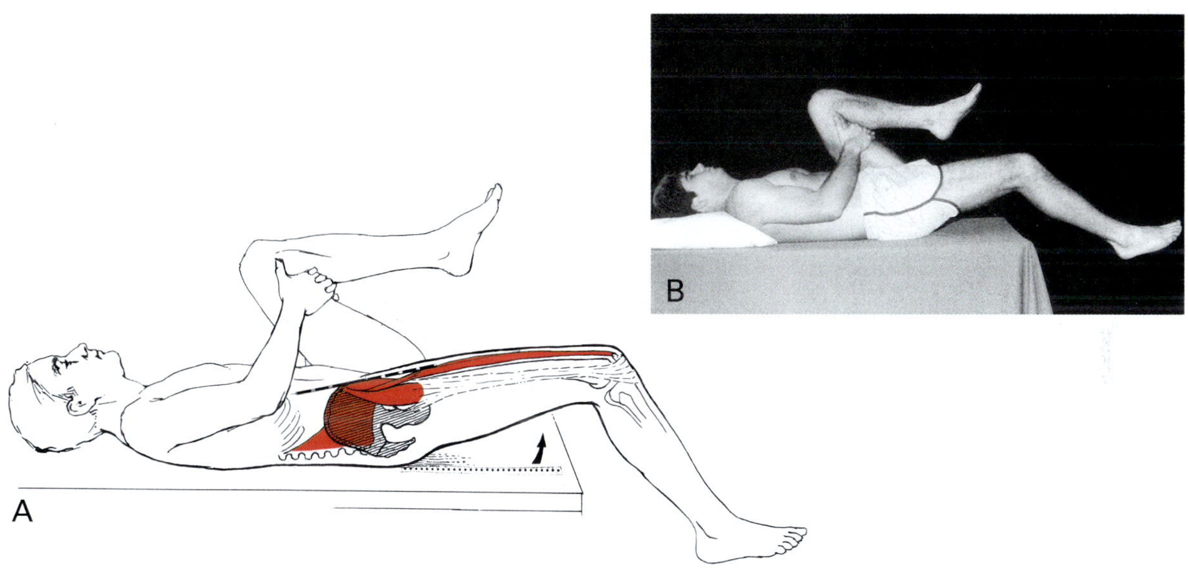

FIGURE 7-35: A, B, Shortness of both one-joint and two-joint hip flexors.

Normal Length of Hip Flexors

With the low back and sacrum flat on the table, the posterior thigh touches the table, and the knee passively flexes approximately 80°. In Figure 7-34, the pelvis is shown in 10° of posterior tilt. This is equivalent to 10° of hip joint extension and, with the thigh touching the table, represents normal length of the one-joint hip flexors. In addition, the knee flexion (about 80°) indicates that the rectus femoris is normal in length and that the tensor fasciae latae probably is normal. To maintain the pelvis in posterior tilt with the low back and sacrum flat on the table, one thigh is held toward the chest while testing the length of the opposite hip flexors.

Shortness of Both One-Joint and Two-Joint Hip Flexors

With the low back and sacrum flat on the table, the posterior thigh does not touch the table, and the knee extends. Figure 7-35 shows shortness of both one-joint and two-joint muscles. If the hip remains in 15° of flexion with the knee extended, the *one-joint* hip flexors lack 15° of length. If the knee will flex to only 70°, the *two-joint* muscles lack 25° of length (15° at the hip plus 10° at the knee).

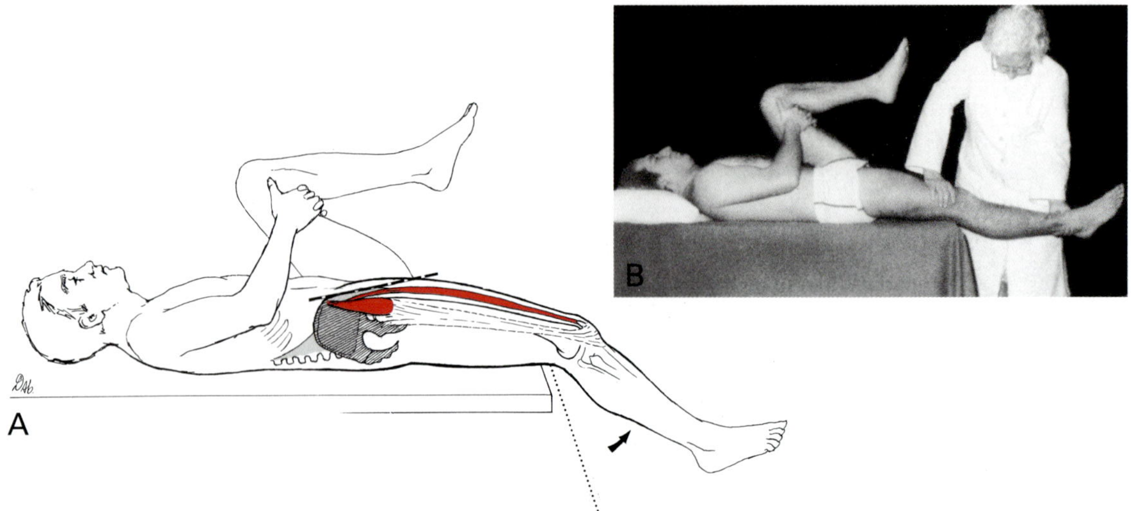

FIGURE 7-36: **A**, **B**, Normal length of one-joint and shortness of two-joint hip flexors.

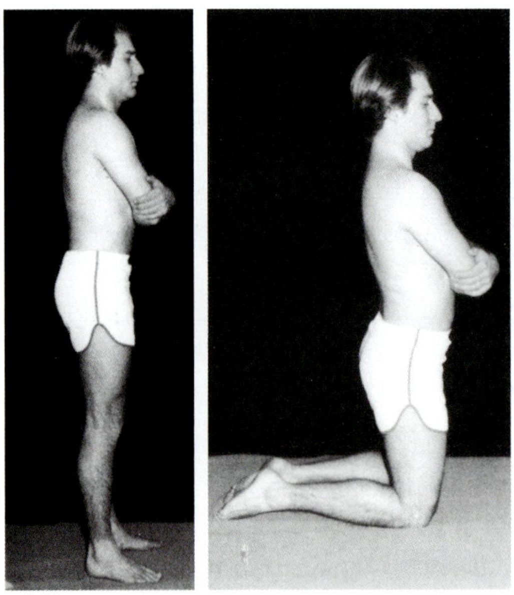

FIGURE 7-37: Testing for length of hip flexor muscles, standing and kneeling.

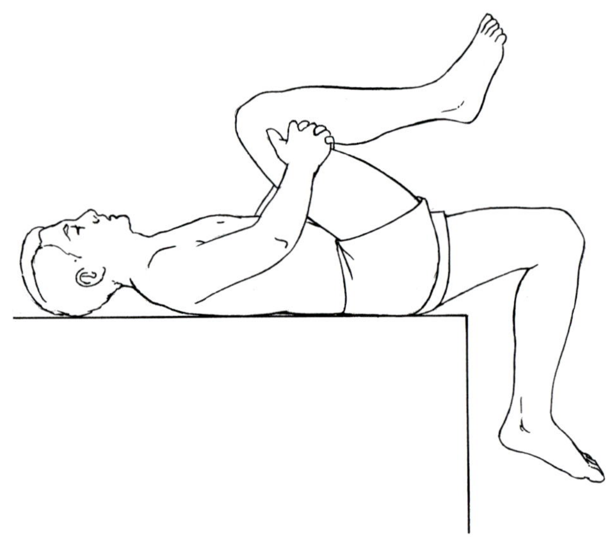

FIGURE 7-38: Shortness of one-joint and no shortness of two-joint hip flexors.

Normal Length of One-Joint and Shortness of Two-Joint Hip Flexors

With the low back and sacrum flat on the table and the knee in extension, the posterior thigh touches the table. Shortness of the two-joint muscles is determined by holding the thigh in contact with the table and allowing the knee to flex. The angle of knee flexion (i.e., number of degrees less than 80°) determines the degree of shortness. Figure 7-36 shows a subject in whom the hip joint can be extended if the knee joint is allowed to extend. This means that the one-joint hip flexors are normal in length, but that the rectus femoris is short.

In standing, the subject does not have a hyperlordosis. This indicates that the shortness is *not* in the one-joint hip flexors.

A kneeling position puts a stretch on the short rectus femoris and tensor fasciae latae over both the hip joints and knee joints, causing these muscles to pull the pelvis into anterior tilt and back into a lordotic position (Figure 7-37).

Shortness of One-Joint and No Shortness of Two-Joint Hip Flexors

The posterior thigh does not touch the table, and the knee can be flexed as many degrees beyond 80° as the hip is flexed. In Figure 7-38, the thigh is flexed 15° and the knee 95°.

Excessive Hip Flexor Length

The subject is tested with the low back flat, hip joint at the end of the table, and knee straight. That the thigh drops below table level is evidence of excessive length in the one-joint hip flexors (Figure 7-39).

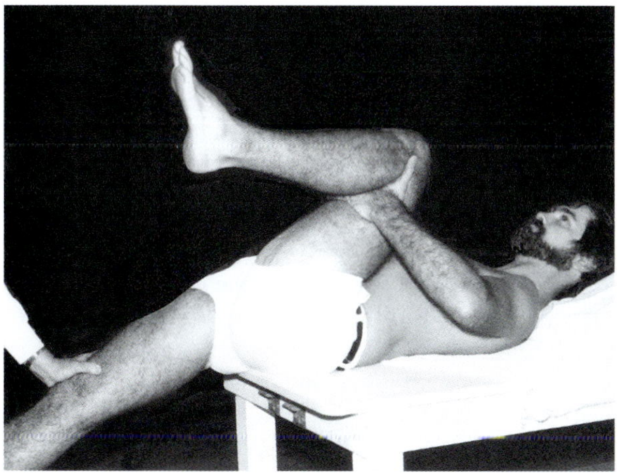

FIGURE 7-39: Excessive hip flexor length.

Shortness of Sartorius

During the hip flexor length test, a combination of three or more of the following indicates tightness of the sartorius: abduction of the hip, flexion of the hip, external rotation of the hip, and flexion of the knee (Figure 7-40).

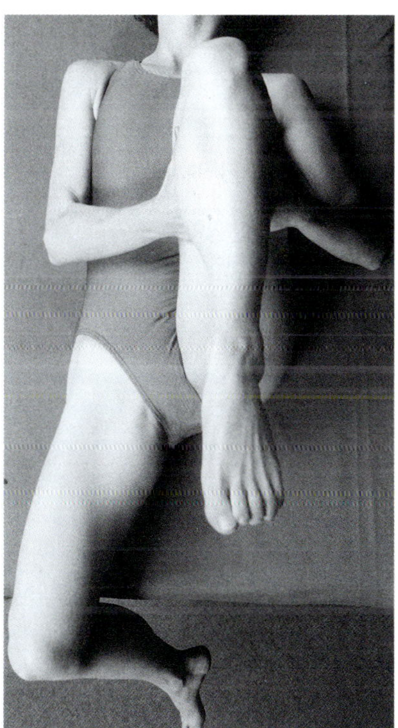

FIGURE 7-40: Shortness of sartorius.

Shortness of Tensor Fasciae Latae During Hip Flexor Length Testing

The following variations noted during hip flexor length testing indicate shortness of the tensor fasciae latae but do not constitute a length test for this muscle:

Abduction of the hip as it extends: Occasionally, the hip joint can be fully extended along with abduction. This finding indicates shortness of the tensor fasciae latae but not in the iliopsoas.

Lateral deviation of the patella: If the hip is not allowed to abduct during extension, there may be a strong lateral pull on the patella because of tensor fasciae latae shortness. It may also occur even if the hip abducts.

Extension of the knee if the thigh is prevented from abducting or is passively adducted as the hip is extended.

Internal rotation of the hip.

External rotation of the knee (open kinetic chain).

Shortness of Tensor Fasciae Latae and Sartorius: Similarities and Differences

Table 7-2 lists how the shortness of tensor fasciae latae and sartorius are similar and different.

TABLE 7-2 Shortness of Tensor Fasciae Latae and Sartorius: Similarities and Differences

Tensor Fasciae Latae	Joint	Sartorius
Abducts	Hip	Abducts
Flexes	Hip	Flexes
Internally rotates	Hip	Externally rotates
Extends	Knee	Flexes

PROBLEMS ASSOCIATED WITH HAMSTRING LENGTH TESTING

There are only two variables in the forward bending hamstring length test: the knee joint and the hip joint. Movement at the knee is controlled by maintaining the knee in extension during the movement of hip flexion. Hip flexion is obtained by movement of the pelvis toward the thigh. This test is not valid when there is a significant difference in length between right and left hamstrings, in which case the straight-leg-raising test should be used.

There are three variables in the straight-leg-raising test: the low back, hip joint, and knee joint. The knee joint is controlled by maintaining it in extension. The pelvis is controlled by maintaining the low back and sacrum flat on the table. The position of the pelvis and low back must be controlled. If the pelvis is in anterior tilt and the low back is hyperextended, the hip joint is already in flexion. The hamstrings will appear shorter than they actually are when measured by the angle of the lower extremity with the table because this measurement does not include the amount of hip joint flexion due to the anterior pelvic tilt.

Hip flexor shortness is the chief cause of anterior pelvic tilt in the supine position, and the degree of shortness varies from one individual to another. In order to stabilize the pelvis with the low back and sacrum flat on the table, one must accommodate the tight hip flexors by passive flexion using pillows or a towel-roll under the knees, but *only* as much as is necessary to obtain the required position of the pelvis.

If the hip and knees are flexed to allow approximately 40° of hip flexion, the position will assure that there will be no anterior pelvic tilt to interfere with testing, but it will not prevent excessive posterior tilting. Standardizing the hip and knee position will not ensure that the position of the low back and pelvis will be standardized.

The hamstrings will appear longer than their actual length if the pelvis is in posterior tilt with the low back in excessive flexion. When the straight-leg-raising test is done starting with one knee and hip flexed and the foot resting on the table as the other leg is raised, the pelvis is free to move in the direction of posterior tilt. An individual with as little as 45 of straight-leg-raising can appear to have as much as 90° of length.

TESTS FOR LENGTH OF HAMSTRING MUSCLES

Straight-Leg Raising

Equipment:

Table or floor.
Folded blanket may be used, but not soft padding. (The examiner cannot confirm that the low back and sacrum are flat if they are on a soft pad.)
Goniometer to measure the angle between the straight leg and the table.
Pillow or towel roll (in case of hip flexor shortness).
Chart to record findings.

Starting Position: Supine with the lower extremities extended and the low back and sacrum flat on the table. (Standardization of the test requires that the knee be in extension, and that the low back and pelvis have a fixed position to control the variables created by excessive anterior or posterior pelvic tilt.) When the low back and sacrum are flat, *hold* one thigh firmly down, making use of passive restraint by the hip flexors to prevent excessive posterior pelvic tilt before starting to raise the other leg in the straight-leg-raising test.

Test Movement: With the low back and sacrum flat on the table and one leg held firmly down, have the subject raise the other leg with the knee straight and the foot relaxed.

Reasons: The knee is kept straight for standardization of test performance. The foot is kept relaxed to avoid gastrocnemius involvement at the knee. (If the gastrocnemius is tight, dorsiflexion of the foot will cause the knee to flex, thereby interfering with the test of the hamstrings.) If the knee starts to bend, lower the leg slightly and have the subject fully extend the knee and again raise the leg until some restraint is felt and the subject feels slight discomfort.

This straight-leg-raising test (Figure 7-41), with the lower back flat on the table, shows normal length of the hamstring muscles, which permits flexion of the hip to an angle of approximately 80° up from the table.

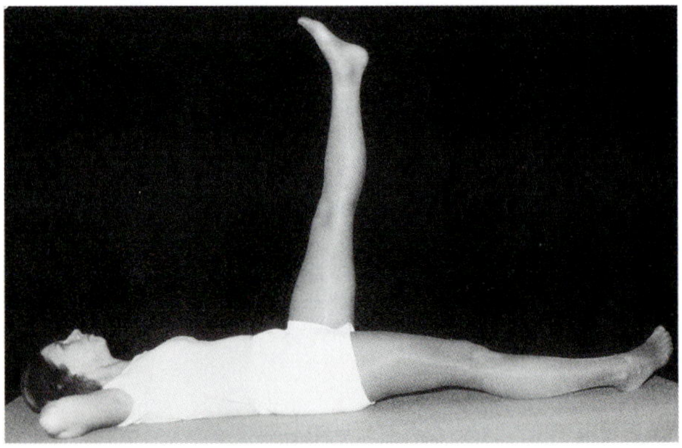

FIGURE 7-41: Straight leg raising test for length of hamstring muscles.

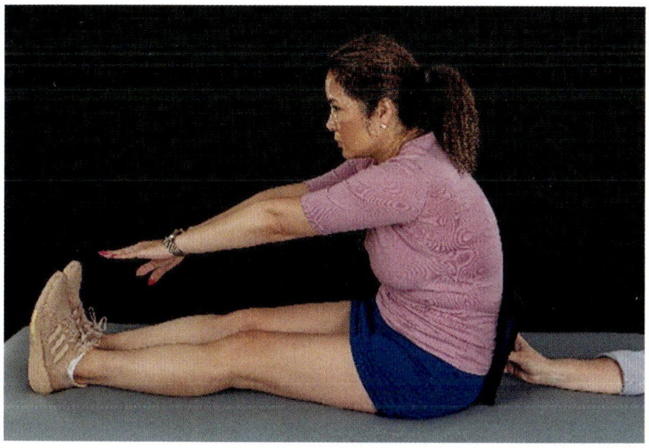

FIGURE 7-42: Measuring arc of motion.

FIGURE 7-43: Forward bending test for length of hamstring muscles.

Sit and Reach (Forward Bending)

Equipment:

Table (not padded) or floor.

Board (3 inches wide, 12 inches long, and approximately ¼ inch thick) to place flat against the sacrum.

Goniometer to measure the angle between the sacrum and the table.

Chart to record findings.

Starting Position: Sitting with the hips flexed and the knees fully extended (long-sitting). Allow the feet to be relaxed and avoid dorsiflexion.

Reasons: Keeping the knee straight maintains a fixed elongation of the hamstrings over the knee joint, eliminating movement at the knee as a variable. Avoiding dorsiflexion of the foot prevents the knee flexion that may occur if the gastrocnemius is tight.

Test Movement: Have the subject reach forward, as far as possible, in the direction of trying to touch the fingertips to the toes or beyond.

Reasons: The subject will tilt the pelvis forward, toward the thighs, flexing the hip joints to the limit allowed by the hamstring length.

Measuring Arc of Motion: Place the board with the 3-inch side on the table and the 12-inch side pressed against the sacrum when the hamstring length appears to be normal or excessive (Figure 7-42). Place the board with the 12-inch side on the table and the 3-inch side pressed against the sacrum when the hamstrings are tight. Measure the angle between the upright board and the table.

Normal Range of Motion: The pelvis tilts anteriorly toward the thigh to the point that the angle between the sacrum and the table is approximately 80° (Figure 7-43), that is, the same angle as that between the leg and the table in the straight-leg-raising test.

In forward bending, normal hamstring length permits tilting of the *pelvis toward the thigh* (i.e., hip joint flexion) as illustrated in Figure 7-43.

Hip Flexion at 90° with Passive or Active Knee Extension in Supine

Sometimes an effort is made to determine hamstring length by ascertaining the number of degrees lacking in knee joint extension. In supine, one lower extremity is placed in approximately 40° of hip flexion, with the knee flexed and the foot resting on the table. The thigh of the opposite leg is raised to a position perpendicular to the table (which may—or may not—be 90° of true hip joint flexion). The knee is then moved in the direction of extension. The length of hamstrings is stated in the number of degrees the knee joint *lacks* in extension.

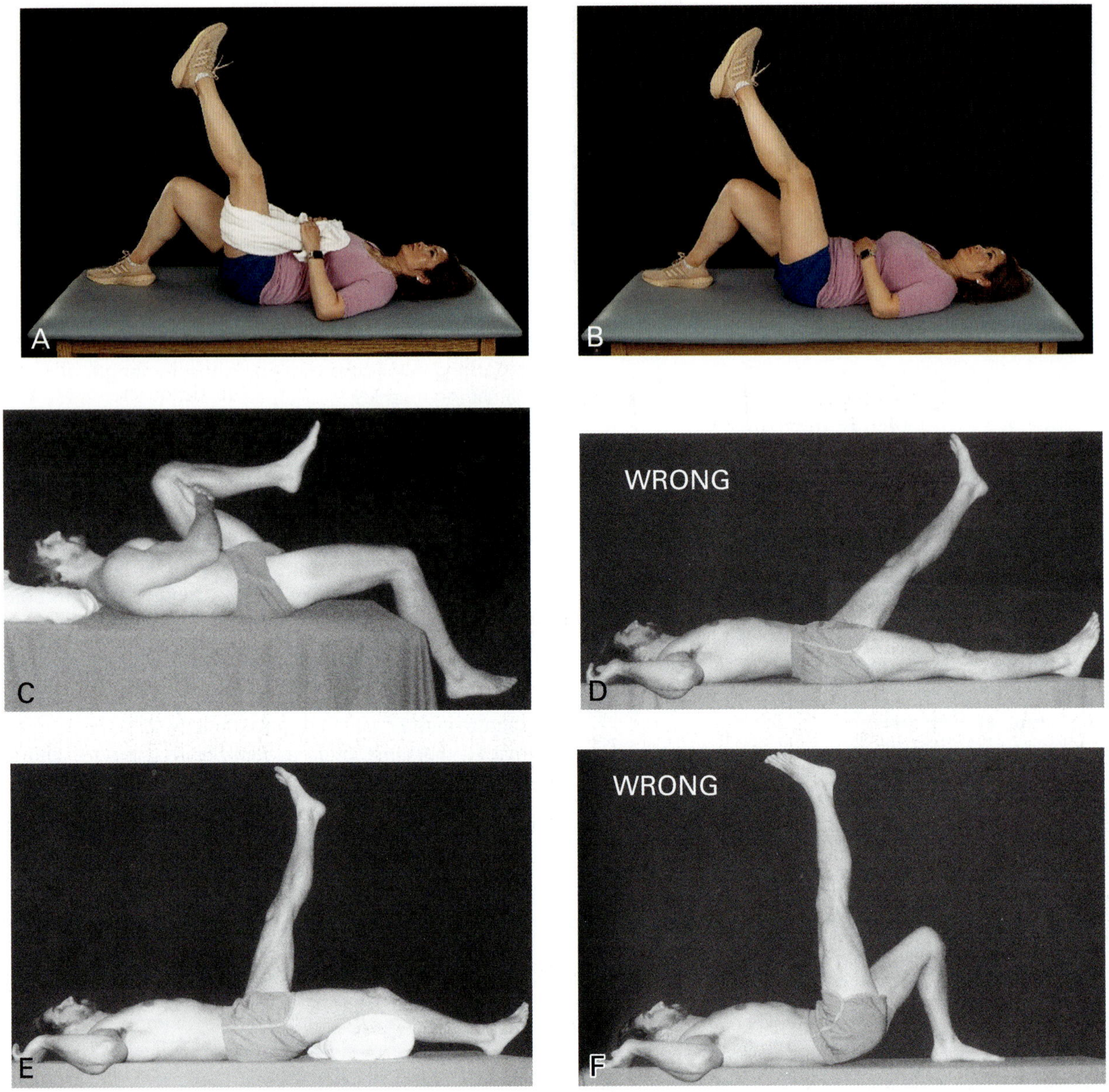

FIGURE 7-44: **A–F**, Hamstring length tests as related to the effect of hip flexor shortness.

EFFECT OF HIP FLEXOR SHORTNESS ON HAMSTRING LENGTH TESTS

Assisted with a towel, holding the position (Figure 7-44A).

Holding the position without assistance (Figure 7-44B).

A test for length of hip flexors confirms shortness of these muscles (Figure 7-44C) (see hip flexor length tests).

The hamstrings appear to be short. This test is not accurate, however, because the low back is not flat on the table. Shortness of the hip flexors on the side of the extended leg holds the back in hyperextension (Figure 7-44D).

To accommodate for hip flexor shortness and allow the low back to flatten, the thigh is *passively* flexed by a pillow under the knee, *not actively* held in flexion by the subject. With the back flat, the test accurately shows the hamstrings to be normal in length (Figure 7-44E).

In testing for hamstring length and in exercising to stretch short hamstrings, *avoid* placing one hip and knee in the flexed position (as illustrated) while raising the other. Otherwise, flexibility of the lower back is added to the range of hip flexion, making the hamstrings appear to be longer than they are. Not infrequently, a subject has excessive back flexibility along with hamstring shortness (Figure 7-44F).

FIGURE 7-45: Forward bending and hip flexor shortness.

Flexion of the pelvis toward the thigh (i.e., hip flexion) appears to be normal in forward bending. Because both hips are in flexion in forward bending, hip flexor shortness does not interfere with movement of the pelvis toward the thigh, as occurs when one leg is extended in the supine position (Figure 7-45).

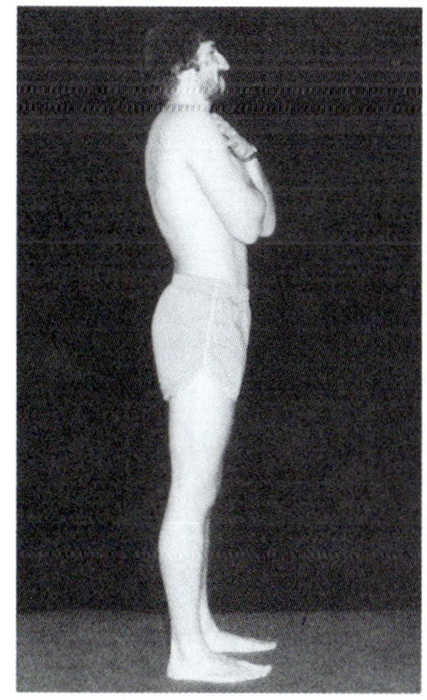

FIGURE 7-46: Lordosis while standing.

The lordosis in standing is evidence of the one-joint hip flexor shortness in this subject (Figure 7-46).

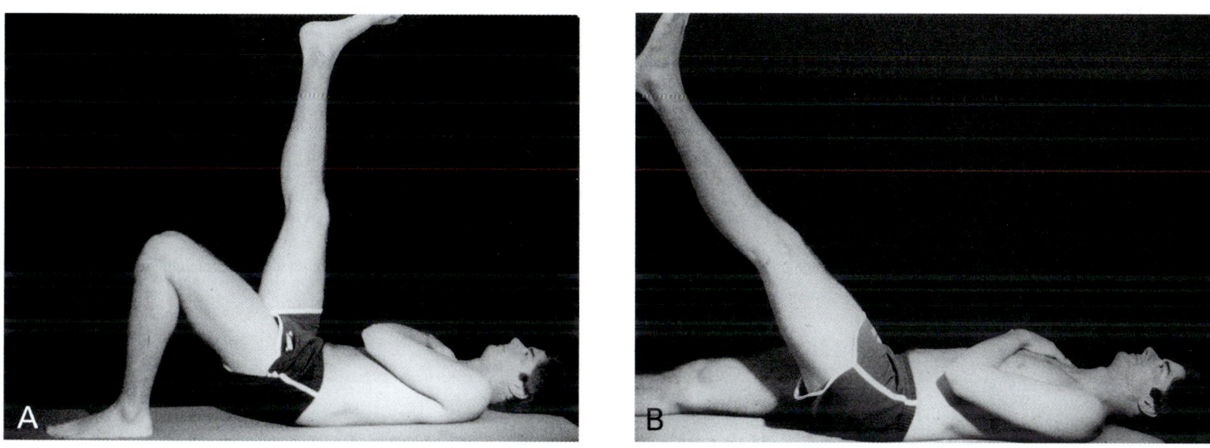

FIGURE 7-47: **A**, Error in testing hamstring length; **B**, Correct test for hamstring length.

ERRORS IN TESTING HAMSTRING LENGTH

Error in Testing

When the straight-leg-raising test is done starting with one knee and hip flexed and the foot resting on the table as the other leg is raised, the pelvis is free to move in the direction of excessive posterior tilt, with the sacrum no longer flat on the table. Depending on the amount of flexibility of the back, the hamstring length will appear to be longer than actual, because back flexion is added to hip flexion. An individual with as little as 45° of actual hamstring length can appear to have as much as 90°, as seen in Figure 7-47.

No Standardization of Low Back and Pelvis

If the hip and knee are flexed to allow approximately 40° of hip flexion, the position will ensure enough slack in the hip flexors that they will not cause anterior pelvic tilt. This will not, however, insure against excessive posterior tilting. Standardization of the amount of hip and knee flexion will not standardize the position of the low back and pelvis, which must be standardized. Hip flexor shortness is the chief cause of anterior pelvic tilt in the supine position, and the degree of shortness varies from one individual to another. To stabilize the pelvis with the low back and sacrum flat on the table, one must "give in" to the tight hip flexors by using a pillow or towel roll under the knees, but only by as much as is necessary to obtain the required position of the pelvis.

THE OBER AND MODIFIED OBER TEST

HISTORICAL NOTE

In the *Journal of the American Medical Association,* May 4, 1935, there appeared an article by Frank Ober of Boston entitled "Back Strain and Sciatica" (7). In it, he discussed the relationship of a contracted tensor fasciae latae and iliotibial band to low back and sciatic pain. The test for tightness was described, but Ober did not mention anything about avoiding hip flexion or internal rotation as the thigh is allowed to drop in adduction.

After the article appeared, Henry O. Kendall, then a physical therapist at Children's Hospital School in Baltimore, expressed concern about the test to his medical director, George E. Bennett (8, 9). The concern was that allowing the thigh to drop in flexion and internal rotation would "give in" to the tight tensor and not accurately test it for length. At some point in late 1935 or early 1936, Dr. Ober visited Children's Hospital School, and Mr. Kendall expressed concern about the test to him personally.

In the *Journal of the American Medical Association,* August 21, 1937, another article appeared in which Dr. Ober again described his test but, this time, he cautioned the examiner to avoid hip flexion and internal rotation as the thigh is allowed to adduct (10).

Apparently, some people who have described the test had access to the first article but not to the second. A well-known text describes positioning the leg in abduction, with the hip in a neutral position and the knee flexed 90°, and then *releasing* the abducted leg (11). The text also states that the normal iliotibial band will allow the thigh to drop to the adducted position (as illustrated by the knee touching the other leg or the table). A tensor fascia lata of normal length will not permit the thigh to drop to table level unless the hip goes into some internal rotation and flexion.

In the first article, Ober stated, "The thigh is abducted and extended in the coronal plane of the body." With respect to what should be considered a "normal" range of motion in the direction of adduction, this article stated, "If there is no contraction present, the thigh will adduct beyond the median line." It must be noted that this statement referred to the test in which no reference was made about preventing flexion and internal rotation.

In the second article, Ober did not specifically refer to the coronal plane, but he did state, "The thigh is allowed to drop toward the table in this plane." By the description, Ober was referring to the coronal plane. Maintaining the thigh in the coronal plane prevents hip joint flexion.

The second article made no mention of how far the thigh should drop toward the table (see below for further discussion about normal range of motion in adduction).

Before deciding what may be considered a normal range of adduction in the Ober test, it is necessary to review normal range of motion of the hip joint. Contrary to the information in several books (12–16), the normal range of hip joint adduction from the anatomical position (i.e., in the coronal plane) is—and should be—limited to approximately 10°.

If adduction is limited to 10°, then in the side-lying position, with the pelvis in a neutral position, the extended extremity should not drop more than 10° below the horizontal if kept in the coronal plane. In flexion and internal rotation, the range in adduction is greater, but *such a position is no longer a test for length of the tensor fasciae latae.* The action of the muscle is abduction, flexion, and internal rotation of the hip as well as assisting in extension of the knee. By "giving in" to flexion and internal rotation, the muscle is *not being lengthened.*

Limitation of range of motion provides stability by preventing excessive motion. Limitation of knee joint extension prevents hyperextension. Limitation of hip joint extension prevents the pelvis from swaying forward abnormally in standing. Limitation of hip joint adduction provides stability for standing on one leg at a time.

In the 1937 article, Ober also stated that "when the maximum amount of fascial contracture is on the side and in front of the femur, the spine is held in lordosis, and that if the contracture is posterolateral, the lumbar curve is flattened. The former condition is common; the latter is rare. Either condition may be associated with pain low in the back and sciatica. Unilateral contracture may produce lateral curvature of the spine" (10).

To test for tightness of the posterolateral iliotibial band, the hip is slightly flexed and medially rotated along with the adduction. Tightness of this band can be a factor in a straight-leg-raising test for hamstring length.

Three-fourths of the gluteus maximus inserts into the iliotibial band, but the fibers are oblique to the band and do not have the direct line of pull as does the tensor fasciae latae. Furthermore, the gluteus maximus is seldom tight.

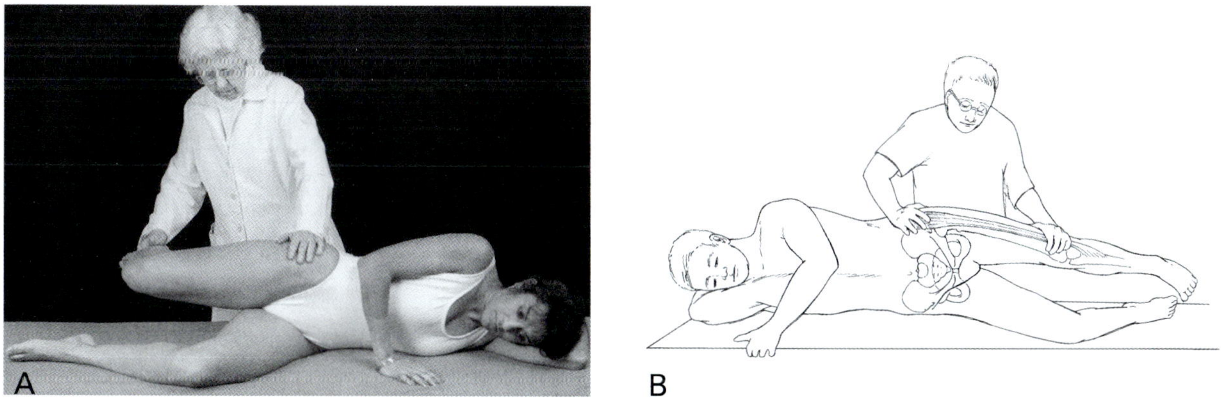

A B

FIGURE 7-48: Normal length: **A**, Ober test; **B**, Modified Ober test.

Ober Test

Below is the description of the test (which Ober called "The Abduction Test") quoted directly from the 1937 article to provide the reader with the author's exact description: (10)

The Abduction Test

1. The patient lies on their side on a table, the shoulder and pelvis being perpendicular to the table.
2. The leg on which the patient is lying is flexed at the knee, and the hip is flexed and kept flexed to flatten the lumbar curve.
3. If the patient is on their left side, the examiner places their left hand over the patient's hip in the region of the trochanter to steady him.
4. The right leg is flexed to a right angle at the knee and is grasped just below the knee with the examiner's right hand, the leg and ankle being allowed to extend backward under this forearm and elbow.
5. The right thigh is abducted widely, then hyperextended in the abducted position, the lower extremity being kept level and care being taken to keep the hip joint in a neutral position as far as rotation is concerned.
6. The examiner slides his right hand backward along the leg until it grasps the ankle lightly but with enough tension to keep the hip from flexing.
7. The thigh is allowed to drop toward the table in this plane. (Caution: Do not bear down on the leg.) If the fascia lata and the iliotibial band are tight, the lower extremity will remain more or less permanently abducted. If the hip is allowed to flex or internally rotate, the iliotibial band becomes relaxed and the lower extremity falls from its own weight.
8. The same procedure for the opposite side is followed in every case.

Ober Test, Normal Length: With the knee maintained at a right angle, the thigh drops *slightly* below horizontal (Figure 7-48A).

Modified Ober Test

A modification of the Ober test was first recommended by the Kendalls in *Posture and Pain* (6). The reasons for modifying the test are as follows: to include less strain medially in the area of the knee joint, less tension on the patella, and less interference by a tight rectus femoris. Additionally, for a muscle with multiple actions, like the tensor fasciae latae, it is not necessary to stretch in the reverse of all actions when testing for length (17).

Place the subject in a side-lying position, with the underneath lower extremity flexed at the hip and knee to flatten the low back, thereby stabilizing the pelvis against anterior pelvic tilt. Anterior pelvic tilt is the equivalent of hip flexion and is to be avoided because it "gives in" to possible tightness of the tensor fasciae latae.

The pelvis must also be stabilized to prevent lateral pelvic tilt downward on the tested side. Downward lateral tilt is the equivalent of hip joint abduction, and such a movement of the pelvis would "give in" to a tight tensor. For most people, the lateral trunk will be in contact with the table in the side-lying position. People with wide hips and narrow waists will be the exceptions.

On the tested side, the examiner places one hand laterally on the subject's pelvis, just below the iliac crest, and pushes upward enough to stabilize the pelvis and keep the lateral trunk in contact with the table. The examiner does not externally rotate the hip but, instead, keeps it from internally rotating and brings it back in extension. If the tensor is tight, it will be necessary to abduct the lower extremity to bring it into extension. Keep the lower extremity extended in line with the trunk (i.e., in the coronal plane), and allow it to drop in adduction toward the table.

In Figure 7-48B, the pelvis is in neutral position, the hip is neutral between medial and lateral rotation, and the lower extremity is in the coronal plane and allowed to drop in adduction. In this case, it drops 10° below the horizontal, which may be considered a normal length for the tensor fasciae latae.

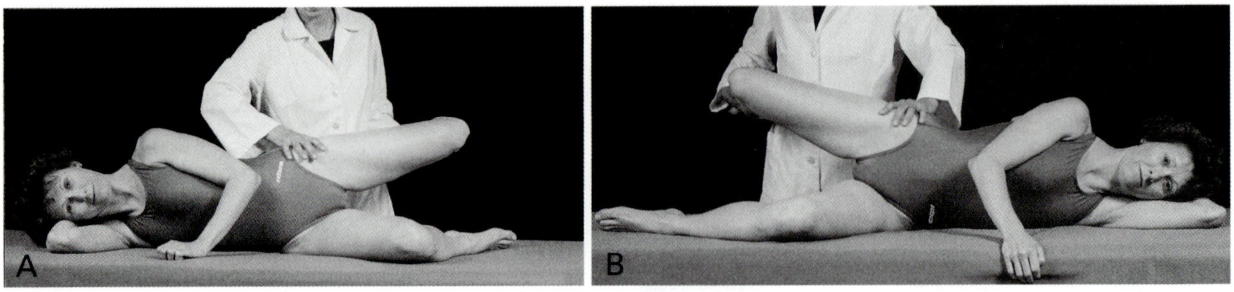

FIGURE 7-49: **A**, **B**, Positive Ober test; bilateral tightness of tensor fasciae latae.

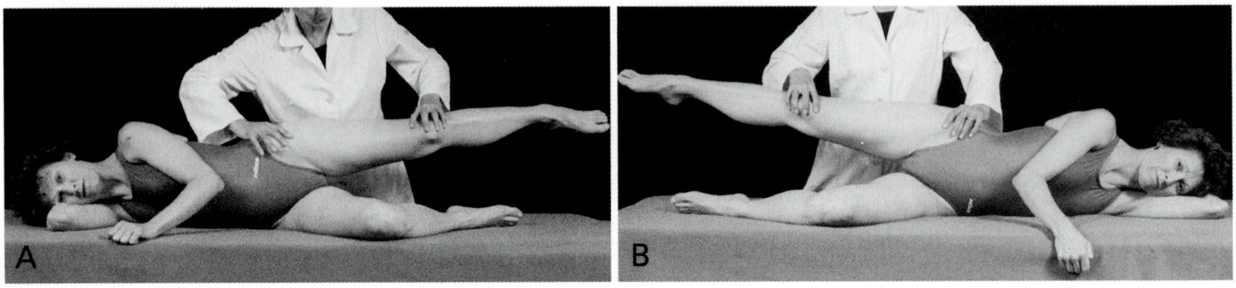

FIGURE 7-50: **A**, **B**, Modified Ober test (knee extended: bilateral tightness of tensor fasciae latae).

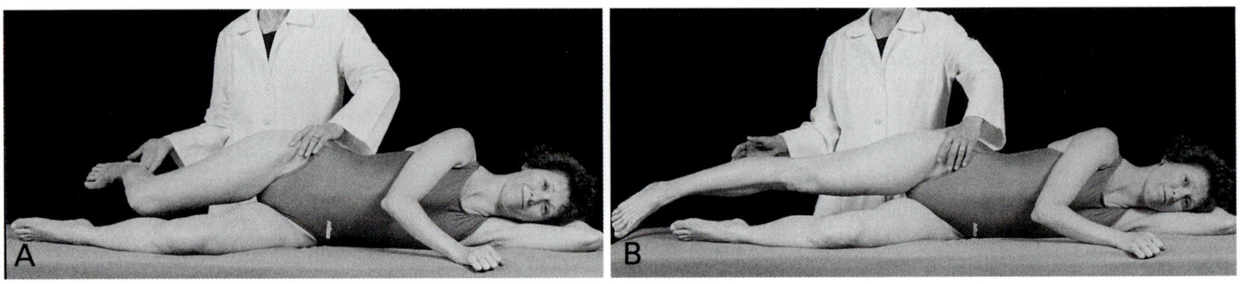

FIGURE 7-51: **A**, **B**, Errors in testing for tightness of tensor fasciae and iliotibial band.

Bilateral Tightness of Tensor Fasciae Latae: Positive Ober Test

The range of motion in adduction may be considered as normal if the lower extremity drops slightly below horizontal with the thigh in neutral rotation in the coronal plane and the knee flexed 90°. This subject's thighs remain in marked abduction because of bilateral tightness of the tensor fasciae latae and iliotibial band (Figure 7-49).

Bilateral Tightness of Tensor Fasciae Latae: Modified Ober Test (Knee Extended)

The range of motion in adduction may be considered as normal if the lower extremity drops 10° below the horizontal with the thigh in neutral rotation in the coronal plane and the knee extended. In this test, this subject's lower extremities do not drop to the horizontal because of tightness in the tensor fasciae latae and iliotibial band (Figure 7-50).

Errors in Testing for Tightness of Tensor Fasciae Latae and Iliotibial Band

According to one reference, the lower extremity, with the knee bent, is maneuvered into the correct Ober test position *and then released* (7). As seen in Figure 7-51A and B, the hip internally rotates and flexes when not controlled by the examiner. The thigh must be kept in the coronal plane and prevented from internally rotating to test accurately for tightness of the tensor fasciae latae and iliotibial band.

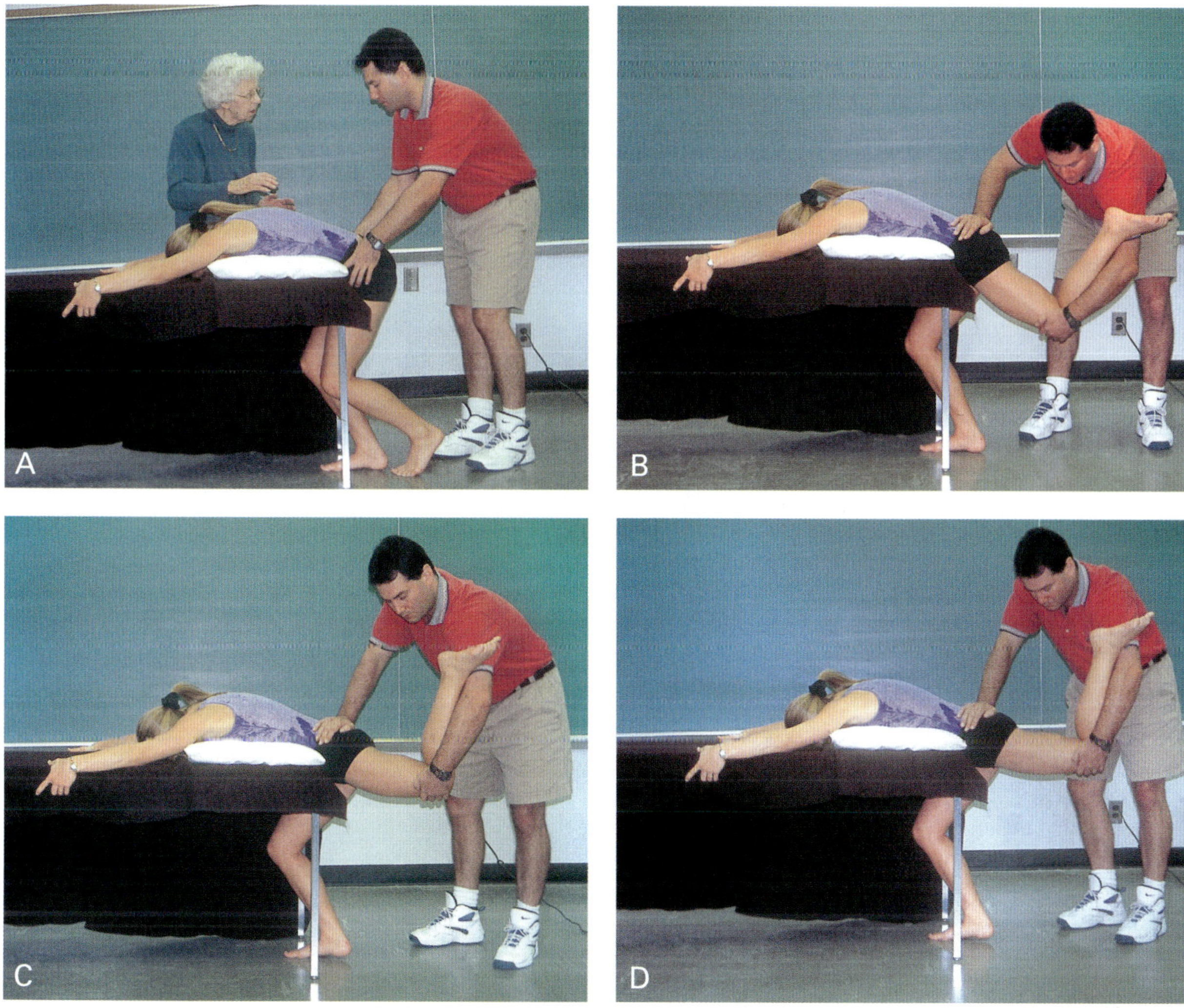

FIGURE 7-52: A–D, Modified Ober test: trunk prone.

MODIFIED OBER TEST: TRUNK PRONE

Equipment: Treatment table. If the table is not padded, place a folded towel or thin pillow at the end of the table as a cushion. For this test, a table that can be raised or lowered is preferable to accommodate for the height of the subject. (Except for this test, treatment tables that have thick padding and are hinged in the middle are not suitable for length and strength tests of most hip joint muscles and trunk muscles.) Adjust the height of the table as needed for the subject to be able to place both feet on the floor with the knees slightly bent.

Starting Position: The subject stands at the end of the table, in contact with the table, and bends forward to rest the trunk prone on the table. For the trunk to rest fully on the table, the knees are bent, and the feet are placed forward under the table as much as necessary. The subject extends both arms overhead and grasps the sides of the table (Figure 7-52).

Reasons: With the trunk prone, the lower back will be flat; keeping the arms fully extended overhead tends to prevent any lateral tilting of the pelvis. This prone position meets the requirements of the Ober test and is more stable than the side-lying position.

Test Movement: To test for length of the left tensor fasciae latae and the iliotibial band, the examiner stands in position to grasp, with the left arm, the subject's left thigh and lower leg, holding the knee bent at a right angle. With the right hand, the examiner holds the pelvis firmly down on the table. Keeping the knee bent, the examiner moves the leg to the completion of hip abduction and then upward in extension. Maintaining the hip joint at the completion of extension, the examiner then moves it in the direction of adduction. (Reverse the instructions for testing the subject's right leg.)

Normal Range of Motion: Moving the thigh to a position of zero adduction (i.e., comparable to horizontal in the side-lying position). (If the hip cannot be fully extended, there will be slightly more adduction.)

LEG-LENGTH MEASUREMENTS

So-called "actual leg length" is a measurement of length from the anterosuperior spine of the ilium to the medial malleolus. Obviously, such a measurement is not an absolutely accurate determination of leg length, because the points of measurement are from a landmark on the pelvis to one on the leg. Because it is impossible to palpate a point on the femur under the anterosuperior spine, it is necessary to use the landmark of the pelvis. It becomes necessary, therefore, to fix the alignment of the pelvis in relation to the trunk and legs before taking measurements to ensure the same relationship of both extremities to the pelvis. Pelvic rotation or lateral tilt will change the relationship of the pelvis to the extremities enough to make a considerable difference in measurement. To obtain as much accuracy as possible, the patient lies supine on a table, with the trunk, pelvis, and legs in straight alignment and close together. The distance from the anterosuperior spine to the umbilicus is measured on the right and on the left to check against lateral pelvic tilt or rotation. If a difference in measurements is found, the pelvis is leveled and any rotation corrected so far as possible before leg-length measurements are taken.

"Apparent leg length" is a measurement from the umbilicus to the medial malleolus. This type of measurement is more often a source of confusion than an aid in determining differences of length for the purpose of applying a lift to correct pelvic tilt. The confusion arises because the picture in standing is the reverse of that in lying and occurs when the pelvic tilt is caused by muscle imbalance rather than by an actual difference in leg length.

In *standing*, a fault in alignment will result when a weak muscle fails to provide adequate support for weight-bearing. For example, a weakness of the right gluteus medius allows the pelvis to deviate toward the right and seemingly elevate because the left side drops, giving the appearance of a *longer* right leg. If the postural fault has been of long standing, there is usually an associated imbalance in the lateral trunk muscles, in which the right lateral trunk muscles are shorter and stronger than the left.

In *lying*, a fault in alignment will more often result from the pull of a strong muscle. In the supine position, an individual with the type of imbalance described above (i.e., a weak right gluteus medius and strong right lateral trunk muscles) will tend to lie with the pelvis higher on the right, pulled upward by the stronger lateral abdominal muscles. This position, in turn, draws the right leg up so that it appears to be *shorter* than the left.

The need for an elevation on a shoe should be determined by measurements in standing rather than the lying position. Boards of various thicknesses are used for this purpose (see also apparent leg-length discrepancy caused by muscle imbalance).

FIGURE 7-53: A–D, Hip flexor length tests and tightness in the tensor fasciae latae.

HIP FLEXOR LENGTH TESTING

A tight one-joint muscle will limit the range of motion in the direction opposite to its action. A muscle that crosses two or more joints may exhibit tightness at only one joint if the other joint (or joints) are maintained in a position of normal elongation of the muscle.

Figure 7-53A: The subject is supine with the legs abducted. The low back is flat on the table, that is, normal flexion of the low back. The pelvis is in posterior tilt, and the hip joint is extended. There is no apparent hip flexor shortness.

Figure 7-53B: The legs are in neutral position, neither adducted nor abducted. The low back is no longer flat on the table, the pelvis is in anterior tilt. Because of the anterior pelvic tilt, the hip joint is in flexion.

Figure 7-53C: The subject is kneeling with the knees flexed about 90°, and the thighs are abducted. The pelvis and femur are in good alignment.

Figure 7-53D: The subject is kneeling with the thighs in neutral position (neither abducted nor adducted). The alignment of the trunk has shifted forward. The low back extension (arching) has increased, providing evidence of hip flexor tightness.

Conclusion: The tightness is in the muscle that both flexes and abducts the hip joint, namely the Tensor fasciae latae.

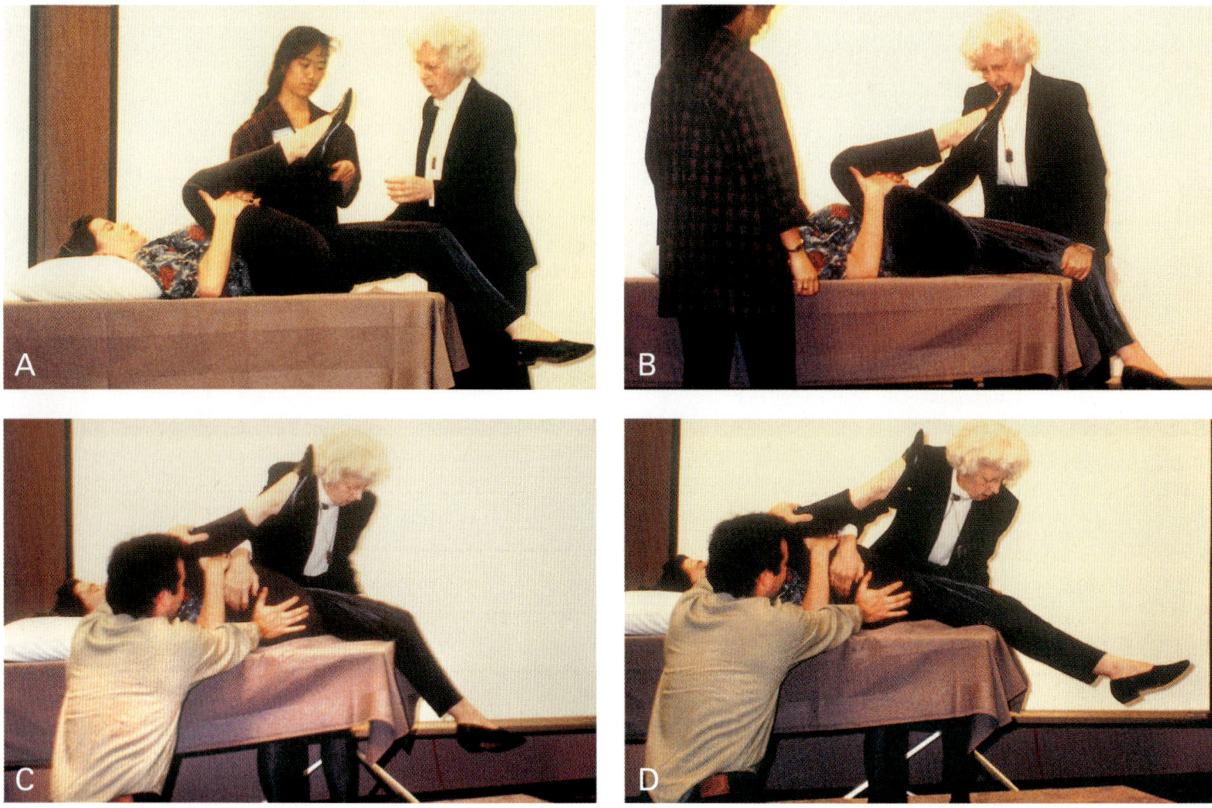

FIGURE 7-54: **A–D**, Hip flexor length tests performed on the same patient by the same examiner five years after the tests shown in Figure 7-53.

Differential Diagnosis

Flexion of the hip joint may be performed by flexion of the thigh toward the pelvis, or by anterior tilting of the pelvis toward the thigh. Hip flexors consist of the following (the sartorius is omitted here because it acts to flex and externally rotate the hip joint and flex the knee joint):

1. The one-joint iliopsoas that flexes the hip joint.
2. The two-joint rectus femoris that flexes the hip joint and extends the knee joint.
3. The two-joint tensor fasciae latae that flexes, abducts, and internally rotates the hip joint, and assists in extension of the knee.

Hip Flexor Length Tests

These photographs show tests for differential diagnosis of hip flexor tightness. The same subject and the same examiner appear in Figures 7-53 and 7-54, shown for comparison.

Figure 7-54A: Starting position for hip flexor length tests. The low back is flat on the table and maintained in that position by holding the right knee toward the chest as the left lower extremity is tested. There is evidence of left hip flexor shortness by the fact that the thigh does not touch the table.

Figure 7-54B: The lower extremity has been moved into a position of hip joint abduction. The thigh now touches the table, providing evidence that there is no tightness in the iliopsoas muscle. The degree of knee flexion indicates that there is little or no tightness in the rectus femoris.

Figure 7-54C: The thigh has been maintained in contact with the table (to keep the iliopsoas at its normal length). The pelvis has been stabilized to prevent any lateral movement of the pelvis as the lower extremity has been moved back (against a fair amount of resistance by the tensor) from the abducted position to zero position.

Figure 7-54D: Normal length of the tensor fasciae latae will permit knee flexion along with the hip extension and adduction. There is undeniable evidence of tightness in the tensor fasciae latae as exhibited by the extended position of the knees, especially evident at the time of the first test.

SECTION IV
MUSCLE STRENGTH TESTS

CHART FOR ANALYSIS OF MUSCLE IMBALANCE: LOWER EXTREMITY

Name:..*Date: 1st. Ex.-*..................*2nd. Ex.-*......................

Diagnosis:..*Onset:*...........................*Exam. of*..............*extremity*

		2nd EX.	1st. EX.	1st. EX.	2nd. EX.		
	ILIOPSOAS SARTORIUS TENSOR FAS. LAT. RECTUS FEMORIS } HIP FLEXORS					GLUTEUS MAXIMUS	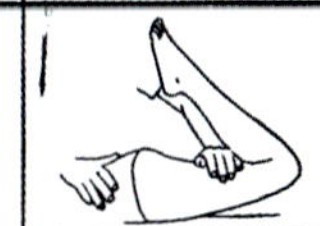
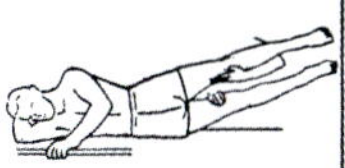	HIP ADDUCTORS					GLUTEUS MEDIUS	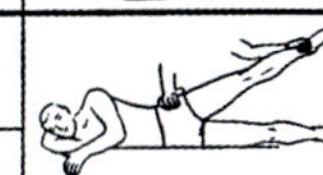
						GLUTEUS MINIMUS	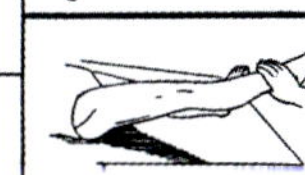
						TENSOR FASCIAE LATAE	
	HIP LATERAL ROTATORS					HIP MEDIAL ROTATORS	
	QUADRICEPS					MEDIAL	
						HAMSTRINGS LATERAL	
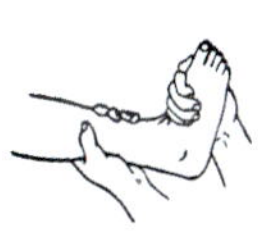	TIBIALIS ANTERIOR					SOLEUS	
						GASTROCNEMIUS & SOLEUS	
						PERONEUS LONGUS & BREVIS	
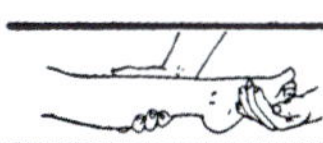	TIBIALIS POSTERIOR					PERONEUS TERTIUS	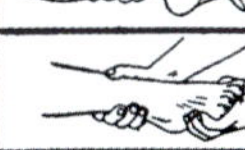
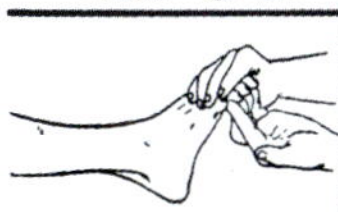	FLEXOR DIGITORUM LONGUS 1					1 DISTAL INTER-PHALANGEAL JOINT EXTENSORS	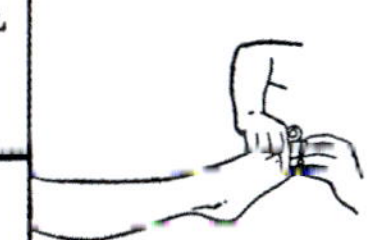
	2					2	
	3					3	
	4					4	
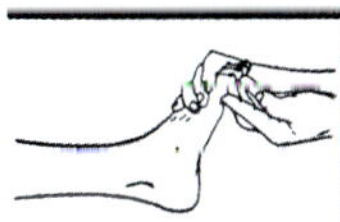	FLEXOR DIGITORUM BREVIS 1					1 PROXIMAL INTER-PHALANGEAL JOINT EXTENSORS	
	2					2	
	3					3	
	4					4	
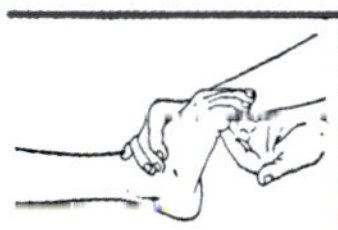	LUMBRICALES & INTEROSSEI 1					1 EXT. DIGITORUM LONGUS & BREVIS	
	2					2	
	3					3	
	4					4	
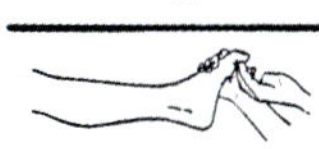	FLEXOR HALLUCIS LONGUS					EXTENSOR HALLUCIS LONGUS & BREVIS	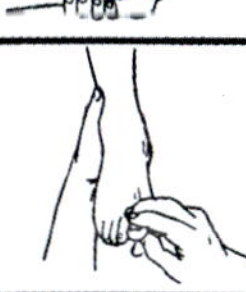
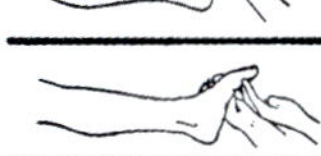	FLEXOR HALLUCIS BREVIS						
	ABDUCTOR HALLUCIS					ADDUCTOR HALLUCIS	

FIGURE 7-55: Chart for analysis of lower extremity muscle imbalance.

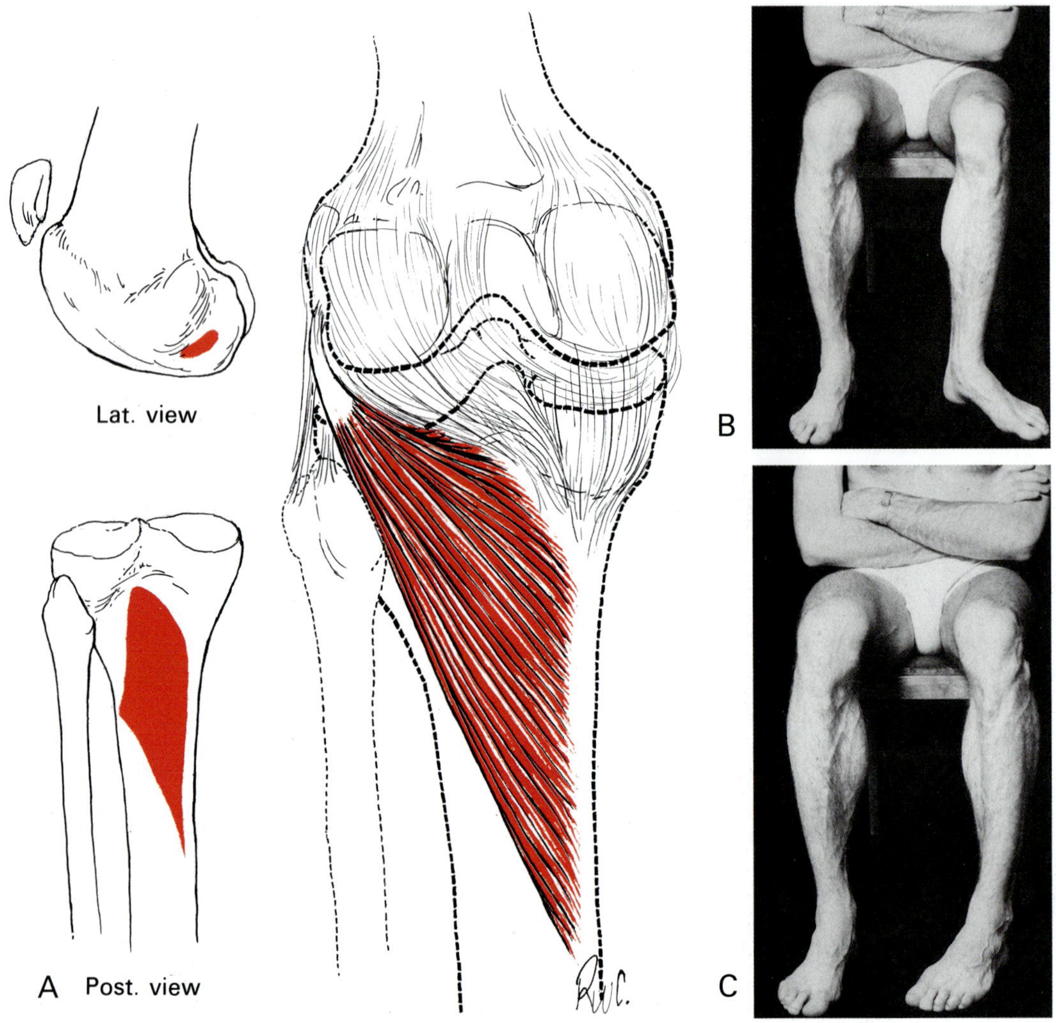

FIGURE 7-56: Popliteus: **A**, Lateral and posterior views; **B**, Testing the popliteus; starting position, left leg; **C**, Completed test position, left leg.

Popliteus

Origin: Anterior part of the groove on the lateral condyle of the femur and oblique popliteal ligament of the knee joint (Figure 7-56A).

Insertion: Triangular area proximal to the soleal line on the posterior surface of the tibia and fascia covering the muscle.

Action: In non-weight-bearing (i.e., *with the origin fixed*), the popliteus medially rotates the tibia on the femur and flexes the knee joint. In weight-bearing (i.e., *with the insertion fixed*), it laterally rotates the femur on the tibia and flexes the knee joint. This muscle helps to reinforce the posterior ligaments of the knee joint.

Nerve: Tibial, L4, **5**, **S1**.

Patient: Sitting, with the knee flexed at a right angle and with leg in lateral rotation of tibia on femur.

Fixation: None necessary.

Test Movement: Medial rotation of the tibia on the femur (Figure 7-56C).

Resistance: Seldom is resistance or pressure applied, because the movement is not used as a test for the purpose of grading the popliteus but, rather, merely to indicate whether the muscle is active.

Weakness: May result in hyperextension of the knee and lateral rotation of the tibia on the femur. Weakness is usually found in instances of imbalance between the lateral and medial hamstrings in which the medial hamstrings are weak and the lateral hamstrings are strong.

Shortness: Results in slight flexion of the knee and medial rotation of the tibia on the femur.

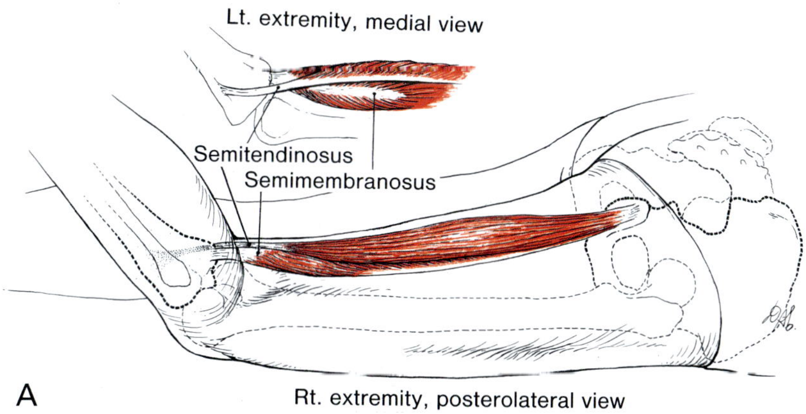

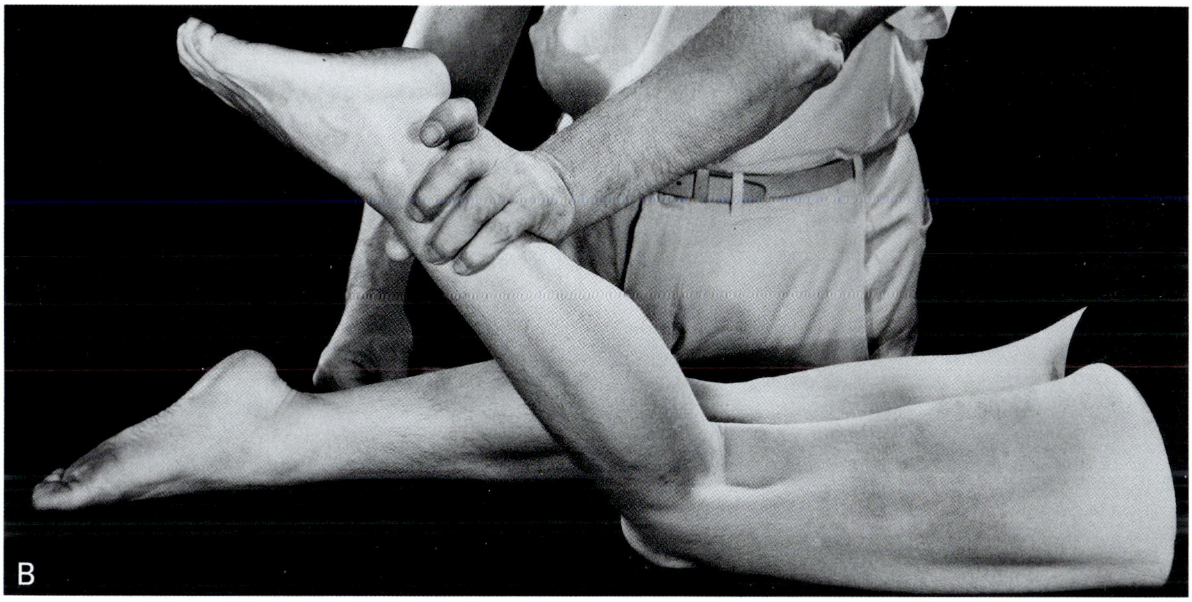

FIGURE 7-57: **A**, Medial hamstrings; **B**, Testing semimembranosus and semitendinosus.

MEDIAL HAMSTRINGS: SEMITENDINOSUS AND SEMIMEMBRANOSUS

Semitendinosus

Origin: Tuberosity of the ischium by the tendon common with the long head of the biceps femoris (Figure 7-57A).

Insertion: Proximal part of the medial surface of the body of the tibia and deep fascia of the leg.

Action: Flexes and medially rotates the knee joint. Extends and assists in medial rotation of the hip joint.

Nerve: Tibial, L4, **5**, S**1**, **2**.

Patient: Prone.

Fixation: The examiner should hold the thigh down firmly on the table (Figure 7-57B). (To avoid covering the muscle belly of the medial hamstrings, fixation is not illustrated.)

Semimembranosus

Origin: Tuberosity of the ischium, proximal and lateral to the biceps femoris and the semitendinosus (Figure 7-57A).

Insertion: Posteromedial aspect of the medial condyle of the tibia.

Action: Flexes and medially rotates the knee joint. Extends and assists in medial rotation of the hip joint.

Nerve: Sciatic (tibial branch), L4, **5**, **S1**, 2.

Test: Flexion of the knee between 50° and 70°, with the thigh in medial rotation and the leg medially rotated on the thigh (Figures 7-57B).

Pressure: Against the leg, proximal to the ankle, in the direction of knee extension. Do not apply pressure against the rotation component.

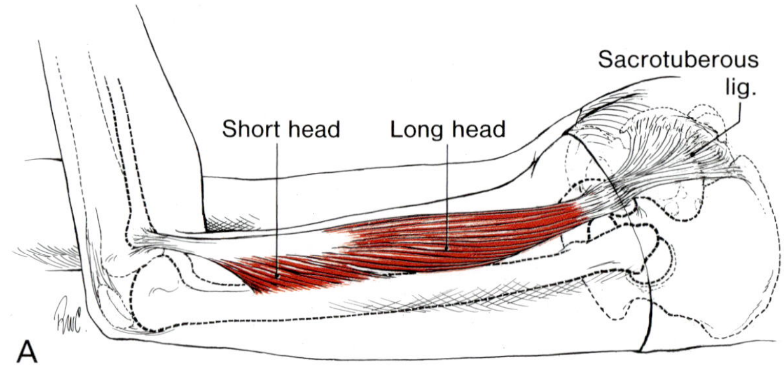

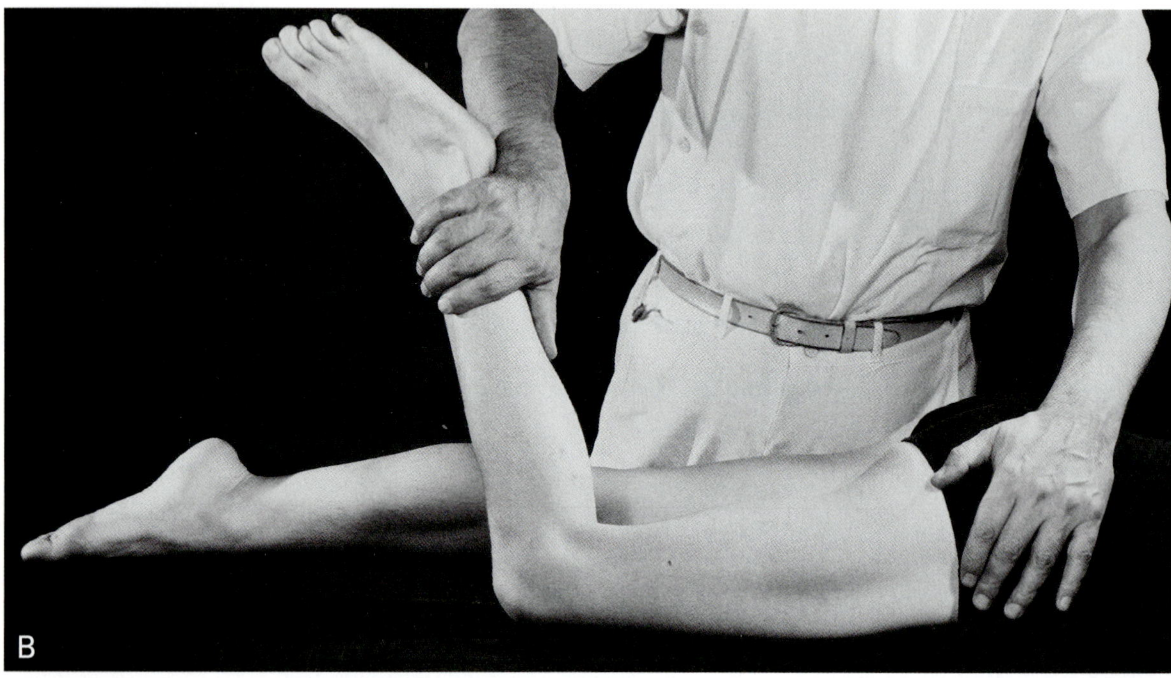

FIGURE 7-58: **A**, Lateral hamstrings; **B**, Testing biceps femoris.

LATERAL HAMSTRINGS: BICEPS FEMORIS

Biceps Femoris

Origin of Long Head: Distal part of the sacrotuberous ligament and posterior part of the tuberosity of the ischium.

Origin of Short Head: Lateral lip of the linea aspera, proximal ⅔ of the supracondylar line, and lateral intermuscular septum.

Insertion: Lateral side of the head of the fibula, lateral condyle of the tibia, deep fascia on the lateral side of the leg.

Patient: Prone.

Fixation: The examiner should hold the thigh firmly down on the table (Figure 7-58). (Not illustrated to avoid covering the muscles.)

Action: The long and short heads of the biceps femoris flex and laterally rotate the knee joint. In addition, the long head extends and assists in lateral rotation of the hip joint.

Nerve to Long Head: Tibial, L5, S**1**, **2**, 3.

Nerve to Short Head: Common Fibular, L**5**, S**1**, **2**.

Test: Flexion of the knee between 50° and 70°, with the thigh in slight lateral rotation and the leg in slight lateral rotation on the thigh.

Pressure: Against the leg, proximal to the ankle, in the direction of knee extension. Do not apply pressure against the rotation component.

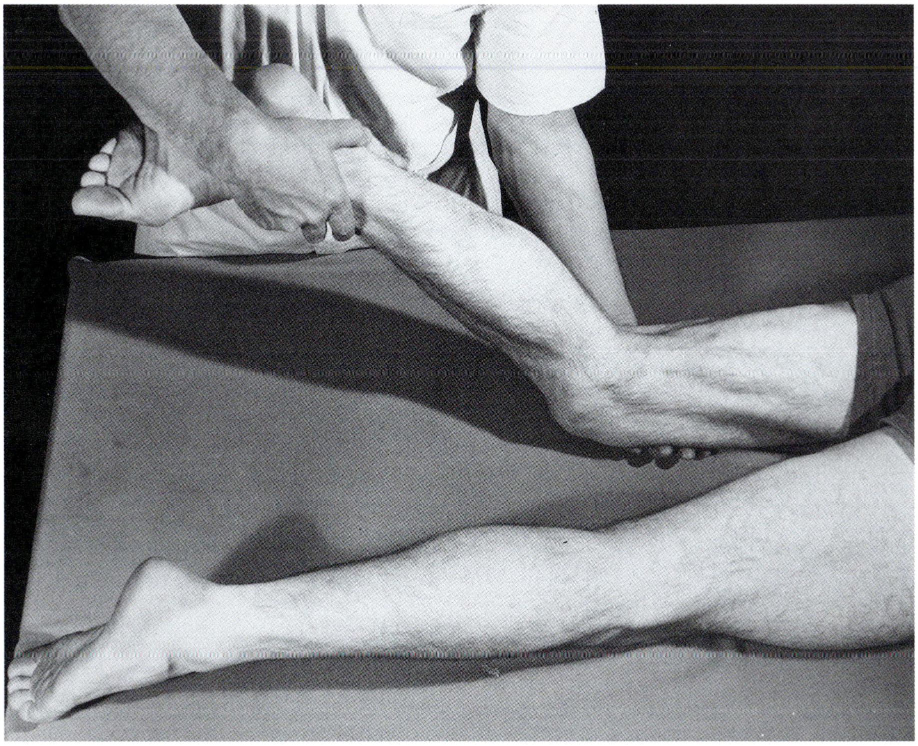

FIGURE 7-59: Action of the gracilis as a knee flexor.

HAMSTRINGS AND GRACILIS

Weakness: Evidence of slight weakness of either the medial or lateral hamstrings is based on the subject's inability to maintain the rotation when asked to hold the test position. Weakness of both the medial and lateral hamstrings permits hyperextension of the knee. When this weakness is bilateral, the pelvis may tilt anteriorly, and the lumbar spine may assume a lordotic position. When this weakness is unilateral, a pelvic rotation may result. Weakness of the lateral hamstrings causes a tendency toward loss of lateral stability of the knee, allowing a thrust in the direction of a bowleg position in weight-bearing. Weakness of the medial hamstrings decreases the medial stability of the knee joint and permits a genu valgum position, with a tendency toward lateral rotation of the tibia on the femur.

Contracture: Contracture of both the medial and lateral hamstrings results in a position of knee flexion, and, if the contracture is extreme, a posterior tilting of the pelvis and lumbar spine flexion.

Shortness: Restriction of knee extension when the hip is flexed, or restriction of hip flexion when the knee is extended. Shortness of the hamstrings does not cause a posterior pelvic tilt, but a posterior pelvic tilt and a flattening of the lumbar spine (lumbar spine flexion) often are seen in subjects with hamstring shortness.

NOTE: *Ordinarily, the hip flexors act to safeguard the hamstrings during knee flexion. Do not expect the subject to hold full knee flexion or to hold against the same amount of pressure with the hip extended in the prone position that could be resisted with a hip flexed in sitting. The frequent occurrence of muscle cramping during the hamstring test results from the muscle being in too short a position and attempting to hold against strong pressure. To test the hamstrings in full knee flexion, the hip must be flexed to take up some of the slack. However, there will be assistance from the sartorius in both hip and knee flexion when the hamstrings are tested with the hip flexed.*

Weakness of the popliteus and gastrocnemius may interfere with initiating knee flexion. Substitution of sartorius action will appear in the form of hip flexion as the knee flexion is initiated. A short rectus femoris, limiting the range of motion of knee flexion, will cause hip flexion as the motion of knee flexion is completed. (Hip flexion in the prone position is seen as an anterior tilt of the pelvis with lumbar spine hyperextension.) Assistance from the gastrocnemius in flexing the knee will be seen as an effort to dorsiflex the ankle, elongating the gastrocnemius over the ankle to make it more effective in knee flexion.

Action of the gracilis as a knee flexor is illustrated (Figure 7-59). The muscle is brought into action by the test position and pressure as used for the medial hamstrings. The gracilis has its origin on the pubis, and the medial hamstrings arise from the ischium.

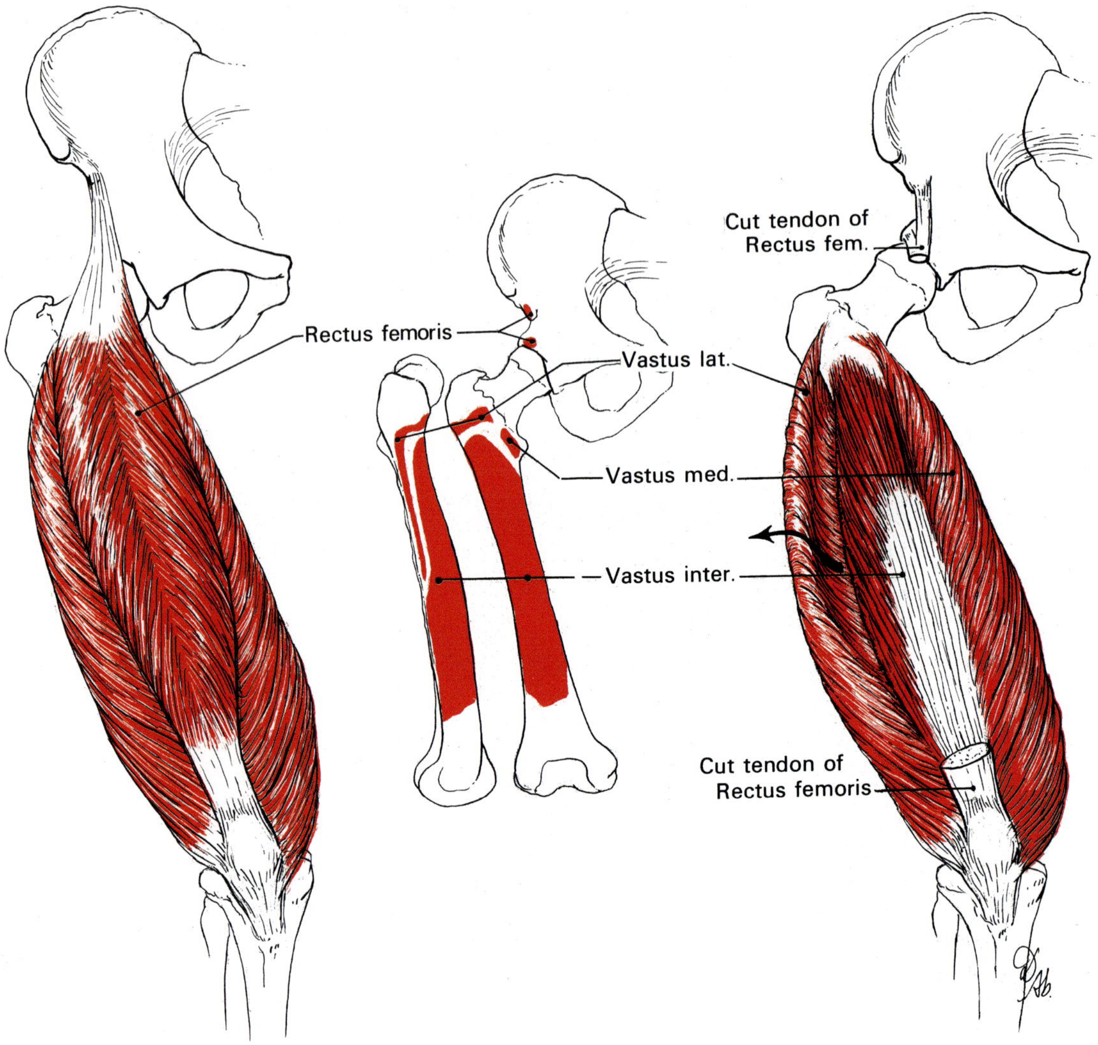

FIGURE 7-60: Quadriceps femoris.

Quadriceps Femoris

Origin of Rectus Femoris:

Straight head: From anteroinferior iliac spine (Figure 7-60).

Reflected head: From groove above rim of acetabulum.

Origin of Vastus Lateralis: Proximal part of intertrochanteric line, anterior and inferior borders of greater trochanter, lateral lip of the gluteal tuberosity, proximal half of lateral lip of linea aspera, and lateral intermuscular septum.

Origin of Vastus Intermedius: Anterior and lateral surfaces of the proximal ⅔ of the body of the femur, distal half of the linea aspera, and lateral intermuscular septum.

Origin of Vastus Medialis: Distal half of the intertrochanteric line, medial lip of the linea aspera, proximal part of the medial supracondylar line, tendons of the adductor longus and adductor magnus and medial intermuscular septum.

Insertion: Proximal border of the patella and through the patellar ligament to the tuberosity of the tibia.

Action: The quadriceps extends the knee joint, and the rectus femoris portion flexes the hip joint.

Nerve: Femoral, L**2**, **3**, **4**.

The **articularis genus** is a small muscle that may be blended with the vastus intermedius but is usually distinct from it. (Not shown in drawing.)

Origin: Anterior surface of the distal part of the body of the femur.

Insertion: Proximal part of the synovial membrane of the knee joint.

Action: Draws the articular capsule proximally.

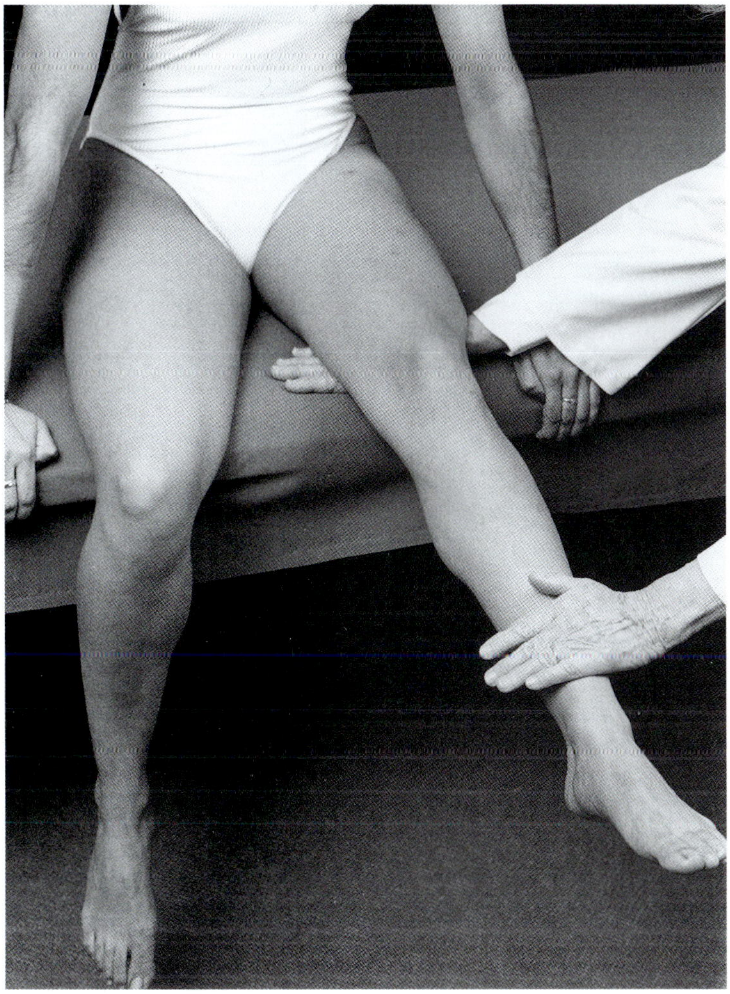

FIGURE 7-61: Testing quadriceps femoris.

Nerve: Branch of the nerve to the vastus intermedius.

Patient: Sitting, with the knees over the side of the table and holding on to the table (Figure 7-61).

Fixation: The examiner may hold the thigh firmly down on the table. Alternatively, because the weight of the trunk is usually sufficient to stabilize the patient during this test, the examiner may put a hand under the distal end of the thigh to cushion that part against table pressure.

Test: Full extension of the knee joint, without rotation of the thigh.

Pressure: Against the leg, above the ankle, in the direction of flexion.

NOTE: *Inclining the body backward may be evidence of an attempt to release hamstring tension when those muscles are contracted. When the tensor fasciae latae is being substituted for the quadriceps, it medially rotates the thigh and exerts a stronger pull if the hip is extended. If the rectus femoris is the strongest part of the quadriceps, the patient will lean backward to extend the hip, thereby obtaining maximum action of the rectus femoris.*

Weakness: Results in knee hyperextension, not by permitting a posterior knee position, but instead by requiring the patient to lock the knee joint in slight hyperextension when walking. Functionally, this weakness interferes with stair climbing, walking up an incline, and getting up and down from a sitting position.

Contracture: Knee extension.

Shortness: Restriction of knee flexion. Shortness of the rectus femoris part of the quadriceps results in restriction of knee flexion when the hip is extended or restriction of hip extension when the knee is flexed (see tests for length of hip flexor muscles earlier in the chapter).

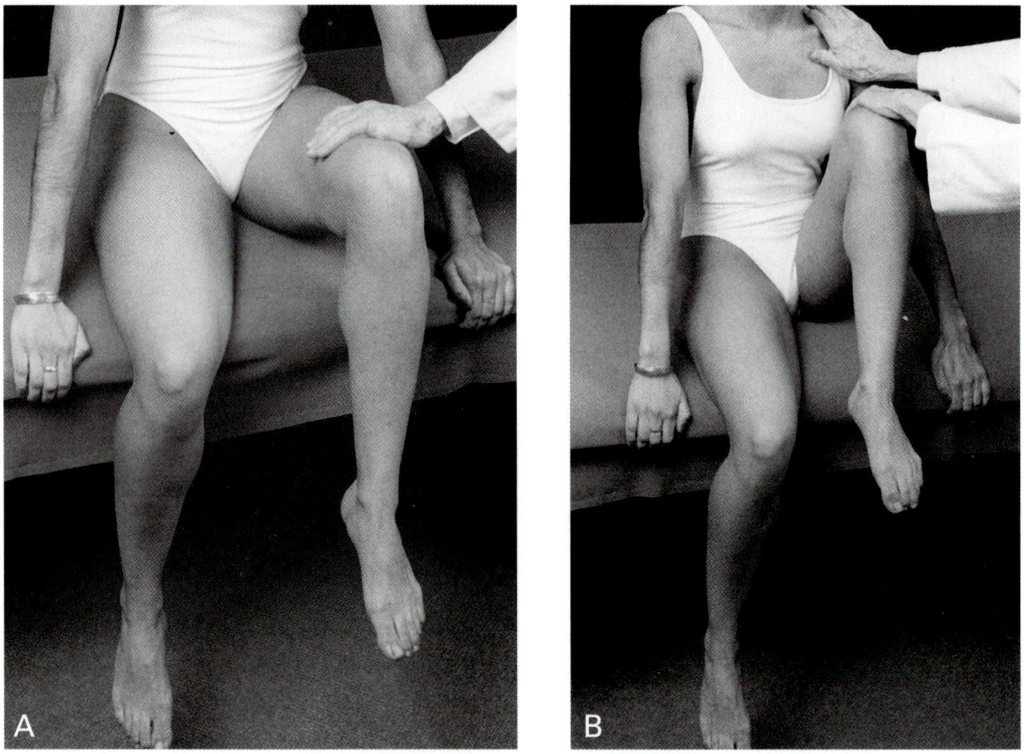

FIGURE 7-62: Hip flexors: **A**, Test of hip flexors as a group; **B**, Test of iliopsoas.

Hip Flexors

Patient: Sitting upright, with the knees bent over the side of the table. Hold on to the table to prevent leaning backward to obtain assistance by two-joint hip flexors.

Fixation: The weight of the trunk may be sufficient to stabilize the patient during this test, but holding on to the table gives added stability. If the trunk is weak, place the patient in the supine position during the test.

Test for Hip Flexors as a Group: (Figure 7-62A) Hip flexion with the knee flexed, raising the thigh a few inches from the table.

Pressure: Against the anterior thigh, in the direction of extension.

Test for Iliopsoas: (Figure 7-62B) Full hip flexion with the knee flexed. This test emphasizes the one-joint hip flexor by requiring completion of the arc of motion. The grade is based on the ability to hold the completed position. With weakness of the iliopsoas, the fully flexed position cannot be held against resistance, but as the thigh drops to the position assumed in the group test, the strength may grade normal. This test is used to confirm the findings of the supine test, which follows and is described under Iliopsoas.

Pressure: One hand against the anterior shoulder area gives counterpressure and the other applies pressure against the thigh, in the direction of hip extension.

NOTE: *Lateral rotation with abduction of the thigh as pressure is applied generally is evidence of sartorius strength or of a tensor fasciae latae that is too weak to counteract the pull of the sartorius. Medial rotation of the thigh shows the tensor fasciae latae as stronger than the sartorius. If adductors are primarily responsible for the flexion, the thigh will be adducted as it is flexed. If the anterior abdominals do not fix the pelvis to the trunk, the pelvis will tilt anteriorly to flex on the thighs, and the hip flexors may hold against strong pressure but not at maximum height.*

Weakness: Decreases the ability to flex the hip joint, resulting in marked disability in stair climbing, walking up an incline, getting up from a reclining position, and bringing the trunk forward in the sitting position prior to rising from a chair. With marked weakness, walking is difficult because the leg must be brought forward by pelvic motion (produced by anterior or lateral abdominal muscle action) rather than by hip flexion. The effect of hip flexor weakness on posture is shown in Chapter 2, Section II.

Contracture: Bilaterally, hip flexion deformity with increased lumbar lordosis (see Chapter 2 for images of kyphotic-lordotic posture). Unilaterally, hip position of flexion, abduction, and lateral rotation.

Shortness: In the standing position, shortness of the hip flexors is seen as a lumbar lordosis with an anterior pelvic tilt.

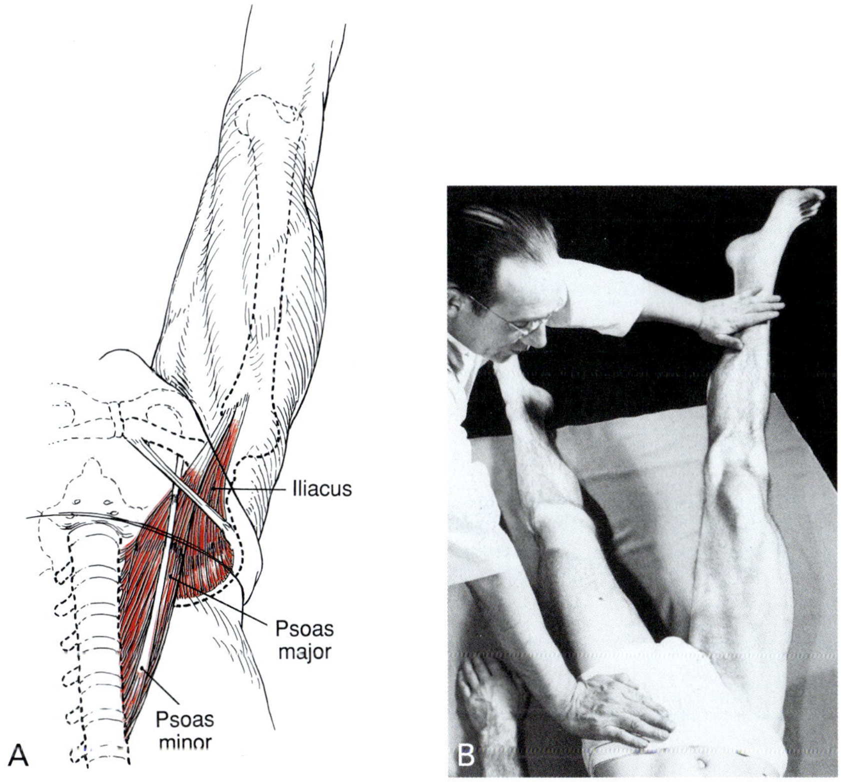

FIGURE 7-63: **A**, Psoas major, psoas minor, and iliacis; **B**, Test of iliopsoas, with emphasis on psoas major.

ILIOPSOAS AND PSOAS MINOR

Psoas Major

Origin: Ventral surfaces of the transverse processes of all lumbar vertebrae, sides of the bodies and corresponding intervertebral disks of the last thoracic and all lumbar vertebrae, and the membranous arches that extend over the sides of the bodies of the lumbar vertebrae (Figure 7-63A).

Insertion: Lesser trochanter of the femur.

Nerve: Lumbar plexus, L1, **2**, **3**, 4.

Iliacus

Origin: Superior ⅔ of the iliac fossa, internal lip of the iliac crest, iliolumbar and ventral sacroiliac ligaments, and ala of the sacrum.

Insertion: Lateral side of the tendon of the psoas major and just distal to the lesser trochanter.

Nerve: Femoral, L(1), **2**, **3**, 4.

Iliopsoas

Action: With the origin fixed, flexes the hip joint by flexing the femur on the trunk, as in supine alternate-leg raising, and may assist in lateral rotation and abduction of the hip joint. With the insertion fixed and acting bilaterally, flexes the hip joint by flexing the trunk on the femur, as in a sit-up from the supine position. The psoas major, acting bilaterally with the insertion fixed, will increase the lumbar lordosis; when acting unilaterally, it will assist in lateral flexion of the trunk toward the same side.

Iliopsoas (with Emphasis on Psoas Major)

Patient: Supine.

Fixation: The examiner stabilizes the opposite iliac crest. The quadriceps stabilize the knee in extension.

Test: Hip flexion in a position of slight abduction and slight lateral rotation. The muscle is not seen in the photograph above because it lies deep beneath the sartorius, the femoral nerve, and the blood vessels contained in the femoral sheath (Figure 7-63B).

Pressure: Against the anteromedial aspect of the leg, in the direction of extension and slight abduction, directly opposite to the line of pull of the psoas major from the origin of the lumbar spine to the insertion on the lesser trochanter of the femur.

Weakness and **Contracture:** See the discussion of hip flexors earlier. Weakness tends to be *bilateral* in cases of lumbar kyphosis and sway-back posture and *unilateral* in cases of lumbar scoliosis.

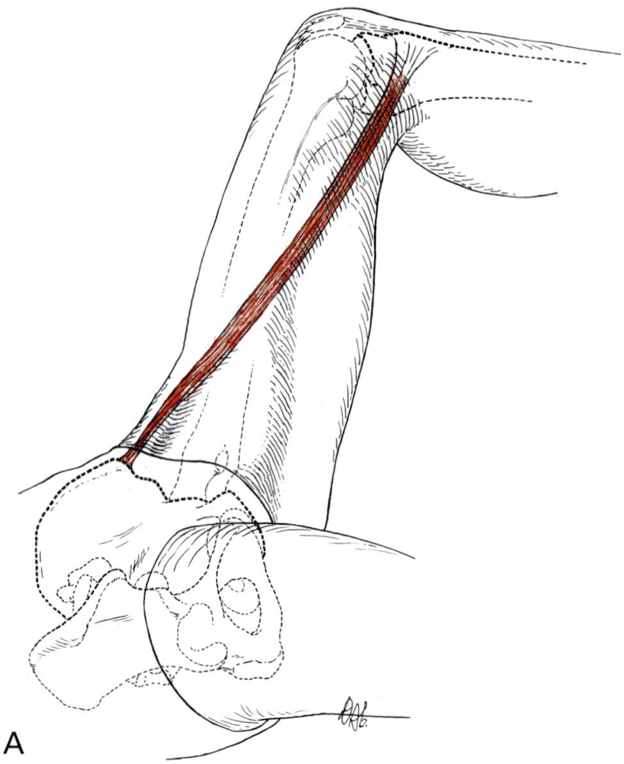

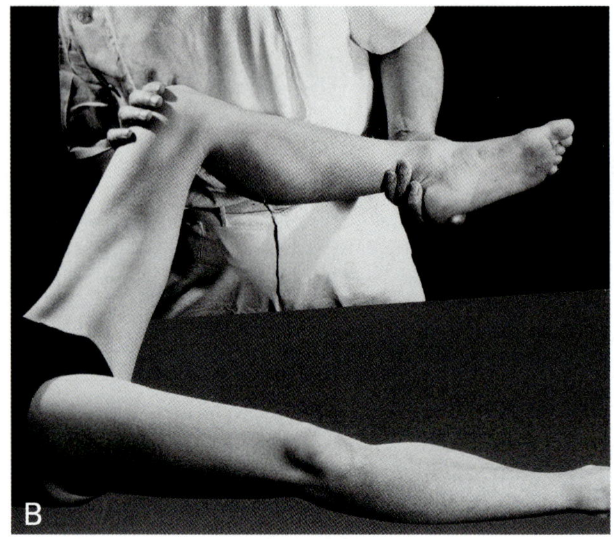

FIGURE 7-64: **A**, Sartorius; **B**, Testing sartorius.

Psoas Minor

This muscle is not a lower extremity muscle because it does not cross the hip joint. It is relatively unimportant and not always present. Its distal attachment to the iliac fascia may allow it to play a role as a stabilizer for the iliopsoas tendon (18).

Origin: Sides of the bodies of the 12th thoracic and first lumbar vertebrae and from the intervertebral disk between them (Figure 7-63A).

Insertion: Iliopectineal eminence, arcuate line of the ilium, and iliac fascia.

Action: Flexion of pelvis on lumbar spine, and vice versa.

Nerve: Lumbar plexus, L**1**, **2**.

Sartorius

Origin: Anterosuperior iliac spine and superior half of the notch just distal to the spine (Figure 7-64A).

Insertion: Proximal part of the medial surface of the tibia near the anterior border.

Action: Flexes, laterally rotates, and abducts the hip joint. Flexes and assists in medial rotation of the knee joint.

Nerve: Femoral, L**2**, **3**, (4).

Patient: Supine.

Fixation: None necessary by the examiner. The patient may hold on to the table.

Test: Lateral rotation, abduction, and flexion of the thigh, with flexion of the knee (Figure 7-64B).

Pressure: Against the anterolateral surface of the distal thigh, in the direction of hip extension, adduction, and medial rotation, and against the leg, in the direction of knee extension. The examiner's hands are in a position to resist the lateral rotation of the hip joint by pressure and counterpressure (as described for the hip lateral rotator test, Figure 7-72).

Weakness: Decreases the strength of hip flexion, abduction, and lateral rotation. Contributes to anteromedial instability of the knee joint.

Contracture: Flexion, abduction, and lateral rotation deformity of the hip, with flexion of the knee.

Error in Testing Sartorius

The position of the leg, as illustrated in Figure 7-65, resembles the sartorius test position in its flexion, abduction, and lateral rotation. The ability to hold this position is essentially a function of the hip adductors, however, and requires little assistance from the sartorius.

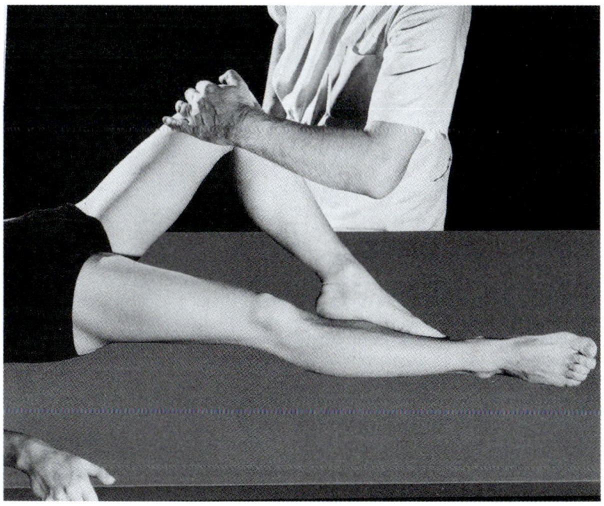

FIGURE 7-65: Error in testing sartorius.

Tensor Fasciae Latae

Origin: Anterior part of the external lip of the iliac crest, outer surface of the anterosuperior iliac spine, and deep surface of the fascia lata (Figure 7-66A).

Insertion: Into the iliotibial tract of the fascia lata at the junction of the proximal and middle thirds of the thigh.

Action: Flexes, medially rotates, and abducts the hip joint. Tenses the fascia lata. May assist in knee extension.

Nerve: Superior gluteal, L**4**, **5**, S**1**.

Shortness: The effect of shortness of the tensor fasciae latae in standing depends on whether the tightness is bilateral or unilateral. If bilateral, there is an anterior pelvic tilt and sometimes bilateral genu valgum. If unilateral, the abductors of the hip and fascia lata are tight, along with the tensor fasciae latae, and there is an associated lateral pelvic tilt (low on the side of tightness). The knee on that side will tend toward a genu valgus position. If the tensor fasciae latae and other hip flexor muscles are tight, there is an anterior pelvic tilt and a medial rotation of the hip, as indicated by the position of the patella.

Patient: Supine.

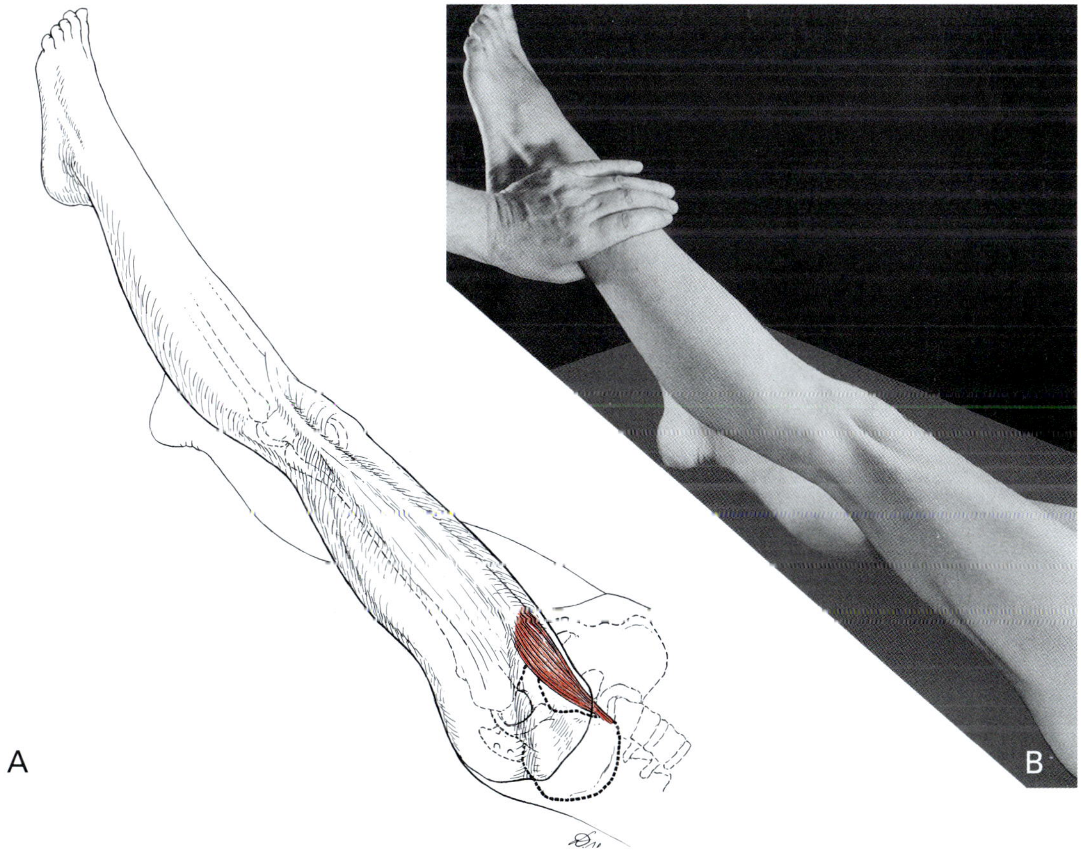

FIGURE 7-66: **A**, Tensor fasciae latae; **B**, Testing tensor fasciae latae.

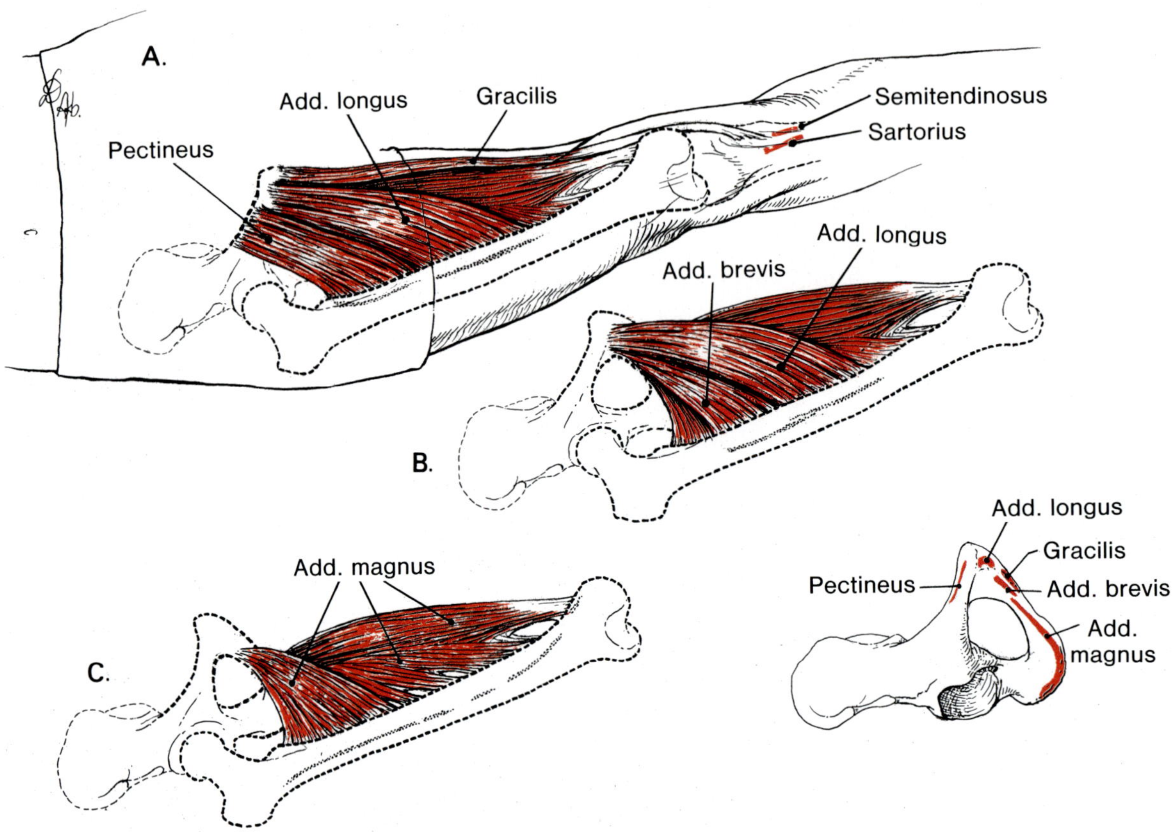

FIGURE 7-67: Hip adductors.

Fixation: The patient may hold on to the table. Quadriceps action is necessary to hold the knee extended. Usually, no fixation is necessary by the examiner, but if there is instability and the patient has difficulty in maintaining the pelvis firmly on the table, one of the examiner's hands should support the pelvis anteriorly on the opposite side.

Test: Abduction, flexion, and medial rotation of the hip, with the knee extended (Figure 7-66B).

Pressure: Against the leg, in the direction of extension and adduction. Do not apply pressure against the rotation component.

Weakness: Moderate weakness is evident immediately by failure to maintain the medially rotated test position. In standing, there is a thrust in the direction of a bowleg position, and the extremity tends to rotate laterally from the hip.

Contracture: Hip flexion and genu valgus position. In a supine or standing position, the pelvis will be anteriorly tilted if the legs are brought into adduction.

HIP ADDUCTORS

In Figure 7-67, dashed lines indicate the borders of the bones. Stippled lines in the figures above indicate muscle attachments located on the posterior surface of the femur.

Pectineus

Origin: Surface of the superior ramus of the pubis, ventral to the pectineal line, between the iliopectineal eminence and the pubic tubercle.

Insertion: Pectineal line of the femur.

Nerve: Femoral and Obturator, L**2**, **3**, 4.

Adductor Magnus

Origin: Inferior pubis ramus, ramus of the ischium (anterior fibers), and ischial tuberosity (posterior fibers).

Insertion: Medial to the gluteal tuberosity, middle of the linea aspera, medial supracondylar line, and adductor tubercle of the medial condyle of the femur.

Nerve: Obturator, L2, **3**, **4**, and Sciatic, L**4**, 5, S1.

Gracilis

Origin: Inferior half of the symphysis pubis and medial margin of the inferior ramus of the pubic bone.

Insertion: Medial surface of the body of the tibia, distal to the condyle, proximal to the insertion of the semitendinosus, and lateral to the insertion of the sartorius.

Nerve: Obturator, L**2**, **3**, **4**.

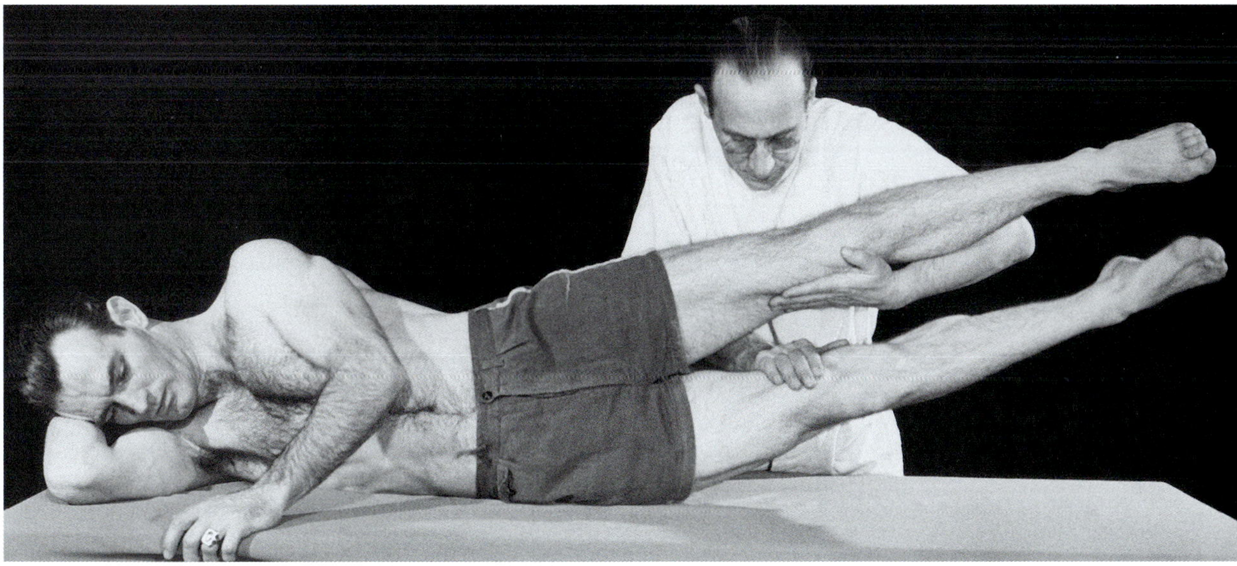

FIGURE 7-68: Testing hip adductors.

Adductor Brevis

Origin: Outer surface of the inferior ramus of the pubis.

Insertion: Distal ⅔ of the pectineal line and proximal half of the medial lip of the linea aspera.

Nerve: Obturator, L**2**, **3**, **4**.

Adductor Longus

Origin: Anterior surface of the pubis at the junction of the crest and the symphysis.

Insertion: Middle ⅓ of the medial lip of the linea aspera.

Nerve: Obturator, L**2**, **3**, 4.

Hip Adductors

Action: All the muscles described here adduct the hip joint. In addition, the pectineus, adductor brevis, and adductor longus flex the hip joint. The anterior fibers of the adductor magnus, which arise from the rami of the pubis and the ischium, may assist in flexion, whereas the posterior fibers that arise from the ischial tuberosity may assist in extension. The gracilis, in addition to adducting the hip joint, flexes and medially rotates the knee joint. All adductor muscles with the exception of the pectineus and posterior fibers of the adductor magnus may assist with medial rotation of the hip although some authors dispute this action (see discussion of rotation action on the hip joint).

Patient: Lying on the right side to test the right adductors (and vice versa), with the body in a straight line and the lower extremities and lumbar spine straight.

Fixation: The examiner holds the upper lower extremity in abduction. The patient should hold on to the table for stability.

Test: Adduction of the involved extremity upward from the table, without rotation, flexion, or extension of the hip or tilting of the pelvis (Figure 7-68).

Pressure: Against the medial aspect of the distal end of the thigh, in the direction of abduction (i.e., downward toward the table). Pressure is applied at a point above the knee to avoid strain of the tibial collateral ligament.

> **NOTE:** Anterior *rotation of the pelvis with extension of the hip joint shows an attempt to hold with the lower fibers of the gluteus maximus. Anterior tilting of the pelvis or flexion of the hip joint (with posterior rotation of the pelvis on upper side) allows substitution by the hip flexors.*
>
> *The adductor longus, adductor brevis, and pectineus aid in hip flexion. If the side-lying position is maintained and the hip tends to flex as the thigh is adducted during the test, it is not necessarily evidence of substitution but, rather, is merely evidence that the adductors that flex the hip are doing more than the rest of the adductors that assist in this movement. Alternatively, it may be evidence that hip extensors are not helping to maintain the thigh in a neutral position.*

Contracture: Hip adduction deformity. In standing, the position is one of lateral pelvic tilt with the pelvis so high on the affected side that it becomes necessary to plantarflex the ipsilateral ankle, holding the foot in equinus so the toes can touch the floor. Alternatively, if the foot is placed flat on the floor, the opposite extremity must be either flexed at the hip and knee or abducted to compensate for the apparent shortness on the adducted side.

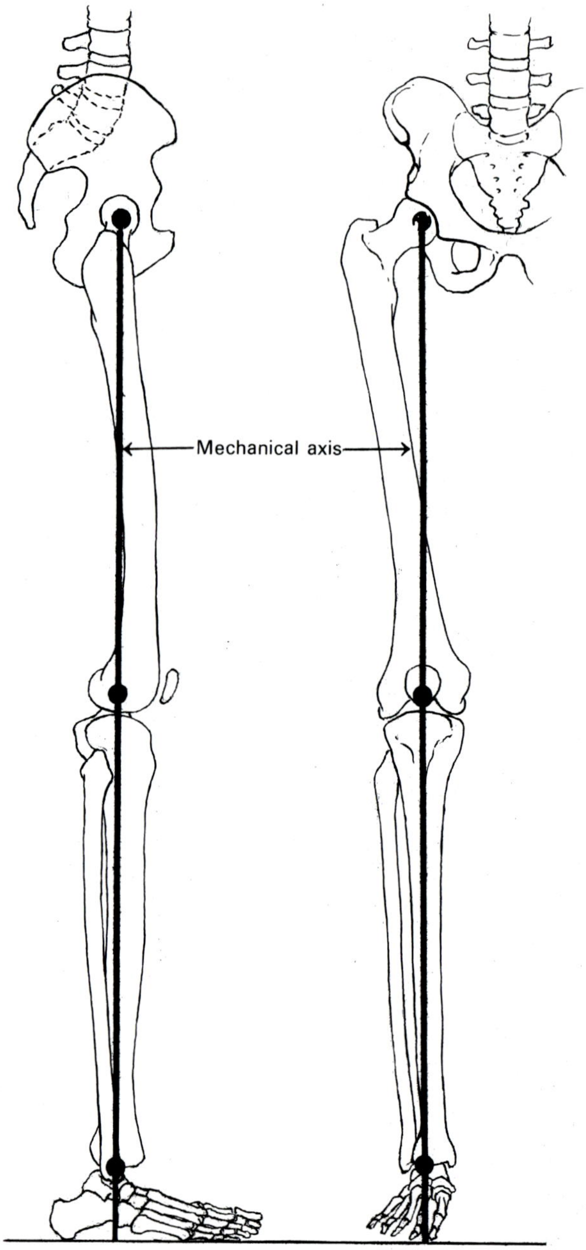

FIGURE 7-69: Mechanical axis of femur and rotation action of adductors.

MECHANICAL AXIS OF FEMUR AND ROTATION ACTION OF ADDUCTORS

The following discussion about the rotator action of the adductors is not an attempt to solve the controversy that appears to exist but, rather, to present some of the reasons why this controversy exists (19).

When considering Figure 7-69, it is important to note that in the anatomical position, from the anterior view, the femur extends obliquely, with the distal end being more medial than the proximal end. From the lateral view, the shaft of the femur is convexly curved in an anterior direction. The *anatomical axis* of the femur extends longitudinally along the shaft. If rotation of the hip took place about this axis, there would be no doubt that the adductors, attached as they are posteriorly along the linea aspera, would be lateral rotators.

Rotation of the hip joint, however, does not occur about the anatomical axis of the femur but, rather, about the *mechanical axis,* which passes from the center of the hip joint to the center of the knee joint and is at the intersection of the two planes represented by the solid black lines in Figure 7-69.

The muscles or major portions of muscles that insert on the part of the femur that is anterior to the mechanical axis will act as medial rotators of the femur (see lateral view). On the other hand, the muscles or major portions of muscles that insert on the part of the femur posterior to the mechanical axis will act as lateral rotators (20).

When the position of the extremity in relation to the pelvis changes from the anatomical position, the actions of the muscles also change. Thus, if the femur is medially rotated, a larger portion of the shaft comes to lie anterior to the mechanical axis, with the result that more of the adductor insertions will be anterior to the axis and, therefore, will act as medial rotators. With increased lateral rotation, more of the adductors will act as lateral rotators.

Besides the change that occurs with movement, normal variations also occur in the bone structure of the femur that tend to make the rotator action of the adductors variable.

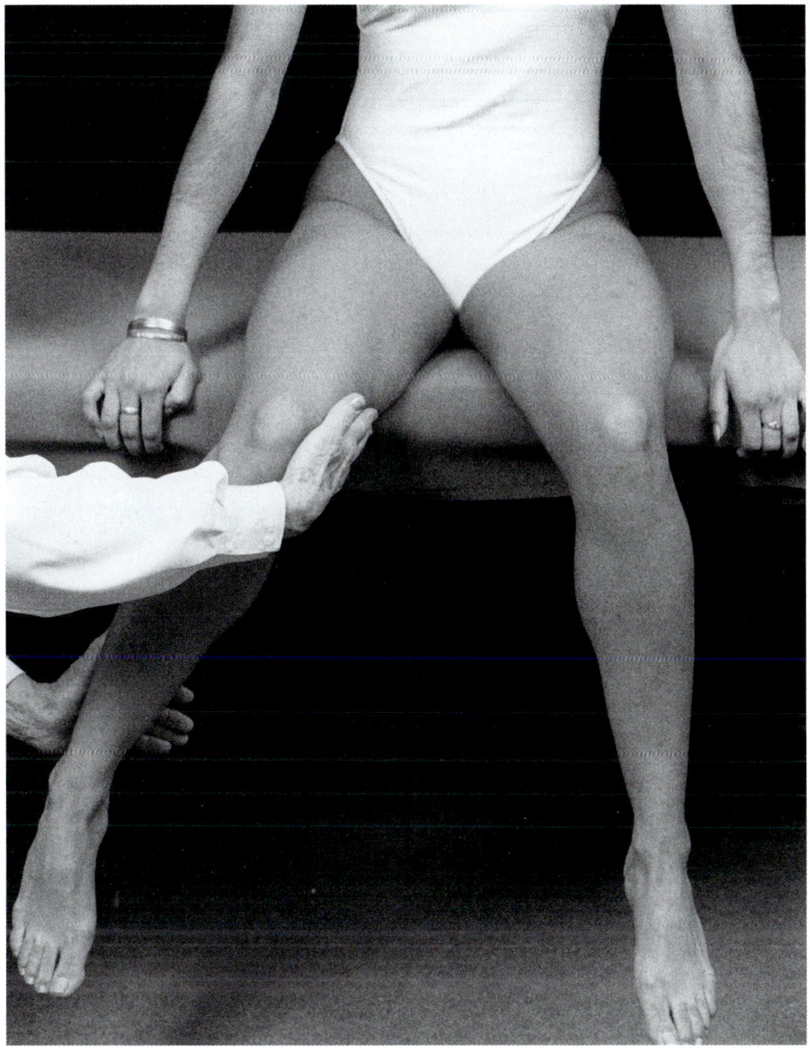

FIGURE 7-70: Testing medial rotators of hip joint.

MEDIAL ROTATORS OF HIP JOINT

The medial rotators of the hip joint consist of the tensor fasciae latae, gluteus minimus, and gluteus medius (anterior fibers).

Patient: Sitting on a table, with the knees bent over the side and the subject holding on to the table.

Fixation: The weight of the trunk stabilizes the patient during this test. Stabilization is also given in the form of counterpressure, as described below under **Pressure**.

Test: Medial rotation of the thigh, with the leg in a position of completion of the outward arc of motion.

Pressure: With one hand, the examiner applies counterpressure at the medial side of the lower end of the thigh. With the other hand, the examiner applies pressure to the lateral side of the leg above the ankle, pushing the leg inward in an effort to rotate the thigh laterally (Figure 7-70).

Weakness: Results in lateral rotation of the lower extremity in standing and walking.

Contracture: Medial rotation of the hip, with in-toeing and a tendency toward genu valgum in weight-bearing.

Shortness: Inability to laterally rotate the hip through the full range of motion, and inability to sit in a cross-legged position (i.e., tailor fashion).

> **NOTE:** *If the rotator test is done in a supine position, the pelvis will tend to tilt anteriorly if much pressure is applied, but this is not a substitution movement. Because of its attachments, the tensor fasciae latae, when contracting to maximum, pulls forward on the pelvis as it medially rotates the hip.*

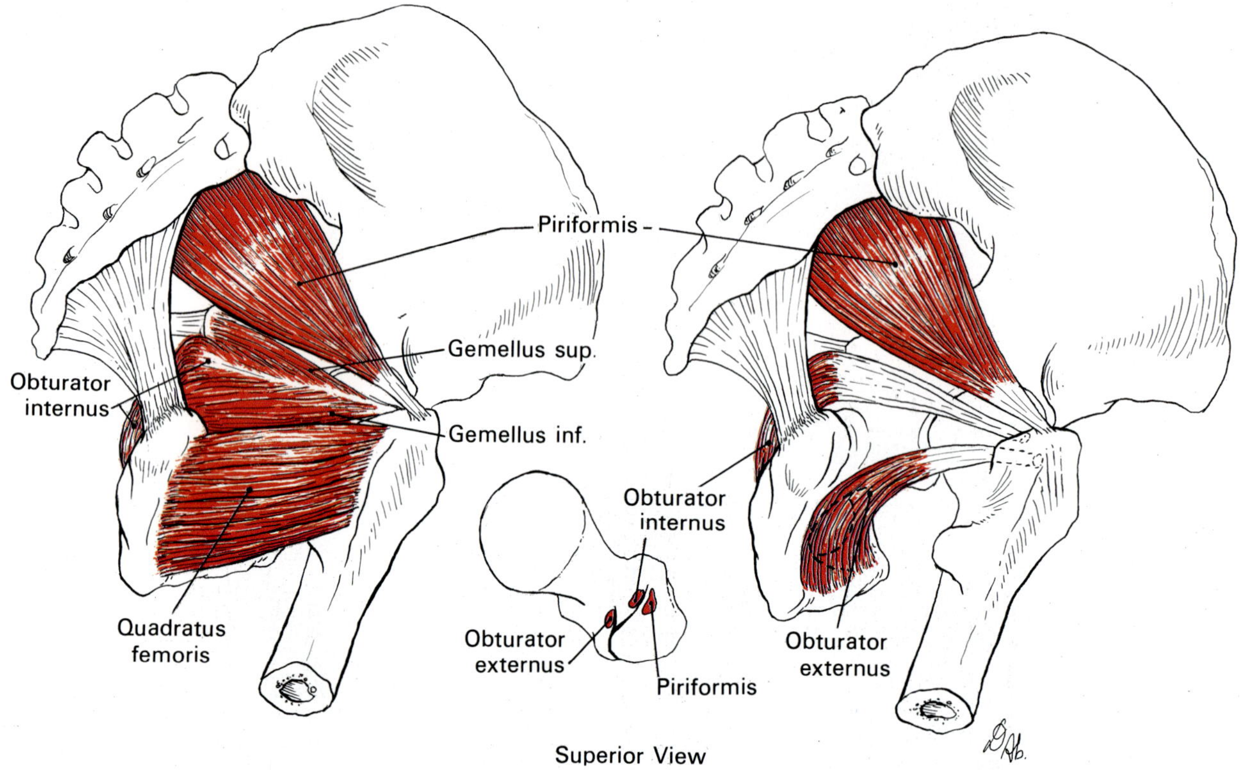

FIGURE 7-71: Superior view of lateral rotators of hip joint.

LATERAL ROTATORS OF HIP JOINT

Piriformis

Origin: Pelvic surface of the sacrum between (and lateral to) the first through fourth pelvic sacral foramina, margin of the greater sciatic foramen, and pelvic surface of the sacrotuberous ligament.

Insertion: Superior border of the greater trochanter of the femur (Figure 7-71).

Nerve: Sacral plexus, L(5), S**1**, **2**.

Quadratus Femoris

Origin: Proximal part of the lateral border of the tuberosity of the ischium.

Insertion: Proximal part of the quadrate line, extending distally from the intertrochanteric crest.

Nerve: Sacral plexus, L**4**, **5**, S**1**, (2).

Obturator Internus

Origin: Internal or pelvic surface of the obturator membrane and margin of the obturator foramen, pelvic surface of the ischium posterior and proximal to the obturator foramen, and to a slight extent the obturator fascia.

Insertion: Medial surface of the greater trochanter of the femur, proximal to the trochanteric fossa.

Nerve: Sacral plexus, L**5**, S**1**, **2**.

Obturator Externus

Origin: Rami of the pubis and ischium, and external surface of the obturator membrane.

Insertion: Trochanteric fossa of the femur.

Nerve: Obturator, L**3**, **4**.

Gemellus Superior

Origin: External surface of the spine of the ischium.

Insertion: With the tendon of the obturator internus, into the medial surface of the greater trochanter of the femur.

Nerve: Sacral plexus, L**5**, S**1**, **2**.

Gemellus Inferior

Origin: Proximal part of the tuberosity of the ischium.

Insertion: With the tendon of the obturator internus, into the medial surface of the greater trochanter of the femur.

Nerve: Sacral plexus, L**4**, **5**, S**1**, (2).

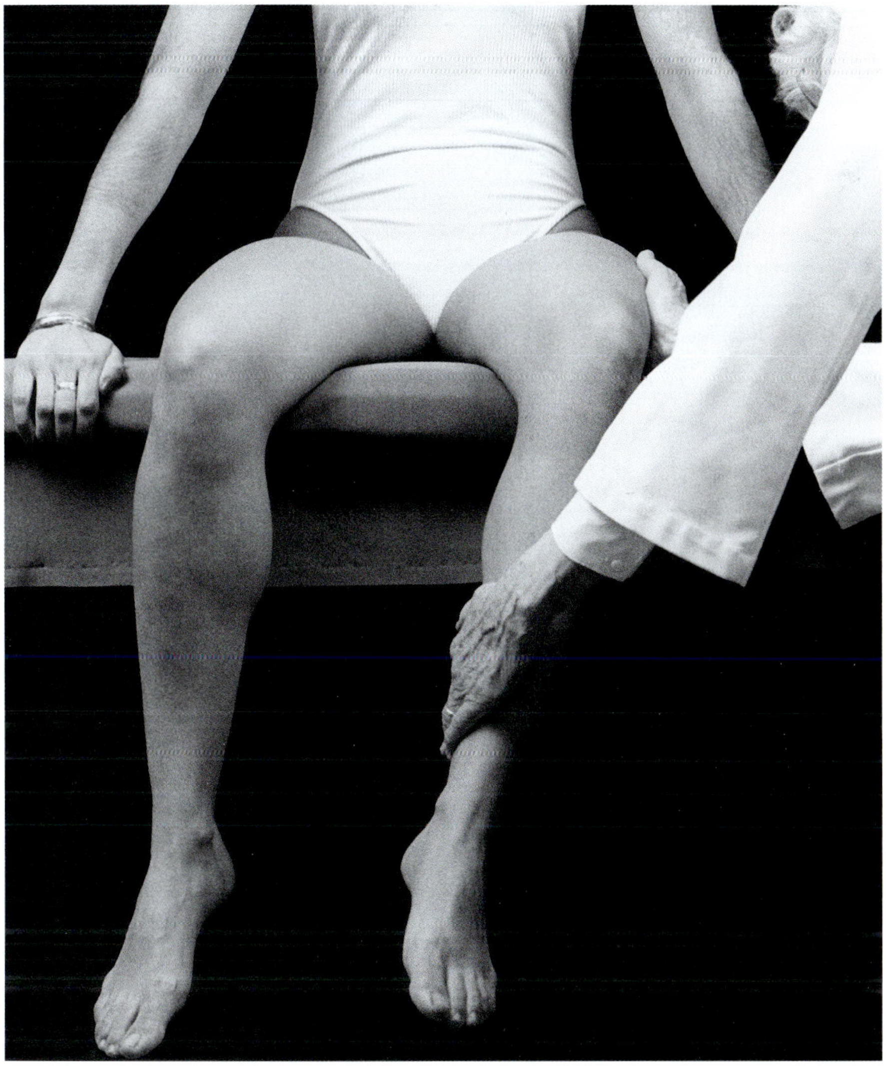

FIGURE 7-72: Testing lateral rotators of hip joint.

Lateral Rotators of Hip Joint

All the muscles cited here laterally rotate the hip joint. In addition, the obturator externus may assist in adduction of the hip joint and the piriformis, obturator internus, and gemelli may assist in abduction when the hip is flexed. The piriformis may assist in extension.

Patient: Sitting on a table, with the knees bent over the side and the subject holding on to the table.

Fixation: The weight of the trunk stabilizes the patient during this test. Stabilization is also given in the form of counterpressure, as described below under **Pressure.**

Test: Lateral rotation of the hip, with the leg in a position of completion of the inward arc of motion.

Pressure: With one hand, the examiner applies counterpressure at the lateral side of the lower end of the thigh. With the other hand, the examiner applies pressure to the medial side of the leg, above the ankle, pushing the leg outward in an effort to rotate the hip medially (Figure 7-72).

Weakness: Usually, medial rotation of the hip accompanied by pronation of the foot and a tendency toward a genu valgus position.

Contracture: Lateral rotation of the hip, usually in an abducted position.

Shortness: The range of medial rotation of the hip will be limited. (Frequently, excessive range of lateral motion is noted.) In the standing posture, a lateral rotation of the hip and out-toeing are observed.

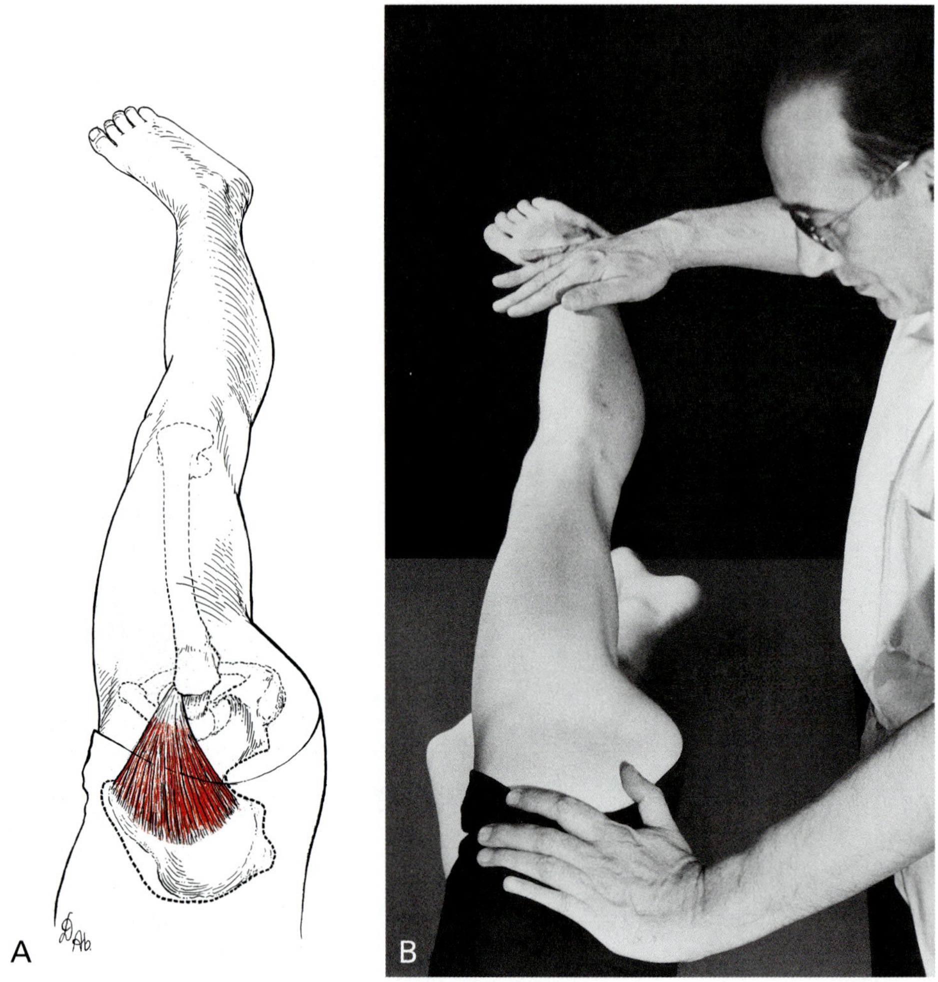

FIGURE 7-73: **A**, Gluteus minimus; **B**, Testing gluteus minimus.

Gluteus Minimus

Origin: External surface of the ilium, between the anterior and inferior gluteal lines and margin of the greater sciatic notch (Figure 7-73A).

Insertion: Anterior border of the greater trochanter of the femur and hip joint capsule.

Action: Abducts, medially rotates and may assist in flexion of the hip joint.

Nerve: Superior gluteal, L**4**, **5**, S**1**.

Patient: Side-lying.

Fixation: The examiner stabilizes the pelvis (see *Note*).

Test: Abduction of the hip in a position neutral between flexion and extension and neutral in regard to rotation (Figure 7-73B).

Pressure: Against the leg, in the direction of adduction and very slight extension.

Weakness: Lessens the strength of medial rotation and abduction of the hip joint.

Contracture and **Shortness:** Abduction and medial rotation of the hip. In standing, lateral pelvic tilt, low on the side of shortness, plus medial rotation of the hip.

NOTE: *In tests of the gluteus minimus and medius or of the abductors as a group, stabilization of the pelvis is necessary but often difficult. It requires a strong fixation by many trunk muscles aided by stabilization on the part of the examiner. Flexion of the hip and knee of the underneath lower extremity aids in stabilizing the pelvis against anterior or posterior tilt. The examiner's hand attempts to stabilize the pelvis to prevent the tendency to roll forward or backward, the tendency to tilt anteriorly or posteriorly, and if possible, any unnecessary hiking or dropping of the pelvis laterally. Any one of these six shifts in position of the pelvis may result primarily from trunk weakness; alternatively, such shifts may indicate an attempt to substitute anterior or posterior hip joint muscles or lateral abdominals in the movement of leg abduction. When the trunk muscles are strong, it is not very difficult to maintain good stabilization of the pelvis, but when trunk muscles are weak, the examiner may need the assistance of a second person to hold the pelvis steady.*

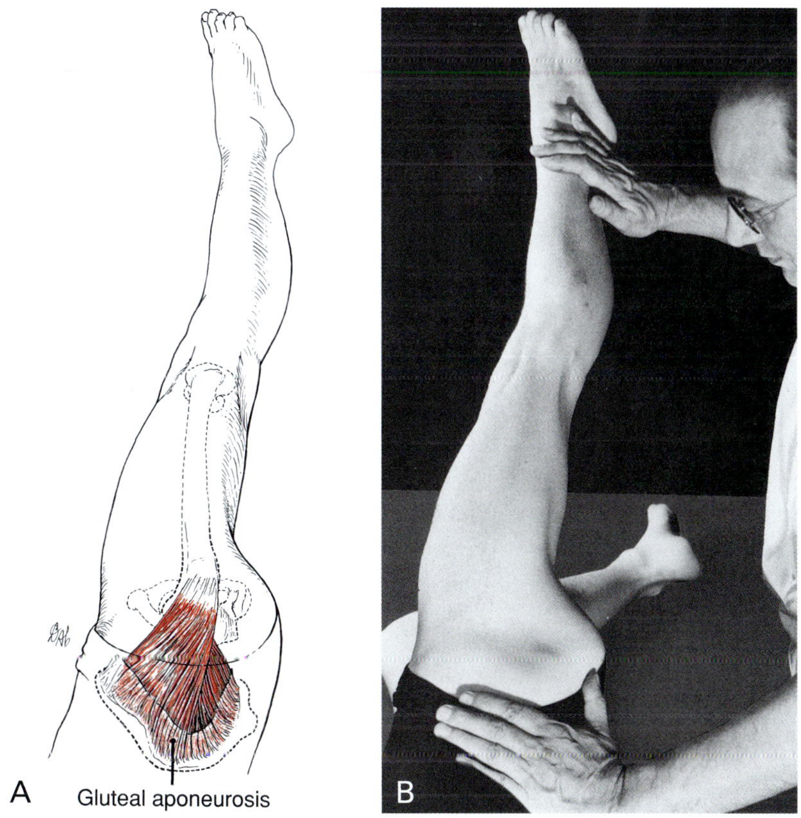

FIGURE 7-74: **A**, Gluteus medius; **B**, Testing gluteus medius.

Gluteus Medius

Origin: External surface of the ilium, between the iliac crest and posterior gluteal line dorsally and the anterior gluteal line ventrally and gluteal aponeurosis (Figure 7-74A).

Insertion: Oblique ridge on the lateral surface of the greater trochanter of the femur.

Action: Abducts the hip joint. The anterior fibers medially rotate and may assist in flexion of the hip joint; the posterior fibers laterally rotate and may assist in extension.

Nerve: Superior gluteal, L**4**, **5**, S**1**.

Patient: Side-lying, with the underneath leg flexed at the hip and knee and the pelvis rotated slightly forward to place the posterior gluteus medius in an antigravity position.

Fixation: The muscles of the trunk and the examiner stabilize the pelvis (see *Note* on facing under gluteus minimus).

Test (Emphasis on Posterior Portion): Abduction of the hip, with slight extension and slight external rotation. The knee is maintained in extension. Differentiating the posterior gluteus medius is very important. Hip abductors, when tested as a group, may be normal in strength, even though a precise test of the gluteus medius may reveal appreciable weakness (Figure 7-74B).

When external rotation of the hip joint is limited, do not allow the pelvis to rotate posteriorly in the transverse plane to obtain the appearance of hip joint external rotation. With posterior rotation of the pelvis, the tensor fasciae latae and gluteus minimus become active in abduction. Even though pressure may be applied properly against the gluteus medius, the specificity of the test is greatly diminished. Weakness of the gluteus medius may become apparent immediately because of the subject's inability to hold the precise test position, the tendency for the muscle to cramp, or an attempt to rotate the pelvis posteriorly to substitute with the tensor fasciae latae and the gluteus minimus.

Pressure: Against the leg, near the ankle, in the direction of adduction and slight flexion; *do not* apply pressure against the rotation component. The pressure is applied against the leg for the purpose of obtaining a long lever. To determine normal strength, strong force is needed, and this force can be obtained by the examiner with the added advantage of a long lever. There is relatively little danger of injuring the lateral knee joint, because it is reinforced by the strong iliotibial tract.

Weakness: See the following discussion regarding weakness of the gluteus medius and abductors.

Contracture and **Shortness:** An abduction deformity that, in standing, may be seen as a lateral pelvic tilt, low on the side of tightness, along with some abduction of the extremity.

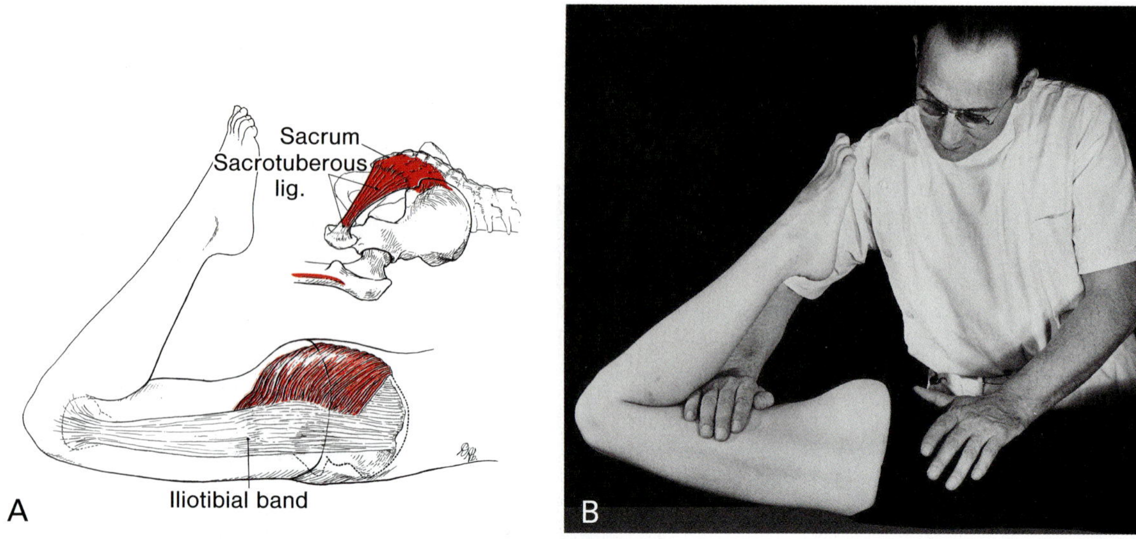

FIGURE 7-75: **A**, Gluteus maximus; **B**, Testing gluteus maximus.

Gluteus Maximus

Origin: Posterior gluteal line of the ilium and portion of the bone superior and posterior to it, posterior surface of the lower part of the sacrum, side of the coccyx, aponeurosis of the erector spinae, sacrotuberous ligament, and gluteal aponeurosis (Figure 7-75A).

Insertion: Larger proximal portion and superficial fibers of the distal portion of the muscle into the iliotibial tract of the fascia lata. Deep fibers of the distal portion into the gluteal tuberosity of the femur.

Action: Extends and laterally rotates the hip joint. Lower fibers assist in adduction of the hip joint; upper fibers assist in abduction. Through its insertion into the iliotibial tract, helps to stabilize the knee in extension.

Nerve: Inferior gluteal, L**5**, S**1**, **2**.

Patient: Prone, with knee flexed 90° or more. (The more the knee is flexed, the less the hip will extend because of restricting tension of the rectus femoris anteriorly.)

Fixation: Posteriorly, the back muscles; laterally, the lateral abdominal muscles; and anteriorly, the *opposite* hip flexors fix the pelvis to the trunk.

Test: Hip extension, with the knee flexed (Figure 7-75B).

Pressure: Against the distal posterior thigh, in the direction of hip flexion.

Weakness: Bilateral marked weakness of the gluteus maximus makes walking extremely difficult and may necessitate the aid of an assistive device. The individual bears weight on the extremity in a position of posterolateral displacement of the trunk over the femur. Raising the trunk from a forward-bent position requires action of the gluteus maximus, and in cases of weakness patients must push themselves to an upright position using their arms.

NOTE: *It is important to test for strength of the gluteus maximus before testing strength of back extensors, and in cases of coccyalgia.*

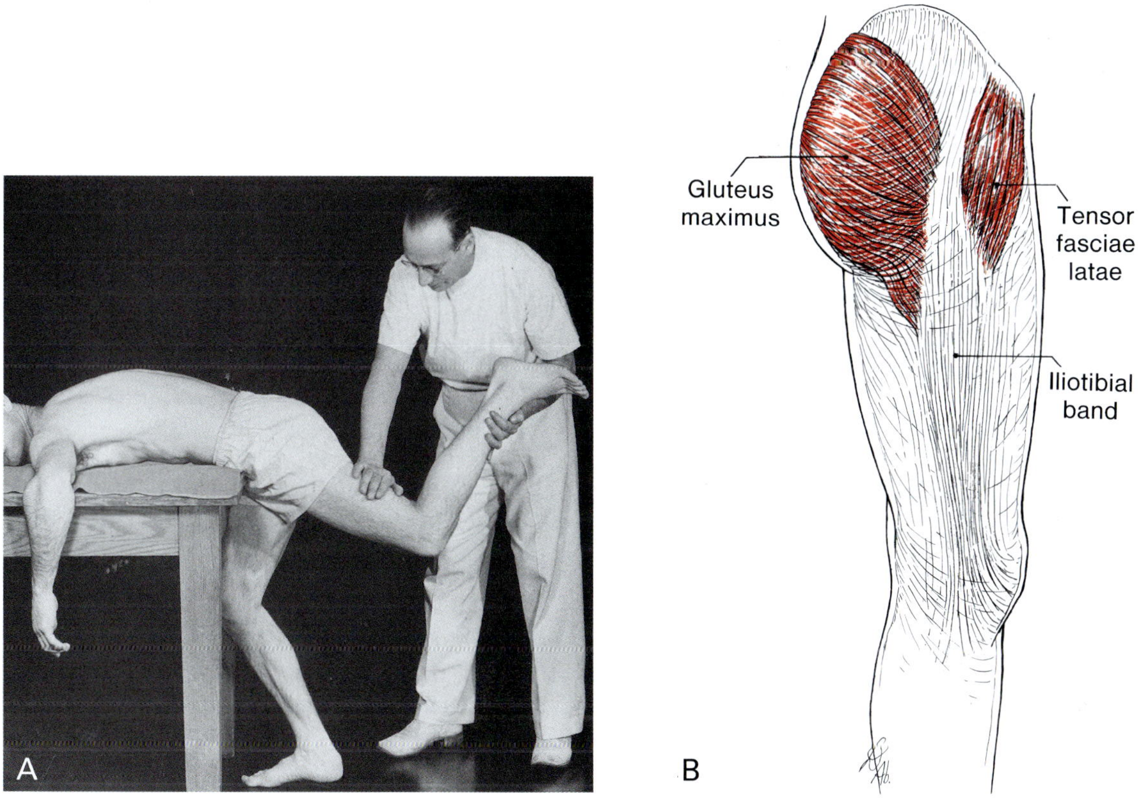

FIGURE 7-76: **A**, Modified test of gluteus maximus; **B**, Muscles inserting to the iliotibial band.

Modified Test for Gluteus Maximus

When the back extensor muscles are weak or the hip flexor muscles are tight, it is often necessary to modify the gluteus maximus test. Figure 7-76A shows the modified test.

Patient: Trunk prone on the table, and legs hanging over the end of the table.

Fixation: The patient usually needs to hold on to the table when pressure is applied.

Test: Extension of the hip, either with the knee passively flexed by the examiner, as illustrated, or with the knee extended, permitting hamstring assistance.

Pressure: This test presents a rather difficult problem regarding application of pressure. If the gluteus maximus is to be isolated as much as possible from the hamstrings, it requires that knee flexion be maintained by the examiner; otherwise, the hamstrings will unavoidably act in maintaining the antigravity knee flexion. Trying to maintain knee flexion passively and applying pressure to the thigh makes it difficult to obtain an accurate test.

If this test is used because of marked hip flexor tightness, it may be impractical to flex the knee, thereby increasing the rectus femoris tension over the hip joint.

SECTION V
CLINICAL FINDINGS

DEFORMITIES OF FOOT AND ANKLE

In the following list, foot deformities are defined in terms of the positions of the involved joints. In severe deformities, the position of the joint is beyond the normal range of joint motion.

Talipes valgus: Foot everted and accompanied by flattening of the longitudinal arch.

Talipes varus: Foot inverted and accompanied by an increase in the height of the longitudinal arch.

Talipes equinus: Ankle joint plantar flexed.

Talipes equinovalgus: Ankle joint plantar flexed and foot everted.

Talipes equinovarus: Ankle joint plantar flexed and foot inverted (i.e., clubfoot).

Talipes calcaneus: Ankle joint dorsiflexed.

Talipes calcaneovalgus: Ankle joint dorsiflexed and foot everted.

Talipes calcaneovarus: Ankle joint dorsiflexed and foot inverted.

Talipes cavus: Ankle joint dorsiflexed and forefoot plantar flexed, resulting in a high longitudinal arch. With the change in position of the calcaneus, the posterior prominence of the heel tends to be obliterated, and weight-bearing on the calcaneus shifts posteriorly.

FOOT PROBLEMS

The foot has two longitudinal arches that extend lengthwise from the heel to the ball of the foot. The *medial longitudinal arch* is made up of the calcaneus, talus, navicular, three cuneiform, and three medial metatarsal bones. The outer or *lateral longitudinal arch* is made up of the calcaneus, cuboid, and two lateral metatarsal bones. The lateral arch is lower than the medial arch, and it tends to disappear in weight-bearing. Any references to the longitudinal arch will therefore mean the medial arch.

There are two transverse metatarsal arches, one across the midsection and one across the ball of the foot. The *posterior metatarsal arch* is at the proximal end (or base) of the metatarsal bones. It is a structural arch with wedge-shaped bones at the apex of the arch. The *anterior metatarsal arch* is at the distal ends (or heads) of the metatarsals.

Painful foot conditions may be roughly divided into three groups:

- Those dealing with longitudinal arch strain.
- Those dealing with metatarsal arch strain.
- Those dealing with faulty positions of the toes.

Examination of faulty and painful feet should include the following steps:

1. Examine the overall postural alignment for evidence of superimposed strain on the feet, such as occurs in cases of postural faults in which the body weight is borne too far forward over the balls of the feet.
2. Check the alignment of the feet in standing, both with and without shoes.
3. Observe the manner of walking, both with and without shoes.
4. Test for muscle weakness or tightness of the toe and foot muscles.
5. Check regarding unfavorable occupational influences.
6. Examine the shoes for overall fit, and check for places of wear on the shoe. Faulty weight distribution in standing or walking is often revealed by excessive wear on certain parts of the shoe. Common sites of wear are the heel and collar of the shoe.

FAULTY AND PAINFUL FOOT CONDITIONS AND TREATMENT INDICATIONS

There is a familiar saying, "If your feet hurt, you hurt all over." For those whose occupation requires constant standing or those engaged in activities that place great stress on the feet, the statement is especially applicable.

In older people, feet may become painful because of the loss of normal padding on the soles of the feet (21). *Insoles* that cushion the foot markedly improve comfort and function. The insole must be thin enough to fit in the shoe without crowding the foot, but thick enough to offer a firm, resilient cushion.

To the extent that the foot pain or discomfort is relieved, the insole may indirectly help alleviate the discomfort elsewhere that has resulted from a painful foot condition.

Pronation without Flatness of the Longitudinal Arch

This type of fault is most often found among women who wear high heels. In weight-bearing, some symptoms of foot strain may occur in the longitudinal arch, but more often, the pronation causes strain medially at the knee. In the foot itself, the anterior arch is subjected to more strain than the longitudinal arch.

Occasionally, the longitudinal arch is higher than average. This situation may require use of an arch support that is higher than usual so that the support may conform to the foot and provide a uniform base of support.

Pronation with Flatness of the Longitudinal Arch

This position of the foot is comparable to a position of dorsiflexion and eversion. In weight-bearing, the position of pronation with flatness of the longitudinal arch is usually accompanied by an out-toeing of the forefoot. Excessive tension is exerted on the muscles and ligaments on the inner side of the foot that support the longitudinal arch. Undue compression is exerted on the outer side of the foot, in the region of the talocalcaneonavicular joint.

The tibialis posterior and abductor hallucis are usually weak. Toe extensor muscles and the flexor brevis digitorum also may be weak. The fibular muscles tend to be tight if pronation is marked.

Supinated Foot

A supinated foot is a very uncommon postural fault. It is essentially the reverse of a pronated foot—the arch is high, and the weight is born on the outer side of the foot. Likewise, shoe corrections are essentially the opposite of those applied to a pronated foot. An outer wedge on the heel, a reverse modified Thomas heel, and an outer sole wedge usually are indicated.

If the genu valgum is associated with supination of the foot, shoe corrections as described above may increase the deformity of the knee. Give careful consideration to any associated faults.

Hammer Toes

The position of hammer toes (as illustrated in Figure 7-77) is one in which the toes are extended at the metatarsophalangeal and distal interphalangeal joints and are flexed at the proximal interphalangeal joints. Usually, calluses are found under the ball of the foot and corns on the toes as a result of pressure from the shoe. Shoes that are too short or too narrow can contribute to the problem.

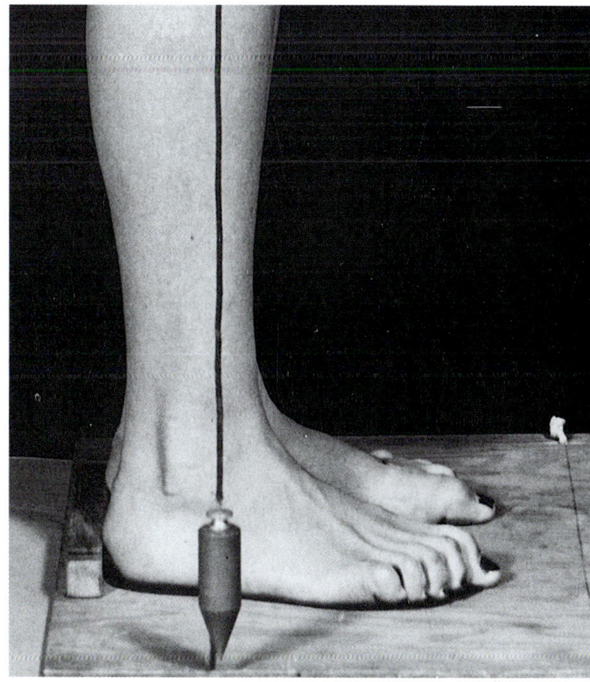

FIGURE 7-77: Hammer toes.

Metatarsal Arch Strain

This type of strain is usually the result of wearing high heels or of walking on hard surfaces in soft-soled shoes. It also may result from an unusual amount of running, jumping, or hopping.

In cases of metatarsal arch strain, the lumbricals, adductor hallucis (transverse and oblique), and flexor digiti minimi are most noticeably weak. If asked to flex the toes and cup the front part of the foot, the patient can only flex the end joints of the toes; little or no flexion of the metatarsophalangeal joints occurs.

Hallux Valgus

A hallux valgus is a position of faulty alignment of the big toe in which the end of the toe deviates toward the midline of the foot, sometimes to the point of overlapping the other toes. The abductor hallucis muscle is stretched and weakened, and the adductor hallucis muscle is tight.

Such cases may require surgery if the fault cannot be corrected or the pain alleviated by conservative means. In the early stages, however, it may be possible to achieve considerable correction.

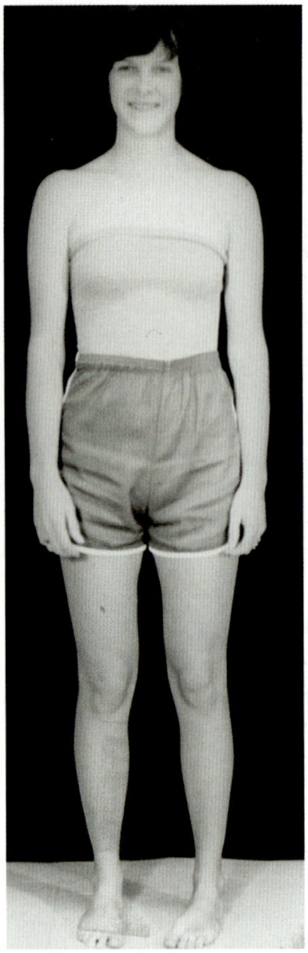

FIGURE 7-78: Anterior foot varus and in-toeing left foot.

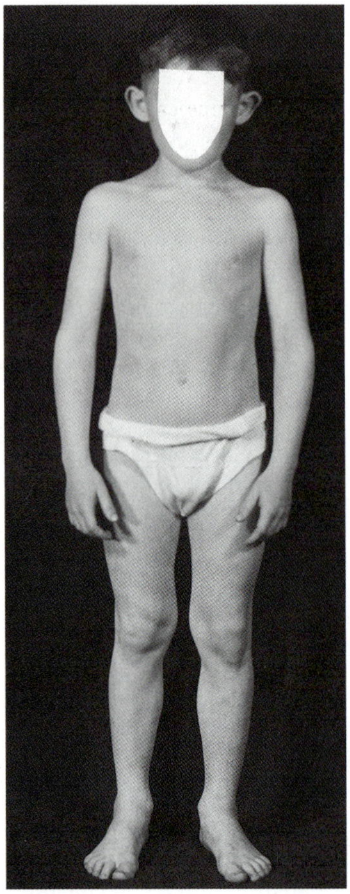

FIGURE 7-79: External rotation of hips and out-toeing of feet.

In-Toeing Position of the Foot

An in-toeing position of the feet, like the out-toeing position, may be related to faults at various levels. The term *pigeon-toes* may be considered to be synonymous with in-toeing.

If the legs are internally rotated at hip level, the patellae face inward, the feet point inward, and usually, pronation of the feet occurs. With in-toeing related to medial torsion of the tibia, the patellae face forward, and the feet point inward. If the problem is within the foot itself, the hips and knees may be in good alignment, but anterior foot varus (i.e., adduction of the forefoot) may be found (Figure 7-78).

Out-Toeing Position of the Feet

Out-toeing may be the result of (a) external rotation of the entire extremity from hip level (b), tibial torsion, in which the shaft of the tibia has developed lateral rotation (c), or a fault of the foot itself in which the forefoot abducts in relation to the posterior part of the foot (Figure 7-79).

Shoes

The protection and support given by shoes are important considerations with regard to the postural alignment in standing. Various factors predispose toward faulty alignment and foot strain and create the need for adequate shoe support. The flat, unyielding floors and sidewalks of our environment, use of heels that decrease the stability of the foot, and prolonged periods of standing, as required in some occupations, are several of the causes contributing to foot problems.

A number of factors relating to the size, shape, and construction of a shoe need to be considered. That said, clinicians should be mindful of overcorrection as it too can contribute to faulty stressors on the foot, ankle, and posture. It is beyond the scope of this text to go into the possibilities for shoe correction.

KNEE PROBLEMS

The habitual position of the knee in standing indicates which areas are subjected to undue compression and tension. The postural faults may appear separately or in various combinations. For example, postural genu varum results from the combination of hyperextension of the knees, medial rotation of the hips, and pronation of the transverse and subtalar joints. Medial rotation of the hips and slight knee valgus are frequently seen in combination. Lateral rotation of the hips is often seen with severe genu valgum.

This text does not deal with the treatment of congenital or acquired deformities of the feet and knees. An excellent reference for such treatment is found in the chapter by Joseph H. Kite in *Basmajian's Therapeutic Exercise* (best in the third edition, 22).

Genu Varum (Bowlegs)

A position of genu varum (bowlegs) may be either actual or apparent (i.e., structural or postural). An actual bowing is of the shaft (femur, tibia, or both) and may be caused by rickets or osteomalacia. An apparent bowing occurs as a result of a combination of joint positions that permit faulty alignment without any structural defect in the long bones. It results from a combination of medial rotation of the hip, hyperextension of the knee, and pronation of the foot. Hyperextension alone does not result in a position of postural bowlegs; the medial rotation component is required. Medial rotation of the hip plus pronation of the foot do not result in bowing unless accompanied by hyperextension. Thus, on testing, the apparent postural bowing will disappear in non-weight-bearing or in standing if the knees are held in neutral extension.

Knee Hyperextension

Hyperextension of the knee joint results in undue compression anteriorly and undue tension on muscles and ligaments posteriorly. Pain may occur in either area. Pain in the popliteal space is not uncommon in adults who have stood with the knees in hyperextension.

Hyperextension may cause further problems if not corrected. The popliteus is a short (one-joint) muscle that acts somewhat as a broad posterior knee joint ligament. Its action is to flex the knee and to rotate the leg medially on the thigh. If it is stretched by knee hyperextension, it allows the lower leg to rotate laterally on the femur in flexion or in hyperextension.

Genu Valgum (Knock-Knees)

Tension on the medial knee ligaments and compression on the lateral surfaces of the knee joint are present in **genu valgum**. These both can cause discomfort and pain that is slow to develop before potentially becoming debilitating. Tightness of the tensor fasciae latae and iliotibial band is frequently seen in conjunction with genu valgum.

Knee Flexion

Knee flexion is a less common finding than the three abovementioned problems, but it is fairly common among older people. Habitually standing with the knees flexed (Figure 7-80) can cause problems at the knee and along the quadriceps muscle. It is a position that requires constant muscular effort to keep the knees from flexing further. Pain is most often associated with muscle strain of the quadriceps or with the effect of traction by the quadriceps (through its patellar tendon insertion) on the tibia.

Sometimes a position of knee flexion is assumed to ease a painful low back that is otherwise pulled into a lordotic curve by tight hip flexors. There may also be actual shortness of the popliteus and the one-joint hamstring, namely the short head of the biceps femoris. If the hip flexors and knee flexors are tight, institute appropriate stretching exercises are indicated.

Effect on Posture: Unilateral knee flexion creates concerns beyond the area of the knee. The effect on posture may be seen in Figure 7-80. With the left knee flexed, the right foot is more pronated than the left, the right thigh is medially rotated, the pelvis tilts down on the left, the spine curves convexly toward the left, the right hip is high, and the right shoulder is low.

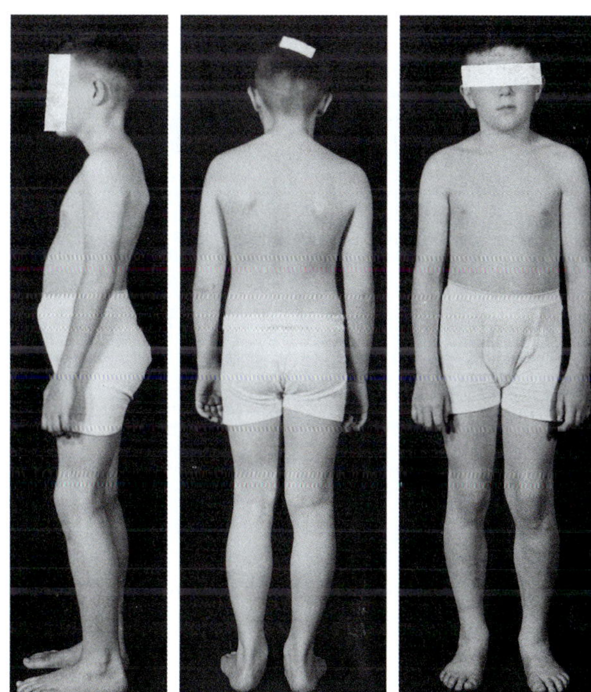

FIGURE 7-80: Knee flexion.

FIGURE 7-81: Reverse tailor or "W" position.

Medial Rotation of Hip and Pronation of Feet

The position of the knees in which the patellae face slightly inward results from medial rotation at the hip joints. As a functional or apparent (i.e., not structural) malalignment, it is usually accompanied by pronation of the feet. The initial problem may be at the hip or at the foot, and it may result from weakness of the hip external rotators or of the muscles and ligaments that support the longitudinal arches of the feet. Whichever predisposes to the fault, the end result is usually that both conditions exist if the initial problem is not corrected. A tight tensor fasciae latae may be a contributing cause, and sitting in a reverse tailor or "W" position may predispose toward faulty hip, knee, and foot position (see Figure 7-81).

There may be a structural malalignment with a lateral tibial torsion accompanying the hip medial rotation. In either event, there tends to be pronation of the subtalar and transverse tarsal joints, but with tibial torsion, there is more out-toeing of the foot. The malalignment affects the knee joint adversely, causing ligamentous strain anteromedially and joint compression laterally.

Reverse tailor or "W" position.

Habitual Positions that Predispose to Bilateral Adaptive Shortening

Sitting in a "W" or reverse-tailor position favors tensor fasciae latae shortness; sitting in a tailor or yoga position favors sartorius shortening. The habit of sitting with one leg—and always the same one—in one of these positions is conducive to unilateral shortness. Changing postural habits is an important part of treatment.

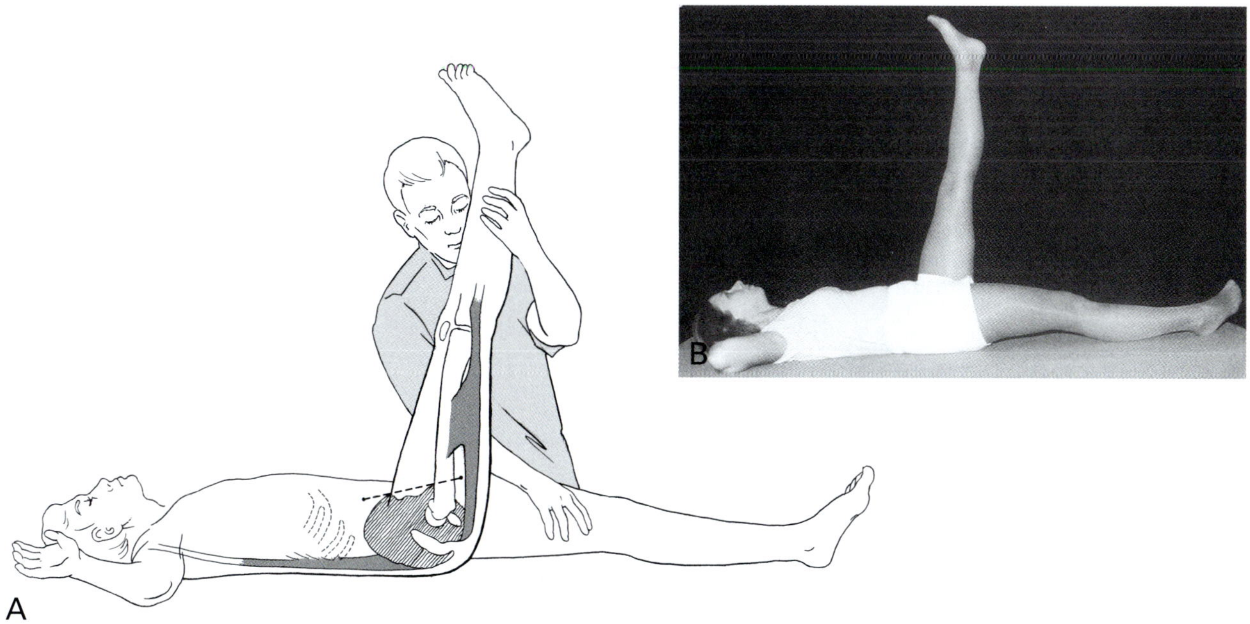

FIGURE 7-82: **A**, **B**, Normal hamstring length.

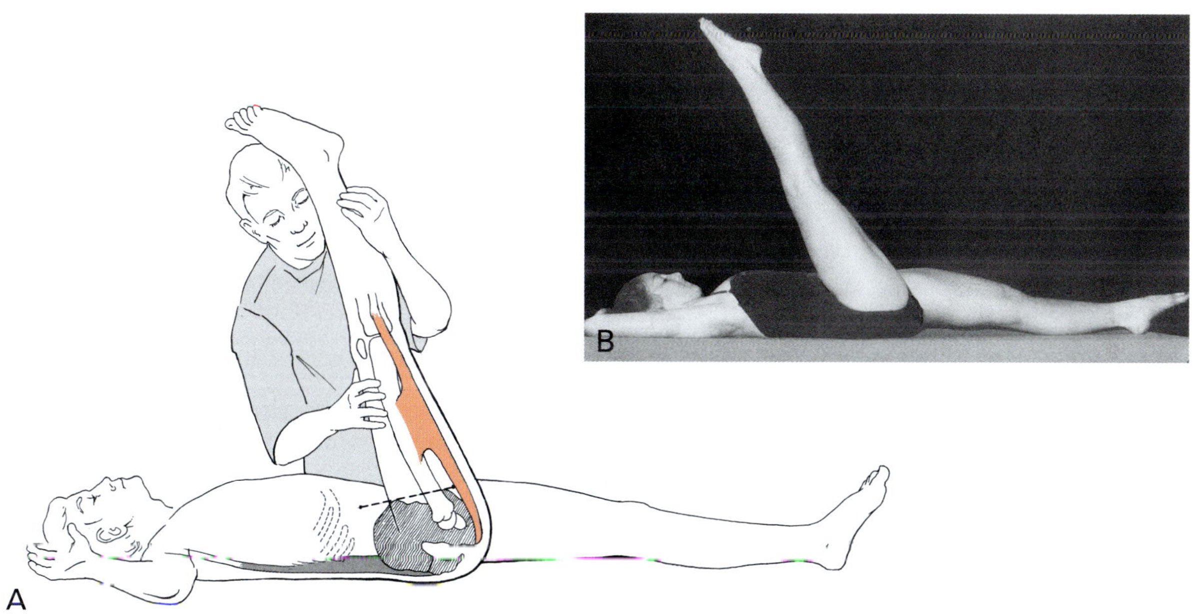

FIGURE 7-83: **A**, **B**, Excessive hamstring length.

Normal Hamstring Length

No Hip Flexor Shortness: Straight-leg raising, with the subject in supine position, the low back and sacrum flat on the table, and the other leg either extended by the subject or held down by the examiner. An angle of approximately 80° between the table and the raised leg is considered to be a normal range of hamstring length (Figure 7-82).

Excessive Hamstring Length

Excessive hamstring length (an angle greater than 80 degrees between the table and raised lower extremity) can be reliably concluded only if the low back and sacrum are maintained flat on the table and the opposite lower extremity is extended to the hip's zero position.

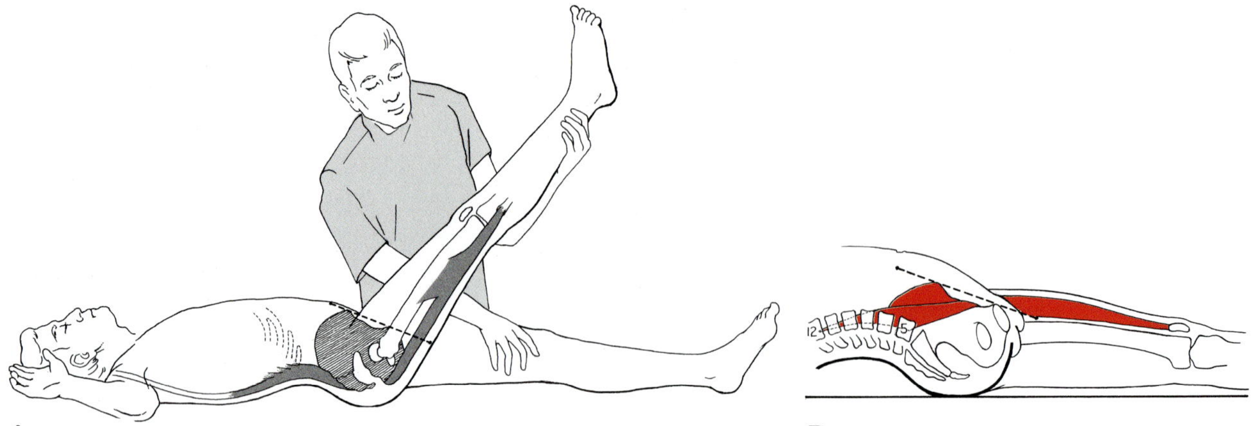

A B

FIGURE 7-84: **A**, Apparently short, actually normal hamstring length; **B**, Short hip flexors hold back to hyperextension and hip joint flexion.

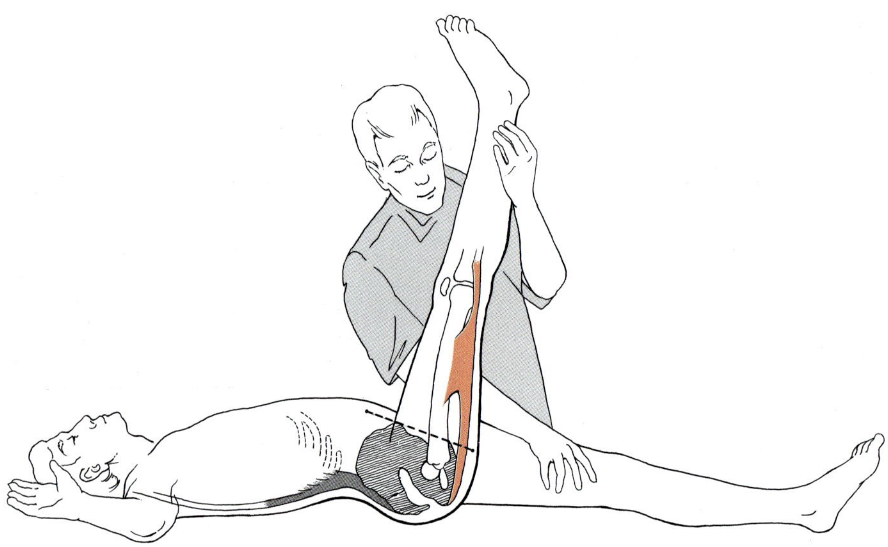

FIGURE 7-85: Apparently normal, actually excessive hamstring length.

SHORTNESS OF HAMSTRINGS

Hamstring Length: Apparently Short, Actually Normal

In the supine position with the legs extended, low back hyperextended, and pelvis in anterior tilt, the hip joint is already in flexion. If the straight-leg-raising test is performed with the low back and pelvis in this position, hamstrings of normal length will appear to be short (Figure 7-84A).

With few exceptions, the position of anterior tilt results from shortness of the one-joint hip flexors, and the amount of flexion varies with the amount of hip flexor shortness. If it were possible to determine how many degrees of hip flexion exist by virtue of the pelvic tilt, this number could be added to the number of degrees of straight-leg raise in determining hamstring length. It is not possible, however, to measure that amount of flexion. Hence, the low back and pelvis must be flat on the table. To get the low back and sacrum flat in a subject with hip flexor shortness, the hips must be flexed, but only by the amount necessary to obtain the desired position (Figure 7-84B).

Hamstring Length: Apparently Normal, Actually Excessive

The actual hamstring length is the same as that in Figure 7-83A (Figure 7-85).

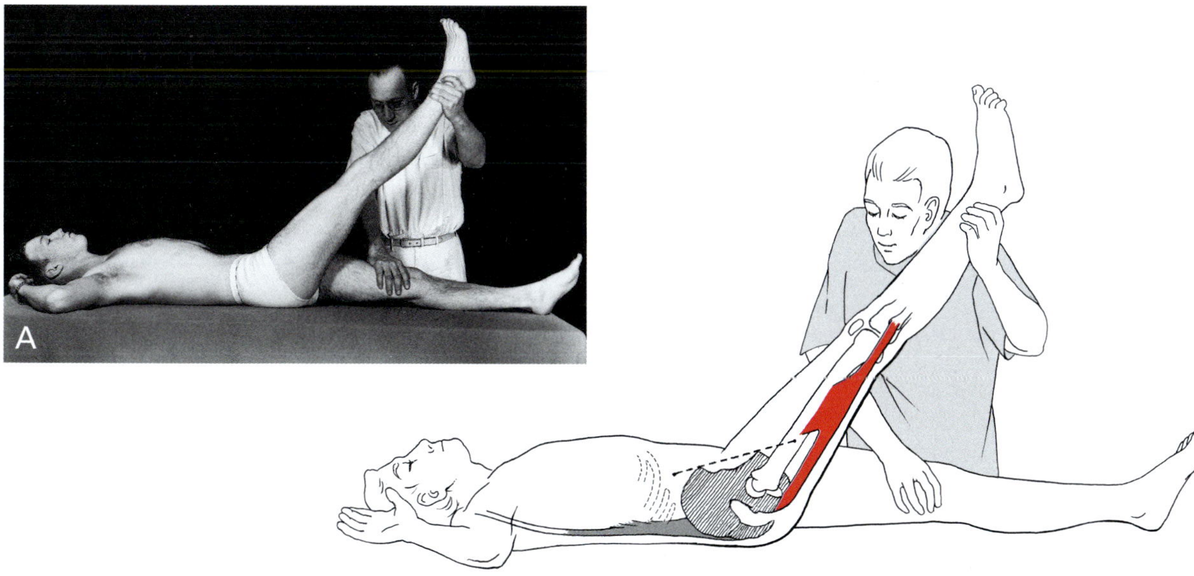

B

FIGURE 7-86: **A**, **B**, Short hamstrings.

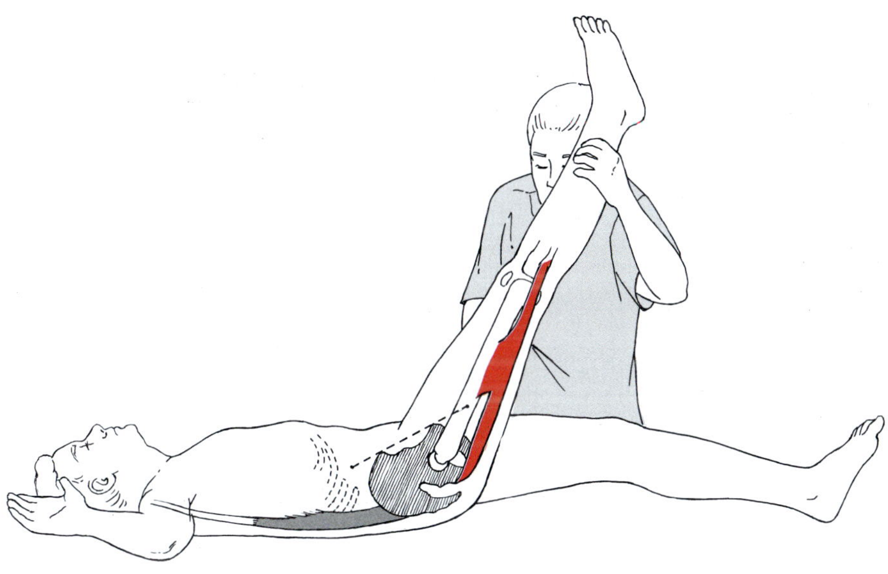

FIGURE 7-87: Apparent hamstring length greater than actual length.

Short Hamstrings

When hip joint flexion has reached the limit of hamstring length in the straight-leg raise, the hamstrings exert a downward pull on the ischium in the direction of posteriorly tilting the pelvis (Figure 7-86). To prevent excessive posterior pelvic tilt and excessive flexion of the back, stabilize the pelvis with the low back in the flat position by holding the opposite leg firmly down. (If there is shortness of hip flexors and a roll or pillow must be put under the knees to get the back flat, then one leg must be held firmly down on the pillow to prevent excessive posterior tilt.)

Apparent Hamstring Length Greater than Actual Length

In Figure 7-87, excessive posterior tilt of the pelvis allows the leg to be raised slightly higher here than shown in Figure 7-86A and B even though the hamstring length is the same in both instances. With the opposite leg held firmly down, excessive posterior tilt will not occur except in subjects with excessive length in the hip flexors, which is not common.

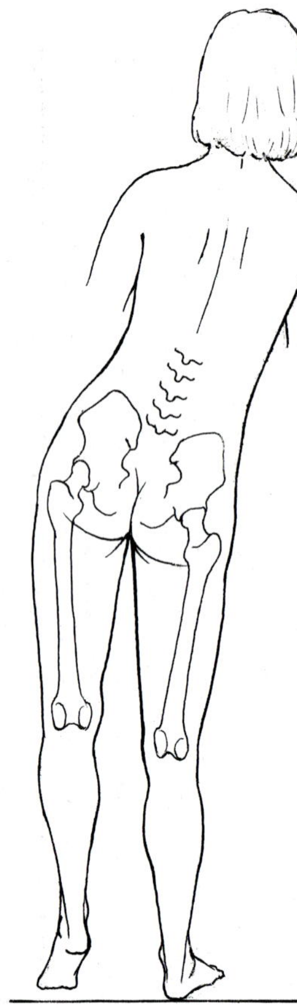

FIGURE 7-88: Right gluteus medius weakness.

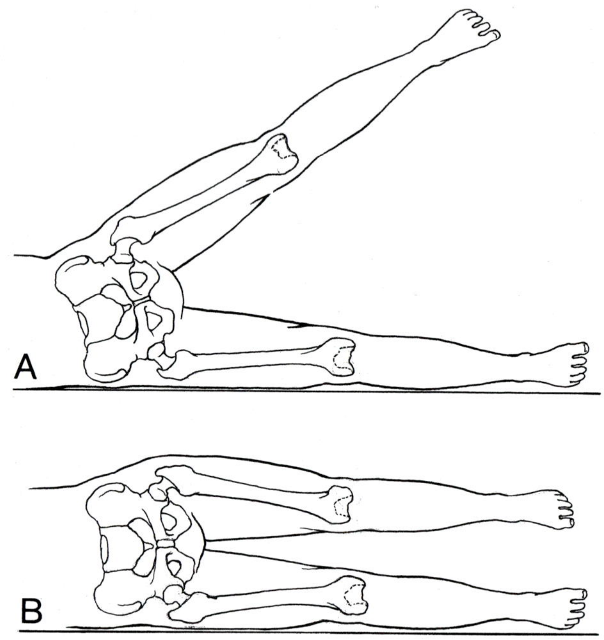

FIGURE 7-89: A, B, Hip joint abduction.

GLUTEUS MEDIUS WEAKNESS

Paralysis or Marked Weakness of Right Gluteus Medius: Paralysis or marked weakness of the gluteus medius can result in observable gait deviation (Figure 7-88). This consists of displacement of the trunk laterally, toward the side of weakness, shifting the center of mass in such a way that the body can be balanced over the extremity with minimal muscular support at the hip joint.

Hip Joint Abduction: *Actual* abduction of the hip joint is accomplished by the hip abductors, with normal fixation by the lateral trunk muscles, as shown in Figure 7-89A. When the hip abductors are weak, *apparent* abduction may occur by substitution action of the lateral trunk muscles. In this case, the leg drops into adduction, the pelvis is hiked up laterally, and the leg is raised upward from the table, as shown in Figure 7-89B.

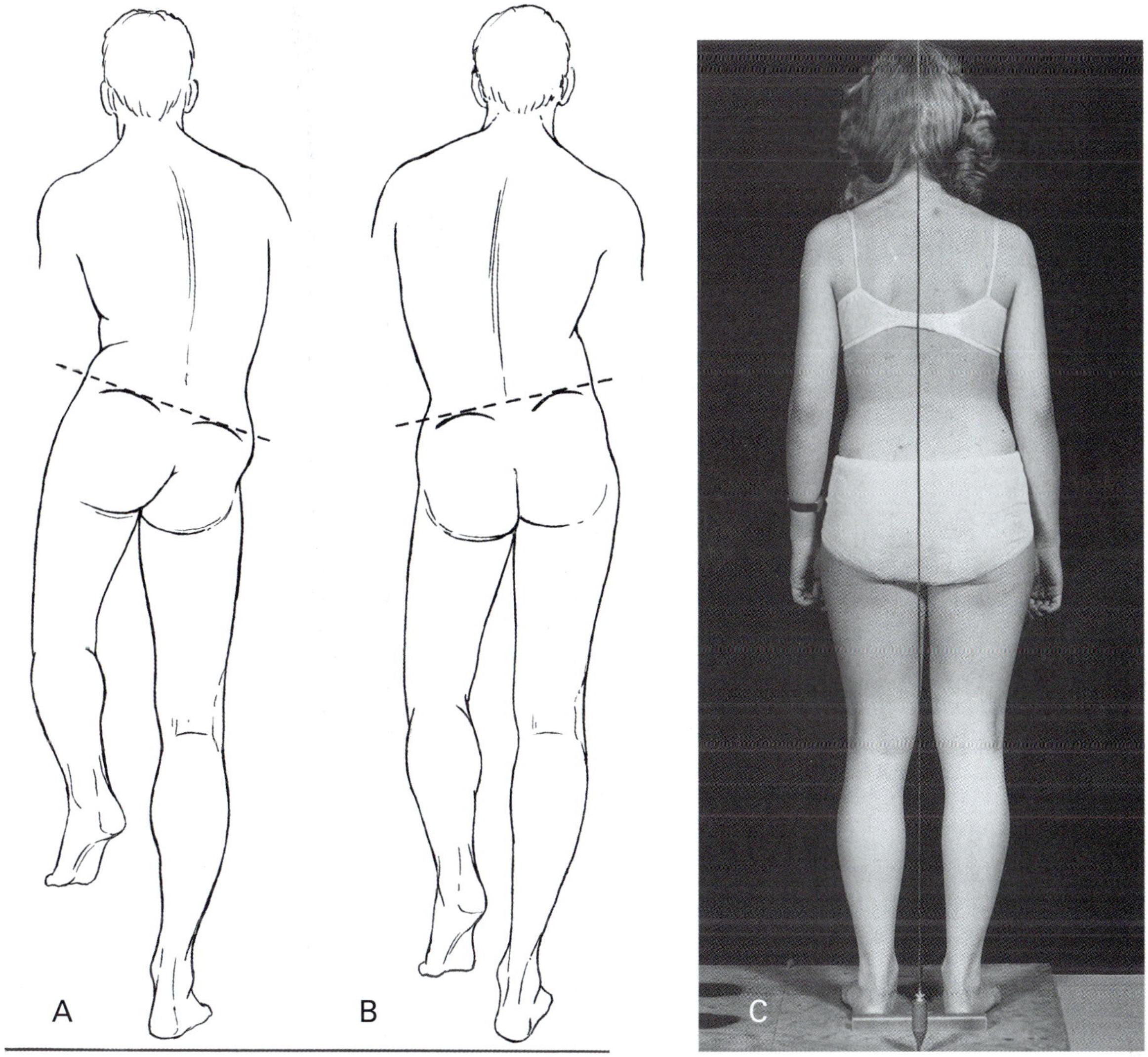

FIGURE 7-90: Trendelenburg sign and hip abductor weakness: **A**, Hip joint abduction; **B**, Hip joint adduction; test for the Trendelenburg sign; **C**, Mild weakness of right hip abductors.

TRENDELENBURG SIGN AND HIP ABDUCTOR WEAKNESS

When body weight is supported alternately on one leg, such as in walking, the body must be stabilized on the weight-bearing leg during each step. By reverse action (i.e., origin pulled toward the insertion), strong hip abductors can stabilize the pelvis on the femur in hip joint *abduction*, as shown in Figure 7-90A. The lateral trunk flexors on the left also act by pulling upward on the pelvis.

Figure 7-90B shows a position of hip joint *adduction* that results when hip abductors are too weak to stabilize the pelvis on the femur. The pelvis drops downward on the opposite side. Strong lateral trunk flexors on the left cannot raise the pelvis on that side, in standing, without the opposite abductors providing a counter-pull on the right. This is also the test used to elicit the Trendelenburg sign.

The *Trendelenburg gait* is one in which the affected hip goes into hip joint adduction during each weight-bearing phase of the gait. The femur rides upward, because the acetabulum is too shallow to support the head of the femur. If the problem is bilateral, a waddling gait is observed.

Figure 7-90C illustrates a relaxed postural position in an individual with mild weakness of the right hip abductors. The gluteus medius is the chief abductor, and a test that emphasizes the posterior gluteus medius often demonstrates more weakness than the test for hip abductors as a group. Often, this weakness of the gluteus medius is found in association with other weaknesses in the handedness patterns.

Testing the strength of the gluteus medius is important in cases of pain in the region of this muscle or of low back pain associated with lateral pelvic tilt.

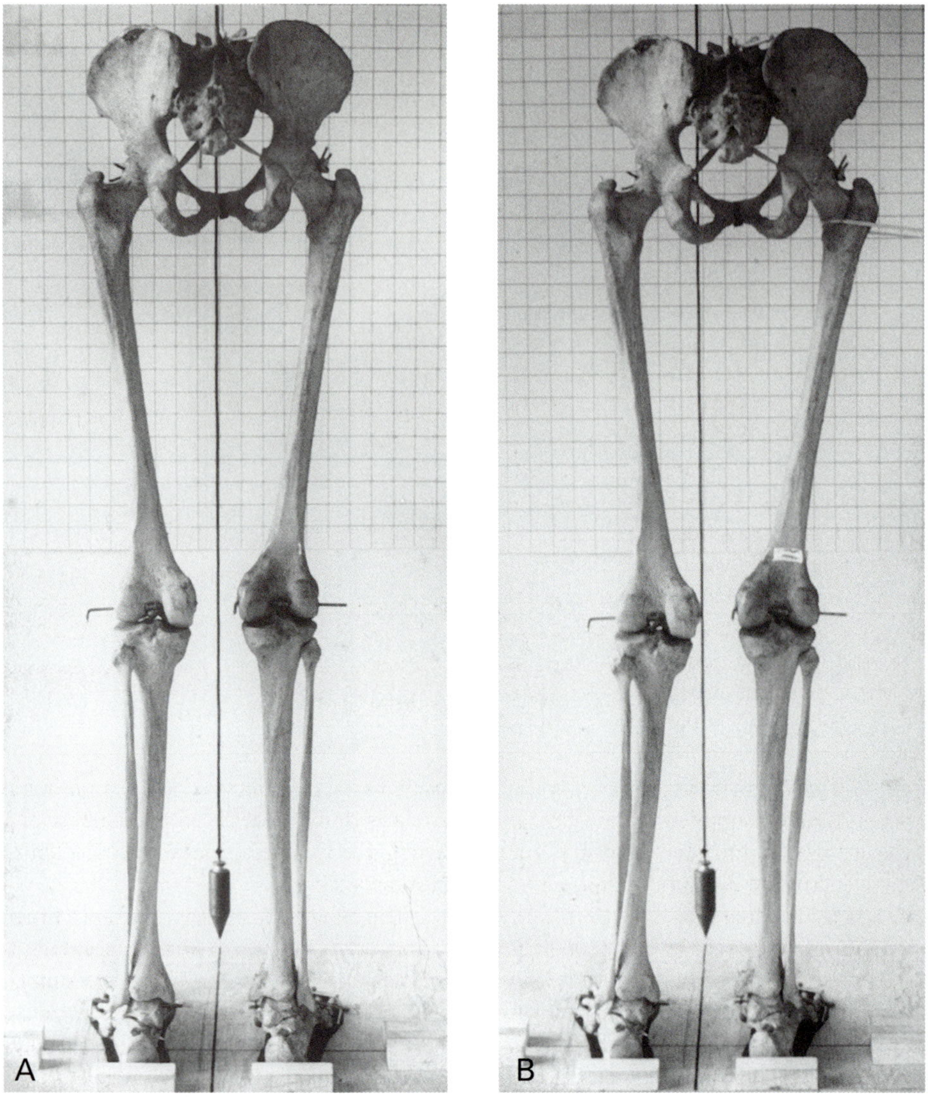

FIGURE 7-91: Apparent leg length discrepancy caused by **A**, Pelvic lateral tilting, and **B**, Pelvic lateral displacement.

APPARENT LEG-LENGTH DISCREPANCY CAUSED BY MUSCLE IMBALANCE

Without any actual difference in leg length, subjects have an appearance of a longer leg on the high side when the pelvis is tilted laterally (Figure 7-91A). In Figure 7-91B, this appearance has been created by displacing the pelvis laterally. (The feet were anchored to the floor.)

If tightness develops in the tensor fasciae latae and iliotibial band on one side, the pelvis will be tilted downward on that side. With gluteus medius weakness on one side, the pelvis will ride higher on the side of the weakness.

Legs are equal in length.
Pelvis is level.
Both hip joints are neutral between adduction and abduction.
Length of abductors is equal.
As the pelvis sways sideways, the pelvis is higher on the right.
The right hip joint is adducted.
The left hip joint is abducted.
The right hip abductors are elongated.
The left hip abductors and fascia lata are in a shortened position.

LEG PAIN

The conditions discussed here include pain associated with a tight tensor fasciae latae and iliotibial band, stretched tensor fasciae latae and iliotibial band, and sciatica associated with a protruded intervertebral disk or a stretched piriformis.

Tight Tensor Fasciae Latae and Iliotibial Band

A condition sometimes mistakenly diagnosed as sciatica is that of pain associated with a tight tensor fasciae latae and iliotibial band. The dermatome area of cutaneous distribution corresponds closely with the area of pain. Pain may be limited to the area covered by the fascia along the lateral surface of the thigh or it may extend upward over the buttocks, involving the gluteal fascia as well.

Palpation over the full length of the fascia lata, from its origin on the iliac crest to the insertion of the iliotibial band into the lateral condyle of the tibia, may elicit pain or tenderness. There is tenderness especially along the upper margin of the trochanter and at the point of insertion near the head of the tibia.

Painful symptoms may be limited to the area of the thigh or may appear in the area supplied by the fibular nerve. A review of the anatomy of the lateral aspect of the knee shows the relationship of the common fibular nerve to the muscles and fascia in this area (Figure 7-92).

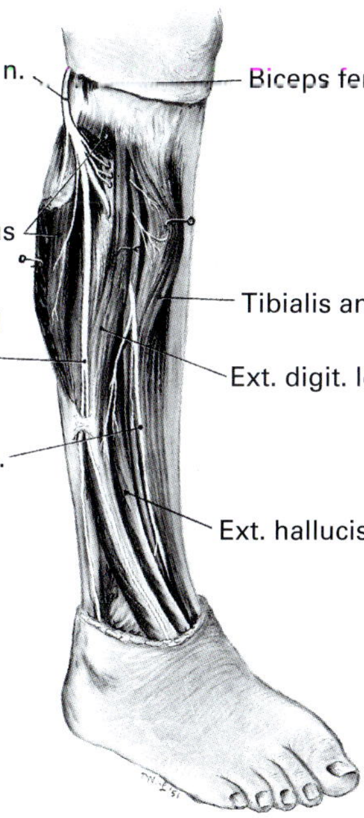

FIGURE 7-92: Anatomy of the lateral aspect of the leg.

The fibular branch of the sciatic nerve passes obliquely forward over the neck of the fibula and crosses directly under the fibers of origin of the fibularis longus muscle. Any prolonged pressure over this area, even if only slight, must be avoided because of the danger of fibular nerve paralysis. Even in the application of adhesive traction to the lower leg, one must avoid either pressure over the nerve or *excessive* traction on the soft tissue at that point.

The mechanism by which the fibular nerve is irritated in cases of tightness of the iliotibial band may be explained by the effect of pressure by the rigid bands of fascia or by the effects of traction on this part. When the fascia is drawn taut, as in movements of walking or on testing for tightness, the fascia is often observed to be extremely rigid.

The effect of traction is often seen in acute cases. With the patient side-lying and the affected leg uppermost, the mere dropping of the foot into inversion (i.e., downward toward the table) puts tension on the muscle and fascial band. Symptoms of nerve irritation in the area supplied by the fibular nerve may be elicited by this simple movement of the foot. When the side-lying position is assumed for sleeping or treatment and pillows are placed between the legs to keep the leg in abduction, the foot should also be supported to prevent it from dropping into inversion. Failure to recognize the peripheral cause of this fibular nerve irritation has often resulted in rather obscure explanations of this problem.

SECTION VI
INTERVENTION

TREATMENT OF MUSCLE LENGTH PROBLEMS

If muscle length is excessive, *avoid* stretching exercises and postural positions that maintain elongation of the already stretched muscles. Work to correct the faulty posture. Because the stretched muscles are usually weak, strengthening exercises are indicated. However, for active individuals, strength may improve simply through avoidance of overstretching.

Supports are indicated to prevent excessive range if the problem cannot be controlled through positioning and corrective exercise. For example, marked knee hyperextension, if unavoidable in weight-bearing, should be prevented by an appropriate support to allow the posterior knee joint ligaments and muscles to shorten.

When muscle shortness exists and stretching exercises are indicated, they must be done with precision to ensure that the tight muscles are the ones that are actually being stretched and to avoid adverse effects on other parts of the body.

TREATMENT OF FOOT PROBLEMS

Treatment may be considered corrective or palliative. Ideally, treatment should be corrective, but considering that painful foot conditions occur in many older people who have bony, ligamentous, and muscular structures that cannot adjust to corrective measures, it is necessary to use measures designed to obtain relief with minimal correction.

Hallux Valgus

The patient should wear shoes with a straight inner border and avoid shoes with cut-out toe space. A "toe-separator," which is a small piece of rubber, is inserted between the big toe and the second toe aids in holding the big toe in more normal alignment. As a pure palliative procedure for the relief of pain caused by pressure, a bunion guard is often useful.

Because excessive pronation often is the cause of the hallux valgus, prevention or correction require that the arch be supported. "Excessive" means marked relaxation of the supporting arch structures that require firm support; rigid orthoses are needed in such instances.

Hammer Toes

Massage and stretching may aid in correcting the faulty alignment of the toes in the early stage, and benefit may be obtained from use of a metatarsal bar. An inside metatarsal bar may be more effective, but an outside metatarsal bar may be more comfortable (Figure 7-93).

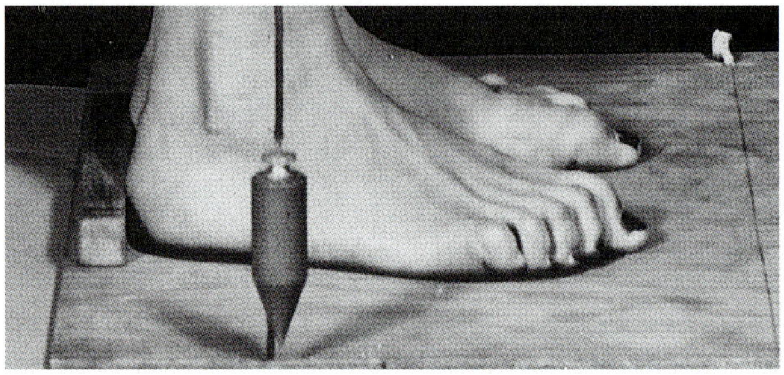

FIGURE 7-93: Hammer toes.

Treatment of Pronation without Flatness of the Longitudinal Arch

Treatment of pronation consists of using an inner heel wedge or an orthosis that provides the same type of correction. Generally, patients should be discouraged from wearing a high heel if they have symptoms of foot or knee pain. Recommending shoes with little or no heel may be inadvisable, however, because the foot tends to pronate more in a flat-heeled shoe. With a medium heel, the longitudinal arch is increased, and a heel wedge or arch support will help to correct pronation.

Regarding shoe correction, on a heel of medium height, a 1/16-inch inner wedge is usually used whereas a 1/8-inch wedge is the usual adjustment on a low heel. A high heel cannot be altered by use of an inner wedge without interfering with the subject's stability.

Treatment of Pronation with Flatness of the Longitudinal Arch

Supportive treatment consists of using an inner-heel wedge and a longitudinal arch support. When the heel has a wide base, a wedge of 1/8-inch thickness is most often used. When the fault is severe, the patient should be discouraged from wearing a shoe without a heel. This type of fault is more prevalent among men and children than among women.

Corrective Foot Exercises for Pronated Feet

Lying on Back:

1. Curl the toes downward, and hold while pulling the foot upward and inward.
2. With the legs straight and together, try to touch the soles of the feet together.

Sitting in Chair:

3. With the left knee crossed over the right, move the left foot in a half-circle downward, inward, and upward, and then relax. (Do not turn the foot outward.) Repeat with the right foot.
4. With the knees apart, place the soles of the feet together and hold while bringing the knees together.
5. Place a towel on the floor. With the feet parallel and approximately 6 inches apart, grip the towel with the toes and pull inward (in adduction) with both feet, bunching the towel between the feet.
6. With a small ball (~1¼ to 1½ inches in diameter) cut in half and placed under the anterior arch of the foot, grip the toes downward over the ball.

Standing:

7. With the feet straight ahead or slightly out-toeing, roll weight to the outer borders of the feet by pulling upward under the arches.

Walking:

8. Walk along a straight line on the floor, pointing the toes straight ahead and transferring weight from the heel along the outer border of the foot to the toes.

TREATMENT OF SUPINATED FOOT

Shoe corrections are essentially the opposite of those applied to a pronated foot. An outer wedge on the heel, a reverse modified Thomas heel, and an outer sole wedge usually are indicated.

Metatarsal Arch Strain

Stretching of the toe extensors is indicated if tightness exists. Supportive treatment consists of use of a metatarsal pad or a metatarsal bar. If calluses are under the heads of the second, third, and fourth metatarsals, a pad is usually indicated; if calluses are under the heads of all the metatarsals, a bar is indicated.

In-Toeing Position of the Foot

Generally, children do not exhibit muscle tightness. It is not uncommon, however, to find that the tensor fasciae latae, which is an internal rotator, is tight in children who exhibit medial rotation from hip level. Stretching of the tensor may be indicated, but this should be done carefully.

FIGURE 7-94: Child in "W" tailor position, which results in in-toeing.

Children who develop this medial rotation from hip level often sit in a reverse tailor or "W" position (Figure 7-94). Encouraging the child to sit in a cross-legged position tends to offset the effects of the other position.

The shoe correction used in cases of in-toeing associated with internal rotation of the extremity is a small semicircular patch, placed on the outer side of the sole at about the base of the fifth metatarsal. To mark the area for the patch, the shoe is held upside down and bent sharply at the sole in the same manner that it bends in walking. The patch extends about equally forward and backward from the apex of the bend.

The patch is of a given thickness (either 1/8 or 3/16 inch, depending on the size of the shoe) along the outer border. It tapers off to zero toward the front, center, and back of the sole.

In-toeing associated with internal rotation of the extremity tends to be more marked in walking than in standing, and the shoe correction helps to change the walking rather than the standing pattern. The effect of changing the walking pattern, in turn, helps to correct the standing position.

The patch, by its convex shape, pivots the foot outward as the sole of the shoe is brought in contact with the floor during the usual transfer of weight forward. Before marking the shoe for alteration, a leather patch may be taped to the sole of the shoe and tested for position by observing the child's walk.

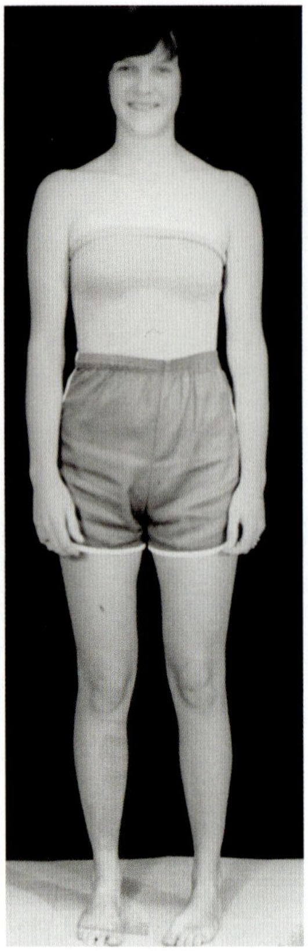

FIGURE 7-95: Anterior foot varus and in-toeing left foot.

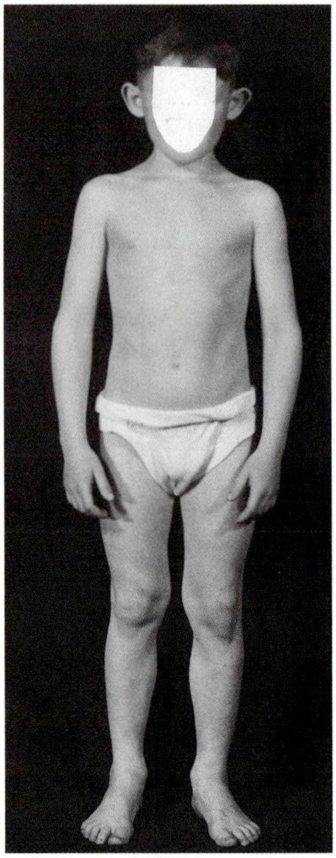

FIGURE 7-96: External rotation of hips and out-toeing of feet.

An in-toeing position caused by malalignment of the forefoot in relation to the rest of the foot is similar to a mild clubfoot, without equinus or supination of the heel. As a matter of fact, there may be pronation of the heel along with the adduction (Figure 7-95).

Inflare shoes may be comfortable, but they will not be corrective. The child should be fitted with shoes that have been made on a straight last. A stiff inner counter, extending from the base of the first metatarsal to the end of the great toe, should be added to the shoe. The outer counter should be stiff from the heel to the cuboid.

Out-Toeing Position of the Feet

For young children in whom the problem is from hip level (Figure 7-96), a *twister* may be used. Usually, results are obtained within a relatively short period of time (i.e., several months) (Figure 7-97).

The external rotation of the extremity (see Figure 7-96) does not automatically cause difficulty in standing. Walking in an out-toeing position, however, tends to put strain on the longitudinal arch as the weight is transferred from the heel to the toes.

If tibial torsion is an established fault in an adult, no effort should be made to have the individual walk with the feet straight ahead. Such "correction" of the foot position would result in a faulty alignment of the knees and hips.

Abduction of the forefoot is the result of a breakdown of the longitudinal arch. In children, measures that correct the arch position will help to correct the out-toeing. Wearing corrective shoes may be advisable because they typically have an inflare last. In adults with an established fault, however, corrective shoes do not change the alignment of the foot and cause undue pressure on the foot. Usually, it is necessary to have the patient wear shoes that have been made over a straight, or even an outflare, last. The patient can tolerate some arch support and inner wedge alterations if these are indicated, but the alignment of the shoe must necessarily conform with that of the foot to avoid pressure.

Toeing out in walking may result from tightness of the tendo achillis, in which case stretching of the plantar flexor muscles is indicated.

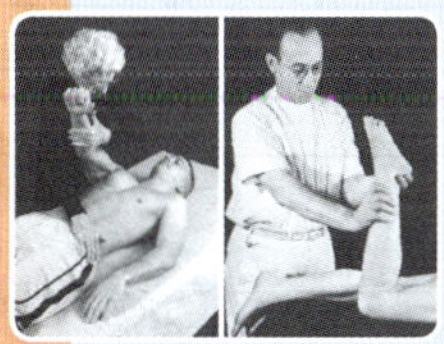

CLASSIC KENDALL

The "Twister"

This elastic rotation leg control device, the Twister, is designed to exert a force of counterrotation on the legs and feet to correct excessive internal or external rotation. This appliance is recommended for children with mild to moderate rotation problems and is frequently combined with other forms of treatment, such as shoe corrections and ankle braces. The simple fitting procedure of lacing the shoe hooks to the shoes, securing the pelvic belt with its Velcro fastener, extending the elastic straps as shown in Figure 7-97, and adjusting the strap tension for the position desired produces an effective rotation control that usually requires only a short adjustment period by the patient (Courtesy C.D. Denison Orthopaedic Appliance Corp, 23).

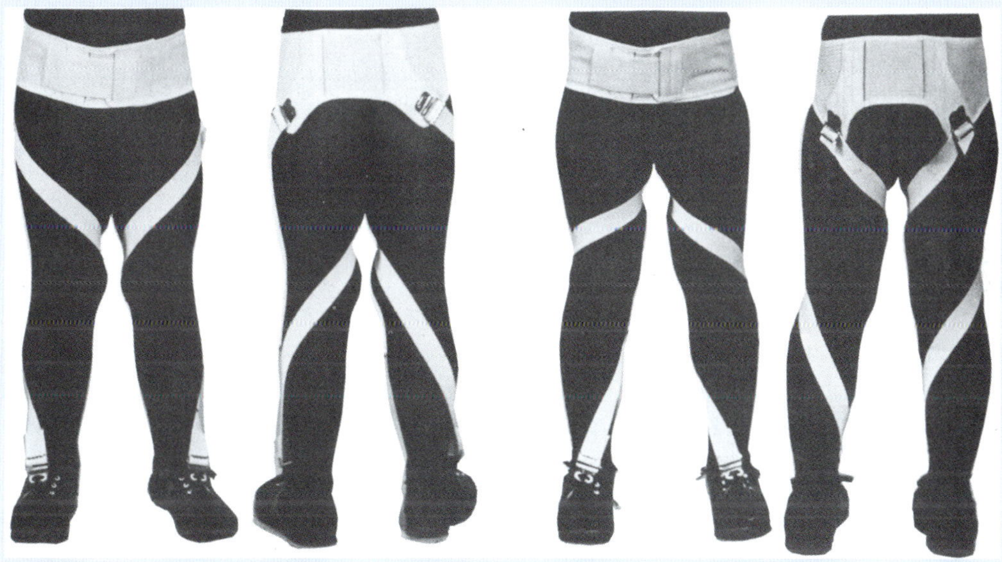

FIGURE 7-97: "Twister" elastic rotation leg control device.

SHOES AND SHOE CORRECTIONS

Shoes

The protection and support provided by shoes are important considerations for postural alignment in standing. Various factors predispose toward faulty alignment and foot strain and create the need for adequate shoe support. The flat, unyielding floors and sidewalks of our environment, use of heels that decrease the stability of the foot, and prolonged periods of standing required in some occupations are several of the causes contributing to foot problems.

A number of factors relating to the size, shape, and construction of a shoe need to be considered. That said, clinicians should be mindful of overcorrection as it too can contribute to faulty stressors on the foot, ankle, and posture. *It is beyond the scope of this text to go into the possibilities for shoe correction.*

GENU VARUM

Correction depends on use of appropriate shoe corrections, exercises to correct pronation, exercises to strengthen hip lateral rotators, and cooperation by the subject in avoiding a position of knee hyperextension.

In some instances, postural bowing and hyperextension are compensatory for knock-knees, as described in Chapter 1. Paradoxically, correction of this type of postural bowing must be based on correction of the underlying genu valgum.

Correction of *structural* bowing depends chiefly on timely intervention and effective bracing. An outer wedge on the heel or sole usually is not indicated because there is a tendency for the foot to pronate as the legs bow outward.

KNEE HYPEREXTENSION

Prevention or correction of hyperextension is based on instruction in good postural alignment and cooperation by the subject in avoiding positions of knee hyperextension in standing. Specific exercises for knee flexors may be indicated. Bracing may be required in cases that do not respond otherwise and in severe cases.

GENU VALGUM

Heat, massage, and stretching of the muscle and fascia lata often are needed, along with shoe corrections to bring about a realignment.

In treatment of early, *mild* genu valgum, an inner border wedge on a shoe tends to realign the extremity, thus relieving the strain medially and the compression laterally. There is danger, however, in using too high an inner wedge, because overcorrection of the foot may be overcompensated for by an increase in genu valgum. A 1/8- to 3/16-inch inner heel wedge is usually adequate. A *moderate* degree of genu valgum may benefit from a knee support in addition to shoe corrections. The support should have lateral steel uprights, with a joint at the knee. *Severe* genu valgum may require bracing or surgery.

HIP FLEXOR STRETCHING

Begin in the supine position, with the low back held flat by keeping one knee toward the chest with the other leg extended. The subject should contract the gluteals to actively extend the hip joint, bringing the thigh down toward the table *without* arching the back. (*Note:* If no table is available, this is the only hip flexor stretching exercise that can be done in the supine position. The stretching will affect the one-joint hip flexors only.)

The test position may be used to stretch both one-joint and two-joint hip flexors. If there is much tightness, take care to progress *gradually* with stretching. A little bit of stretching can cause soreness that may be felt more the next day. Also, remember that the psoas muscle is attached to the bodies, transverse processes, and intervertebral disks of the lumbar spine, and vigorous stretching can create or aggravate a problem with the low back.

The prone position on a table is unsatisfactory for stretching hip flexors because the low back, which is already in a hyperlordosis, cannot be held flat or controlled in any fixed position. If a table is available, the subject may lie with the trunk prone at the end of the table and the legs hanging down, with the knees bent as necessary and the feet on the floor. Have the subject raise one leg in hip extension, high enough to put a stretch on the hip flexors, with the knee straight for a one-joint stretch and the knee bent approximately 80° for a one-joint and two-joint stretch.

When two-joint hip flexors are short, avoid the kneeling lunge. (The kneeling lunge may be used to stretch the one-joint muscles, providing that the two-joint hip flexors are not tight.) Be cautious in use of the kneeling lunge because of potential strain on the sacroiliac joint as well as on the low back.

When one-joint hip flexors are short, *avoid* the lunge (Figure 7-98). Because the low back is not stabilized, tight hip flexors pull it into a hyperlordosis. In the supine position, the low back is held flat, and tightness appears at the hip joint.

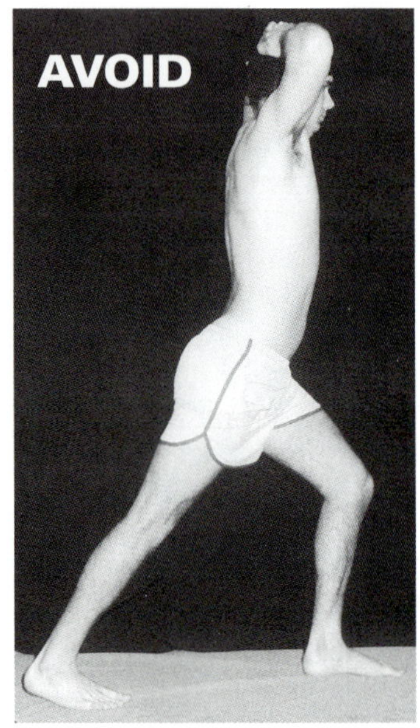

FIGURE 7-98: The lunge is to be avoided when one-joint hkp flexors are tight.

Exercise to stretch the one-joint hip flexors (Figure 7-99A). Contract the gluteus maximus to pull the thigh toward the table, maintaining the knee in extension and *keeping the back flat.*

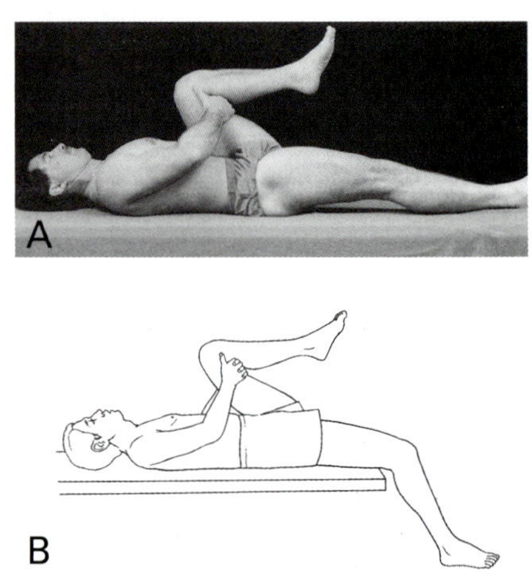

FIGURE 7-99: Exercise to stretch **A**, the one-joint hip flexors, and **B**, two-joint hip flexors.

To stretch the one-joint and two-joint hip flexors on the right, lie on the back with the right lower leg hanging over the end of the table (Figure 7-99B). Pull the left knee toward the chest just enough to flatten the low back and sacrum on the table. With hip flexor tightness, the thigh will be up from the table. *Keeping back flat and the knee bent,* press the right thigh down toward the table by pulling with the buttock muscle. If stretching one-joint hip flexors only, passive extension of the knee is permitted. Reverse the procedure to stretch the left hip flexors.

An effective stretch of one-joint hip flexors can be done standing by a door frame. Place one leg forward to help brace the body against the door frame, and place the other leg back to extend the hip joint. In the starting position (Figure 7-100A), the low back will be arched because of hip flexor tightness. Keep the hip extended, and pull upward and inward with the lower abdominal muscles to tilt the pelvis posteriorly and stretch the hip flexors (Figure 7-100B). This exercise requires a *strong* pull by the abdominals and is useful in building up the strength of these muscles, which are direct opponents of the hip flexors in standing.

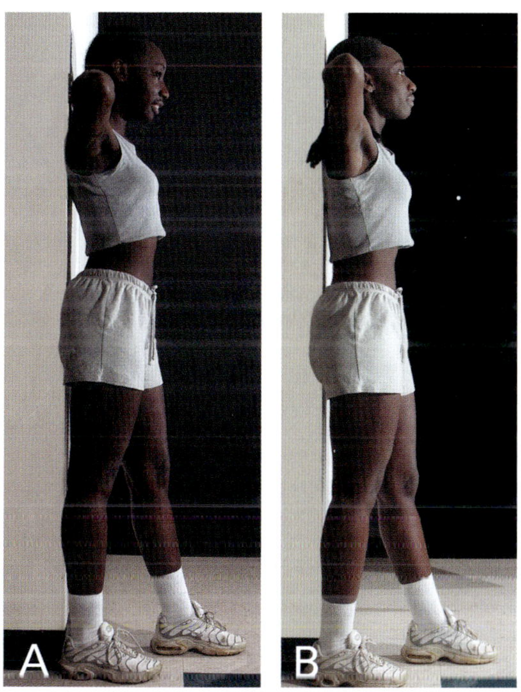

FIGURE 7-100: Exercise to stretch the one-joint hip flexors in standing: **A**, Starting position; **B**, Stretching the hip flexors by tilting the pelvis posteriorly.

HAMSTRING STRETCHING

Straight-Leg Raising

As illustrated by Figure 7-101, hamstring stretching may be performed as a passive exercise or as an active, assisted exercise. It may be performed as an active exercise if not contraindicated because of tightness of the hip flexors.

To stretch the right hamstring, lie on the table with the legs extended. Have an assistant hold the left leg down and gradually raise the right leg with the knee straight (or strap the left leg down and raise the right leg actively). Repeat on the left side.

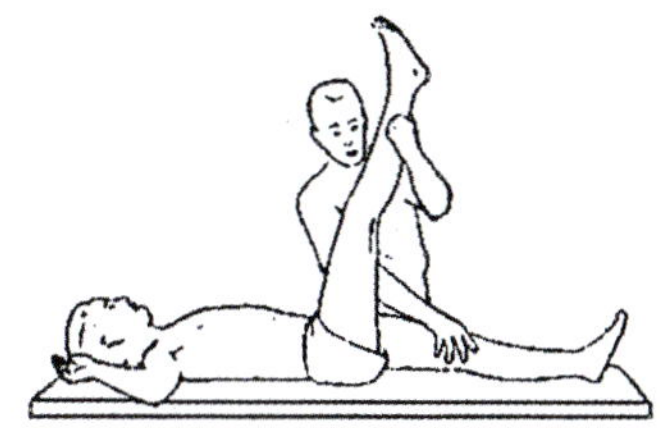

FIGURE 7-101: Straight leg raising to stretch the hamstrings.

The exercise also may be performed by putting the leg in a position that places a stretch on hamstrings, such as supine on the floor, with one leg extended, the other leg raised, and the heel resting on the back of a chair, or lying in an open doorway area, with one leg extended and the other raised and the heel resting against the wall (Figure 7-102). To increase the stretch, move the body closer to the chair or wall. *Avoid* placing both legs in the raised position at the same time, because the low back will be stretched instead of the hamstrings. Keeping one leg extended prevents excessive posterior tilt of the pelvis and excessive low back flexion.

Lie on the floor behind a sturdy chair.

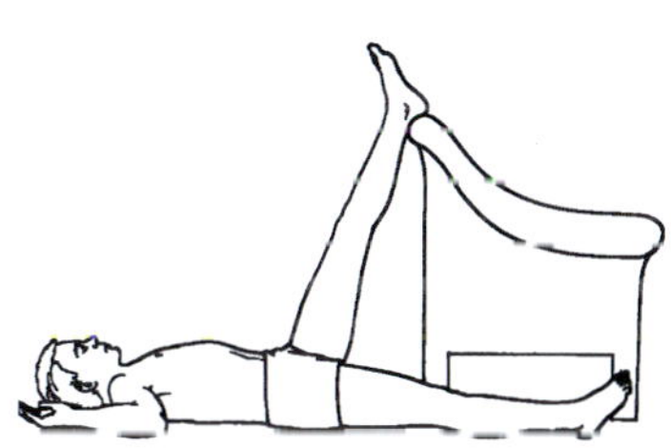

FIGURE 7-102: Stretch the hamstrings while supine on the floor behind a chair.

Knee Extension in Sitting Position

Sit with the back against a wall, as illustrated by Figure 7-103. With the back kept straight and the buttocks touching the wall, raise one leg, extending the knee as much as possible.

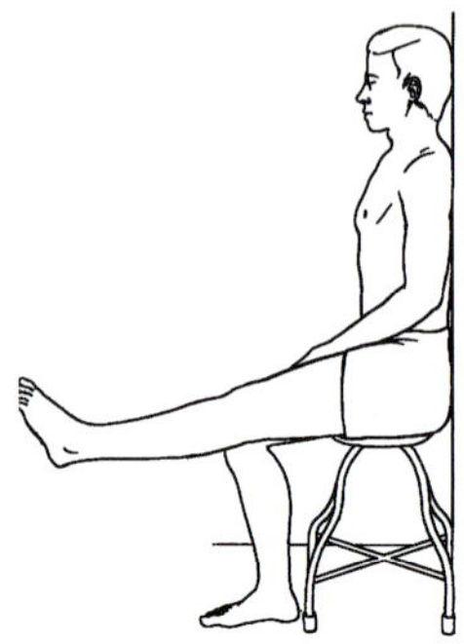

FIGURE 7-103: Knee extension in a sitting position to stretch the hamstrings.

Positions to Avoid

Avoid the standing position, with one heel on a stool or table and forward bending (Figure 7-104). For patients with pain or disability, this is a risky position. It also makes it impossible to control the pelvic position to ensure proper hamstring stretching. Furthermore, the exercise has an adverse effect on anyone with a kyphosis of the upper back. Exercise should be localized to stretching the hamstrings.

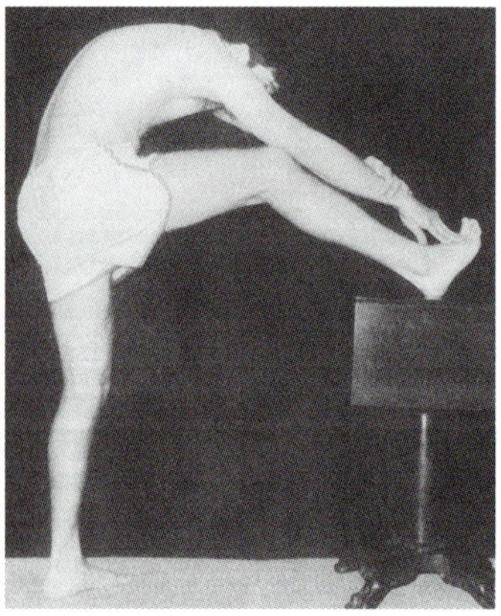

FIGURE 7-104: Avoid propping the foot in standing in an effort to stretch the hamstrings.

Avoid the "hurdler's position" for stretching the hamstrings. Excessive strain is placed on the bent knee and the low back is excessively stretched (Figure 7-105).

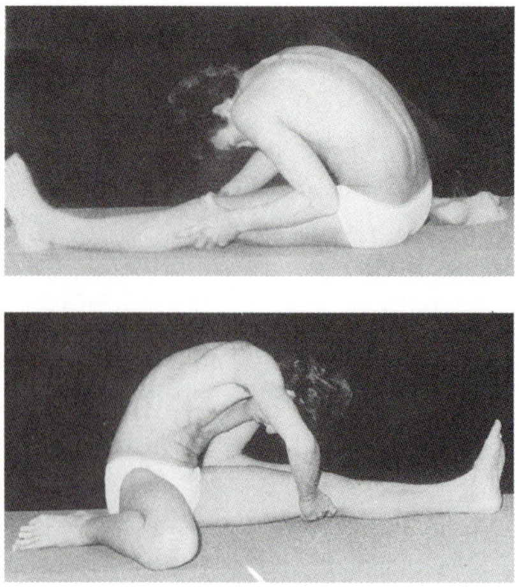

FIGURE 7-105: Avoid the "hurdler's" position while stretching the hamstring.

Avoid forward bending to stretch the hamstrings in cases with excessive flexion of the back, as seen in Figure 7-106.

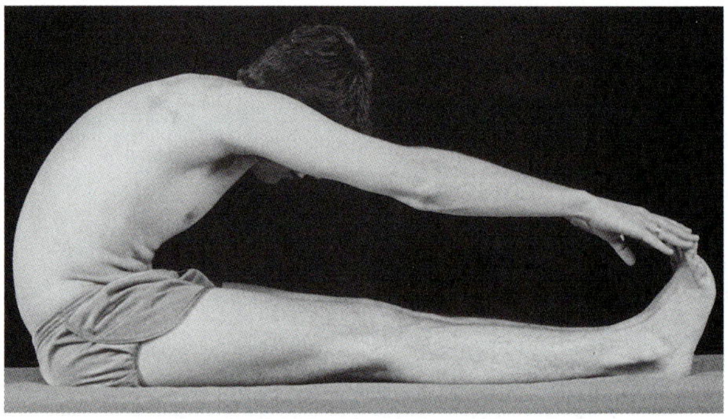

FIGURE 7-106: Avoid forward bending to stretch the hamstrings in cases with excessive flexion of the back.

MEDIAL ROTATION OF HIP AND PRONATION OF FEET

Treatment consists of shoe alterations and/or orthoses that support the longitudinal arch, exercises for foot inverters, strengthening exercises for hip lateral rotators, and stretching of the tensor fasciae latae if tight.

APPARENT LEG-LENGTH DISCREPANCY CAUSED BY MUSCLE IMBALANCE

The habit of standing with the weight mainly on one leg and the pelvis swayed sideways weakens the abductors, especially the gluteus medius on that side. If tightness of the tensor fasciae latae on one side and weakness of the gluteus medius on the other is mild, treatment may be as simple as breaking the habit and standing evenly on both feet. If the imbalance is more marked, treatment may involve stretching of the tight tensor fasciae latae and iliotibial band and use of a heel lift on the low side. The lift will help stretch the tight tensor and relieve strain on the opposite gluteus medius.

TENSOR FASCIAE LATAE STRETCHING

Tightness or even contracture of the iliotibial band is frequently seen. The relationship to painful conditions is discussed in Section IV. The following discussion concerns exercises to stretch the tensor fasciae latae and the anterolateral iliotibial band (24).

The tensor fasciae latae abducts, flexes, and internally rotates the hip joint, and it assists in knee extension. When a muscle has multiple actions, it is not necessary to elongate the muscle in all the directions opposite to its actions to stretch it. An exercise may only need to include two or three movements in the direction of stretching. Most importantly, the stretching should be specifically directed to the area in need of stretch. Some commonly prescribed exercises are not meeting this requirement.

Standing with the legs crossed puts the hip joints in adduction (Figure 7-107A). However, in this position, the hips are usually in internal rotation and in some degree of flexion by virtue of the pelvis being tilted anteriorly. If, besides standing in a position of adduction, the person sways sideways toward a wall or a table (Figure 7-107B), the stretch will often affect the posterior gluteus medius more than the tensor fasciae latae.

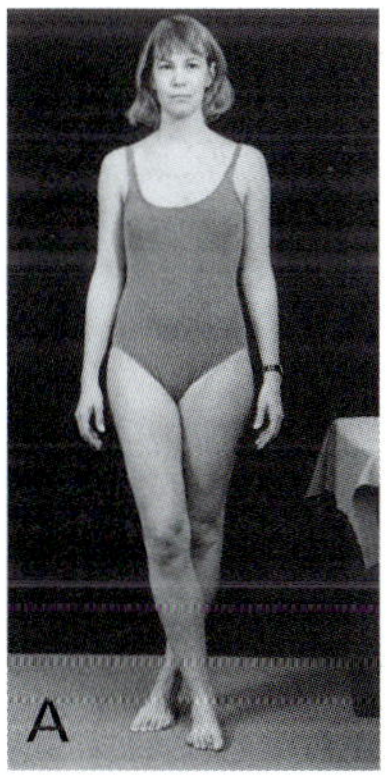

FIGURE 7-107: **A**, **B**, Apparent leg length discrepancy as a result of the habit of standing with the weight mainly on one leg and the pelvis swayed sideways.

Better control and more precision in stretching can be obtained by moving the pelvis in relation to the femur. To understand this mechanism, it is necessary to describe the effect of pelvic tilt on the hip joints.

When the legs are of equal length and the pelvis is level in standing, both hip joints are neutral as far as adduction and abduction are concerned. If the person shifts weight to one side, however, the position of the hip joints changes. Shifting weight toward the left results in adduction of the left hip joint. Likewise, if a lift is placed under the left foot (Figure 7-108), the left side of the pelvis will be elevated and the left hip joint will also be in adduction.

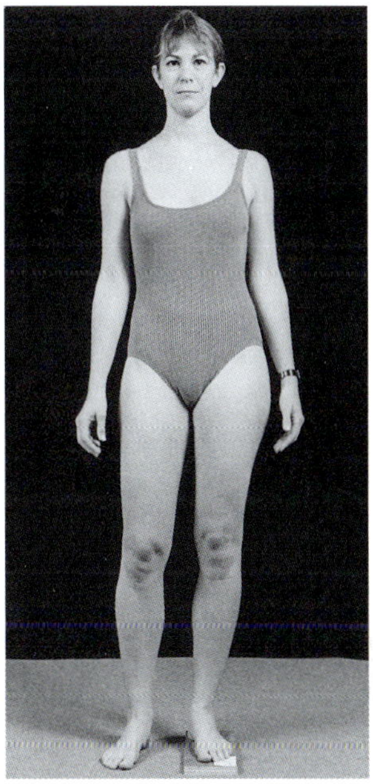

FIGURE 7-108: A lift under the left foot elevates the left side of the pelvis, creating left hip adduction.

To stretch a tight left tensor and anterior iliotibial band, stand with a board, book, or magazine under the left foot; the thickness of such a raise should be determined by the amount tolerated. Keep the weight on both feet, and keep the feet and knees (i.e., femurs) in good alignment (i.e., the feet out-toeing approximately 8°–10° on each side and the patellae facing straight ahead). Then, attempt to tilt the pelvis posteriorly. This posterior pelvic tilt results in extension of the hip joint. The range of motion will be slight, but the stretch should be felt very specifically in the area of the left tensor fasciae latae. The tensor will be stretched by adduction and extension of the hip joint without allowing internal rotation. Additionally, the stretching can be done by removing the right shoe (if the heel is not too high) instead of putting a lift under the left foot.

For bilateral tightness, place the lift alternately under the left and right, or alternately remove one shoe, and hold the stretch position for a comfortable length of time (e.g., 1–2 minutes).

When tightness is unilateral, a lift (¼-inch heel pad) in the shoe on the side of tightness will passively stretch the tensor. Make sure this lift is worn in all shoes and bedroom slippers and that the person avoids any bad habit of standing on the opposite leg. *A lift will not do any good unless the person stands with the weight distributed evenly over both feet.* (For assisted stretching of a tight tensor fasciae latae, see the section on Leg Pain.)

Stretched Tensor Fasciae Latae and Iliotibial Band

Although pain associated with a contracted tensor fasciae latae is more common, there are instances of *strain* on the high side of the pelvis. When a lower extremity is in a position of postural adduction, there is continuous tension on the abductors of the ipsilateral thigh. Symptoms of pain may become quite acute. If present, they are treated by relief of strain—that is, by leveling the pelvis and correcting any opposing muscle tightness that may be causing the persistent tension. Because the chief opponent is the opposite tensor, this problem may sometimes be resolved by treatment of the contracted muscles and fascia on the low side, even though symptoms of strain are present on the side that is higher.

There are instances in which the tensor and the fascia lata are stretched by a fall sideways or a sideways thrust in which the pelvis moves laterally on the fixed extremity, thrusting the hip joint into adduction.

On several occasions, adhesive taping has been successfully used in a way that limits the adduction. Figures 7-109 and 7-110 and the information below explain the procedure.

Adhesive tape, preferably 1½ inches wide, is cut in lengths that will extend from the area of the anterosuperior spine of the pelvis to just below the lateral knee joint.

The subject removes the shoe on the affected side, or if both shoes are removed, a raise of approximately half an inch is placed under the unaffected side. The subject stands with the feet apart to place the affected leg in some abduction. It is not expected that the tape will hold that same degree of abduction—there is always some give in the tape.

It is very important that patients be checked for skin sensitivity to adhesive tape, particularly if the weather is hot. Tincture of benzoin has been used on the skin each time the taping procedure has been used.

The tape is directed from anterolateral pelvis to posterolateral area of the knee in such a manner that hip and knee flexion will not be restricted in the sitting position (see the Brief Case History later in this chapter).

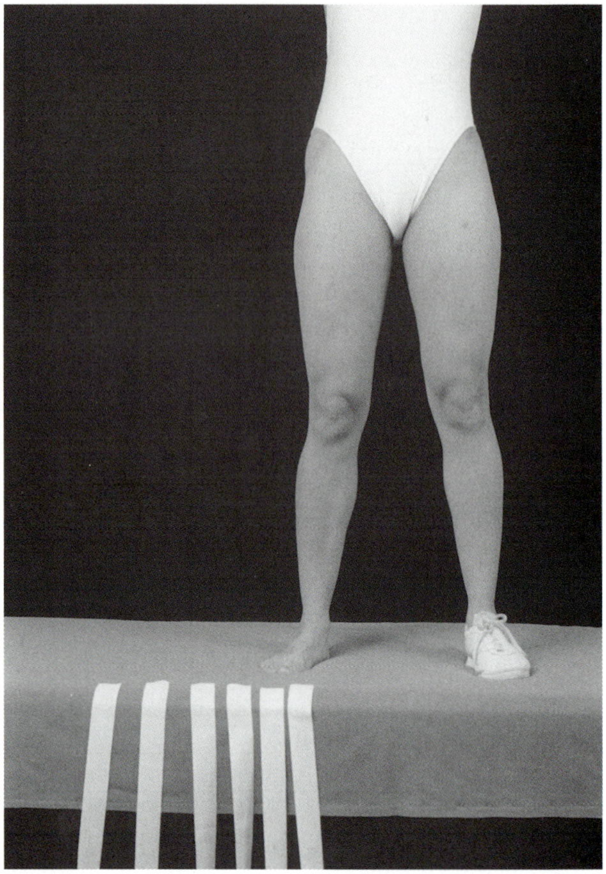

FIGURE 7-109: Stretch the tensor fasciae latae by use of an iliotibial band.

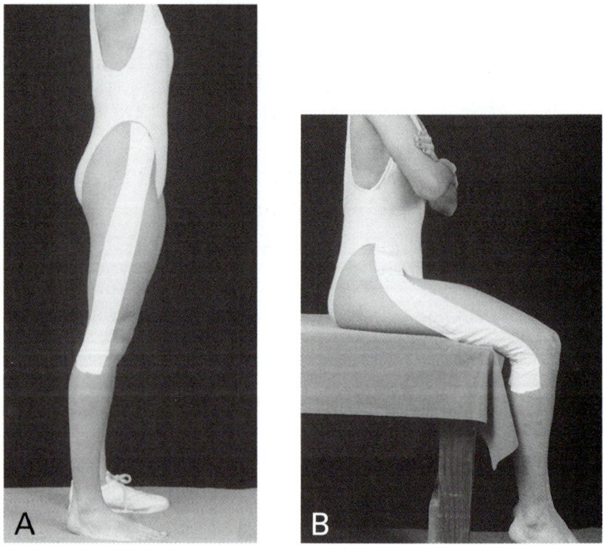

FIGURE 7-110: **A**, **B**, Adhesive taping to limit adduction while stretching the tensor fasciae latae.

Abductor Exercises

The normal range of hip joint abduction is approximately 45° and that of adduction is approximately 10°. When abductors are too weak to raise the leg in abduction against gravity in a side-lying position, *avoid* exercises in that position. A subject can learn to substitute by hiking the pelvis up laterally and bringing the lower extremity into *apparent* abduction, but doing so actually stretches and strains the abductors rather than shortening and strengthening them. Substitution also can take place in the supine position, but this can be prevented and an appropriate exercise can be done.

On a table or firm bed, the *unaffected* leg is moved in abduction to completion of the range of motion. This position will block any effort to hike the pelvis up on the *affected* side, thereby preventing substitution. Movement of the thigh in abduction will require true hip joint motion—not just a sideways movement of the extremity. Whatever assistance is appropriate may be used: Manually assist, or assist with some apparatus or adaptive measures, such as a smooth or powdered board or roller skate.

Treatment Indications for Symptoms of Acute Leg Pain

Heat may be applied to the lateral aspect of the thigh while the patient is in a position that gives in to the tightness. This is done by abducting the lower extremity in either supine or side-lying. To support the lower extremity in abduction in side-lying, firm pillows are placed between the thighs and lower legs, making sure that the foot is also supported. A pillow at the back or abdomen helps to balance the patient comfortably in this side-lying position. As soon as the patient can tolerate it, which may be during the first treatment or may be 2 or 3 days later, massage may be started. Massage should be *firm* but not deep. A superficial stroking may be more irritating than a firm, gentle massage. Massaging downward may be more effective than an upward stroke. Patients frequently describe their reaction to the massage as "a hurt that feels good." They are aware of a feeling of tightness and describe "wishing they could make the muscle let go" or that "it would feel good if somebody stretched it." Patients should avoid exposure to cold or drafts, because even the slightest exposure often causes an increase in the pain.

The almost immediate relief of symptoms that occurs in some instances indicates that the condition is basically one of tight muscles and fascia. (These treatment reactions differ from those in sciatica. The same procedures applied to the painful area along the hamstring muscles in cases of sciatic irritation would give rise to increased pain.)

For Subacute Stages: As acute pain subsides, succeeding treatments should be directed toward stretching the tight fascia. The position and movement for assisted stretching is illustrated in Figure 7-111.

Self-stretching in the standing position (as first described by Frank Ober (7)) may be done if the hip does not internally rotate or flex, but this is hard to control. More precise stretching should be used instead, as described and illustrated below.

To stretch the left tensor fasciae latae, have subject lie on right side with right hip and knee bent. Relax left lower extremity on pillows placed between thighs and lower legs. Apply heat and massage to left lateral thigh. Remove the pillows. Bend right hip and knee enough to flatten low back. Stabilize the pelvis firmly with one hand, draw the thigh slightly back, and press gently (on thigh, not leg) downward toward the table, stretching the muscles and fascia between the hip and knee. (The knee should not be allowed to rotate inward, and care should be taken to avoid strain at the knee joint.) To stretch right tensor, have subject lie on left side and reverse the procedure.

For mild to moderate unilateral tensor fasciae latae tightness, place a heel lift of 1/8- to 3/16-inch thickness in the shoe on the side of the tightness to level the pelvis and provide a gradual stretch in the standing position.

The shoe correction indicated for treatment of the lateral pelvic tilt associated with tensor fasciae latae

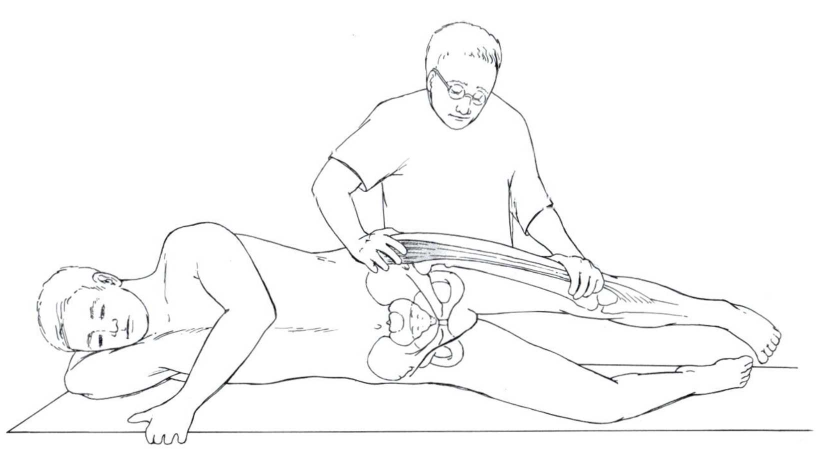

FIGURE 7-111: Position and movement for assisted stretching.

tightness also aids in gradual stretching of the tight fascia. For this reason, such shoe alterations may not be tolerated until acute symptoms have subsided and until some active treatment, in the form of heat, massage, and stretching, has been instituted to relax and stretch the tight fascia.

Brief Case History

The subject caught her right heel on the edge of a step and averted a fall down a long flight of stairs by suddenly stepping down three steps with the left leg.

At first, pain was felt in the left hip. Two days later, the left knee gave way. The left knee continued to be painful.

Four days after the injury, the patient was seen by an orthopedist who obtained a radiograph of the left knee. Five days later, the patient was seen by another orthopedist who obtained radiographs of the hip and knee.

Two weeks after the injury, the patient was seen by a neurosurgeon who recommended disk surgery.

Four days later, the patient was referred to a physical therapist. The pertinent findings were:

1. When lying supine, the patient could not let the knee extend without severe pain.
2. When the patient was placed in a sitting position and tested for quadriceps strength, the knee extended fully, with no pain of any consequence.
3. When the patient was again put in a supine position and the thigh was supported to keep the hip in flexion, the patient extended the knee without pain.
4. Any attempt to extend the knee in the supine position while also extending the hip resulted in severe pain at the knee.
5. The test for strength of the tensor fasciae latae was painful.
6. On palpation, the tensor fasciae latae muscle appeared to be in spasm.

Impression: The site of injury appeared to be in the tensor fasciae latae muscle, with pain referred to the lateral knee via the fascia lata (i.e., the muscle in spasm placing tension on the iliotibial tract whenever the hip was extended).

Following the examination, the patient was given moist heat and massage (stroking downward) to the tensor fasciae latae. The patient felt considerable relief of pain in the lying position, but pain was felt in standing.

The anterolateral aspect of the left thigh was strapped from the crest of the ilium to just below the knee (in such a way as not to interfere with hip or knee flexion). The patient felt much relief of symptoms after strapping. (Nonallergic adhesive was used.)

Two days later (and again 6 days after that), the strapping was checked to be sure there was no irritation and to reinforce it with more tape.

Three days later, no skin irritation was found, and new strapping was applied.

Six days after that visit, the patient removed the strapping and was walking without a cane.

Approximately 5 weeks after the patient removed the strapping, a note received from her doctor stated, "The examination of [the patient's] leg assures me that she is well and there has been no residual. I feel that we can discharge her, and she can assume her general duties."

The procedure for the taping was the same as illustrated in Figure 7-110A, B.

Protruded Intervertebral Disk

The basic concepts regarding flexion and extension of the spine in relation to disk protrusion play an important role in determining treatment. The following quotes are pertinent to this topic.

Nordin and Frankel state, "The forward inclination of the spine makes the disk bulge on the concave side. Hence, when the spine is flexed the disk protrudes anteriorly and is retracted posteriorly" (25). Pope et al. record the findings of Brown et al. and Roaf (26). Brown et al. reported disk bulging anteriorly during flexion, posteriorly during extension, and toward the concavity of the spinal curve during lateral bend (27). Roaf stated that the bulging of the annulus is always on the concave side of the curve and that, during flexion and extension, the nucleus does not change in shape or position (28). This information is contrary to what many people believe or have been taught. In the analysis of low back problems and sciatica, however, this concept is important.

Strong back muscles are essential for both posture and function. Although low back muscles are seldom weak, back extension exercises are frequently prescribed. Overemphasis on back extension can contribute to an increase in a lordotic position. Quoting again from Nordin and Frankel, "The erector spinae muscles are intensely activated by arching the back in the prone position. Loading the spine in extreme positions such as this one produces high stresses on spine structures, so this hyperextended position should be avoided" (25).

Good strength in the abdominal muscles is also important to counterbalance the back muscles and to stabilize the trunk in good postural alignment and during activities such as lifting. Unfortunately, abdominal muscles are often weak, especially the lower abdominals, and not enough attention is paid to appropriate exercises.

If a disk has ruptured and is pressing on a nerve root with intractable pain and no relief has been obtained from conservative measures, there may be no alternative to surgery. However, there are many cases of sciatica in which clinical findings suggest a disk lesion, but the fluctuation of symptoms suggest that the protrusion is not constant. Conservative treatment of many such cases has brought about effective relief of symptoms without surgery. In instances when the patient declines operation or the doctor does not elect to perform surgery, conservative treatment becomes the necessary alternative.

The rationale for conservative treatment is based on the premise that any bending, torsional loads, or compressive force—whether caused by muscle spasm, tightness of back muscles, or stress of superimposed weight on the lumbar spine— may be factors in causing the disk protrusion.

Two measures provide effective conservative treatment: First, immobilization of the back for relief of acute muscle spasm and for restriction of motion; second, use of an hourglass type of support that acts to transmit the weight of the thorax to the pelvis and relieves stress on the lumbar spine (in much the same manner as a cervical collar is used to relieve pressure on the cervical spine).

To treat by immobilization and for relief of superimposed body weight, a fitted support is reinforced with strong lateral and posterior stays. Following relief of acute symptoms, therapeutic measures may be instituted to correct any underlying muscle imbalance or faults in alignment.

Acute sciatic symptoms associated with protrusion of a ruptured disk often occur as a result of a sudden twist and extension of the spine from a forward-bent position, such as twisting the trunk while lifting a weight. That such a type of stress should be related to this type of lesion is not surprising in view of the fact that "rotation of the lumbar spine takes place at the intervertebral disk" (29).

Sciatic symptoms that have been acute or subacute often cause the body to be drawn into faulty alignment such that secondary symptoms of compression and muscle strain are added to the original problem. These secondary symptoms may, on occasion, persist after the original underlying problems have subsided.

Piriformis Muscle and Its Relation to Sciatic Pain

Albert Freiberg described the piriformis muscle and its relation to sciatic pain and furnished an interesting explanation for a possible cause of sciatic symptoms (30). Although there may be numerous cases in which sciatic pain is associated with a *contracted* piriformis, as he described, it was the opinion of previous authors that irritation of the sciatic nerve by the piriformis muscle is often associated with a *stretched* piriformis.

The piriformis arises with a broad origin from the anterior aspect of the sacrum and inserts into the superior border of the greater trochanter. This muscle has three functions in standing. It acts as an external rotator of the femur, aids slightly in tilting the pelvis down laterally, and aids in tilting the pelvis posteriorly by pulling the sacrum downward toward the thigh.

In a faulty position with a leg in postural adduction and internal rotation in relation to an anteriorly tilted pelvis, there is marked stretching of the piriformis along with other muscles that function in a similar manner. The mechanics of this position are such that the piriformis muscle and the sciatic nerve are thrust into close contact. Figure 7-112 shows the relationship of the sciatic nerve to the piriformis muscle.

Evaluation: The following points should be considered in the diagnosis of sciatic pain associated with a stretched piriformis (Figure 7-112):

1. Do the sciatic symptoms diminish or disappear in non-weight-bearing?
2. Do internal rotation and adduction of the thigh in the flexed position, with the patient supine, increase sciatic symptoms?
3. Do the symptoms diminish in standing if a straight raise is placed under the opposite foot?
4. Does the patient seek relief of symptoms by placing the leg in external rotation and abduction in both the lying and standing positions?

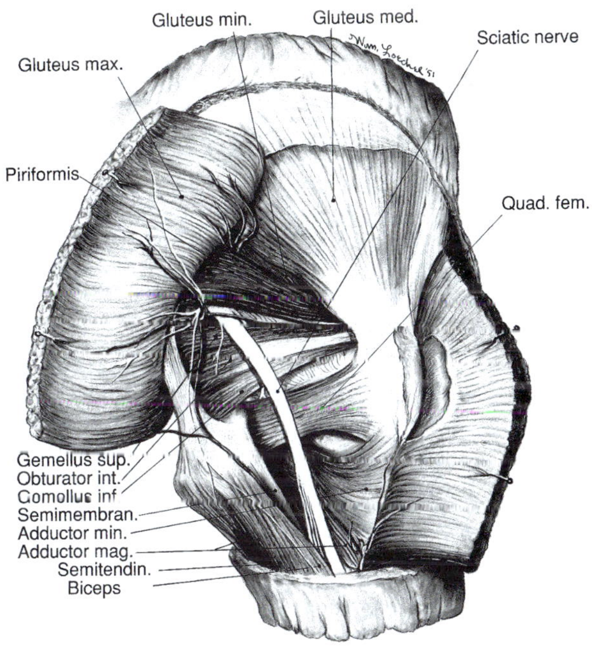

FIGURE 7-112: Relationship of the sciatic nerve to the piriformis muscle.

The test movement to place the piriformis on maximum stretch (see point 2 above) is done in the following manner: The patient is supine on a table. The knee and hip of the affected leg are flexed to right angles. Flexion of the knee rules out any confusion with pain due to irritation of the hamstring muscles. The examiner then internally rotates and adducts the thigh passively.

In regard to point 3 above, it has been a frequent clinical observation that during the course of examination, a lift applied under the foot of the affected side increases symptoms, whereas a lift placed under the foot of the unaffected side gives some immediate relief in the affected leg.

Shoe corrections for cases of irritation resulting from a stretched, rather than a contracted, piriformis, consist of a straight raise (usually ⅛ to ¼ inch) on the heel of the *unaffected* side to relieve tension on the abductors of the affected side as well as an inner wedge on the heel on the *affected* side to correct the internal rotation of the leg. Heat, massage, and stretching of the low back muscles if they are contracted, abdominal muscle exercise if abdominal weakness is present, and correction of the faulty position of the pelvis in standing are used as indicated.

Sciatica

Sciatica refers to a neuritic type of pain along the course of the sciatic nerve. Pain extends down the posterior thigh and lower leg to the sole of the foot and along the lateral aspect of the lower leg to the dorsum of the foot.

Sciatica may occur in connection with various infections or inflammatory disease processes, or it may be caused by mechanical compression or tension.

Symptoms may originate from a lesion of one or more of the nerve roots that later join through a plexus to form the sciatic nerve. A protruded intervertebral disk is an example of mechanical irritation at the level where the nerve roots emerge from the spinal canal. Distribution of pain tends to extend from the root origin to the terminal nerve endings, with the result that pain is quite widespread. An L5 lesion, for example, may give rise not only to symptoms down the course of the sciatic nerve but also to pain in the region of the posterior and lateral thigh supplied by the inferior and superior gluteal nerves.

Symptoms of sciatica may arise from irritation anywhere along the course of the sacral plexus, the sciatic nerve trunk, or its peripheral nerve branches. Sciatica may arise as reflex pain from irritation of peripheral nerve endings. A lesion along the course of the nerve or its branches often may be distinguished from a root lesion by the localization of pain to the distribution below the level of the lesion.

Other than the root, there are two commonly recognized sites of lesions giving rise to sciatic pain: The sacroiliac region, where the spinal nerves emerge through the sacral foramen, and at the level of the piriformis muscle where the sciatic nerve trunk emerges through the sciatic notch and passes either through or under the piriformis muscle.

This discussion about sciatica is concerned with faulty body mechanics in relation to disk protrusion and with sciatic symptoms associated with the piriformis syndrome. There will be no discussion of sciatica in relation to the sacroiliac strain other than to suggest that the faulty mechanics causing this strain may put tension on the sacral plexus because of the close association of the involved structures in this area.

NEUROMUSCULAR PROBLEMS

The use of manual muscle testing and the accurate recording of the results aid in establishing a diagnosis, as illustrated by the six cases illustrated in Figures 7-113 through 7-118. For figures in Cases 1 through 3, test results are recorded in the left-hand column, and the corresponding dots to the right are circled to indicate the nerves involved.

CASE 1: FIBULAR NERVE INJURY

Trunk and Lower Extremity

Name Date

		Muscle strength grade	Left leg muscle	D. T1-12, L1-5, S1-3	V. Spinal segment T1, 2, 3, 4	V. T5, 6	V. T7, 8	V. T9, 10, 11, 12	V. Iliohypogastric T12 L1	V. Ilioinguinal T(12) L1	V. Lumb. Plex. T(12) L1, 2, 3, 4	P. Femoral L(1) 2, 3, 4	A. Obturator L(1) 2, 3, 4	P. Sup. Glut. L4, 5, S1	P. Inf. Glut. L5, S1, 2	V. Sac. Plex. L4, 5, S1, 2, 3	P. Sciatic L4, 5, S1, 2	A. Sciatic L4, 5, S1, 2, 3	P. C. Peroneal L4, 5, S1, 2	A. Tibial L4, 5, S1, 2, 3	Sensory
			ERECTOR SPINAE	•																	
Thoracic Nerves			SERRATUS POST SUP		•																
			TRANS THORACIS		•	•	•														
			INT INTERCOSTALS		•	•	•	•													
			EXT INTERCOSTALS		•	•	•	•													
			SUBCOSTALES		•	•	•	•													
			LEVATOR COSTARUM		•	•	•	•													
			OBLIQUUS EXT ABD			(•)	•	•													
			RECTUS ABDOMINIS			•	•	•													
			OBLIQUUS INT ABD				•	•	•	(•)											
			TRANSVERSUS ABD				•	•	•	(•)											
			SERRATUS POST INF					•													
Lumb. Plexus			QUAD LUMBORUM								•										
			PSOAS MINOR								•										
			PSOAS MAJOR								•										
Femoral			ILIACUS									•									
			PECTINEUS									•	(•)								
			SARTORIUS									•									
			QUADRICEPS									•									
Obturator	Ant.		ADDUCTOR BREVIS										•								
			ADDUCTOR LONGUS										•								
			GRACILIS										•								
	Post		OBTURATOR EXT										•								
			ADDUCTOR MAGNUS										•					•			
Gluteal	Sup		GLUTEUS MEDIUS											•							
			GLUTEUS MINIMUS											•							
			TENSOR FAS LAT											•							
	In.		GLUTEUS MAXIMUS												•						
Sacral Plexus			PIRIFORMIS													•					
			GEMELLUS SUP													•					5 1 2
			OBTURATOR INT													•					5 1 2
			GEMELLUS INF													•					4 5 1 (2)
			QUADRATUS FEM													•					4 5 1 (2)
Sciatic	P		BICEPS (SHORT H)														•				5 1 2
	Tibial		BICEPS (LONG H)															•			5 1 2 3
			SEMITENDINOSUS															•			4 5 1 2
			SEMIMEMBRANOSUS															•			4 5 1 2
Common Peroneal	Deep	3	TIBIALIS ANTERIOR																(•)		4 5 1
		0	EXT HALL LONG																(•)		4 5 1
		0	EXT DIGIT LONG																(•)		4 5 1
		0	PERONEUS TERTIUS																(•)		4 5 1
		0	EXT DIGIT BREVIS																(•)		4 5 1
	Sup	2	PERONEUS LONGUS																(•)		4 5 1
		2	PERONEUS BREVIS																(•)		4 5 1
Tibial	Tibial	—	PLANTARIS																	•	4 5 1 (2)
		10	GASTROCNEMIUS																	•	1 2
		—	POPLITEUS																	•	4 5 1
		10	SOLEUS																	•	5 1 2
		7	TIBIALIS POSTERIOR																	(•)	(4) 5 1
		0	FLEX DIGIT LONG																	(•)	5 1 (2)
		0	FLEX HALL LONG																	(•)	5 1 2
	Med Pl	10	FLEX DIGIT BREVIS																	•	4 5 1
		—	ABDUCTOR HALL																	•	4 5 1
		10	FLEX HALL BREVIS																	•	4 5 1
		8	LUMBRICALIS I																	•	4 5 1
	Lat Plant	—	ABD DIGITI MIN																	•	1 2
		—	QUAD PLANTAE																	•	1 2
		—	FLEX DIGITI MIN																	•	1 2
		—	OPP. DIGITI MIN																	•	1 2
		—	ADDUCTOR HALL																	•	1 2
		—	PLANT INTEROSSEI																	•	1 2
		—	DORSAL INTEROSSEI																	•	1 2
		8	LUMB II,III,IV																	•	(4) (5) 1 2

KEY
D Dorsal Primary Ramus
V Ventral Primary Ramus
A Anterior Division
P Posterior Division

SENSORY

The patient, on whom muscle and sensory tests were done 6 weeks after onset, had fallen through a glass door and sustained a laceration injury of the left leg. Muscle test findings indicated the following:

- Involvement of the nerve branches to the flexor digitorum longus and flexor hallucis longus, without involvement of the tibial nerve and its terminal branches.
- Involvement of the superficial fibular nerve and of the deep fibular nerve, probably below the level of a proximal branch to the tibialis anterior.

The weakness of the posterior tibial muscle may have been caused by trauma of the muscle rather than by nerve involvement because it made a complete recovery within 3.5 months of onset. By that time, the flexor digitorum longus and flexor hallucis longus had made a good recovery, and by the end of 6 months they had made a complete recovery. Progress was slow, and muscle weakness remained in all muscles supplied by the deep and superficial fibular nerves.

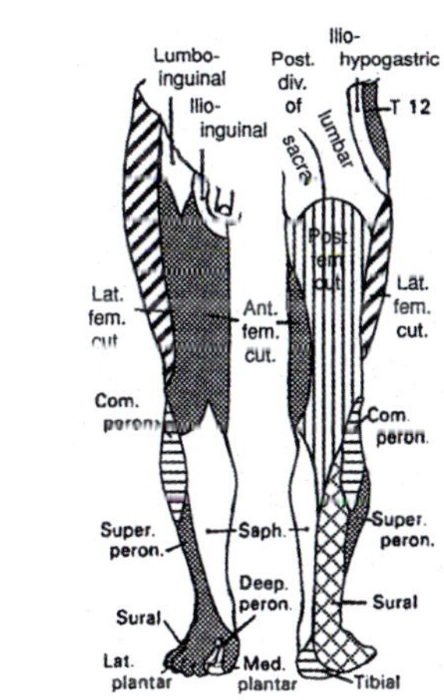

Dermatomes redrawn from Keegan and Garrett Anat Rec 102. 409. 437. 1948
Cutaneous Distribution of peripheral nerves redrawn from *Gray's Anatomy of the Human Body.* 28th ed

FIGURE 7-113: Case 1: Peripheral nerve involvement of the common fibular nerve.

CASE 2: LESION INVOLVING LUMBOSACRAL NERVES

Diagnostic Chart for Nerve Lesions: Trunk and Lower Extremity

Name ____________________ Date ____________________

MOTOR — *Left leg*

	MUS. STREN. GRADE	MUSCLE	T 1-12, L1-5, S1-5, Dorsal	VENTRAL DIVISION: T 2-6	T 5-6	T 7-11	T 9-11	Iliohypogastric (L1)	Ilioinguinal (L1)	Lumb. Plex. (T12, L1-4) V.	Obturator (L2,3,4) Vent.	Femoral (L2,3,4) Dorsal	Sup. Gluteal (L4,5, S1) D.	Inf. Gluteal (L5, S1,2) D.	Sac. Plex. (L4,5 S1,2,3)	Sciatic (L4,5,S1,2,3) D.&V.	Tibial (Int. Popliteal) V.	Peroneal (Ext. Popliteal) D.
	100	ERECTOR SPINAE	x															
THORACIC NERVES		INT. INTERCOSTALS		x														
		EXT. INTERCOSTALS		x														
		SUBCOSTALES		x														
		LEVATOR COSTARUM		x														
		SERRATUS POST. SUP.		x														
		TRANS. THORACIS		x														
		EXTERNAL OBLIQUE			x													
		RECTUS ABDOMINIS				x												
		DIAPHRAGM				x												
		INTERNAL OBLIQUE				x		x	(x)									
		TRANSVERSUS ABD.				x		x	(x)									
		SERRATUS POST. INF.					x											
LUMBAR PLEXUS		PSOAS MINOR								x								
	100	PSOAS MAJOR								x								
		ILIACUS								x								
		QUADRATUS LUMBORUM								x								
OBTURATOR		GRACILIS									x							
		ADDUCTOR BREVIS									x							
	100	ADDUCTOR LONGUS									x							
		ADDUCTOR MAGNUS									x							
		OBTURATOR EXTERNUS									x							
		PECTINEUS									x	x						
FEM.	100	SARTORIUS										x						
	100	QUADRICEPS										x						
SUP. GLUT.	70	GLUTEUS MEDIUS											(x)					
	70	GLUTEUS MINIMUS											(x)					
	70	TENSOR FASCIA LAT.											(x)					
INF. GL.	70	GLUTEUS MAXIMUS												(x)				
SACRAL PLEXUS		PIRIFORMIS													(x)			
		QUADRATUS FEMORIS													(x)			
	60	GEMELLUS SUPERIOR													(x)			
		GEMELLUS INFERIOR													(x)			
		OBTURATOR INTERNUS													(x)			
SCIATIC — TIBIAL	70	SEMI-MEMBRANOSUS														(x)		
	70	SEMI-TENDINOSUS														(x)		
	60	BICEPS (LONG HEAD)														(x)		
SCIATIC — P.	60	BICEPS (SHORT HEAD)														(x)		
TIBIAL (INTERNAL POPLITEAL) — TIBIAL	?	GASTROCNEMIUS	*Tendon has been lengthened*															
	—	PLANTARIS															x	
	—	POPLITEUS															(x)	
	?	SOLEUS	*Tendon has been lengthened*														(x)	
	0	TIBIALIS POSTICUS															(x)	
	0	FLEX. DIGIT. LONG.															(x)	
	0	FLEX. HALL. LONG.															(x)	
MED. PLANT.	0	FLEX. DIGIT. BREVIS															(x)	
	0	FLEX. HALL. BREVIS															(x)	
	0	ABDUCTOR HALLUCIS															(x)	
	—	LUMBRICALES (1 & 2)															—	
LAT. PLANTAR	—	LUMBRICALES (3 & 4)															—	
	—	DORSAL INTEROSSEI															—	
	—	PLANTAR INTEROSSEI															—	
	—	QUADRATUS PLANTAE															—	
	—	FLEX. DIGITI QUINTI															—	
	—	ABD. DIGITI QUINTI															—	
PERONEAL (EXT. POP.) — SUP.	(60)	PERONEUS LONGUS																x
	(60)	PERONEUS BREVIS																x
DEEP	(60)	PERONEUS TERTIUS																x
	10	TIBIALIS ANTICUS																x
	10	EXT. DIGIT. LONG.																x
	0	EXT. HALL. LONG.																x
	0	EXT. DIGIT. BREVIS																x

SPINAL SEGMENT

MUSCLE	T 2,3,4	T 5,6	T 7,8	T 9,10,11	T 12	L 1	L 2	L 3	L 4	L 5	S 1	S 2	S 3
ERECTOR SPINAE	x	x	x	x	x	x	x	x	x	x	x	x	x
INT. INTERCOSTALS	x	x											
EXT. INTERCOSTALS	x	x											
SUBCOSTALES	x	x											
LEVATOR COSTARUM	x	x											
SERRATUS POST. SUP.	x	x											
TRANS. THORACIS	x	x											
EXTERNAL OBLIQUE		x											
RECTUS ABDOMINIS			x	x									
DIAPHRAGM			x	x									
INTERNAL OBLIQUE			x	x		x							
TRANSVERSUS ABD.			x	x		x							
SERRATUS POST. INF.				x									
PSOAS MINOR						x							
PSOAS MAJOR						(x)	X	X	(x)				
ILIACUS						(x)	X	X	(x)				
QUADRATUS LUMBORUM					x	x	x						
GRACILIS							X	X	x				
ADDUCTOR BREVIS							X	X	x				
ADDUCTOR LONGUS							X	X	x				
ADDUCTOR MAGNUS							X	X	x				
OBTURATOR EXTERNUS								x	x				
PECTINEUS							x	x	x				
SARTORIUS							X	x	x				
QUADRICEPS							x	X	X				
GLUTEUS MEDIUS									X	X	X	(x)	
GLUTEUS MINIMUS									X	X	X	(x)	
TENSOR FASCIA LAT.									X	X	X	(x)	
GLUTEUS MAXIMUS										X	X	x	
PIRIFORMIS											x	x	
QUADRATUS FEMORIS									(x)	x	x		
GEMELLUS SUPERIOR										(x)	x	x	x
GEMELLUS INFERIOR									(x)	x	x		
OBTURATOR INTERNUS										(x)	x	x	x
SEMI-MEMBRANOSUS									x	X	X	(x)	(x)
SEMI-TENDINOSUS									(x)	X	X	x	(x)
BICEPS (LONG HEAD)									(x)	X	X	x	x
BICEPS (SHORT HEAD)									(x)	X	X	x	x
GASTROCNEMIUS											x	X	
PLANTARIS									x	x	x		
POPLITEUS									x	x	x		
SOLEUS											x	x	
TIBIALIS POSTICUS									X	x	x		
FLEX. DIGIT. LONG.										x	x	x	
FLEX. HALL. LONG.									(x)	x	x	(x)	
FLEX. DIGIT. BREVIS									x	x	x	x	(x)
FLEX. HALL. BREVIS									x	x	x	x	(x)
ABDUCTOR HALLUCIS									x	x	x	x	(x)
LUMBRICALES (1 & 2)									x	x	x	x	(x)
LUMBRICALES (3 & 4)									x	x	x	x	(x)
DORSAL INTEROSSEI									x	x	x	x	(x)
PLANTAR INTEROSSEI									x	x	x	x	(x)
QUADRATUS PLANTAE									x	x	x	x	(x)
FLEX. DIGITI QUINTI									x	x	x	x	(x)
ABD. DIGITI QUINTI									x	x	x	x	(x)
PERONEUS LONGUS									x	X	X	(x)	
PERONEUS BREVIS									x	X	X	(x)	
PERONEUS TERTIUS									x	x	x	(x)	
TIBIALIS ANTICUS									X	x	x	(x)	
EXT. DIGIT. LONG.									x	X	x	(x)	
EXT. HALL. LONG.									x	X	x	(x)	
EXT. DIGIT. BREVIS									x	x	X	(x)	

Note: Lesion involves lumbo-sacral nerves, (L4+5, S1,2,3) on left side only, with slightly more severe involvement of ventral than of dorsal division.)

SENSORY

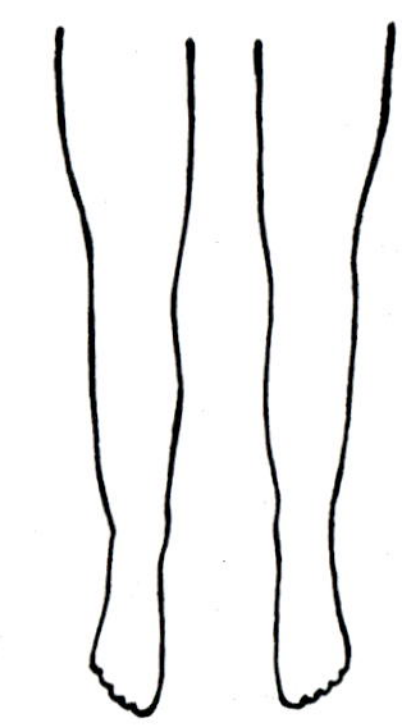

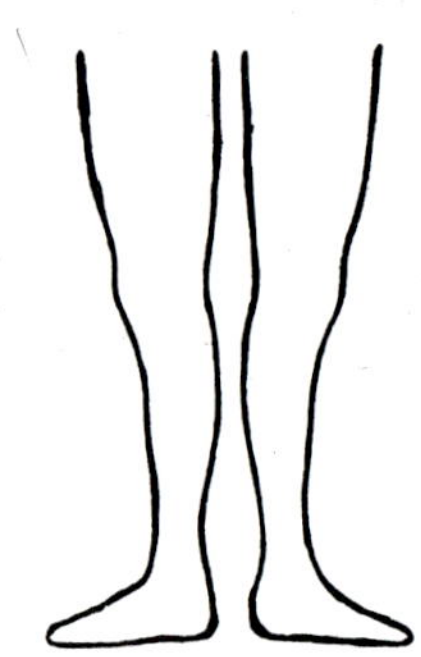

FIGURE 7-114: Case 2: Involvement of the dorsal and ventral divisions of L4, 5, and S1, 2, 3 on one side only. (The other side essentially normal.)

Name ______________________ Date ______________________

MOTOR

	Mus. Stren. Grade	Muscle (*Right leg*)	T 1-12, L1-5, S1-5, Dorsal	Ventral Division: T 2-6	Ventral Division: T 5-6	Ventral Division: T 7-11	Ventral Division: T 9-11	Iliohypogastric (L1)	Ilioinguinal (L1)	Lumb. Plex. (T12, L1-4) V.	Obturator (L2,3,4,) Vent.	Femoral (L2,3,4,) Dorsal	Sup. Gluteal (L4,5, S1) D.	Inf. Gluteal (L5, S1,2) D.	Sac. Plex. (L4,5 S1,2,3)	Sciatic (L4,5,S1,2,3) D.&V.	Tibial (Int. Popliteal) V.	Peroneal (Ext. Popliteal) D.
	100	ERECTOR SPINAE	x															
THORACIC NERVES		INT. INTERCOSTALS		x														
		EXT. INTERCOSTALS		x														
		SUBCOSTALES		x														
		LEVATOR COSTARUM		x														
		SERRATUS POST. SUP.		x														
		TRANS. THORACIS		x														
		EXTERNAL OBLIQUE			x													
		RECTUS ABDOMINIS				x												
		DIAPHRAGM				x												
		INTERNAL OBLIQUE				x		x	(x)									
		TRANSVERSUS ABD.				x		x	(x)									
		SERRATUS POST. INF.					x											
LUMBAR PLEXUS	100	PSOAS MINOR								x								
	100	PSOAS MAJOR								x								
	100	ILIACUS								x								
	—	QUADRATUS LUMBORUM								x								
OBTURATOR	100	GRACILIS									x							
	100	ADDUCTOR BREVIS									x							
	100	ADDUCTOR LONGUS									x							
	100	ADDUCTOR MAGNUS									x							
	100	OBTURATOR EXTERNUS									x							
	100	PECTINEUS									x	x						
FEM.	100	SARTORIUS										x						
	100	QUADRICEPS										x						
SUP. GLUT.	60	GLUTEUS MEDIUS											x					
	60	GLUTEUS MINIMUS											x					
	80	TENSOR FASCIA LAT.											x					
IN. GL.	100	GLUTEUS MAXIMUS												x				
SACRAL PLEXUS	70	PIRIFORMIS													x			
	70	QUADRATUS FEMORIS													x			
	70	GEMELLUS SUPERIOR													x			
	70	GEMELLUS INFERIOR													x			
	70	OBTURATOR INTERNUS													x			
SCIATIC — TIBIAL	100	SEMI-MEMBRANOSUS														x		
	100	SEMI-TENDINOSUS														x		
	100	BICEPS (LONG HEAD)														x		
SCIATIC — P.	100	BICEPS (SHORT HEAD)														x		
TIBIAL (INTERNAL POPLITEAL) — TIBIAL	100	GASTROCNEMIUS															x	
	—	PLANTARIS															x	
	—	POPLITEUS															x	
	100	SOLEUS															x	
	100	TIBIALIS POSTICUS															x	
	100	FLEX. DIGIT. LONG.															x	
	100	FLEX. HALL. LONG.															x	
MED. PLANT.	100	FLEX. DIGIT. BREVIS															x	
	100	FLEX. HALL. BREVIS															x	
	—	ABDUCTOR HALLUCIS															x	
	100	LUMBRICALES (1 & 2)															x	
LAT. PLANTAR	100	LUMBRICALES (3 & 4)															x	
	—	DORSAL INTEROSSEI															—	
	—	PLANTAR INTEROSSEI															—	
	—	QUADRATUS PLANTAE															—	
	—	FLEX. DIGITI QUINTI															—	
	—	ABD. DIGITI QUINTI															—	
PERONEAL (EXT. POP.) — SUP.	100	PERONEUS LONGUS																x
	100	PERONEUS BREVIS																x
DEEP	100	PERONEUS TERTIUS																x
	100	TIBIALIS ANTICUS																x
	100	EXT. DIGIT. LONG.																x
	100	EXT. HALL. LONG.																x
	100	EXT. DIGIT. BREVIS																x

SPINAL SEGMENT

Muscle	T 2,3,4	T 5,6	T 7,8	T 9,10,11	T 12	L 1	L 2	L 3	L 4	L 5	S 1	S 2	S 3
ERECTOR SPINAE	x	x	x	x	x	x	x	x	x	x	x	x	x
INT. INTERCOSTALS	x	x											
EXT. INTERCOSTALS	x	x											
SUBCOSTALES	x	x											
LEVATOR COSTARUM	x	x											
SERRATUS POST. SUP.	x	x											
TRANS. THORACIS	x	x											
EXTERNAL OBLIQUE		x											
RECTUS ABDOMINIS			x	x									
DIAPHRAGM			x	x									
INTERNAL OBLIQUE			x	x		x							
TRANSVERSUS ABD.			x	x		x							
SERRATUS POST. INF.				x									
PSOAS MINOR						x							
PSOAS MAJOR						(x)	X	X	(x)				
ILIACUS						(x)	X	X	(x)				
QUADRATUS LUMBORUM					x	x	x						
GRACILIS							X	X	x				
ADDUCTOR BREVIS							X	X	x				
ADDUCTOR LONGUS							X	X	x				
ADDUCTOR MAGNUS							X	X	x				
OBTURATOR EXTERNUS								x	x				
PECTINEUS							x	x	x				
SARTORIUS							X	x	x				
QUADRICEPS							x	X	X				
GLUTEUS MEDIUS									X	X	X	(x)	
GLUTEUS MINIMUS									X	X	X	(x)	
TENSOR FASCIA LAT.									X	X	X	(x)	
GLUTEUS MAXIMUS										X	X	x	
PIRIFORMIS											x	x	
QUADRATUS FEMORIS									(x)	x	x		
GEMELLUS SUPERIOR										(x)	x	x	x
GEMELLUS INFERIOR									(x)	x	x		
OBTURATOR INTERNUS										(x)	x	x	x
SEMI-MEMBRANOSUS									x	X	X	(x)	(x)
SEMI-TENDINOSUS									(x)	X	X	x	(x)
BICEPS (LONG HEAD)									(x)	X	X	x	x
BICEPS (SHORT HEAD)									(x)	X	X	x	x
GASTROCNEMIUS											x	X	
PLANTARIS									x	x	x		
POPLITEUS									x	x	x		
SOLEUS											x	x	
TIBIALIS POSTICUS									X	x	x		
FLEX. DIGIT. LONG.										x	x	x	
FLEX. HALL. LONG.									(x)	x	x	(x)	
FLEX. DIGIT. BREVIS									x	x	x	x	(x)
FLEX. HALL. BREVIS									x	x	x	x	(x)
ABDUCTOR HALLUCIS									x	x	x	x	(x)
LUMBRICALES (1 & 2)									x	x	x	x	(x)
LUMBRICALES (3 & 4)									x	x	x	x	(x)
DORSAL INTEROSSEI									x	x	x	x	(x)
PLANTAR INTEROSSEI									x	x	x	x	(x)
QUADRATUS PLANTAE									x	x	x	x	(x)
FLEX. DIGITI QUINTI									x	x	x	x	(x)
ABD. DIGITI QUINTI									x	x	x	x	(x)
PERONEUS LONGUS									x	X	X	(x)	
PERONEUS BREVIS									x	X	X	(x)	
PERONEUS TERTIUS									x	x	x	(x)	
TIBIALIS ANTICUS									X	x	x	(x)	
EXT. DIGIT. LONG.									x	X	x	(x)	
EXT. HALL. LONG.									x	X	x	(x)	
EXT. DIGIT. BREVIS									x	x	X	(x)	

Note: The abductor and external rotator weakness in this right leg is undoubtedly weight-bearing weakness, secondary to the involvement of the left leg.

SENSORY

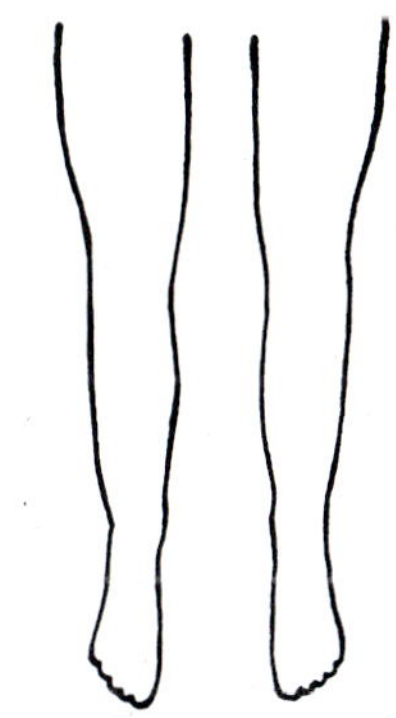

Post.: Left or Ant.: Right   Post.: Right or Ant.: Left

Lat.: Left or Med.: Right   Lat.: Right or Med.: Left

FIGURE 7-114: *(Continued)*

CASE 3: POSSIBLE L5 LESION

Trunk and Lower Extremity

Muscle test findings indicate a possible L5 lesion. Numerous muscles that receive innervation from L4 were normal in strength, leading to the assumption that L4 was not involved. The patient was able to stand on one foot at a time and rise on the toes without any difficulty, hence the normal grade for the gastrocnemius. With the innervation to this muscle from S1 and S2, the grade of normal rules out a disk below L5.

Subsequent examination by a neurologist confirmed a probable disk lesion, and the patient had a complete recovery.

Name Date

KEY: D Dorsal Primary Ramus; V Ventral Primary Ramus; A Anterior Division; P Posterior Division

		MUSCLE STRENGTH GRADE	RIGHT LEG MUSCLE	D. T1-12, L1-5, S1-3	V. T1, 2, 3, 4	V. T5, 6	V. T7, 8	V. T9, 10, 11, 12	V. Iliohypogastric T12 L1	V. Ilioinguinal T(12) L1	V. Lumb. Plex. T(12) L1, 2, 3, 4	P. Femoral L(1) 2, 3, 4	A. Obturator L(1) 2, 3, 4	P. Sup. Glut. L4, 5, S1	P. Inf. Glut. L5, S1, 2	V. Sac. Plex. L4, 5, S1, 2, 3	P. Sciatic L4, 5, S1, 2	A. Sciatic L4, 5, S1, 2, 3	P. C. Peroneal L4, 5, S1, 2	A. Tibial L4, 5, S1, 2, 3	L1	L2	L3	L4	L5	S1	S2	S3
			ERECTOR SPINAE	•																								
Thoracic Nerves			SERRATUS POST SUP		•																							
			TRANS THORACIS		•	•	•																					
			INT INTERCOSTALS		•	•	•	•																				
			EXT INTERCOSTALS		•	•	•	•																				
			SUBCOSTALES		•	•	•	•																				
			LEVATOR COSTARUM		•	•	•	•																				
			OBLIQUUS EXT ABD			(•)	•	•																				
			RECTUS ABDOMINIS			•	•	•																				
			OBLIQUUS INT ABD				•	•																				
			TRANSVERSUS ABD				•	•																				
			SERRATUS POST INF					•																				
Lumb.	Plexus	—	QUAD LUMBORUM																									
		—	PSOAS MINOR																									
		10	PSOAS MAJOR																									
Femoral		10	ILIACUS																									
		—	PECTINEUS																									
		10	SARTORIUS																									
		10	QUADRICEPS																									
Obturator	Ant.		ADDUCTOR BREVIS										•									2	3	4				
			ADDUCTOR LONGUS										•									2	3	4				
		10	GRACILIS										•									2	3	4				
	Post		OBTURATOR EXT										•										3	4				
			ADDUCTOR MAGNUS										•					•				2	3	4	5	1		
Gluteal	Sup	4	GLUTEUS MEDIUS											•										4	5	1		
		4	GLUTEUS MINIMUS											•										4	5	1		
		6	TENSOR FAS LAT											•										4	5	1		
	In.	6	GLUTEUS MAXIMUS												•										5	1	2	
Sacral	Plexus		PIRIFORMIS													•									(5)	1	2	
			GEMELLUS SUP													•									5	1	2	
		7	OBTURATOR INT													•									5	1	2	
			GEMELLUS INF													•								4	5	1	(2)	
			QUADRATUS FEM													•								4	5	1	(2)	
Sciatic	P	7	BICEPS (SHORT H)														•								5	1	2	
	Tibial		BICEPS (LONG H)															•							5	1	2	3
		7	SEMITENDINOSUS															•						4	5	1	2	
			SEMIMEMBRANOSUS															•						4	5	1	2	
Common Peroneal	Deep	4	TIBIALIS ANTERIOR																•					4	5	1		
		8	EXT HALL LONG																•					4	5	1		
		8	EXT DIGIT LONG																•					4	5	1		
		8	PERONEUS TERTIUS																•					4	5	1		
		8	EXT DIGIT BREVIS																•					4	5	1		
	Sup	7	PERONEUS LONGUS																•					4	5	1		
		7	PERONEUS BREVIS																•					4	5	1		
Tibial	Tibial	—	PLANTARIS																	•				4	5	1	(2)	
		10	GASTROCNEMIUS																	•						1	2	
		—	POPLITEUS																	•				4	5	1		
		10	SOLEUS																	•					5	1	2	
		7	TIBIALIS POSTERIOR																	•				(4)	5	1		
		6	FLEX DIGIT LONG																	•					5	1	(2)	
		7	FLEX HALL LONG																	•					5	1	2	
	Med Pl	7	FLEX DIGIT BREVIS																	•				4	5	1		
		—	ABDUCTOR HALL																	•				4	5	1		
		7	FLEX HALL BREVIS																	•				4	5	1		
		8	LUMBRICALIS I																	•				4	5	1		
	Lat Plant	—	ABD DIGITI MIN																	•						1	2	
		—	QUAD PLANTAE																	•						1	2	
		—	FLEX DIGITI MIN																	•						1	2	
		—	OPP. DIGITI MIN																	•						1	2	
		—	ADDUCTOR HALL																	•						1	2	
		—	PLANT INTEROSSEI																	•						1	2	
		—	DORSAL INTEROSSEI																	•						1	2	
		6	LUMB II,III,IV																	•				(4)	(5)	1	2	

SENSORY

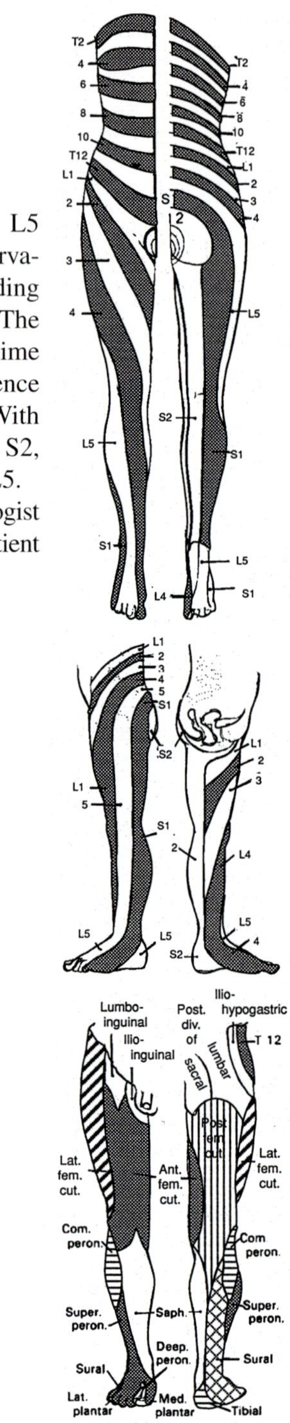

Dermatomes redrawn from Keegan and Garrett Anat Rec 102. 409. 437. 1948
Cutaneous Distribution of peripheral nerves redrawn from *Gray's Anatomy of the Human Body.* 28th ed

FIGURE 7-115: Case 3: Protruded intervertebral disc at L5.

CASE 4: GUILLAIN-BARRÉ SYNDROME 1

PATIENT'S NAME CLINIC No. 6694

LEFT MUSCLE CHART, No. 3 RIGHT

					6-7-47 HOK		6-7-47 HOK					
					70	Anterior Neck	70					
					100	Posterior Neck	100					
					100	Back	100					
					—	Quadratus Lumborum	—					
						Rectus Abdominis						
						External Oblique						
						Internal Oblique						
						Lateral Abdominals						
					55	Gluteus Maximus	55					
					60	Gluteus Medius	60					
					70	Inner Hamstrings	60					
					70	Outer Hamstrings	70					
					65	Internal Rotators	70					
					60	External Rotators	60					
					70	Hip Flexors	80					
					60	Sartorius	80					
					60	Hip Abductors	60					
					70	Hip Adductors	60					
					60	Tensor Fascia Lata	80					
					70	Quadriceps	70					
					100	Soleus	100					
					weak	Gastrocnemius	weak					
					20	Longus Longus	55					
					20	Brevis Peroneals Brevis	55					
					10	Tertius Tertius	30					
					30	Tibialis Posticus	20					
					20	Tibialis Anticus	10					
					0	Extensor Proprius Hallucis	0					
					55	Flexor Longus Hallucis	60					
					70	Flexor Brevis Hallucis	70					
					0	1 Extensor Longus Digitorum 1	0					
					0	2 Extensor Longus Digitorum 2	0					
					0	3 Extensor Longus Digitorum 3	0					
					0	4 Extensor Longus Digitorum 4	0					
					0	1 Extensor Brevis Digitorum 1	0					
					0	2 Extensor Brevis Digitorum 2	0					
					0	3 Extensor Brevis Digitorum 3	0					
					0	4 Extensor Brevis Digitorum 4	0					
					60	1 Flexor Longus Digitorum 1	70					
					55	2 Flexor Longus Digitorum 2	60					
					50	3 Flexor Longus Digitorum 3	60					
					20	4 Flexor Longus Digitorum 4	60					
					0	1 Flexor Brevis Digitorum 1	55					
					0	2 Flexor Brevis Digitorum 2	(60)					
					0	3 Flexor Brevis Digitorum 3	60					
					0	4 Flexor Brevis Digitorum 4	60					
					0	1 Lumbricales 1	0					
					0	2 Lumbricales 2	0					
					0	3 Lumbricales 3	0					
					0	4 Lumbricales 4	0					
						Length						
						Calf						
						Thigh						

	Contractions and Deformities	
	Neck	
	Back	
	Hip	
	Knee	
	Ankle	
	Foot	

FIGURE 7-116: Case 4: Guillain Barré Syndrome 1. Diagnosis confirmed on the basis of the symmetry of strengths and of weaknesses of the right and left extremities on the basis of one examination.

CASE 5: GUILLAIN BARRÉ SYNDROME 2

PATIENT'S NAME CLINIC No.

LEFT						MUSCLE CHART, No. 3						RIGHT
12-1-47 HOK	8-11-47 HOK	7-15-47 HOK	5-22-47 HOK	5-3-47 HOK	4-30-47 HOK		4-30-47 HOK	5-3-47 HOK	5-22-47 HOK	7-15-47 HOK	8-11-47 HOK	12-1-47 HOK
100	80	85	60	40	30	Anterior Neck	30	40	60	85	80	100
100	100	100	100	80	80	Posterior Neck	80	80	100	100	100	100
100	90	100	100	70	70	Back	70	70	100	100	90	100
—	—	—	—	—	—	Quadratus Lumborum	—	—	—	—	—	—
						Rectus Abdominis						
						External Oblique						
						Internal Oblique						
						Lateral Abdominals						
100	100	100	70	50	50	Gluteus Maximus	60	60	70	100	100	100
90	60	80	60	45	50	Gluteus Medius	50	50	60	75	60	90
100	100	100	100	65	60	Inner Hamstrings	60	65	100	100	100	100
100	100	100	90	65	60	Outer Hamstrings	60	65	100	100	100	100
100	80	100	70	60	60	Internal Rotators	60	60	80	70	90	100
100	90	90	70	60	60	External Rotators	60	60	90	85	70	100
100	90	100	60	(50)	(50)	(1) Hip Flexors (1)	(50)	(50)	60	100	100	100
100	100	100	100	80	70	Sartorius	70	80	100	100	100	100
90	60	80	60	45	50	Hip Abductors	50	50	60	75	60	100
100	90	80	65	55	50	Hip Adductors	50	55	65	80	90	100
100	90	100	60	60	60	Tensor Fascia Lata	60	60	60	100	80	100
100	100	100	70	60	70	Quadriceps	70	60	70	100	100	100
100	100	100	100	100	60	Soleus	80	90	85	100	100	100
100	90	L) 100	L) 100	L) 90	L) 80	Gastrocnemius	L) 80	L) 90	L) 100	L) 100	100	100
100	100	100	100	80	70	Longus Longus	90	100	100	100	100	100
100	100	100	100	80	70	Brevis Peroneals Brevis	90	100	100	100	100	100
100	100	100	100	80	70	Tertius Tertius	100	100	100	100	100	100
100	100	100	100	100	80	Tibialis Posticus	90	100	90	100	100	100
100	100	100	100	100	80	Tibialis Anticus	100	100	100	100	100	100
100	90	100	70	60	80	Extensor Proprius Hallucis	70	60	70	100	90	100
100	100	100	80	70	70	Flexor Longus Hallucis	90	100	100	100	100	100
100	100	100	100	100	100	Flexor Brevis Hallucis	100	100	100	90	100	100
100	100	100	100	70	90	1 Extensor Longus Digitorum 1	90	80	100	100	100	100
						2 Extensor Longus Digitorum 2						
/	/	/	/	/	/	3 Extensor Longus Digitorum 3	/	/	/	/	/	/
						4 Extensor Longus Digitorum 4						
100	100	100	100	70	90	1 Extensor Brevis Digitorum 1	90	80	90	100	100	100
						2 Extensor Brevis Digitorum 2						
/	/	/	/	/	/	3 Extensor Brevis Digitorum 3	/	/	/	/	/	/
						4 Extensor Brevis Digitorum 4						
100	100	100	100	80	80	1 Flexor Longus Digitorum 1	100	100	100	100	100	100
						2 Flexor Longus Digitorum 2						
/	/	/	/	/	/	3 Flexor Longus Digitorum 3	/	/	/	/	/	/
						4 Flexor Longus Digitorum 4						
100	100	90	80	70	60	1 Flexor Brevis Digitorum 1	80	80	80	90	100	100
						2 Flexor Brevis Digitorum 2						
/	/	/	/	/	/	3 Flexor Brevis Digitorum 3	/	/	/	/	/	/
						4 Flexor Brevis Digitorum 4						
100	100	100	100	80	90	1 Lumbricales 1	100	100	100	100	100	100
						2 Lumbricales 2						
/	/	/	/	/	/	3 Lumbricales 3	/	/	/	/	/	/
						4 Lumbricales 4						
						Length						
						Calf						
						Thigh						

Notes	Contractions and Deformities	
(1) 4-30-47, 5-3-47 } Masked hamstring tightnes s.	Neck	
L) Lying dow n	Back	
	Hip	
	Knee	
	Ankle	
	Foot	

FIGURE 7-117: Case 5: Guillain Barré Syndrome 2. The symmetry of right and left sides based on six examinations over a 7-month period.

CASE 6: POLIO

PATIENT'S NAME CLINIC No.

LEFT MUSCLE CHART, No. 3 RIGHT

4-18-45 HOK	3-8-45 HOK	1-22-45 HOK	12-18-49 HOK	10-18-44 HOK	9-21-44 HOK		9-21-44 HOK	10-18-44 HOK	12-18-44 HOK	1-22-45 HOK	3-8-45 HOK	4-18-45 HOK
	50	40	40	30	10	Anterior Neck	10	30	40	40	40	
	60		60		20	Posterior Neck	20		60	60	60	70
70			40		20	Back	20		40			70
						Quadratus Lumborum						
						Rectus Abdominis						
						External Oblique						
						Internal Oblique						
						Lateral Abdominals						
100	100	60	40	30	10	Gluteus Maximus	50	100	100	100	100	100
60	80	40	30	20	10	Gluteus Medius	30	40	60	70	70	60
60	60	30	30	20	10	Inner Hamstrings	100	100	100	100	100	100
60	40	30	30	5	0	Outer Hamstrings	100	100	100	100	100	100
70	70	60	60	60	20	Internal Rotators		80	80	90	90	
90	90	70	70	60	60	External Rotators		80	80	100	100	
90	90	60	60	40	10	Hip Flexors	30	40	70	90	90	
80	80	80	60	40	10	Sartorius	60	90	100	100	100	100
60	80	60	30	20	10	Hip Abductors	30	60	60	60	70	60
100	90	80	80	60	60	Hip Adductors	60	60	90	90	100	100
80	100	80	80	40	20	Tensor Fascia Lata	50	60	70	80	100	100
100	100	70	60	60	10	Quadriceps	80	70	70	80	100	100
10	5	5	5	0	5	Soleus	90	100	90	100	100	100
0	0	0	5	0	5	Gastrocnemius	100	100	100	100	100	100
5	5	5	5	5	0	Longus Longus	100	100	100	100	100	
5	5	5	5	5	0	Brevis Peroneals Brevis	100	100	100	100	100	
0	0	0	0	0	0	Tertius Tertius	100	100	100	100	100	
100	5	5	5	0	0	Tibialis Posticus	100	100	90	100	90	100
20	20	20	20	5	0	Tibialis Anticus	60	100	100	100	100	
0	0	0	0	0	0	Extensor Proprius Hallucis	100	100	90	100	100	
20	20	0	0	0	0	Flexor Longus Hallucis	60	80	100	100	100	
60	70	20	5	5	5	Flexor Brevis Hallucis	100	100	100	100	100	
0	0	5	0	0	5	1 Extensor Longus Digitorum 1	80	100	100	100	100	
0	0	5	0	0	5	2 Extensor Longus Digitorum 2	80	100	100	100	100	
0	0	5	0	0	5	3 Extensor Longus Digitorum 3	80	100	100	100	100	
0	0	5	0	0	5	4 Extensor Longus Digitorum 4	80	100	100	100	100	
0	0	0	0	0	0	1 Extensor Brevis Digitorum 1	100	100	100	100	100	
0	0	0	0	0	0	2 Extensor Brevis Digitorum 2	100	100	100	100	100	
0	0	0	0	0	0	3 Extensor Brevis Digitorum 3	100	100	100	100	100	
0	0	0	0	0	0	4 Extensor Brevis Digitorum 4	100	100	100	100	100	
40	0	0	0	0	0	1 Flexor Longus Digitorum 1	70	60	90	100	90	
40	0	0	0	0	0	2 Flexor Longus Digitorum 2	70	60	90	100	100	
60	60	0	0	0	0	3 Flexor Longus Digitorum 3	70	60	90	100	100	
60	60	0	0	0	0	4 Flexor Longus Digitorum 4	70	60	90	100	100	
70	60	60	60	60	50	1 Flexor Brevis Digitorum 1	40	70	60	60	80	
60	70	60	60	60	50	2 Flexor Brevis Digitorum 2	40	70	60	70	80	
70	70	60	60	60	40	3 Flexor Brevis Digitorum 3	40	70	60	70	80	
70	70	60	60	60	40	4 Flexor Brevis Digitorum 4	60	70	60	70	80	
70	70	60	40	40	10	1 Lumbricales 1	70	80	90	90	100	
70	70	60	40	40	10	2 Lumbricales 2	70	80	90	90	100	
60	60	60	40	40	10	3 Lumbricales 3	70	60	90	70	100	
60	60	60	40	40	10	4 Lumbricales 4	70	60	90	70	100	
						Length Calf						
						Thigh						

	Contractions and Deformities	
	Neck	
	Back	
	Hip	
	Knee	
	Ankle	
	Foot	

FIGURE 7-118: Case 6: A poliomyelitis case. In this example, the left leg has rather extensive involvement and the right leg is essentially normal. Poliomyelitis cases do not show patterns of weakness.

CORRECTIVE EXERCISES: LOWER EXTREMITY

Exercises in the lying position should be done on a firm surface (e.g., a board on the bed, a treatment table, or the floor, with a thin pad or folded blanket placed on the hard surface for comfort) (Figure 7-119).

Stretching exercises should be preceded by gentle heat and massage to help relax tight muscles. (Avoid using heat on weak, overstretched muscles.) Stretching should be done gradually with a conscious effort to relax. Continue until a firm, but tolerable "pull" is felt, breathing comfortably while holding the stretch, then return *slowly* from the stretched position.

Strengthening exercises should also be done slowly, with an effort to feel a strong "pull" by the muscles being exercised. Hold the completed position for several seconds, then relax and repeat the exercise the number of times indicated by your therapist.

Active Hamstring Stretching
To stretch right hamstrings, lie on table with legs extended. Hold left leg down and gradually raise right leg with knee straight. (Reverse the procedure to stretch left Hamstrings.)

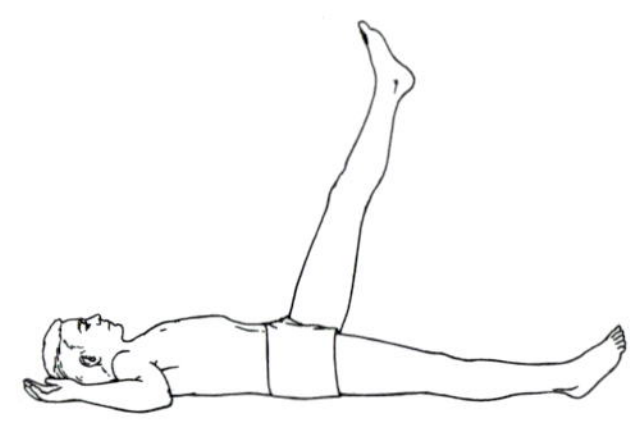

A

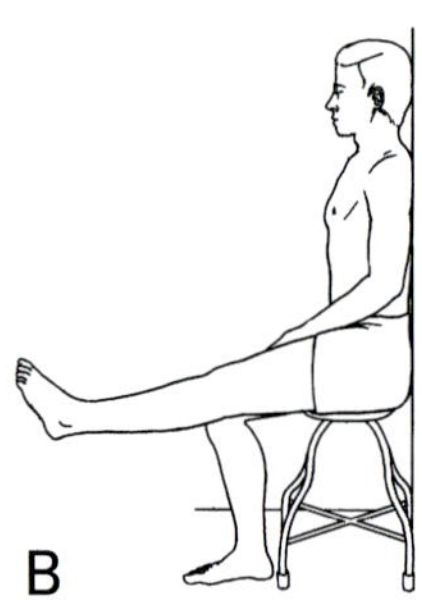

B

Sit on a stool with back against wall. Keep one knee bent and straighten other leg. A stretch should be felt under the knee and along Hamstring muscles.

Passive Hamstring Stretch in Doorway

Lie on floor by an indoor doorway. Place one leg out straight on the floor inside the doorway with the other in a position of straight leg raise, resting the heel against the doorframe. As muscles relax, move closer to the doorframe, raising the leg higher and giving an added stretch to the Hamstrings.

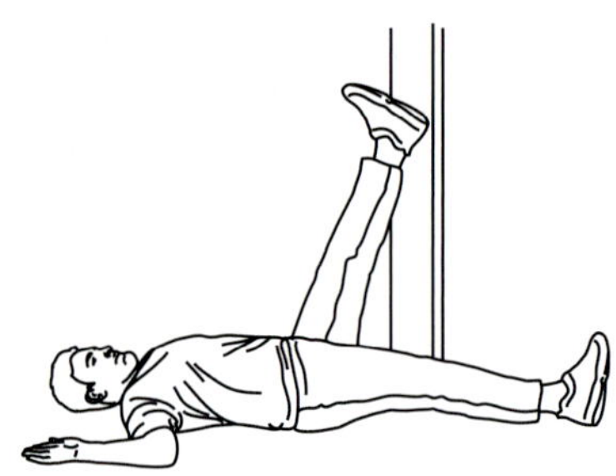

D

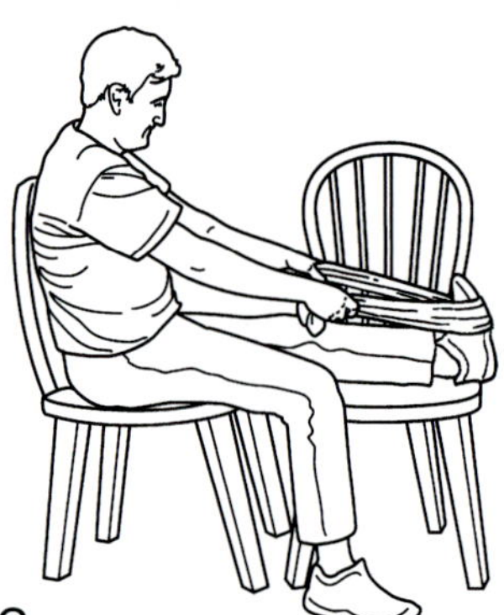

C

Passive Seated Hamstring and Calf Stretch (with towel assist)
Sitting in chair, place one leg on stool or chair seat of same height, keeping knee supported. You will feel a "pull" in the back of your thigh. (To add a stretch to the calf muscle, place a towel or strap around the ball of the foot and slowly pull foot toward you.) Hold for ___ seconds. Repeat ___ times.

E

Active Hamstring Stretch (with towel assist)
Lying on back on firm, but padded surface, use towel to pull thigh to slightly less than vertical position (80°), keeping upper arms resting supported at your side. Straighten knee until you feel a "pull" in the back of your thigh and knee. Hold for ___ seconds. Repeat ___ times.

FIGURE 7-119: A–K, Corrective exercises for the lower extremity. © 2005 Florence P. Kendall and Patricia G. Provance.

One-Joint Hip Flexor Stretching
In back-lying position, pull one knee toward chest until low back is flat on table. *Keeping back flat*, press other leg, with knee straight, down toward the table by tightening the buttock muscle.

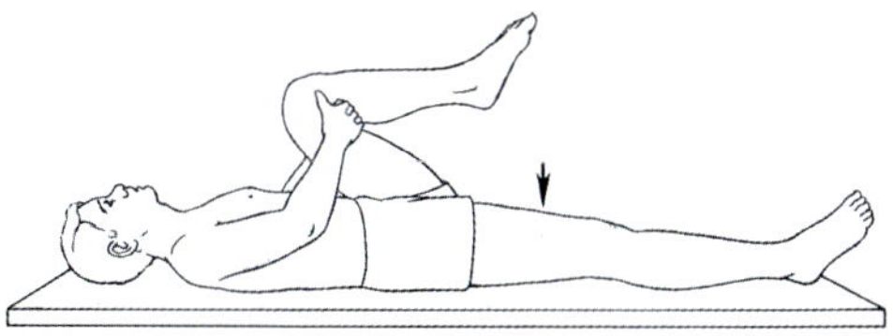

F

Two-Joint Hip Flexor Stretching and Hip Extensor Strengthening
To stretch right hip flexors, lie on back with right lower leg hanging over end of a *sturdy* table. Pull left knee toward chest just enough to flatten low back on table. (When there is hip flexor tightness, the right thigh will come up from table.) *Keeping back flat*, stretch right hip flexors by pulling thigh downward with the right buttock muscle, trying to touch thigh to table. Keeping thigh down toward table, try to bend knee until a firm "pull" is felt in front of the right thigh (no more than 80°).

To stretch left hip flexors, pull right knee toward chest and apply the stretch to left thigh, as described above. (Note: This can be done at the top of a flight of stairs if no sturdy table is available.)

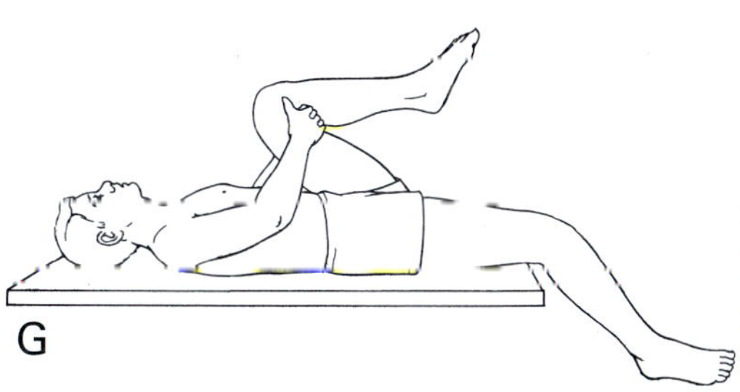

G

Hip Abduction Strengthening, Back-Lying
Lying on back with your hands on your hips, slide the (right) (left) leg out to the side and keep it in that position without hiking the hip up on that side. Slowly slide the other leg out as far as possible. Return to midline. Repeat ___ times.

H

I

Dynamic Single Leg Stand (for balance, Gluteal, and Quadriceps strengthening)

Use hand support as needed for balance and safety.

Balance on one foot with posture erect, keeping pelvis level and abdomen and buttocks firm, and other leg lifted forward off the floor. Keeping weight over supporting leg, slowly bend knee as though stepping off a curb with the other foot. Keep back straight and avoid tilting the pelvis forward, backward, or sideways.

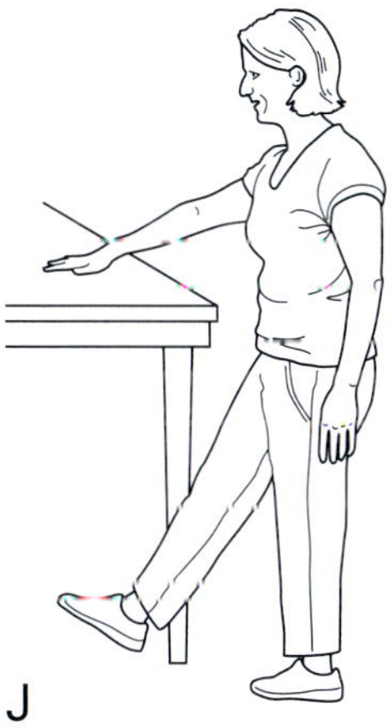

J

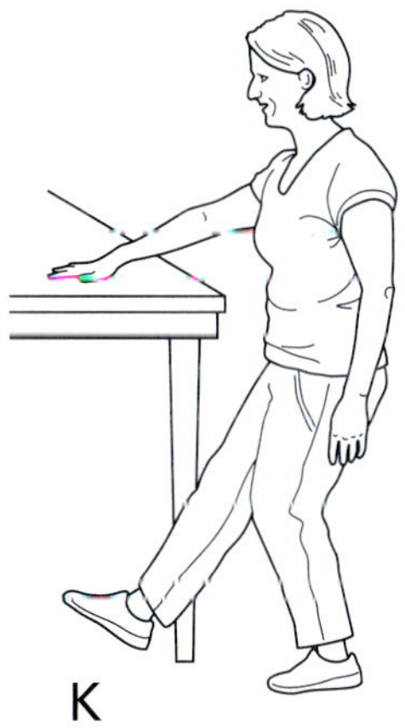

K

FIGURE 7-119: *(Continued)*

REFERENCES

1. Reddy RS, Alahmari KA. Effect of lower extremity stretching exercises on balance in geriatric population. Int J Health Sci (Qassim). 2016; 10(3): 389–395.
2. Reiman MP, Thorborg K. Clinical examination and physical assessment of hip joint-related pain in athletes. Int J Sports Phys Ther. 2014; 9(6): 737–755.
3. Agur AMR. Grants Atlas of Anatomy, 9th ed. Baltimore: Williams and Wilkins; 1991: 263.
4. Monteiro DP, Britto RR, Fregonezi GAF, Dias FAL, Silva MGD, Pereira DAG. Reference values for the bilateral heel-rise test. Braz J Phys Ther. 2017; 21(5): 344–349. doi:10.1016/j.bjpt.2017.06.002.
5. Hébert-Losier K, Wessman C, Alricsson M, Svantesson U. Updated reliability and normative values for the standing heel-rise test in healthy adults. Physiotherapy. 2017; 103(4): 446–452. doi:10.1016/j.physio.2017.03.002.
6. Kendall HO, et al. Posture and Pain. Baltimore: Williams & Wilkins; 1952.
7. Ober FR. Back strain and sciatica. JAMA. 1935; 104(18): 1580–1581.
8. Kendall HO, Kendall FP. Muscles, Testing and Function, 1st ed. Baltimore: The Williams and Wilkins Company; 1949.
9. Kendall HO, Kendall FP, Wadsworth GE. Muscles, Testing and Function, 2nd ed. The Williams and Wilkins Company; 1971.
10. Ober FR. Relation of the fascia lata to conditions of the lower part of the back. JAMA. 1937; 109(8): 554–555.
11. Hoppenfeld S. Physical examination of the spine and extremities. East Norwalk: Appelton-Century-Crofts; 1976: 167.
12. Rothstein J, Roy S, Wolf S. The rehabilitation specialist's handbook. Philadelphia: FA Davis; 1991: 64–65
13. Guides to the evaluation of permanent impairment. Chicago: American Medical Association; 1984.
14. Daniels L, Worthingham C. Muscle Testing-Techniques of Manual Examination, 5th ed. Philadelphia: WB Saunders; 1986: 54.
15. Palmer M, Epler M. Clinical assessment procedures in physical therapy. Philadelphia: JB Lippincott; 1990: 247–248.
16. Norkin CC, White DJ. Measurement of Joint Motion: A Guide to Goniometry. Philadelphia: FA Davis; 1985: 139.
17. Willett GM, Keim SA, Shostrom VK, Lomneth CS. An anatomic investigation of the Ober test. Am J Sports Med. 2016 Mar; 44(3): 696–701. doi: 10.1177/0363546515621762. Epub 2016 Jan 11. PMID: 26755689.
18. Neumann DA, Garceau LR. A proposed novel function of the psoas minor revealed through cadaver dissection. Clin Anat. 2015; 28(2): 243–252. doi:10.1002/ca.22467.
19. Reimann R, Sodia F, Klug F. Die umstrittene Rotationswirkung ausgewählter Muskeln im Hüftgelenk [Controversial rotation function of certain muscles in the hip joint]. Ann Anat. 1996; 178(4): 353–359.
20. Leighton RD. A functional model to describe the action of the adductor muscles at the hip in the transverse plane. Physiother Theory Pract. 2006; 22(5): 251–262. doi:10.1080/09593980600927385.
21. Yi TI, Lee GE, Seo IS, Huh WS, Yoon TH, Kim BR. Clinical characteristics of the causes of plantar heel pain. Ann Rehabil Med. 2011; 35(4): 507–513. doi:10.5535/arm.2011.35.4.507.
22. Kite JH. Exercise in foot disabilities. In: Basmajian JV, ed. Therapeutic Exercise, 3rd ed. Baltimore: Williams and Wilkins; 1978: 485–513.
23. Denison CD. Orthopaedic Appliance Corporation, 220 W. 28th St. Baltimore, MD.
24. Bae HI, Kim DY, Sung YH. Effects of a static stretch using a load on low back pain patients with shortened tensor fascia lata. J Exerc Rehabil. 2017; 13(2): 227–231. Published 2017 Apr 30. doi:10.12965/jer.1734910.455.
25. Nordin M, Frankel V. Basic Biomechanics of the Musculoskeletal System, 2nd ed. Philadelphia: Lea and Feibiger; 1989: 193, 201.
26. Pope M, Wilder D, Booth J. The biomechanics of low back pain. In: White AA, Gordon SL, eds. Symposium on Idiopathic Low Back Pain. St. Louis, MO: CV Mosby; 1982.
27. Brown T, Hanson R, Yorra A. Some mechanical tests on the lumbosacral spine with particular reference to the intervertebral disc. J Bone Joint Surg [AM]. 1957; 39-A: 1135.
28. Roaf R. A study of the mechanics of spinal injuries. J Bone Joint Surg [Br]. 1960; 42-B: 810.
29. Goss CM, ed. Gray's Anatomy of the Human Body, 28th ed. Philadelphia: Lea & Febiger; 1966: 311.
30. Freiberg AH Vinke TH. Sciatica and sacro-iliac joint. J Bone Joint Surg. 1934;16:126–136.

APPENDIX

A

POSTURE OF CHILDREN

OUTLINE

INTRODUCTION

This appendix introduces a variety of concepts dealing with the development of postural habits in the growing individual and a variety of influences that affect such development. No attempt is made to give the various concepts either exhaustive or equal treatment. The Kendalls hoped this material would be useful from the standpoint of prevention and that it would create, through a recognition of the factors involved in postural development, a more positive approach toward providing, within available limits, the best possible environment for good posture.

Good posture is not an end in itself; it is a part of general well-being. Ideally, posture instruction and training should be a part of general experience rather than a separate discipline. To the extent that parents and teachers are able to recognize the influences and habits that help to develop good or faulty posture, they will be able to contribute to this aspect of well-being in the daily life of growing individuals. Nevertheless, posture instruction and training should not be neglected in a good program of health education; attention should be paid to observable faults. When instruction is given, it should be simple and accurate; while it must not be neglected, neither should it be overemphasized. It should be given in such a manner as to capture the interest and cooperation of the child.

FACTORS INFLUENCING CHILDREN'S POSTURE

Nutritional Factors

Good postural development is dependent on good structural and functional development of the body, which in turn is highly dependent on adequate nutrition. The influence of nutrition on the proper structural development of skeletal and muscular tissues is particularly significant. Rickets, for example, which is often responsible for severe skeletal deformities in children, is a disease of vitamin D deficiency.

After growth is completed, poor nutrition is less likely to cause structural faults that directly affect posture. At this stage, deficiencies are more likely to interfere with physiological function and to be represented posturally in a position of fatigue. The body uses food not only for growth but also for fuel, transforming it into heat and energy. If the fuel is insufficient, energy output decreases and so does general physiological efficiency. Nutritional deficiencies in the adult are most likely to occur when unusual physiological demands are made on the individual over time.

Defects, Diseases, and Disabilities

Certain physical defects, diseases, and disabilities have associated postural problems. These conditions can be roughly divided into three groups regarding the importance of attention to posture in their treatment.

The first group consists largely of physical defects in which the postural aspects are more potential than actual during the initial stages, then become a problem only if the defect cannot be completely corrected by medical or surgical means. These defects may be visual, auditory, skeletal (e.g., clubfoot or dislocation of the hip), neuromuscular (e.g., brachial plexus injury), or muscular (e.g., wry neck).

The second group includes conditions that are in themselves potentially disabling but in which continuing attention to posture from the early stages can minimize the disabling effects. In an arthritic condition of the spine (e.g., Marie-Strümpell), if the body can be kept in good functional alignment during the time that fusion of the spine is taking place, the individual may have little obvious deformity and only moderate disability when the fusion is complete. If the postural aspect is disregarded, however, the trunk is usually in marked flexion when fusion of the spine is complete. This is a position of severe deformity and associated severe disability.

The third group contains conditions in which a degree of permanent disability exists as a result of injury or disease but in which added postural strain can greatly increase the disability. The amputation of a lower extremity, for example, throws an unavoidable extra burden on the remaining weight-bearing structures. A postural alignment that minimizes (as much as possible) the mechanical strains of position and motion does much to keep these structures from breaking down.

Environmental Factors

A number of environmental factors influence the development and maintenance of good posture. These environmental influences should be made as favorable to good posture as is practical. When no major adjustment is possible, small adjustments will often contribute considerably. The following discussion takes into account factors such as chairs, desks, and beds because they illustrate environmental influences on posture in the sitting and lying positions. After children start school, the amount of time they spend in the sitting position increases considerably. The school seat is an important factor affecting posture.

Both the chair and the desk should be adjusted to fit the child. The child should be able to sit with both feet flat

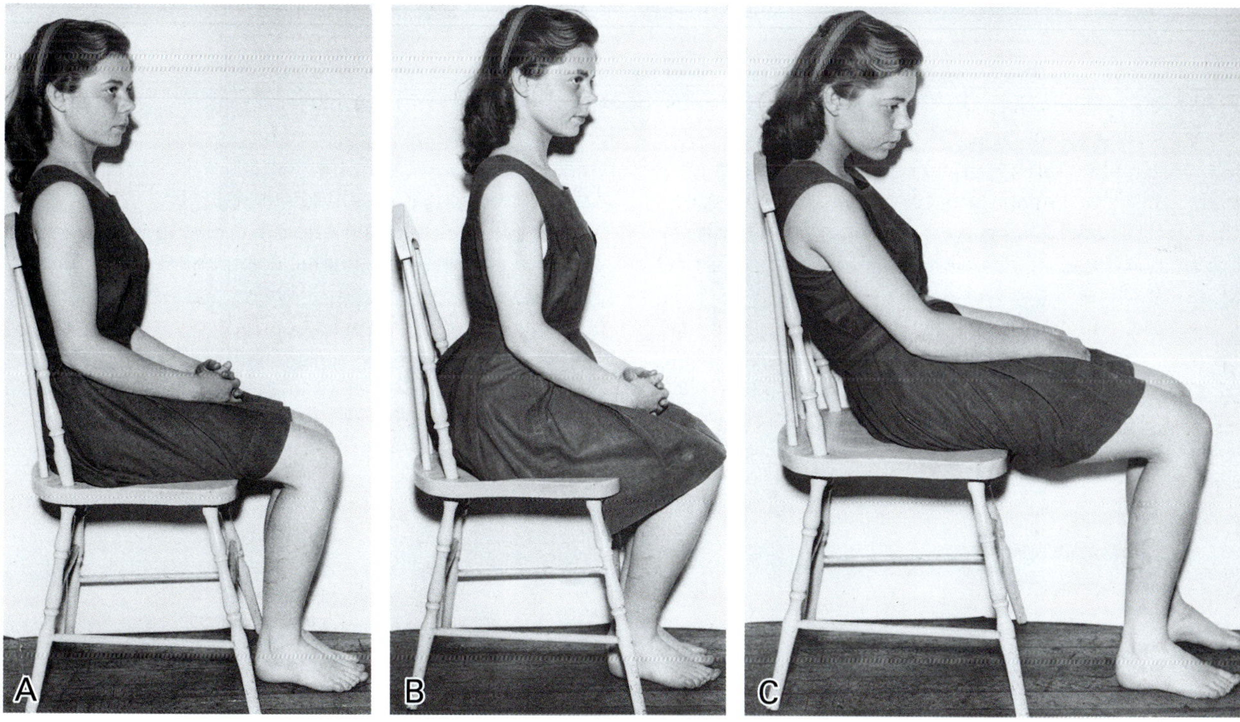

FIGURE APP-A-1: Sitting posture and alignment: **A**, Good alignment; **B**, The low back in lordosis; **C**, Slumped position.

on the floor and with knees bent to about a right angle. If the chair is too high, there will be a lack of support for the feet. If the chair is too low, the hips and knees will bend in too much flexion. The seat of the chair should be deep enough from front to back to support the thighs adequately, but the depth should not interfere with bending of the knees. The back of the chair should support the child's back. It should also incline backward a few degrees so that the child can relax against it (Figure App-A-1).

The top of the desk should be at about elbow level when the child is sitting in a good position, and it may be slightly inclined. The desk should be close enough that the arms can rest on it without the need to lean too far forward or to sit forward on the seat of the chair.

Developmental Factors

It is important to recognize marked or persistent postural deviations in the growing individual, but it is equally important to recognize that children are not expected to conform to an adult standard of alignment. This is true for a variety of reasons, but primarily because the developing individual exhibits much greater mobility and flexibility than the adult.

Most postural deviations in the growing child fall into the category of developmental deviations; when patterns become habitual, they may result in postural faults. Developmental deviations are those that appear in many children at approximately the same age and that improve or disappear without any corrective treatment, sometimes even despite unfavorable environmental influences (1). Whether a deviation in a child is becoming a postural fault should be determined by repeated or continued observation, not by a single examination. If the condition remains static or the deviation increases, corrective measures are indicated. Severe faults need treatment as soon as they are observed, regardless of the age of the individual.

A young child is not likely to have habitual faults and can actually be harmed by unneeded corrective measures. Overcorrection may lead to atypical faults that are more harmful and difficult to deal with than the ones that caused the original concerns.

Some of the differences between children and adults result because in the years between birth and maturity, the structures of the body grow at varying rates. In general, body structures grow rapidly at first, then at a gradually reduced rate. An example of this is the increase in size of the bones. Associated with increased overall length of the skeleton is a change in the proportionate lengths of its various segments. This change in proportions occurs as first one part of the skeleton and then another has the most rapid rate of growth (2, 3). The gradual tightening of ligaments and fascia as well as the strengthening of muscles are significant developmental factors. Their effect is to gradually limit the range-of-joint motion toward the range that is typical of maturity. The increase in stability that results is advantageous because it decreases the danger of strain from handling heavy objects or from other strenuous activities. Normal joint range for adults should provide an effective balance between motion and stability. A joint that is either too limited in range or not sufficiently limited is vulnerable to strain.

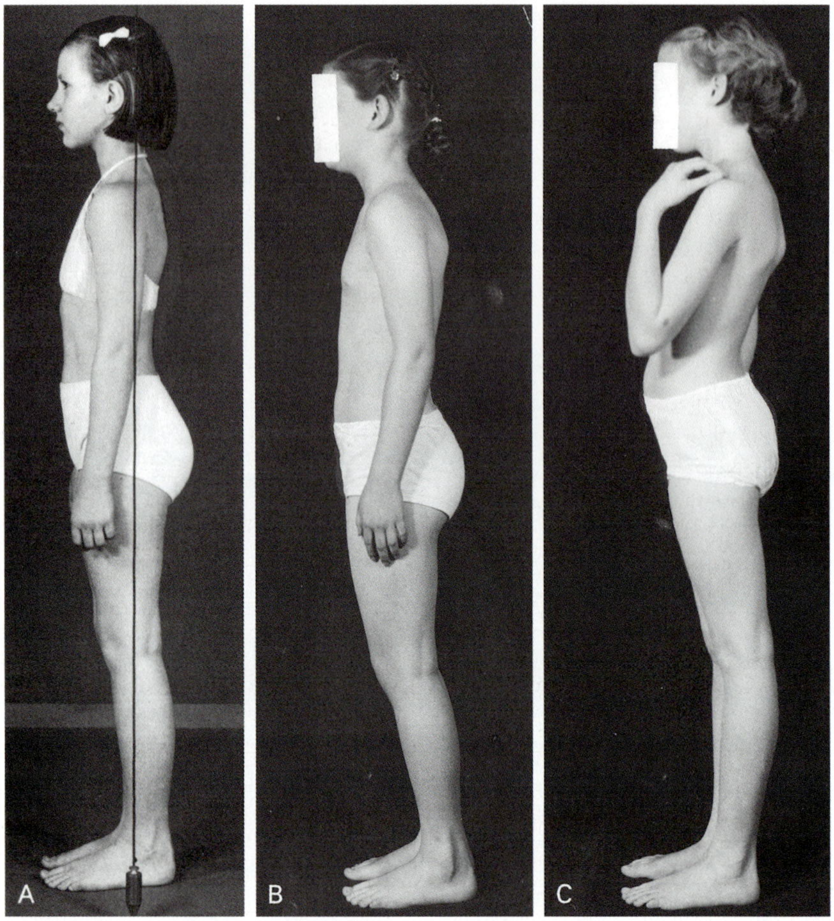

FIGURE APP-A-2: Standing posture and alignment: **A**, Good alignment; **B**, Average for 9 year old **C**, Faulty posture.

Figure App-A-2, A shows a 10-year-old child who has very good posture for this age. The posture resembles the normal adult posture more than that of a younger child. The curves of the spine are nearly normal, and the scapulae are less prominent. It is characteristic of small children to have a protruding abdomen, but there is a noticeable change at approximately 10–12 years of age, when the waistline becomes relatively smaller and the abdomen no longer protrudes.

Figure App-A-2, B shows a 9-year-old child whose posture is about average for this age.

Figure App-A-2, C shows a 11-year-old child whose posture is very faulty, with forward head, kyphosis, lordosis, anterior pelvic tilt, and hyperextended knees.

The child's greater range-of-joint motion makes possible momentary and habitual deviations in alignment that would be considered distortions in the adult. At the same time, the flexibility serves as a protection against developing *fixed* postural faults.

As early as 8 or 10 years of age, handedness patterns related to posture may appear. The slight deviation of the spine to the side opposite to the higher hip makes an early appearance. There also tends to be a compensatory low shoulder on the side of the higher hip. In most cases, the low shoulder is a less significant factor. Usually, shoulder correction tends to follow correction of lateral pelvic tilt, but the reverse does not occur. No attempts should be made to raise the shoulder into position by constant muscular effort.

Activities that are rather neutral in their effect on posture are games or sports in which walking or running predominates. Sports that exert an influence toward muscle imbalance are predominantly one-sided, such as those involving use of a racket or a bat.

The play activities of young children usually are varied enough that no problem of muscle imbalance or habitual alignment fault is present. However, when a child becomes old enough to engage in competitive athletics, a point may be reached at which further development of skill through intensive practice requires a sacrifice of some degree of good muscle balance and skeletal alignment. Although seemingly unimportant at the time, the faults acquired may progress until a painful condition results.

Specific exercises may be needed to maintain range-of-joint motion and to strengthen certain muscles if opposing muscles are being overdeveloped by the activity. These exercises must be specific for the part in question and therapeutic for the body as a whole.

GOOD AND FAULTY POSTURE OF CHILDREN

A consideration of both normal and abnormal variations in the posture of children can be discussed from the standpoints of the overall posture and the deviations of the various segments. Variations in overall posture of children at approximately the same age are illustrated in Figure App-A-2 and Figure App-A-3A.

Feet

When a small child is beginning to stand and walk, the foot is normally flat. The bones are in a formative stage, and the arch structure is incomplete. The arch develops gradually along with development of the bones and with strengthening of muscles and ligaments. By the age of 6 or 7 years, one may expect good arch formation. Footprints taken at regular intervals help to gauge the amount of change that has occurred in the arch. These can be taken with a podograph; if this is not available, the sole of the foot can be painted with Vaseline and a footprint made on paper. As the arch increases in height, less of the sole in the area of the arch will be seen in the footprint.

Flat longitudinal arches may persist as a fixed fault, or they may recur because of foot strain at any age. Improper shoes or a habit of standing and walking with the feet in an out-toeing position may cause such strain. If a child's foot is very flat, is pronated, and toes-out in a manner that allows the body weight to be borne constantly on the inner side of the foot, it may be necessary to use a slight correction, such as an inner heel-wedge or small longitudinal pad in the shoe quite soon after the child begins to stand and walk. In most cases, however, it is advisable to institute corrective measures only after a period of observation. Some individuals fail to develop a longitudinal arch and have what is termed a *static flat foot*. In these cases, the alignment of the foot is not faulty in regard to pronation or out-toeing, and no symptoms of foot strain are observed. The corrective measures usually indicated for flat arches are not indicated in such cases.

Knees

Hyperextension is a fairly common fault, usually associated with lack of firm ligamentous support. It tends to disappear as the ligaments tighten, but if it persists as a postural habit, a corrective effort should be made by postural training.

A degree of *genu valgum* (knock-knees) is common in children and usually first observed when the child begins to stand. The height and build of the child must be taken into consideration when judging whether the deviation is a fault, but in general, it may be said that a fault exists if the ankles are more than 2 inches apart when the knees are touching. The genu valgum should show definite improvement and be nonexistent by the age of 6 or 7 years (see Figure App-A-3, Part A).

In some cases, children with genu valgum may stand with one knee slightly flexed and the other slightly hyperextended so that the knees overlap to keep the feet together. Genu valgum may persist, and in adults, it is more prevalent among women than among men.

Records of the change in the degree of genu valgum can be kept by drawing an outline of the legs on paper while the child is standing with the knees touching each other. Mild to moderate genu valgum conditions are usually treated by shoe corrections, but bracing or even surgery may be required for the more severe cases.

Genu varum (bowlegs) is an alignment fault in which the knees are separated when the feet are together. It may be a postural or a structural fault. Postural bowing is a deviation associated with knee hyperextension and hip medial rotation. As the posterior ligaments tighten and hyperextension decreases, this type of fault tends to become less pronounced. If it persists as a postural habit, the child should be given instruction to correct the alignment faults. This fault is less easy to correct as the individual approaches maturity, although some degree of correction may be obtained in young adults who are very flexible.

Postural genu varum may be compensatory for genu valgum. If a child with genu valgum stands with the knees thrust back into hyperextension, the resultant postural genu varum will let the feet be brought together without having the knees overlap. In this position, the genu valgum fault may be obscured, but it will become obvious if the lower extremities are brought into a neutral position of knee extension.

Postural genu varum usually disappears when an individual is recumbent, whereas structural genu varum does not. Structural genu varum requires early treatment; in later stages, it may require surgery.

Drawings to record the change in *structural* genu varum can be made while the child is in a supine position with the feet together. Because *postural* genu varum shows up only in standing, the drawing must be made in standing. This can be done by placing the paper on a wall behind the standing child.

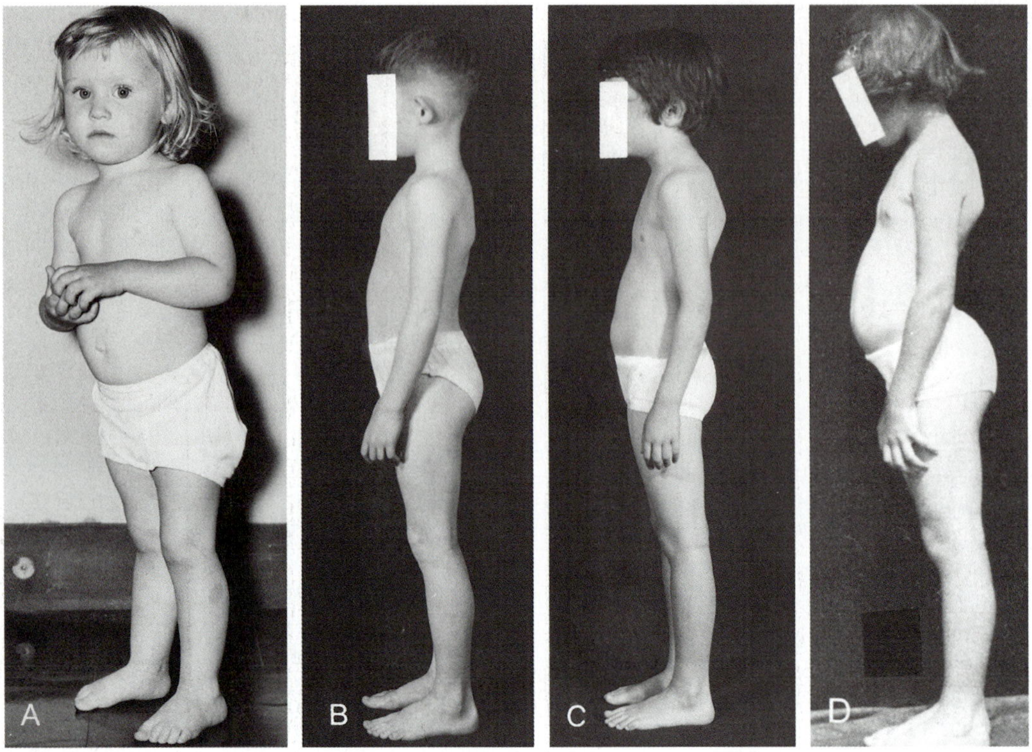

FIGURE APP-A-3: Standing posture and alignment: **A**, 18 month old; **B**, 7 year old with poor posture; **C**, 6 year old; **D**, 8 year old with poor posture.

Figure App-A-3, A shows the posture of an 18-month-old child. The flexed hips and wide stance suggest the uncertain balance associated with this age. Although it is not very evident in the picture, the subject had a mild degree of genu valgum. (This deviation gradually decreased without any corrective measures, and by the age of 6 years, the child's lower extremities were in good alignment.) The development of the longitudinal arch is very good for a child of this age.

Figure App-A-3, B shows a 7-year-old child who has very good posture for his age.

Figure App-A-3, C shows poor posture in a 6-year-old child. There is forward head, kyphosis, depressed chest, and a tendency toward sway-back posture. Prominence of the scapulae is evident in side view.

Figure App-A-3, D shows hyperlordosis in an 8-year-old child. A corset to hold the back in good alignment and to support the abdomen is needed, along with therapeutic exercises when alignment is this faulty.

Neck and Trunk

Beginning in infancy, there is a persistent imbalance between the strength of the anterior and posterior muscles of the trunk and neck. The greater strength of the posterior muscles permits the child to raise the head and trunk backward long before being able to raise either one forward without assistance. Although the abdominal and neck flexor muscles never match the strength of their opponents, their relative strength is much greater in the adult than in the child. Thus, in this regard, individuals should not be expected to conform to the adult standard until they are approaching maturity.

It is characteristic of small children to have a protruding abdomen. For the most part, the contour of the abdominal wall changes gradually, but a noticeable change occurs at approximately 10–12 years of age, when the waistline becomes relatively smaller and the abdomen no longer protrudes.

The posture of the back varies somewhat with the age of the child. A small child may stand bent slightly forward at the hips and with the feet apart for better balance (see Figure App-A-3, A). Children of early school age appear to have a typical deviation of the upper back in which the shoulder blades are quite prominent. Beginning at approximately 9 years of age, there seems to be a tendency for increased forward curve or lordosis of the low back. The deviations should become less pronounced as the child grows older (1, 4).

Normal range of motion for lumbar flexion and extension has been shown to decrease with increasing age in both children and adults (5–7).

NORMAL FLEXIBILITY ACCORDING TO AGE LEVEL

The ability to touch the toes with the fingertips may be considered normal for young children and adults. Between the ages of 11 and 14 years, however, many individuals who show no signs of muscle or joint tightness are unable to complete this movement. The proportionate length of the trunk and lower extremities is different in this age group compared with that of younger and older age groups.

The five illustrations in Figure App-A-4 are representative of the majority of individuals in each of the following age groups: Figure App-A-4, A, 1–3 years; Figure App-A-4, B, 4–7 years; Figure App-A-4, C, 8–10 years; Figure App-A-4, D, 11–14 years; and Figure App-A-4, E, 15 years and older.

Reaching to touch the fingertips to the toes while sitting with lower extremities extended shows interesting and significant variations according to age. Figures App-A-5 through A-8 indicate the variations in normal accomplishment of this movement at different age levels (8).

The change from the apparently extreme flexibility of the youngest child to the apparently limited flexibility of the child in Figure App-A-4, D occurs gradually, over a period of years, as the lower extremities become proportionately longer in relation to the trunk. Standards of performance for children that involve forward bending should take into consideration normal variations in the ability to complete the range of this movement (9).

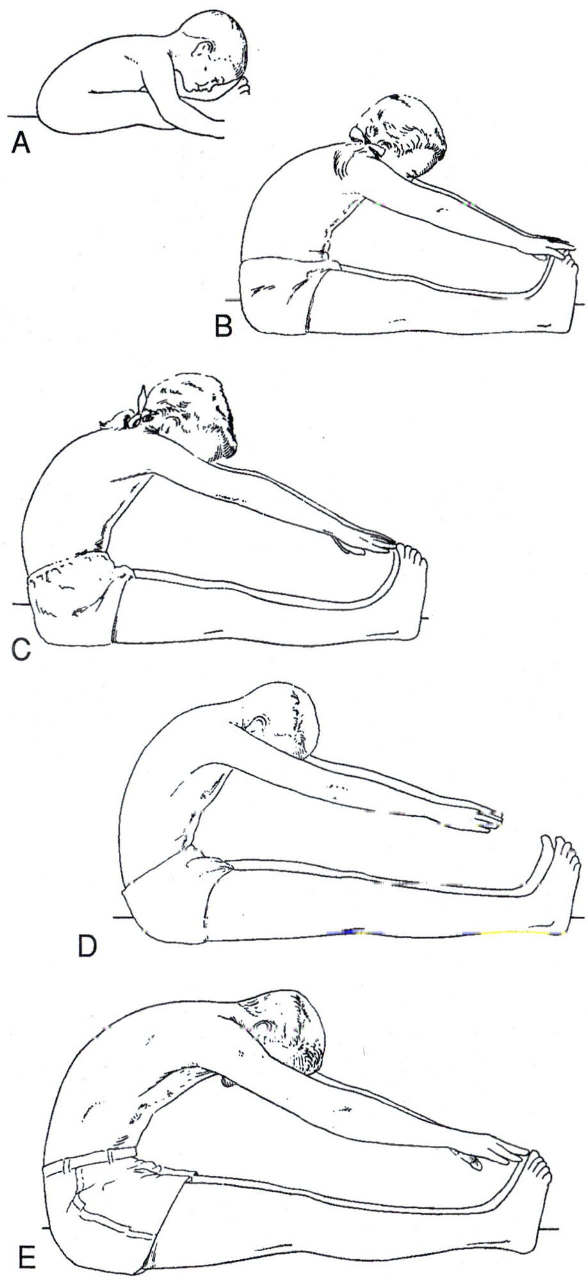

FIGURE APP-A-4: Normal flexibility according to age level.

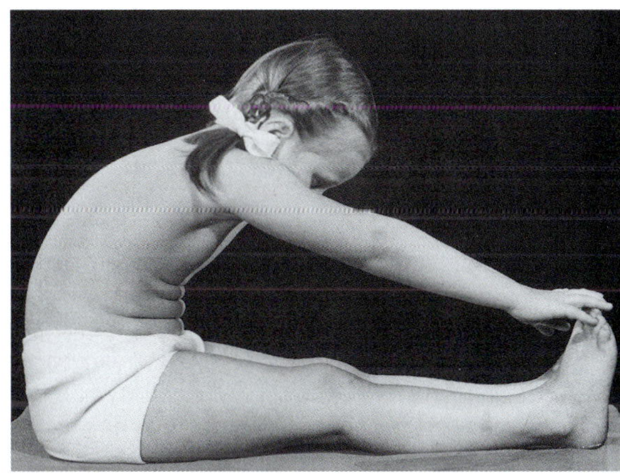

FIGURE APP-A-5: Six-year-old girl touching toes in long sitting.

This 6-year-old girl touches her toes easily. There is good contour of the back and normal hamstring length.

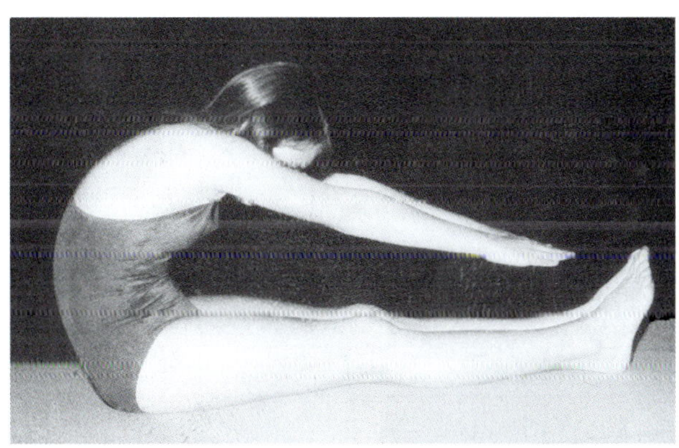

FIGURE APP-A-6: Twelve-year-old girl unable to touch toes.

This subject is a 12-year-old girl. The inability to touch the toes is typical of this age. Sometimes, leg length is the determining factor, and sometimes, as in this case, there is slight shortness of the hamstrings at this age.

FLEXIBILITY TESTS: CHARTS

Flexibility Test 1: Touching Fingertips to Toes

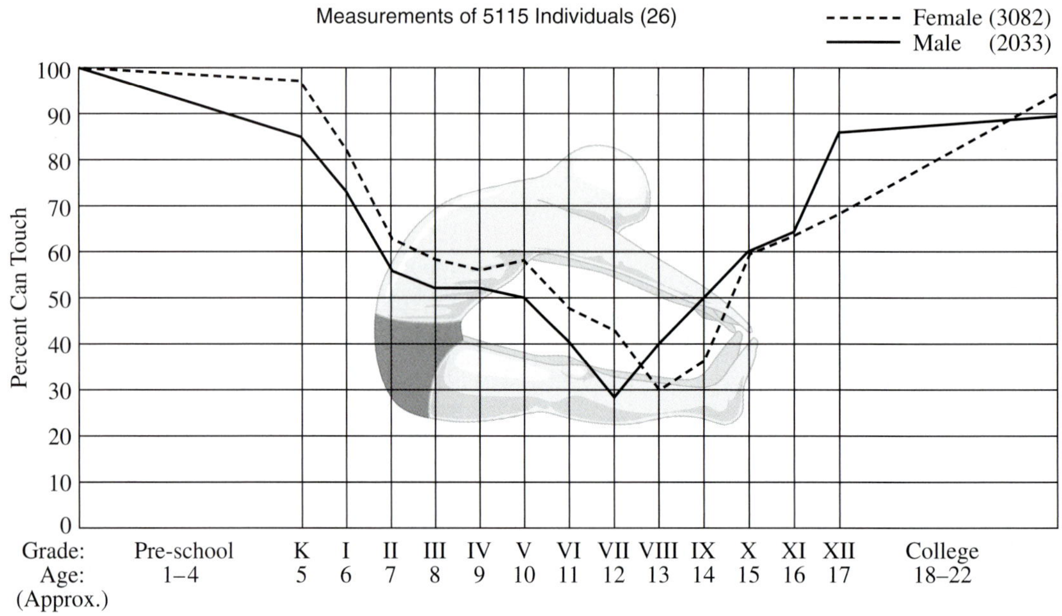

A

FLEXIBILITY TEST #I: TOUCHING FINGER TIPS TO TOES
Measurements of 5115 Individuals

MALE					FEMALE			
Range of Limitation	Mean	% Can Touch	Total Ex'd.	Grade Age	Total Ex'd.	% Can Touch	Mean	Range of Limitation
1/2″–9″	2 3/4″	86	102	K 5	102	98	3 3/4″	3 1/2″–4″
1″–10″	4″	74	125	I 6	108	83	3″	1/2″–4″
1/2″–10 1/2″	3″	56	147	II 7	152	63	3 1/2″	1/2″–10 1/2″
1/2″–9 1/2″	3 1/2″	52	150	III 8	192	59	4″	2″–8 1/2″
1/2″–10 1/2″	4 1/2″	52	150	IV 9	158	57	4 1/2″	1″–13 1/2″
1″–10″	4 1/2″	50	158	V 10	174	59	4″	1/2″–8″
1″–11 1/2″	4 1/4″	41	140	VI 11	156	49	4 1/2″	1/2″–10″
1/2″–9 1/2″	4″	28	100	VII 12	100	43	6″	1/2″–11 1/2″
1 1/2″–13″	4 1/2″	40	151	VIII 13	115	30	5″	1/2″–10″
1/2″–10″	4 1/2″	50	222	IX 14	108	37	5 1/2″	2″–13″
1/2″–12 1/2″	3 1/2″	60	100	X 15	498	59	5″	1/2″–12″
1/2″–12 1/2″	5″	64	100	XI 16	507	64	5″	1″–12″
1″–12″	3″	87	113	XII 17	405	69	5″	1″–14″
1″–11″	4″	90	275	18–22	307	95	3″	1″–6 1/2″
Total number tested:			2033		3082	:Total number tested		

B

FIGURE APP-A-7: Flexibility test: touching fingers to toes.

Flexibility Test 2: Touching Forehead to Knees

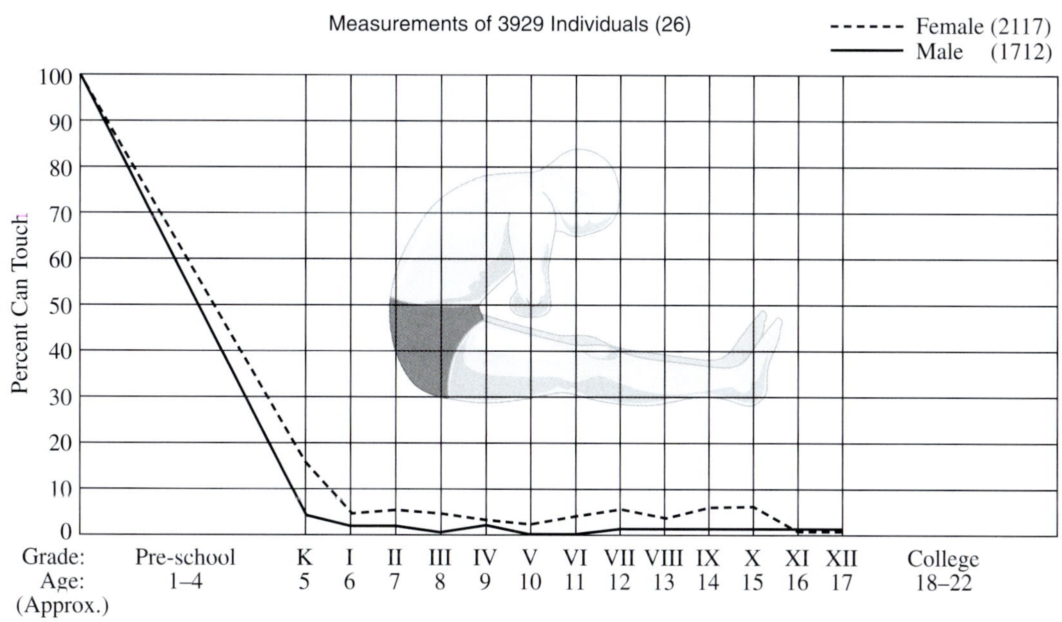

A

FLEXIBILITY TEST #2: TOUCHING FOREHEAD TO KNEES
Measurements of 3929 Individuals

MALE					FEMALE			
Range of Limitation	Mean	% Can Touch	Total Ex'd.	Grade Age	Total Ex'd.	% Can Touch	Mean	Range of Limitation
½″–10″	5″	5	102	K 5	102	16	4″	½″–7½″
2″–11½″	7″	2	125	I 6	108	5	6″	½″–10½″
3″–13″	7½″	2	147	II 7	152	6	7″	1″–13½″
½″–11″	6½″	1	150	III 8	192	5	6″	1″–11½″
4″–14″	9″	2	150	IV 9	158	3	7½″	1″–12½″
1″–12½″	7″	0	158	V 10	174	2	6″	1″–10½″
1½″–15″	7½″	0	140	VI 11	156	4	6¾″	2″–11½″
3½″ 13½″	9″	1	100	VII 12	100	5	6″	½″–11½″
1″–18″	8″	1	112	VIII 13	116	4	7″	1½″–20″
2″–19″	10″	1	215	IX 14	129	6	7″	½″–12″
1½″–19″	9″	1	100	X 15	173	6	8″	1″–18½″
2½″–23½″	11″	1	100	XI 16	277	0	8″	1″–18½″
½″–18″	8″	1	113	XII 17	281	1	8″	1½″–20″
Total number tested:			1712		2117	:Total number tested		

B

FIGURE APP-A-8: Flexibility test: touching forehead to knees.

PROBLEMS WITH PHYSICAL FITNESS TESTS

Many tests have been designed to evaluate the physical fitness of schoolchildren, armed services personnel, athletic teams, and countless others engaged in health and fitness programs. The same movements also have been used as exercises to build strength, endurance, and flexibility. Awards, promotions, and accolades are given or withheld on the basis of these test results.

In spite of long-standing and widespread use, three tests in particular need to be reevaluated:

- Knee-bent sit-ups
- Push-ups
- Sit-and-reach

The usefulness of these tests depends on their accuracy and ability to detect deficiencies. Unfortunately, these tests have become an evaluation of the performance rather than a measure of the physical fitness of the performer (9, 10). Emphasis is on excesses—speed of performance, number of repetitions, and extent of stretching—rather than on quality and specificity of movement.

The Kendalls decided to discuss these tests in this book because of the need to correct misleading information and because of the adverse effects of these tests and their results on both children and adults.

Knee-Bent Sit-Ups with Feet Held Down

The knee-bent sit-up test requires that a person perform as many sit-ups as possible in a period of 60 seconds. The stated purpose of the test is to measure endurance and strength of the abdominal muscles. The test does not fulfill that purpose, however. Instead, it measures strength and endurance of the hip flexor muscles, aided in their performance by stabilization of the feet.

The sit-up movement requires flexion of the hip joints, and this movement can be performed *only* by hip flexors. Abdominal muscles do not cross the hip joint, so they cannot assist in the hip flexion movement. Abdominal muscles flex the spine (i.e., curl the trunk), and to test the strength of these muscles, the trunk must be curled. If these muscles can *hold the trunk curled* as the movement of hip flexion is performed, this indicates good upper abdominal muscle strength.

The problem with using the sit-up movement as a test or an exercise lies in the failure to differentiate between a "curled-trunk sit-up" and an "arched-back sit-up." The former involves strong contraction of the abdominal muscles to hold the trunk curled; the latter puts a stretch on the abdominal muscles and a strain on the low back. This strain may be felt by both children and adults when they are required to perform as many sit-ups as possible in the time allotted.

Many will start the sit-up with the trunk curled. The endurance of the abdominal muscles will not be sufficient to maintain the curl, however, and as the test progresses, the back will arch increasingly. Some will not have the strength even to curl the trunk initially, and the test will be done with the back arched for the entire 60 seconds. The problem is that those with weak abdominal muscles can pass this so-called "abdominal muscle test" with a high score.

The test, as advocated, requires speed of performance. For accurate testing of abdominal muscle strength, however, the test must be done slowly, making sure that the trunk curls *before* hip flexion starts and that the *curl is maintained* both when hip flexion starts and while moving to the sitting position. To have validity, the test should require that credit be given only for the number of sit-ups that can be performed with the trunk curled. Currently, the test has no such requirement. Furthermore, the test cannot be done rapidly if the position of the trunk is to be observed closely (see Chapter 5 for extensive coverage of the sit-up movement and testing of upper and lower abdominal muscle strength).

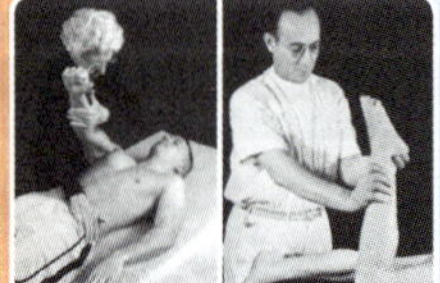

CLASSIC KENDALL

The people most in danger of being adversely affected by repeated sit-ups with the knees bent are children and youths because they start with more flexibility than adults. Those adults who have low back pain associated with excessive low back flexibility also may be adversely affected by this exercise. An interesting phenomenon noted by the Kendalls: Some subjects who have done a great number of knee-bent sit-ups show excessive flexion in sitting or in forward bending but a lordosis in standing.

It is unfortunate that the ability to do a certain number of sit-ups, regardless of how they are performed, is used as a measure of physical fitness. Along with push-ups, these two exercises probably are stressed more than any others in fitness programs. Done to excess, however, these exercises tend to increase—or even produce—postural faults.

When, how, and to what extent the knee-bent position should be used is discussed in Chapter 5.

Push-Ups

When a push-up is performed properly, the scapulae abduct as the trunk is pushed upward. The scapulae move forward to a position that is comparable to that of reaching the arms directly forward. When the serratus anterior muscle is weak, the push-up movement still can be performed, but the scapulae do not move into the abducted position as in a properly performed push-up.

If the primary purpose of the push-ups is to test strength and endurance of the arm muscles, it accomplishes that purpose, but in the presence of serratus weakness, it does so at the expense of the serratus muscle. Evidence for this is seen in the winging of the scapulae and in the inability to complete the range of scapular motion in the direction of abduction.

When push-ups are done at the expense of the serratus muscle, the activity can no longer be considered an index for the physical fitness of the person being tested.

Sit-and-Reach

Sitting with knees extended, this test is done by reaching forward to touch fingertips to toes. For *young* children and *most* adults, touching the toes in this position may be considered a normal accomplishment. Reaching beyond the toes usually denotes excessive flexibility of the back, excessive length of the hamstrings, or both. The stated purpose of the sit-and-reach test is to evaluate the flexibility of the low back and hamstrings. Scoring is based on how many inches *beyond* the toes the individual can reach. Ostensibly, the distance beyond equates with good, better, or best flexibility of the back and hamstrings, with emphasis on "the more, the better."

This test fails to address important variables that affect test results. Variations in "normal" occur according to age group, and limitations result from imbalances between the length of back and hamstring muscles.

This inability to touch toes—much less to reach beyond them—is normal for many youths between the ages of 10 and 14 years. These children are at a stage of growth when the lower extremities are long in relation to the trunk, and they should not be forced to touch their toes (9).

Limited back flexibility can go undetected if the hamstrings are stretched. Individuals with this imbalance may "pass" the test, whereas many children with normal flexibility for their age will "fail." *It would be more accurate to say that the test has failed these children than to say that these children have failed the test.*

In addition to being told that they have "failed," many young people are then given exercises to increase flexibility of the spine and/or stretch the hamstrings when such exercises are unnecessary or even contraindicated.

Adults will demonstrate numerous variations in length of the hamstrings and back muscles. Like adolescents, those adults whose lower extremities are long in relation to the trunk may have normal flexibility of the back and hamstrings yet be unable to touch their toes.

The extensive use of physical fitness tests and the importance placed on their results make it imperative that these tests be carefully scrutinized and performed.

REFERENCES

1. Nissinen M. Spinal posture during pubertal growth. Acta Paediatr 1995; 84: 308–312.
2. Buschang P. Differential long bone growth of children between two months and eleven years of age. Am J Phys Anthropol 1982; 58: 291–295.
3. Nissinen M, Heliovaara M, Scitsamo J, Kononen M, Hurmerinta K, Poussa M. Development of trunk asymmetry in a cohort of children ages 11 to 22 years. Spine 2000; 25(5): 570–574.
4. Willner S, Johnson B. Thoracic kyphosis and lumbar lordosis during the growth period in children. Acta Paediatr Scand 1983; 72: 873–878.
5. Einkauf DK, Gohdes ML, Jensen GM, Jewell MJ. Changes in spinal mobility with increasing age in women. Phys Ther 1987; 67(3): 370–375.
6. Hein V. A method to evaluate spine and hip range of motion in trunk forward flexion and normal values for children at age 8–14 years. Med Sport 1996; 49: 379–385.
7. Widhe T. Spine: Posture, mobility, and pain. A longitudinal study from childhood to adolescence. Eur Spine J 2001; 10: 118–123.
8. Kendall HO, Kendall FP. Normal flexibility according to age groups. J Bone Joint Surg [AM] 1948; 30: 690–694.
9. Cornbleet SL, Woolsey NB. Assessment of hamstring muscle length in school-aged children using the sit-and-reach test and the inclinometer measure of hip joint angle. Phys Ther 1996; 76: 850–855.
10. Kendall F. A criticism of current tests and exercises for physical fitness. Phys Ther 1965; 45: 187–197.

APPENDIX

B

CONSIDERATIONS FOR SPECIALTY DIAGNOSES/POPULATIONS

OUTLINE

SCOLIOSIS

The normal spine has curves in both anterior and posterior directions, but a curve in a lateral direction is considered to be abnormal. Scoliosis is a lateral curvature of the spine. Because the vertebral column cannot bend laterally without also rotating, scoliosis involves both lateral flexion and rotation.

There are many known causes of scoliosis. It may be congenital or acquired; it may result from disease or injury. Some of the causes involve changes in bony structure, such as wedging of a vertebral body, and some relate to neuromuscular problems directly affecting the musculature of the trunk. Still others relate to impairment of an extremity, such as shortness of one leg, or impairment of vision or hearing (1).

Many cases of scoliosis, however, have no known cause. These cases are referred to as **idiopathic**. In spite of the battery of tests that are available to help establish a cause, a high percentage of cases fall into this category. This section deals chiefly with idiopathic scoliosis. Muscle imbalance that exists as a result of disease, such as poliomyelitis, is readily recognized as a cause of scoliosis when it affects the musculature of the trunk. However, muscle imbalance also is present in so-called "normal" individuals but often goes unrecognized except by those who employ muscle testing when examining cases of faulty posture. A basic problem in the management of idiopathic scoliosis is failure to accept the fact that muscle imbalance, which can exist without a known cause, plays an important role in the etiology.

The following discussion focuses on one segment of this subject that deserves attention—that is, the care of patients with early scoliosis for whom proper exercises and supports can make a difference in outcome.

In examining patients with scoliosis, it is especially important to observe the relationship of the overall posture to the plumb line. Suspending a plumb line in line with the seventh cervical vertebra or the buttocks crease (as is frequently done) may be useful in ascertaining the curvature of the spine itself. It does not, however, reveal the extent to which the spine may be compensating for a lateral shift of the pelvis or other postural faults that contribute to the lateral pelvic tilt and associated spinal deviations. Analysis of postural alignment appears in Chapter 2.

HISTORICAL NOTE
Exercise Programs

Throughout the years, elaborate exercise programs have been instituted in response to the treatment needs of patients with scoliosis. The creeping exercises advocated by Klapp were discarded when problems with children's knees forced the discontinuance of such a program (2). Exercises that overemphasized flexibility created problems by making the spine more vulnerable to collapse. When treating patients with S-curves, one must avoid exercises that adversely affect one of the curves while attempting to correct the other.

It is not surprising, therefore, that the usefulness of exercises in cases of scoliosis has been questioned. For many years, the attitude has been that exercises have little or no value. This idea is not new. The following statement was made years ago by Risser:

> It was customary at the scoliosis clinic at New York Orthopedic Hospital, as late as 1920–1930, to send new patients with scoliosis to the gymnasium for exercises. Invariably the patients who were 12 to 13 years of age showed an increase of the scoliosis … it was therefore assumed that exercises and spinal motion made the curve increase (3). (Reprinted with permission from Risser JC: Scoliosis Past and Present. In: Basmajian JV, ed. *Therapeutic Exercise*. 4th ed. Baltimore: Williams & Wilkins; 1984:469.)

Except in some isolated instances, exercise programs for patients with scoliosis continued to be looked on with skepticism. The American Academy of Orthopedic Surgeons 1985 lecture series included this statement:

> Physical therapy cannot prevent a progressive deformity, and there are those who believe specific spinal exercise programs work in a counterproductive fashion by making the spine more flexible than it ordinarily would be and by so doing making it more susceptible to progression (4).

However, intervention strategies such as Schroth exercises and the Functional Individual Therapy of Scoliosis (FITS) method have demonstrated that a supervised and guided exercise regimen can be beneficial to this patient population (5, 6).

The overemphasis on flexibility was wrong. Adequate musculoskeletal evaluation has been lacking. As a result, there has been little scientific basis on which to justify the selection of therapeutic exercises. Scoliosis is a problem of asymmetry, and to restore symmetry requires the use of asymmetrical exercises along with appropriate support. Stretching of tight muscles is desirable, but overall flexibility of the spine is not. It is better to have stiffness in the best attainable position than to have too much flexibility of the back (7).

Scoliosis Resulting from Neuromuscular Disease

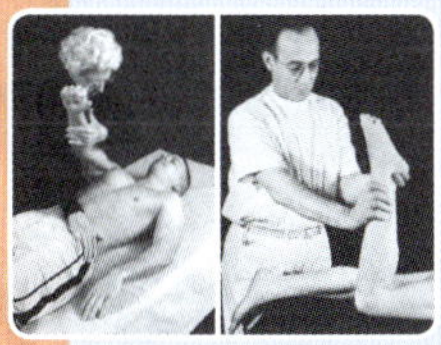

CLASSIC KENDALL

Treatment for Polio

The lessons learned from treating patients with polio were easily understood because of the obvious effects of the disease on the functions of muscles. People who treated these patients appreciated that deformities could develop where muscle imbalance existed. They saw the devastating effects of muscle weakness and the subsequent tightness or contracture in opposing muscles, not the least of which were the effects on the spine. Some potentially severe problems were helped by appropriate intervention.

Figure App-B-1 shows the marked weakness of the right abdominal musculature and the associated lateral curve. This patient had polio at the age of 1 year and 4 months but was not admitted to a hospital for treatment until the age of 8 years and 8 months. She was placed on a flexed frame to relax the abdominal muscles with a pullover strap pulling in the direction of the right external oblique. Specific exercises were given to the weak muscles of the trunk in addition to the support from the pullover strap. Seven months after treatment was started, the strength of the abdominal muscles had improved, with the right external oblique showing an increase from a poor minus to a good grade.

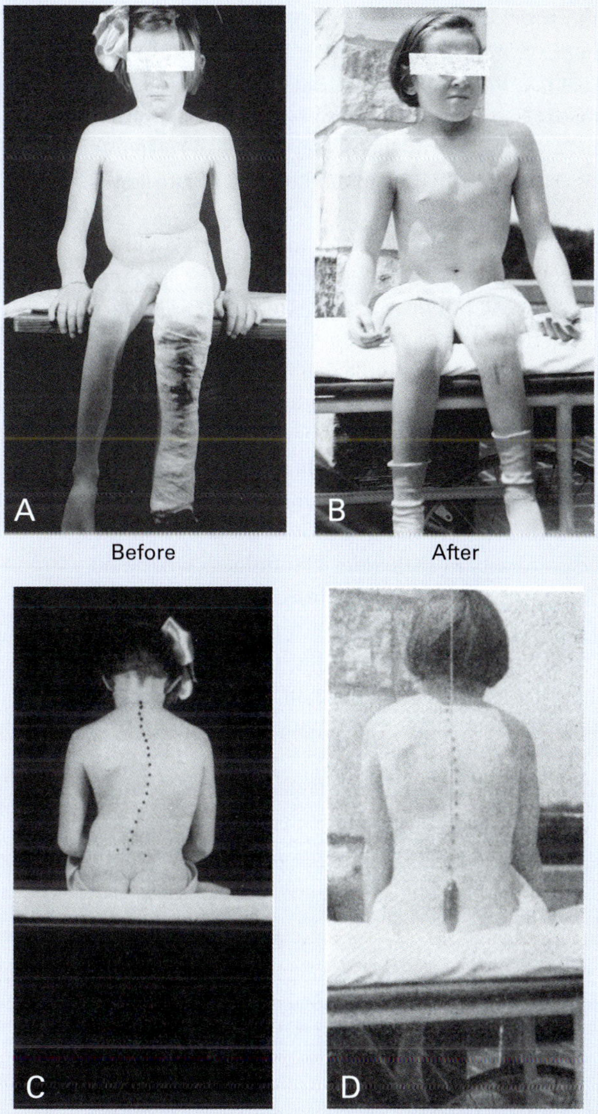

FIGURE APP-B-1: Photos of a child with polio: **A**, Before treatment; **B**, After treatment; **C**, Before treatment; **D**, After treatment.

In treating patients with polio, it became obvious in many instances that weakness caused by stretching had been superimposed on the initial weakness caused by the disease. As in the case illustrated here, the muscles were not reinnervated by relieving stretch and strain on them, innervation existed as a latent factor. The stretched muscles were incapable of response until the stretch and strain were relieved by adequate support, and until the weak muscles were stimulated by proper exercises.

Postural Examination

Instead of abandoning the use of exercises in the treatment of scoliosis, attention should be focused on a more scientific approach toward the evaluation and selection of appropriate exercises. Musculoskeletal evaluation should include alignment and muscle tests.

Postural *alignment* tests, both plumb line and segmental, in back, side, and front views should be included.

Muscle *length* tests should include, but not be limited to, hip flexor, hamstring, forward bending for contour of the back and length of the posterior muscles, tensor fasciae latae and iliotibial band, and teres and latissimus dorsi.

Muscle *strength* tests should include back extensors, upper and lower abdominals, lateral trunk, oblique abdominal muscles, hip flexors, hip extensors, hip abductors and gluteus medius, hip adductors, and in the upper back, the middle and lower trapezius.

An essential part of examination is observation of the back *during movement*. The examiner stands behind the subject, and the subject bends forward and then returns *slowly* to the upright position. If there is a *structural* curve, some fullness (prominence) will be noted on the side of the convexity of the curve. The fullness will be on one side only if there is a single curve, (i.e., C-curve). In a double curve (i.e., S-curve), as in a right thoracic, left lumbar, there will be fullness on the right in the upper back and on the left in the low back area. In a *functional* curve, however, there may be no evidence of rotation in forward bending. This is especially true if the *functional* curve is caused by lateral pelvic tilt that results from hip abductor or abdominal muscle imbalance.

Rotation of the spine or thorax, as seen in scoliosis cases, is observed with the patient bending forward (Figure App-B-2).

FIGURE APP-B-2: Mild right thoracic curve.

For most people, the curves in the spine are functional; they do not become fixed or structural. When curves do become fixed, they also tend to change and become compensatory—that is, change from a single C-curve to an S-curve. Usually, a single curve toward the left stays as a left curve in the low back and changes to a right curve in the upper back.

In an ordinary C-curve, the shoulder is low on the side of the high hip. If the shoulder is high on the same side as the high hip, there probably is an S-curve. In some cases, faulty alignment appears to be limited to the spine. Figure App-B-3 shows a simple C-curve in which overall plumb alignment of the body is good. Segmentally, the right shoulder is low along with the C-curve.

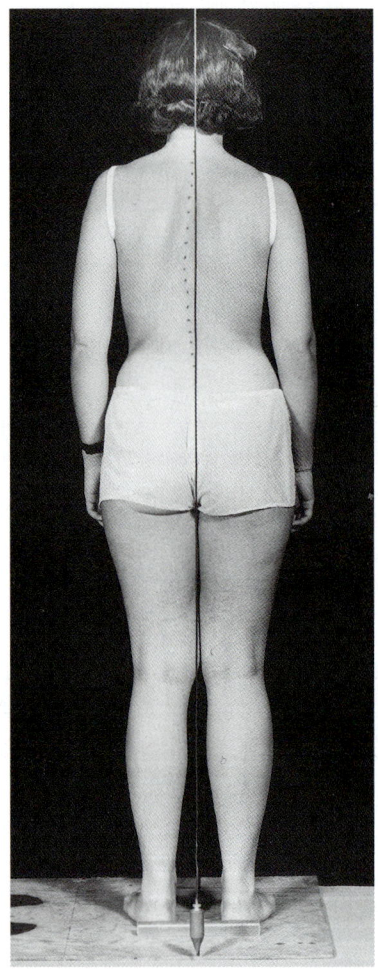

FIGURE APP-B-3: Mild left thoracolumbar curve (C-Curve).

For this patient, a shoe lift is not indicated because the pelvis is level. Exercise is indicated for the right internal oblique and left external oblique by shifting the upper trunk toward the right without any lateral movement of the pelvis.

Figures App-B-4 and App-B-5 show faulty alignment, weakness of the lower abdominal muscles, and error in testing for hamstring length and normal hamstring length.

Name .... Doctor .... —
Diagnosis .... *Faulty posture, mild scoliosis* .... Date of 1st Ex. .... —
Onset .... Date of 2nd Ex. .... —
Occupation .... *Student* .... Height .... — Weight .... —
Handedness .... *Right* .... Age ... *17* .... Sex .... Leg length: Left .... — Right .... —

PLUMB ALIGNMENT

Side view: Lt. .... Rt. ....

Back view: Deviated lt. .... Deviated rt. ....

SEGMENTAL ALIGNMENT

	X	Hammer toes		Hallux valgus		Low ant. arch		Ant. foot varus
Feet	*L*	Pronated >		Supinated		Flat long. arch		Pigeon toes
	B	Med. rotat. >		Lat. rotat.	*B*	Knock-knees *slight*		
Knees		Hyperext. >	*B*	Flexed *L>R*		Bowlegs		Tibial torsion
Pelvis	*R*	Leg in postural add.		Rotation	*Ant.*	Tilt	*Ant.*	Deviation
Low back	*X*	Lordosis *marked*		Flat		Kyphosis		Operation
Up. back	*X*	Kyphosis		Flat	*B*	Scap. abducted *R>L*		Scap. elevated
Thorax		Depressed chest		Elevated chest		Rotation	*Post.*	Deviation *slight*
Spine		Total curve	*L*	Lumbar *-Thoracic*		Dorsal	*R*	Cervical *-Thoracic*
Abdomen	*X*	Protruding *slight*		Scars		....		....
Shoulder		Low		High	*B*	Forward	—	Med. rotated
Head	*X*	Forward		Torticollis		....		Rotation ....

TESTS FOR FLEXIBILITY AND MUSCLE LENGTH

Forward bending ..... *Limited 7"* .... Bk. ... *(1)* ..... H.S. .. *N (2)* ... G.S. *Sl. tight*
Arm overhead elevation: Lt. .. *Sl. limited* .... Rt. .. *Normal length* ....
Hip flexors: Lt. .. *Tight* .... Rt. .. *Tight* ....
Tensor fas. lata.: Lt. .. *Sl. tightness* .... Rt. .. *Normal length* ....
Trunk extension: .. *Normal range* ....
Trunk lat. flex.: To lt. .. *slightly limited* .... To rt. .. *Normal range* ....

MUSCLE STRENGTH TESTS

L		R
G-	Mid. trapezius	*G+*
F+	Low. trapezius	*F+*
N	Back extensors	*N*
N	Glut. medius	*G-*
N	Glut. maximus	*N*
N	Hamstrings	*N*
N	Hip flexors	*N*
G	Tib. posterior	*N*
Weak	Toe flexors	*Weak*

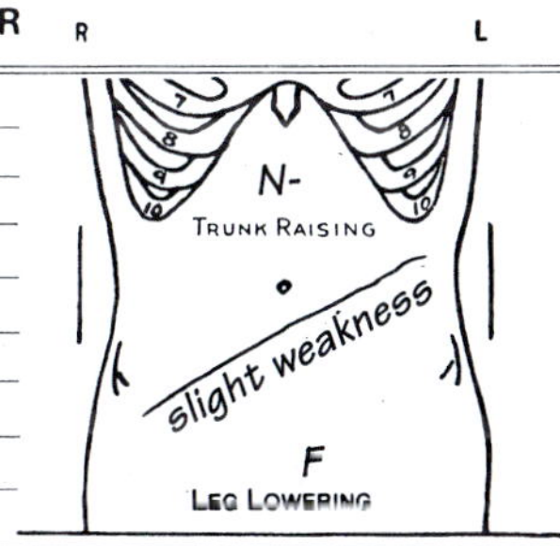

SHOE CORRECTION

Left		Right
1/8"	(Wide Heel) Inner wedge (Narrow heel)	
3/16"	Level heel raise	
medium bar	Metatarsal support	*medium bar*
	Longitudinal support	

NOTES: *(1) Back flexibility limited slightly in lower thoracic area*
(2) Hamstrings normal in forward bending (i.e., angle of sacrum with thigh.)
Hs. appear tight in leg raising due to tight hip flexors keeping pelvis in anterior tilt.

TREATMENT

Knees tend to flex slightly, left>right (probably due to hip flexor shortness).

Exercises:

Bk. Lying	Pel. tilt and breath.	*X*
	Pel. tilt and leg sl.	*X*
	Head and sh. raising	*(omit)*
	Shoulder add. stretch	*X*
	Straight leg-raise	*(omit)*
	Hip flex. stretch	*X*
Sd. Lying	Stretch .. *left* .. tensor	*X*
Sitting	Forward bending	
	To stretch low bk.	—
	To stretch h. s.	—
	Wall-sitting	
	Middle trapezius	*X*
	Lower trapezius	*X*
Standing	Foot and knee ex.	*X*
	Wall-standing	*X*

Other Exercises: ....
stretch toe extensors
In standing, with pelvis stabilized, shift upper trunk slightly toward the right (using left external oblique and right internal oblique abdominal muscles).

Support: ....

FIGURE APP-B-4: Postural examination chart.

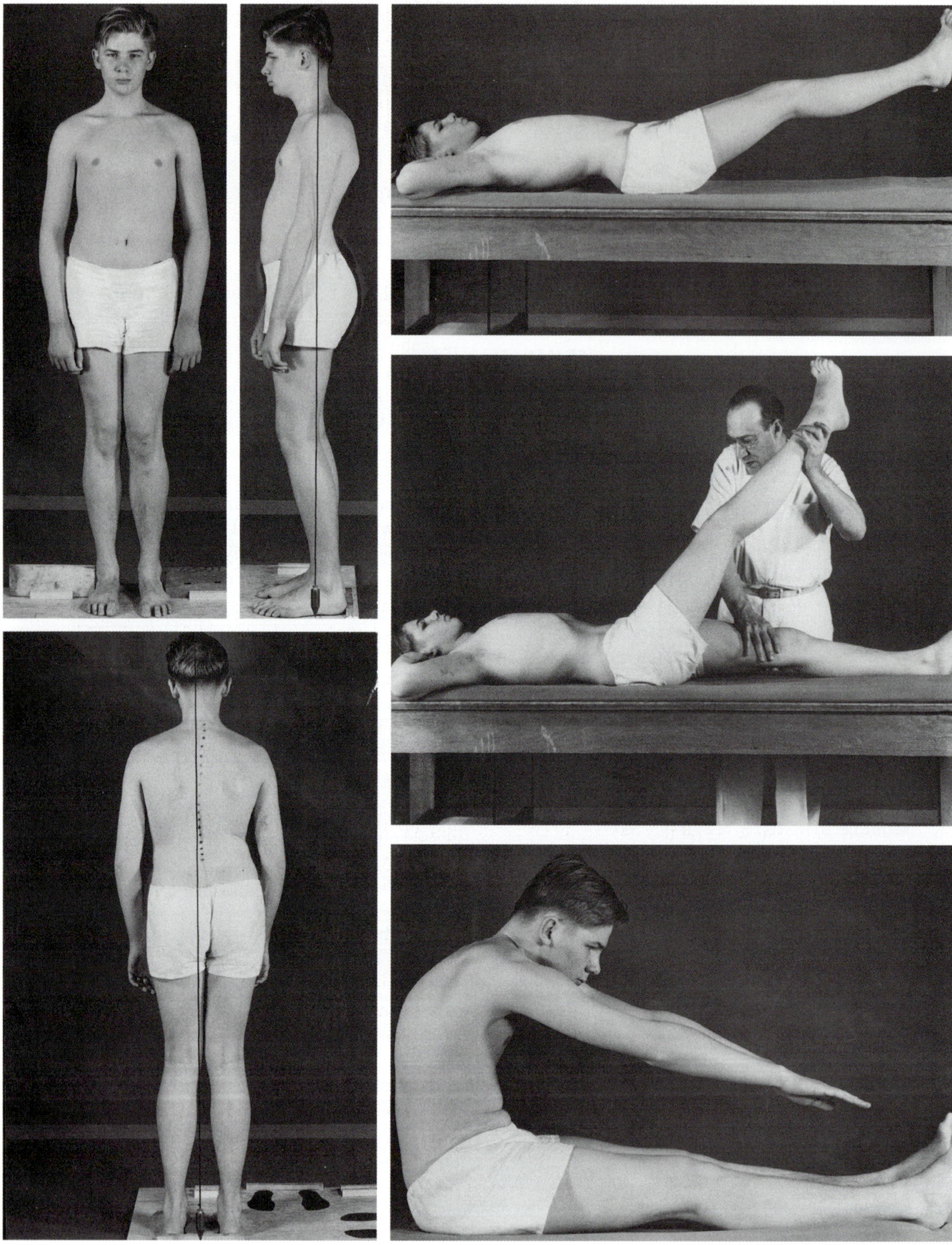

FIGURE APP-B-5: Faulty alignment, weakness of lower abdominal muscles, and error in testing for hamstring length and normal hamstring length.

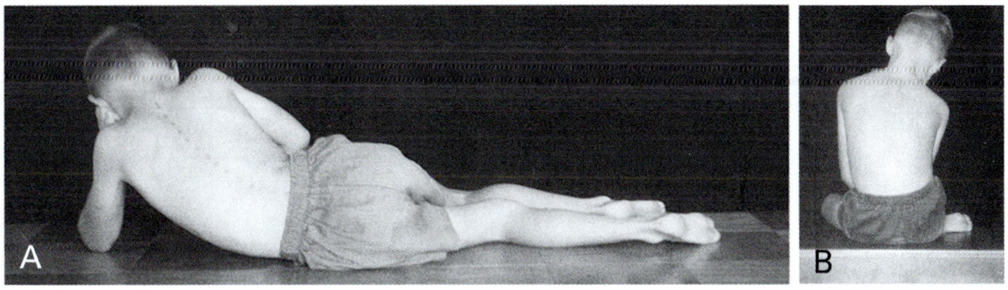

FIGURE APP-B-6: Faulty postural habits.

Functional Scoliosis

Scoliosis and Lateral Pelvic Tilt

If the pelvis tilts laterally, the lumbar spine moves with the pelvis into a position of lateral curve, convex toward the low side. An actual leg-length difference causes a lateral tilt in standing, low on the side of the shorter leg. A temporary position of lateral tilt can be demonstrated by standing with a lift under one foot.

An example of a muscle problem that was recognized as a contributing cause of scoliosis among patients with polio is *unilateral tightness* of the tensor fasciae latae and iliotibial band. The effect of such tightness is to produce a lateral tilt of the pelvis, low on the side of the tightness. The existence of unilateral tightness of these structures is not limited to persons with some known etiology; it is common among so-called "normal" individuals.

Less understood but equally important is the fact that *unilateral weakness* can result in a lateral pelvic tilt. Weakness of the right hip abductors as a group or, more specifically, of the right posterior gluteus medius will allow the pelvis to ride upward on the right side, tilting downward on the left side. Likewise, weakness of the left lateral trunk muscles will allow the left side of the pelvis to tilt downward. These weaknesses may be present separately or in combination, but they occur more often in combination.

In the sitting position, lateral pelvic tilt accompanied by a lateral curve in the spine will result from unilateral weakness and atrophy of the gluteus maximus muscle.

Handedness in Relation to Scoliosis

Seen frequently among right-handed individuals who also exhibit a functional left curve: pronation of the left foot, *tightness* of the left iliotibial band and *weakness* of the right gluteus medius, left hip adductors, and left lateral abdominals. Most people do not develop a scoliosis, but among those who do, there is a predominance of right thoracic, left lumbar curves. There is also a predominance of right-handed people in our society, and many activities and postural positions predispose these people to problems of muscle imbalance that are only discovered by precise and adequate manual muscle testing. Among left-handed individuals, the patterns tend to be the opposite. However, they occur with somewhat less frequency, probably because these people must conform to so many activities or positions that are designed for right-handed use. (Muscle imbalance as related to handedness is illustrated in Chapter 2.)

Faulty Postural Habits

It is important to be cognizant of the postural habits of a child in the various positions of the body in standing, sitting, and lying. For a right-handed individual seated at a desk to write, the position is one in which the body (or upper body) is turned slightly counterclockwise, the paper is turned diagonally on the desk, and the right shoulder is slightly forward.

Children sometimes assume a side-lying position on the floor or bed to do their homework. A right-handed person will lie on the left side so that the right hand is free to write or turn pages in a book. Such a position places the spine in a left curve (Figure App-B-6-A).

Sitting on one foot, such as the left foot, will cause the pelvis to tilt downward on the left and upward on the right because the right buttock is raised by resting on the left foot. The spine then curves to the left (Figure App-B-6-B).

If a back pack is carried by a strap over the left shoulder and the child keeps that shoulder raised to keep the strap from slipping off, there will be a tendency for the spine to curve toward the left.

Children who engage in repetitive, asymmetrical activities, whether vocational or recreational, are prone to develop muscle imbalance problems that can lead to lateral deviations of the spine (8).

When the spine habitually curves toward the same side in the various postural positions, it becomes a matter of concern with respect to correction or prevention of early scoliosis. Not to be overlooked are problems associated with pronation of one foot with one knee slightly bent, if it is always the same knee that is bent. Logically, the imbalance in hip musculature and faulty foot or leg positions, which result in lateral pelvic tilts, are more closely related to primary lumbar or thoracolumbar curves than to primary thoracic curves.

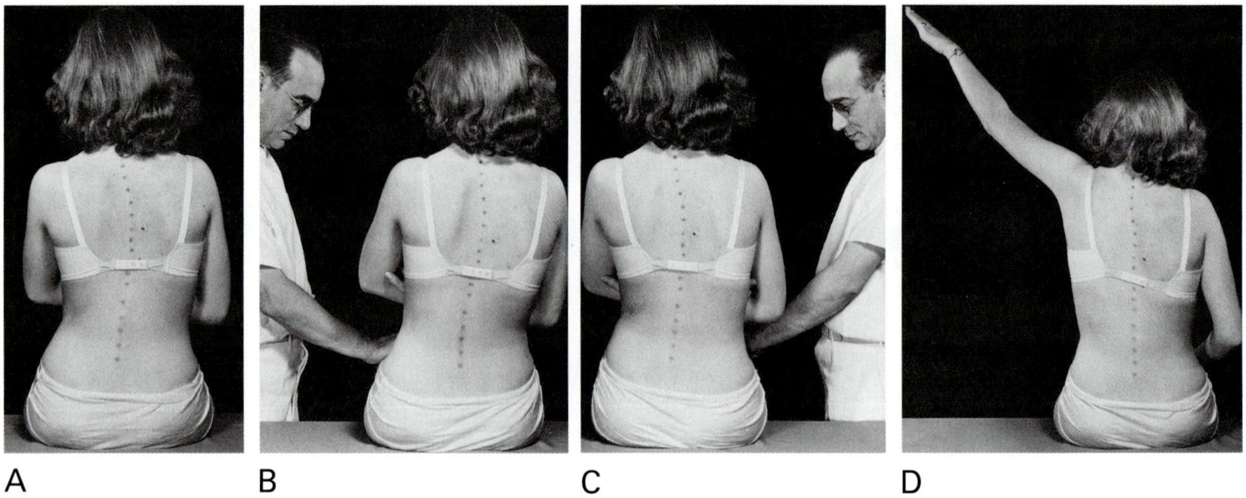

FIGURE APP-B-7: **A**, In a sitting position, a right thoracic and slight left lumbar curve; **B**, The adverse effects of exercising the left iliopsoas; **C**, The correction that takes place with exercise of the right iliopsoas; **D**, The overall correction when the appropriate exercise to correct the thoracic curve is added.

Exercises and Supports

Exercises

Exercises should be carefully selected on the basis of examination findings. There must be adequate instruction to ensure that the exercises will be performed with precision. If possible, a parent or other individual in the home should monitor performance until the child becomes capable of doing the exercises without supervision. The object is to use asymmetrical exercises to bring about optimal symmetry.

It has been determined in Figure App-B-7 that the right iliopsoas is weak. The subject is a dancer. One of the stretching exercises she performs is a split in which one leg is forward and the other is back. Routinely, the left leg has been forward and the right leg back. There is a left lateral curve in the lumbar region and a right curve in the thoracic area.

Because the psoas muscle attaches to the lumbar vertebrae, transverse processes, and the intervertebral disks, this muscle can pull directly on the spine. If the spine is flexible, it can be influenced by exercises, carefully performed, that help to correct the lateral deviation. The exercise is done sitting at the side of a table with the knees bent and the legs hanging down. (It is not done in a supine position.) A strong effort is made as if to lift the right thigh in flexion, but enough resistance is applied (by an assistant or the subject) to prevent movement of the thigh. By so doing, the force is not dissipated by movement of the thigh but is exerted on the spine, pulling it toward the right (Figure App-B-7-C).

The person who monitors this exercise should stand behind the subject while the exercise is being performed to ensure that both curves are being corrected simultaneously. Because curves vary greatly, close monitoring is necessary to avoid emphasis on the correction of one curve at the expense of the other.

In a right thoracic, left lumbar scoliosis, there is often weakness of the posterolateral part of the right external oblique muscle and shortness of the upper anterior part of the left external oblique. In the supine position, the subject places the right hand on the right lateral chest wall and the left hand on the left side of the pelvis. Keeping the hands in position, the object of the exercise is to bring the two hands closer by contracting the abdominal muscles, but without flexing the trunk. It is as if the upper part of the body shifts toward the left and the pelvis shifts toward the right. By not allowing trunk flexion and by contracting the posterolateral fibers of the external oblique, there will be a tendency toward some counterclockwise rotation of the thorax in the direction of correcting the thoracic rotation that accompanies a right thoracic curve.

It is of particular importance that girls between the ages of 10 and 14 years have periodic examination of the spine. More spinal curvatures occur in girls than in boys, and it usually appears between these ages.

Regarding the thoracic curve correction: Sitting tall with the spine in as good an anteroposterior alignment as possible, the subject reaches in a diagonally upward direction, slightly forward from the coronal plane. The aim is to practice holding the corrected position to develop a new kinesthetic sense of what is straight. The faulty position has become so habitual that the straight position feels abnormal.

All too often, early cases of lateral curvature are "treated" merely by observation, with radiographs obtained at specified intervals. Early tendencies toward a lateral curvature are potentially more serious than the anteroposterior deviations seen in the usual faulty postures. Instruction in good body mechanics and appropriate postural exercises, plus the necessary shoe alteration to mechanically assist in the correction of alignment, constitutes more rational treatment than mere observation.

Correction of lateral pelvic tilt associated with a lateral curvature can be helped by proper heel lifts (9). The cooperation by the subject is of utmost importance. The lifts need to be used in all footwear. No amount of lift can help if the subject continues to stand with the weight predominantly on the leg with the higher hip and with the knee flexed on the side of the lift. For use of a lift in connection with a tight tensor fasciae latae and iliotibial band, see Chapter 7. For use of a lift in the heel of the opposite shoe to relieve strain on a weak gluteus medius, see Chapter 7.

Along with the use of appropriate exercises, it is important to avoid those exercises that would have an adverse effect. Increasing the overall flexibility of the spine carries an inherent danger. Gains in flexibility in the direction of correcting the curves are indicated, provided that strength is also increased to maintain the corrections. If the subject has the potential for gaining strength and is dedicated to a strict program of strengthening exercises and wearing a support, exercises that increase flexibility can have a desirable end result.

A subject who is developing a kyphoscoliosis along with a lordosis should not do back extension exercises from a prone position because in an effort to obtain better extension in the thoracic spine, the low back problem increases. Extension of the thoracic spine may be done while sitting on a stool with the back against a wall, but the lumbar spine must not arch in an effort to make it appear that the thoracic spine is straight. In this same instance, upper abdominal exercises by trunk curls or situps should be avoided even if the upper abdominals are weak. The exercise would be counterproductive, because curling the trunk is flexing the thoracic spine. If there is a developing kyphoscoliosis, such an exercise would increase the kyphotic curve. Exercise of the lower abdominals in the form of pelvic tilt or of pelvic tilt and leg sliding, emphasizing the action by the external oblique, however, would be strongly indicated.

The significance of muscle imbalance and overall faulty posture as etiological factors in idiopathic scoliosis should not be overlooked. Scoliosis is a complex postural problem. As such, it calls for thorough evaluation procedures to determine any weakness or tightness of muscles that results in distortion of alignment. Verification can come only from repeated testing, but the testing must be done with precision. There must be adherence to the principles on which manual muscle testing is founded. Using a long lever whenever appropriate is vitally important to distinguish differences in strength of some of the large muscles (e.g., hip abductors) when comparing one side with the other.

Supports

In addition to exercise and proper shoe corrections, many patients with early scoliosis need some support. It may be that only a corset type of support is needed or, as in more advanced cases, a more rigid support. The Kendalls made many of these rigid supports.

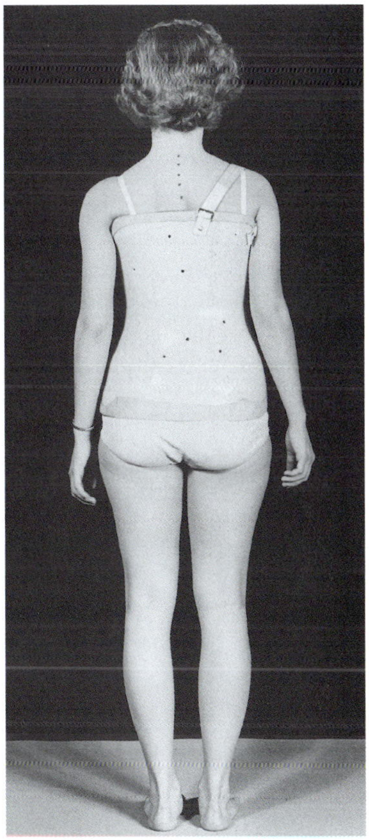

FIGURE APP-B-8: The subject is shown wearing a removable cellulose jacket, which was often used for scoliosis cases.

In Figure App-B-8, the subject is shown wearing a removable cellulose jacket, which was often used for scoliosis cases. The procedure for making this jacket requires the subject to be placed in a standing position with head traction from a Sayre head sling. A heel raise was used to level the pelvis, and straps of adhesive tape or moleskin were placed diagonally from the rib cage to the opposite iliac crest to obtain the best possible correction of the trunk position before the original plaster case was made. For young females, a brassiere with small extra padding would be put on under the stockinet to allow room for development of the breasts.

After the positive plaster mold was poured and allowed to dry, further adjustments were made by shaving down slightly on the side of convexity and adding an equal amount of plaster at places of concavity at the same level to maintain the necessary circumference measurements. The jacket was then made over the plaster mold.

Today, newer materials provide greater versatility and ease of handling, but the basic principles for use of supports have changed little: Obtain the best possible alignment, allow for expansion in the area of concavity, and apply pressure in the area of convexity to the extent tolerated without adverse effects or discomfort.

Early Intervention

Intervention in the very early stages of a lateral curve does not mean getting involved in a vigorous, active program of exercises. Rather, it means prescribing a few carefully selected exercises that help establish a kinesthetic sense of good alignment. It means providing good instruction to the patient and parents or guardians on how to avoid habitual positions or activities that clearly are conducive to increasing the curvature.

It may mean taking a picture of the child's back in the usual sitting or standing position, and then another in a corrected position, so that the child can see the effect of the exercise on posture. It also means providing incentives to help keep the person interested and cooperative, because achieving correction is an ongoing project.

For those in whom the curve has become more advanced, in many instances, it is necessary and advisable to provide some kind of a support to help maintain the improvement in alignment that has been gained through an exercise program.

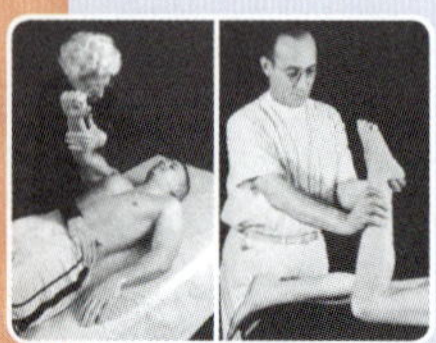

CLASSIC KENDALL

Kendall's Work in Scoliosis

Henry O. Kendall was the first physical therapist at the Children's Hospital in Baltimore, beginning work there in June 1920. The following is a quote from some handwritten notes made by him in the early 1930s regarding scoliosis:

> Symmetrical exercises should not be attempted. A careful muscle examination should be made and muscles graded according to their strength. If one group or one muscle is too strong for its antagonist, that muscle or group should be stretched and the weaker antagonist built up to sufficient strength to compete with it.
>
> In examination of more than one hundred cases of lateral curvature, I have yet to find a case with weak erector spinae muscles, each and every case was able to hyperextend the spine against gravity and in most cases against resistance as well.
>
> The muscle weakness was almost always found in the lateral abdominals, anterior abdominals, pelvic, hip and leg muscles. This weakness caused the body to deviate from either the lateral median plane or the anterior-posterior median plane, causing the patient to compensate for the deviation by substituting other muscles in order to maintain equilibrium. In doing the substituting, the patient invariably develops muscles which cause lateral rotatory movements and it is easy to see why we have lateral curvature with rotation.
>
> By correcting muscle imbalance we get at the primary cause of many cases of lateral curvature.

Corrective Exercises: Posture

The exercises below are designed to help correct some common postural faults. Elsewhere in this text, there are additional corrective exercises. Specific exercises are done to improve muscle balance and restore good posture (Figure App-B-9). To be effective, they should be done every day for a period of weeks, plus daily practice in assuming and maintaining good posture until it becomes a habit.

While working to correct muscle imbalance, it is usually advisable to avoid the following exercises:

- Lying on the back and raising both lower extremities at the same time;
- Lying on the back and coming up to a sitting position with the feet held firmly down;
- Lying on the back with most of the weight resting on the upper back and doing "bicycling" exercise;
- Standing or sitting with knees straight, reaching forward to touch toes;
- The exercise of raising the trunk to arch the back from a face-lying position (for those who have an increased forward curve in the low back).

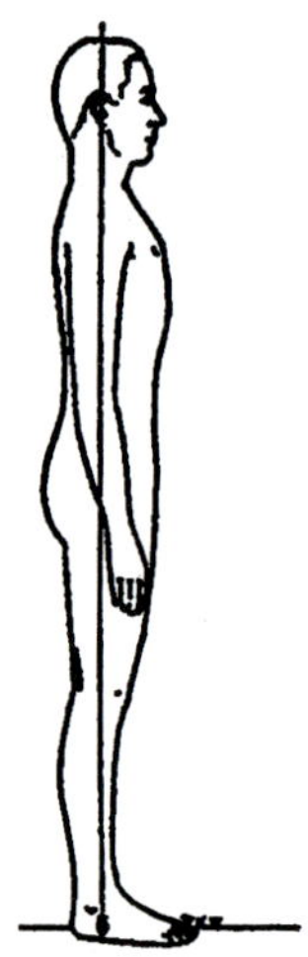

FIGURE APP-B-9: An ideal postural position.

Posterior Neck Stretch

In back-lying position, bend knees and place feet flat on floor (Figure App-B-10-A). With elbows bent and hands up beside head, tilt pelvis to flatten low back. Press head back, with chin down and in, trying to flatten neck.

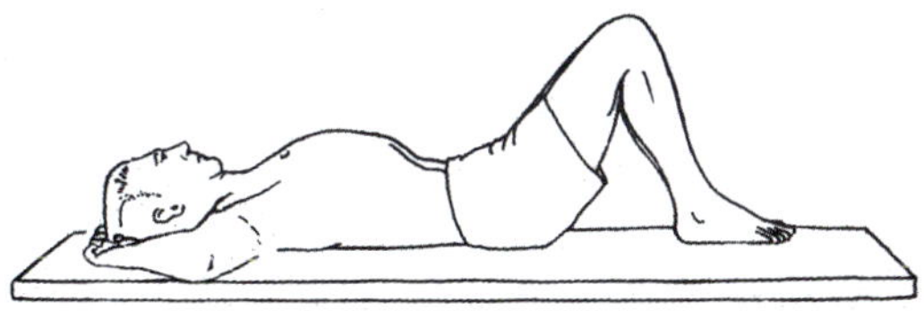

A

Shoulder Adductor Stretching

With knees bent and feet flat on floor, tilt pelvis to flatten low back. Hold the back flat, place both arms overhead, and try to reach arms to the table with elbows straight (Figure App-B-10-B). Bring upper arms as close to sides of head as possible. (Do NOT allow the back to arch.)

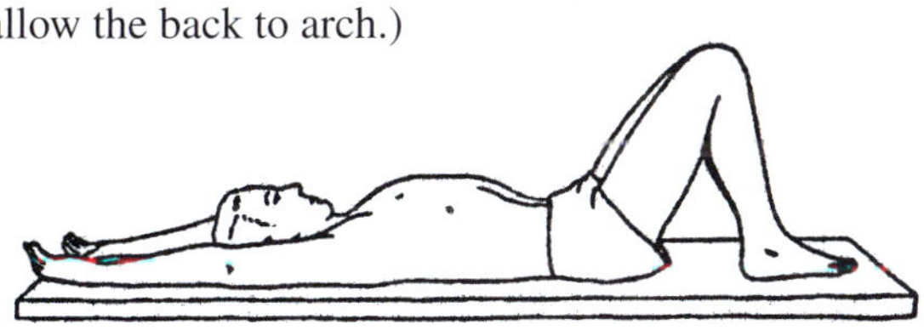

B

FIGURE APP-B-10: **A**, **B**, Corrective exercise for the shoulder adductors.

Wall-Sitting Postural Exercise

Sit on a stool with back against wall (Figure App-B-12). Place hands up beside head. Straighten upper back, press head back with chin down and in, and pull elbows back against wall. Flatten low back against wall by pulling up and in with lower abdominal muscles. Keep arms in contact with wall and slowly move arms to a diagonally overhead position.

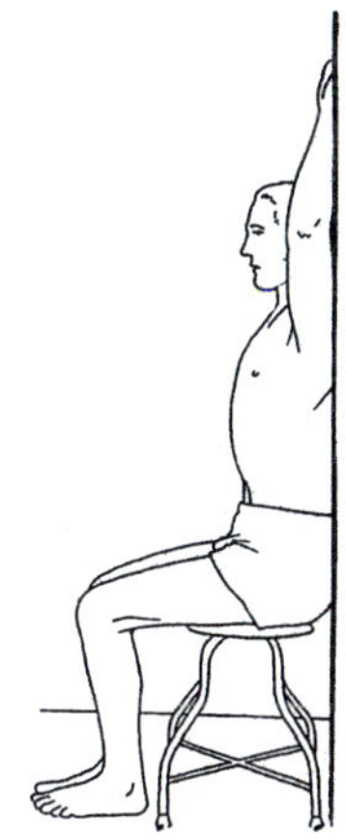

FIGURE APP-B-12: Corrective exercise while sitting in a chair against a wall.

Wall-Standing Postural Exercise

Stand with back against a wall, heels about 3 inches from wall. Place hands up beside head with elbows touching wall. If needed, correct feet and knees as in Figure App-B-11, then tilt to flatten low back against the wall by pulling up and in with the lower abdominal muscles. Keep arms in contact with the wall and move arms slowly to a diagonally overhead position.

FIGURE APP-B-11: Corrective exercise while standing against a wall.

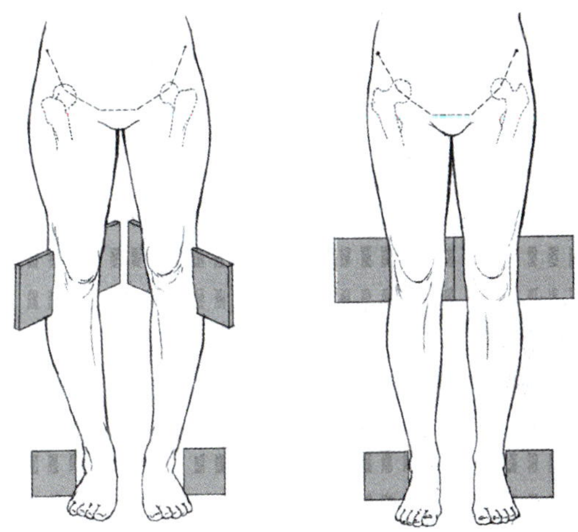

FIGURE APP-B-13: Corrective exercise for pronation, hyperextension, and internal rotation.

Correction of Pronation, Hyperextension, and Internal Rotation

Stand with feet about 4 inches apart and toeing-out slightly (Figure App-B-13). Relax knees into an "easy" position, that is, neither stiff nor bent. Tighten buttock muscles to rotate legs slightly outward (until kneecaps face directly forward). Tighten muscles that lift the arches of the feet, rolling the weight slightly toward outer borders of feet.

POLIO

Factors Influencing Treatment

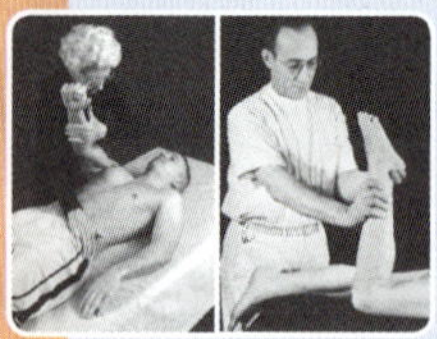

CLASSIC KENDALL

Orthopedic and Physical Therapy Objectives in Poliomyelitis Treatment

When someone asks "How do you treat poliomyelitis?" there is no specific answer because every patient requires a different approach in treatment. In considering the treatment of the individual case, the following questions must be answered:

How long since onset?
How old is the patient?
What is the extent of involvement at present?
What was the extent of original involvement?
How much improvement has there been in individual muscles?
What kind of treatment has the patient had thus far?

Time since onset is important because treatment varies in many respects with the stage of the disease. The relation between duration of the disease and improvement aids in determining prognosis.

Age of the patient is very important because underlying bony and ligamentous structural variations must be considered in relation to treatment.

A comparison between the extent of the *original and present involvement* is important in considering further course of treatment.

The kind of treatment to date is important for various reasons:

- Treatment which has allowed unnecessary contractures to develop distorts the poliomyelitis picture. Such contractures create a superimposed stretch weakness on opposing muscles. Neither an accurate diagnosis nor prognosis can be made until such secondary superimposed factors are corrected.
- Treatment which causes stretch and relaxation of joint structures superimposes an even more serious problem than muscle contracture. It is more difficult to restore normal tightness to stretched ligaments than it is to restore joint motion in cases of some muscle tightness. One cannot determine accurately the power of a muscle if the joint is so relaxed that the muscle does not have a stable joint on which to act.
- Early excessive and prolonged heat treatment tends to distort the degree of involvement. The basic reason for the use of the heat, in itself, explains this phenomenon. Heat is used to relax muscles and to act as a general sedative. When continuous and prolonged heat is applied, muscles lack their normal contractility. We observed an unusual situation in many 1944 poliomyelitis patients transferred from an isolation hospital where heat and stretching movements were employed in the early stage of the disease. Upon admission to the orthopedic hospital, more rest and less active treatment was instituted. There occurred a sudden, unexpected amount of improvement that was not at all typical of the usual course of progress in poliomyelitis muscles. Our explanation for this rapid improvement is that on top of poliomyelitis weakness, which was not pronounced, there was a superimposed weakness due to too much heat and manipulation, which cleared up when such treatment was stopped.
- Prolonged immobilization, which permits joint stiffness or unnecessary disuse atrophy to develop, also distorts the poliomyelitis picture and prolongs recovery.

Polio and Postpolio Muscle Tests

Functional manual muscle tests are necessary parts of diagnostic procedures in the field of neuromuscular disorders. They were essential tools in the early evaluation of patients with polio. The pattern of muscle weakness enabled the examiner to determine the type and location of a neuromuscular lesion. Weakness of specific muscles helped to indicate which spinal motor neurons were involved.

Although polio has been eradicated in most parts of the world, it remains endemic in some countries and poses a serious health threat. In 2003, a polio outbreak in Nigeria spread to neighboring countries and put 15 million children at risk (10). In the first six months of 2004, five times as many children in central and west Africa were paralyzed by polio than during the same period in 2003 (11).

Also of great concern is the emergence in the western hemisphere of West Nile Virus (WNV). According to the Centers for Disease Control and Prevention, 9,006 cases of WNV infection were reported in 2003. This was more than double the 4,156 cases reported in 2002 (12). WNV can cause a polio-like syndrome of muscle weakness and paralysis, because it attacks the same motor cells in the spinal cord as are attacked in polio (13). Richard Bruno states that "nearly 1% of those affected with WNV have paralysis, almost the same percentage as in those affected with polio viruses" (14).

Late Complications of Polio: Postpolio Syndrome

Although most doctors in the United States today have never seen the neuromuscular weakness and paralysis of acute polio, many of them are now confronted with former polio patients who are experiencing new muscle weakness, pain, fatigue, and decreased endurance. Referred to as the postpolio syndrome (PPS), these aftereffects of polio can appear anywhere from 10 to 40 years after the initial attack (15). "The World Health Organization estimates that 10 to 20 million polio survivors are alive worldwide, and some estimates suggest that 4 to 8 million of them may get PPS" (16).

Nearly 2 million North Americans alive today had polio 50 years ago (14). Estimates vary, but as many as 50% of these people may be affected by PPS (17). Many of these former patients had recovered good strength and mobility and considered themselves to be cured. The recurrence of old weaknesses and the appearance of new ones can be extremely challenging and difficult, both psychologically and physically, for the patient. Many people who had felt they had conquered the disease when they were able to discard their braces, supports, canes, and wheelchairs are now being told these same aids may once again be necessary to protect and preserve existing strength.

Unlike earlier treatments, however, the goal may no longer be the return of muscle strength. Instead, the weakness associated with PPS is often the result of long-term overuse and substitution of muscles. To restore muscle balance and preserve strength, treatment usually entails some limitation or curtailment of activity and the use of protective supports.

Diagnosis of PPS

The diagnosis of PPS is established by excluding other neuromuscular disorders. People who suffered polio many years ago may present with a wide variety of symptoms. This constellation of symptoms can mimic or overlap those of other disorders, such as multiple sclerosis, amyotrophic lateral sclerosis, Guillain-Barré syndrome, fibromyalgia, and osteoarthritis. Lauro Halstead, MD, believes that new weakness is the hallmark symptom of PPS. When new weakness appears in muscles previously affected by the polio and/or muscles thought to be unaffected originally, it may or may not be accompanied by other symptoms. This is a crucial point to recognize—a patient can have PPS even if new weakness is the only symptom (18).

Some controversy exists regarding the exact role and value of manual muscle tests in the evaluation of patients with PPS. Debate centers on the argument that such a test measures only the strength at the moment of testing, whereas the problem for the patients may be not so much a loss in strength but a loss of ability to maintain strength after exercise or exertion. There is increased muscle fatigue, which leads to more frequent episodes of weakness or gradual progressive weakness.

A single test may show a muscle that grades as normal even though it has lost 50% of the motor neurons that originally supplied it (19, 20). In other words, half a muscle's reserve may be lost before clinical symptoms of weakness are observed (21, 22).

In addition to the loss of reserve, there may be motor unit dysfunction. Patients experience a return of old weaknesses when motor units can no longer sustain the increased workload of their adopted muscle fibers.

To develop the best treatment plan, it would be helpful for the clinician to know whether the weakness experienced is in the muscles that were originally involved, or is a "new" weakness occurring in muscles that were not previously affected but are now weakened because of years of overuse and substitution, or is it a combination of the two.

Manual muscle tests can help to define the problem, especially when previous test results are available. A comparison of the data may show the return of old weaknesses as well as the emergence of new weaknesses. In most cases, however, the original test results are no longer available, and in many cases too few muscles were originally tested for a valid comparative analysis to be performed.

The absence or inadequacy of records covering a long period of time has made it difficult to determine with any degree of accuracy the relationship between current and past symptoms. By providing the results of manual muscle tests done over a span of 50 years, this text contains a long-term case study that should be of interest to all clinicians and persons affected by postpolio sequelae.

Determining both the nature and the extent of weakness is essential. What is needed in addition to baseline manual muscle tests (or muscle tests done every few months) are tests of selected muscles after periods of exercise or exertion. Data from specific protocols for serial muscle testing of normal individuals and patients with PPS immediately after exercise will enable the clinician to design more appropriate and individualized treatment plans. Findings from such serial tests will help to answer the question of whether exercise should be prescribed and, if so, how much and what kind. James Aston, MD, has suggested the following:

> "Any muscle being considered for exercise should be strong enough to withstand more than the force of gravity and should be retested two or three times after the patient takes a 1- to 2-minute walk. Any decline in strength after the walk indicates that the muscle has no reserve and should not be exercised (21)." (Reprinted from Paul Tornetta III, Mark W. Pagnano, *Instructional course lectures* vol. 61 2012 American Academy of Orthopaedic Surgeons. Price: AAOS member: $175, Resident: $130, 624 pages, Orthopaedics and Trauma, with permission from Elsevier.)

Treatment of the patients with PPS is clinically very challenging. Patricia Andres has summarized the role of the physical therapist as follows:

> "Physical therapy management of the PPS patient should focus on restoring postural alignment by (1) use of orthotics &/or assistive devices and (2) exercises that stretch tight, overworked muscles combined with nonfatiguing exercises of weak, overstretched muscles in the shortened range (23)."

Clinicians should also refer to the earlier Classic Kendall box. Although it was written specifically for patients with polio, it is applicable to patients with PPS and to anyone—including those with WNV infection—suffering from weakness or paralysis because of anterior horn cell involvement.

Explanation of Polio/Postpolio Charts

The compilation of six upper extremity, trunk, and lower extremity muscle tests on one patient with polio over a 50-year period by only two examiners gives a rare, comprehensive picture regarding the history of this disease in one individual. Six of nine recorded tests are presented in this text. At the time of the first exam, this male patient was 17 years old. He was 67 when last tested.

Not all of the same muscles were tested during each exam. The decision of which muscles to test was based on the patient's particular complaint or pain at the time of testing, previous test results, and the examiner's discretion. The neck, trunk, and upper and lower extremities were all affected by polio. The lower extremity was more involved than the upper extremity.

Upper Extremity

Examiners and test dates:

HOK: 10/18/49, 2/21/50, 8/30/50

FPK: 2/5/90, 2/21/92, 10/7/99

During the initial examination, nine muscles demonstrated marked weakness. Less than a year later, only three muscles remained in that category.

Two muscles never regained adequate strength, and what little strength they had diminished to zero or trace levels 50 years later.

Only 22 of 84 muscles tested showed good to normal strength (i.e., score of 8–10) when first tested. Less than a year later, 59 of 67 muscles had regained good to normal strength.

Forty-two years after the first test, with 94 muscles having been tested, 4 muscles remained only moderately strong and 2 muscles significantly weak. Only one muscle, the right opponens pollicis, tested as weaker than when it was first examined.

Fifty years later, the left lower trapezius, right opponens pollicis, and right abductor pollicis brevis had lost strength. The latter two tested weaker than they had on the initial exam. Loss of regained strength was noted in five muscles of the left upper extremity. Other than the two with marked weakness, the right upper extremity muscles were not tested at this time (Figure App-B-14).

Lower Extremity

Examiners and test dates:

HOK: 10/18/49, 2/21/50, 4/31/51, 5/16/68

FPK: 1/26/90, 10/7/99

During the initial examination, 87 muscles were tested. Marked weakness (i.e., score of less than 5) was found in 17 muscles and some weakness (i.e., score of 5–7) was found in 70 muscles. Good strength (i.e., score of 8–10) was found in only three muscles.

Six months later, marked weakness remained in only 1 muscle, some weakness in 16 muscles, and good strength in 63 muscles.

Nineteen years after the first test, marked weakness persisted in the tibialis anterior, with some weakness in 24 muscles and good strength in 58 muscles.

Fifty years later, marked weakness was noted in four muscles and good strength in 46 muscles (Figure App-B-15).

> **NOTE:** *Grading was originally done on a scale of 0–100 to indicate the percentage of strength in a muscle. These numbers have been converted to a scale of 0–10 in keeping with the grading system presented in this text. Conversion to a scale of 0–5 can be made by referring to the Key to Grading Symbols (Chapter 1).*

PATIENT'S NAME CLINIC No.

LEFT 10-7-99 FPK	2-21-92 FPK	2-5-90 FPK	8-30-50 HOK	2-21-50 HOK	10-18-49 HOK		10-18-49 HOK	2-21-50 HOK	8-30-50 HOK	2-5-90 FPK	2-21-92 FPK	RIGHT 10-7-99 FPK
						Facial						
						Tongue						
						Deglutition						
						Speech						
	10	10		6	3	*Anterior Deltoid*	8	10		10	10	
	10	9		6	3	*Middle Deltoid*	8	10		10	10	
	10	9		5	3	*Posterior Deltoid*		10		10	10	
	10	10				*Upper Trapezius*				10	10	
	6	6	5	4	POOR	*Middle Trapezius*	POOR	7	6	8	7	
4	5	6	5	3		*Lower Trapezius*		7	7	7	6	
	8	6				*Serratus Magnus*		10		8	10	
	10	10	10	9		*Rhomboids*		8	10	10	10	
	8	8		10		*Latissimus Dorsi*		9		8	8	
	10	10	10	10	7	*Pectoralis Major*	7	9	10	10	10	
		10		8	6	*Pectoralis Minor*	9	10		10	10	
		10		10	6	*Internal Rotators*	7	10		10	10	
				10	7	*External Rotators*	7	10		10	10	
10	10	10	10	10	7	*Biceps*	7	10	10	10	10	
10	10	10	10	10	6	*Triceps*	4	6		10	10	
10	10	10	10	10	7	*Brachio Radialis*	7	10	10	10	10	
7	10	10	10	/	7	*Supinators*	8	10		9	9	
7	10	10	10	/	7	*Pronators*	7	/		7		
10	10	10	10	/	7	*Flexor Carpi Radialis*	7	/	10	10	10	
10	10		/	/	7	*Flexor Carpi Ulnaris*	6	/	/	10	/	
10	/	10	/	/	7	*Extensor Carpi Radialis*	8	/	/	10	/	
10	/	10	/	/	7	*Extensor Carpi Ulnaris*	7	/	/	10	/	
10	10	10	10	10	10	*1 Flexor Profundus Digitorum 1*	7	10	10	10	10	
/	/	/	/	/	/	*2 Flexor Profundus Digitorum 2*	10	/	/	/	/	
/	/	/	/	/	/	*3 Flexor Profundus Digitorum 3*	10	/	/	/	/	
/	/	/	/	/	/	*4 Flexor Profundus Digitorum 4*	10	/	/	/	/	
10	10	10	10	/	10	*1 Flexor Sublimis Digitorum 1*	7	10	10	10	10	
/	/	/	/	/	/	*2 Flexor Sublimis Digitorum 2*	10	10	/	/	/	
/	/	/	/	/	/	*3 Flexor Sublimis Digitorum 3*	/	/	/	/	/	
/	/	/	/	/	/	*4 Flexor Sublimis Digitorum 4*	/	/	/	/	/	
10	10	10	10	10	7	*1 Finger Extensors 1*	7	9	10	10	10	
/	/	/	/	/	/	*2 Finger Extensors 2*	/	/	/	/	/	
/	/	/	/	/	/	*3 Finger Extensors 3*	/	/	/	/	/	
/	/	/	/	/	/	*4 Finger Extensors 4*	/	/	/	/	/	
10	10	10	10	10	7	*1 Lumbricales 1*	6	10	9	10	10	
/	/	/	/	/	/	*2 Lumbricales 2*	/	/	/	/	/	
/	/	/	/	/	/	*3 Lumbricales 3*	/	/	/	/	/	
/	/	/	/	/	/	*4 Lumbricales 4*	/	/	/	/	/	
10	10	10	10		6	*Dorsal Interossei*	7	9	10	10	10	
10	10	10	10	10	7	*Palmar Interossei*	7	10	10	10	10	
7	10	7	6	5	4	*Opponens Pollicis*	2	4	3	1	1	0
10	10	8	10	10	7	*Adductor Pollicis*	7	10	10	6	10	
10	10	10	10	9	8	*Adductor Pollicis—Longus*	6	10		10	10	
8	10	9	4	3	6	*Adductor Pollicis—Brevis*	3	4	3	(5)	4	1
	10	10	10	10	10	*Thumb Flexors*	7			10	10	
	10	7	8	10	7	*Thumb Extensors*	8	10	10	10	10	
	10	10				*Abductor Minimi Digiti*	7			10	10	
	10	10				*Opponens Minimi Digiti*	5			10	10	
	10	10				*Flexor Minimi Digiti*	7			9	10	

	Contractions and Deformities	
Key: 4 and under 5–7 8–10	*Shoulder*	
( ) Range of motion limited	*Elbow*	
	Forearm	
Poor = 2	*Wrist*	
	Fingers	
	Thumb	

FIGURE APP-B-14: Upper extremity muscle chart for the 1949–1999 Polio and Postpolio Muscle Tests.

PATIENT'S NAME CLINIC No.

LEFT 10-7-99 FPK	1-26-90 FPK	5-15-68 HOK	4-3-50 HOK	2-21-50 HOK	10-18-49 HOK	EXAMINER DATE	RIGHT 10-18-49 HOK	2-21-50 HOK	4-31-51 HOK	5-15-68 HOK	1-26-90 FPK	10-7-99 FPK
				9	3	Neck Flexors	3	10				
					weak	Neck Extensors	weak					
		8	9	9		Back Extensors	3	9	9	8		
						Quadratus Lumborum						
weak	5	5	6	6	4	Trunk-raising: Rectus abdominis, Internal oblique / Leg-lowering: Rectus abdominis, External oblique	3	4	5	6	7	weak
10	7	9	10	10	5	Gluteus Maximus	5	10	10	10	9	10
10	7	9	10	10	4	Gluteus Medius	5	8	9	8	7	10
10	8	9			4	Hip Abductors	5			8	10	10
	7	7	10	4	2	Hip Adductors	4	10	10	10	9	
	9	9			6	Hip Medial Rotators	6			10	9	
	7	9			8	Hip Lateral Rotators	7			10	10	
6	7	6	10	6	3	Hip Flexors	7	10	10	10	10	10
5	7	10	10	9	6	Tensor Fasciae Latae	5	10	10	10	10	10
8	8	10	10	10	6	Sartorius	7	10	10	10	10	10
10	9	10	10	10	6	Medial Hamstrings	7	10	10	10	10	10
7	7	10	10	10	6	Lateral Hamstrings	7	10	10	10	8	7
5	(6) A.	8	10	7	6	Quadriceps	8	10	10	10	10	10
6 C.	B.	10		10		Gastrocnemius	7	10	10	10	B.	C.
		10		10	7	Soleus	7					
10	10	10	10	9	6	Peroneus Longus	7	10	10	10	10	6
10	10	10	10	9	6	Peroneus Brevis	7	/	/	/	/	6
10	10	10	10	9	4	Peroneus Tertius	7	/	/	/	/	10
4	6	6	6	6	3	Tibialis Posterior	4	6	8	9	9	5
0	0	1	3	3	3	Tibialis Anterior	4	6	10	10	8	6
10	10	10	9	7	6	Extensor Hallucis Longus	6	7	10	10		10
8	9	7	10	10	6	Flexor Hallucis Longus	6	10	9	8	10	10
(5)	8	6	(9)	10	8	Flexor Hallucis Brevis	7	9	10	8	10	9
7	8	6	6	7	5	1 Extensor Digitorum Longus 1	6	8	10	8	9	(7)
8	8	8	7	7	5	2 Extensor Digitorum Longus 2	6	/	/	10	/	(7)
4	6	8	7	7	7	3 Extensor Digitorum Longus 3	7	/	/	10	/	(6)
(5)	6	8	7	7	7	4 Extensor Digitorum Longus 4	7	/	/	8	/	?
9	10	10	9	7	6	1 Extensor Digitorum Brevis 1	7	8	10	10	10	10
/	/	/	10	/	/	2 Extensor Digitorum Brevis 2	/	/	/	/	/	/
/	/	/	/	/	/	3 Extensor Digitorum Brevis 3	/	/	/	/	/	/
/	/	/	/	/	/	4 Extensor Digitorum Brevis 4	/	/	/	/	/	/
2	8	7	10	7	?	1 Flexor Digitorum Longus 1	6	6	9	10	9	10
6	/	6	/	7	?	2 Flexor Digitorum Longus 2	/	6	/	7	/	/
8	/	7	/	8	5	3 Flexor Digitorum Longus 3	/	8	/	/	/	/
8	/	7	/	8	5	4 Flexor Digitorum Longus 4	/	8	/	/	/	/
7	8	6	9	7	6	1 Flexor Digitorum Brevis 1	5	6	8	8	9	8
/		/	10	/	/	2 Flexor Digitorum Brevis 2	5	6	8	8	/	/
/		/	/	/	/	3 Flexor Digitorum Brevis 3	6	8	9	8	/	/
/		/	/	8	/	4 Flexor Digitorum Brevis 4	6	8	9	8	/	/
5	5	5	6	7	7	1 Lumbricalis 1	5	8	7	9	9	8
/	/	/	/	/	/	2 Lumbricalis 2	5	/	/	9	9	8
/	/	/	/	/	/	3 Lumbricalis 3	6	/	/	7	7	6
/	/	/	/	/	/	4 Lumbricalis 4	6	/	/	7	7	6
						Leg Length						
						Thigh Circumference						
						Calf Circumference						

NOTES: KEY: ■ 4 and below ■ 5–7 ■ 8–10 ( ) = Range of motion limited **weak** = 5 or 6

A. 1-26-90 Quadriceps - cannot extend last 15∞ of knee in sitting

B. 1-26-90 Cannot rise on toes one foot at a time but can come up when rising on both at the same time but with some forward displacement of the body

C. 10-7-99 Cannot rise on toes one foot at a time

FIGURE APP-B-15: Neck, trunk, and lower extremity muscle chart. © 2003 Elizabeth E. McCreary and Florence P. Kendall.

Late Complications of Polio

Figure App-B-16 illustrates possible causes of late complications of polio.

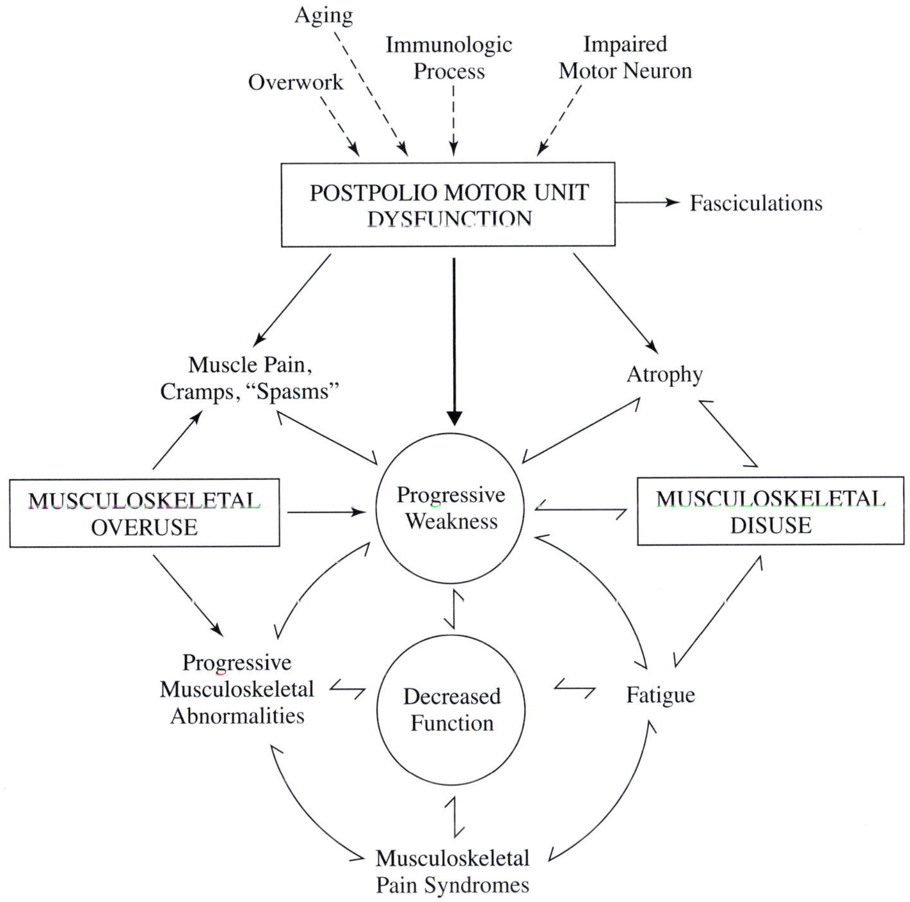

FIGURE APP-B-16: Schematic model showing three possible causes for late neuromuscular and musculoskeletal complications of polio and their interactions. Reprinted from Halstead L. Late complications of poliomyelitis. In: Goodgold J, ed. *Rehabilitative Medicine.* St. Louis: CV Mosby; 1988: 328–342 with permission from Elsevier (24).

REFERENCES

1. Nissinen M, Heliovaara M, Seitsamo J, Poussa M. Trunk asymmetry, posture, growth, and risk of scoliosis. Spine 1993; 18(1): 8 13.
2. Licht S. History. In: Basmajian J, ed. Therapeutic Exercises. 4th ed. Baltimore: Williams & Wilkins; 1984: 30.
3. Risser JC. Scoliosis past and present. In: Basmajian JV, ed. Therapeutic Exercise. 4th ed. Baltimore: Williams & Wilkins; 1984: 469.
4. American Academy of Orthopedic Surgeons (AAOS). Instructional Course Lectures. St. Louis: CV Mosby; 1985.
5. Kuru T, Yeldan İ, Dereli EE, Özdinçler AR, Dikici F, Çolak İ. The efficacy of three-dimensional Schroth exercises in adolescent idiopathic scoliosis: A randomised controlled clinical trial. Clin Rehabil 2016 Feb; 30(2): 181–190. doi: 10.1177/0269215515575745. Epub 2015 Mar 16. PMID: 25780260.
6. Białek, Marianna PT. Mild angle early onset idiopathic scoliosis children avoid progression under FITS method (functional individual therapy of scoliosis). Medicine May 2015; 94(20): e863. doi: 10.1097/MD.0000000000000863
7. Kocaman H, Bek N, Kaya MH, Büyükturan B, Yetiş M, Büyükturan Ö. The effectiveness of two different exercise approaches in adolescent idiopathic scoliosis: A single-blind, randomized-controlled trial. PLoS One 2021; 16(4): e0249492. Published 2021 Apr 15. doi:10.1371/journal.pone.0249492

8. Perrone M, Orr R, Hing W, Milne N, Pope R. The impact of backpack loads on school children: A critical narrative review. Int J Environ Res Public Health 2018; 15(11): 2529. Published 2018 Nov 12. doi:10.3390/ijerph15112529
9. Menez C, L'Hermette M, Coquart J. Orthotic insoles improve gait symmetry and reduce immediate pain in subjects with mild leg length discrepancy. Front Sports Act Living 2020; 2: 579152. Published 2020 Dec 16. doi:10.3389/fspor.2020.579152
10. 15 Million children threatened by polio outbreak. [http://www.unicef.org/UK/press]. Accessed 1/22/04.
11. http://www.newscientist.com. Polio. [http://www.newscientist.com]. Accessed 6/29/04.
12. West Nile Virus statistics, surveillance, and control. [http://www.cdc.gov/nci-dod/dvfbid/westnile/index.htm]. Accessed 1/21/04.
13. West Nile Virus can cause polio-like symptoms. [http://sciencedaily.com/releases/2003/04/030401074409.htm]. Accessed 1/21/04.
14. Bruno R. Polio by any other name. West Nile Virus, postpolio syndrome, chronic fatigue syndrome, and a double standard of disbelief. [http://www.ChronicFatigueSupport.com/library.print.cfm/ID=3938]. Accessed 7/28/03.
15. Postpolio syndrome fact sheet. National Institute of Neurological Disorders and Stroke. [http://www.ninds.nih.gov/health_and_medical/pubs/post-polio.htm]. Accessed 1/31/04.
16. Report on postpolio syndrome in Australia, Canada, France, Germany, Japan, UK, and USA: Disability World; 2001.
17. Mayo Clinic Staff. Postpolio syndrome. [http://www.mayoclinic.com/invole.cfm/id=Ds00494]. Accessed 1/31/04.
18. Polio experts grapple with the complexities of postpolio syndrome. [http://www.post-polio.org/task/expertsa.html]. Accessed 2/17/04.
19. Halstead L. Postpolio syndrome. Sci Am 1998; 278(4): 36–44.
20. Bollenbach E. Polio biology X. A Lincolnshire Post-Polio Library Publication. [http://www.ott.zynet.co.uk/polio/lincolnshire/library/bollenbach/biology10.html]. Accessed 5/14/03.
21. Aston J. Postpolio syndrome. An emerging threat to polio survivors. Postguard Med 1992; 92: 249–256.
22. Anderson W, Oregon tMABotPPPESo. An approach to the patient with suspected postpolio syndrome. [http://www.pke.com/pps/ppspamoh.htm]. Accessed 5/10/03.
23. Andres P. Rehabilitative principles and the role of the physical therapist. In: Munsat T, ed. Postpolio Syndrome. Stoneham, MA: Butterworth–Heinemann; 1991. [http://polio.dyndns.org/polio/documentslibrary]. Accessed 4/27/03.
24. Halstead L. Late complications of poliomyelitis. In: Goodgold J, ed. Rehabilitative Medicine. St. Louis: CV Mosby; 1988: 328–342.

SUGGESTED READINGS

Berry P. West Nile Virus, polio-like symptoms. *The Clarion-Ledger*. September 24, 2002. http://www.shadetreephysics.com/vel/wnv-pol.htm

Bruno RL. *The Polio Paradox*. Warner Books, Inc.; 2002.

Dalakas MC, Elder G, Hallet M, et al., A long-term follow-up study of patients with post-poliomyelitis neuromuscular symptoms. *New Engl J Med*. 1986;314(15):959–963.

Dean E. Clinical decision making in the management of the late sequelae of poliomyelitis. *Phys Ther*. 1991;71(10):752–761.

Gawns AC, Halstead LS. *Evaluation of the Post-Polio Patient*. The Lincolnshire Post-Polio Library. 2004; 1–4.

Gross MT, Schuch CP. Exercise programs for patients with post-polio syndrome: A case report. *Phys Ther*. 1989;69(1):172–175.

Halstead LS. The residual of polio in the aged. *Top Geriatr Rehabil*. 1988;3(4):9–26.

Kendall HO, Kendall FP. Care during the recovery period in paralytic poliomyelitis. *JAMA*. 1938;111(5):472.

Kendall HO, Kendall FP. Orthopedic and physical therapy objectives in poliomyelitis treatment. *Physiotherap Rev*. 1947;27(3):159–165.

Krivickas LS. *Breathing Problems Caused by Post-Polio Syndrome*. Greater Boston Post-Polio Association; 2003.

Post-Polio Health International. *Management of PPS, About Polio and PPS-Monograph*; 2003.

Maynard FM. *The Post-Polio Syndrome and Re-Rehabilitation*. 2003.

Mense S, Simons DG. *Muscle Pain: Understanding Its Nature, Diagnosis, and Treatment*. Lippincott Williams & Wilkins; 2001.

Perry J, Fontaine JD, Mulroy S. Findings in post-poliomyelitis syndrome. Weakness of muscles of the calf as a source of late pain and fatigue of muscles of the thigh after poliomyelitis. *J Bone Jt Surg*. 1995;77-A(8):1148–1153.

Polio and the Era of Fear, *The Mission*. 1994.

Polio and Post-Polio Fact Sheet. Post-Polio Health International; 2004. http://www.post-polio.org/ipn/fact.html

Sharrad WJW. Muscle recovery in poliomyelitis. *J Bone Jt Surg* 1955;37B(1):63–79.

Swensrud G. Post polio syndrome, aging with a disability. Oakland Kaiser Conference, September 19, 2003.

APPENDIX

C

SELECTED MUSCLE STRENGTH, PERFORMANCE, AND LENGTH TESTS

In this era of time-limited treatments, advancing technology, and the need to justify the initiation or continuation of care by practitioners, it is essential for practitioners to choose and effectively perform tests that aid in solving problems, whether to provide a differential diagnosis, establish or change treatment procedures, improve function, or relieve pain. No matter the reason for implementing a certain test, objectivity is essential so that clinical reasoning can be both accurate and meaningful. Following are details pertaining to certain muscle strength, performance, and length tests referenced throughout this text.

Test	Handheld Dynamometry	Grip Strength	Heel Raise	30 Second Sit to Stand*	Sit and Reach
Purpose	To ascertain muscle strength pre and post intervention	• Grip strength • Possible functional recovery prognosis indicator	Ascertain strength of plantar flexor muscle group	Test leg strength and endurance	Ascertain hamstring flexibility
Target population	Variety of patient populations			Adults/older adults	Variety of patient populations
Advantage	Provides objective measure for return to activity	Age- and sex-related normative values	Age- and sex-related normative vales	Normative value for older adults	Age- and sex-related normative values
Disadvantage	• Variable reliability • Lack of standardized positioning of device and patient	Preset device handle adjustments not compatible for maximized positioning of all hand sizes	Challenged by weight-bearing and ROM limitations	Requires appropriate ROM and sitting balance	Compensation through lumbar and thoracic spine
Equipment	Handheld dynamometer; plinth or chair for LE and UE testing	Handgrip dynamometer	Plinth for stability (no significant weight-bearing on hand)	Chair without arm rests and 17 inches in height (seated surface from floor)	Measuring device, firm surface, tape (sit and reach box if available)
Patient	Prone, supine, or sitting (dependent on muscle being tested)	Sitting	Single-leg stance for tested lower extremity	• Seated in chair • Upper extremities crossed with fingertips at shoulder	Long sitting
Fixation	Proximal to muscle tested unless specified	Maintain position of 0° of shoulder flexion and abduction, elbow 90° flexion, wrist and forearm in neutral position, wrist extension allowed during activation of grip	Maintain form through full ROM	Achieve full extension of knees and hips	• Control pelvis/lumbar spine interval if possible • Knees to remain in full extension
Test procedures (abridged)	• Performed as a **Make test** or **Break test**. • Handheld dynamometer placed as indicated in muscle test procedures outlined in previous chapters • Up to three may be performed	• Set handle to second width setting • Dial of device calibrated to zero • Subject instructed to squeeze device as hard as possible • Repeat for three trials • Compare performance to opposite upper extremity	Raising of heel via ankle plantar flexion, with knee in full extension, to bear weight on metatarsal heads and phalanges of digits 1–5	• Seated in chair with arms crossed • Complete as many sit to stand tasks completely • Timed for 30 seconds	• Measuring device placed between lower extremities • 15-inch mark placed at heel • Inhale • Reach by flexing at hips, lumbar, and thoracic spine while exhaling, keeping upper extremities parallel to floor and symmetrical • Record, measure distance achieved • Repeat for two trials
Scoring	Record highest value achieved in lbs, kgs, or N	Record highest value achieved in lbs or kgs	Record number of completed heel raises	Record number of completed sit to stands	Record best of two trials
Source(s)	1, 2	3–5	6	7, 8	9

*Component of Senior Fitness Test (4).
ROM, range of motion; LE, lower extremity; UE, upper extremity; lbs, pounds; kg, kilograms; N, newtons

REFERENCES

1. Chamorro C, Armijo-Olivo S, De la Fuente C, Fuentes J, Chirosa LJ. Absolute reliability and concurrent validity of hand held dynamometry and isokinetic dynamometry in the hip, knee and ankle joint: Systematic review and meta-analysis. Open Med 2017; 12(1): 359–375. https://doi.org/10.1515/med-2017-0052.
2. Arnold CM, Warkentin KD, Chilibeck PD, Magnus CRA. The reliability and validity of handheld dynamometry for the measurement of lower-extremity muscle strength in older adults. J Strength Cond Res 2010; 24(3): 815–824. https://doi.org/10.1519/JSC.0b013e3181aa36b8.
3. Savino E, Martini E, Lauretani F, Pioli G, Zagatti AM, Frondini C, et al. Handgrip strength predicts persistent walking recovery after hip fracture surgery. Am J Med 2013; 126(12): 1068–1075.
4. Lee SC, Wu LC, Chiang SL, Lu LH, Chen CY, Lin CH, Ni CH, Lin CH. Validating the capability for measuring age-related changes in grip-force strength using a digital hand-held dynamometer in healthy young and elderly adults. BioMed Res Int 2020 Apr 21; 2020.
5. Rodacki AL, Moreira NB, Pitta A, Wolf R, Melo Filho J, Rodacki CD, Pereira G. Is handgrip strength a useful measure to evaluate lower limb strength and functional performance in older women?. Clin Interv Aging 2020; 15: 1045.
6. Lunsford BR, Perry J. The standing heel-rise test for ankle plantar flexion: Criterion for normal. Phys Ther 1995; 75(8): 694–698. https://doi:10.1093/ptj/75.8.694.
7. Miotto JM, Chodzko-Zajko WJ, Reich JL, Supler MM. Reliability and validity of the Fullerton Functional Fitness Test: An independent replication study. J Aging Phys Act 1999; 7(4): 339–353, accessed Apr 27, 2022, https://doi.org/10.1123/japa.7.4.339.
8. Rikli RE, Jones CJ. Development and validation of criterion-referenced clinically relevant fitness standards for maintaining physical independence in later years. Gerontologist 2013; 53(2): 255–267. https://doi:10.1093/geront/gns071.
9. ACSM's Resource Manual for Guidelines for Exercise Testing and Prescription/American College of Sports Medicine; [senior editor, Deborah Riebe; associate editors, Jonathan K. Ehrman, Gary Liquori, Meir Magal]. 10 ed. Pages 104–105. Philadelphia: Wolters Kluwer Health/Lippincott Williams & Wilkins, c2018.

APPENDIX

D

SPINAL SEGMENT DISTRIBUTION TO NERVES AND MUSCLES

For anatomists and clinicians, the determination of spinal segment distribution to the peripheral nerves and muscles has been a difficult task. The pathway of the spinal nerves is obscured by the intertwining of the nerve fibers as they pass through the nerve plexuses. It is almost impossible to trace the course of an individual nerve fiber through the maze of its plexus, so information regarding spinal segment distribution has been derived mainly from clinical observation. The use of this empirical method has resulted in a variety of findings regarding the segmental origins of these nerves and the muscles that they innervate. An awareness of possible variations is important in establishing the diagnosis and identifying the location of a nerve lesion. To focus attention on the range of variations that exist, the Kendalls tabulated information from six well-known sources.

Figure App-D-1 shows the spinal segment distribution to the nerves. Figure App-D-2 shows the distribution to the muscles. The compilations derived from these charts became part of the *Spinal Nerve and Muscle Charts.*

The symbols used in tabulating the reference material are as follows:

- A large X to denote a major distribution.
- A small x to denote a minor distribution.
- A parenthetical (x) to denote a possible or infrequent distribution.

For Figure App-D-1, T2 was included in the brachial plexus by all the sources. Separate columns for T2 were not added to the upper extremity portion of the figure, however, because T2 contains only cutaneous sensory fibers. The information in the compilation columns in Figure App-D-1 has been converted from X symbols to numbers in the right column. This information regarding spinal segment distribution to the nerves appears at the top of Figures 1-10 and 1-11 (the upper- and lower-extremity *Spinal Nerve and Muscle Charts* under the heading *Peripheral Nerves.*

In the Kendall compilation of spinal segment supply to the muscles as it appears in the last column on the right of the tabulation (Figure App-D-1), the x symbols represent an arithmetical summary. As a general rule, the symbols were chosen as follows:

- If five or six authorities agreed that a spinal segment was distributed to a given muscle, then the nerve supply was indicated by a large X.
- If three or four authorities agreed, a small x was used.
- If only two authorities agreed, a small x in parentheses was used.
- If only one authority mentioned the given distribution, it was disregarded (see the triceps tabulation as an example).

TRICEPS

	C6	C7	C8	T1
Gray (1)		X	X	
deJong (2)	X	X	X	(x)
Cunningham (3)	X	X	X	
Spalteholz (4)	x	X	X	(x)
Foerster and Bumke (5)	(x)	X	X	x
Haymaker and Woodhall (6)		X	X	x
Total	4	6	6	4
Kendall compilation of triceps innervation	x	X	X	x
	C6	C7	C8	T1

When one of the six sources did not specify the spinal segment, agreement among four or five sources was indicated by a large X. This occurred for the popliteus and for some intrinsic muscles of the foot.

The tabulation of data focuses attention on the range of variations that exist among these sources, but the arithmetical summary indicates the extent of their agreement. Only in the case of three thumb muscles (opponens, abductor brevis, and superficial head of the flexor brevis) were the six authorities divided in their opinion, resulting in an apparent overstatement of the number of roots of origin. The method used in compiling the information resulted in all segments being listed with small x symbols (i.e., C6, 7, 8, and T1), without major emphasis on any one segment.

In most instances, the arithmetical summary preserved the major emphasis on the spinal segments that provide innervation to the muscles. When the summary did not, exceptions were made. For example, all sources included C3, 4, and 5 innervations to the diaphragm. All placed emphasis on C4, however, so only C4 was given a large X symbol. All sources also included the following spinal segment innervations:

- C5 for the supinator.
- C8 for the extensor carpi radialis longus and brevis.
- L4 for the adductor longus.
- L4 as a component of the sacral plexus.

All sources represented these innervations by a small x symbol, indicating a minor distribution, and so the compilation preserved the lesser emphasis. All sources included T(12) innervation to the lumbar plexus but indicated that it was a minimal supply, so T(12) remained in parentheses in the compilation.

Innervation was omitted in the compilation in two instances, because a discrepancy existed between the spinal segment innervation to the *muscle* and that to the *peripheral nerve* supplying the muscle. C8 innervation, which was mentioned by two of the sources as supplying the subscapularis, was omitted because no indication was observed that the upper or lower subscapular nerve received C8 innervation. Likewise, C(4), which was included by two sources for the teres minor, was omitted because no indication was found that the axillary nerve received C4 innervation. In two other instances, innervation was added in the compilation. C6 and C7 were added to the medial pectoral nerve. Above the communicating loop, the medial pectoral nerve is composed of C8 and T1 fibers. Below the loop, C7 and possible C6 fibers (branching from the lateral pectoral nerve) join the medial pectoral nerve. The medial cord of the plexus is derived from C8 and T1, but the ulnar nerve, as the terminal branch of this cord, is listed as having a C7 component in addition to C8 and T1. Numerous anatomists (2–4) record this information, and some (7–9) indicate that the C7 component is variable.

The compilation was modified regarding spinal segment distribution to the upper and lower portions of the pectoralis major. In the muscle sections of the books used as references for the compilation, only one text (3) divided the pectoralis major muscle into upper and lower portions and listed the spinal segment innervation to each. Gray (1), however, in the description of the lateral and medial pectoral nerves, indicated that the lateral pectoral supplied the more cranial part of the muscle, whereas the medial pectoral, joined by two or three branches from the lateral, supplied the more caudal part. In addition, several other references (3, 6, 10) differentiated the peripheral supply to the upper and lower parts. In certain lesions of the cervical region of the spinal cord, it has been clinically noted that the upper part of the pectoralis major has had normal strength while the lower part has been paralyzed. This observation suggests a difference in spinal segment innervation to the parts of the muscle. Based on the above information, the compilation distinguishes between the upper and lower parts of the pectoralis major in regard to spinal segment distribution.

The results of the compilation have been used in the spinal segment column on the nerve–muscle charts. The X symbols have been converted to numbers that indicate the specific spinal segment. In the nerve–muscle charts, the major emphasis, as designated in the compilation by large X symbols, has been obtained by using numbers in bold type while those used for minor emphasis are not bold. Possible or infrequent innervation has been obtained by numbers in parentheses.

A

Spinal Segment Distribution to Nerves: Neck, Diaphragm, and Upper Extremity

NERVE	CUNNINGHAM[3]									GRAY[1]								
	C1	2	3	4	5	6	7	8	T1	C1	2	3	4	5	6	7	8	T1
Cervical Plex.	X	X	X	X						X	X	X	X					
Brachial Plex.				(x)	X	X	X	X	X				(x)	X	X	X	X	X
Phrenic			(x)	X	(x)							X	X	X				
Long Thoracic					X	X	X							X	X	X		
Dorsal Scapular					X									X				
N. to Subclavius				X	X	X								X	X			
Suprascapular				(x)	X	X								X	X			
Upp. Subscap.				(x)	X	X	(x)							X	X			
Thoracodorsal						X	X	X						X	X	X		
Low. Subscap.					X	X								X	X			
Lat. Pectoral					X	X	X							(Not listed)				
Med. Pectoral								X	X								X	X
Axillary					X	X								X	X			
Musculocutan.				(x)	X	X	X							X	X	X		
Radial					X	X	X	X	X					X	X	X	X	X
Median					(x)	X	X	X	X						X	X	X	X
Ulnar							(x)	X	X								X	X

NERVE	MORRIS[10]									SPALTEHOLZ[4]								
	C1	2	3	4	5	6	7	8	T1	C1	2	3	4	5	6	7	8	T1
Cervical Plex.	X	X	X	X						X	X	X	X					
Brachial Plex.				(x)	X	X	X	X	X				(x)	X	X	X	X	X
Phrenic			(x)	X	(x)							x	X	x				
Long Thoracic					X	X	X							x	X	X	(x)	
Dorsal Scapular				(x)	X								(x)	x				
N. to Subclavius					X	(x)								X	(x)			
Suprascapular				(x)	X	X							(x)	X	X			
Upp. Subscap.				(x)	X	X								x	x	x		
Thoracodorsal					x	X	X	X							(x)	X	(x)	
Low. Subscap.				(x)	X	X								x	x	x		
Lat. Pectoral					X	X	X							(Not listed)				
Med. Pectoral								X	X					(Not listed)				
Axillary					X	X								X	X	(x)		
Musculocutan.				(x)	X	X	(x)							X	X	(x)		
Radial				(x)	(x)	X	X	X	(x)					X	X	X	X	X
Median					(x)	X	X	X	X					X	X	X	X	X
Ulnar								X	X							X	X	X

NERVE	deJONG[2]									HAYMAKER & WOODHALL[6]								
	C1	2	3	4	5	6	7	8	T1	C1	2	3	4	5	6	7	8	T1
Cervical Plex.					(Not listed)					X	X	X	X					
Brachial Plex.					(Not listed)								(x)	X	X	X	X	X
Phrenic			X	X	X							X	X	X				
Long Thoracic					X	X	X	x						X	X	X		
Dorsal Scapular				X	X									X				
N. to Subclavius					X	X								X	X			
Suprascapular					X	X								X	X			
Upp. Subscap.					X	X								X	X			
Thoracodorsal						X	X	X							x	X	X	
Low. Subscap.					X	X	(x)							X	X			
Lat. Pectoral					(Not listed)									X	X	X		
Med. Pectoral							X	X	X								X	X
Axillary					X	X								X	X			
Musculocutan.					X	X	X							X	X	X		
Radial					X	X	X	X	(x)					X	X	X	X	(x)
Median					(x)	X	X	X	X					x	X	X	X	X
Ulnar							(x)	X	X								X	X

NERVE	COMPILATION BY KENDALLS									SPINAL SEGMENTS USED FOR SPINAL NERVE & MUSCLE CHART
	C1	2	3	4	5	6	7	8	T1	
Cervical Plex.	X	X	X	X						Cervical Plex. **C 1,2,3,4**
Brachial Plex.				(x)	X	X	X	X	X	Brach. Plex. C(4),**5,6,7,8,T1**
Phrenic			x	X	x					Phrenic C3,**4**,5
Long Thoracic					X	X	X	(x)		Long. Thor. C**5,6,7**,(8)
Dorsal Scapular				x	X					Dor. Scap. C4,**5**
N. to Subclavius					X	X				N. to Subclavius C**5**, **6**
Suprascapular				x	X	X				Suprascap. C4, **5**, **6**
Upp. Subscap.				(x)	X	X	(x)			U. Subscap. C(4),**5**,**6**,(7)
Thoracodorsal					(x)	X	X	X		Thoracodor. C(5), **6**, **7**, **8**
Low. Subscap.				(x)	X	X	(x)			L. Subscap. C**5**,**6**,(7)
Lat. Pectoral					X	X	X			Lat. Pect. C**5**,**6**,**7**
Med. Pectoral								X	X	Med. Pect.* C(6),7,**8**, T**1**
Axillary					X	X				Axillary C**5**,**6**
Musculocutan.				(x)	X	X	X			Musculocutan. C(4),**5**,**6**,**7**
Radial					X	X	X	X	x	Radial C**5**, **6**, **7**, **8** T1
Median					x	X	X	X	X	Median C5, **6**, **7**, **8** T**1**
Ulnar							x	X	X	Ulnar C7,**8** T**1**

*See innervation to pectoral muscles, pp. 406–407.

B

Spinal Segment Distribution to Nerves: Trunk and Lower Extremity

	CUNNINGHAM[3]									GRAY[1]								
	T	LUMBAR					SAC			T	LUMBAR					SAC		
	12	1	2	3	4	5	1	2	3	12	1	2	3	4	5	1	2	3
Iliophypogastric	(x)	X								(x)	X							
Ilioinguinal	(x)	X									X							
Lumb. Plex.	(x)	X	X	X	X					(x)	X	X	X	X				
Femoral			X	X	X							X	X	X				
Obturator			X	X	X							X	X	X				
Sup. Gluteal					X	X	X							X	X	X		
Inf. Gluteal						X	X	X							X	X	X	
Sac. Plex.					x	X	X	X	X					x	X	X	X	X
Sciatic					X	X	X	X	X					X	X	X	X	X
Common Peroneal					X	X	X	X						X	X	X	X	
Tibial					X	X	X	X	X					X	X	X	X	X

	MORRIS[10]									SPALTEHOLZ[4]								
	T	LUMBAR					SAC			T	LUMBAR					SAC		
	12	1	2	3	4	5	1	2	3	12	1	2	3	4	5	1	2	3
Iliophypogastric	x	X								X	X							
Ilioinguinal	(x)	X									X							
Lumb. Plex.	(x)	X	X	X	X					(x)	X	X	X	x				
Femoral		(x)	X	X	X						x	x	x	x				
Obturator		(x)	X	X	X						(x)	x	X	x				
Sup. Gluteal					X	X	X							x	X	x	(x)	
Inf. Gluteal						X	X	X						(x)	x	X	x	
Sac. Plex.					x	X	X	X	X					x	X	X	X	X
Sciatic					X	X	X	X	X					X	X	X	X	X
Common Peroneal					X	X	X	X						X	X	X	X	
Tibial					X	X	X	X	X					X	X	X	X	X

	deJONG[2]									HAYMAKER & WOODHALL[6]								
	T	LUMBAR					SAC			T	LUMBAR					SAC		
	12	1	2	3	4	5	1	2	3	12	1	2	3	4	5	1	2	3
Iliophypogastric		(x)									X							
Ilioinguinal		(x)									X							
Lumb. Plex.					(Not listed)					(x)	X	X	X	X				
Femoral			X	X	X							X	X	X				
Obturator			X	X	X							X	X	X				
Sup. Gluteal					X	X	X							X	X	X		
Inf. Gluteal						X	X	X							X	X	X	
Sac. Plex.					(Not listed)									x	X	X	X	X
Sciatic					X	X	X	X	X					X	X	X	X	X
Common Peroneal					X	X	X	X						X	X	X	X	
Tibial					X	X	X	X	x					X	X	X	X	X

	COMPILATION BY KENDALLS									SPINAL SEGMENTS USED FOR SPINAL NERVE & MUSCLE CHART
	T	LUMBAR					SAC			
	12	1	2	3	4	5	1	2	3	
Iliophypogastric	x	X								Iliohypogastric **T**12, **L1**
Ilioinguinal	(x)	X								Ilioinguinal **T**(12), **L1**
Lumb. Plex.	(x)	X	X	X	X					Lumb. Plex. **T**(12), **L1**, **2**, **3**, **4**
Femoral		(x)	X	X	X					Femoral **L**(1), **2**, **3**, **4**
Obturator		(x)	X	X	X					Obturator **L**(1), **2**, **3**, **4**
Sup. Gluteal					X	X	X			Sup. Glut. **L4**, **5**, **S1**
Inf. Gluteal						X	X	X		Inf. Glut. **L5**, **S1**, **2**
Sac. Plex.					x	X	X	X	X	Sac. Plex. **L**4, **5**, **S1**, **2**, **3**
Sciatic					X	X	X	X	X	Sciatic **L4**, **5**, **S1**, **2**, **3**
Common Peroneal					X	X	X	X		C. Peroneal **L4**,**5**,**S1**,**2**
Tibial					X	X	X	X	X	Tibial **L4**, **5**, **S1**, **2**, **3**

FIGURE APP-D-1: Spinal Segment Distribution to Nerves. **A**, Neck, diaphragm, and upper extremity; **B**, Trunk and lower extremity.

MUSCLE	GRAY[1]									deJONG[2]									CUNNINGHAM[3]								
	SPINAL SEGMENT									SPINAL SEGMENT									SPINAL SEGMENT								
	C1	C2	C3	C4	C5	C6	C7	C8	T1	C1	C2	C3	C4	C5	C6	C7	C8	T1	C1	C2	C3	C4	C5	C6	C7	C8	T1
HEAD & NECK EXTENSORS	X	X	X	X	X	X	X	X	X	X	X	X	X	X	X	X	X		X	X	X	X	X	X	X	X	X
INFRAHYOID MUSCLES	X	X	X																X	X	X						
RECTUS CAP. ANT. & LAT.	X	X								X	X								X	X							
LONGUS CAPITIS	X	X	X							X	X	X	X						X	X	X	X					
LONGUS COLLI		X	X	X	X	X	X				X	X	X	X	X					X	X	X	X	X	X	X	
LEVATOR SCAPULAE			X	X	(x)							X	X	X							X	X	X				
SCALENI (A.M.P.)			X	X	X	X	X	X					X	X	X	X	X				X	X	X	X	X	x	
STERNOCLEIDOMASTOID		X	X							(x)	X	X								X							
TRAPEZIUS (U.M.L.)			X	X							(x)	X	X								X	X					
DIAPHRAGM			x	X	x							x	X	x							(x)	X	X				
SERRATUS ANTERIOR					X	X	X							X	X	X	X						X	X	X		
RHOMBOIDS, MAJ. & MINOR					X								X	X							x	X	X				
SUBCLAVIUS					X	X							(x)	X	X								X	X			
SUPRASPINATUS					X								(x)	X	X								X	X			
INFRASPINATUS					X	X								X	X								X	X			
SUBSCAPULARIS					X	X								X	X								X	X			
LATISSIMUS DORSI						X	X	X							X	X	X							X	X	X	
TERES MAJOR					X	X								X	X	(x)							X	X			
PECTORALIS MAJ. (UPPER)			}		X	X	X	X	X			}		X	X	X	X	X					X	X			
PECTORALIS MAJ. (LOWER)																							X	X	X	X	X
PECTORALIS MINOR								X	X							X	X	X							X	X	X
TERES MINOR					X								(x)	X	X								X	X			
DELTOID					X	X								X	X								X	X			
CORACOBRACHIALIS						X	X								X	X								X	X		
BICEPS					X	X								X	X								X	X			
BRACHIALIS					X	X								X	X								X	X			
TRICEPS							X	X							X	X	X	(x)						X	X	X	
ANCONEUS							X	X								X	X								X	X	
BRACHIALIS (SMALL PART)					X	X								X	X								X	X			
BRACHIORADIALIS					X	X								X	X								X	X			
EXT. CARPI RAD. L. & B.						X	X							(x)	X	X	(x)						X	X	X	X	
SUPINATOR						X								X	X	X							X	X			
EXT. DIGITORUM						X	X	X							X	X	X							X	X	X	
EXT. DIGITI MINIMI						X	X	X							X	X	X							X	X	X	
EXT. CARPI ULNARIS						X	X	X							X	X	X							X	X	X	
ABD. POLLICIS LONGUS						X	X								X	X	X							X	X	X	
EXT. POLLICIS BREVIS						X	X								X	X	X							X	X	X	
EXT. POLLICIS LONGUS						X	X	X							X	X	X							X	X	X	
EXT. INDICIS						X	X	X							X	X	X							X	X	X	
PRONATOR TERES						X	X								X	X								X	X		
FLEX. CARPI RADIALIS						X	X								X	X	(x)							X	X	X	
PALMARIS LONGUS						X	X								(x)	X	X								X	X	X
FLEX. DIGIT. SUPERFICIALIS							X	X	X							X	X	X							X	X	X
FLEX. DIGIT. PROF. I & II								X	X							X	X	X							X	X	X
FLEX. POLLICIS LONGUS								X	X						X	X	X	X							X	X	X
PRONATOR QUADRATUS								X	X							X	X	X							X	X	X
ABD. POLLICIS BREVIS						X	X								X	X										X	X
OPPONENS POLLICIS						X	X								X	X										X	X
FLEX. POLL. BREV. (SUP. H.)						X	X								X	X										X	X
LUMBRICALES I & II						X	X								X	X	X	X								X	X
FLEX. CARPI ULNARIS								X	X							(x)	X	X								X	X
FLEX. DIGIT. PROF. III & IV								X	X							X	X	X								X	X
PALMARIS BREVIS								X								(x)	X	X								X	X
ABD. DIGITI MINIMI								X	X							(x)	X	X								X	X
OPPONENS DIGITI MINIMI								X	X							(x)	X	X								X	X
FLEX. DIGITI MINIMI								X	X							(x)	X	X								X	X
PALMAR INTEROSSEI								X	X								X	X								X	X
DORSAL INTEROSSEI								X	X								X	X								X	X
LUMBRICALES III & IV								X									X	X								X	X
ADDUCTOR POLLICIS								X	X								X	X								X	X
FLEX. POLL. BREV. (DEEP H.)								X	X								X	X								X	X

FIGURE APP-D-2: Spinal segment distribution to muscles.

SPALTEHOLZ[4]									FOERSTER & BUMKE[5]									HAYMAKER & WOODHALL[6] (modified after Bing)									COMPILATION by Kendalls								
SPINAL SEGMENT									SPINAL SEGMENT									SPINAL SEGMENT									SPINAL SEGMENT								
C1	C2	C3	C4	C5	C6	C7	C8	T1	C1	C2	C3	C4	C5	C6	C7	C8	T1	C1	C2	C3	C4	C5	C6	C7	C8	T1	C1	C2	C3	C4	C5	C6	C7	C8	T1
x	x	x	x	x	x	x	x	x	X	X	X	X	X	X	X	X	X		X	X	X	X	X	X	X	X	X	X	X	X	X	X	X	X	X
x	x	x							(x)	X	X	(x)						X	X	X							X	X	X						
x	x								X	x								X									X	X							
x	x	x							X	X	X							X	X	X							X	X	X	(x)					
	x	x	x	x	x						X	X	X	x					X	X	X	X	X					X	X	X	X	X	(x)		
	(x)	x	x	x							x	X	x							X	X	X							X	X	X				
			(x)	x	x	x	x					X	X	X	X	x				X	X	X	X	X	X				x	X	X	X	X	X	
	x								(x)	X	X								X								(x)	X	x						
	x	x	x						X	X	X	x							X	X	X							x	X	X					
		x	X	x						(x)	x	X	x							x	X	x							x	X	x				
				x	X	X	(x)						x	X	X	(x)						X	X	X							X	X	X	x	
				x								x	x									X	x							x	X				
				x	x								x	X	X							X	X								X	X			
			(x)	x	(x)							x	X									X	x							x	X	x			
			(x)	X	x							x	X	x								X	x							(x)	X	X			
				x	x	(x)							x	X	X	x						X	X	X	X						X	X	x		
					(x)	X	x							x	X	x							x	X	X							X	X	X	
					x	x								x	X							X	X								x	X	x		
		}		(x)	X	X	X	x			}		x	X	X	X	x			}		X	X	X	X	X					X	X	X		
																																X	X	X	X
						x	x							x	X	x							X	X	X							(x)	X	X	x
				x	X	(x)						(x)	X	x								X	x								X	X			
				X	x							(x)	X	X								X	x								X	X			
					x	X								X	x	(x)							x	X								X	X		
				x	x	(x)							X	X								X	X								X	X			
				x	x	(x)							x	X								X	X								X	X			
					x	X	X	(x)						(x)	X	X	x							X	X	x						x	X	X	x
					(x)	X	x	(x)					(not listed)											X	X								X	X	
				x	x	(x)							x	X								X	X								X	X			
				x	x								x	X								x	X								X	X			
				(x)	x	x	x							x	X	(x)							x	X	x						x	X	X	x	
				(x)	x	x	x						x	X								x	X								x	X	(x)		
					(x)	x	X							x	X	(x)								X	x							X	X	X	
						x	x						(not listed)											X	x							x	X	X	
					(x)	x	X								x	X								X	x							x	X	X	
					(x)	x	x								x	X								X	X							x	X	X	
					(x)	x	x									x	X							X	x							x	X	X	
					(x)	x	x								x	X								X	x							x	X	X	
					(x)	x	x									X	x							X	x							x	X	X	
					X	x								x	X								x	X								X	X		
					x	x	(x)							(x)	X	(x)							x	X	x							X	X	x	
						x	x	x								X	x							x	X	x						(x)	X	X	x
					(x)	(x)	X	x							x	X	X							x	X	X							X	X	X
					x	x	X	x								X	X							x	X	X							x	X	X
					x	x	x									X	X							x	X	X						(x)	x	X	X
					x	x	X	x								X	x							x	X	X							x	X	X
					x	X	(x)										X								X	x						x	x	x	x
					x	x	(x)	(x)								x	X								X	x						x	x	x	x
					x	X	x									x	X								X	X						x	x	x	x
						x	x	x								x	X							x	X	X						(x)	x	X	X
							x	x							X	x							x	X	x								x	X	x
					x	x	X	x								X	X							x	X	X							x	X	X
						(x)	x	(x)								x	X								x	X							(x)	X	X
						(x)	x	(x)								x	X								x	X							(x)	X	X
						(x)	x	(x)								x	X								x	X							(x)	X	X
						(x)	x	(x)								x	X								x	X							(x)	X	X
						(x)	x	x								x	X								x	X								X	X
						(x)	x	x								x	X								x	X								X	X
						x	x	x								x	X							x	X	X							(x)	X	X
						x	x	x								x	X								x	X								X	X
					x	X	x									x	X								X	X								X	X

FIGURE APP-D-2: *(continued)*

MUSCLE	GRAY[1]												
	SPINAL SEGMENT												
	THORACIC					LUMBAR					SACRAL		
	T1, 2, 3, 4	T5, 6	T7, 8	T9, 10, 11	T12	L1	L2	L3	L4	L5	S1	S2	S3
ERECTOR SPINAE	X	X	X	X	X	X	X	X	X	X	X	X	X
SERRATUS POST. SUP.	X												
TRANS. THORACIS	X	X	X	X									
INT. INTERCOSTALS	X	X	X	X									
EXT. INTERCOSTALS	X	X	X	X									
SUBCOSTALES	X	X	X	X									
LEVATOR COSTARUM	X	X	X	X									
OBLIQUUS EXT. ABD.			X	X	X								
RECTUS ABDOMINIS			X	X	X								
OBLIQUUS INT. ABD.			X	X	X	X							
TRANSVERSUS ABD.			X	X	X	X							
SERRATUS POST. INF.				X	X								
QUAD. LUMBORUM					X	X							
PSOAS MINOR						X							
PSOAS MAJOR							X	X					
ILIACUS							X	X					
PECTINEUS							X	X	X				
SARTORIUS							X	X					
QUADRICEPS							X	X	X				
ADDUCTOR BREVIS								X	X				
ADDUCTOR LONGUS								X	X				
GRACILIS								X	X				
OBTURATOR EXT.								X	X				
ADDUCTOR MAGNUS								X	X				
GLUTEUS MEDIUS									X	X	X		
GLUTEUS MINIMUS									X	X	X		
TENSOR FAS. LAT.									X	X	X		
GLUTEUS MAXIMUS										X	X	X	
PIRIFORMIS											X	X	
GEMELLUS SUPERIOR										X	X	X	
OBTURATOR INTERNUS										X	X	X	
GEMELLUS INFERIOR									X	X	X		
QUADRATUS FEMORIS									X	X	X		
BICEPS (LONG HEAD)											X	X	X
SEMITENDINOSUS										X	X	X	
SEMIMEMBRANOSUS										X	X	X	
BICEPS (SHORT HEAD)										X	X	X	
TIBIALIS ANTERIOR									X	X	X		
EXT. HALL. LONG.									X	X	X		
EXT. DIGIT. LONG.									X	X	X		
PERONEUS TERTIUS									X	X	X		
EXT. DIGIT. BREVIS										X	X		
PERONEUS LONGUS									X	X	X		
PERONEUS BREVIS									X	X	X		
PLANTARIS									X	X	X		
GASTROCNEMIUS											X	X	
POPLITEUS									X	X	X		
SOLEUS											X	X	
TIBIALIS POSTERIOR										X	X		
FLEX. DIGIT. LONG.										X	X		
FLEX. HALL. LONG.										X	X	X	
FLEX. DIGIT. BREVIS									X	X			
ABDUCTOR HALLUCIS									X	X			
FLEX. HALLUCIS BREVIS									X	X	X		
LUMBRICALIS I									X	X			
ABD. DIGITI MINIMI											X	X	
QUAD. PLANTAE											X	X	
FLEX. DIGITI MINIMI											X	X	
OPP. DIGITI MINIMI											X	X	
ADDUCTOR HALLUCIS											X	X	
PLANT. INTEROSSEI											X	X	
DORSAL INTEROSSEI											X	X	
LUMBRICALES II, III, IV											X	X	

MUSCLE	deJONG[2]												
	SPINAL SEGMENT												
	THORACIC					LUMBAR					SACRAL		
	T1, 2, 3, 4	T5, 6	T7, 8	T9, 10, 11	T12	L1	L2	L3	L4	L5	S1	S2	S3
ERECTOR SPINAE	X	X	X	X	X	X	X	X	X	X	X	X	X
SERRATUS POST. SUP.	X												
TRANS. THORACIS	X	X	\										
INT. INTERCOSTALS	X	X	X	X	X								
EXT. INTERCOSTALS	X	X	X	X	X								
SUBCOSTALES	(not listed)												
LEVATOR COSTARUM			\	X									
OBLIQUUS EXT. ABD.			X	X	X								
RECTUS ABDOMINIS		X	X	X	X								
OBLIQUUS INT. ABD.			X	X	X	X							
TRANSVERSUS ABD.			X	X	X	X							
SERRATUS POST. INF.				X	X								
QUAD. LUMBORUM	(not listed)												
PSOAS MINOR						X	X						
PSOAS MAJOR						(X)	X	X	X				
ILIACUS							X	X	X				
PECTINEUS							X	X	X				
SARTORIUS							X	X	X				
QUADRICEPS							X	X	X				
ADDUCTOR BREVIS							X	X	X				
ADDUCTOR LONGUS							X	X	X				
GRACILIS							X	X	X				
OBTURATOR EXT.							X	X	X				
ADDUCTOR MAGNUS							X	X	X	X			
GLUTEUS MEDIUS									X	X	X		
GLUTEUS MINIMUS									X	X	X		
TENSOR FAS. LAT.									X	X	X		
GLUTEUS MAXIMUS										X	X	X	
PIRIFORMIS											X	X	
GEMELLUS SUPERIOR										X	X	X	
OBTURATOR INTERNUS										X	X	X	
GEMELLUS INFERIOR									X	X	X		
QUADRATUS FEMORIS									X	X	X		
BICEPS (LONG HEAD)											X	X	X
SEMITENDINOSUS										X	X	X	
SEMIMEMBRANOSUS									X	X	X		
BICEPS (SHORT HEAD)										X	X	X	
TIBIALIS ANTERIOR									X	X	X		
EXT. HALL. LONG.									X	X	X		
EXT. DIGIT. LONG.									X	X	X		
PERONEUS TERTIUS									X	X	X		
EXT. DIGIT. BREVIS									X	X	X		
PERONEUS LONGUS									X	X	X		
PERONEUS BREVIS									X	X	X		
PLANTARIS										X	X		
GASTROCNEMIUS											X	X	
POPLITEUS										X	X		
SOLEUS											X	X	
TIBIALIS POSTERIOR									X	X	X		
FLEX. DIGIT. LONG.										X	X		
FLEX. HALL. LONG.									X	X	X		
FLEX. DIGIT. BREVIS									X	X	X		
ABDUCTOR HALLUCIS									X	X	X		
FLEX. HALLUCIS BREVIS									X				
LUMBRICALIS I									X	X	X		
ABD. DIGITI MINIMI											X	X	
QUAD. PLANTAE											X	X	
FLEX. DIGITI MINIMI												X	
OPP. DIGITI MINIMI	(not listed)												
ADDUCTOR HALLUCIS										X	X	X	
PLANT. INTEROSSEI											X	X	
DORSAL INTEROSSEI											X	X	
LUMBRICALES II, III, IV									X	X	X	X	

MUSCLE	CUNNINGHAM[3]												
	SPINAL SEGMENT												
	THORACIC					LUMBAR					SACRAL		
	T1, 2, 3, 4	T5, 6	T7, 8	T9, 10, 11	T12	L1	L2	L3	L4	L5	S1	S2	S3
ERECTOR SPINAE	X	X	X	X	X	X	X	X	X	X	X	X	X
SERRATUS POST. SUP.	X												
TRANS. THORACIS	X	X	X	X									
INT. INTERCOSTALS	X	X	X	X									
EXT. INTERCOSTALS	X	X	X	X									
SUBCOSTALES	X	X	X	X									
LEVATOR COSTARUM	X	X	X	X									
OBLIQUUS EXT. ABD.			X	X									
RECTUS ABDOMINIS			X	X									
OBLIQUUS INT. ABD.			X	X	X	X							
TRANSVERSUS ABD.			X	X	X	X							
SERRATUS POST. INF.				X									
QUAD. LUMBORUM					X	X	X	X					
PSOAS MINOR						X	(X)						
PSOAS MAJOR						X	X	X	(X)				
ILIACUS							X	X	X				
PECTINEUS							X	X					
SARTORIUS							X	X					
QUADRICEPS							X	X	X				
ADDUCTOR BREVIS							X	X	x				
ADDUCTOR LONGUS							X	X	x				
GRACILIS							X	X	x				
OBTURATOR EXT.								X	X				
ADDUCTOR MAGNUS								X	X	X	x		
GLUTEUS MEDIUS									X	X	X		
GLUTEUS MINIMUS									X	X	X		
TENSOR FAS. LAT.									X	X	X		
GLUTEUS MAXIMUS										X	X	X	
PIRIFORMIS											X	X	
GEMELLUS SUPERIOR										X	X	X	
OBTURATOR INTERNUS										X	X	X	
GEMELLUS INFERIOR									X	X	X		
QUADRATUS FEMORIS									X	X	X		
BICEPS (LONG HEAD)									X	X	X	X	X
SEMITENDINOSUS									X	X	X	X	
SEMIMEMBRANOSUS									X	X	X	X	
BICEPS (SHORT HEAD)										X	X	X	
TIBIALIS ANTERIOR									X	X	X		
EXT. HALL. LONG.									X	X	X		
EXT. DIGIT. LONG.									X	X	X		
PERONEUS TERTIUS									X	X	X		
EXT. DIGIT. BREVIS									X	X	X		
PERONEUS LONGUS									X	X	X		
PERONEUS BREVIS									X	X	X		
PLANTARIS									X	X	X		
GASTROCNEMIUS											X	X	
POPLITEUS									X	X	X		
SOLEUS										X	X	X	
TIBIALIS POSTERIOR										X	X		
FLEX. DIGIT. LONG.										X	X		
FLEX. HALL. LONG.										X	X	X	
FLEX. DIGIT. BREVIS									x	X	X		
ABDUCTOR HALLUCIS									x	X	X		
FLEX. HALLUCIS BREVIS									x	X	X		
LUMBRICALIS I									x	X	X		
ABD. DIGITI MINIMI											X	X	
QUAD. PLANTAE											X	X	
FLEX. DIGITI MINIMI											X	X	
OPP. DIGITI MINIMI											X	X	
ADDUCTOR HALLUCIS											X	X	
PLANT. INTEROSSEI											X	X	
DORSAL INTEROSSEI											X	X	
LUMBRICALES II, III, IV											X	X	

FIGURE APP-D-2: *(continued)*

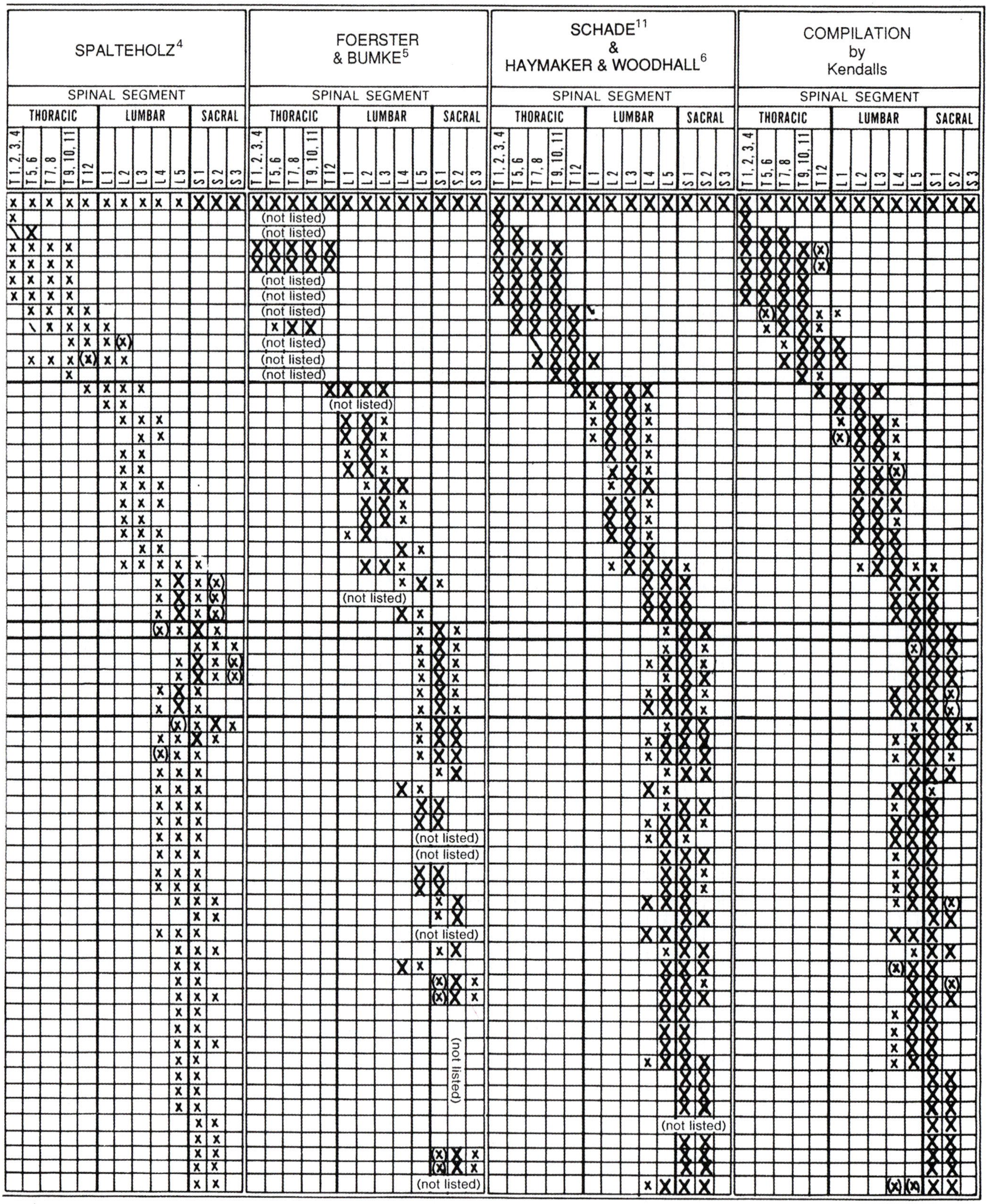

FIGURE APP-D-2: *(continued)*

REFERENCES

1. Goss CM, ed. Gray's Anatomy of the Human Body. 28th ed. Philadelphia: Lea & Febiger, 1966.
2. deJong RN. The Neurologic Examination. 3rd ed. New York: Harper & Row, 1967.
3. Romanes GJ, ed. Cunningham's Textbook of Anatomy. 10th ed. London: Oxford University Press, 1964.
4. Spalteholz W. Hand Atlas of Human Anatomy, Vol II, III. 6th ed. In English. London: JB Lippincott.
5. Foerster O, Bumke O. Handbuch Der Neurologie. Volume 5. Berlin: J Springer, 1936.
6. Haymaker W, Woodhall B. Peripheral Nerve Injuries. 2nd ed. Philadelphia: WB Saunders, 1953.
7. Brash JC, ed. Cunningham's Manual of Practical Anatomy. Vol 1. 11th ed. New York: Oxford University Press, 1948.
8. Hollinshead WH. Functional Anatomy of the Limbs and Back. 3rd ed. Philadelphia: WB Saunders, 1969.
9. Tavores AS. L'Innervation des muscles pectoraux. Acta Anat 1954; 21: 132–141.
10. Anson BJ, ed. Morris Human Anatomy. 12th ed. New York: McGraw-Hill, 1966.
11. Schade JP. The Peripheral Nervous System. New York: American Elsevier, 1966.

GLOSSARY

Abdomen. Region of the trunk that extends from the thorax to the pelvis.

Abduction. Movement about a sagittal axis in the coronal plane, away from the midsagittal plane of the body. For fingers and toes, *abduction* is movement away from midline of the hand or foot. For the thumb, *abduction* is a perpendicular movement away from the palm of the hand.

Active insufficiency. The inability of a class III or class IV two-joint (or multi-joint) muscle to generate an effective force when placed in a fully shortened position. The same meaning is implied by the expression "the muscle has been put on a slack."

Adaptive shortening. Tightness that occurs as a result of a muscle remaining in a shortened position.

Adduction. Movement about a sagittal axis in the coronal plane, toward the midsagittal plane of the body. For fingers and toes, *adduction* is movement toward midline of the hand or foot. For the thumb, *adduction* is a perpendicular movement toward the palm of the hand.

Agonist. A contracting muscle whose action is opposed by another muscle (antagonist).

Alignment. The arrangement of body segments as seen in various postural positions. See *ideal alignment* and *faulty alignments*.

Anatomical position. The position assumed when a person is standing erect with arms at the sides and the palms facing forward. Used as starting reference for all movement, positional, and directional terms.

Antagonist. A muscle that works in opposition to another muscle (agonist); opponent.

Anterior. Toward the front or ventral surface.

Anterior pelvic tilt. Pelvic tilt in which the vertical plane through the anterior-superior spines is anterior to the vertical plane through the symphysis pubis.

Aponeurosis. A flat fibrous sheet of connective tissue that attaches muscle to bone or other tissue.

Articulation. A joint; the site of close approximation of two or more bones.

Assessment. An appraisal of objective data from tests and examinations.

Asymptomatic. Having no subjective complaints; presenting no symptoms of disease or dysfunction.

Axes. Lines, real or imaginary, about which movement takes place. There are three basic types of axes at right angles to each other: Coronal axis (medial/lateral), Longitudinal axis (superior/inferior), and Sagittal (anterior/posterior).

Axial line. A line of reference in the hand or foot. *In the hand*, the axial line extends in line with the third metacarpal and the third digit. *In the foot*, the axial line extends in line with the second metatarsal and the second digit.

Body mechanics. The science concerned with static and dynamic forces acting on the body; the efficient or inefficient use of these forces in relation to body positions and movements.

Bowlegs (genu valgum). Legs curve outward.

Break test. A muscle strength test used to elicit the maximal effort exerted by a subject who is performing an isometric contraction as the examiner applies a gradual buildup of pressure to the point that the effort by the subject is overcome, i.e., the "breaking point." The break test is applicable for grading muscle strength of fair+ (6) through good+ (9) but not for grades of fair or below, nor for the grade of normal.

Capsular. Pertaining to sheath or continuous enclosure around an organ or structure.

Caudal. Downward, away from the head (toward the tail).

Center of gravity. That point on a body; freely acted upon by the earth's gravity, about which the body is in equilibrium; the point at which the three midplanes of the body intersect. In an ideally aligned posture, it is considered to be slightly anterior to the first or second sacral segment.

Circulation. Movement in a regular or circular course such as blood through the vasculature.

Circumduction. A circular (conical) movement that results from a combination of flexion, extension, abduction, adduction, and rotation.

Clockwise rotation. Used in describing rotation of the thorax or pelvis. With the transverse plane as a reference and 12 o'clock at midpoint anteriorly, rotation forward on the left is clockwise rotation. (Also described as facing toward the right.)

Compression. The force (or stress) that tends to shorten a body or squeeze it together. See **Tension, 2** for opposite meaning.
Concentric contraction. A shortening contraction; an isotonic contraction.
Connective tissue. Tissue that supports and connects other tissues and parts of the body. Types include areolar, adipose, fibrous, elastic, reticular, cartilage, bone, and blood.
Contractility. The property of a muscle that enables it to generate an effective force (produce tension). See **Tension, 1**.
Contraction. An increase in muscle tension, with or without change in overall length.
Contracture. A marked decrease in muscle length; range of motion in the direction of elongating the muscle is markedly limited. An irreversible contracture is resistant to treatment because elastic tissue has been replaced by inelastic tissue.
Contraindication. A sign or symptom that indicates that a particular treatment or procedure is not appropriate.
Contralateral. On the opposite side.
Coronal (frontal or lateral) plane. A vertical plane, extending from side to side, dividing the body into an anterior and a posterior portion.
Counterclockwise rotation. Used in describing rotation of the thorax or pelvis. With the transverse plane as a reference and 12 o'clock at midpoint anteriorly, rotation forward on the right is counterclockwise rotation. (Also described as facing toward the left.)
Cranial. Upward, toward the head.
Criteria. Standards upon which a decision can be based; established rules or principles for any given test.
Depression (motion). The lowering of a body part or region.
Diagnosis. The identification and classification of a disease, injury, or dysfunction based on examination findings.
Diagnostic. Useful in determining a diagnosis; pertaining to the art and science of distinguishing one injury, disease, or dysfunction from another.
Distal. Farther from the center or median line, or from the trunk.
Dorsal. Pertaining to the dorsum or back of the body or segment.
Dorsal primary rami. A branch of the spinal nerve that contains efferent and afferent axons that communicate with muscles of the deep back and the skin overlying the region of the deep back posterior neck and head.
Dorsiflexion. Ankle joint extension; opposite of plantar flexion (often mistakenly called flexion).
Dysfunction. Inability to function properly, functional impairment, or disability.
Eccentric contraction. A lengthening contraction.
Elevation (motion). The raising of a body part or region.
Evaluation. Interpretation of objective and subjective data for the purpose of determining a musculoskeletal diagnosis and the appropriate course of treatment.
Eversion. A combination of pronation and forefoot abduction; talipes valgus. (Eversion is more free in dorsiflexion than in plantar flexion.)
Examination. A procedure that includes more than one type of test; e.g., a postural examination that includes a variety of tests.
Extensibility. The property of muscle that permits it to lengthen or be elongated.
Extension. The act of straightening or being straight about a joint resulting in increased angle between the bones of a limb or body segment in the sagittal plane. A return to the anatomical position.
External rotation (lateral rotation). Turning the anterior surface of the extremity away from the midline of the body.
Extracapsular. Outside a capsule.
Fascia. A fibrous membrane covering, supporting, and separating muscle (deep) or connecting muscle to skin (superficial).
Fasciculi. Plural form of a small bundle of nerve or muscle fibers.
Fixation. Includes stabilization, support, and counterpressure; implies holding firm.
Flexibility. The ability to readily adapt to changes in position or alignment; may be expressed as normal, limited, or excessive.
Flexion. The act of bending or being bent about a joint resulting in decreased angle between the bones of a limb in the sagittal plane. Moving away from the anatomical position.
Frontal (coronal) plane. A vertical plane, extending from side to side, dividing the body into an anterior and a posterior portion.
Fusiform. Tapering at both ends. Spindle-shaped.
Genu valgum. Knock-knees.
Genu varum. Bowlegs.
Goniometer. An instrument for measuring angles and determining range of joint motion.
Gravity line. A vertical line through the center of gravity: a line analogous to the intersection of the midsagittal and midcoronal planes.
Horizontal abduction. A shoulder movement of the arm about a longitudinal axis in the transverse plane away from the midline. A movement of shoulder extension in an abducted position.
Horizontal adduction. Movements of the arm about a longitudinal axis in the transverse plane toward the midline. A movement of flexion in an adducted position.

Hyperextension. **1.** *Movement* beyond the normal range of joint motion in extension. **2.** A *position* of extension that is greater than normal postural alignment but not beyond the normal range of joint motion. It is seen as a lordotic position of the cervical spine in a typical forward-head posture, as a lordosis of the lumbar spine along with anterior pelvic tilt, and as hip joint extension in a sway-back posture.

Hyperkyphosis. An abnormal posterior curve, usually found in the thoracic region of the spine. As such, it is an *exaggeration of the normal posterior curve*.

Ideal alignment. The alignment used as a standard when evaluating posture.

Idiopathic. Pertaining to illnesses whose cause is either uncertain or unknown.

Impingement. An encroachment on the space occupied by soft tissue, such as nerve or muscle. In this text, impingement refers to nerve irritation (i.e., from pressure or friction) associated with muscles.

Indication. A sign or symptom that points out that a particular treatment or procedure is appropriate.

Internal (medial) rotation. Turning the anterior surface of the extremity toward the midline of the body.

Intervening muscles. Muscles that hold an adjacent part (usually an arm or a leg) firmly fixed to the bone of insertion, thereby providing a longer lever for the purpose of testing and grading muscle strength. Examples: posterior deltoid in trapezius test and shoulder flexors in serratus anterior test.

Intracapsular. Within a capsule.

Inversion. A combination of supination and forefoot adduction; talipes varus. (Inversion is more free in plantar flexion than in dorsiflexion.)

Ipsilateral. On the same side.

Isometric contraction. Increase in tension without change in muscle length.

Isotonic contraction. Increase in tension with change in muscle length (in the direction of shortening); concentric contraction.

Joint. A skeletal, bone to bone connection, held together by fibrous, cartilaginous, or synovial tissue. Joints are named according to the bones that are held together.

Knock-knees. Knees touch with feet apart; genu valgum.

Kyphosis. **1.** An abnormal posterior curve, usually found in the thoracic region of the spine. As such, it is an *exaggeration of the normal posterior curve*. If used without any modifying word, it refers to a thoracic kyphosis. In the low back, there is, occasionally, a lumbar kyphosis which is a *reversal of the normal anterior curve*. **2.** Normal curve in thoracic and sacrococcygeal regions of the spine.

Kyphotic. Pertaining to a position of kyphosis.

Lateral. Directional term referencing away from midline of the body or body segment.

Lateral flexion. Side bending; movement in which the body bends toward the side of concavity while the spine curves convexly toward the opposite side. (Curves of the spine are named according to the convexity; a curve to the right is lateral flexion toward the left.)

Lateral pelvic tilt. Pelvic tilt in which the crest of the ilium is higher on one side than on the other.

Lateral (external) rotation. Turning the anterior surface of the extremity away from the midline of the body.

Ligament. A band of fibrous connective tissue connecting bones, cartilage, and other structures and serving to support or attach fascia or muscles.

Lordosis. An abnormal anterior curve, usually found in the lumbar region, and as such is an *exaggeration of the normal anterior curve* (avoid use of the term "normal lordosis"); often called "hollow back." It is accompanied by anterior pelvic tilt and hip joint flexion. If used without any modifying word, it refers to a lumbar lordosis. In the thoracic region, occasionally, there is a slight lordosis which is a *reversal of the normal posterior curve*. In a typical forward head position, the neck is in a position of extension that is greater than the normal anterior curve and as such resembles a lordosis.

Lordotic. Pertaining to a position of lordosis.

Measurable test. A test that is quantifiable, based on a standard. One of the criteria for muscle length and strength tests.

Medial. Directional term referencing toward midline of the body or body segment.

Medial (internal) rotation. Turning the anterior surface of the extremity toward the midline of the body.

Median (sagittal, midsagittal, or anterior-posterior) plane. A vertical plane, extending from front to back. A plane that divides the body into right and left halves.

Mobility. Ability to move freely.

Muscle. A type of tissue composed of contractile cells or fibers that effects movement of an organ or body part. Three types include skeletal, smooth, and cardiac.

Muscle balance. A state of equilibrium that exists when there is a balance of strength of opposing muscles acting on a joint, providing ideal alignment for movement and optimal stabilization.

Muscle imbalance. Inequality in strength of opposing muscles; a state of muscle imbalance exists when a muscle is weak and its antagonist is strong; leads to faults in alignment and inefficient movement.

Muscle length. The extent to which a muscle can be elongated.

Musculoskeletal diagnosis. The identification and classification of musculoskeletal dysfunction.

Nerve. A bundle of neurons that transmit electrochemical impulses between the central nervous system and body tissues.

Neutral position of the pelvis. One in which anterior-superior spines are in the same transverse plane, and the anterior-superior spines and the symphysis pubis are in the same vertical plane.

Normal. Conforming to a standard. See normal alignment; normal flexibility according to age; normal range of motion; and normal strength.

Normal flexion of lumbar spine. Straightening or flattening of the lumbar spine.

Ober test. A test for tightness of the tensor fasciae latae and the iliotibial band.

Objective. Pertaining to findings evident to the examiner. See **Sign**.

Opposition. The ability to move the thumb or 5th finger (digit) into contact with other fingers (digits) by means of rotation at their designated carpometacarpal joints.

Optimal strength test position. Completed range of motion for one-joint muscles and class II two-joint muscles; a position within the mid-range of the overall length for class III and class IV two-joint or multi-joint muscles.

Optimal test position. Completed range of motion for one-joint muscles; a position within the mid-range of the overall length for *two-joint* muscles.

Overstretch. Stretch beyond the normal range of muscle length.

Overstretch weakness. Weakness in a two-joint (or multi-joint) muscle resulting from repetitive movements or habitual positions that *elongate the muscle beyond normal range of muscle length.*

Palmar (volar). Pertaining to the palm of the hand.

Passive insufficiency. Shortness of a two-joint (or multi-joint) muscle; the length of the muscle is not sufficient to permit *normal elongation* over both joints simultaneously, e.g., short hamstrings.

Passive range of motion. Movement through available, pain-free range of motion, performed by another individual without participation by the subject.

Pelvic tilt. An anterior (forward), a posterior (backward), or a lateral (sideways) tilt of the pelvis from neutral position (see also **Neutral position of the pelvis**).

Pelvis (osteologic). The bony compartment comprised of the innominate bone, the sacrum, and the coccyx. Joined at the sacroiliac joints, sacrococcygeal joint, and the pubic symphysis.

Pennate (muscle). A muscle in which fibers extend at an angle from point of origin.

Planes. Two-dimensional, flat surfaces, real or imaginary, at right angles to each other.

Plantar. Pertaining to the sole of the foot.

Plantar flexion. Ankle joint flexion; opposite of dorsiflexion (often mistakenly called extension).

Plexus. A network of nerves, or of blood or lymph vessels.

Plumb line. A line (piece of cord) to which is attached a plumb bob (a small lead weight). When suspended, it represents a vertical line. When used for analyzing standing posture, it must be suspended in line with fixed points, namely midway between heels in posterior view, and just anterior to the lateral malleolus in a lateral view.

Position. The manner in which the body is arranged by oneself, another individual, or other external means.

Posterior. Toward the back or dorsal surface.

Posterior pelvic tilt. Pelvic tilt in which the vertical plane through the anterior-superior spines is posterior to the vertical plane through the symphysis pubis.

Postural bowing. An *apparent bowing* that results from a combination of pronation of the feet, hyperextension of the knees, and medial rotation of the hips.

Posture. Attitude or position of the body.

Practical test. A test that is relatively easy to perform and requires a minimum of equipment. One of the criteria for muscle length and strength tests.

Pressure. In muscle testing, the force applied by the examiner to elicit the strength of a muscle holding in *test position* (pertains to muscles grading fair+ (6) or better).

Pronation. A rotation movement. *Pronation of the forearm* occurs when the distal end of the radius moves from the anatomic lateral position (supination) to a medial position, causing the hand to face posteriorly. *Pronation of the foot* occurs when the foot rotates so that the sole of the foot faces in a somewhat lateral direction. In standing, weight is on the inner side of the foot.

Prone. Lying face downward; face-lying.

Protective muscle spasm. A reflex muscle spasm by which nature "splints" or immobilizes a part to avoid movement that would cause further irritation of the injured structure.

Protrusion. The state or condition of being thrust forward.

Proximal. Nearer to the center or median line, or to the trunk.

"Put on a slack." Placing the muscle in a shortened position in which it is incapable of developing enough tension to exert an effective force. (Applies to class III and IV two-joint muscles, but not to one-joint or class II two-joint muscles. See **Active insufficiency**.)

Radiograph. A photographic film produced as a result of taking an x-ray.

Range of motion. The range, usually expressed in degrees, through which a joint can move or be moved.

Referred pain. Pain that is felt at some distance from its source.

Reliable test. A test that produces the same results on successive trials. One of the criteria for muscle length and strength tests.

Resistance. A force tending to hinder motion; in muscle testing, refers to resistance by the examiner or by gravity during *test movements*.

Retrusion. The state or condition of being thrust backward.

Reverse action. With the insertion fixed, the muscle contracts to move the origin toward the insertion.

Rotation. Movement about a longitudinal axis in a transverse plane.

Round shoulders. Forward shoulders.

Round upper back. A kyphosis.

Sagittal (midsagittal, median, or anterior-posterior) plane. A vertical plane, extending from front to back that divides the body into right and left halves.

Scoliosis. Lateral curvature of the spine. The spine may curve toward one side only or may have compensatory curves. A lateral curve convex toward the right is a right curve and vice versa.

Shortness. Tightness; denotes a slight to moderate decrease in muscle length; movement in the direction of elongating the muscle is limited.

Sign. Indication of an abnormality, related to disease or dysfunction, that is evident to the examiner, i.e., objective evidence. Compare with **Symptom**.

Sit-up. The movement of coming from a supine to a sitting position by flexing at the hip joints. ("Trunk curl," which is spine flexion, should not be called a partial sit-up.)

Spasm. An involuntary muscle contraction.

Spinal cord. Is part of the central nervous system and is a collection of nervous tissue that extends from the medulla oblongata to approximately the second lumbar vertebrae.

Spinal nerve. Is a mixed nerve containing afferent and efferent axons that is formed by the dorsal and ventral roots emanating from the spinal cord.

Sprain. Injury to a joint with possible rupture of ligaments or tendons, but without dislocation.

Stability. Capacity to provide support; firmness in position.

Stabilization. Fixation; implies holding steady or holding down.

Strain. The effect of an injurious tension.

Stress. Any force that tends to distort a body. It may be in the direction of either pulling apart or pressing together.

Stretch. Elongate; increase in length. The implied meaning is that it is not beyond the normal length of the muscle. See also **Overstretch**.

Stretch weakness. Weakness that results from muscles remaining in an elongated condition, however slight, beyond the neutral physiological rest position, but *not* beyond the normal range of muscle length. The concept relates to the duration of the faulty alignment rather than to the severity of it.

Structural bowing. Involves *actual bowing* of the bones of the lower extremities; genu varum.

Subjective. Perceived by the individual; not evident to the examiner. See **Symptom**.

Substitution. The action of muscles in attempting to function in place of other muscles that fail to perform because of weakness or pain.

Supination. A rotation movement. *Supination of the forearm* occurs when the distal end of the radius moves from a position of medial rotation (pronation) to the anatomic lateral position, causing the palm of the hand to face anteriorly. *Supination of the foot* occurs when the foot rotates so that the sole of the foot faces in a somewhat medial direction. In standing, weight is borne on the outer side of the foot.

Supine. Lying face upward; back-lying.

Sway-back posture. A faulty postural alignment in which there is a posterior displacement (swaying back) of the upper trunk and an anterior displacement (swaying forward) of the pelvis. There is a long kyphosis extending into the upper lumbar region, and a flattening of the low lumbar region. The pelvis is in posterior tilt and the hip joints are extended. The head and neck are in a forward head position.

Symptom. An abnormality in function or sensation, perceived by the patient, and indicative of disease or dysfunction; i.e., subjective evidence. Compare with **Sign**.

Syndrome. A group of signs and symptoms that appear together as characteristic of a disease, lesion, or dysfunction.

Taut. Firm when fully elongated; not slack. Muscles become taut at the end of the available range of motion permitted by the muscle length, i.e., when stretched to their limit.

Tendon. A fibrous connective tissue serving for the attachment of muscle to bone and other structures.

Tension. 1. As *applied to muscles*: the effective force generated by a muscle. **2.** As *applied to body mechanics*: the force (or stress) that tends to lengthen a body. Compression and tension are opposite in meaning. **3.** As *applied to headaches*: tightness of the posterior neck muscles.

Test movement. A movement of the part in a specified direction and through a specified arc of motion.

Test position. The position in which the part is placed by the examiner and held (if possible) by the patient.

Test. A procedure for obtaining measurements to be interpreted according to a standard; e.g., a muscle length, muscle strength, range of motion, or alignment test.

Thomas test. Definition from Jones and Lovett: "The Thomas flexion test is founded upon our inability to extend a diseased hip without producing a lordosis. If there is flexion deformity, the patient is unable to extend the thigh on the diseased side, and it remains at an angle" (61).

Thorax. A body region that extends from base of the neck to the diaphragm. A region component of the trunk.

Tight. Short, limiting the range of motion; i.e., the muscle *is* tight. Firm on palpation when drawn taut, i.e., the muscle *feels* tight (may be true of a short or a stretched muscle).

Tightness. Shortness; denotes a slight to moderate decrease in muscle length; movement in the direction of lengthening the muscle is limited.

Tilt. Rotation about a transverse axis. See **Pelvic tilt**.

Torticollis. A stiff neck associated with muscle spasm classically resulting in lateral flexion and rotation of the cervical spine. Types include congenital or acquired.

Transverse plane. A horizontal plane, dividing the body into upper (cranial) and lower (caudal) portions.

Trendelenburg sign. Indication of hip abductor weakness as evidenced by the hip going into *adduction* when standing with full weight on the affected leg with the other foot off the floor. Initially, the Trendelenburg test was used in diagnosing a dislocated hip. The Trendelenburg gait is one in which the affected hip goes into *adduction* during each weight-bearing phase of the gait. This is in contrast to the *abducted* position of the hip joint in the gait associated with paralysis of hip abductors.

Trunk. A region of the body exclusive of the head and limbs.

Useful test. A test that provides information of value for determining the proper course of treatment. One of the criteria for muscle length and strength tests.

Valgus. Knee (genu v.): knock-knees. Foot (talipes v.): pronation with forefoot abduction. Big toe (hallux v.): adduction of big toe (toward midline of foot), associated with a bunion.

Valid test. One that measures, quantitatively and qualitatively, what it purports to measure. One of the criteria for muscle length and strength tests.

Varus. Knee (genu v.): bowlegs. Foot (talipes v.): supination with forefoot adduction.

Ventilation. The movement of air in and out of the lungs.

Ventral. Front or anterior, as anterior surface of the body.

Ventral primary rami. A branch of the spinal nerve that contains efferent and afferent axons that communicate with muscles and skin (excluding muscle of the deep back and the skin overlying the region of the deep back).

Ventral root. A nerve branch comprised of ventral rootlets extending from the ventral root of the spinal cord.

SUGGESTED READINGS

Adams MA, Hutton WC. Prolapsed intervertebral disc a hyperflexion injury. Spine 1982; 7: 3.

Andersson GBJ, Ortengren R, Nachemson AL, et al. Lumbar disc pressure and myoelectric back muscle activity during sitting. Scand J Rehabil Med 1974; 6: 104.

Andersson GBJ, Ortengren R, Nachemson AL, et al. The sitting posture: An electromyographic and discometric study. Orthop Clin North Am 1975; 6: 105.

Andersson GBJ, Ortengren R, Herberts P. Quantitative electromyographic studies of back muscle activity related to posture and loading. Orthop Clin North Am 1977; 8: 85.

Ardran GM, Kemp FH. The mechanism of the larynx. II. The epiglottes and closure of the larynx. Br J Radiol 1967; 40: 372.

Arnold GE. Physiology and pathology of the cricothyroid muscle. Laryngoscope 1961; 71: 687.

Atkinson M, Dramer P, Wyman SM, et al. The dynamics of swallowing. I. Normal pharyngeal mechanisms. J Clin Invest 1957; 36: 581.

Barun N, Arora N, Rochester D. Force-length relationship of the normal human diaphragm. J Appl Physiol 1982; 53(2): 4405–412.

Basmajian JV. Electromyography of two-joint muscles. Anat Rec 1957; 129: 371.

Basmajian JV. Electromyography of iliopsoas. Anat Rec 1958; 132: 127.

Basmajian JV. Grant's Method of Anatomy. 9th ed. Baltimore: Williams & Wilkins, 1975.

Basmajian JV, Travill A. Electromyography of the pronator muscles in the forearm. Anat Rec 1961; 139: 45–49.

Basmajian JV, Wolf SL. Therapeutic Exercise. 5th ed. Baltimore: Williams & Wilkins, 1990.

Battı'e MC, Bigos SJ, Sheehy A, Wortley MD. Spinal flexibility and individual factors that influence it. Phys Ther 1987; 67: 5.

Beattie P, Rothstein JM, Lamb RL. Reliability of the attraction method for measuring lumbar spine backward bending. Phys Ther 1987; 67: 364–368.

Bender JA, Kaplan HM. The multiple angle testing method for the evaluation of muscle strength. J Bone Joint Surg [Am] 1963; 45-A: 135.

Black SA. Clinical applications in muscle testing. Rehab Man 1990; 3(1): 30, 32, 61.

Blackburn SE, Portney LG. Electromyographic activity of back musculature during Williams' flexion exercises. Phys Ther 1981; 61: 878.

Blakely WR, Garety EJ, Smith DE. Section of the cricopharyngeus muscle for dysphagia. Arch Surg 1968; 96: 745.

Blankenship KL. Industrial rehabilitation-seminar syllabus. Stress and lift-pull indexes (Ch. 9). Proper lifting techniques (Ch. 10). American Therapeutics, Inc., 1989.

Blanton PL, Biggs NL, Perkins RC. Electromyographic analysis of the buccinator muscle. J Dent Res 1970; 49: 389.

Bohannon RW. Cinematographic analysis of the passive straight-leg-raising test for hamstring muscle length. Phys Ther 1982; 62(9): 1269–1274.

Bohannon RW. Grip Strength: An Indispensable Biomarker For Older Adults. *Clin Interv Aging*. 2019;14: 1681–1691. Published 2019 Oct 1. doi:10.2147/CIA.S194543.

Bohannon RW, Gajdosik RL. Spinal nerve root compression-some clinical implications. Phys Ther 1987; 67: 3.

Bohannon RW, Gajdosik RL, LeVeau BF. Contribution of pelvic and lower limb motion to increases in the angle of passive straight leg raising. Phys Ther 1985; 65(4): 474–476.

Borello-France DF, Handa VL, Brown MB, et al. Pelvic-floor muscle function in women with pelvic organ prolapse. *Phys Ther*. 2007;87(4):399–407. doi:10.2522/ptj.20060160.

Bosma JF. Deglutition: Pharyngeal stage. Physiol Rev 1957; 37: 275.

Bouman HD, ed. An exploratory and analytical survey of therapeutic exercise: Northwestern University Special Therapeutic Exercise Project. Am J Phys Med 1967; 46: 1.

Bourn J, Jenkins S. Postoperative respiratory physiotherapy: Indications for treatment. Physiotherapy 1992; 78(2): 80–85.

Bourne MN, Timmins RG, Opar DA, et al. An Evidence-Based Framework for Strengthening Exercises to Prevent Hamstring Injury. *Sports Med*. 2018;48(2): 251–267. doi:10.1007/s40279-017-0796-x.

Brand PW, Beach RB, Thompson DE. Relative tension and potential excursion of muscles in the forearm and hand. J Hand Surg [Am] 1981; 6: 209.

Breig A, Troup JDG. Biomechanical considerations in the straight-leg-raising test. Spine 1979; 4(3): 242–250.

Brunnstrom, S. Clinical Kinesiology. 3rd ed. Philadelphia: FA Davis, 1972.

Bullock-Saxton J. Normal and abnormal postures in the sagittal plane and their relationship to low back pain. Physiother Pract 1988; 4(2): 94–104.

Bunnell's Surgery of the Hand. 4th ed. Boyes JH, ed. Philadelphia: JB Lippincott, 1964.

Campbell EJM. The Respiratory Muscles and the Mechanics of Breathing. Chicago: Year Book, 1958.

Campbell EJM, Agostini E, Davis JN. The Respiratory Muscles: Mechanisms and Neural Control. 2nd ed. Philadelphia: WB Saunders, 1970.

Capuano-Pucci D, Rheault W, Aukai J, Bracke M, Day R, Pastrick M. Intratester and intertester reliability of the cervical range of motion device. Arch Phys Med Rehabil 1991; 72: 338–340.

Carmen DJ, Blanton PL, Biggs NL. Electromyographic study of the anterolateral abdominal musculature utilizing indwelling electrodes. Am J Phys Med 1972; 51: 113.

Cash JE, ed. Chest, Heart and Vascular Disorders for Physiotherapists. Philadelphia: JB Lippincott, 1975.

Cassella MC, Hall JE. Current treatment approaches in the nonoperative and operative management of adolescent idiopathic scoliosis. Phys Ther 1991; 71: 12.

Chusid JG. Correlative Neuroanatomy and Functional Neurology. 15th ed. Los Altos, CA: Lange Medical Publications, 1973.

Clapper MP, Wolf SL. Comparison of the reliability of the orthoranger and the standard goniometer for assessing active lower extremity range of motion. Phys Ther 1988; 68(2): 214–218.

Clayson SJ, Newman IM, Debevec DF, et al. Evaluation of mobility of hip and lumbar vertebrae of normal young women. Arch Phys Med Rehabil 1962; 43: 1.

Close JR. Motor Function in the Lower Extremity. Springfield, IL: Charles C Thomas, 1964.

Close JR, Kidd CC. The functions of the muscles of the thumb, the index and long fingers. J Bone Joint Surg [Am] 1969; 51-A: 1601.

Close RI. Dynamic properties of mammalian skeletal muscles. Physiol Rev 1972; 52: 129.

Cohen-Sobel E, Levitz SJ. Torsional development of the lower extremity. J Am Podiatr Med Assoc 1991; 81(7): 344–357.

Cole TM. Goniometry: The measurement of joint motion. In: Krusen FH, Kottke F, Elwood PM, eds. Handbook of Physical Medicine and Rehabilitation. 2nd ed. Philadelphia: WB Saunders, 1971.

Cooperman JM. Case studies: Isolated strain of the tensor fasciae latae. J Orthop Sports Phys Ther 1983; 5(4): 201–203.

Crommert ME, Bjerkefors A, Tarassova O, Ekblom MM. Abdominal Muscle Activation During Common Modifications of the Trunk Curl-up Exercise. *J Strength Cond Res*. 2021;35(2):428–435. doi:10.1519/JSC.0000000000002439.

Cunningham DP, Basmajian JB. Electromyography of genioglossus and geniohyoid muscles during deglutition. Anat Rec 1969; 165: 401.

Currier DP. Maximal isometric tension of the elbow extensors at varied positions. Phys Ther 1972; 52: 1265.

Currier DP. Positioning for knee strengthening exercises. Phys Ther 1977; 57: 148.

Cyriax J. Textbook of Orthopaedic Medicine. Vol 1: Diagnosis of Soft Tissue Lesions. 7th ed. London: Bailliere-Tindall, 1978.

Cyriax J, Cyriax P. Illustrated Manual of Orthopaedic Medicine. London: Butterworth, 1983.

DeJong RN. The Neurological Examination. 4th ed. New York: Harper & Row, 1979.

DeLuca CJ, Forrest WJ. Force analysis of individual muscles acting simultaneously on the shoulder joint during isometric abduction. J Biomech 1973; 6: 385.

DeRosa C, Porterfield JA. The sacroiliac joint. Postgraduate advances in the evaluation and treatment of low back dysfunction. Forum Medicum 1989.

Des Jardins TR. Cardiopulmonary Anatomy and Physiology. Albany, NY: Delmar, 1988.

DeSousa OM, Furlani J. Electromyographic study of the m. rectus abdominis. Acta Anat 1974; 88: 281.

DeSousa OM, Demoraes JL, Vieira FL. Electromyographic study of the brachioradialis muscle. Anat Rec 1961; 139: 125.

DeSousa OM, Berzin F, Berardi AC. Electromyographic study of the pectoralis major and latissimus dorsi during medial rotation of the arm. Electromyography 1969; 9: 407.

Dickson RA, Lawton JL, Archer IA, Butt WP. The pathogenesis of idiopathic scoliosis. J Bone Joint Surg [Br] 1984; 66-B(1): 8–15.

Donelson R, Silva G, Murphy K. Centralization phenomenon-its usefulness in evaluating and treating referred pain. Spine 1990; 15(3): 211–213.

DonTigny RL. Anterior dysfunction of the sacroiliac joint as a major factor in the etiology of idiopathic low back pain syndrome. Phys Ther 1990; 70(4): 250–265.

Dostal WF, Soderberg GL, Andrews JG. Actions of hip muscles. Phys Ther 1986; 66(3): 351–361.

Downer AH. Physical Therapy Procedures. 3rd ed. Springfield, IL: Charles C Thomas, 1978.

Duval-Beaupere G. Rib hump and supine angle as prognostic factors for mild scoliosis. Spine 1992; 17: 1.

Eaton RG, Littler JW. A study of the basal joint of the thumb. J Bone Joint Surg [Am] 1969; 51-A: 661.

Ekholm J, Arborelius U, Fahlcrantz A, et al. Activation of abdominal muscles during some physiotherapeutic exercises. Scand J Rehabil Med 1979; 11: 75.

Elftman H. Biomechanics of muscle. J Bone Joint Surg [Am] 1966; 48-A: 363.

Escamilla RF, Lewis C, Pecson A, Imamura R, Andrews JR. Muscle Activation Among Supine, Prone, and Side Position Exercises With and Without a Swiss Ball. Sports Health. 2016;8(4):372–379. doi:10.1177/1941738116653931Pelvic Floor.

Eyler DL, Markee JE. The anatomy and function of the intrinsic musculature of the fingers. J Bone Joint Surg [Am] 1954; 36-A: 1.

Farfan HF. Mechanical Disorders of the Low Back. Philadelphia: Lea & Febiger, 1973.

Farfan HF. Muscular mechanism of the lumbar spine and the position of power and efficiency. Orthop Clin North Am 1975; 6: 135.

Farrell C, Kiel J, Seemann L, Pujalte GGA. Popliteus Tendon Injuries [published online ahead of print, 2022 Jul 25]. *Orthopedics*. 2022;1–6. doi:10.3928/01477447-20220719-10.

Fast A. Low back disorders: Conservative management. Arch Phys Med Rehabil 1988; 69: 880–891.

Fenn WO, Rahn H. Handbook of Physiology. Section 3: Respiration. Vol 1. Washington, DC: American Physiological Society, 1964: 377–384.

Fischer FJ, Houtz SJ. Evaluation of the function of the gluteus maximus muscle. Am J Phys Med 1968; 47: 182.

Fishman AP, ed. Pulmonary Diseases and Disorders. 2nd ed. New York: McGraw-Hill, 1988.

Flint MM. Abdominal muscle involvement during performance of various forms of sit-up exercise. Am J Phys Med 1965; 44: 224.

Flint MM. An electromyographic comparison of the function of the iliacus and the rectus abdominis muscles. J Am Phys Ther Assoc 1965; 45: 248.

Francis RS. Scoliosis screening of 3,000 college-aged women: The Utah Study-Phase 2. Phys Ther 1988; 68(10): 1513–1516.

Franco AH. Pes cavus and pes planus. Phys Ther 1987; 67(5): 688–693.

Frank JS, Earl M. Coordination of posture and movement. Phys Ther 1990; 70(12): 855–863.

Frese E, Brown M, Norton BJ. Clinical reliability of manual muscle testing-middle trapezius and gluteus medius muscles. Phys Ther 1987; 67(7): 1072–1076.

Fujiwara M, Basmajian JV. Electromyographic study of two-joint muscles. Am J Phys Med 1975; 54: 234.

Gajdosik R, Lusin G. L Hamstring muscle tightness. Phys Ther 1983; 63(7): 1085–1090.

Girardin Y. EMG action potentials of rectus abdominis muscle during two types of abdominal exercises. In: Cerquigleni S, Venerando A, Wartenweiler J, eds. Biomechanics III. Baltimore: University Park Press, 1973.

Gleeson PB, Pauls JA. Obstetrical physical therapy-review of the literature. Phys Ther 1988; 68(11): 1699–1702.

Glennon TP. Isolated injury of the infraspinatus branch of the suprascapular nerve. Arch Phys Med Rehabil 1992; 73: 201–202.

Godfrey KE, Kindig LE, Windell EJ. Electromyographic study of duration of muscle activity in sit-up variations. Arch Phys Med Rehabil 1977; 58: 132.

Goldberg CJ, Dowling FE. Idiopathic scoliosis and asymmetry of form and function. Spine 1991; 16(1): 84–87.

Gose JC, Schweizer P. Iliotibial band tightness. J Orthop Sports Phys Ther 1989; 9(4): 399–406.

Gowitzke BA, Milner MM. Understanding the Scientific Basis of Human Motion. 2nd ed. Baltimore: Williams & Wilkins, 1980.

Gracovetsky S, Farfan HF, Lamy C. The mechanism of the lumbar spine. Spine 1981; 6: 249.

Gray ER. The role of leg muscles in variations of the arches in normal and flat feet. J Am Phys Ther Assoc 1969; 49: 1084.

Grieve GP. The sacro-iliac joint. Physiother 1976; 62: 384.

Guffey JS. A critical look at muscle testing. Clin 1991; 11(2): 15–19.

Halpern A, Bleck E. Sit-up exercise: An electromyographic study. Clin Orthop Relat Res 1979; 145: 172.

Halski T, Ptaszkowski K, Słupska L, Dymarek R, Paprocka-Borowicz M. Relationship between lower limb position and pelvic floor muscle surface electromyography activity in menopausal women: a prospective observational study. *Clin Interv Aging*. 2017;12:75–83. Published 2017 Jan 4. doi:10.2147/CIA.S121467.

Hart DL, Stobbe TJ, Jaraiedi M. Effect of lumbar posture on lifting. Spine 1987; 12(2): 1023–1030.

Hasue M, Fujiwara M, Kikuchi S. A new method of quantitative measurement of abdominal and back muscle strength. Spine 1980; 51: 143.

Haymaker W. Bing's Local Diagnosis in Neurological Diseases. 15th ed. St. Louis: CV Mosby, 1969.

Hicks JH. The three weight-bearing mechanisms of the foot. In: Evans FG, ed. Biomechanical Studies of the Musculoskeletal System. Springfield, IL: Charles C Thomas, 1961.

Hirano M, Koike Y, von Leden H. The sterno-hyoid muscle during phonation. Acta Otolaryngol 1967; 64: 500.

Houtz SJ, Lebow MJ, Beyer FR. Effect of posture on strength of the knee flexor and extensor muscles. J Appl Physiol 1957; 11: 475.

Hsieh C, Walker JM, Gillis K. Straight-leg-raising test. Phys Ther 1983; 63(9): 1429–1433.

Ingher RS. Iliopsoas myofascial dysfunction: A treatable cause of "failed" low back syndrome. Arch Phys Med Rehabil 1989; 70: 382–385.

Itoi E. Roentgenographic analysis of posture in spinal osteoporotics. Spine 1991; 16(7): 750–756.

Johnson JTH, Kendall HO. Localized shoulder girdle paralysis of unknown etiology. Clin Orthop 1961; 20: 151–155.

Joint Motion, Method of Measuring and Recording. Chicago: American Academy of Orthopaedic Surgeons, 1965.

Jonsson B, Olofsson BM, Steffner LCH. Function of the teres major, latissimus dorsi and pectoralis major muscles. Acta Morph Neerl Scand 1972; 9: 275.

Kamonseki DH, Gonçalves GA, Yi LC, Júnior IL. Effect of stretching with and without muscle strengthening exercises for the foot and hip in patients with plantar fasciitis: A randomized controlled single-blind clinical trial. *Man Ther*. 2016;23:76–82. doi:10.1016/j.math.2015.10.006.

Kaplan EB. Functional and Surgical Anatomy of the Hand. 2nd ed. Philadelphia: JB Lippincott, 1965.

Karabay D, Emük Y, Özer Kaya D. Muscle Activity Ratios of Scapular Stabilizers During Closed Kinetic Chain Exercises in Healthy Shoulders: A Systematic Review. J Sport Rehabil. 2019;29(7):1001–1018. Published 2019 Dec 19. doi:10.1123/jsr.2018-0449.

Keagy RD, Brumlik J, Bergan JJ. Direct electromyography of the psoas major muscle in man. J Bone Joint Surg [Am] 1966; 48-A: 1377.

Keller RB. Nonoperative treatment of adolescent idiopathic scoliosis. In: Barr JS, ed. The Spine-Instructional Course Lectures. Vol 30. 1989: 129.

Kendall HO. Some interesting observations about the after care of infantile paralysis patients. J Excep Children 1937; 3: 107.

Kendall HO. Watch those T.V. exercises. TV Guide 1963; II-31: 5.

Kendall HO, Kendall FP. Study and Treatment of Muscle Imbalance in Cases of Low Back and Sciatic Pain. Pamphlet. Baltimore: privately printed, 1936.

Kendall HO, Kendall FP. Care during the Recovery Period of Paralytic Poliomyelitis. U.S. Public Health Bulletin No 242. Washington, DC: U.S. Government Printing Office, 1939.

Kendall HO, Kendall FP. Gluteus medius and its relation to body mechanics. Physiother Rev 1941; 21: 131.

Kendall HO, Kendall FP. The role of abdominal exercise in a program of physical fitness. J Health Phys Ed 1943; 480.

Kendall HO, Kendall FP. Unpublished report on the Posture Survey at U.S. Military Academy, West Point, 1945.

Kendall HO, Kendall FP. Physical Therapy for Lower Extremity Amputees. War Department Technical Manual TM-8-293:14/42 and 58/65, Washington, DC: U.S. Government Printing Office, 1946: 12–42.

Kendall HO, Kendall FP. Orthopedic and physical therapy objectives in poliomyelitis treatment. Physiother Rev 1947; 27: 159.

Kendall HO, Kendall FP. Functional muscle testing. In: Bierman W, Licht S, eds. Physical Medicine in General Practice. New York: Paul B Hoeber, 1952: 339–384.

Kendall HO, Kendall FP. Posture, Flexibility, and Abdominal Muscle Tests (Leaflet). Baltimore: Waverly Press, 1964.

Kendall HO, Kendall FP. Developing and maintaining good posture. J Am Phys Ther Assoc 1968; 48: 319.

Kendall HO, Kendall FP, Boynton DA. Posture and Pain. Baltimore: Williams & Wilkins, 1952. Reprinted Melbourne, Florida: Robert E Krieger, 1971.

Kendall FP. Range of Motion. The Correlation of Physiology with Therapeutic Exercise. New York: American Physical Therapy Association, 1956.

Kendall FP. A criticism of current tests and exercises for physical fitness. J Am Phys Ther Assoc 1965; 45: 187–197.

Kisner C, Colby LA. Therapeutic Exercise: Foundations and Techniques. 2nd ed. Philadelphia: FA Davis, 1990.

Kleinberg S. Scoliosis: Pathology, Etiology, and Treatment. Baltimore: Williams & Wilkins, 1951.

Klousen K, Rasmussen B. On the location of the line of gravity in relation to L5 in standing. Acta Physiol Scand 1968; 72: 45.

Koes BW, Bouter LM, vanMameren H, Essers AHM, Verstegen GMJR, Hofhuizen DM, Houben JP, Knipschild PG. The effectiveness of manual therapy, physiotherapy, and treatment by the general practitioner for nonspecific back and neck complaints. Spine 1992; 17(1): 28–35.

Kotby MN. Electromyography of the laryngeal muscles. Electroencephalog Clin Neurophysiol 1969; 26: 341.

Kraus H. Effects of lordosis on the stress in the lumbar spine. Clin Orthop 1976; 117: 56.

LaBan M, Raptou AD, Johnson EW. Electromyographic study of function of iliopsoas muscle. Arch Phys Med 1965; 46: 676–679.

Lieb FJ, Perry J. Quadriceps function. J Bone Joint Surg [Am] 1971; 53-A: 749.

Lilienfeld AM, Jacobs M, Willis M. A study of the reproducibility of muscle testing and certain other aspects of muscle scoring. Phys Ther Rev 1954; 34(6): 279–290.

Lindahl O. Determination of the sagittal mobility of the lumbar spine. Acta Orthop Scand 1966; 37: 241.

Lindahl O, Movin A. The mechanics of extension of the knee joint. Acta Orthop Scand 1967; 38: 226.

Lindstrom A, Zachrisson M. Physical therapy for low back pain and sciatica. Scand J Rehabil Med 1970; 2: 37.

Lipetz S, Gutin B. Electromyographic study of four abdominal exercises. Med Sci Sports 1970; 2: 35.

Loebl WY. Measurement of spinal posture and range of spinal movement. Ann Phys Med 1967; 9: 103.

Long C. Intrinsic-extrinsic muscle control of the fingers. J Bone Joint Surg [Am] 1968; 50-A: 973.

Loptata M, Evanich MJ, Lourenco RV. The electromyogram of the diaphragm in the investigation of human regulation of ventilation. Chest 1976; 70(Suppl): 162S.

Loring SH, Mead J. Action of the diaphragm on the rib cage inferred from a force-balance analysis. J Appl Physiol 1982; 53; 3: 756–760.

Low JL. The reliability of joint measurement. Physiother 1976; 62: 227.

Mann R, Inman VT. Phasic activity of intrinsic muscles of the foot. J Bone Joint Surg [Am] 1964; 46-A: 469.

McCreary EK. The control of breathing in singing. [Research paper for Physiology Department] John A. Burns School of Medicine, Honolulu, Hawaii, 1982.

Mayhew TP, Norton BJ, Sahrmann SA. Electromyographic study of the relationship between hamstring and abdominal muscles during a unilateral straight leg raise. Phys Ther 1983; 63(11): 1769–1775.

Mendez-Rebolledo G, Morales-Verdugo J, Orozco-Chavez I, Habechian FAP, Padilla EL, de la Rosa FJB. Optimal activation ratio of the scapular muscles in closed kinetic chain shoulder exercises: A systematic review. J Back Musculoskelet Rehabil. 2021;34(1): 3–16. doi:10.3233/BMR-191771.

Michelle AA. Iliopsoas. Springfield, IL: Charles C Thomas, 1962.

Mines, AH. Respiratory Physiology. New York: Raven Press, 1981.

Moller M, Ekstrand J, Oberg B, Gillquist J. Duration of stretching effect on range of motion in lower extremities. Arch Phys Med Rehabil 1985; 66: 171–173.

Moore KL. Clinically Oriented Anatomy. Baltimore: Williams & Wilkins, 1980.

Moore ML. Clinical assessment of joint motion. In: Licht S, ed. Therapeutic Exercise. 2nd ed. Baltimore: Waverly Press, 1965.

Mulligan E. Conservative management of shoulder impingement syndrome. Athl Train 1988; 23(4): 348–353.

Nachemson A. Electromyographic studies on the vertebral portion of the psoas muscle. Acta Orthop Scand 1966; 37: 177.

Nachemson A. Physiotherapy for low back pain patients. Scand J Rehabil Med 1969; 1: 85.

Nachemson A. Towards a better understanding of low back pain: A review of the mechanics of the lumbar disc. Rheumatol Rehabil 1975; 14: 129.

Nachemson A. A critical look at the treatment for low back pain. Scand J Rehabil Med 1979, 11. 143.

Nachemson A, Lindh M. Measurement of abdominal and back muscle strength with and without low back pain. Scand J Rehabil Med 1969; 1: 60.

Nagler W, Pugliese G. Facet syndrome (letter to the editor). Arch Phys Med Rehabil 1989; 70.

Okubo Y, Kaneoka K, Hasebe K, Matsunaga N, Imai A, Hodges PW. Differential activation of psoas major and rectus femoris during active straight leg raise to end range. J Electromyogr Kinesiol. 2021;60:102588. doi:10.1016/j.jelekin.2021.102588.

Oliva-Lozano JM, Muyor JM. Core Muscle Activity During Physical Fitness Exercises: A Systematic Review. Int J Environ Res Public Health. 2020;17(12):4306. Published 2020 Jun 16. doi:10.3390/ijerph17124306.

Ouaknine G, Nathan H. Anastomotic connections between the eleventh nerve and the posterior root of the first cervical nerve in humans. J Neurosurg 1973; 38: 189.

Padua L, Coraci D, Erra C, et al. Carpal tunnel syndrome: clinical features, diagnosis, and management. Lancet Neurol. 2016;15(12):1273-1284. doi:10.1016/S1474-4422(16)30231-9.

Paré EB, Schwartz JM, Stern JT. Electromyographic and anatomical study of the human tensor fasciae latae muscle. In: Proceedings of the 4th Congress of the International Society of Electrophysiological Kinesiology. Boston: Published by the organizing committee, 1979.

Partridge MJ, Walters CE. Participation of the abdominal muscles in various movements of the trunk in man. Phys Ther Rev 1959; 39: 791–800.

Patton NJ, Mortensen OA. A study of some mechanical factors affecting reciprocal activity in one-joint muscles. Anat Rec 1970; 166: 360.

Pearsall DJ, Reid JG, Hedden DM. Comparison of three non-invasive methods for measuring scoliosis. Phys Ther 1992; 72: 9.

Pearson AA, Sauter RW, Herrin GR. The accessory nerve and its relation to the upper spinal nerves. J Anat 1964; 114-A: 371.

Pennal CF, Conn GS, McDonald G, et al. Motion studies of the lumbar spine. J Bone Joint Surg [Br] 1972; 54-B: 442.

Physical Therapy, Journal of the American Physical Therapy Association. Special issues:

Pain. 1980; 60: 1. (Lister MJ, ed.)

Respiratory care. 1980; 60: 12. (Lister MJ, ed.)

Muscle biology. 1982; 62: 12. (Lister MJ, ed.)

Biomechanics. 1984, 64. 12. (Lister MJ, ed.)

Shoulder complex. 1986; 66: 12. (Lister MJ, ed.)

Clinical measurement. 1987; 67: 12. (Lister MJ, ed.)

Foot and ankle. 1988; 68: 12. (Rose SJ, ed.)

Clinical decision making. 1989; 69: 7. (Rose SJ, ed. em.)

Hand management in physical therapy. 1989; 69: 12. (Rothstein JM, ed.)

Physiotherapy. Journal of the Chartered Society of Physiotherapy. Special issues:

The hand. 1977, 63: 9. (Whitehouse J, ed.)

Update in respiratory care. 1992; 78: 2. (Whitehouse J, ed.)

Pruijs JEH, Keessen W, van der Meer R, van Wieringen JC, Hageman MAPE. School screening for scoliosis: Methodologic considerations – Part 1: External measurements. Spine 1992; 17(4): 431–435.

Ptaszkowski K, Zdrojowy R, Ptaszkowska L, Bartnicki J, Taradaj J, Paprocka-Borowicz M. Electromyographic evaluation of synergist muscles of the pelvic floor muscle depending on the pelvis setting in menopausal women: A prospective observational study. *Gait Posture.* 2019;71:170–176. doi:10.1016/j.gaitpost.2019.04.024.

Ralston HJ, Todd FN, Inman VT. Comparison of electrical activity and duration of tension in the human rectus femoris muscle. Electromyogr Clin Neurophysiol 1976; 16: 271.

Ramsey GH, Watson JS, Gramiak R, et al. Cinefluorographic analysis of the mechanism of swallowing. Radiology 1955; 64: 498.

Riddle DL, Finucane SD, Rothstein JM, Walker ML. Intrasession and intersession reliability of hand-held dynamometer measurements taken on brain-damaged patient. Phys Ther 1989; 69(3): 182–194.

Roberts RH, ed. Scoliosis. CIBA Found Symp 1972; 24: 1.

Rodgers MM, Cavanagh PR. Glossary of biomechanical terms, concepts, and units. Phys Ther 1984; 64(12): 1886–1902.

Root ML, Orien WP, Weed JH. Normal and Abnormal Function of the Foot. Los Angeles: Clinical Biomechanics Corp, 1977: 95–107.

Salminen JJ, Maki P, Oksanen A, Pentti J. Spinal mobility and trunk muscle strength in 15-year-old schoolchildren with and without low-back pain. Spine 1992; 17(4): 405–411.

Salter N, Darcus HD. The effect of the degree of elbow flexion on the maximum torques developed in pronation and supination of the right hand. J Anat 1952; 86-B: 197.

Saunders JB deCM, Davis C, Miller ER. The mechanism of deglutition. Ann Otol Rhinol Laryngol 1951; 60: 897.

Schuit D, Adrian M, Pidcoe P. Effect of heel lifts on ground reaction force patterns in subjects with structural leg-length discrepancies. Phys Ther 1989; 69(8): 663–670.

Schultz JS, Leonard JA Jr. Long thoracic neuropathy from athletic activity. Arch Phys Med Rehabil 1992; 73: 87–90.

Scoliosis: An anthology. (Articles reprinted from *Physical Therapy*). Alexandria, VA: American Physical Therapy Association, 1984.

Shaffer T, Wolfson M, Bhutani VK. Respiratory muscle function, assessment, and training. Phys Ther 1981; 61: 12.

Sharf M, Shvartzman P, Farkash E, Horvitz J. Thoracic lateral cutaneous nerve entrapment syndrome without previous lower abdominal surgery. J Fam Pract 1990; 30: 2.

Sharp JT, Draz W, Danon J, et al. Respiratory muscle function and the use of respiratory muscle electromyography in the evaluation of respiratory regulation. Chest 1976; 70(Suppl): 150S.

Sharrard WJW. The segmental innervation of the lower limb muscles in man. Ann R Coll Surg Engl 1964; 35: 106.

Shelton RL, Bosma JF, Sheets BV. Tongue, hyoid and larynx displacement in swallow and phonation. J Appl Physiol 1960; 15: 283.

Simonds AH, Abraham K, Spitznagle, T. Executive summary of the clinical practice guidelines for pelvic girdle pain in the postpartum population. Journal of Women's Health Physical Therapy: January/March 2022; 46 (1): 3-10 doi: 10.1097/JWH.0000000000000235.

Slonim NB, Hamilton LH. Respiratory Physiology. St. Louis: CV Mosby, 1981.

Smidt GL, Rogers MW. Factors contributing to the regulation and clinical assessment of muscular strength. Phys Ther 1982; 62(9): 1283–1289.

Smith JW. Muscular control of the arches of the foot in standing: An electromyographical assessment. J Anat 1954; 88-B: 152.

Smith RL, Brunolli J. Shoulder kinesthesia after anterior glenohumeral joint dislocation. Phys Ther 1989; 69(2): 106–112.

Soderberg GL, Dostal WF. Electromyographic study of three parts of the gluteus medius muscle during functional activities. Phys Ther 1978; 58(6): 691–696.

Southwick WO, Keggi K. The normal cervical spine. J Bone Joint Surg [Am] 1964; 46-A(8): 1767–1777.

Speakman HGB, Weisberg J. The vastus medialis controversy. Physiother 1977; 63: 8.

Spitzer WO et al. Scientific approach to the assessment and management of activity-related spinal disorders: A monograph for clinicians-report of the Quebec Task Force on Spinal Disorders. Spine [European Edition] 1987; 12: 7s.

Stoff MD, Greene AF. Common peroneal nerve palsy following inversion ankle injury. Phys Ther 1982; 62(10): 1463–1464.

Stokes IAF, Abery JM. Influence of the hamstring muscles on lumbar spine curvature in sitting. Spine 1980; 5(6): 525–528.

Stone B, Beekman C, Hall V, Guess V, Brooks HL. The effect of an exercise program on change in curve in adolescents with minimal idiopathic scoliosis. Phys Ther 1979; 59(6): 759–763.

Straus WL, Howell AB. The spinal accessory nerve and its musculature. Rev Biol 1936; 11: 387.

Sullivan MS. Back support mechanisms during manual lifting. Phys Ther 1989; 69(1): 38–45.

Suzuki N. An electromyographic study of the role of muscles in arch support of the normal and flat foot. Nagoya Med J 1972; 17: 57.

Thomas HO. Diseases of the Hip, Knee and Ankle Joints. (Reproduction of 2nd ed, 1876.) Boston: Little, Brown, 1962.

Travell JG, Simons DG. Myofascial Pain and Dysfunction. Baltimore: Williams & Wilkins, 1983.

Trief PM. Chronic back pain: A tripartite model of outcome. Arch Phys Med Rehabil 1983; 64: 53–56.

Truex RC, Carpenter MG, eds. Strong and Elwyn's Human Neuroanatomy. 6th ed. Baltimore: Williams & Wilkins, 1969.

Urban LM. The straight-leg-raising test: A review. J Orthop Sports Phys Ther 1981; 2(3): 117–133.

Urquhart DM, Hodges PW. Differential activity of regions of transversus abdominis during trunk rotation. *Eur Spine J*. 2005;14(4):393–400. doi:10.1007/s00586-004-0799-9.

Vander AJ, Sherman JH, Luciano DS. Human Physiology: The Mechanism of Body Function. 3rd ed. New York: McGraw-Hill, 1980.

Wadsworth CT, Krishnan R, Sear M, Harrold J, Nielsen DH. Intrarater reliability of manual muscle testing and handheld dynametric muscle testing. Phys Ther 1987; 67(9): 1342–1347.

Walters CE, Partridge MJ. Electromyographic study of the differential action of the abdominal muscles during exercise. Am J Phys Med 1957; 36: 259.

Warfel JH. The Head, Neck and Trunk. 5th ed. Philadelphia: Lea & Febiger, 1985.

Watkins MA, Riddle DL, Lamb RL, Personius WJ. Reliability of goniometric measurements and visual estimates of knee range of motion obtained in a clinical setting. Phys Ther 1991; 71(2): 90–97.

Weiss HR. The effect of an exercise program on vital capacity and rib mobility in patients with idiopathic scoliosis. Spine 1991; 16: 1.

Wells KF. Kinesiology. 4th ed. Philadelphia: WB Saunders, 1966.

White A, Panjabi M. Clinical Biomechanics of the Spine. Philadelphia: JB Lippincott, 1978.

Williams M, Lissner HR. Biomechanics of Human Motion. Philadelphia: WB Saunders, 1962.

Williams M, Stutzman L. Strength variation through the range of joint motion. Phys Ther Rev 1959; 39: 145.

Williams PC. The Lumbosacral Spine. New York: McGraw-Hill, 1965.

Wolf S. Normative data on low back mobility and activity levels. Am J Phys Med 1979; 58: 217.

Youdas JW, Carey JR, Garrett TR. Reliability of measurements of cervical spine range of motion-comparison of three methods. Phys Ther 1991, 71(2): 98–106.

Zimny N, Kirk C. A comparison of methods of manual muscle testing. Clin Man 1987; 7(2): 6–11.

INDEX

Page numbers followed by "t" denote tables.

C

O